SURFACE EFFECTS IN ADHESION, FRICTION, WEAR, AND LUBRICATION

TRIBOLOGY SERIES

Vol. 1 Tribology — A Systems Approach to the Science and Technology of Friction, Lubrication and Wear (Czichos)
Vol. 2 Impact Wear of Materials (Engel)
Vol. 3 Tribology of Natural and Artificial Joints (Dumbleton)
Vol. 4 Tribology of Thin Layers (Iliuc)
Vol. 5 Surface Effects in Adhesion, Friction, Wear, and Lubrication (Buckley)

TRIBOLOGY SERIES, 5

SURFACE EFFECTS IN ADHESION, FRICTION, WEAR, AND LUBRICATION

DONALD H. BUCKLEY

National Aeronautics and Space Administration, Lewis Research Center, Cleveland, Ohio USA

ELSEVIER SCIENTIFIC PUBLISHING COMPANY
Amsterdam – Oxford – New York 1981

ELSEVIER SCIENTIFIC PUBLISHING COMPANY
1, Molenwerf, 1014 AG Amsterdam
P.O. Box 211, 1000 AE Amsterdam, The Netherlands

Distributors for the United States and Canada:

ELSEVIER NORTH-HOLLAND INC.
52, Vanderbilt Avenue
New York, N.Y. 10017

Library of Congress Cataloging in Publication Data

Buckley, Donald H.
Surface effects in adhesion, friction, wear, and lubrication.

(Tribology series ; 5)
Includes bibliographies and indexes.
1. Surfaces (Technology) 2. Tribology. 3. Adhesion. I. Title. II. Series.
TA418.7.B76 621.8'9 81-2331
ISBN 0-444-41966-7 AACR2

ISBN: 0-444-41966-7 (Vol. 5)
ISBN: 0-444-41677-3 (Series)

Printed in The Netherlands

Preface

Lubrication has emerged as a science principally since World War II. Prior to that, a number of fundamental contributions had been made by DeVinci, Coulomb, Reynolds, Amontons, Hardy, Bowden, and Tabor. From a scientific viewpoint, however, the most significant contributions, as well as the greatest number of contributions, have been made since World War II.

The war effort itself was responsible for initiating a number of efforts to try and understand the fundamental nature of surfaces and their interactions in sliding, rolling, and rubbing contact. An example of this was the development during World War II of adjuvants for carbon bodies to improve their wear resistance in high altitude aircraft generator applications. Basic researchers found that moisture in the carbon was critical to its lubrication. Therefore, the presence of moisture on the surface of the carbon was important. With it present, the carbon lubricated very effectively and very low wear was observed. In its absence, however, extremely high wear occurred and what was commonly called dusting of the carbon took place. With a fundamental study of the surface behavior of the carbon it was discovered that something had to be substituted for the moisture that was imparting the good lubricating characteristics to the carbon. The importance of surfaces became readily apparent.

After World War II, the need in the aircraft industry for liquid lubricants to substitute for the conventionally used mineral oils dictated the studying of synthetic materials. Again, the fundamentals and the basic structure of organic compounds became important, and studies began on the basic mechanism of the molecular structure in the lubrication of solid surfaces. As a result of those studies, synthetic lubricants were developed for aircraft applications; currently, most aircraft use synthetic lubricants for lubricating aircraft engine components. The study of the synthetic lubricants necessitated understanding the fundamental interactions of the synthetic molecular structure with the solid surface so that the lubrication of solid surfaces could be understood.

Prior to the late 1950's it was extremely difficult to gain any fundamental understanding of or to study solid surfaces because of the absence of a good vacuum environment in which to characterize and carefully control the surface. The interaction of various environmental constituents with the solid surface prevented such studies. However, with the advent of Sputnik at the end of the 1950's, a considerable amount of research resulted in the extensive development of vacuum technology in the United States and elsewhere. As a result of this development and the advances in vacuum technology, it was possible in the early 1960's to obtain clean vacuum systems which could achieve pressures to 10^{-10} torr. At these pressures it is possible to obtain and maintain atomically clean surfaces.

Shortly after the development of vacuum science and technology in the early 1960's, surface analytical tools became available for analyzing and characterizing surfaces. Some of these tools include (1) LEED (low energy electron diffraction), (2) field ion microscopy, (3) atom probe, (4) Auger emission spectroscopy , and more recently (5) photospectroscopy.

With the combination of a good vacuum system and analytical surface tools, it is now possible to characterize and analyze surfaces in tribological systems very carefully. In earlier years it was taken for granted that when one cleaned a solid surface with ordinary organic solvents the surface was, in essence, clean. Today we know that is not the case. Even very carefully cleaned surfaces inside a vacuum chamber, where such techniques as ion bombardment are used to remove oxides and adsorbed layers, may not necessarily be clean. The contaminant may not come from the environment if the vacuum system is a good one with pressures of 10^{-10} torr or below. They can come from within the bulk of the material and impart a surface film to the solid. For example, small concentrations of impurities such as sulfur and phosphorus in materials like iron, nickel, and copper can, with mild heating or straining of the solid, migrate to the surface and contaminate an essentially clean surface.

In the 17th century, the philospher John Granville recognized the importance of understanding the fundamental nature of materials. He wrote, "Iron seemeth a simple metal, but in its nature are many mysteries, and men who bend to them, their minds shall, in arriving days, gather therefrom greater profit, not to themselves alone but to all of mankind." Since John Granville wrote those words a considerable amount of research has gone into understanding iron and iron chemistry. In the field of tribology, iron-based alloys are one of the most commonly used materials. We are currently trying to understand its surface; even elemental iron in a state of high purity is a virtual chemical factory that can bring to its surface from the bulk surface contaminants such as carbon and sulfur which can impart surface films that alter markedly its adhesion, friction, and wear characteristics. Thus, it has become apparent in recent years that to work with and understand the nature of solid surfaces it is extremely important to have surface analytical characterizing tools so that one knows and can fully appreciate those films that are present on the surface and appreciate their role in tribological behavior of materials in solid-state contact.

The objective of this text, then, is to introduce some recent advances in the development of analytical surface tools for studying the tribological surface by using a combination of the two factors referred to before: (1) a vacuum system capable of holding and maintaining clean surfaces and (2) analytical surface tools. Throughout this text, attempts are made to understand the nature and character of solid surfaces that are used in tribological systems. The primary emphasis is on the use of these tools as they relate to tribology, both in situ type and post mortem analysis. The emphasis herein, however, is on the tribological implications rather than on the instrumentation. The nature and the basic mechanism of operating the analytical tools are reviewed very briefly in chapter 2.

Contents

CHAPTER 1

Introduction

Tribology and its importance are always present. While you are reading this text, your fingers as they touch the pages are virtually a tribological factory. The skin on the fingers has furrows and ridges which give a fingerprint pattern on paper. If the fingers are very carefully washed with organic solvents, very high friction coefficients are measured—in fact, as high as 1.8. With such high friction values, sliding the finger causes discomfort and even pain. Normally, with the secretion of body fluid (sebum, which contains fats, fatty acids, and hydrocarbons) to the fingertips, the friction coefficient measured is about 0.5. With a friction coefficient of 0.5 in the presence of these substances very little discomfort or pain is observed, and presumably this surface contamination, the basic lubricant if you will, prevents undue wear and damage to the skin (ref. 1).

The fingertips even have a mechanism for dissipating an excessive excretion of sebum so that the friction coefficient of the fingertip does not become too low and cause a failure to grip objects. The small furrows between ridges in the fingerprint pattern act as channels for carrying the fluid away from the points of contact of the finger with a solid object. When the finger presses down on a solid object the excess fluid in the contact zone is forced through the channels and away from the actual point of contact to improve or increase the static friction coefficient between the furrows and the object being gripped so that solid objects can be gripped.

The fingers in contact with a solid surface are a good example of a fundamental tribological system involving the importance of surfaces. The two solids in contact are (1) the finger and some solid object and (2) the lubricant; in this case, sebum is secreted from the body to produce a lubricating film. Although we need some lubrication to reduce wear, we do not need an excessive amount or friction would be reduced to the point where the finger would lose its basic function—namely, to grip objects.

When the sebum is removed by an organic solvent the friction coefficient reaches an extremely high value. This is comparable to the observations made for material surfaces in solid-state contact such as metal to metal contacts, metals and polymers, and metals and carbons. When the surfaces are very clean and brought into solid-state contact the friction is higher than it was in the presence of contaminating organic hydrocarbons or other surface films.

Surface Geometry

Just as the finger contains ridges and furrows, the surfaces of all other solids (in general) contain irregularities or hills and valleys. These surface irregularities, commonly called asperities, are indicated schematically in figure 1-1(a). The surfaces of most solids that are prepared either in the laboratory or in the machine shop have surfaces which contain these asperities (ref. 2). Figure 1-1(a) shows an exaggeration of the steepness of these surface irregularities. Typically surface irregularities have an angle of approximately 15° from the surface (ref. 3). Thus, they are fairly flat hills or peaks which lie on the solid surface. These irregularities are found on metals, polymers, ceramics, and carbon bodies.

With metals, in addition to the presence of surface irregularities or asperities, the solid surface itself is covered with films. For example, on the outermost surface there may be a layer of adsorbate, which is water vapor

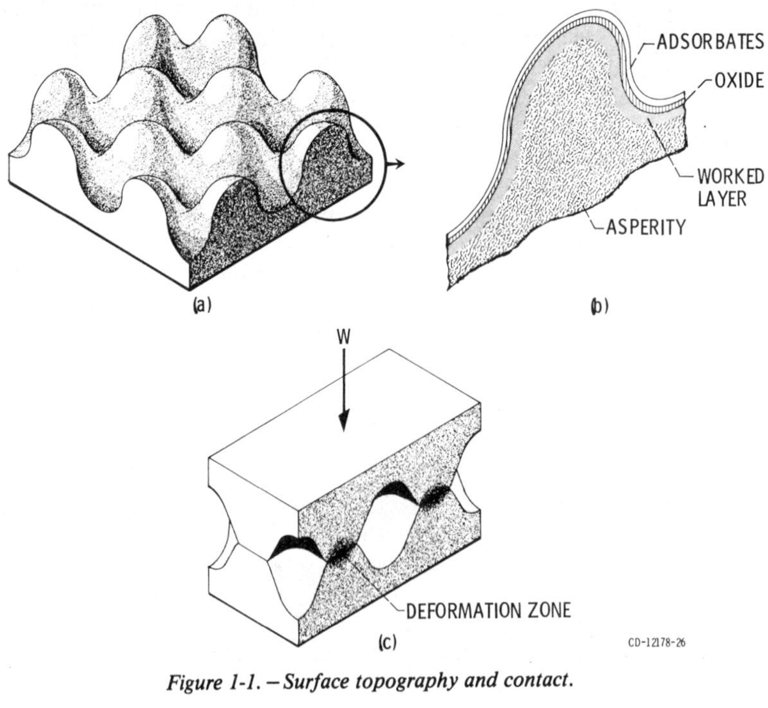

Figure 1-1. – Surface topography and contact.

or hydrocarbons from the environment that may have condensed and become physically adsorbed to the solid surface. This is shown schematically in figure 1-1(b). On metal surfaces or alloys, beneath this layer of adsorbate is generally a layer of metal or alloy oxide. In the case of an elemental metal, the oxide layer may be one of the oxides of the elemental metal. So, for example, on iron it may be iron oxide (Fe_2O_3). Or the film may contain a mixture of oxides such as Fe_2O_3, Fe_3O_4, and FeO. With some metals, such as copper, the oxide layer consists of more than one layer of oxide film. With copper, the layer closest to the metal is the lower oxide of copper (Cu_2O), and the outermost layer is the higher oxide of copper (CuO), for that portion of the surface which is exposed directly to oxygen in the environment. With alloys, the surface oxides may consist of a mixture of oxides. For example, on stainless steels the oxides may be a mixture of iron oxide and chromium oxides, principally chromium oxide (Cr_2O_3). With some alloys, the oxides may be combined into compounds called spinels; that is, there is a mixed oxide system on the solid surface.

The thickness of the oxide layer depends on two factors: (1) the nature of the substrate metal which has been oxidized and (2) the environment. With some metals, the oxides that are formed are very tenacious, very thin films form on the metal, and the surface becomes passivated with no further oxidation taking place. Good examples of this are aluminum and titanium surfaccs. With other metals, however, the oxide can continue to grow; for example, Fe_2O_3 continues to grow in a humid air environment.

Beneath the oxide layer on alloy surfaces is a region of the surficial material which may be highly worked or deformed as a result of the forming process with which the metal surface was prepared. For example, in grinding, honing, machining, or polishing, the surface layers (the outermost layers of the solid surface) become highly strained. This strain is reflected in what is called the worked layer, which lies subsurface to the oxide and is an integral part of the metal itself in the surface region. The amount of this material present and the degree of deformation that occurs are functions of two factors: (1) the amount of work or energy that was put into the deformation process, and (2) the nature of the metal. Some metals are much more prone to deformation and work hardening than are others. This, of course, would be reflected in these surface layers.

These layers are extremely important because their properties, from a surface chemistry point of view, can be entirely different from the annealed bulk metal or alloy. Likewise, their mechanical behavior is also influenced by the amount and the depth of deformation of the surface layers.

With ceramic materials the oxide layer may or may not be present. For example, on aluminum oxide, oxygen is an integral part of the structure so an oxide surface layer is not expected. However, adsorbates may still be present on the surface from the environment, and these include water vapor and hydrocarbons. In addition, with carbon bodies (mechanical carbons that are very frequently used), adsorbates play a very large role. Even though an oxide may not be present because the oxides of the carbon (carbon monoxide and carbon dioxide) are volatile, adsorbed species (oxygen and water vapor as well as hydrocarbons) play a very heavy role in the mechanical as well as the chemical behavior of carbon solid surfaces in tribological systems.

With polymers, much as with ceramics, the outermost surface layer may not be an oxide as it is in the case of metals or alloys. But adsorbates are certainly present on the solid surface, and again they may include water vapor and hydrocarbons from the environment.

When two solid surfaces are placed in contact, the actual contact takes place over a very small area, actually at the tips of the asperities or surface irregularities, as indicated in figure 1-1(c). These asperity regions initially deform elastically and, if the load is sufficiently high, they deform plastically until the load can be supported; that is, the real contact area continues to increase with deformation until the contact area is sufficient to support the load applied to the two surfaces in solid-state contact. With deformation of the surfaces, the adsorbed layers, oxides, and worked layers generally deform with the material. In some cases, depending on the mechanical properties of the surface films, they are completely compliant with the surface and deform with it. In other cases, they become disrupted or dislodged, and solid-state clean material contact can occur at the asperity junctions through the films because of the breakup of these surface films. When that occurs, clean, solid-state contact occurs. At this point the basic material properties of the solids themselves become extremely important in the adhesion, friction, and wear behavior of the materials.

The surface irregularities or asperities indicated in figure 1-1 are gross surface geometric characteristics of solids. In addition to these there are many other minor or smaller surface irregularities that can occur on the surface of solids. For example, it was mentioned earlier that most practical solid surfaces contain these irregularities or asperities. There are, however, situations where the surface is free of these surface defects. For example, although brittle inorganic crystals which are cleaved have atomically smooth surfaces, they contain surface cleavage steps (fig. 1-2). Between the cleavage steps are atomically smooth or flat surfaces; in other words, a complete absence of surface asperities. This occurs, for example, in the

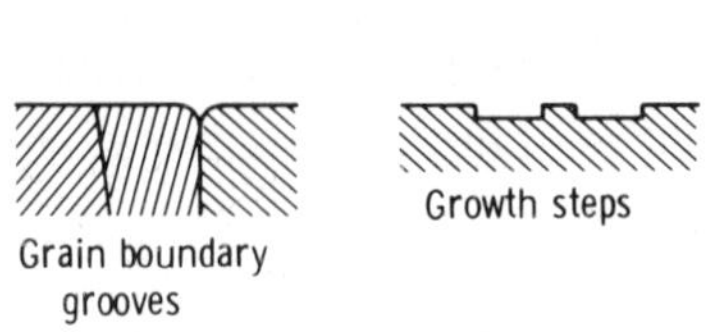

Figure 1-2. – Schematic drawing of a few modes of small geometrical surface alterations.

cleavage of a brittle material like lithium fluoride, sodium chloride, or aluminum oxide along cleavage planes. In addition, some metals can be cleaved in this fashion at cryogenic temperatures; for example, zinc can be cleaved along the basal plane at liquid nitrogen temperatures.

In addition to cleavage steps, other small alterations in surface geometry can occur as shown in figure 1-2. These include, for example, growth steps that develop during the growth of crystalline solids. They can also develop when materials solidify from the liquid state. There are deposits that may form on a solid surface that serve as surface geometric irregularities. These could include deposits of solid wear particles, particles that adhere in a particular manner to the surface, and deposits of materials to the surface from the environment. With all crystalline materials there are also grain boundaries. Grain boundaries are high energy sites on a solid surface; they generally contain or act as an irregularity in the surface because of a cusp that usually forms at the region where the two adjacent grains meet in the boundary. The resulting cusp at the surface, then, is a defect or alteration in the surface geometry. Almost all surfaces that are prepared by mechanical techniques contain scratches in addition to the foregoing. Scratches can be generated by the rubbing of one hard surface particle against a softer surface or by the entrapment of small, hard particles between two solid surfaces.

Another type of surface defect that can be found is the etch pit when the environment or a constituent of a lubricant interacts with the surface in a reactive manner. The high energy sites (e.g., grain boundaries and dislocation sites) react chemically much more rapidly than the bulk of the surface. This can result in surface defects which are commonly called etch pits. The minor defects revealed in figure 1-2 can then be superimposed on the larger surface defects or asperities seen in figure 1-l so that, in any real surface, there may be a combination of these particular geometric irregularities. That is, there may be asperities in addition to, for example, growth steps, scratches, or etch pits. The presence of all of these then comprise the real surface geometry of a solid. When examining a surface it thus becomes apparent that a number of considerations are important. Surface chemistry is important from the viewpoint of the surface films (such as the oxides) that are adsorbed on the solid surface. Metallurgy is important from the standpoint of the layers that may develop as a result of deformation in the worked layers. Physics is important because of the nature of the bonding or adhesion of the solid surfaces in solid-state contact. Similarly, mechanics is important to understand the deformation mechanisms when two solid surfaces are brought into solid-state contact.

Surface Chemistry

If we take an analytical surface tool such as Auger emission spectroscopy analysis (described in detail in chapter 2) and analyze the elements that are present in the surface layers shown in figure 1-1(b), we can determine the chemical composition of those layers for various materials. In figure 1-3 such an anlysis was conducted on an iron (011) single crystal surface. We

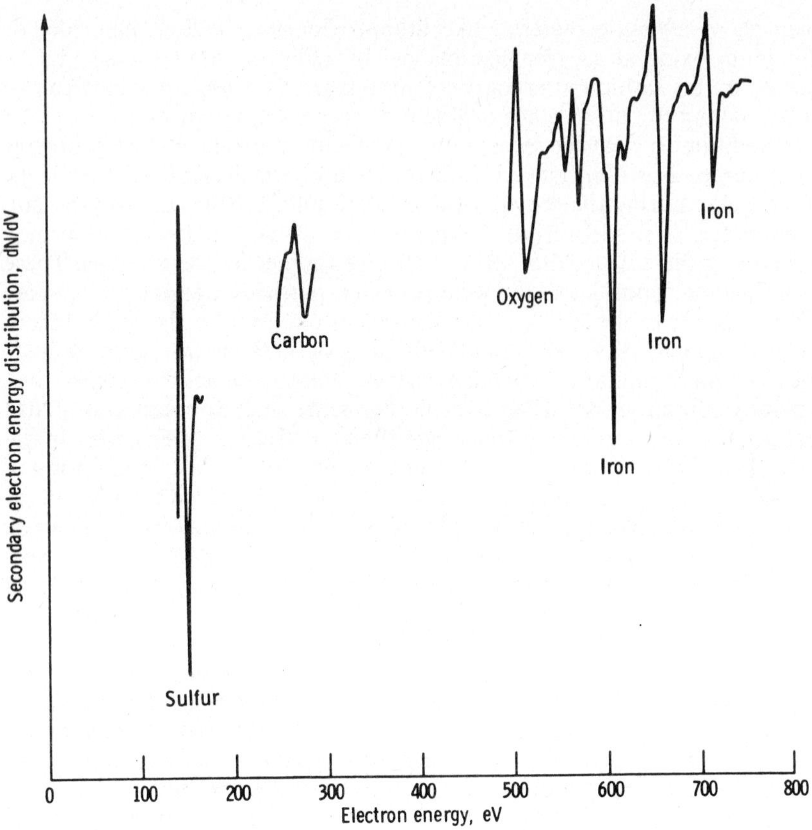

Figure 1-3. — Auger electron spectrometer analysis of iron (011) surface.

plotted the secondary electron energy distribution as a function of electron energy for the specific elements involved and found, by analysis of the Auger spectra, that a number of elements are present on the single crystal of the iron surface. It should be noted that the single crystal was present in a vacuum system before any attempts were made to clean the surface. In the spectra, sulfur, carbon, oxygen, and iron are present. The sulfur can arrive at the surface from within the iron itself by segregation from the bulk, or it can arrive at the surface either by itself or combined with other species such as oxygen and adsorb on the surface as an adsorbed surface layer.

The carbon can also originate from two sources. It can come from the bulk iron itself as a result of diffusion to the surface (even though this particular iron sample is a triple zone refined, high purity iron containing only parts per million of carbon). It takes very small concentrations of a carbon contaminant to appear on the surface. A second source of the carbon, like the sulfur, can be the environment. Carbon can arrive at the surface as carbon monoxide or carbon dioxide which is generally found to be physically or chemically adsorbed to most solid metal surfaces.

In addition to sulfur and carbon, oxygen is present on the surface (fig.

1-3). The oxygen is present in two forms: (1) combined with the iron in the form of iron oxide, and (2) present on the surface as an adsorbate by itself or in combination with either carbon or sulfur. The iron sample examined in figure 1-3 was cleaned with a solvent before it was inserted in the vacuum chamber, which was evacuated to a pressure of 11^{-11} torr with intermediate bakeout at 251° C. Despite these conditions and the environment, the surface contains a number of contaminants (fig. 1-3). Thus, any examination of an iron surface such as that represented by the data of figure 1-3 is not in fact an iron surface but rather an iron surface covered by oxide and adsorbed films such as those indicated in figure 1-1(b).

Low energy electron diffraction (LEED) is a surface analytical tool used to analyze the structural arrangement of atoms in the outermost atomic layer of solid surfaces. The mechanism for operating with LEED is described in detail in chapter 2. At this point, however, it is sufficient to indicate that LEED can indicate the arrangement of atoms in the outermost atomic layer of a solid surface. If a LEED pattern is obtained from the surface shown in figure 1-3, the structural arrangements are as shown in the photomicrograph in figure 1-4. The iron surface before cleaning showing sulfur and carbon monoxide present on the surface as well as iron oxides is indicated as a fractioned pattern to the left (fig. 1-4). The pattern contains a number of white spots. These spots reflect the diffraction from the surface and various species present on the solid surface.

When the iron surface is very clean—that is, if a technique such as argon iron bombardment is used to remove these surface contaminants and only clean iron is left—a LEED pattern such as that shown on the right in figure 1-4 is obtained. The clean surface is represented by only four white diffraction spots in a rectangular array. These four diffraction spots are the characteristic pattern for the iron (011) clean crystal surface. All the additional diffraction spots on the left pattern in figure 1-4, prior to cleaning, are due to the contaminants (including oxygen, sulfur, and carbon) on the solid surface. Thus, from an elemental analysis of the solid surface (fig. 1-3) and a structural pattern of that surface (fig. 1-4), it becomes readily apparent that what might normally be considered a clean surface is in reality quite heavily contaminated with surface films. A clean surface such as that indicated on the right in figure 1-4 can only be maintained in a good vacuum

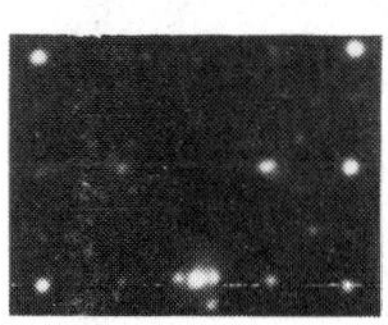

Fe (011) SURFACE BEFORE CLEANING. SULFUR AND CARBON MONOXIDE PRESENT ON SURFACE

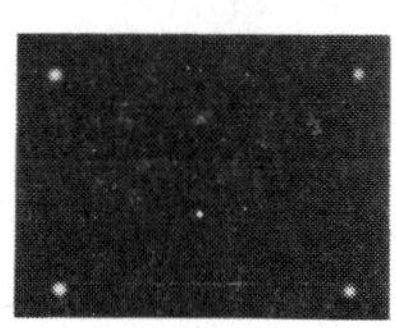

CLEAN Fe (011) SURFACE

Figure 1-4. – LEED patterns obtained before and after cleaning of iron (011) surface.

environment at a pressure, for example, of 10^{-10} torr. At 10^{-6} torr the surface would be contaminated in a matter of seconds with the constituents of the environment.

The clean surface in figure 1-4 is extremely reactive and highly energetic. All types of reactions and interactions can take place with that surface because it is in an unstable, highly energetic state. The surface atoms are only bounded in the bulk and not above the free surface; thus, there are free electrons available for interaction with species from the environment and also from within the bulk of the solid itself.

One of the most common types of surface interactions that can take place with a clean surface is the physical adsorption of species on that solid surface. For example, the admission of an inert gas, such as argon, to the surface can produce the physical adsorption of the argon to the clean surface. There would be a lack of electron interaction of the argon or a sharing of electrons between the metal and the adsorbate.

The physical adsorption process is a relatively weak process and is depicted in figure 1-5. Oxygen can also physically adsorb to the surface containing the normal residual contaminants (as shown in fig. 1-4). For example, in addition to sulfur and carbon monoxide on the surface, there can be free oxygen.

The molecule depicted in figure 1-5 for physical adsorbtion, bonding itself to the surface, is shown as a diatomic molecule such as might occur in oxygen (O_2). In such a case, both oxygen atoms of the diatomic molecule can bond to the already contaminated surface. However, it takes very little energy to remove physically adsorbed species from a solid surface, and almost all surfaces that are examined in a vacuum environment of 10^{-10} torr are already free of physically adsorbed species. The physical adsorption process typically involves van der Waal's forces. It has frequently been said that, if the interaction involves less than 10 kilocalories per mole, the pro-

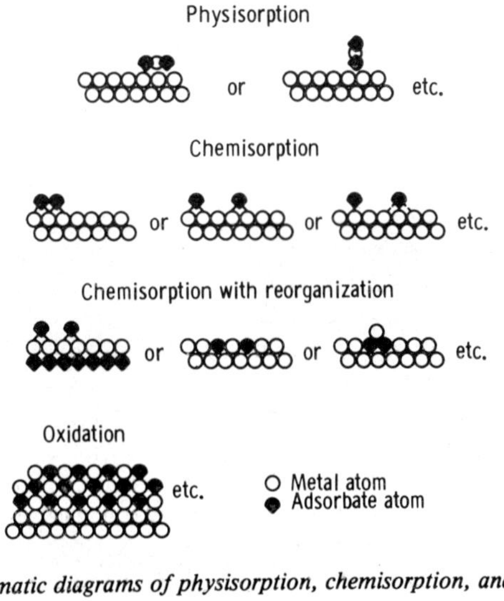

Figure 1-5. – Schematic diagrams of physisorption, chemisorption, and oxidation.

cess is one of physical adsorption. If, however, the energy involves in excess of 10 kilocalories per mole, the process is similar to chemisorption.

Chemisorption is also depicted in figure 1-5. In chemisorption, in contrast to physical adsorption, there is an actual sharing of electrons or electron interchange between the chemisorbed species and the solid surface. In chemisorption the solid surface very strongly bonds to the adsorbing species; it therefore requires a great deal of energy to remove the adsorbed species, the energy being a function of the solid surface to which the adsorbing species attaches itself and the character of the adsorbing species as well. For example, while oxygen may chemisorb very strongly to iron or titanium, it may chemisorb very weakly to a metal such as one of the noble metals—for example, copper or silver. Thus, if oxygen is admitted to the vacuum system containing the clean iron surface seen in figure 1-4, the first thing that occurs is chemisorption of the oxygen to the iron surface. This alters the LEED pattern seen in figure 1-4 because the presence of oxygen adds diffraction spots to the original diffraction pattern seen for the clean iron surface. Also, the oxygen can be detected on the solid surface with an Auger emission spectroscopy analysis.

In chemisorption, the chemisorbing species, while chemically bonding to the surface, retains its own individual identity so that we can, by proper treatment of the surfaces, recover the initial adsorbing species. This is a distinction between chemisorption and chemical reaction and sets chemisorption apart from chemical reaction per se. Once the surface is covered with a layer (e.g., a clean metal surface is covered with a layer of oxygen), chemisorption ceases; any subsequent layer formation is either by physical adsorption or chemical reaction.

Chemisorption is primarily a monolayer process. For example, once a monolayer of oxygen is formed on the solid surface of iron (fig. 1-4), the oxygen need not stay in the position or initial sites at which it adsorbs. If the energy situation or condition at the surface is such that the surface is not in the lowest energy state, rearrangement or, as it is commonly called, reconstruction can take place at the surface and bring about a change in the ordering or arrangement of iron to oxygen atoms on the solid surface. This is depicted schematically in figure 1-5 where reorganization is shown. The oxygen and iron atoms can switch positions until such time as the surface species, the iron and the oxygen, achieve the lowest energy state; at this time the reorganization or reconstruction of the solid surface ceases.

Another surface process that can take place is chemical reaction or interaction of the other surface species with the solid surface itself. For example, with oxygen adsorbing on the surface of iron, oxidation can take place at the iron surface if the concentration of oxygen in the environment is sufficiently high or if the temperature of the surface is sufficiently high. That is, the chemisorbed oxygen can begin to react with the iron surface to form iron oxides, and this phenomena, depicted in figure 1-5, indicates the oxidation process of the surface of the metal. Surface oxides are true chemical compounds, and one does not normally recover the oxygen (as is possible in simple chemisorption) by supplying energy to desorb the adsorbing species. With oxidation, true chemical compounds are formed.

If the surface of figure 1-1(b) is denuded or cleaned of the adsorbates and oxides, any one of the interactions or reactions depicted in figure 1-5 can occur with the solid surface. In all these interactions or reactions, one of the constituents is the solid surface and the other is a species which comes from the environment.

Bulk Chemistry

If a clean surface is generated as a result of the deformation process depicted in figure 1-1(c), the clean surface can interact with the environment, as indicated in figure 1-5, or bulk chemistry can play a part in the behavior of the solid surface. For example, if the clean iron surface referred to in figure 1-4 is generated in vacuum, heating or straining the iron can cause carbon to diffuse from the bulk of the material to the surface and produce a structure of carbon on the solid surface (fig. 1-6). The four bright diffraction spots in figure 1-6 indicate the iron basic pattern that was seen in figure 1-4 for the clean iron surface. The additional diffraction spots in a ring structure which encompass and include the four diffraction spots for the clean iron are associated with carbon. Auger emission spectroscopy analysis of the iron surface revealed that the contaminant was carbon. The source of this carbon is the bulk iron, which contains 10 parts per million of carbon in the bulk. This is sufficient to diffuse to the surface when heating or straining the iron so as to produce the surface structure seen in figure 1-6. In addition to carbon, other species have been observed to diffuse from the bulk of metals to the surfaces; these include sulfur, nitrogen, boron, and oxygen.

In addition, for various metals and binary alloys, the solute dissolved in the solvent in small concentrations has been observed to diffuse to the surface and produce surface-rich layers of the solute on the solvent. This role of bulk alloying elements in diffusing and segregating on the surface markedly alters surface chemistry and surface behavior in adhesion, friction, wear, and lubrication. A common technique for obtaining clean metal surfaces is to ion bombard the surface with positive argon ions. When this is done the surface contaminants, such as the carbon seen in figure 1-6, can be

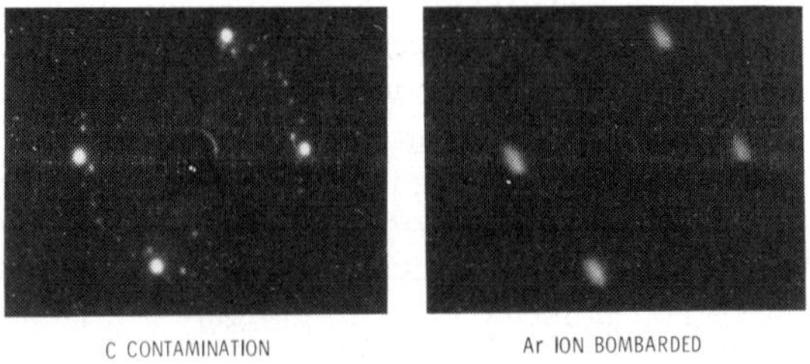

Figure 1-6. – LEED patterns of iron (011) surface with carbon present and after argon ion bombardment.

removed from the iron. When removing the carbon, however, the incoming argon ions strike the surface with sufficient energy to produce a strain in the crystal lattice. And the diffraction spots for the iron take on an elongated or irregular shape (fig. 1-6). This is a strained iron surface in the surficial layers. A modest heating to 211° C is sufficient to produce an annealing effect in the surface layers and to obtain, once again, the sharp diffraction spots that were observed in figure 1-4 for the clean iron solid surface.

Thus far we have discussed the presence of adsorbates and oxides of figure 1-1(b) on the solid surfaces. In addition to these, there is what is referred to as the worked layer. The polishing, grinding, machining, or cutting of a solid surface produces the worked layer. The worked layer can consist of (1) recrystallized material, (2) highly deformed or strained crystallites, or (3) a textured surface produced by the rubbing of the solid surface. This may generate a reorientation of the individual crystals or grains in the surficial layers so that they become oriented in a preferred direction. These surface changes also produce a change in the properties of adhesion, friction, and wear for two solid surfaces in contact.

Worked Layer

The metallurgical properties of the surface layer of a metal or alloy can vary markedly from the bulk of the material. This effect is depicted in figure 1-7, which shows a tapered section of a ground zinc metal surface (ref. 4). At the base the grains are very large; up near the surface, however, the grains are extremely small. The small grain size has resulted from the recrystallization of the grains at the surface. The properties of the small grains are different from those of the large grains. Furthermore, beneath the recrystallization zone (where insufficient energy was available to produce recrystallization) there is sufficient energy to bring about twinning in the individual grains of the zinc. Therefore, a high concentration of twins is seen in a band or region beneath that where the recrystallized layer of small grains appears. Thus, the energy in the grinding process is dissipated in the surface region by recrystallization (where the energy is the greatest), and it is dissipated into the bulk of the solid (where less energy is available for recrystallization) by twinning, which is a very common occurrence in the deformation of hexagonal metals. Taper sectioning, which is a very effective technique for showing the surficial layers in solids, is discussed in more detail in chapter 2.

At the very surface of figure 1-7, the individual crystallite or grains of the zinc can, with rubbing or grinding, take on a preferred orientation. That is, the crystals can orient themselves at the surface with a preferred crystallographic slip plane oriented directly onto the surface or near the surface. Each of these crystallographic orientations that may arise at the surface or develop on the surface has its own properties. The various orientations, for example, in different crystal systems have different reaction rates because the number of free bonds available for interaction with species from the environment varies with different orientations.

Figure 1-7. – Taper section of ground zinc surface showing recrystallized surface layer and zone containing deformation twins. Taper ratio, 16.2 (ref. 4).

This orientation effect relative to interaction with the environment is demonstrated in the data of table 1-I for a germanium surface in an oxygen-rich water environment (ref. 5). There are three orientations of germanium presented in table 1-I: {100}, {110}, and {111}. The number of free bonds associated with each of these surfaces, which conceivably are present at the interface of solid surface and vacuum, varies with surface orientation. The number of free bonds is greatest with the {100} orientation and the least with the {111} orientation. Because there are a greater number of bonds available

TABLE 1-I. – DENSITY OF FREE BONDS ON GERMANIUM SURFACES AND DISSOLUTION RATES IN OXYGEN-SATURATED WATER[a]

Orientation	Free bonds/cm^2	Relative free bond density	Relative dissolution rate
{100}	1.25×10^{15}	1.00	1.00
{110}	8.83×10^{14}	.71	.89
{111}	7.22×10^{14}	.58	.62

[a]Reference 5.

for interaction with environmental species on the {100} surface, that particular surface is much more reactive with the water environment. It is dissolved at a much higher rate than is either the {110} or the {111} orientation; this is indicated by the relative dissolution rates of the various orientations for germanium in table 1-I.

The results in table 1-I indicate that, for polycrystalline materials, the orientations of the individual grains in the solid surface can produce localized variations in reactivity with variations in the environment. This localized alteration in chemical reactivity can take place if the environment is a gaseous one or if the surface is covered with a lubricant. Thus, for example, where additives are placed in a conventional oil to interact with a solid surface, the reactivity or reaction rates of different orientations that may be exposed to the solid surface varies. In many tribological systems, however, the sliding, rolling, or rubbing contact helps to promote or generate a specific surface orientation or texture of nearly all the grains. They orient themselves with one particular plane exposed near the surface so that the relative reactivity rates of adjacent ones may be fairly comparable. The grain boundaries, however, always retain a different and higher energy condition than the bulk surface of the grains themselves, and as a consequence, they react at a much higher rate. The boundaries are zones of high defect densities, great concentrations of vacancies, and dislocations; as a consequence, they are all high energy sites and thus sites for greater reactivity (ref. 6).

Surface Effects in Tribology

The importance of surface films, even fractions of a monolayer, on the behavior of two solid surfaces in contact is depicted in the data of figure 1-8 (from ref. 7) where the static friction coefficient is plotted as a function of the adsorbate concentration from fractions of monolayer to a full monolayer for the adsorbates chlorine and oxygen on various metal surfaces including copper, iron, and steel.

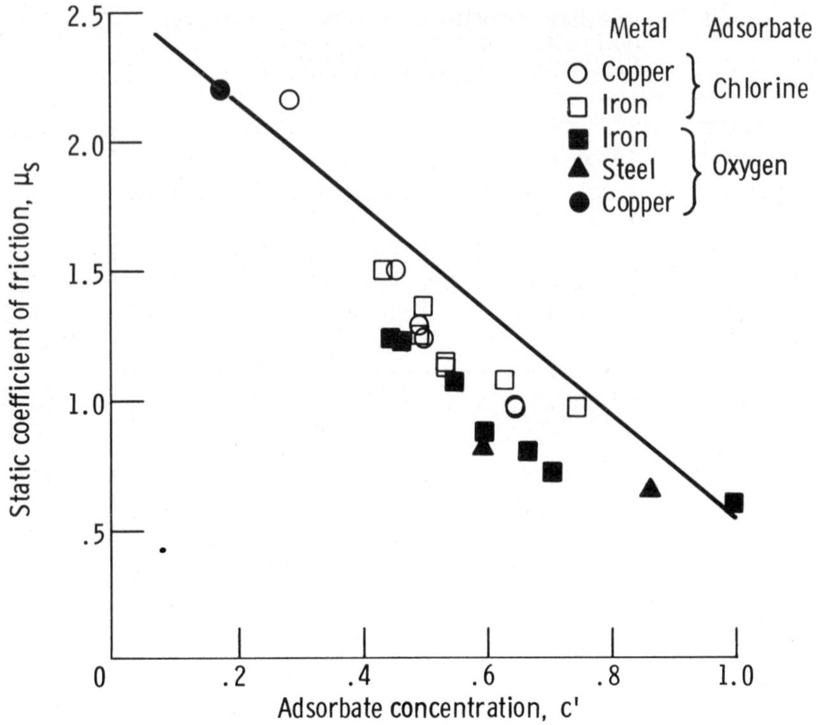

Figure 1-8. – Static coefficient of friction as function of adsorbate concentration (ref. 7).

An examination of the data in figure 1-8 indicates that the static friction coefficient for all three materials (copper, iron, and steel) decreases with increasing concentration of adsorbate up to one monolayer. From the data it is obvious that even fractions of a monolayer of an adsorbed film on the surface of a solid can markedly alter static friction behavior. Adhesion in these same experiments decreased appreciably with the presence of even fractions of a monolayer of adsorbates on the solid metal surfaces. For these data the adsorbates were present as chemisorbed species as opposed to a reaction product.

It is important to note that, while two different adsorbates and two different base metals (viz., copper and iron) are involved, the static friction coefficient appears to be insensitive to the difference in the adsorbing species and the differences in the metals. These are polycrystalline metal samples of small grain size. If, however, one examines very carefully the influence of orientation on the solid surface, it can be established that the orientation makes a significant difference in the friction behavior of metals in contact in the presence of adsorbates. In addition to the particular orientation of the metal, the adsorbate species that is present on the solid surface also makes a difference in the measured friction behavior. Numerous experiments have been conducted with a variety of different metals, in the single crystal form as well as and large-grain polycrystalline form, to determine the influence of orientation and adsorbate on friction behavior. In general, the friction responds not only to the orientation effects but also to the particular species that may be present on the solid surface. This orientation and adsorbate specificity are demonstrated by the data of table 1-II.

Table 1-II presents friction data for a large-grained tungsten sample containing crystallites of various orientations on the solid surface. This was one large disk specimen containing various orientations. Three grains are a {110} surface, a {110} surface, and a {100} surface, with various chemisorbed gases present including hydrogen, oxygen, carbon dioxide, and hydrogen sulfide. The data of table 1-II indicate that, in the absence of any adsorbate, the friction coefficient varies with orientation, friction being highest on the {100} surface and least on the {110} surface. Friction coefficients for metals (in general, body centered cubic, face centered cubic, and close packed hexagonal) are usually lowest on the highest atomic density, low surface energy planes in the metal. For the body-centered-cubic system, this generally is the {110} surface; for the face-centered-cubic system, the {111} surface; and for the close-packed-hexagonal system, the {0001} surface. These surfaces are the high atomic density, low surface energy planes in their respective crystal systems, and they accordingly exhibit the lowest friction characteristics.

It is of interest to note in table 1-II that even the presence of a gas such as hydrogen, normally considered to be a reducing gas adsorbed to the surface of the tungsten, produces a reduction in the friction coefficient on each of the three tungsten planes. The friction coefficient on the {100} surface is reduced to nearly half of what it was in the absence of the adsorbate. The difference between the clean tungsten surface and the hydrogen covered surface is less for the {210} and the {110} surfaces, but there is still a reduction of friction with adsorption of hydrogen.

Table 1-II shows that the most effective adsorbed gaseous species for reducing the friction coefficient for all three planes of tungsten is oxygen. There are two tribological implications of the data of figure 1-8 and table 1-II. First, extremely small concentrations of species on a solid metal surface can markedly influence a tribological property such as friction

TABLE 1-II. - INFLUENCE OF VARIOUS CHEMISORBED GASES ON FRICTION COEFFICIENT OF TUNGSTEN IN VACUUM

[Rider specimen, (100) atomic plane of tungsten; load, 50 g; sliding velocity, 0.001 cm/sec; temperature, 20° C; pressure, 10^{-10} torr (1.33×10^{-8} N/m^2).]

Chemisorbed	Coefficient of friction		
	For (110) plane	For (210) plane	For (100) plane
None	1.33	1.90	3.00
H_2	1.25	1.33	1.66
O_2	.95	1.00	1.30
CO_2	1.15	1.15	1.40
H_2S	1.00	----	1.35

behavior; even fractions of a monolayer can appreciably reduce friction coefficients. Second, the presence of something as simple as hydrogen on a solid surface can influence the friction behavior of materials and solid-state contact. Third, the orientation of the grains on the solid surface causes variations in the friction behavior of the grains. The influence of the adsorbed species and the orientation of the solid surface indicate the extreme sensitivity of tribological properties to surfaces and surface films. This extreme sensitivity also stresses the importance of using surface tools and characterizing solid surfaces in the defining and understanding of tribological mechanisms.

References

1. Spurr, R. T.: Fingerprint Friction Wear, vol. 39 pp. 167–171, 1976.
2. Bowden, Frank Phillip; and Tabor, D.: The Friction and Lubrication of Solids. Oxford University Press (London) 1950.
3. Williamson, J. B. P.: Topography of Solid Surfaces An Interdisciplinary Approach to Friction and Wear. NASA SP-181, 1968, pp. 85–142.
4. Samuels, L. E.: Damaged Surface Layers: Metals. The Surface Chemistry of Metals and Semiconductors, Harry C. Gatos, ed., John Wiley & Sons, Inc., 1960, pp. 82–103.
5. Gatos, Harry C.: The Reaction of Semiconductors with Aqueous Solutions. The Surface Chemistry of Metals and Semiconductors, Harry C. Gatos, ed., John Wiley & Sons, Inc., 1960, pp. 381–406.
6. McLean, Donald: Grain Boundaries in Metals. Clarendon Press (London), 1957.
7. Wheeler, D. R.: The Effect of Adsorbed Chlorine and Oxygen on the Sheer Strength of Iron and Copper Junctions. NASA TN D-7894, 1975.

CHAPTER 2

Surface Analytical Tools

The most effective and universal surface tool available to the tribologist for understanding and studying tribological surface behavior is the naked eye. Very frequently a careful examination of surfaces in sliding, rubbing, or rolling contact with just the naked eye can provide a considerable amount of information and insight into the behavior of the materials and a history of what has transpired to the surfaces.

Optical Techniques

While examination with the naked eye can provide considerable information about the nature of tribological surfaces, it has its limitations. The first and simplest surface tool that can be employed beyond the naked eye is the simple magnifying glass which magnifies the surface and thereby provides more detail of the character of the surface. A simple magnifying glass, or lens, magnifies the image as indicated in figure 2-1. In figure 2-1(a) the ob-

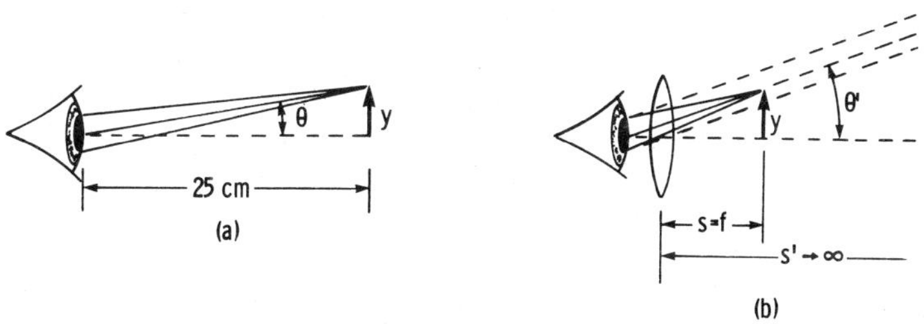

Figure 2-1. – Effect of simple magnifier.

ject is at the near point where it subtends an angle of θ at the eye. In figure 2-1(b) a magnifier lens is placed in front of the eye, and this forms an image at infinity with the angle subtended at the magnifier being θ'. The angular magnification M is defined as the ratio of the angle θ to the angle θ'. Thus, a simple magnifying glass, which provides magnifications of 30 to 40, is frequently used to examine surfaces and provide further detail about the nature of those surfaces that can not be observed with the naked eye.

When a magnification greater than that obtainable with a simple magnifier is desired, it is necessary to use a microscope. Essential elements of a microscope are illustrated in figure 2-2. The object O to be examined is placed just beyond the first focal point F of the objective lens which forms a real and enlarged image I. This image lies just within the first focal point F_1 of the ocular or eyepiece, which forms a virtual image of I and I′. The position of I′ may be anywhere between near and far points of the eye. While

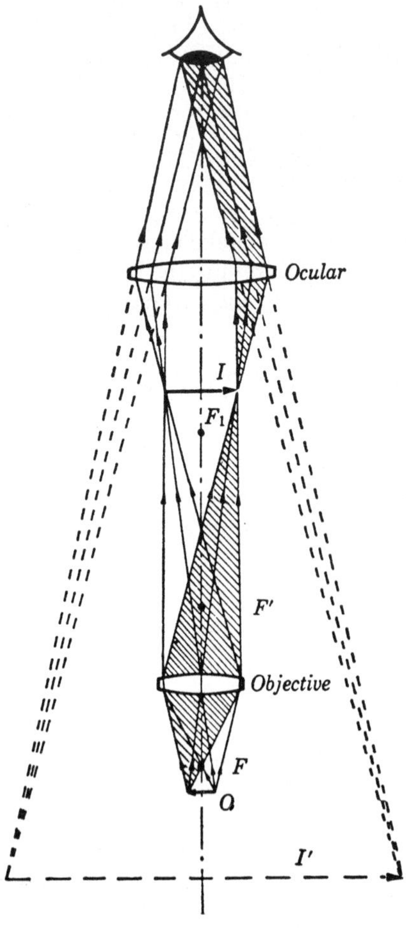

Figure 2-2. – Optical microscope.

both the objective and ocular of an actual microscope are highly corrected compound lenses, for simplicity they are shown in figure 2-2 as simple, thin lenses. Since the objective merely forms an enlarged real image which is examined by the ocular, the overall magnification M in the compound microscope is a product of the lateral magnification M_1 of the objective and the angular magnification M_2 of the ocular. The ordinary optical microscope is an extremely useful tool in tribological studies. It can provide, with oil immerging of the objective, magnifications up to 1000. Thus, for many practical systems, the study of wear and adhesion of surfaces can be very effectively carried out within the range of magnifications provided by the ordinary optical microscope.

The optical microscope has a limitation in that it does not have a great depth of focus; as a consequence, many tribological surfaces (e.g., wear surfaces) and the topography of wear cannot be totally seen. A further limitation of the ordinary optical microscope is the resolution limit, which is about 2000 angstroms; this means that features contained in a surface or on a surface that are smaller than this dimension are not revealed by the ordinary optical microscope.

Figure 2-2 shows that the ordinary microscope consists of nothing more than two lenses: an ocular (or eyepiece) and an objective. The simplest form of an optical microscope is probably the toolmaker's microscope (fig. 2-3). An ocular and an objective are mounted in a tube. The light source to operate the microscope is nothing more than room light as indicated by the arrows in figure 2-3. The light is incident on a surface just beneath the objective. The typical toolmaker's microscope can magnify a surface about 100 times.

In addition to using the optical microscope to examine the surface topography of solid surface at normal incidence to the surface, the optical microscope has been very effectively used to examine surfaces in cross section by using a taper sectioning technique to magnify the surface

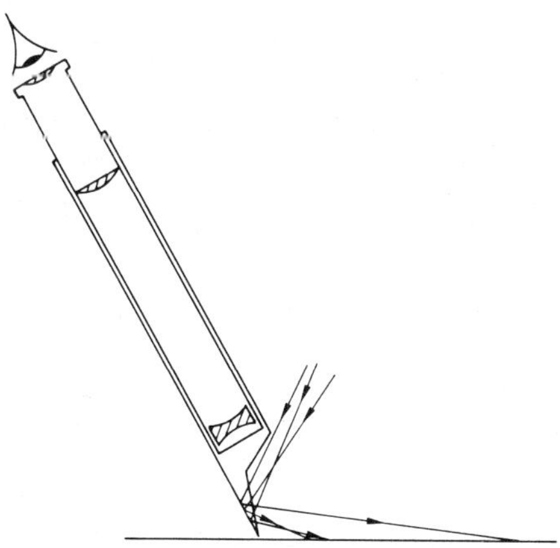

Figure 2-3. – Toolmakers microscope.

topography. The zinc photomicrograph (fig. 1-7) showing the microstructure of zinc at and near the surface is an example of a photomicrograph taken from a tapered section. Taper sectioning is shown schematically in figure 2-4. The normal incidence of the optical microscope occurs along the plane or slice of material shown by plane A. The topography is revealed, but it is not amplified or exaggerated. If, however, a slice of material is cut through the solid surface at an angle θ from the surface, as indicated in figure 2-4, the surface is magnified as indicated in plane B in figure 2-4. The tapered section magnifies the surface to a considerable extent and allows a detailed examination of surface topography. This particular technique has been very effectively used by Bowden and Tabor in their examination of wear of surfaces (ref. 1) and by Samuels in his examination of surfaces undergoing plastic deformation (ref. 2).

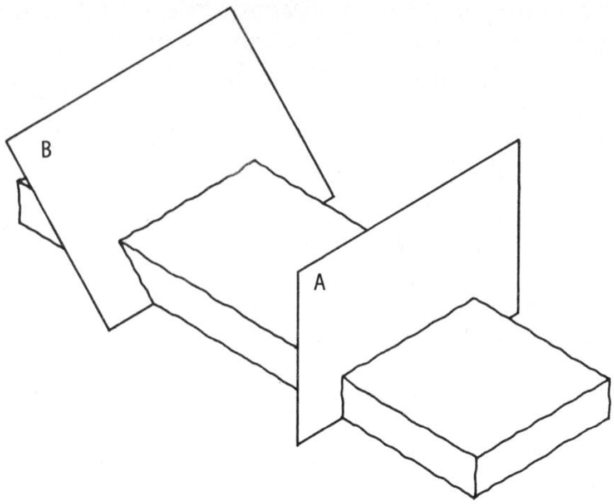

Figure 2-4. – Cross sections of a surface: A at normal incidence and B at an angle to surface.

Optical microscopy can be used to study the details of surface topography both in normal incidence and tapered sections as already indicated. In addition, optical microscopy can be used for in situ study of tribological surface behavior. Sliney, for example, incorporated an optical microscope in a microtribometer to study two solid surfaces in contact in the presence of a liquid oil film (ref. 3). With this technique the interface between two solid surfaces can be examined optically. It is necessary, however, for one of the two surfaces in contact to be transparent.

In Sliney's experiments a disk of sapphire or glass was used in contact with a steel ball. The two specimens are shown schematically in figure 2-5 with the optical microscope. In the upper portion of figure 2-5 the steel ball is shown being loaded against one side of a glass disk surface. The point of contact between the steel ball and the glass disk is viewed by an optical microscope through the glass disk. An examination of the microscope scheme shows a microscope objective (corresponding to that shown in fig.

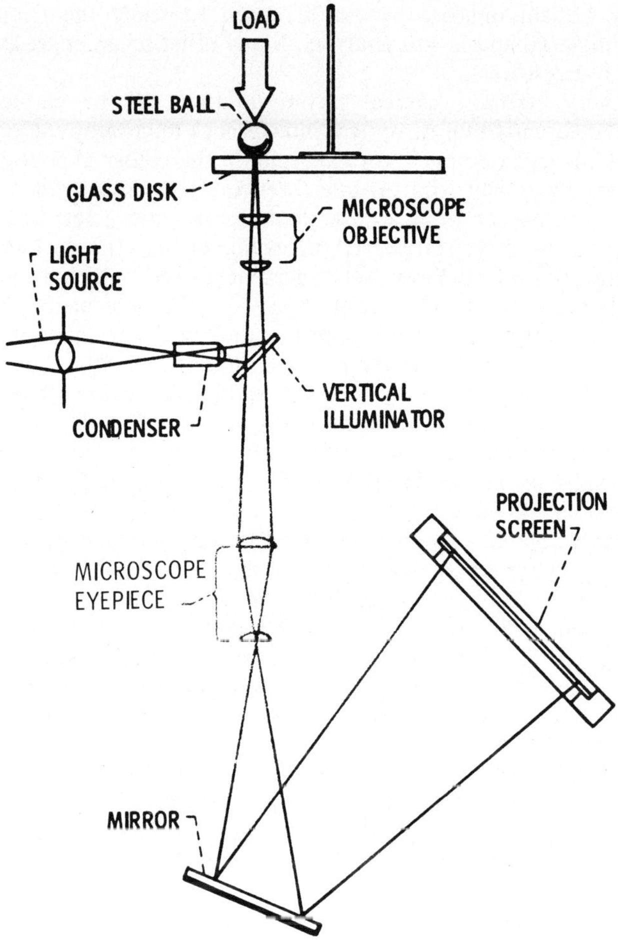

Figure 2-5. – Schematic of optical system (ref. 5).

2-2) and a microscope ocular. In addition, there is a light source in the vertical illuminator (fig. 2-5) to assist in supplying sufficient light to the ball-disk contact so that the image of the contact region is picked up by a mirror surface and then transmitted to a projection screen where it can be viewed.

The disk specimen is mounted on a shaft that can be rotated at relatively slow speeds to allow viewing the ball-disk contact region in sliding or rolling motion. The steel ball can be fixed to permit sliding between the ball and glass surface, or it can be fixed to allow rotation of the ball against the glass disk surface. Such a device can be used to study Hertzian contact, elastrohydronamic lubrication, conventional liquid lubricants, and the effect of foreign particles in the lubricant. The apparatus can also be used to study the behavior of solid film lubricants. This is just one example of how an ordinary optical microscopy can be used to study the tribological behavior of material and in situ analyses. Many other schemes are available to the inventive researcher.

Another very effective optical technique that can be employed in tribological research is that of optical interference microscopy. Optical interference effects are observed, for example, in the colors of soap bubbles seen in the sunlight. This optical interference effect occurs when light is reflected by transparent films whose thickness is only a few angstroms. Such thin films may have refractive indices higher than those of adjoining media (for instance, oil slicks on wet pavement) or they may have lower indices (for example, the air films trapped between the elements of two glass surfaces). Color fringes are readily seen in either case, and they arise by interference of beams reflected from upper and lower film boundaries. Because these films are very thin, the resulting differences in path length or, equivalently, angular phase must be treated coherently. The optical interference microscope can be effectively used in the studies of microtopography as well as in studies of the presence of films and their thickness on solid surfaces.

The optical interference microscope is a relatively simple instrument (fig. 2-6). In figure 2-6 S is a light source, essentially monochromatic for Fuseau fringes or polychromatic for fringes of equal chromatic order, and C is a columnator from which light passes to a beam splitter and falls on the reference surface R. The test surface T is inclined at a small angle for Fuseau fringes or effectively parallel for fringes of equal chromatic order or for interference contrast. A magnifying system or spectrograph is M. For

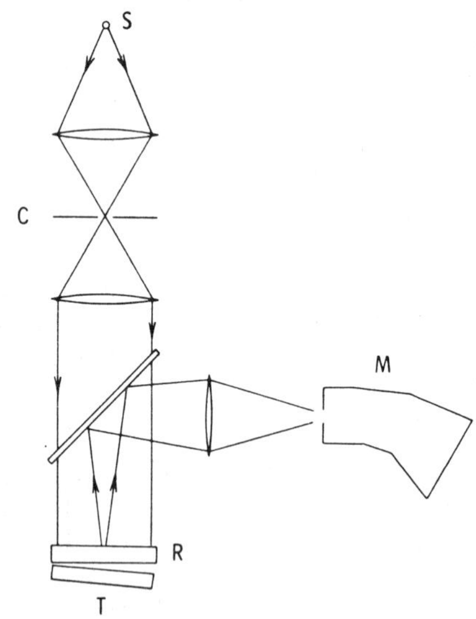

Figure 2-6. – Optical interference microscope.

transmission fringes, the beam splitter is removed and the observation system alined to receive light directly from the test specimen which must, therefore, be transparent to a certain amount of light.

Suppose that R and T are ordinary surfaces inclined to enclose a small wedge of air between them. Then, in accordance with familiar principles associated with interference microscopy, a series of lines or fringes is formed. The successive orders N appearing at wedge thicknesses is given by

$$N\lambda = 2\mu T \cos \theta$$

in which λ is the wavelength, μ the refractive index of the medium, and θ the angle of incidence. At these N values, fringes are dark because of the π change in phase between reflections at R and T, the one at a glass-air interface and the other at an air-glass interface. Thus, we have

$$\mu = 1$$

$$\cos \theta = 1$$

A series of equally spaced linear fringes appear, each neighboring one corresponding to an increase or decrease in T by $\lambda/2$. As to which is the increase and which is the decrease can be assessed by several methods; the common practice is to press down, say with the hand, on the side of R, wherein the fringes move N to the left if R is the thicker side of the wedge. If the contrary is true, the fringes move to the right.

One very good use of the optical interference microscope is to examine defects in the surfaces of solids and to gain some insight into the relative thickness of surface imperfections. This can be seen with the aid of figure 2-7. Take a work piece or a disk specimen with a scratch or a groove in the surface, such as that shown in figure 2-7(a), intersected by a series of inclined planes from the optical interference microscope (i.e., the light

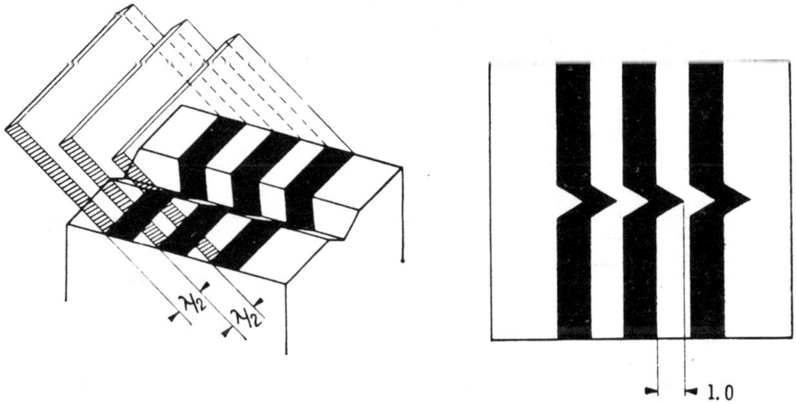

Figure 2-7. – Interference microscropy.

comes in at inclined planes and is separated by one-half of a wavelength). The intersection lines between these planes and the work piece have the same pattern as the interference bands; thus, an image is then seen (fig. 2-7(b)). If there is a scratch or a groove in the surface of the test specimen or work piece, the bands deviate to an amount equal to or directly proportional to its depth. The depth of such a groove is determined by multiplying the band deviation (measured in fractions of the band spacing) by half the wavelength of the light used. Figure 2-7(b) shows the depth of the groove to be 1×10^{-6} centimeter. Practical surfaces always contain scratches, large and small, and interference bands act as profile lines with very high magnification of the depth of these surface imperfections. The spacing and direction of the bands on the surface of the test specimen can be adjusted at will by using two glass wedges as shown in figure 2-6.

The optical interference microscope can also be used to examine concentrated contacts (e.g., a ball in contact with a flat surface) as might be experienced in tribological systems. Sliney, using the apparatus shown in figure 2-5, very effectively used the interference microscope to study concentrated contacts. If the glass surface in figure 2-5 is coated with a thin metallic film (e.g., silver or chromium) to gain the proper amount of light reflectance, interference microscopy can be very effectively achieved. Figure 2-8 shows the view obtained when a steel ball is brought in contact with the glass disk and pressed against the disk surface; the disk-ball contact zone is viewed through the disk from the back side. In this figure the center region, which is black, is the contact zone of the ball with the glass surface. This is the area where elastic deformation under an applied load has taken place between the steel ball and the glass disk surface. The rings out beyond

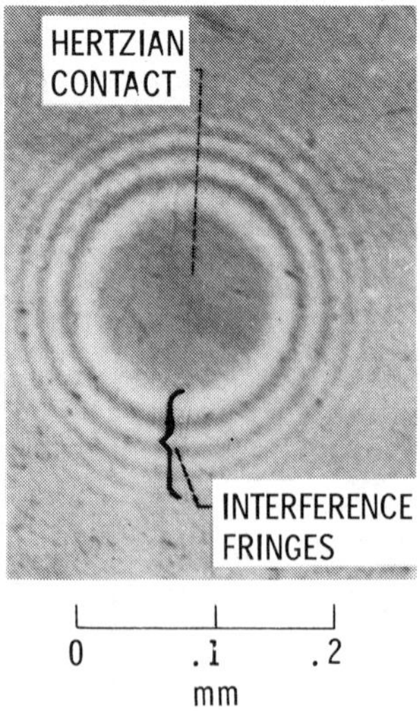

Figure 2-8.—Static contact at 4.4 newton load. Central circle is contact area formed by elastic deformation at contact of tool steel ball on glass flat. Original magnification, 250.

the center, black, circular region are optical interference fringes caused by the divergence of the ball away from the flat surface of the glass disk. These rings, which are called Newton rings, are nothing more than contour map lines for a spherical hill. Such contours in a geographical map are obtained by cutting the Earth's surface features by a succession of equidistant parallel horizontal planes. Similarly, Newton rings can be considered as the contours resulting from cutting the lens' spherical surface by parallel planes $\lambda/2$ apart.

The optical interference microscope can be used to study other characteristics than defects in surfaces and single solid surfaces in contact with a second surface. For example, the interaction of three solids can be examined with the interference microscope (ref. 3). When a third solid enters the contact zone (see fig. 2-8), the particle and its path can be followed with the optical interference microscope. This observation is demonstrated by the photomicrographs in figure 2-9. In figure 2-9(a) the black center spot due to the loaded contact of a steel ball against a glass surface is seen. Beyond the black center circle are the Newton rings, and in the lower left corner (at 7 o'clock), a small black particle (in this particular instance a glass wear particle) is observed to interrupt the Newton rings. If the glass disk is moved slightly, the position of the particle of debris moves; that movement can be followed relative to the solid-state contact with the optical interference microscope. In figure 2-9(b) the particle is shown entering the contact zone between the ball and the glass plate. The glass particle moves through the contact zone and exits on the opposite side, as is indicated in figure 2-9(c).

From the foregoing discussion on optical interference microscopy it is apparent that the optical interference microscope is a very effective tool in tribological studies. It provides a way to examine surface defects such as scratches and surface imperfections including fracture and fatigue cracks; in addition, it provides a way to examine solid-state contact between tribological components and the nature of that contact. Furthermore, it is a tool which can be utilized to study the interaction of three solid surfaces, the two involved in the tribological system and a foreign third surface such as that which may be encountered in abrasive wear. The information realized from the examination of tribological surfaces using the ordinary optical microscope can be enhanced considerably if the microscope is coupled with other surface techniques.

Surface Etching and Chemical Reaction

Ordinary chemical reagents can be used to etch solid surfaces to bring out various features of the solid in the ordinary optical microscope. With metal surfaces, for example, conventional chemical etches like nital (nitric acid and ethyl alcohol) are very effective in bringing out characteristic features of a solid surface such as grain boundaries and other structural characteristics. The chemical etchant or reagent for bringing out various species or structures of a solid surface is somewhat specific to the particular

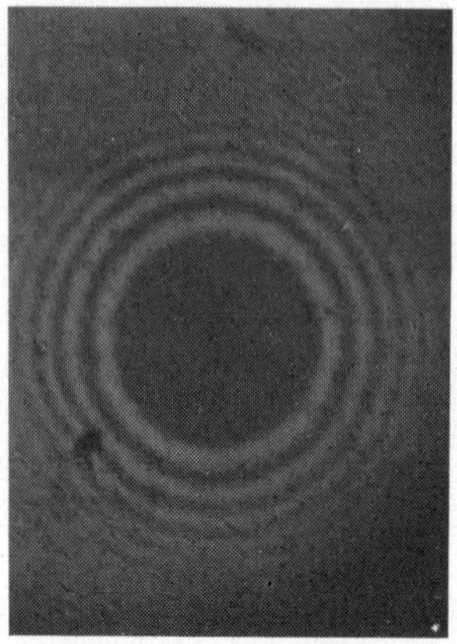

(*a*) *Particle at inlet.*

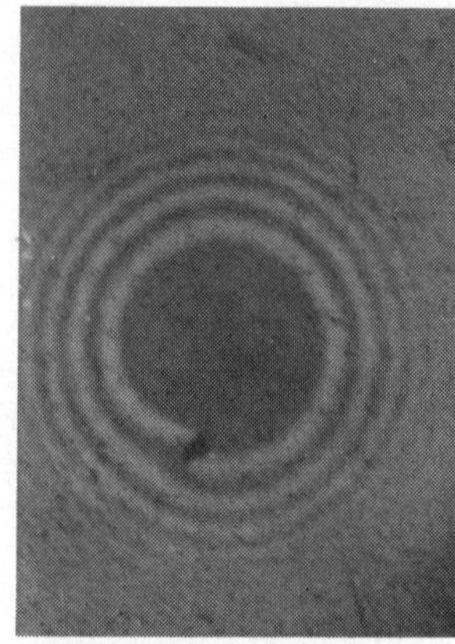

(*b*) *Particle in contact.*

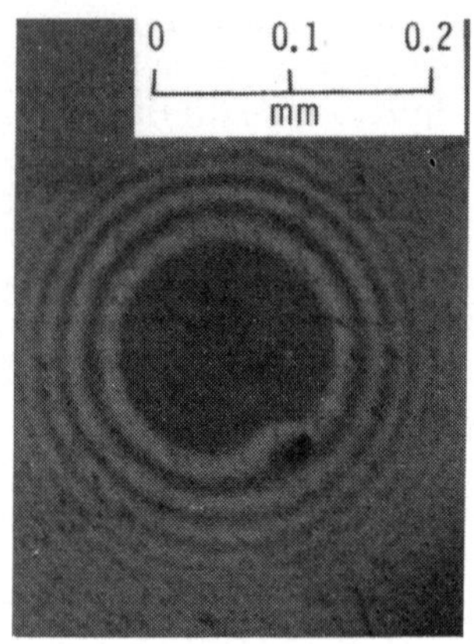

(*c*) *Particle exiting contact.*

Figure 2-9. – Passage of glass wear particle through periphery of concentrated contact. Load, 4.4 newtons; original magnification, 250.

metal or material involved. That is, selective reagents must be used for particular metals or materials to achieve the effect of showing a particular type of surface structure (e.g., grain boundaries, second phases, or subsurface defects). *Smithell's Handbook on Metallurgy* (ref. 4) and the *American Society for Metals Handbook* (ref. 5) are good sources for the desired chemical reagents for obtaining the surface state desired for examination in the optical microscope.

The etching is accomplished because of the differences in the energy states in the solid surface. If an etchant such as an acid is placed on the surface to bring out grain boundaries, which are generally high-energy sites, the acid preferentially reacts with those high-energy sites more rapidly than it reacts with other regions of the solid surface, such as the surface of the grain.

Dislocations are surface defects which exist in all real solids. In tribological systems it is frequently desirable to know the amount of deformation that has taken place in a solid surface. Etch pitting used in conjunction with optical microscopy can bring out the dislocation sites on a solid surface and reflect a change in dislocation behavior in the solid. Dislocations are line defects in the solid and they are sites on the surface of higher energy state. Thus, they interact or react more rapidly with certain chemical agents than do the bulk grain surfaces. As a result, pits or cavities are formed on the solid surface at the dislocation sites; examining a solid surface carefully with dislocation etch pitting can give information about dislocation activity on the solid surface.

It is possible to gain some insight into the total concentration of dislocations in a material from the concentration of dislocations in a particular area of the solid surface. Dislocation movement on the solid surface can also be followed and, furthermore, the effects of deformation of the solid can be seen by examining the dislocation structures. Figure 2-10 shows a single crystal (001) surface of lithium fluoride on which a sapphire ball was dropped. The point of contact between the sapphire ball and the lithium fluoride surface is in the center of the photomicrograph. The small squares throughout the entire photomicrograph are dislocation etch pits brought about by etching the lithium fluoride surface with an etch pit reagent. The

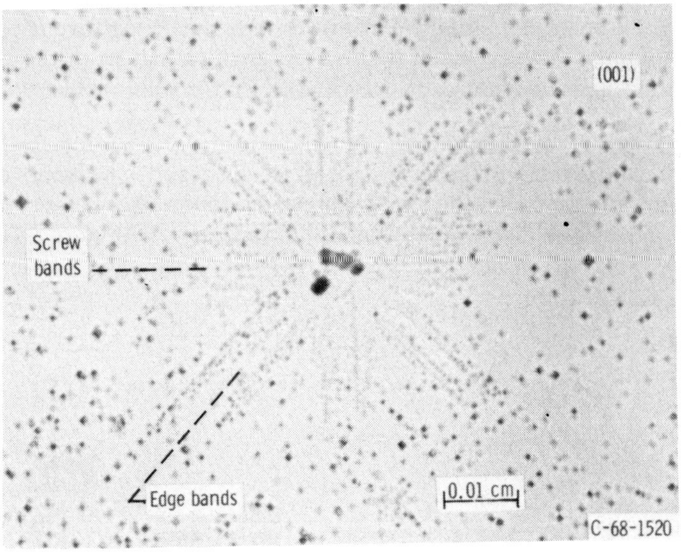

Figure 2-10. – Dislocation band developed by dropping 1.6-millimeter sapphire ball on lithium fluoride (001) surface.

normal distribution of the dislocations in the lithium fluoride crystal can be seen from the concentration of the small black squares throughout the matrix of the single crystal surface.

In the center of the photomicrocraph, where the sapphire ball contacted the lithium fluoride crystal, a few very dark or black spots are indicated; emanating from those points are lines of dislocation etch pits. The surface was etch pitted twice. It was initially etch pitted for background dislocations. A careful examination of the photomicrograph reveals that the background dislocations in the bulk of the crystal surface are twice as large as the new dislocation etch pits generated by the second etching of the solid surface after the sapphire ball contacted that surface. The dislocations on the solid surface generated as a result of the plastic deformation of the lithium fluoride on impact of the sapphire ball form distinct patterns on the surface and can be associated with distinct types or characteristic types of dislocation behavior.

The bands of dislocations, as revealed by the etch pits running in the horizontal direction from the point of impact, are associated with screw dislocations that are present in the crystal surface as a result of plastic deformation. The rows of etch pits which run at a 45° diagonal in the photomicrograph are associated with edge dislocations. Just as the difference in the dislocations that are present in the bulk crystal prior to deformation by impact with the sapphire ball can be detected by the sequential etching of the solid surface (the difference being detected by the size of the dislocation etch pit, the original dislocation etch pits were larger as a result of having received a second etching), the deformation process itself can be followed using an ordinary optical microscope, relatively modest magnification, and sequential etch pitting of the solid surface. When, for example, the surface is initially deformed, one can etch pit for the presence of dislocations, as was done in figure 2-10, giving a pattern of the array of dislocations and the manner of deformation of the solid surface as well as the extent.

On subsequent deformation of the solid surface, the surface can be etch pitted again with etch pit reagent and, if the dislocations have moved with the deformation process, the movement can be followed because the size of the etch pit changes. This is indicated in the schematics in figure 2-11 (from

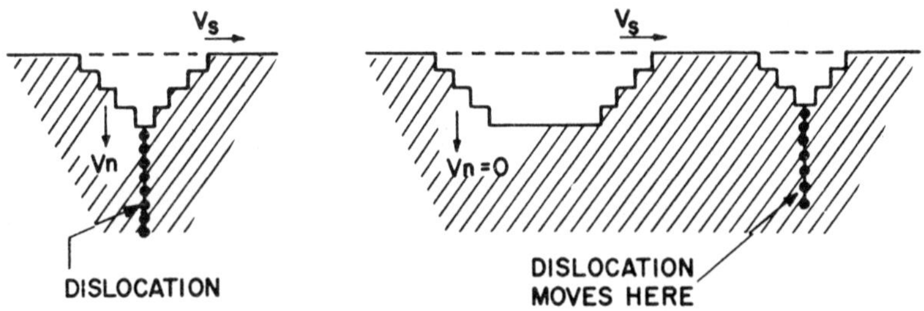

(*a*) *Initial dislocation position.* (*b*) *Final dislocation position.*

Figure 2-11. – Etch pitting to follow dislocation movement (*ref. 6*).

ref. 6). The original dislocation etch pit is shown in figure 2-11(a). The pit forms as a result of the chemical reagent attacking the surface at the dislocation site, a higher energy site than the bulk grain, resulting in a tear-shaped pit or step. When the surface is deformed further, the dislocation moves. When the dislocation moves, a new dislocation site is revealed, as indicated in figure 2-11(b), as a new, small dislocation pit. The former location of this pit is now revealed as a much larger etch pit; the larger size of the older etch pit occurred as a result of the second etching of the surface. Careful examination of the etch pit with the optical microscope can indicate the actual orientation of the planes on which the dislocations originate. This can be arrived at from the symmetry of the dislocation etch pit on the solid surface. And if the orientation of the solid surface is known, then the orientations of other planes relative to the surface plane are known. From this information and the orientation of the etch pit, the source of the dislocations can be derived (ref. 7).

In figure 2-12 the asymmetry of etch pits due to inclination of the etch pit with the solid surface are depicted schematically. When the dislocation lies normal to the solid surface, the etch pit is fairly symmetrical, indicating that this location is on a plane normal to the solid surface. When the dislocations are inclined at some angle other than normal to the solid surface, dislocation etch pits are asymmetrical, and the asymmetry is oriented in a particular direction depending on the planes from which dislocations arise. When the dislocations lie on more than a single set of slip planes, the dislocation etch pits are oriented in different directions, as indicated in figure 2-12(c). In figure 2-12(c) the pits form at the emergence of a hexagonal grid of dislocations. The pits are asymmetrical and successive pits are oriented differently.

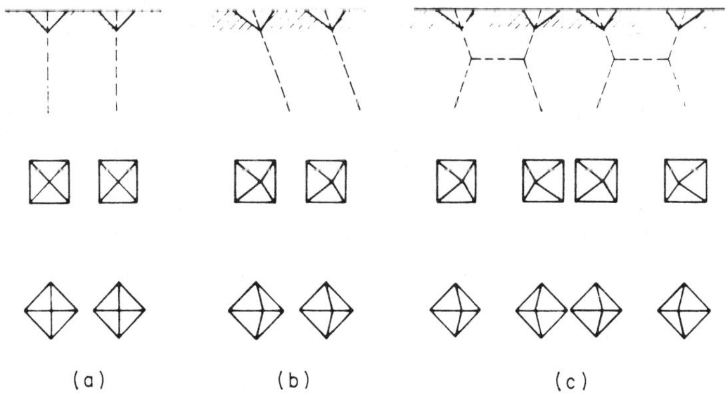

(a) Parallel lines perpendicular to surface produce symmetrical pits.

(b) Parallel lines inclined with respect to surface, as in a tilt boundary, produce asymmetrical pits all oriented same way.

(c) Pits formed at emergence points of hexagonal grid of dislocation; pits are asymmetrical, and successive pits are oriented differently.

Figure 2-12. – Asymmetry of etch pits due to inclination with respect to surface of dislocation lines giving rise to them (ref. 7).

There are other techniques available for examining surfaces for dislocation behavior such as electron microscopy, X-ray techniques, and decoration, but the etch pit has an advantage in its simplicity. All it requires is a simple chemical reagent and an ordinary optical microscope, and these are available to most individuals interested in examining tribological surfaces. Etch pitting is compared with other techniques for estimating dislocation densities in table 2-I. In this table the techniques include electron microscopy, X-ray transmission, X-ray reflectance, decoration, and etch pitting. Notice that with all techniques other than etch pitting the specimen thickness is critical; with etch pitting, however, there is no limit to the specimen thickness. This is very important, of course, in tribology. Many times the solid surfaces to be examined are fairly thick and cannot be destroyed or sectioned. Thus, etch pitting in these instances has a distinct advantage over the other techniques available. The width the image can be when using the etch pitting technique is actually the limit resolution of the etch pit itself; it happens to be approximately 0.5 micrometer.

TABLE 2-I. – METHODS FOR ESTIMATING DISLOCATION DENSITIES[a]

Technique	Specimen thickness	Image width	Maximum practical density, per cm^2
Electron microscopy	>1000 Å	~100 Å	10^{11} - 10^{12}
X-ray transmission	0.1 - 1.0 mm	5μ	10^4 - 10^5
X-ray reflection	<2μ (min.) - 50μ (max.)	2μ	10^6 - 10^7
Decoration	~10μ (depth of focus)	0.5μ	2×10^7
Etch pits	No limit	[b]0.5μ	4×10^8

[a]Based on W. G. Johnston. Prog. Ceramic Sci. 2, 1 (1961).
[b]Limit of resolution of etch pits.

The real limitation in the etch pitting technique is its inability to detect high concentrations of dislocation densities. It is limited to observing dislocation densities (per cm^2) of approximately 10^8. With electron microscopy, however, dislocation concentrations of the order of 10^{11} and 10^{12} can be effectively studied. The etch pitting technique is limited to 10^8 because when etching the solid surface the dislocation concentration becomes so high that the etch pits begin to run together and it is very difficult to distinguish and count the separate individual etch pits in a solid surface, even with the highest magnifications of optical microscopy.

The use of dislocation etch pitting for following the deformation behavior of solid surfaces and solid-state contact in tribological systems is of sufficient importance that a complete table of etchants which can be used for etch pit dislocation detection is presented in an appendix to this chapter. There are many different materials listed therein which can be used for etching various surfaces including metals, nonmetals, and semiconductors.

The ordinary light microscope is a very effective tool for examining the solid surface orientations in a relatively crude fashion on the surfaces of solids. Particularly with single crystals, the various orientations on the solid surface can be determined to some extent by simply oxidizing to a limited degree the metal surface in an oxidizing environment and then examining the surface in the optical microscope. On a single crystal metal surface, for example, the high atomic density areas or planes (i.e., the low surface energy planes) oxidize at a much slower rate than do the low atomic density, high surface energy planes. This is because the high atomic density, low surface energy surfaces are more stable and more resistant to interaction with the environment. As a consequence, the surface oxidizes in a patchy fashion. Those areas which have low atomic densities and high surface energies contain relatively thick oxide, and those areas containing relatively high atomic density, low surface energy planes have a very thin oxide layer.

With controlled oxidation of the solid surface and with information such as that available in Kubaschewski and Hopkins (ref. 8) on oxidation, it is possible to gain some information as to the relative thickness of the oxides in these various areas by a color comparison. Figure 2-13 is a stereographic projection of a single crystal surface showing some of the orientations; the diagonal cross hatching indicates some of the areas of high oxidation and those of much lower oxidation. The stereographic plot in figure 2-13 is for a copper single crystal surface (ref. 9). Since copper is a face-centered-cubic metal, it is anticipated that the {111} planes (these are the high atomic density, low surface energy planes) would contain the least amount of oxide on their solid surface. In confirmation, these are represented in figure 2-13 as the clear (starlike) regions, four areas surrounding the central portion of the figure at the (001) location.

The regions of maximum oxidation would occur on the {011} and {110} areas. These are the cross-hatched areas in figure 2-13. There are four such areas located in and around the central point of the (001) surface. They are indicated as irregular diamond-shaped regions or patches. In addition, there are somewhat triangular regions near the periphery in four areas. The oxidation of the (001) surface, the central region of the figure (a somewhat clover leaf pattern) would be intermediate between that of the {111} surfaces, which have the least amount of oxidation, and the {011} surfaces, which have the maximum amount of oxidation on the surface.

A mild oxidation by heating in air can be used to look at the orientations on a polycrystalline surface as well. The high atomic density grains, those grains which have high atomic density and low surface energy, oxidize much more slowly than those grains having low atomic density and a high surface energy. Thus, the surface of the polycrystalline solid when oxidized may well appear to be spotted or patchy as a result of the variations in orientation on the solid surface; the variation and thickness of the oxide on the individual grains is the determinant. This technique can be used for examining the amount of texturing that has occurred on a metal surface as a result of deformation. It provides a relatively simple, inexpensive way of examining texturing by the amount and uniformity of the oxide that forms on a metal surface in a contact region; it is therefore unnecessary to resort to more expensive tools such as X-ray techniques.

One can become a little more sophisticated and develop the X-ray technique to the point where the environment is very carefully selected for the interaction with the specific metal surface. This can be done to bring about surface colorations which will give an indication of gradations in the degree of thickness of the films formed on a solid surface, thereby giving a relative rating of the orientation of the solid surface grains in a polycrystalline

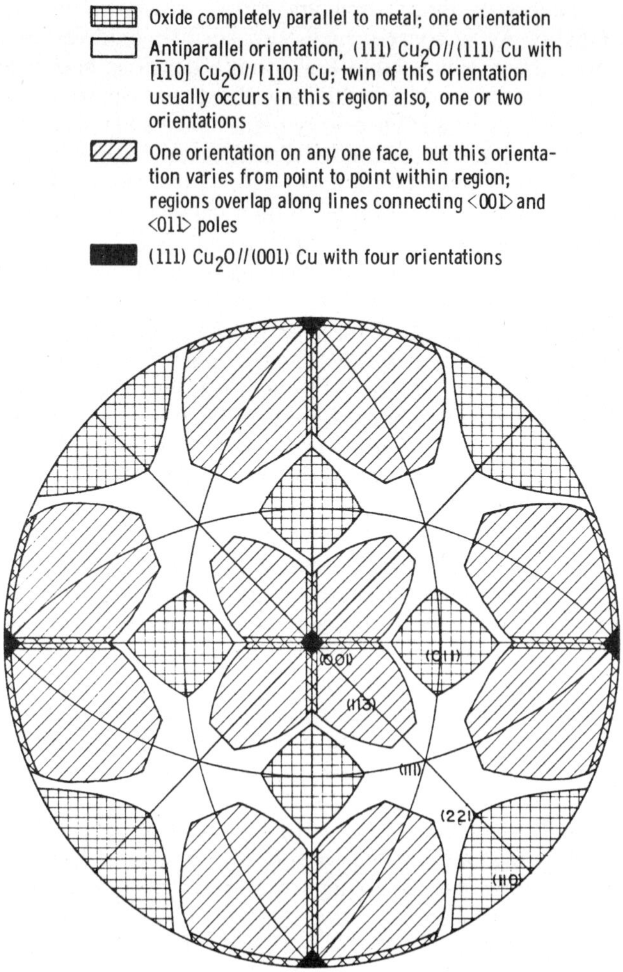

Figure 2-13.—Stereographic plot of cuprous oxide orientation regions on copper single crystal (ref. 9).

material. Gold reacts with chlorine to form a range of gold chlorides having variations in colors with thickness from a pale yellow to a dark brown with intermediate orange and green layers. By controlled chlorination of a polycrystalline gold surface, one can characterize the surface by the colors produced in the various grains.

Mechanical Surface Topographical Devices

The most widely used tool for analyzing the topography of surfaces today is, in all probability, the ordinary surface profilometer. The surface profilometer is an electromechanical device wherein a stylus passes across the surface and follows the contour of the surface to trace the irregularities. This is fed into an amplification system and then to a device that records the amplitude of the surface irregularities. The stylus tracing these irregularities then gives a profile of the topography of the solid surface. In the process of amplification, the vertical and horizontal magnifications of these irregularities in topography are magnified many times but at different rates (vertical magnification greater than horizontal). Generally, the stylus is a diamond that is attached to a motorized arm which moves at a constant speed across the solid surface to trace its profile. A typical stylus trace across a dent in a metal surface is shown in figure 2-14. The crater or dent in the center of the photomicrograph is the result of the diamond having followed the surface contour and traced out the surface irregularity produced by the indentation of an abrasive particle on the solid surface. Figure 2-14 shows that the magnifications are different in the horizontal and vertical directions; in the horizontal direction the magnification is one-fourth of that in the vertical direction (1 unit = 20 μm versus 1 unit = 0.5 μm).

With most of these surface profilometric mechanical devices, this difference in magnification in the horizontal and vertical directions generally exists. The traces or profiles of the surface, therefore, are not a true picture of the actual shapes of the surface irregularities or asperities because they become distorted by this variation in magnification. One must correct for this variation by interpreting the actual shapes of the asperities or peaks and valleys on the solid surface.

There are standards that have been prepared by those organizations that manufacture surface profilometers for reference purposes so that the

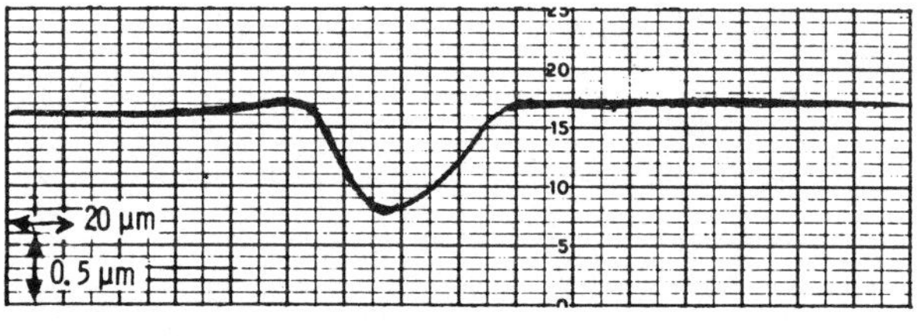

Figure 2-14. – Stylus trace across dent center.

researcher or the engineer can have standards for reference to his experimental surface. Some of these standards are presented in figure 2-15 (ref. 10). There are essentially three magnifications of a solid surface. The horizontal magnification is held constant in all three traces in figure 2-15. The vertical magnification, however, is varied. Notice that in figure 2-15(a) the vertical magnification is 50 000, in figure 2-15(b) it is 10 000, and in figure 2-15(c) it is only 5000. A considerable amount of information about the true topography of solid surfaces can be obtained by utilizing such surface profilometer techniques.

Williamson (ref. 11) has done an outstanding job in characterizing solid surfaces by tracing the solid surface in many directions with a surface profilometer and then feeding that information into a computer to generate a surface topographical map. He has done this with a number of different solid surfaces and then generated contour maps very analogous to those one might find for geographical data. As a result of his work it has become apparent that, contrary to earlier thinking, the true nature of the solid surface asperities for most metals is not sharp peaks or spikes but rather relatively

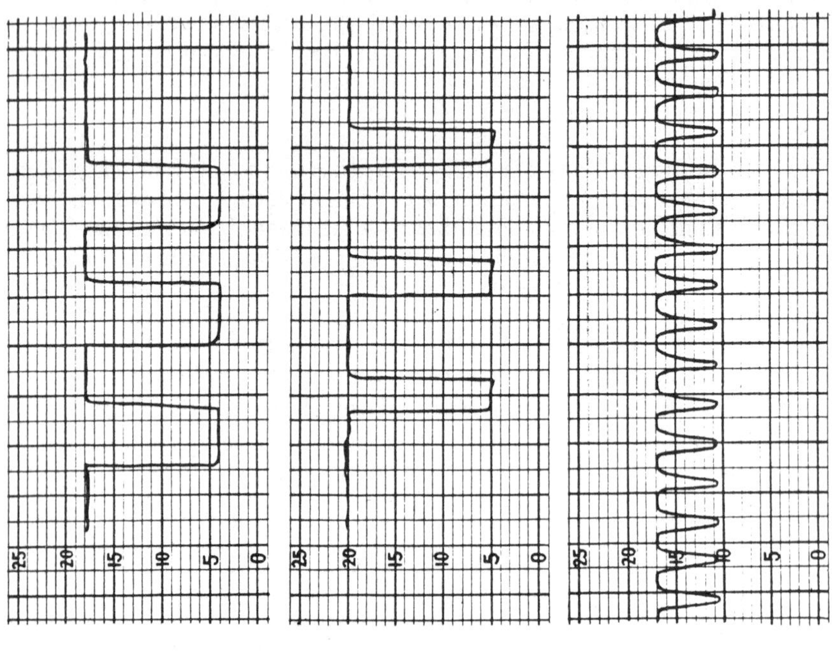

(a) (b) (c)

(*a*) *Vertical magnification, 50 000; one division represents 0.04 micrometer.*

(*b*) *Vertical magnification, 10 000; one division represents 0.2 micrometer.*

(*c*) *Vertical magnification, 5000; one division represents 0.4 micrometer; centerline average, 0.89 micrometer.*

Figure 2-15. – Surface profile standards. Horizontal magnification, 100 (ref. 10).

broad based hills with angles of inclinations from the base line of the solid surface of approximately 15°. The real surface then can be represented more correctly by a series of bumps on the surface as opposed to adjacent spikes.

There are some inherent limitations in using the mechanical stylus technique to measure the profile of solid surfaces. For example, the stylus has a radius and thus, for very sharp surfaces, it is very difficult to follow the true surface profile. As a consequence, some modifications for the existence of the stylus technique must be incorporated in following the topography of very rough surfaces (containing very irregular surface spikes or peaks). An example of such a surface might be a grinding wheel or an abrasive paper. The profile of that surface is very irregular and would be very difficult to follow with a conventional profilometer. Techniques have been developed, however, for characterizing such abrasive surfaces using a profile measuring system coupled with a computer. In one such scheme the stylus oscillates while the surface below moves incrementally so that the surface is stationary when contacted by the stylus. Both stylus and surface motion are controlled by a digital computer. Measurements of surface elevation are the input to the computer for digital processing and the various surface characteristics are computed. Examples of the types of profile that can be obtained are presented in figure 2-16. The profiles in figure 2-16 are for

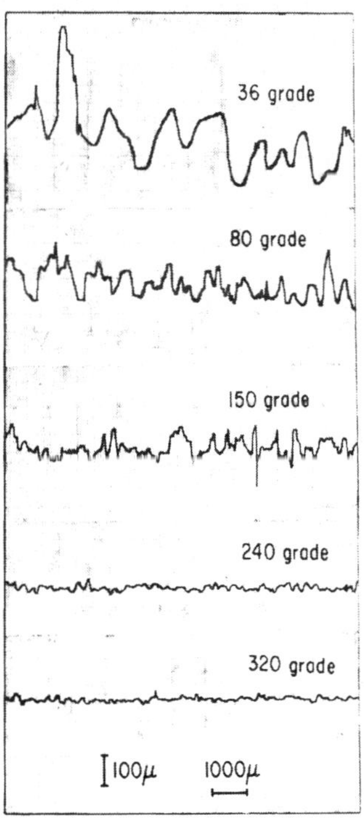

Figure 2-16. – Profiles of coated abrasive samples (ref. 12).

coated abrasive samples containing various grade abrasives (ref. 12). The figure shows progressively (from top to bottom) 36, 80, 150, 240, and finally 320 grade abrasive surfaces. As the surface topography reduces in roughness, that is, as the abrasive paper becomes finer, the surface topography becomes much smoother with less and less surface irregularities. Note again, however, the variation in magnification. The magnification in the vertical direction is 10 times that in the horizontal direction (1 unit = 100 μm versus 1 unit = 1000 μm), as indicated by the scales in figure 2-16, so that we again have a distorted picture of the true nature of the solid surface. Despite the assistance of the computer in helping to increase the utility of the mechanical surface profilometer, it still has inherent limitations because of a mechanical stylus tracing across a solid surface.

An interesting device for measuring surface topography, particularly of ultrafine surface topography without solid-state contact of a stylus with the surface, is the one developed by the National Bureau of Standards and frequently referred to as the field emission ultra micrometer (ref. 13). This micrometer is an arrangement of field emission electrodes wherein the field emission pin tip (which is very analgous to the stylus) is placed above, but not in direct contact with, the solid surface in a well evacuated chamber. The solid surface as well as the emission tip are connected to a constant current electrical circuit, which indicates a voltage that is directly related to the distance between emission tip and solid surface. A more sensitive but much more limited device would apply a fixed emitter anode voltage to measure the dependence of emission current on the emission tip to solid surface spacing. This instrument can operate in spacings as small as a few hundred angstroms. With suitable calibration, distance measurements in the range of 10 to 10^{-2} micrometer can be reproduced to one part in 10^5 and can be expected to have accuracies limited only by available calibration techniques. The instrument is most useful as a null or differential distance measuring device. It has the unusual property that resolution improves over several orders of magnitude as the null point is approached. Since the instrument contains no optical or mechanical lever systems or delicately balanced bridges, it has inherent long-term stability.

A surface topographical map obtained with a field emission ultramicrometer is presented in figure 2-17 (Vedam, ref. 17). The series of contour lines (white lines on a black background) indicates the surface profile or surface contour of the solid.

There are other devices for examining the profile of solid surfaces, and these include the recording optical lever (ref. 14), the capitance measuring devices (ref. 15), and the magnetic flux sensors (ref. 16). These latter devices, however, suffer from one inherent limitation not experienced with the field emission ultramicrometer; they require physical contact with the surfaces to be measured.

Figure 2-17.—Topographic map of 180-line-per-millimeter diffraction-grating replica, obtained with "topografiner," a noncontacting field-emission probe (ref. 17)

Radioisotope Techniques

The introduction of nuclear reactors in the early 1950's has made radioistopes readily available, and this, in turn, has led to extensive use of radio tracers and characterization techniques with them. The extension to surface studies is a natural one and these techniques have been widely applied to solid surfaces. In the field of tribology, Rabinowicz and Tabor (ref. 18) have very effectively used radio tracers to study friction and wear surfaces. The basic principal involved in the radio tracer technique is that the radioactivity of a substance is a nuclear property and as such has a negligible effect on the outer valence electrons of the atom. Since these outer electrons control the chemical properties of an atom, a radioactive isotope (e.g., Fe_{59}) behaves chemically in exactly the same manner as stable isotopes of that element (e.g., Fe_{54}, Fe_{56}, Fe_{57}, and Fe_{58}). If the separability of nuclear and chemical properties is used, it is possible to trace the path of a particular element through a chemical system by following the radioactive isotope of that element.

The further ability to control the number of radioactive atoms in a given isotopic mixture provides a powerful quantitative tool for the researcher. This property is referred to as specific activity and is generally described in units of activity per weight—for example, counts per minute per microgram (cpm/μg) or millicuries per gram (mCi/g). The sensitivity or minimum detectable quantity can be controlled by regulating the amount of radioactivity used in the experiment. This is accomplished by the controlled addition of radioactive species; for example, although the addition of 10^6 cpm of activity to either 1 gram or 1 milligram of a stable species changes the sensivity by a factor of 1000, the total amount of activity is the same in both cases. Certain radioisotopes can be obtained "carrier free," or essentially free of any stable isotopes. These carrier free radioisotopes provide radio tracer sensitivity on the atomic scale; however, experimental problems are very difficult because of the few atoms in the system.

The techniques used in the radio chemical characterization of the surface utilize all of the standard radio chemical technology. In brief, the process involves three steps. The first is adding the radioactive isotopes to the experiment and ensuring that, chemically, calibration has been established between the active and stable isotopes. The second is to conduct the experiment itself, and the third is to perform a radio assay (e.g., beta, gamma, or alpha counting) to locate and measure the amount of radioactivity at any point in the experiment.

Friedlander, Kennedy, and Miller published their book *Nuclear and Radio Chemistry* in 1955 with a second edition in 1964 (ref. 19). Today this book remains a classic text in the techniques of radio chemistry. Oberman and Clark cover much of the experimental detail required for studies using radioisotopes (ref. 20). Radioisotope techniques also include the use of autoradiography, which is a service analytical technique that can complement other radioisotope techniques. It can be used very effectively in conjunction with radio tracer studies to obtain information on the topographic distribution of the radioactivity across the surface. This technique normally gives semiquantitive information. Under favorable conditions, however, quantitative information on radio tracer species may be obtained.

The advantage of autoradiography is that a picture is obtained of the distribution of the radioactive species on the solid surface; thus, it is a particularly useful method for surface analysis. Autoradiography is very simple in principal. However, as will be discussed, the technology tends more to be an art with highly specialized emulsions and dark room techniques. In principal, the technique is shown schematically in figure 2-18 where a radioactive sample is brought into intimate contact with a photographic emulsion. The ionizing radiations in the specimen interact with the bromide salts in the emulsion. In the presence of a developing agent the activated sites catalyze the complete conversion of the silver bromide crystals to metallic silver. Unexposed silver bromide crystals are dissolved out of the emulsion by the developer. The entire technique depends on an efficient interaction between the ionization radiation and the photographic emulsion.

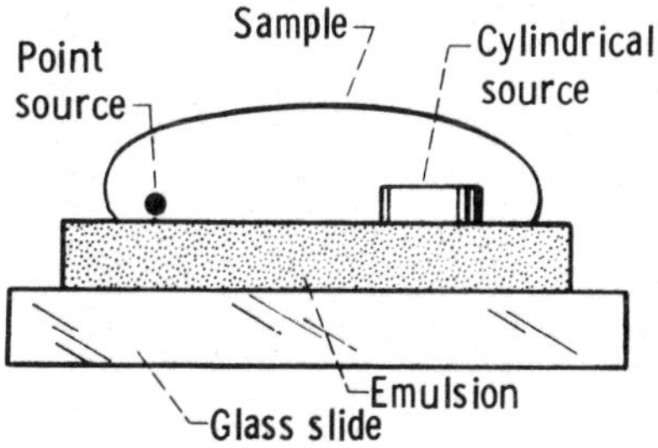

Figure 2-18. – Schematic illustrating technique of autoradiography.

The type of radiation (alpha, beta, gamma, etc.) has a profound effect on both the nature and clarity of the autoradiograph, as well as the technique used to obtain the best exposure. There is generally a compromise between resolution and exposure time. Unfortunately, the two criteria are mutually exclusive for any given radioisotope since large grain emulsions are required to minimize exposure times, while very fine grain emulsions are necessary to obtain maximum resolution. Virtually all autoradiograms are made from beta emitting radioisotopes. The use of radioisotopes in studying a wide variety of friction and wear problems has been conducted by Cohen (ref. 21). Since both friction and wear involve the rubbing of surfaces in solid-state contact, it is a particularly fertile area for the use of radioisotopes in autoradiography. Burwell and Strang (ref. 22), as well as Merchant and Ernst (ref. 23), recognized the usefulness of these techniques in friction and wear work in the very early 1950's and effectively used the same in studying surfaces.

X-ray Techniques

In general, X-ray techniques or methods are not thought of as being surface sensitive, since they penetrate the surface layers and move into the bulk of the solid. Today, tools such as Auger, XPS, and LEED are considered true surface analytical devices, whereas X-ray is thought of more as a bulk material analytical device. There are, however, some surface techniques that have been developed using ordinary X-rays. Some of these techniques are very useful in studying surface topographic characteristics. Probably the simplest of all is based on the Laue back reflection technique. This particular method is extremely surface sensitive, especially for high absorbing materials, since most of the information is derived from longer wavelength components of a polychromatic source. Many different geometrical arrangements using polychromatic radiation have been used but the Laue back reflection technique developed by Swink is probably of greatest interest here (ref. 24). The simple geometry of the Laue X-ray back reflection technique is shown schematically in figure 2-19(a). The polychromatic X-rays from a point source P are collimated to a 6 to 12 millimeter beam C and impinge on the specimen S at a 90° incident angle. The diffracted images I are then formed by back reflections in the same manner as in pinhole

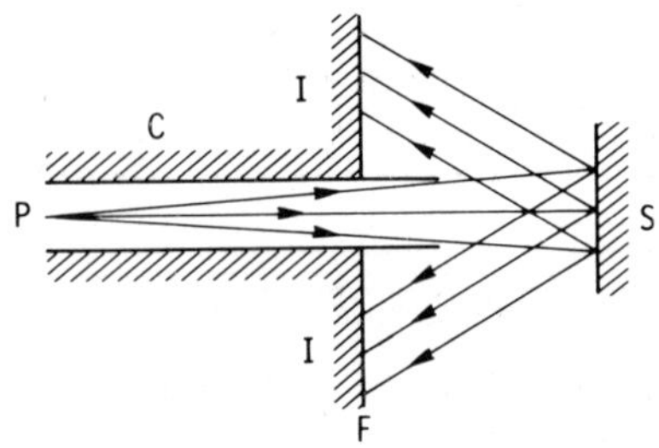

(a) Backreflection topography geometry.

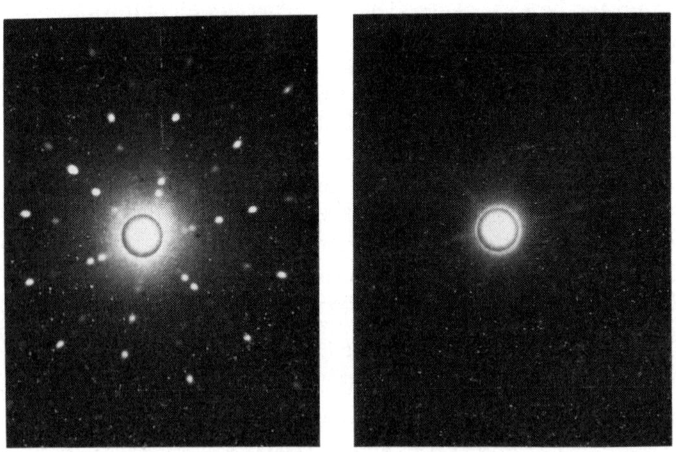

(b) Patterns for diamond (111) surface and silicon carbide (0001) surface.

Figure 2-19. – Laue technique.

orientation devices. Each set of {hkl} planes which diffract have a characteristic back reflection angle; this results in the {hkl} diffraction images separating on the film plane. Since any grain tilts, twins, or crystal deformations can be regarded as having an independent set of {hkl} planes, they are recorded on different portions of the film. Dislocation aggregates which do not constitute a grain mismatch sufficient to be separated in the more perfect parts appear as intensity contrasts in each image similar to monochromatic topographs.

Figure 2-19(b) contains two Laue patterns, the one on the left for a diamond (111) surface and that on the right for a silicon carbide (0001) surface. From the symmetry of the diffraction spots and the size and shape of these individual spots, both crystalline surfaces are nearly perfect, indicating very few defects or strain in the solids. The Laue X-ray technique not only provides information on the orientation of a particular single crystal surface or a grain in a polycrystalline sample but it can also give additional information relative to the amount of strain or deformation that has taken place in the solid surface. If the surficial layers have undergone strain or deformation to a small degree, the diffraction spots for a metal, analogous to those for the nonmetals diamond and silicon carbide (fig. 2-19(b)), would become arc shaped. That is, the diffraction spots would no longer be sharp spherical dots on the film but rather would become arc shaped and begin to form circular patterns about the central beam spot on the film.

Initially, with small degrees of strain or small amounts of surface strain, the diffraction spots would elongate to a small degree. As the amount of strain or deformation increases, the length of the arc of the diffraction spot increases. With fairly high degrees of strain the individual arcs would tend to join, forming nearly complete circles. If sufficient strain is put into the surface and the recrystallization temperature is sufficiently low so as to bring about recrystallization of the solid surface layers (resulting in the formation of a polycrystalline surface), the diffraction pattern would then no longer consist of individual spots from the single crystal surface but would instead consist of a series of rings joining the spots moving out from the center X-ray source. Thus, from a simple Laue back reflection photograph, one can obtain information on (1) orientation of single crystal surfaces, (2) grains in a polycrystalline sample, (3) degree of deformation of the surface or crystal perfection of the solid surface and, (4) information about the recrystallization process (if recrystallization has taken place).

In addition to conventional Laue back reflection topography techniques, there are also other devices which can use the Laue geometry. One that has been developed for polycrystalline thin films (of the order of 1000 to 10 000 Å thick) is the low angle Laue method (ref. 25). From this particular technique, the grain size and preferred orientation can be inferred in thin films. The geometry of this method is shown schematically in figure 2-20. The method is applicable to thin films since the low incidence angle φ gives a much longer path length with the beam in the film. A depth study of the film can be made by changing the angle φ. If the substrate is sufficiently thin, the diffracted rays from the film and film substrate interface are recorded on the lower half of the photographic film, whereas the informa-

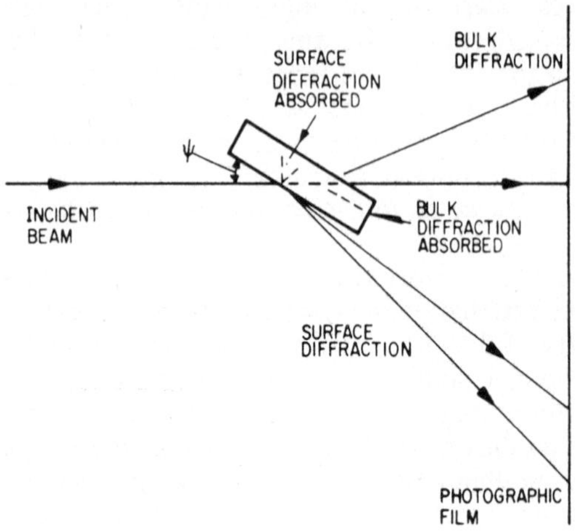

Figure 2-20.–Low-angle Laue geometry (ref. 25).

tion from the back of the substrate is recorded on the upper half. Hence, if the substrate is a single crystalline material, the Laue spots can be compared, and information on how the film is affecting the substrate can be obtained.

Most real crystalline solids can be considered as nearly ideally perfect materials with local perturbations of the perfection of the crystal lattice created by the defects present in the solid. One technique which photographically records the intensity differences between the perfect regions and the regions containing defects is referred to as X-ray topography. X-ray topography has a resolution of 1 to 10 micrometers which, for low dislocation density materials, results in a defect distribution map. For high dislocation densities (e.g., greater than 10^5), X-ray topography cannot resolve individual dislocations, but it is still useful for studying the size, locations, and relative orientation of grains, the existence, position, and habit of twins, and the direction and magnitude of crystal deformation.

The X-ray transmission and reflection techniques indicated in table 2-I as methods for estimating dislocation densities are examples of X-ray topography. You can see from reexamining table 2-I that, with X-ray transmission, the maximum practical dislocation density which can be observed is of the order of 10^4 to 10^5 dislocations per square centimeter, whereas with the X-ray reflection technique somewhat higher sensitivity can be achieved, of the order of 10^6 to 10^7 dislocations.

One of the most sensitive X-ray topographic techniques that has been developed for analyzing the surface topography of crystalline solids is that originally devised by Berg in 1931 (ref. 26). It was later modified by Barrett in 1945 (ref. 27) and further refined by Newkirk in 1959 (ref. 28). This particular technique gives resolutions of about 1 micrometer. The method is relatively simple and is depicted schematically in figure 2-21. It consists of

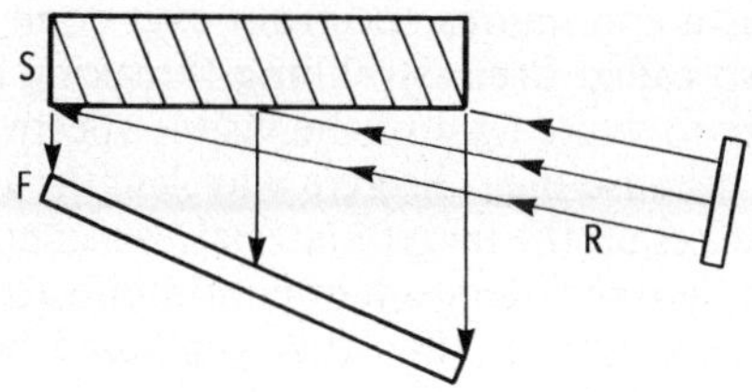

Figure 2-21. – Berg-Barrett geometry.

holding the specimen S in a roughly collimated X-ray beam R so that a set of {hkl} planes makes the proper Bragg angle for diffraction. The diffraction image is recorded on a high-resolution, fine-grained photographic plate F that is held as close as possible to the specimen without intercepting the incident beam R.

X-ray Fluorescence

While conventional X-ray techniques are not, strictly speaking, surface sensing devices, X-ray spectral chemical analysis is a surface technique in the sense that the X-rays being emitted (and measured) originate in a surface layer of finite thickness. The question is whether or not the thickness of the layer being studied is such that it constitutes a surface or a surface layer. Generally there is no hard and fast definition for what constitutes a surface. It depends on the concept of the individual examiner.

Since Moseley conducted his classic experiments on the relation of X-ray wavelength to atomic number back in 1913, X-ray fluorescence as an analytical technique has seen considerable use. With modern instrumentation and the use of soft X-rays, fairly thin surface layers can be probed, and it is now practical to make measurements in the 10 to 100 angstrom wavelength region. These results make X-ray fluorescence analysis a valuable tool despite the valence effects which trouble normally used soft X rays.

X-rays are electromagnetic radiation of specific wavelength, usually between and 0.1 and 250 angstroms, although these are not absolute limits. X-rays result from the interaction of high-energy photons or particles (electrons, protons, alpha particles, etc.) with the atoms of a material. The X-rays, as well as all electromagnetic radiation have an energy equal to the product of Planck's constant H and the frequency V ($E = HV$). Substituting for all of the constants in the equation shows the wavelength λ and the energy have the simple relationship $\lambda = 12\ 398/E$, where λ is in angstroms and E in electron volts. Thus, a 2-angstrom X-ray photon has an energy of 6199 electron volts.

When a particle of sufficiently high energy interacts with a target material such as in an X-ray tube, two kinds of X-rays are produced—continuum and characteristic X-ray lines. Continuum X-rays are so called because the

X-rays are present as a continuous spectrum over some finite wavelength. These X-rays are also called Bremsstrahlung (breaking radiation) or white radiation (by analogy to white light of the visible spectrum). Characteristic X-rays are so called because their discrete wavelengths are characteristic of the element which makes up the target material. No continuum is generated by photon excitation, and the quantum of continuum generated by particles heavier than electrons is smaller than that generated by electrons (by approximately the square of their mass ratios).

There are basically two experimental techniques for X-ray fluorescence analysis: wavelength dispersion and energy dispersion. In both an X-ray photon source is used to excite a sample which emits its own characteristic X-rays. The wavelength or energy of these emitted photons is measured in terms of what elements are present in the sample (qualitative analysis), and their intensity is a measure of how much of each element makes up the specimen (quantitative analysis). Practically, the recognition of the elements is based on identifying the peaks present in the spectrum. Quantitative analysis, on the other hand, is a more complex problem because of what is referred to as a matrix effect. The matrix effect is generally a problem which is solved by using comparison standards, which are chemically analyzed samples similar in composition to the unknown sample. When the unknown sample is analyzed, comparison with the standards is made.

There are a number of different spectrometers that are used for X-ray fluorescence analysis. These include the flat crystal spectrometer, the curved crystal spectrometer, and the grading spectrometers. Probably the simplest and most widely used is the flat crystal spectrometer, a schematic of which is shown in figure 2-22. As shown, an X-ray tube excites a few square centimeters of a sample material. The specimen is located fairly close to the window of the X-ray tube to provide as high an incident flux of X-rays to the surface as possible. The emitted radiation is directed by a collimator to the analyzing crystal where it is diffracted to the detector set at the appropriate angle. Detector and crystal are coupled so that they provide twice as much angular movement to the detector as to the crystal. With the crystal set at the angle θ, the detector is then at an angle of 2θ.

The spectrometer may be fixed to measure the intensity of a single X-ray line or it may be a scanning model that can be manually or motor driven through a range of Bragg angles to measure a range of wavelengths. In addi-

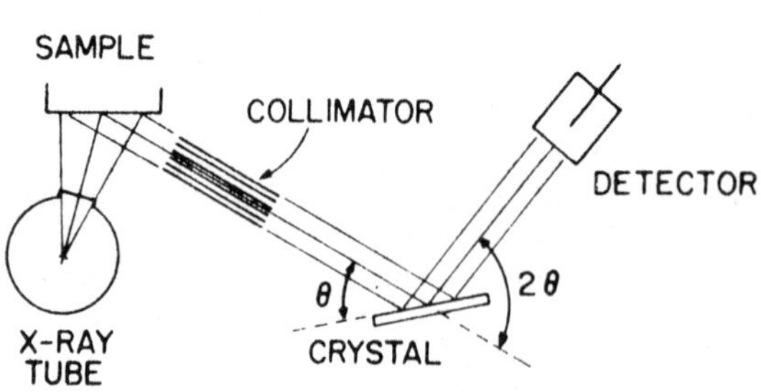

Figure 2-22. – Schematic of flat-crystal X-ray spectrometer.

tion, it may be programmed to move at a relatively rapid speed from one angle to another, just stopping long enough at each position to measure the intensity of the radiation at that location. A modern multichannel X-ray fluorescence instrument may be designed to produce analytical controls with as many as 24 fixed spectrometers to measure that many elements at the same time. These spectrometers may be used with either flat or curved focusing crystals. The automatic machines may have programmable scanning spectrometers incorporated as well. The sequential analyzers are of the programmable scanning type, while research instruments are usually manual or motor-driven.

The modern spectrometers are capable of being operated in an atmosphere of air, helium, or other controlled gas as well as in a vacuum environment. When operating the spectrometer in air, the analyst is limited to measuring X-ray wavelengths shorter than 2.5 angstroms because longer ones are severely attenuated. For softer X-rays ($\lambda < 2.5$ Å), which we are interested in for service analysis, an environment of vacuum, hydrogen, or helium is necessary. Of these options, the vacuum system is the most desirable because a very reproducible environment can be achieved by simply using a mechanical pump to obtain the required vacuum level.

Exception can be taken to the statement that X-ray fluorescence analysis is a surface technique, because many individuals would argue that the thickness measured is very large when compared to analytical surface tools such as XPS (X-ray photoelectron spectroscopy) or Auger analysis. However, because X-rays are attenuated in passing through rather thin layers of solid materials, it is necessary for the surface of a sample to be representative of the bulk for analytical results to characterize the whole specimen. The fraction of the generated photon absorbed in passing through a finite layer of some material is a very strong function of the elements present in the material and of the energy (or wavelength) of the emitted X-ray. A rather simple method of determining the thickness of the surface layer which contributes to the X-rays being measured is a consideration of this absorption criterion. Absorption is an exponential process and therefore it is impossible to say that X-rays of a particular wavelength emerge from a layer of a certain thickness and not any deeper. It is possible, however, to say that 95 percent of the information comes from a certain thickness by calculating what that thickness is which absorbs 95 percent of the X-rays.

X-ray Photoelectron Spectroscopy (XPS or ESCA)

A surface analytical tool has been developed recently that shows great promise for use in studying tribological systems; it is the X-ray photoelectron spectroscope (XPS) or ESCA (electron spectroscopy for chemical analysis) as it was formerly called. The XPS or ESCA technique was originated by professor Kye Siegbahn and his coworkers at Upsilon University in Sweden (refs. 29 and 30). A description of their early work in instrumentation can be found in their book describing many possible applications of XPS or ESCA (ref. 29). Because of the versatility in studying surface composition by XPS, this new technique is undergoing very rapid expansion, and the literature is being produced at a prodigious rate. As a consequence, there is a journal specifically devoted to electron spectroscopy for chemical analysis.

While either photons of low energy or electrons can be used as the exciting radiation in X-ray photoelectron spectroscopy, X-rays are generally used by most investigators. The X-ray photoelectron spectrometric technique uses instrumentation having five components basically: (1) the source, (2) a sample-holding container, (3) an energy analyzer for electrons, (4) a detector, and (5) some sort of readout system. The source must produce an intense monoenergetic stable beam of either photons, electrons, or X-rays of sufficient energy to excite the electrons of interest in the solid surface being examined. The beam is directed onto the sample at its surface where it impinges on the material to be studied and causes the ejection of electrons. The electron energy analyzer sorts out the resulting electrons according to their kinetic energies and focuses them on a detector at the output. The detector produces an electrical signal proportional to the intensity of the electron beam, and the readout system translates it into graphic form. The specific choice of source monochrometer or detector is dependent on the purpose of the instrument—for example, the type of electron spectroscopy to be employed.

The quantity ultimately measured by the XPS system is the binding energy of the ejected electrons. In their original work, Siegbahn and coworkers found that the binding energy for a given electron in an atom did not remain constant for a series of compounds containing that atom. Instead, it was markedly affected by chemical environment, and it was roughly a function of the atomic charge (ref. 29). Core electrons of the atoms in the solid surface are subject to a combination of various forces, the results of which are known as the binding energy. From the nucleus, the electrons in the atom experience a strong attractive force in proportion to the magnitude of the nuclear charge or atomic number of the particular atomic species being examined. The core shell electrons exert a repulsive force which in effect screens the outer or valence electrons from the nuclear charge, thus diminishing the nuclear attractive force. The resultant force by which the electron is bound to the atom is E_b.

If an electron is removed from the outer shell, the screening of the inner electrons is reduced by one electron charge, and the core electrons therefore feel an increased force of attraction from the nucleus. The gain of an outer electron (e.g., through radiation) has the opposite effect, effectively increasing the shielding and decreasing E_b. Binding energies are therefore responsive to change in the chemical environment, and as such they represent a new key to the study of the basic chemical structure of materials in solid surfaces. The soft X-rays impinging on the surface in XPS produce the photo-ejection of the core electrons.

The source for the X-rays is an X-ray tube of rather conventional design consisting of a heated cathode and a cooled anode. The power levels are frequently of the order of 5 kilowatts. The precision of the final electron energy measurement is dependent on using the most narrowly defined source of exciting radiation available. Since the line width of X-ray emission is proportional to atomic number, the lightest elements with well-defined intense radiation are the most desirable sources and the most frequently used. Although copper and chromium were used historically, the targets in most common use today are aluminum or magnesium with K alpha lines at 1486.6 and 1253.6 electron volts, respectively.

A variety of sample solid surfaces can be examined with XPS. One of the requirements, however, is that the sample be in a chamber maintained at a pressure of approximately 10^{-6} torr or less. Differential pumping techniques, however, can be used to study the gaseous samples on solid surfaces. In addition, cryogenically cooled surfaces can be used to condense volatile materials for examination. One limitation of this particular technique is that the material being examined must not be adversely affected or be subject to decomposition by the X-ray beam impinging on the solid surface. If it is, the chemistry can be changed by the analytical tool itself.

XPS is a very sensitive technique requiring very small amounts of sample, typically of the order of 10^{-6} gram, for measurements. Spectra have been obtained in samples as small as 10^{-8} gram. Because the ejected electrons do not emerge from depths in excess of about 100 angstroms, XPS is extremely well suited to study solid surface films, particularly tribological films. When this particular technique is to be used for studying solids in the bulk, however, there must be some assurance that the bulk is representative of what is seen on the surface. To study bulk composition, ion depth profiling (used in conjunction with XPS) can provide considerable insight into whether or not the surface layers are comparable in composition to the bulk chemistry of the material. Ion beam sputtering (profiling) or removal of solid surface layers are discussed in more detail later in the section on ion beam techniques (p. 77).

Figure 2-23 is a schematic diagram of the X-ray photoelectron spectrometer. The upper portion of the schematic shows the X-ray tube, which provides the source for X-rays that strike the sample specimen surface; the emitted photoelectrons from the sample surface then pass through an annular slit to a spherical condenser, an electron multiplier, and a counter. Various techniques are used for recording the output, including minicomputers, conventional x-y recorders, and teletype.

An example of the information which can be generated with the XPS system is shown by examining the oxidation of lead data presented in figure 2-24. The spectra are obtained for a clean, evaporated lead film and for this film in various stages of oxidation (ref. 17). The spectra reflect the influence of the binding energy shift of the 4f electrons in lead as an oxide layer is formed on the pure elemental lead sample. The spectrum closest to the abscissa is for the evaporated lead film. Note the electron binding energy in electron volts for the lead peaks, particularly the lead 4f electrons. After the surface is exposed to air for about 4 seconds an oxide begins to form. Notice the flattening in the lead peak (after 4 sec, peak shows a decrease at 136.8 eV) at 136.8 electron volts. With further exposure to air, about 8 minutes, the peak becomes much more singular in shape. With further exposure at a

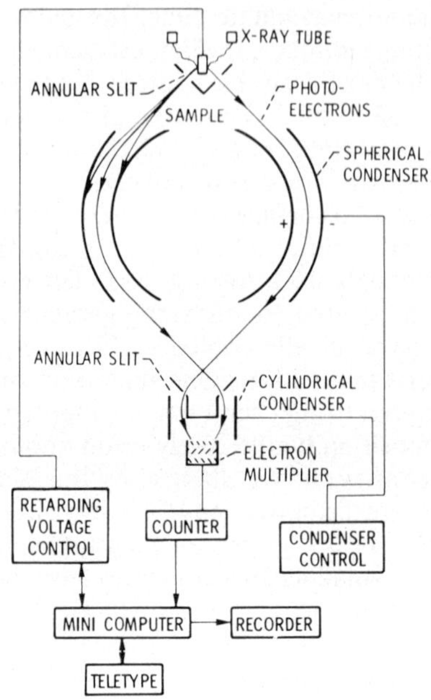

Figure 2-23. – Schematic diagram of X-ray photoelectron spectrometer.

higher temperature, a much more distinct lead oxide (PbO) peak appears, with the complete disappearance of the normal lead peak.

In the second and third curves from the base of the figure, some residual lead peak at 136.8 electron volts is still visible in the emerging lead 4f peak associated with the oxidation of lead. If the surface is exposed to air at 100° C for about 20 hours, the oxidation of the lead not only is complete, but the oxide shifts from the lower oxide of lead (PbO) to the higher oxide of lead (PbO_2), and another marked shift in the lead 4f peak is observed. The elemental lead peak appears at 136.8 electron volts, the PbO at 138.1 electron volts, and the completely oxidized PbO_2 at 139.5 electron volts. These are the kinds of chemical shifts that occur for the various peaks of elements when combined to form compounds that help to fingerprint the environment in which the elemental species finds itself. For example, from the data presented in figure 2-24, it is easy to distinguish the environment in which lead is present. When it is elemental lead, a peak is at 136.8 electron volts; if the lead is oxidized to PbO, a peak is at 138.1 electron volts; and if we have lead at its highest oxidation state—namely, that of PbO_2—the energy peak is at 139.5 electron volts.

This kind of fingerprinting is extremely useful in tribology, particularly since XPS is surface sensitive. It is fairly well known, that extreme pressure additives are placed in oils to react chemically under heavy load conditions. With the presence of these organometallics in the oil (which break down on solid surfaces to liberate surface-active species such as sulfur, chlorine, or phosphorus), reactions with the metal surface occur. The reactions

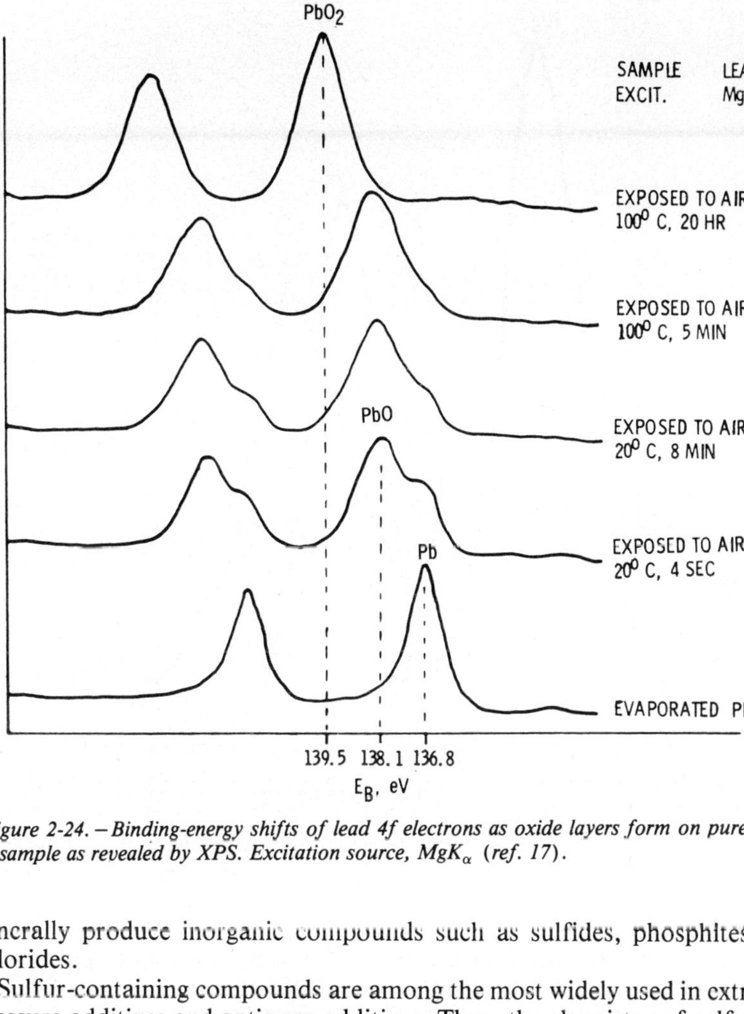

Figure 2-24. – Binding-energy shifts of lead 4f electrons as oxide layers form on pure lead sample as revealed by XPS. Excitation source, MgK_{α} (ref. 17).

generally produce inorganic compounds such as sulfides, phosphites, or chlorides.

Sulfur-containing compounds are among the most widely used in extreme pressure additives and antiwear additives. Thus, the chemistry of sulfur on a solid surface is extremely important in tribology. An example of how XPS can be useful in this area of study is shown in the figure 2-25. In figure 2-25 the sulfur 2p electron binding energies are shown for various representative sulfur types, that is, the environment in which the sulfur finds itself (ref. 31). We can see the binding energy for the elemental sulfur occurring at approximately 162 electron volts; if we ionize the sulfur to a sulfur -2 charge, there is a shift downward in the binding energy from 162 electron volts to approximately 160 electron volts. If, however, this elemental sulfur is oxidized and finds itself as an oxidized species such as SO_3^{-2} there is a shift in the opposite direction. The shift is to a higher binding energy, approximately 166 electron volts. Further oxidation of the sulfur from SO_3 to SO_4 results in yet a further shift in the binding energy to approximately 168 electron volts, indicating the ability to separate the form in which sulfur is com-

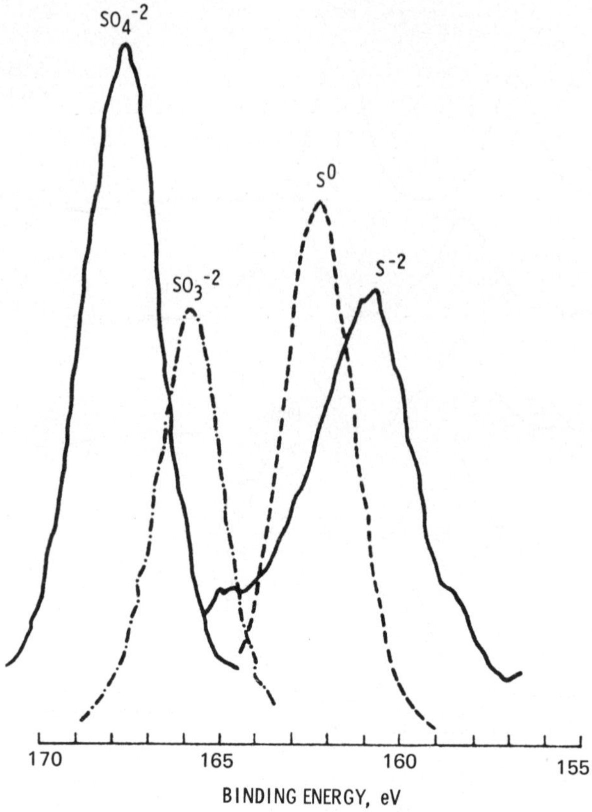

Figure 2-25. – Sulfur (2p) X-ray photoelectron spectra of representative types of sulfur.

bined with oxygen. Thus, XPS can detect the presence of the elemental material (in this case sulfur) and distinguish it from compounds (in this case oxides), but it can also distinguish between different combinations of oxygen with the sulfur as reflected in figure 2-25. Going from SO_3 to an SO_4 structure results in sufficient chemical shift in the binding energies of the photoelectrons emitted that one can separate out and identify these two different species.

Transmission Electron Microscopy

A number of surface analytical tools employ electrons for their source of energy: these electrons are directed at a solid surface to identify it. Probably one of the oldest electron beam sources used in the identification of surfaces is the ordinary transmission electron microscope. In the transmission microscope, the electrons that form the image must pass through the specimen. Thus, the specimen thickness is limited to a few thousand angstroms, or a few microns for very high voltage instruments. To study the surfaces of solids, there are two possible approaches which can be taken. In the first approach, a replica of the surface can be made; for example, a carbon replica may be made by vacuum depositing a 100- to 1000-angstrom film of carbon on the surface, then carefully removing it by some etching technique, and finally mounting it in the microscope. The image obtained from such a replica does not represent the surface topography because it is frequently subject to distortions and artifacts that are often difficult to interpret. Furthermore, the process of replication seriously cuts down the resolution ultimately obtainable with the electron microscope.

A second approach is to plate a suitable material onto the surface of interest and then to section it in thin slices normal to that surface. The section is then mounted for observation on the microscope, and this permits observation of the surface in profile. The resolving power of the instrument can be fully exploited by this method, which is commonly referred to as the profile method. The profile method has the additional advantage of relieving the surface topography in relation to the underlying structure of the material.

Another approach would be to deposit a nonconducting film on the solid surface which one desires to examine—that is, deposit a layer of some material like a polymer on the surface and then electropolish the sample from the opposite side after cutting a thin section very close to the surface. The local thinning near the surface would enable an observer to see the rather large areas of the surface if the nonconducting layer can be dissolved by washing it in a suitable solvent. Thus, in the transmission electron microscope, either sectioning, replication, or thinning must be employed in order to effectively analyze surface layers.

The manner in which replication can be achieved is shown schematically with the aid of figure 2-26. In figure 2-26(a) the metal surface is the surface that one desires to examine in the microscope. The surface is coated with a film of plastic material (ref. 32). After the plastic has set up, it is stripped from the metal surface and backed with Scotch tape. The plastic can then be coated by deposition of a film of carbon; the final carbon replica is available (fig. 2-26(d)) after removing the plastic by washing with a suitable solvent. The carbon film can then be examined directly in the transmission electron microscope to gain some insight into the surface structure. Frequently, a lighter deposit of a shadowing metal is applied during evaporation of carbon (as indicated in fig. 2-26(d)) to bring about better resolution of surface topographical characteristics.

The requirement of either section thinning or replication is the Achilles heel of the transmission electron microscope. It requires a very well developed ability to practice the art of sectioning or thinning samples for transmission electron microscopy examination. Furthermore, the replication technique requires a great degree of skill in preparing the replication of the solid surface. It is frequently very difficult to dissolve the plastic film without causing disintegration of the carbon replica. Before the plastic dissolves it expands, and the strains transmitted to the carbon replica can be destructive. While a carefully prepared replica should accurately reflect

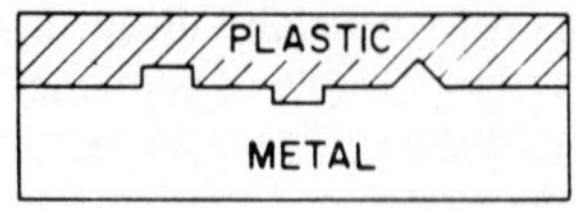

(*a*) *Surface covered with layer of plastic.*

(*b*) *Backed plastic replica after stripping.*

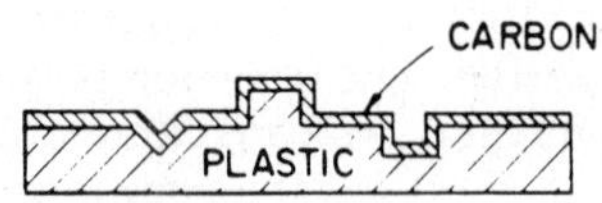

(*c*) *Backed plastic replica after stripping and coating with carbon.*

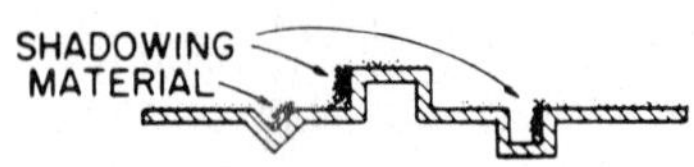

(*d*) *Final carbon replica after removal of plastic by washing* (*with a suitable solvent*) *and shadowing of carbon.*

Figure 2-26. – Two-stage (*plastic/carbon*) *replication method for solid surfaces* (*ref. 32*).

horizontal distances in the surface of interest, considerable doubts may exist as to how well the vertical dimensions are represented by the replica.

Despite the inherent limitations involved in using replication and thinning techniques, the transmission electron microscope is, today, probably still one of the most effective tools for direct observation of crystalline defects in solids. Such defects appear as dislocations, dislocation reactions with impurities, distribution of impurities in solids, grain boundary behavior, twinning, and other metallurgical phenomena that occur in solids.

Scanning Electron Microscopy

The single most useful tool available to the tribologist today interested in studying the behavior of solid surfaces is undoubtedly the scanning electron microscope (SEM). This instrument requires the least amount of technical expertise for interpreting of results and presents an extremely good physical picture of a solid surface. Unlike the conventional transmission electron microscope which requires a special skill in thinning or replication, no such skill is required in the SEM. One simply places the sample to be analyzed into the chamber of the SEM, evacuates the system to the required pressure of operation, and examines the solid surface. Furthermore, unlike the transmission electron microscope, the thickness of the sample is unimportant. Therefore, very thick samples, such as those frequently experienced and found in tribological systems, can be accommodated.

The principle of the SEM is indicated schematically in figure 2-27. Electrons from an emission source or filament are accelerated by a voltage commonly in the range of 1 to 30 kilovolts and directed down the center of an electron optical column consisting of two to three magnetic lenses. These lenses cause a fine electron beam to be focused on the specimen's surface. The scanning coils placed before the final lens cause the electron spot to be scanned or rastered across the specimen surface in the form of a square raster similar to that on a television screen. The currents passing through the scanning coils are made to pass through the corresponding deflection coils of the cathode ray tube so as to produce a similar but larger raster on the viewing screen in an synchronous fashion. The electron beam incident on the specimen surface causes various phenomena of which the emission of secondary electrons is used in the SEM. The emitted electrons strike the collector, and the resulting current is amplified and used to modulate the brightness of the cathode ray tube. Times associated with the emission and collection of secondary electrons are negligible compared to the times associated with the scanning of the incident electron beam across the specimen's surface. Hence, there is a 1:1 correspondence in the number of secondary electrons collected from any particular point on the specimen's surface and the brightness of the analogous point on the screen. Consequently, an image of the surface is progressively built up on the screen.

The SEM has no imaging lenses in the true sense of the word. The image magnification is determined solely by the ratio of the sizes of the rasters on the screen and the specimen's surface. In order to increase the magnification it is only necessary to reduce the currents in the SEM scanning coils. For example, if the image on the cathode ray tube screen is 10 centimeters across, magnifications of 100, 1000, and 10 000 are obtained by scanning the specimen 1, 0.1, and 0.01 millimeter across, respectively.

One consequence of this is that high magnifications are easy to obtain with the SEM while low magnifications are difficult. Thus, for a magnification of 10 it would be necessary to scan a specimen approximately 10

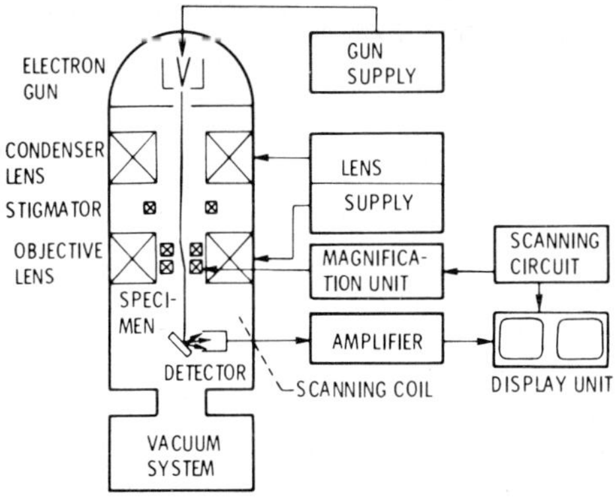

Figure 2-27. – Simplified block diagram of SEM.

millimeters across, and this presents difficulties because of the large deflection angles required. For instance, the electron beam may strike the lens hole or aperture and, at the extremes of the scan, linearity may not be maintained. The completely different operation of the SEM compared with most other microscopes is possible because there are no imaging lenses and any signal that arises from the action of incident electron beam (reflected electrons, transmitted electrons, emitted light, etc.) can be used to form an image on the screen.

An apparatus to study friction and wear has been incorporated in the SEM to observe the wear process and friction behavior of materials in solid-state contact while it is taking place. Figure 2-28 is a schematic diagram of the apparatus. A disk specimen 1.9 centimeters in diameter is mounted on an adapter to the rotary specimen feedthrough. The surface of the disk is inclined at approximately 70° with respect to the electron beam of the SEM. This steep angle allows viewing the interface from the near side. A variable speed small electric motor and a gear train are attached to the external rotary specimen feedthrough to provide rotary motion to the disk specimen at rotating velocities of 0.001 to 5 revolutions per minute. The rotation can be either clockwise or counterclockwise to provide for SEM observation of either the prow or the wake of the rider-disk contact and for direct observation of the friction or wear process while it is occurring. In addition to the foregoing, a side view of the rider-disk contact zone can be obtained.

The stylus (rider) which contacts the disk surface is mounted in an arm which can be moved in and out, up and down, and laterally by means of a bellows and gimbal system. The gimbal system is composed of a precision

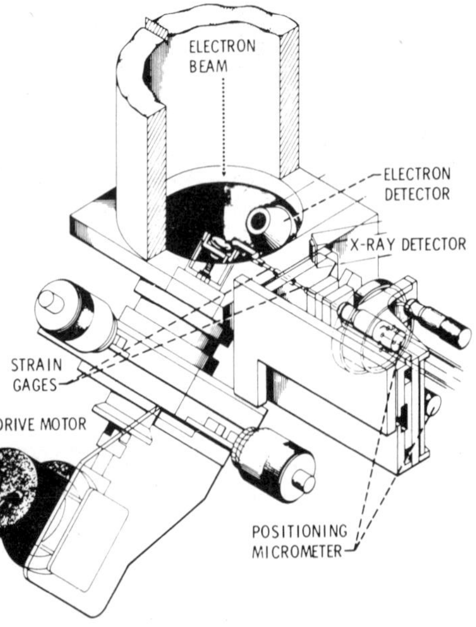

Figure 2-28. – Detailed drawing of wear apparatus mounted on SEM.

optical orienter which is mounted on a translational stage. This stage and the optical orienter are micrometer controlled, thereby allowing very precise positioning of the rider on the disk under the scanning beam. The arm in which the stylus (rider) is mounted contains two flex bands of beryllium copper on which strain gauges are attached. The normal loading is accomplished by allowing the magnets mounted on the optical orienter ring to pull the arm downward to the disk surface. The strain gauge output of the load sensing gauge is amplified and displayed on a digital millivoltmeter which, with suitable calibration, provides a direct reading of load. The friction force gauge is read out on either a strip chart recorder or an oscilloscope to provide direct observation of the friction trace which is more transient in nature.

The entire experiment is viewed on the television monitor of the SEM, and the video signal is recorded along with audio comments on video tape to provide data recording. The tape can be played back in slow motion and with stop action to facilitate interpretation. In addition, kinescopic motion pictures can readily be made from the video tape. To provide for analysis of the wear track, an energy-dispersive X-ray analysis system is mounted on the SEM. The analyzer has 400 channels and a resolution of 170 electron volts. The capability of elemental mapping is also provided in the system. The use of energy-dispersive X-ray analysis in conjunction with the SEM permits the direct elemental analysis of the wear and friction surface as well as an examination of wear debris. For example, the wear debris generated from various specimens can be identified separately by X-ray techniques.

The SEM is ideally suited for studying wear surfaces and general surfaces in topographical tribological studies because of its great depth of focus. Thus, the irregular nature of a worn surface can be seen in detail with thc SEM. Furthermore, wear debris present on the surface can also be visually observed as well as its method of attachment to the solid surface.

Readily apparent in a SEM are those tribological systems where a hard surface is sliding on a relatively softer material and a great degree of plastic deformation occurs to the solid surface. Furthermore, where wear mechanisms involve the generation of surface cracks or fracturing of the surface, the start of the cracks and their propagation, as well as their nature and character, can be readily observed in the SEM.

The depth of field with the SEM is apparent from the photomicrograph presented in figure 2-29. The photomicrograph is for a single crystal of naturally occurring molybdenum disulfide, which has been fractured by breaking it. Examination of a broken edge of a molybdenum disulfide crystal in the SEM reveals, in the upper left corner, the smooth lines generated by fracture of the molybdenum disulfide crystallites. The material shows a larger lattice structure. There are thin sheets of the lamellae, the basal planes of the molybdenum disulfide crystallites being the smooth, flat surfaces visible in the photomicrograph in figure 2-29 and the edge sites being those very thin sections generated by the fracture of the molybdenum disulfide crystallite. The individual lamellae, or crystallites of the molybdenum disulfide crystalline solid sample, appear as sheets much like those observed in mica. If one visualizes that the bonding between these

Figure 2-29. – Molybdenum disulfide crystallites.

adjacent sheets is relatively weak (which it actually is), then it is readily apparent from the photomicrograph in figure 2-29 that these sheets, if they slip or slide over one another readily, would be inherently good lubricating materials.

The SEM has an in-focus depth of field of 1 to 300 times that of the ordinary light microscope; this gives it one of its unique characteristics for use in tribology. The SEM combines some of the best features of the optical microscope, electron microscope, and the electron microprobe into an instrument of outstanding performance, high reliability, and ease of operation or use.

Yet another feature of the SEM which makes it an even more attractive tool for analyzing surfaces in tribology is its ability to provide crystallographic information by an electron diffraction effect, which has been termed electron channeling. By proper incidence of the electron beam on the surface, lines or channels are obtained in a pattern. These patterns were originally interpreted by Booker; he found that they can provide orientation and crystalline perfection information from minute regions of the surface of a specimen (ref. 33). Electron channeling is a very effective tool for studying deformation of solid surfaces, and in tribology, one is concerned with the deformation behavior of surfaces. In its simplest terms, the electron channeling patterns develop as a result of the crystalline nature of materials. The portion of the primary beam which is backscattered is dependent on not only the atomic weight of the sample but also on its crystallographic orientation. If the backscattered electron intensity is measured above the specimen it is found to change with changes in the orientation of the specimen surface.

In the normal mode of operation the essentially parallel primary beam of the SEM experiences slight angular changes with respect to the surface of the specimen as it sweeps from one side to the other. If the entire sweep is within a single crystal, the angle of the beam is changing in relation to the lattice planes of the crystal. These angular variations are, of course, greater with large beam deflections; therefore, electron channeling patterns are frequently observed superimposed on low magnification scanning electron micrographs. Because a large number of Bragg conditions can be satisfied with quite small angular variations, numerous diffraction lines are frequently observed in the scanning process. With a rocking method, the primary beam about a point in the sample surface can be utilized in such a way that the electron channeling patterns are obtained from a very small area. This technique is termed selective area electron channeling patterns. The size of the area from which patterns of this type can be obtained is limited with present instruments to about 10 micrometers. It is, however, anticipated that future instruments should permit electron channeling patterns to be obtained from areas of less than 1000 angstroms.

These patterns are somewhat similar to the Kukuchi patterns obtained in transmission electron microscopy and to the patterns obtained from electron microprobe analysis in the sense that the crystallographic orientations can be determined from the geometry of the patterns and the crystalline perfection information can be obtained from the quality of the patterns. Analyzing the patterns obtained creates some difficulty; therefore, standard patterns are frequently generated from previously oriented samples after which comparisons of the unknown patterns can be made to these standards. Through a knowledge of orientations of solid surfaces, computer patterns can be generated and a comparison to these computer patterns can be made. A technique using this basic concept has recently been developed and worked out. For a more detailed description of the interpretation of these patterns the reader is referred to a paper by Weiss and Hughes (ref. 34). These authors have demonstrated that the amount of cold working or strain that occurs on a metal surface can be effectively followed using electron channeling patterns, and from this an indication of the dislocation density of the material is obtained. The electron channeling technique has been very effectively used by Ruff for looking at the tribological behavior of iron surfaces (ref. 35). Figure 2-30(b) contains a typical electron channeling pattern. The term SACP is used in the figure referring to selected area channeling pattern; it is from one grain of the iron specimen surface, and superimposed over the electron channeling pattern are signal traces which reflect the degree of crystal imperfection in the grain. A series of these traces are presented in figure 2-30(b) across the grain surface along the 220 band of the grain. When a crystalline solid surface loses its perfection, or when it is deformed plastically such as in tribological studies, the sharp lines seen in

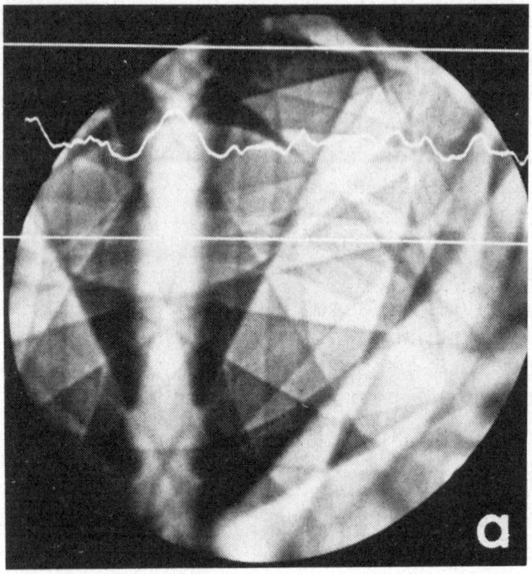

(*a*) *SACP from one grain in iron specimen; 220 band is prominent; one signal trace is superimposed.*

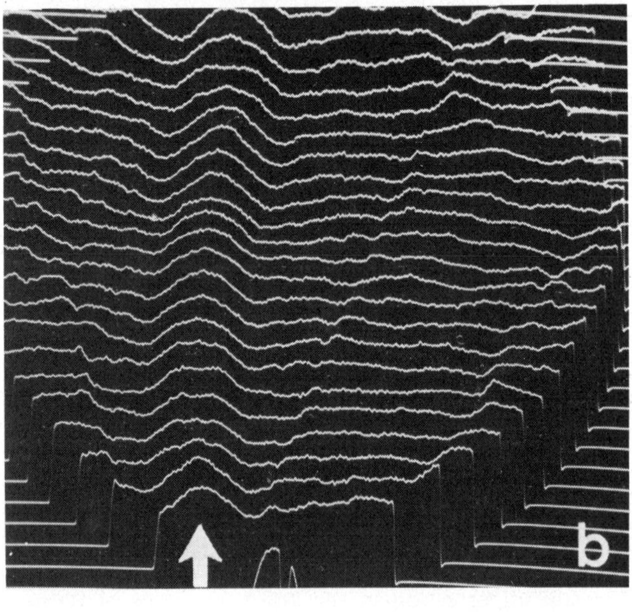

(*b*) *Set of signal traces at different locations on SACP showing variation along 220 band.*

Figure 2-30. – Electron channeling (ref. 35).

figure 2-30(a) are lost and become much more diffuse and difficult to identify with increasing amounts of strain. The signal shown in figure 2-30(a) and the series thereof in figure 2-30(b) become smoother with higher degrees of strain and a loss in the actual channeling lines due to the increase in the deformation of the solid surface. This is discussed in more detail in chapter 7, which addresses itself specifically to wear surfaces.

One of the inherent difficulties in the SEM is that the beam of electrons results in an accumulation of electrons on the surface, and this produces a surface charging effect, which results in degradation of the image. With metals this is not a serious problem because the electrons are dissipated in the conductive solid. With polymeric materials, which are frequently of interest in tribology and ceramic materials, it becomes a difficult task to obtain good surface micrographs or channel patterns without resorting to some auxiliary technique or method of preparing the specimen surface prior to inserting it in the SEM. A frequently and commonly used technique is to paint the specimen surface with a conducting type of film; this tends to dissipate the surface charge from the electron beam as it builds up. There are schemes, however, that can be used which involve modifying the SEM so that films that do charge can be studied directly.

Electron Microprobe

Another surface tool for studying the characteristics of solid surfaces that uses the electron beam is the electron microprobe. The basic principle behind the technique of the electron microprobe is relatively simple and straightforward. The electron beam strikes a sample surface and produces X-rays whose wavelengths and energies are characteristic of the elements present in the solid surface in the region of the impinging electron beam. The intensities of these X-rays are measured and can be compared with the intensities of X-rays from a pure standard of the element and the ratio of intensities measured on the sample to the intensity measured on the specimen can be considered as a measure of the amount of the element present in the sample. In other words, a calibration standard can be used and a comparison made against an unknown. The correlation, however, is not entirely straightforward. There are several factors which tend to complicate the interpretation. Most important of these factors is the influence of other elements present in the sample which will absorb some of the X-rays generated and produce fluorescent radiation characteristics of their own elemental properties. Yet another factor to be considered is the effect of depth of penetration of the electron beam; this is a function of the atomic number of the sample.

In many cases, however, a detailed quantitative chemical analysis is not needed; simple, qualitative, and semi-quantitative analyses are sufficient. While the electron microprobe is not strictly a surface tool as is the Auger analyzer, which is described later, it does have a rather unique feature in that it can analyze various spots on the solid surface and do a point by point analysis of a surface. This facilitates the study of such things as wear tracks of relatively small size, the presence of defects in surfaces, and fatigue and fracture cracks which may be generated in tribological surfaces. Those regions can be studied in detail from an elemental point of view with the electron microprobe because it can pinpoint the location of the analysis on the solid surface.

The essential elements of the electron microprobe analyzer are depicted schematically in figure 2-31. The basic elements of the microprobe consist of an electron beam which is produced by an electron gun in the upper portion of the schematic. This particular gun generally uses a fine tungsten filament. Electrons generated at the tungsten filament are accelerated to a potential in a range of 15 to 50 kilovolts, depending on the sample to be analyzed. On the way to the sample surface, the electrons pass through one or more electron lenses, generally a condenser and an objective which produce a finely focused electron beam on the specimen surface. The X-rays generated from the impinging electron beam are then collected by one of two types of spectrometers. These include either a wavelength dispersive spectrometer in which the wavelengths of the emitted X-rays may be analyzed or an energy dispersive spectrometer, which is frequently called a solid-state detector. The wavelength dispersive spectrometer is selective for one wavelength only. It is generally a linear spectrometer intended to maintain a constant takeoff angle in the sample, and it is usually fully focusing. In the case of the energy dispersive spectrometer, a good semiquantitative concept of the total composition of the sample may be obtained very rapidly, say, in approximately 20 seconds, while for more accurate quantitative work, which can be done within the instrument, longer times are required.

The display of the distribution of elements that the electron beam sees is obtained by scanning the beam over a small area in the sample surface in a

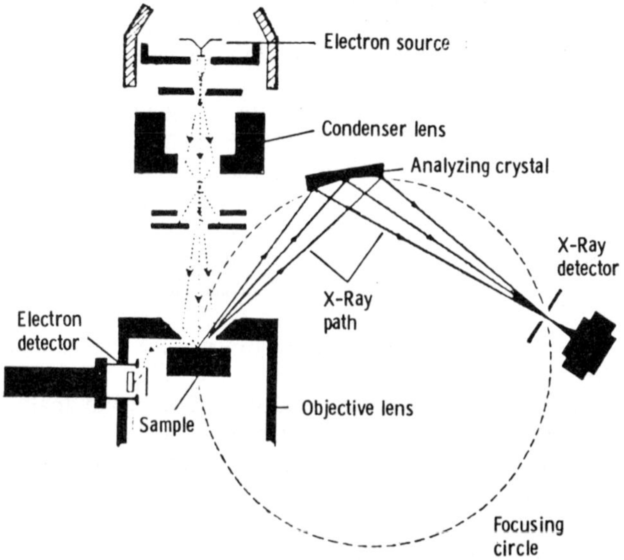

Figure 2-31. – Schematic drawing of elecron microprobe analyzer (ref. 36).

rastered pattern; a cathode ray tube is scanned in synchronization with this. The brightness of the cathode ray tube is modulated by the detector signal of the spectrometer. Magnification is achieved by lowering the area of scan on the sample since the area of the cathode ray tube is constant.

Location of the beam on the specimen surface is achieved by moving in x,y, or z directions. Observation of the specimen position is generally by a light microscope. The electron microprobe is a very useful tool for studying, for example, the transfer of wear particles from one surface to another. Where adhesive transfer takes place in the wear contact region, if material is transferred from one surface to another and the two solid surfaces contain different elemental compositions, some idea of the location of adhesive transfer can be obtained with the electron microprobe. An electron microprobe is also very effective for showing the distribution of the elements on a sample solid surface. Thus, for example, where an additive in an oil would deposit and interact with the solid surface in various regions in a nonuniform manner, the elemental mapping of that surface can be achieved with the electron microprobe.

Figure 2-32 presents data for the normalized beam intensity as a function of wavelength in angstroms for the silicon K'_β X-ray line together with satellite K'_β (ref. 37). The low energy lines are for the compounds silicon carbide, silicon nitride, and silicon dioxide ($\alpha - SiO_2$). It can be seen from the data of figure 2-32 that each compound yields a K'_β peak position characteristic for each material. Since each K'_β peak can be completely resolved it is easy to use the K'_β intensity ratios to establish the relative population of oxygen, carbon, and nitrogen combined with silicon in each of the situations.

The electron microprobe is especially useful (1) where one wishes to look at surfaces where the elements in the surface layers are different from those of the bulk, (2) in situations where lateral variations in composition on the surface or thickness are important, and (3) where one wishes to determine the valency or coordination of elements in the surface layers. Still another

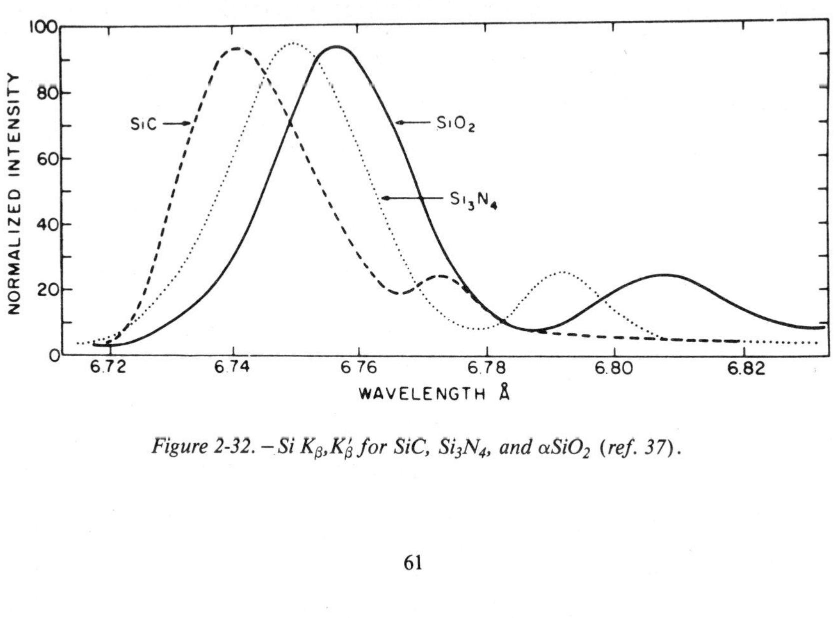

Figure 2-32. – Si K_β,K'_β for SiC, Si_3N_4, and αSiO_2 (ref. 37).

area where the electron microprobe is useful is the situation where an analytical technique that is nondestructive is desired.

Auger Emission Spectroscopy (AES)

One of the surface analytical tools which is extremely versatile and can be very effectively used with little training and interpretation of results is Auger emission spectroscopy (AES) analysis. AES, like the devices described before, employs an electron beam source as its means of excitation of the solid surface. It is a device for chemical analysis of the solid surface. It is truly a surface analytical tool in that it can analyze for all the elements present on the solid surface (with the exception of hydrogen or helium) at surface concentrations of as little as 1/100 of a monolayer. In addition, it has a depth analysis capability of 4 or 5 atomic layers. Thus, it is a true surface analytical tool. The combination of its surface sensitivity and depth of penetration makes it an extremely useful device for studying tribological surfaces. Therefore, a greater indepth description is provided of AES than has been provided for some of the devices described earlier.

In AES, a primary beam of electrons in the voltage range of from 1000 to 3000 electron volts is directed at the solid surface. The incoming beam of electrons strikes the solid surface and penetrates the electron shells of the atoms of the outermost atomic layers (the first four or five atomic layers). In penetrating the electron shells of the atom's outermost layers, the incident electron causes inner level electron shell ionization. That process results in the liberation of energy, and the liberation of energy results in electrons being removed from the electron shells of the atoms in the outermost atomic layer. This particular electron, which is ejected from the solid surface, is referred to as the Auger electron. With a measure of its energy, which is characteristic of the electron shells from which it came, the particular atomic species on the solid surface can be identified. Each electron has a characteristic energy dependent on the particular solid surface from which it originates. The process is depicted schematically in figure 2-33

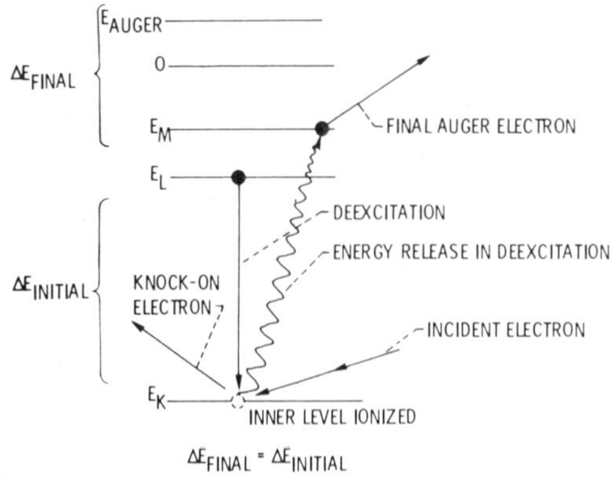

Figure 2-33. – Auger transition diagram for an atom.

where the incident electron from the primary electron beam source is seen to knock out an electron in the E_K energy shell level and thus create a vacancy. This vacancy or absence of an electron from a shell is filled by an electron from an outer shell, the E_L level, dropping down and filling or occupying the vacancy. In the process of doing so, there is an amount of deexcitation energy involved, and that energy results in the knocking out of an electron from the E_M level. This is the Auger electron, which is then liberated from the solid surface and analyzed for its characteristic energy.

One of the earliest ways of using the AES analysis, and one that is still employed today, is to couple the AES system with a LEED system (low energy electron diffraction), which is described later (p. 73). When the two systems are coupled, common components can be used for both LEED and AES analysis. A schematic diagram for one such system is shown in figure 2-34. In figure 2-34, the specimen is indicated as the crystal. For LEED work, the material used is normally a single crystal. With Auger, however, if only the Auger mode of the device is used, it is not necessary that the solid surface be a single crystal because AES analysis can be used to examine polycrystalline as well as single crystal surfaces of metals, nonmetals, or semiconductors. It must be noted, however, that Auger analysis, like other electron devices, does have the associated problem of surface charging if the solid surface being analyzed is a nonmetal (e.g., glass or ceramic). This charging can be adequately taken care of so that AES analysis has been and can be effectively used to examine nonconducting solid surfaces.

The LEED screen in figure 2-34 contains a series of curved grids (dashed lines) which are used for detecting the Auger electrons. The center grids have retarding voltage applied to them and it weaves through the entire energy spectrum range. A small perturbation potential is superimposed on the retarding voltage at a known frequency. This retarding voltage enables differentiation of the signal using phase sensitive detection techniques. The perterbation signal is fed into the reference channel of a lock-in amplifier

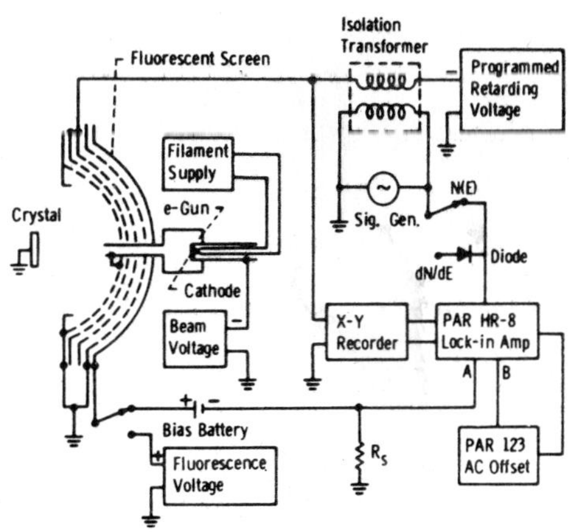

Figure 2-34. – Auger spectrometer using LEED optics.

(shown in fig. 2-34) from the same oscillator to midphase matching. The first derivative is the coefficient of the fundamental frequency as detected by the lock-in amplifier, and the second derivative is the coefficient of second harmonic. The first derivative of the detection current is the secondary electron distribution function. The data are generally taken off the system and plotted on an X-Y recorder. A typical Auger trace for an iron surface containing adsorbed ethylene is shown in figure 2-35.

In figure 2-35 the differentiated derivative energy is plotted as a function of electron energy for the various species present on the solid surface. Four peaks are seen in the spectrum. The one farthest to the left is associated with carbon occurring at approximately 270 electron volts. The other three peaks are associated with iron of the iron single crystal surface. The carbon source is the ethylene molecule which is adsorbed on the iron surface. There is another iron Auger peak which occurs at a very low electron energy. However, this low energy electron iron peak is very sensitive to the presence of surface contaminants. With small amounts of surface contamination, or coverage of the surface by another species (ethylene), that particular peak is not detected. Thus, only three of the iron carbon peaks are seen in figure 2-35.

The electron energies plotted in the abscissa then give us an indication of the elemental species present on a solid surface, and the relative peak heights give some indication of the quantity of the material present on the solid surface. While AES analysis is not strictly a quantitative analytical tool, some semiquantitative information can be obtained by proper calibration and control of the instrument. Chang at Bell Laboratories has done considerable research in using AES as an analytical quantitative tool (see ch. 20, "Characterization of Solid Surfaces," edited by Kane and Larabee, ref. 38). Auger is a very desirable tool for use in tribological studies because it is extremely sensitive to surface elements such as oxygen, carbon, sulfur, phosphorous, and chlorine which are involved in many practical lubricating systems.

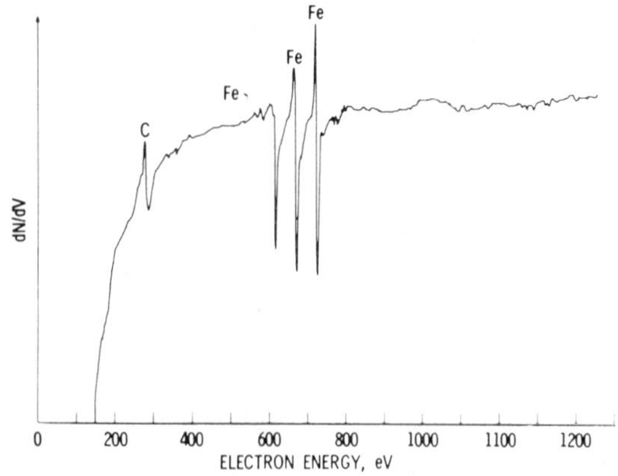

Figure 2-35. – Derivative of electron energy distribution for ethylene absorbed on Fe (001) surface.

In figure 2-35 there are basically only two elements involved: iron from the iron single crystal surface and carbon from the surface contaminant ethylene which is adsorbed on the solid surface. So we see only four Auger peaks. If surfaces that contain a multitude of elements are examined with Auger, very good differentiation results with respect to the source of the individual peaks. For example, figure 2-36 is an Auger emission spectrum for magnesium zinc ferrite which has been sputter cleaned by argon ion bombardment (ref. 39). The Auger spectra shows the presence of a low energy iron peak, a zinc peak, a manganese peak, oxygen, and all the additional associated peaks with each of these particular elements. In addition, a small peak is seen associated with argon as a result of argon becoming embedded in the surface in the process of argon ion bombardment for cleaning purposes. Note that while the lower energy iron, zinc, and manganese peaks are fairly close, they can still be effectively differentiated and identified as to the source element for each peak.

The peaks tell us from their electron energy what particular element is on the solid surface, and the relative height of the peak is some quantitative indication of the amount of material present on the surface. In addition, Auger can supply information about the nature of the species present on the solid surface. For example, a careful, indepth analysis of the peak shapes themselves reveals information as to the form of the element. For example, Grant and Haas (ref. 40) have done a considerable amount of research ex-

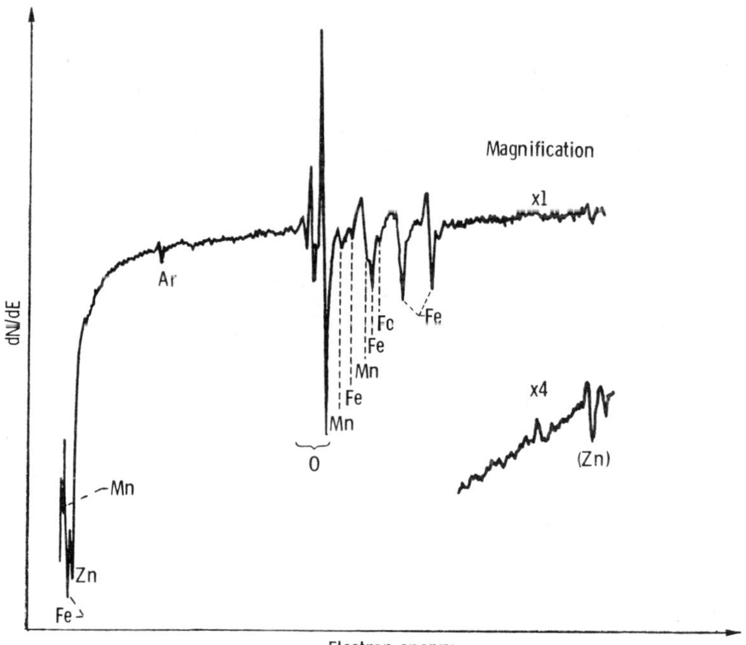

Figure 2-36.—Auger emission spectroscopy spectrum for manganese-zinc ferrite (110) surface after sputter cleaning.

amining the carbon peak and identifying the source of the carbon by the shape of the peak. If the carbon is combined with a metal, such as a carbide, the peak may take one characteristic shape. If the carbon is present as an adsorbed gas on the solid surface, a different shape of the peak may take place, and if the carbon is present as free carbon in the form of graphite, still another shape or form of the peak may exist. Some data obtained by Grant and Haas are shown in figure 2-37. In figure 2-37 we see the carbon peak at approximately 270 electron volts for carbon in various forms.

In the upper Auger peak, the carbon is combined with molybdenum as molybdenum carbide; in the middle peak, carbon is present as carbon monoxide adsorbed on the molybdenum metal surface. If the second peak is compared with the first peak, there is a difference in the shape of the carbon peaks which is a function of the particular species to which the carbon finds itself bonded. Elemental graphite gives the peak shape shown in the lower portion of the figure. It has a distinctly different shape from the other two peaks. Thus, a very careful analysis of peak shapes gives a considerable amount of information and insight into the particular source of the carbon present on the solid surface. While Auger does not have the sensitivity in this regard, and is not as definitive as XPS (already discussed), it can do

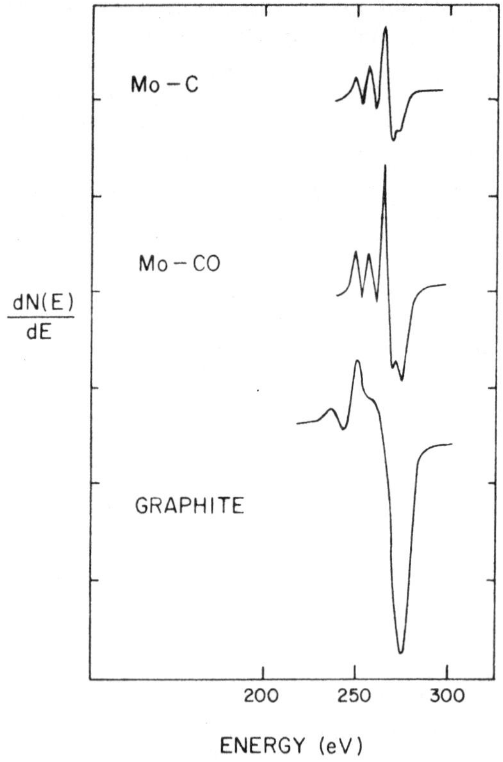

Figure 2-37. – Auger electron spectra of carbon segregated at Mo (110) surface during initial cleaning, in CO on clean Mo (110) surface, and in graphite (ref. 40).

much to help identify the source of elements on the solid surface and how they are combined with various other elements.

When elements combine on the solid surface to form compounds, there is a characteristic energy shift that takes place in the elemental peak. The electron energy may shift in one direction or another depending on the nature of the compound formed. A considerable amount of insight into the particular type of compound that is formed, and, also, the very formation of the compound itself can be learned from an examination of these shifts in the Auger spectra. The use of the chemical shifts is very effective, for example, in studying the oxidation of metal surfaces because the basic metal peak shifts when oxidation takes place on the solid surface. One can also distinguish between adsorption and oxidation. The shifts indicate whether or not the presence of the oxygen on the surface is a result of oxidation or of simple adsorption. Chang et al. have used this technique to study the oxidation of silicon (ref. 41). Figure 2-38 presents an Auger spectra for the elemental silicon peak before oxidation and after silicon oxide is formed on the solid surface. The two peaks are compared in figure 2-38. One can see there is a difference in the shape of the peak and in the size of the peak and, much more importantly, there is a shift in the position of the peak. In the elemental silicon, the peak occurs at approximately 1620 electron volts when the silicon surface oxidizes. In the silicon oxide, the peak shifts downward to an electron energy of approximately 1612 electron volts. These shifts occur with the formation of many compounds, but using the chemical shift to gain basic information is somewhat specific in nature so that, with some particular compounds, there is a relatively marked shift and the formation of the compound can readily be distinguished from the Auger peak associated with the elemental material. With others, however, the shift may be only marginal, so it may be very difficult to give positive identifica-

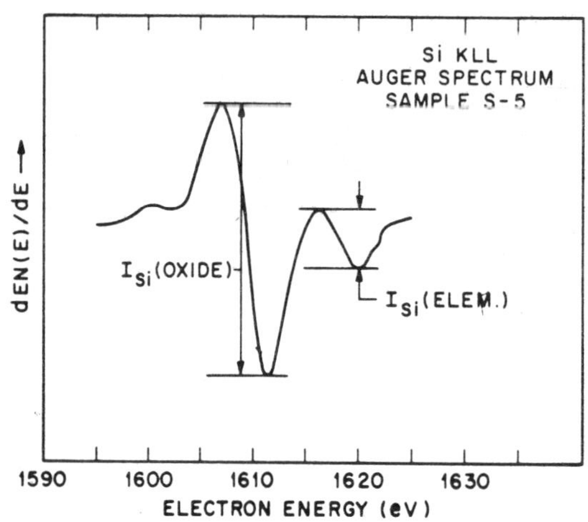

Figure 2-38. – Shift in Auger peak electron energy for silicon with oxidation (ref. 41).

tion to the formation of a compound as the result of the interaction of the element with other species on the solid surface.

Because AES analysis is such a sensitive tool and because it measures the true surface layers, small changes in the surface character can produce marked changes in the Auger spectra observed on the solid surface. For example, some elemental peaks become completely extinguished on a solid surface by the presence of adsorbates or by the formation of reaction products. Thus, it is possible to distinguish between the clean surface and a surface that is contaminated simply by observing the peaks of the elemental solid surface. As mentioned earlier, low energy iron peak can be completely obliterated by the presence of adsorbed ethylene. Copper peaks can be extinguished by the oxidation process. This is demonstrated with the two peaks presented in figure 2-39 (ref. 42). The copper 937-electron-volt Auger peak is presented on the left side of figure 2-39. This is for a clean copper surface. This particular peak has been singled out from the many Auger peaks for copper that are generally observed.

Upon oxidation of the copper surface, however, this particular peak, the 937-electron-volt peak, becomes almost completely extinguished as is observed in the Auger spectrum on the right side of figure 2-39. The copper surface was oxidized by heating to 670° to 700° C for 1 hour in an oxygen atmosphere at 10^{-4} torr. Thus, the rate at which a metal surface (e.g., copper) is being consumed in the oxidation process can be followed by measuring the Auger peak height intensity as a function of time and exposure to oxygen. Thus, the Auger peak can be used to study the kinetics of reactions, and this is very important in tribology.

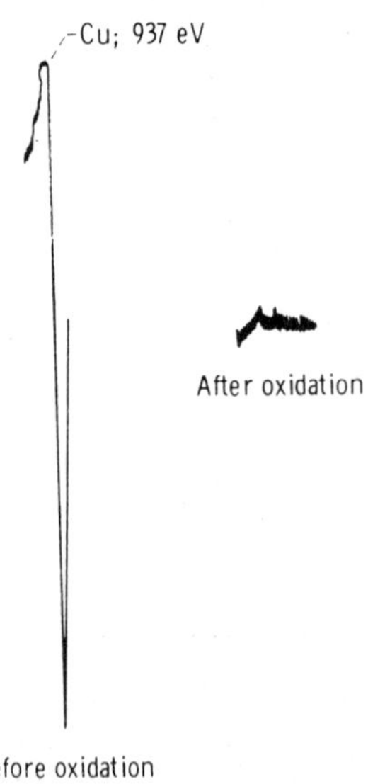

Figure 2-39. – Disappearance of 937-eV copper AES peak as result of oxidation by heating at temperature of 670° to 700° C for 1 hour in oxygen atmosphere of $5x10^4$ torr (ref. 42).

Not only can AES analysis be very effective when studying solid surface chemistry, but it can also give information about the mechanical or deformation characteristics of solid surface. That is, it can show the amount of damage that may have taken place in a relatively perfect solid surface. Chang has demonstrated, for example, that bombarding a solid surface with an energy source produces surface effects which can be detected in an analysis of the Auger spectrum (ref. 38).

In figure 2-40(a) we again see the Auger spectrum for a single crystal graphite solid surface. Note the presence of sharp peaks in figure 2-40(a).

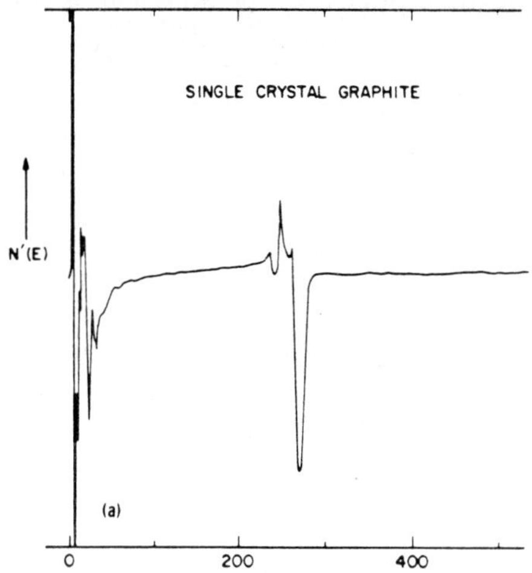

(a) Before electron bombardment.

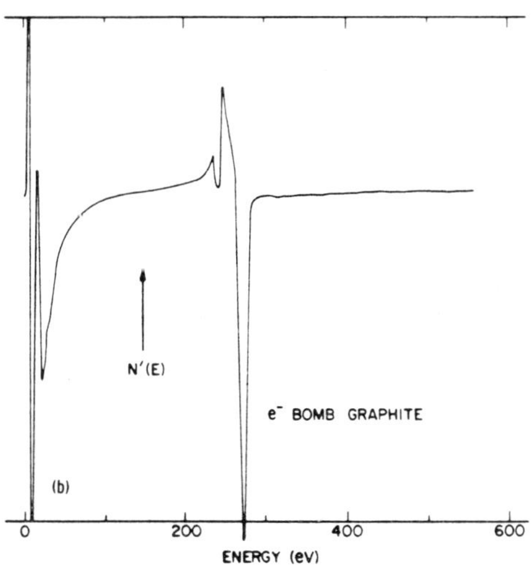

(b) After 15 minutes of electron bombardment (vertical scale expanded 2.5 times); note that most sharp peaks shown in part (a) have disappeared.

Figure 2-40. – Spectra from cylindrical mirror Auger analysis (40 μA, 3.1-keV primaries) of single-crystal graphite.

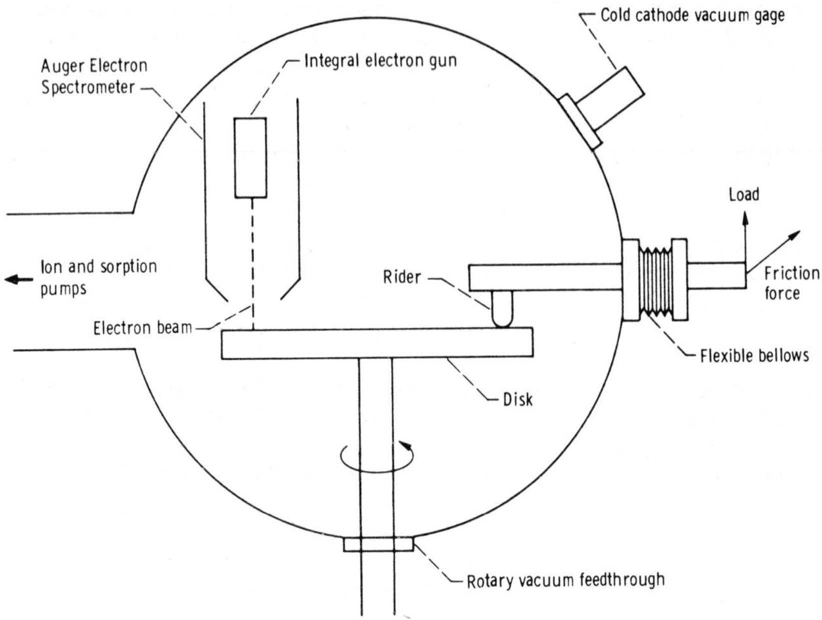

Figure 2-42. – Experimental apparatus with Auger electron spectrometer.

Figure 2-43 shows two Auger spectra obtained with the cylindrical mirror Auger analyzer shown in figure 2-42. Figure 2-43(a) shows a polycrystalline iron disk specimen surface prior to surface cleaning. The spectra appear on the black background because these represent photographs taken from the oscilloscope window. Figure 2-43(a) shows the iron surface before sputter cleaning; the surface contains sulfur, carbon, oxygen, and iron. The same spectra can be obtained whether the disk is stationary or rotating. When the disk specimen is rotating, variations in the peak intensities are observed on the solid surface and all peaks are represented. After this surface has been

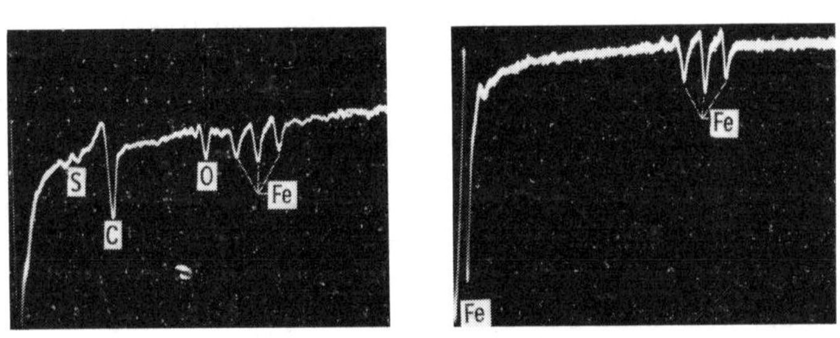

(*a*) *Before sputter cleaning.* (*b*) *After sputter cleaning.*

Figure 2-43. – Auger spectra for iron surface before and after sputter cleaning.

sputter cleaned with argon ion bombardment, the contaminants on the solid surface (including sulfur, carbon, and oxygen) have beem removed by the sputtering process and only the elemental iron peaks (fig. 2-43(b)) are left.

After the sputter cleaning is completed, the Auger peaks presented in figure 2-43(b) include the low energy iron peak, which is very sensitive to surface contaminants, and the three higher energy iron peaks. Note, as discussed earlier, that the low energy iron peak is not visible in the Auger spectra in figure 2-43(a) but is very pronounced in the Auger spectra in figure 2-43(b).

Low Energy Electron Diffraction (LEED)

In the last few sections, the use of electron beams for elemental analysis has been discussed. The electron beam can also be used for examining the structure of solid surfaces by such techniques as low energy electron diffraction (LEED). LEED is a surface analytical technique which can gain structural information about a solid surface. LEED was experimentally demonstrated in 1927 by Davisson and Germer (ref. 43). They showed that, as a result of the wave nature of an electron, electrons could be diffracted from a crystal lattice in a manner similar to X-ray diffraction from a solid surface. After their early work, the technique was pursued as an analytical surface tool by Farnsworth using a Faraday cup to detect the diffracted electrons (ref. 44).

In LEED, as opposed to standard or conventional electron diffraction, the beam of electrons directed at the solid surface are in the voltage range of 0 to 200 electron volts; consequently, they penetrate only the surface layers—that is, the first few atomic layers, of the solid. LEED is, therefore, another analytical surface tool that can be used to study the structural arrangement of atoms in the outermost atomic layer. LEED became a popular surface tool in the late 1950's when it was demonstrated that a diffraction pattern could be displayed on the fluorescent screen by postaccelerating the diffracted electrons. The basic technique is displayed in figure 2-44. A primary beam of incident electrons strikes the crystal surface from the source (an electron gun) as with the Auger analysis, but the incoming electrons are at a much lower energy than that employed in Auger analysis. The two-dimensional crystal lattice, on the outermost solid surface, serves as a diffraction grating for the incoming incidental electron beam. The beam is diffracted from the surface and the electrons strike a phosphorous or a fluorescent screen and form a distinct array of diffraction spots; these spots reflect the orientation of the crystal lattice of the solid surface. The surface crystal lattice is derived from what is referred to as the Ewald sphere.

The Ewald sphere is shown in figure 2-45. The sphere is constructed for a given energy (eV) and for the wavelength of the incident electrons. In LEED, as is the case with ordinary conventional X-ray diffraction, the reciprocal lattice of the solid surface crystallography is observed. The reciprocal lattice for a two-dimensional array is a set of rods; thus, we see

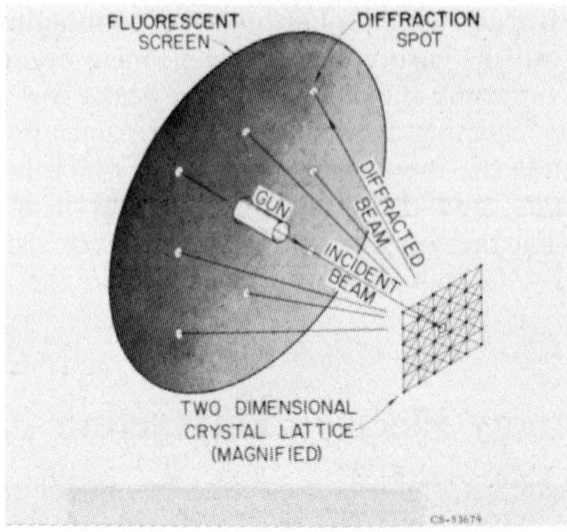

Figure 2-44. – Formation of diffraction patterns.

the reciprocal net rods are shown schematically in figure 2-45. The diffracted electrons then are represented as

$$n\lambda = d \sin \theta$$

A schematical diffraction pattern for a crystal lattice which produces a square net arrangement on a solid surface is shown schematically on the right side of figure 2-45. There are four diffraction spots about the central beam spot arranged in a square array. With LEED the specimen surface examined is generally a single crystal or a polycrystalline sample where large

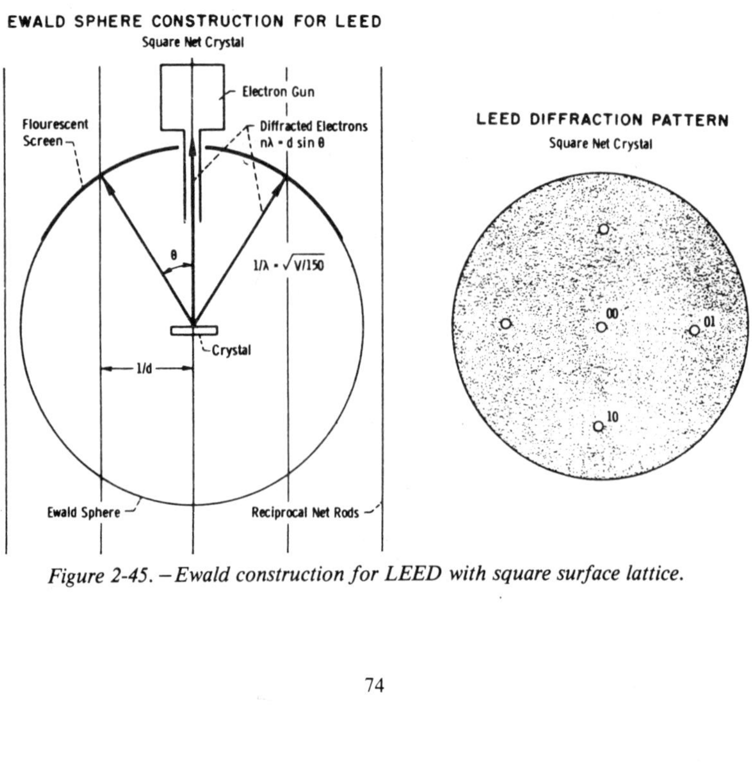

Figure 2-45. – Ewald construction for LEED with square surface lattice.

grains are involved so that the primary electron beam can be directed at a relatively small region or within one of the grains in the polycrystalline surface. LEED, like Auger and the other analytical tools described heretofore, must be operated in a vacuum system.

The schematic arrangement of the system for LEED is presented in figure 2-46. The observer is generally looking through a window at the back of the sample surface (e.g., crystal) being examined. The primary electron beam is generated from an energy source such as a heated filament with a cathode, and the primary beam is directed in a voltage range of 0 to 200 volts at this solid crystal surface from which it is diffracted. On the specimen (e.g., crystal) side of the fluorescent screen, which is depicted schematically in figure 2-46, there are a set of grids (shield grids, suppressor grid, and another shield grid). The first grid is grounded and gives a field free region. The second grid (generally two grids for better resolution in Auger work) repels all the scattered electrons but those that are the primary beam. The final grid is grounded in order to shield the retarding grids from the high voltage and the fluorescent screen. One then views the diffraction pattern on the fluorescent screen through a window and a set of three grids. The grids themselves are a fine mesh screen which do not hamper or impede visual observation of the diffraction spots on the fluorescent screen.

A typical series of LEED patterns from an iron (011) single crystal surface are presented in figure 2-47. The LEED pattern in the upper left is for the iron (011) surface with normal carbon contamination present on the solid surface. The four bright diffraction spots in a rectangular array are associated with the iron single crystal mesh from the iron (011) orientation. The additional small spots arranged in a ring structure about the four bright diffraction spots are spots associated with the diffraction from carbon contamination on the solid surface. This is carbon that has diffused from the bulk of the iron to the solid surface. This is believed to be a graphitic form of the carbon contaminant on the solid surface.

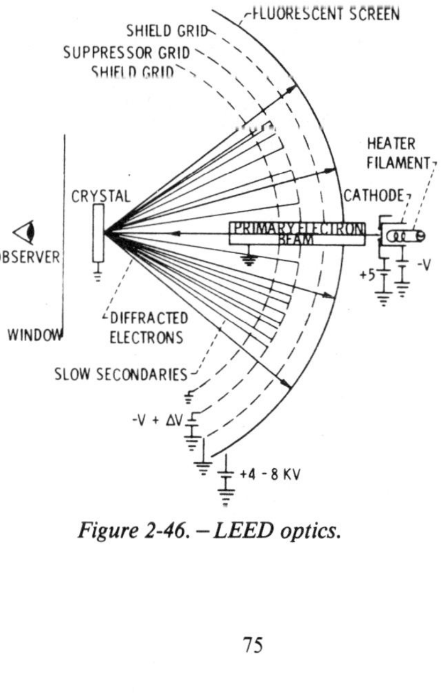

Figure 2-46. – LEED optics.

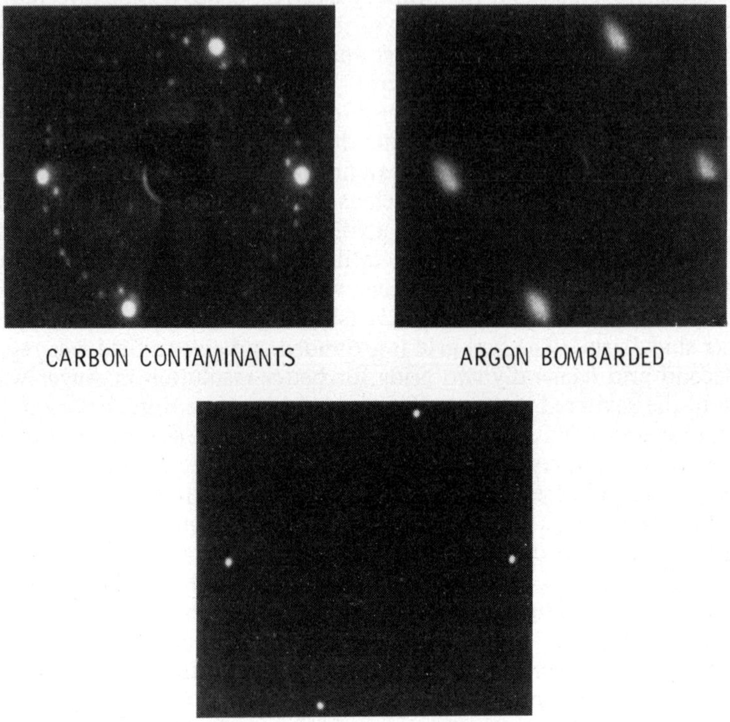

CARBON CONTAMINANTS ARGON BOMBARDED

CLEAN SURFACE (110 V) CS-78-1970

Figure 2-47. – LEED patterns of iron (011) surface.

If that surface is argon ion bombarded, the carbon contaminating layer can be removed and a diffraction pattern is obtained such as that shown in the upper right pattern in figure 2-47. This particular pattern shows four diffraction spots. The spots, however, have become somewhat elongated as a result of the argon ion bombardment. The argon ion bombarding for cleaning the carbon from the surface produces a strain in the crystal lattice in the outermost atomic layers. Since LEED is truly a surface tool and senses only the outermost atomic layer, the strain of those layers is readily reflected in the LEED pattern. So, simple argon ion bombardment at a voltage of 1500 volts is sufficient to produce this crystal lattice strain. A very mild heating at 200° C is sufficient for a short period (approximately 10 to 20 min) to anneal out and remove that strain. When that is done, one obtains the lower diffraction pattern presented in figure 2-47 for the clean iron (011) surface. Note that all that is seen visually are four bright circular diffraction spots in a rectangular array. These are the characteristic diffraction spots for a clean iron (011) surface mesh. The diffraction pattern represents the diffraction from the two-dimensional crystal lattice of the iron (011) solid surface.

Considerable controversy has arisen in recent years in the interpretation of LEED patterns. For a time, some investigators were lending various interpretations to LEED patterns as a result of examining a number of surface conditions. Careful examination of solid surfaces, however, resulted in a serious questioning of some of the interpretations made of these patterns. As a consequence, there has been a stepping back from the interpretation of complex LEED patterns. The basic LEED pattern, however, is very useful in defining a structure, if one is very careful as to how the patterns are obtained and the manner in which they are interpreted. They can be very effectively used to show the structure of the solid surface before and after. For example, in the tribological experiment where two solid surfaces are brought together in solid-state contact in an adhesion experiment, the surface can be examined both before and after contact; as a result, LEED patterns can be obtained under the two conditions, and direct comparisons can be made of these LEED patterns.

This approach or technique can also be used to study the oxidation of solid surfaces or the reactivity of the environmental species with the solid surface. It was used by Bradshaw et al. in their study of the oxidation of the tungsten (100) surface (ref. 45). The (100) surface of tungsten produces a diffraction pattern similar to the square surface lattice indicated in figure 2-45. There are four diffraction spots in a square as opposed to the rectangular array shown for iron in figure 2-47. The (100) surface of any body-centered-cubic metal produces the square diffraction pattern indicated in figure 2-45. That same pattern is reproduced in figure 2-48(a) for the tungsten (100) surface (a clean surface).

If a small amount of oxygen (approximately 0.5 langmuir of oxygen) is admitted to that clean surface, the new pattern shown in figure 2-48(b) is obtained. In addition to the four diffraction spots in a square array, additional smaller diffractions spots appear between the larger spots. This particular pattern can then be characterized as the one obtained when the oxidation of tungsten occurs at a relatively low surface coverage of oxygen on tungsten. If the exposure of the tungsten surface to additional concentrations of oxygen (1.0 langmuir) takes place, the pattern changes from that seen in figure 2-48(b) to that seen in figure 2-48(c).

Heating the crystal surface to 1250 K after 10 langmuirs of exposure of oxygen produces the pattern seen in figure 2-48(d). There are elliptical or elongated diffraction spots located midway between the four diffraction spots characteristic for the tungsten (100) surface. This relatively stable structure is observed on the tungsten surface. If the same surface is heated to 1500 K for the same concentration of oxygen as shown in figure 2-48(d), a pattern such as that in figure 2-48(e) is obtained. Instead of the diffraction spots intermediate between the tungsten diffraction spots and those associated with oxygen being elongated, as they are in figure 2-48(d), the diffraction spots become sharp, which indicates the possible formation of a distinct surface oxide on the tungsten (100) surface. Heating to still higher temperatures with the same concentration of oxygen produces the pattern in figure 2-48(f). Here there are basically four diffraction spots, in addition to the four associated with tungsten, positioned intermediate between the four

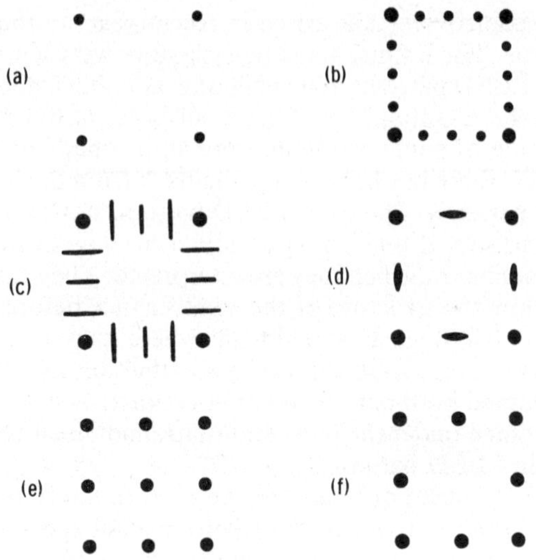

(a) Clean W(100).

(b) 0.5 Langmuir of oxygen.

(c) 1.0 Langmuir of oxygen.

(d) Heated to 1250 K after 10 langmuirs of oxygen.

(e) Heated to 1500 K after 10 langmuirs of oxygen.

(f) Heated to 1900 K after 10 langmuirs of oxygen.

Figure 2-48. – Schematic representation of LEED sequence for oxygen adsorption on W(100) (ref. 45).

tungsten diffraction spots. This kind of stepwise characterization of the surface structure, resulting from different surface treatments, can be a very effective in identifying the particular condition to which a solid surface has been exposed.

Other ways in which the LEED pattern can be used to provide information on the structure of solid surfaces are discussed in chapter 5. A distinction must be made between LEED and high energy diffraction by electrons, frequently referred to as HEED. In LEED, the electron energies are typically of the order of 0 to 200 electron volts; in HEED, electron energies are in the voltage range of 1000 to 100 000 electron volts. Both methods rely on the observation of the elastic electrons backscattered from the surface. With high energy electrons, however, the system is sometimes referred to as RHEED to distinguish the method from the more common transmission electron diffraction techniques, commonly abbreviated TED or THEED, which are used extensively to examine thin films structures.

Appearance Potential Spectroscopy (APS)

Another technique for examining solid surfaces involving the bombardment of the sample surface with electrons is called appearance potential spectroscopy or soft X-ray appearance potential spectroscopy (APS). In APS, the sample is bombarded with electrons of variable energy, and the total X-ray emission intensity is measured as a function of the incident electron energy. Elements present on the solid surface can be identified easily by the energy at which abrupt changes in intensity occur. The energy values correspond to the core level binding energies for the atoms involved. The sensitivity of this technique to chemical effects can be clearly seen from the data presented in figure 2-49 for titanium before and after oxidation (ref. 17). The solid line represents the spectra obtained for the clean titanium polycrystalline surface; the dashed line represents the spectra obtained for the oxidized titanium surface. For the L-2 and L-3 spectra lines, there is a clear and distinct shift in the peaks associated with the oxidation process and the redistribution of valence electrons. This is analogous to what was observed in AES where chemical shifts were observed after oxidation of a metal and semiconductor on nonmetal surfaces. The chemical shift with Auger analysis was seen in the data in figure 2-38 for the oxidation of silicon. The ΔE in figure 2-49 represents the change or the actual chemical shift.

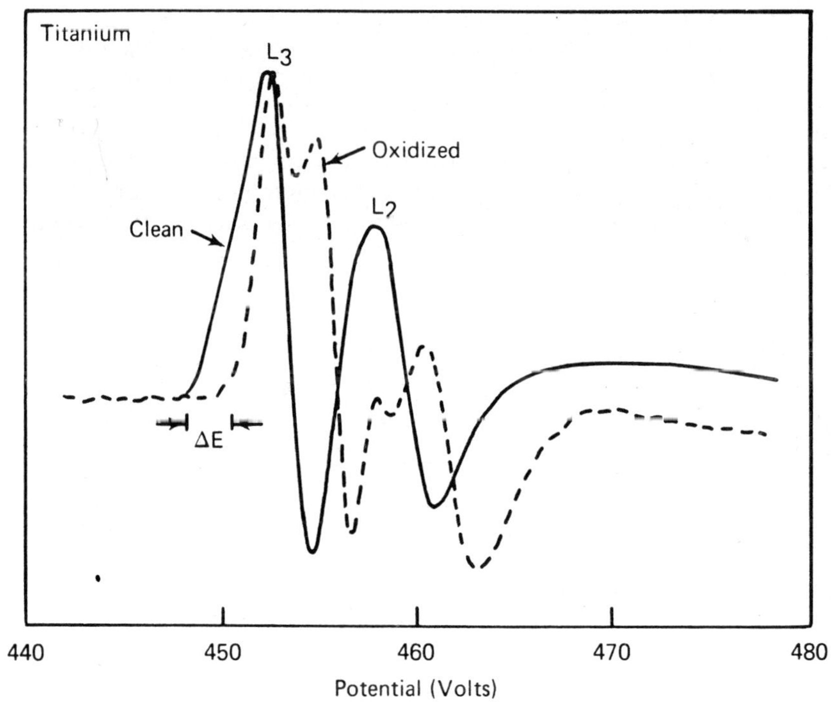

Figure 2-49.—Soft X-ray appearance potential spectrum of clean polycrystalline titanium surface before and after oxidation. Redistribution of valence electrons produces "chemical" shift ΔE in threshold (ref. 17).

Ion Beam Analytical Sources

In 1911, Rutherford and his coworkers detailed the elastic scattering of energetic ions from the solid surface. It was not, however, until 1957 that this particular concept was exploited; that is, nuclear interactions with solid surfaces were used as an analytical surface tool (ref. 46). The basic principle involved in Rutherford scattering is that a beam of monoenergetic ions is made to strike a solid surface; some of these ions are elastically scattered from the surface and lose energy in the amount related to the mass of the atom struck. The number of particles scattered is a function of the number of atoms present on the surface. Hence, if a sample is used as the target of a particle beam and if a way of measuring the energy and counting the number of scattered particles is provided, the identity and number of atoms in the solid can be determined. A schematic of a typical Rutherford scattering analytical system is shown in figure 2-50. A beam of ions is generated with a Van de Graaf generator or some other device and projected through an aperture to the target specimen surface. A detector is provided for detecting the scattered ions and measuring the energy associated with them. The Rutherford backscattering technique is very effective in analyzing or identifying small concentrations of material on a solid surface where the material on the surface has a lower mass than the substrate. The technique is simple, rapid, sensitive, and nondestructive.

The sensitivity of the Rutherford scattering technique is very good. For example, Ball, Buck, and Wheatley (ref. 47) found that they could detect 3 nanograms (10^{-9} g) per square centimeter of oxygen on a graphite surface. The technique is also useful for looking at thin layers on solid surfaces. For example, Rubin et al. (ref. 46) found that they could analyze electrostatically precipitated aerosols on aluminum foils and obtain 5, 0.13, and 0.3 microgram per square centimeter of oxygen, sulfur, and silicon,

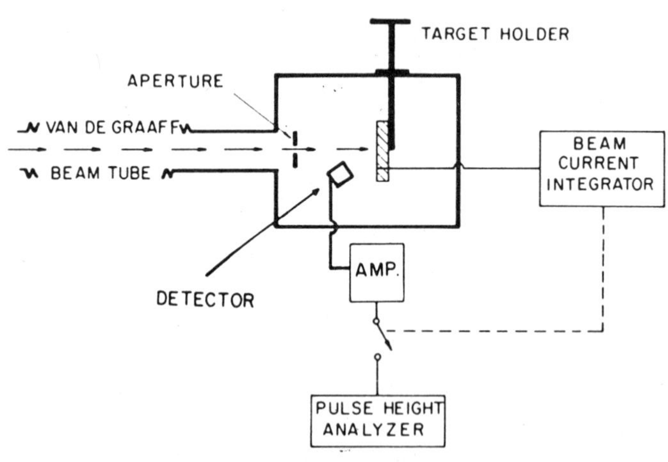

Figure 2-50. – Typical target-chamber arrangements for Rutherford scattering analysis (ref. 46).

respectively, on the aluminum foil surface. Furthermore, Nicolet et al. (ref. 48) found that they could study the films deposited by drawing smog through cellulose ester filters. During their investigation they found that lead was dominant in the spectrum, but they did not report any really quantitative results. When a sample surface has been altered chemically, information on the thickness and composition of the surface layer may be obtained from the Rutherford scattering spectra. The most commonly encountered situations are, for example, oxides and sulfides on all metal substrates.

As seen in figure 2-51, the spectrum for such a sample exhibits a sharp edge at the energy representing scattering from the high mass component at the surface (ref. 49). The sloping plateau is interrupted by a second sharp step which represents scatter from the interface of the substrate and the compound formed on the solid substrate. The count at this point is greater

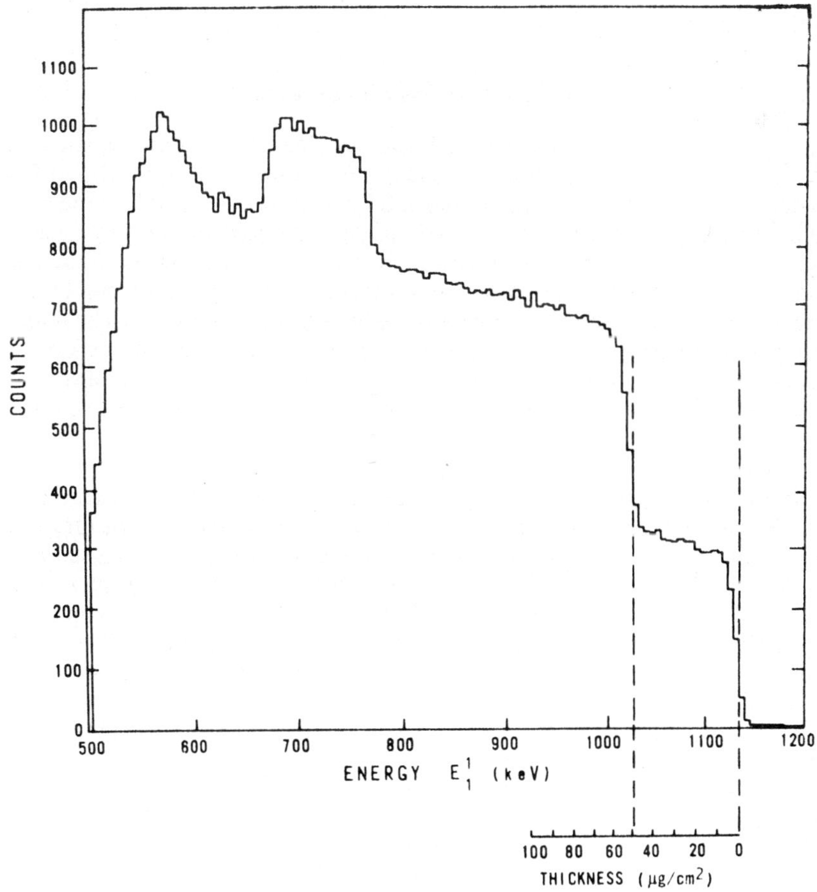

Figure 2-51. – Rutherford scattering spectrum of aluminum with surface film of oxide which is 50 micrograms per square centimeter thick (ref. 49).

because the number of atoms contributing to the yield per unit of depth sharply increases from compound to substrate. Imposed on the plateau resulting from scattering deeper and deeper in the substrate is a peak derived from the low mass component of the surface compound. The area of this peak is difficult to obtain with sufficient precision to calculate the thickness of the surface film. When, however, the stoichiometry of the compound is known so that the rate of energy lost can be calculated from tabulations, the energy of the spectrum can be converted to a depth scale and the position of the interface step indicates the thickness of the compound. The thickness of the film or compound that can be measured as a lower limit is set by the minimum energy differences detectable by the counting system, and an upper limit is set by the requirement that the ions back scattering from the substrate must have sufficient residual energy to emerge from the film surface. Film thicknesses on aluminum surfaces from 10 to 100 micrograms per square centimeter have been measured by Peisach (ref. 50).

Ion Scattering Spectroscopy (ISS)

In ion scattering spectroscopy (ISS) the sample surface is bombarded with an ion beam of low energy ions, usually helium or argon, typically of the order of 1 kilovolt. They impact the solid surface at about 45° to the normal. Those ions reflected at 90° are energy analyzed by an electrostatic analyzer. The energy distribution of the reflected ions is then measured. The identification of the species in the outermost monolayers is provided by the energy of the reflected ions using a simple binary collision model. The intensity of the peaks provides a measure of the abundance of the particular species. ISS is capable of observing between 10^{-2} and 10^{-3} monolayer of material on the solid surface. In addition, isotopes can be observed, but the mass resolution is poor at high masses even when using higher mass probing ions such as argon instead of helium.

In ISS the ion gun is probably the simplest component of the instrument. It uses bigger upper gauge geometry to produce the inner gas ions (argon). These ions are extracted actually from one end and focused onto a sample, as indicated in figure 2-52, by a set of electrostatic pin-hole apertures. Nominal beam diameters are from 10 to 1/2 millimeter, and they are obtained by additional focusing adjustments in the energy range of 300 to 3000 electron volts. This focusing gives a variation in the beam intensity of 1 to 15 microamperes per square centimeter. The energy spread of the primary ion beam is less than 0.15 percent. The beam currents are typically around 100 to 200 nanoamperes. Uniaxial manipulation of the beam is possible by employing electrostatic deflection plates just in front of the sample. One of the better features of the ISS is that there is no need for special preparation of insulating surfaces. The ISS technique is equally effective for metals and nonconductors.

After the inert gas ions from the ion gun strike the sample surface (fig. 2-52) they are reflected or scattered. A 5-centimeter (2-in.) radius, 127°

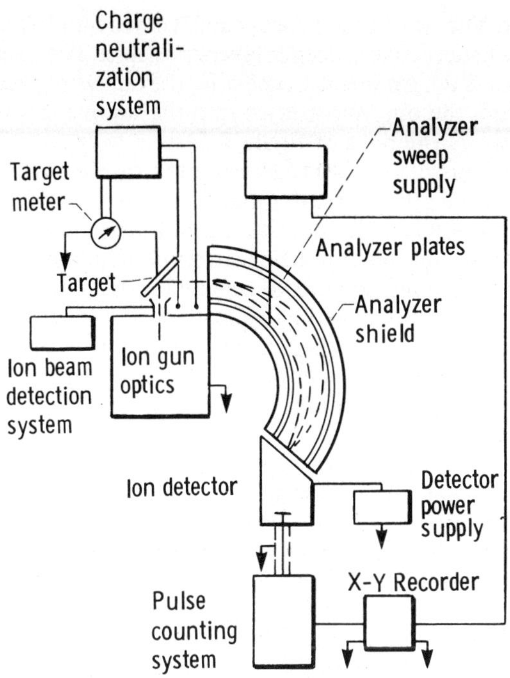

Figure 2-52.—Schematic drawing of ion scattering spectroscopy (ISS) experimental apparatus.

parallel plate electrostatic energy analyzer with a 12.7×10^{-5} meter (5-mil) acceptance aperture is placed 90° from the instant beam to monitor the scattered inert gas ions leaving the sample surface. Ions that pass through the electrostatic analyzer have undergone a simple binary elastic collision with an atom on the surface and thereby have had their energy altered according to the elastic scattering energy considerations. Detecting the scattered ions that have passed through the energy analyzer is done by an electron multiplier with a special cathode. The gain factor of this detector is of the order of 10^8. This type of detector is very stable; it can withstand air cycling without any kind of degradation of the unit. The entire system is housed in a vacuum chamber which should be capable of achieving pressures of the order of 10^{-9} or 10^{-10} torr.

An ion pumping system is desirable to avoid surface contaminants which may arise from other types of pumping systems. The data obtained from the ISS are shown by a plot of intensity as a function of atomic mass, and this is generally plotted on an X-Y recorder. The plot shows the presence of all of the ions that have an atomic mass greater than that of the bombarding gas ion. When helium is the bombarding ion, all elements except hydrogen and helium are analyzed. In addition to the primary ions that are scattered from the surface, sputtering also takes place as a result of the primary beam dislodging surface atoms that achieve enough momentum to leave the surface of the solid. The sputtering process uncovers the underlying layers and makes analysis of successively deeper layers possible. The composition as a function of depth is determined by repeating the analysis several times. The depth profile from the outermost layer into the sample is established (ref. 51). This is shown in the data of figure 2-53.

The depth profiles with monolayer resolution can be extremely useful in studies relating to the properties of a material from its surface to the bulk; for example, in the study of composition as a function of thickness and in the studies of contamination and the penetration of contaminants into solid surfaces. With helium ions, the rate of removal generally ranges from 3 to 5 monolayers per hour, thus enabling a single monolayer to be carefully examined or quickly removed. When surface cleaning is desired at a higher rate, heavier ions can be substituted for helium (e.g., argon or neon in place of helium). The removal rate can be increased by as much as a factor of 10 by using the heavier mass ions.

In figure 2-53, three spectra are presented as a result of ISS analysis of a René solid surface. René 41 is a nickel-aluminum-chromium alloy which is used at high temperatures because of its good oxidation resistance. The upper spectra is for the solid surface after removal of four monolayers of material. It can be seen that a number of elements are present on the solid surface; these include oxygen, aluminum, silicon, calcium, chromium, iron, nickel, molybdenum, and iridium. After the surface sits 1 hour in the system, oxidation takes place on the solid surface and some of the smaller peaks are masked by the process. Note that the oxygen peak grows after 1

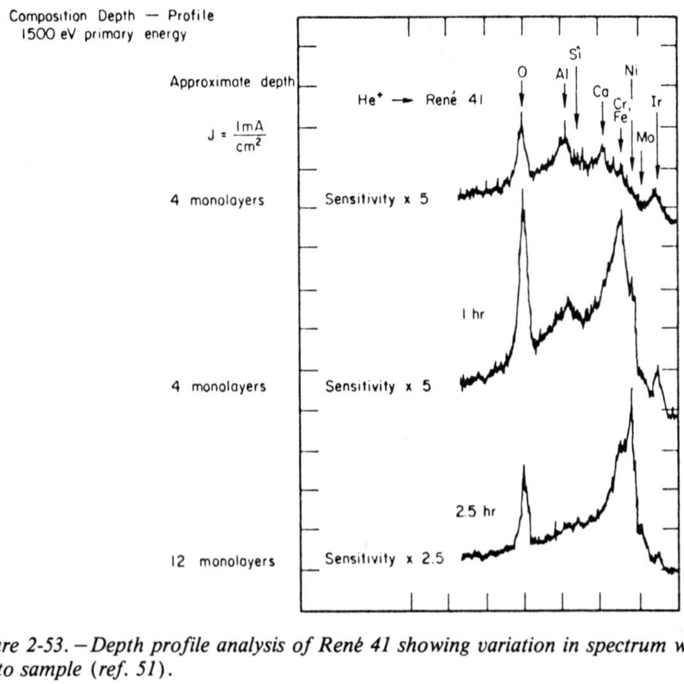

Figure 2-53.—Depth profile analysis of René 41 showing variation in spectrum with depth into sample (ref. 51).

hour, and the middle spectra of figure 2-53 and the other peaks decrease in intensity with the exception of the iron, chromium, and nickel which increase as a result of the formation of the oxides of iron, chromium, and nickel. After twelve monolayers have been removed by sputtering, the principle peaks present are those associated with oxygen (which is decreased in intensity over the peak seen in the middle spectra) and those associated with the transition metals iron, chromium, and nickel.

Because of the direct relationship between the mass and energy in the ISS technique, a given scattering peak is uniquely related to a given mass. This unique mass to energy relationship allows the positive spectrum identification without the confusion introduced by overlapping peaks. Furthermore, since the intensity of a given scattering peak is proportional to the number of scattered ions, it is directly related to the amount of material present in the surface.

Secondary Ion Mass Spectrometry or Ion Microprobe Mass Spectrometer (SIMS)

Secondary ion mass spectrometry or ion microprobe mass spectrometer (SIMS) is another technique that can be used in the characterization of solid surfaces. (These two terms are used interchangeably.) SIMS must be distinguished from ISS. In ISS, the ions from an ion source striking the surface are reflected off the surface and the energies of the reflected ions are measured. In SIMS, the ions which strike the surface cause the uppermost atomic layers to be sputter removed; that is, solid neutral atoms or molecules in surface layers are knocked out by the impinging ions. Although most of the material that leaves the solid surface as a result of this impingement are neutral atoms or molecules, a fraction are ejected as positive or negative ions. The secondary ions are collected and mass analyzed—hence, the term secondary ion mass spectrometry, or as referred to by some authors, ion microprobe mass spectrometer.

Control and localization of the sputtering process permits chemical analysis of diameters as small as 1 micrometer; the examination is of such a nature that fractional portions of monolayers can be detected and an in-depth analysis to approximately 50 to 100 angstroms can be achieved. All the elements can be detected, from hydrogen to uranium including isotopes of these elements with a sensitivity of 10^{-15} to 10^{-19} gram, depending on the element being examined. There is probably no other instrumental analytical technique that can provide such sensitivities. One of the disadvantages of this technique, of course, is that it is destructive to the solid surface.

In the SIMS, the primary spot can be focused to diameters as small as 1 micrometer and can be positioned electrically over the specimen surface. The secondary ions are collected by a high aperture double focusing mass spectrometer and detected by a highly sensitive ion detector. Only one specific mass can be detected at a time. It is impossible to display the ion intensities as the brightness of a spot on an oscilloscope screen. By synchronously sweeping the primary ion spot on the specimen, and a spot

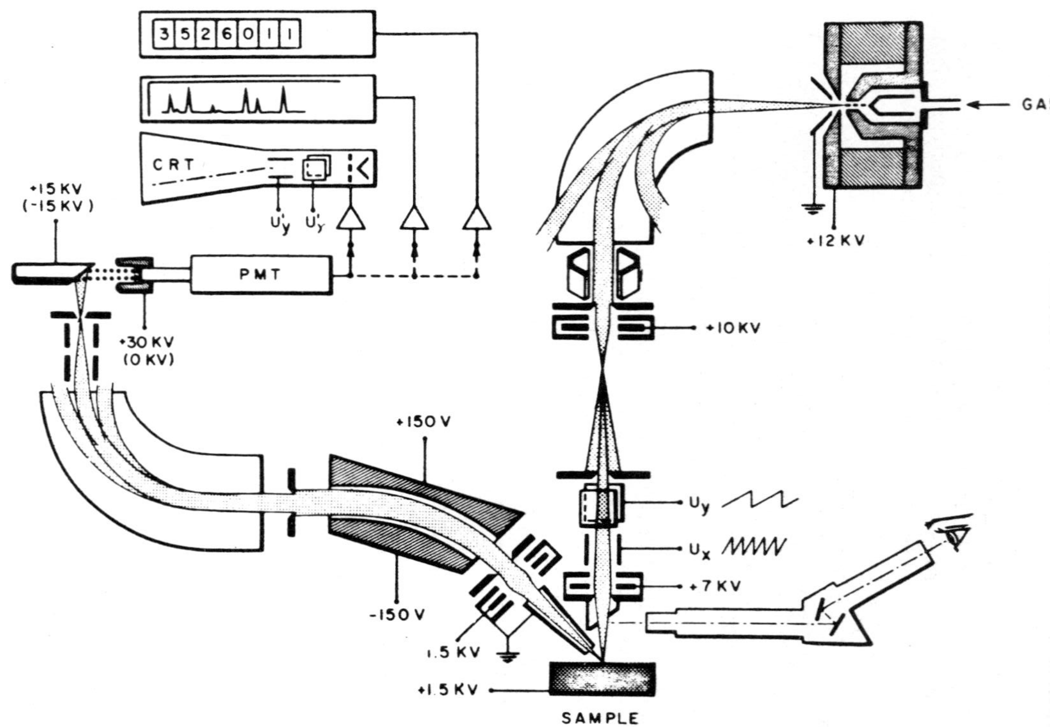

Figure 2-54. –*Schematic diagram of ion microprobe mass spectrometer. U's represent sweep voltages in x and y directions; CRT, a cathode ray tube; and PMT, a photomultiplier tube (courtesy of AAAS).*

representing the ion intensity, a picture of the concentration of that isotope on the surface of the specimen can be obtained. The picture or isotopic concentration map obtained is achieved at a sample consumption rate that may be significant if several iosotopes of several elements are to be mapped because the pictures must be obtained for each element in sequence. This is another inherent limitation in the device.

Basically, the ion microprobe in design is a blend or mixing of two mass spectrometer systems. The primary ion system consists of a high efficiency ion source for gases, often a dual plasmatron with a mass analyzing system for positioning the primary ion beam on the specimen surface as indicated in figure 2-54, which is a diagramatic sketch of the SIMS. Secondary ions are collected from the specimen surface by a second ion optical system that is double focusing in design for a high ion acceptance in transmission. The optical axes of the primary and secondary systems are usually inclined about 45°, but a perpendicular arrangement has also been employed by some investigators. It might be expected that the angular distribution at which the secondary ion is produced would have a great effect on instrument capabilities. A sensitive ion detector is used to determine the ions in the mass resolve secondary ion beam because the ion current at this final stage is extremely low.

Instruments like that depicted schematically in figure 2-54 are commercially available. Some typical spectra obtained with the SIMS are shown in figure 2-55. The spectra are for a silicon carbide surface having contacted

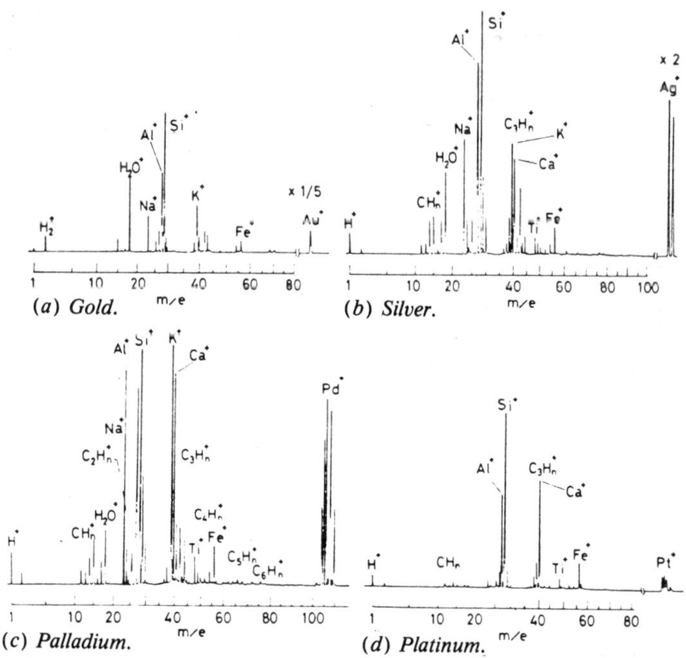

Figure 2-55. – Secondary ion mass spectroscopy (SIMS) spectra of surfaces of SiC abrasive paper used to finish various metals (ref. 52).

four different metals in the abrasion process (ref. 52). The silicon carbide is an integral part of an abrasive paper which rubs the four metals gold, silver, palladium, and platinum.

In figure 2-55(a), the spectra reveal the presence of gold on the surface of the silicon carbide. Likewise, in figure 2-55(b), the presence of silver is detected, and in both spectra the silicon from the silicon carbide is seen as one of the predominant peaks. Figure 2-55(a) shows a relatively small concentration of gold being transferred to the surface whereas with silver a much higher quantity is observed. In figure 2-55(c), palladium is present on the silicon carbide surface in addition to the silicon peaks with a silicon carbide.

With platinum (fig. 2-55(d)), as (fig. 2-55(a)) with gold, very small concentrations of the metal are transferred in the abrasion process to the silicon carbide surface. In contrast, however, with both silver (fig. 2-55(b)) and palladium (fig. 2-55(c)), relatively large concentrations, as indicated in the peak intensity of silver and palladium, are transferred to the silicon carbide; this indicates a difference in the affinity between these two metals and silicon carbide.

Conventional Mass Spectrometer

The conventional gas analyzing mass spectrometer, while normally not thought of as being an analytical surface tool, can be used very effectively as one. For example, adsorbed gaseous species on a solid surface can be desorbed from that solid surface and their chemistry analyzed with the conventional mass spectrometer. Basically, the conventional mass spectrometer can be considered as an apparatus that produces a supply of gaseous ions from a sample surface. It has the function of separating the ions in either space or time according to their mass to charge ratios, and it provides an output record, or display, indicating the intensities of the separate ions liberated from the solid surface. In reality, the ion microprobe is a mass spectrometer.

In conventional mass spectrometery, the ions produced from a sample are generally examined in the gaseous phase and in a vacuum of approximately 10^{-6} torr. A matter of primary importance in the characterization of solid surfaces is the method used to convert sample material to ions in the gaseous phase and the chemical and physical mechanisms involved in the process. The most widely used ion source in mass spectrometery is the electron bombardment source in which a beam of electrons is used to eject electrons from neutral gas molecules and produce positive ions. In order to use this ionization technique to study a solid, the solid must have a vapor pressure great enough to produce the necessary gas for ionization. Furthermore, the vaporization process must be under experimental control if meaningful data are to be obtained.

Conventional mass spectrometery is very useful for the tribologist because organic molecules that may be present on the solid surface can be readily analyzed. For example, naphthalene, a crystalline, white organic compound with a high vapor pressure at room temperature, can be studied easily in a mass spectrometer with an electron bombardment source. The same instrumentation, however, is not effective for studying degassed graphite unless an auxiliary step of selectively is used to convert the graphite to a gaseous product, possibly by treating it with hydrogen or oxygen.

For the analysis of solids, thermal ionization sources for the mass spectrometer are almost exclusively used. For this reason, this particular spectrometer system is described here. The thermal ionization source involves the principle that at high temperatures materials are evaporated and ions are given off in addition to neutral molecules. The ions produced are separated to obtain a mass spectrum. The thermal ionization source is most useful for iosotopic analysis of a number of elements, but it does present difficulties in securing the consistent operation necessary for good quantitative analysis. A considerable amount of mass spectrometer research has been conducted with the 180° magnetic-sector instrument (fig. 2-56)). Gas admitted to the system is ionized at location B and directed at the sample surface C. The ionized species liberated from the solid surface are directed through focusing plates and through the ion trajectory at location F with the 180° sector segment. The directional focusing of the ions is accomplished through the collector slits at G. At H the collector leads to an amplifier which identifies the particular species and the intensity of that species which has been liberated from the solid surface. The system is continuously pumped at location K, and the vacuum is typically of the order of 10^{-6} torr.

A typical spectrum obtained from the analysis of a nickel-hardened gold electroplate on a solid surface is shown in figure 2-57 (ref. 53). The ion in-

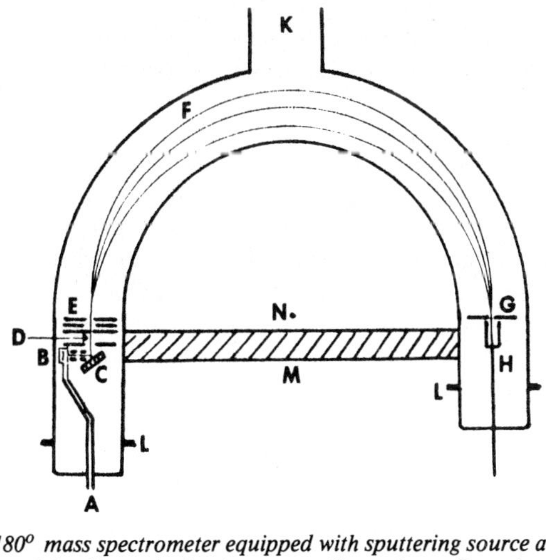

Figure 2-56. – 180° mass spectrometer equipped with sputtering source and Faraday-cup ion collector. A, gas inlet; B, gas-ionization source; C, specimen; D, region of ionizing electron beam for ionization of neutrals; E, focusing plates of secondary-ion source; F, ion trajectories illustrating direction-focusing principle; G, collector slit; H, collector lead to amplifier; K, pumpout tube; L, demountable flanges; M, mechanical brace for rigidity; N, center for central orbit.

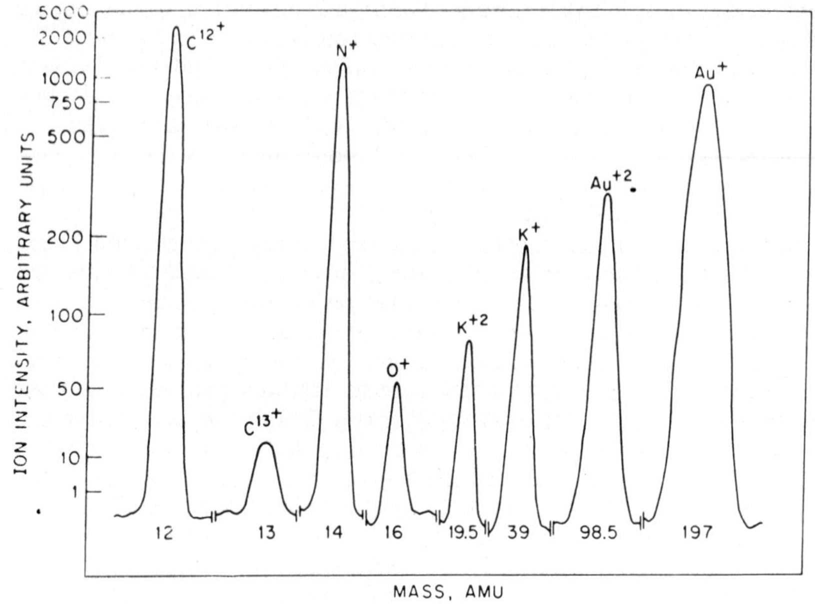

Figure 2-57.—Microphotometer trace of ion-sensitive plate resulting from bulk analysis of nickel-hardened gold electroplate (ref. 53).

tensity in arbitrary units is plotted as a function of atomic mass units (AMU), and the various elements that are present on the solid surface can be readily seen from the spectrum. Note that gold is a very large peak and that there is a considerable amount of carbon and nitrogen trapped in the film. For the tribologist, conventional mass spectrometery is a very useful tool because, with complex organic molecules present on a solid surface, fingerprint patterns (obtained using controls or standards) can give indications of the source of unknown compounds present on the solid surface. Thus, when one has a molecular structure associated with a particular lubricating species, and there is a need to identify that source by analyzing with the mass spectrometer, the spectrum obtained can be compared to that for the particular organic molecule as well as to various known standards. For complex molecular structures, it is probably still one of the simplest and most inexpensive techniques to use in characterizing the presence of these complex organic molecules on solid surfaces.

Field Ion Microscope (FIM)

The field ion microscope (FIM) allows the tribologist to examine each individual atom site on the solid surface. The microscope was originally invented by Erwin Muller of Penn State in the mid-1950's. Its construction is quite simple. A very fine piece of wire is etched on its end to generate a relatively smooth radius of 500 to 1000 angstroms. The piece of wire is mounted in a cryogenically cooled holder inside a vacuum system. At about 10 centimeters from the end of the pin tip is a phosphur-coated fiber optic window, or in earlier times, simply a phosphorescent screen. The system is evacuated by a vacuum pumping technique, and the tip is cryogenically cooled to liquid hydrogen or liquid helium temperatures (for less sophisticated work, ordinary liquid nitrogen is sufficient). Some gaseous species are then admitted to the vacuum system and allowed to condense on the cryogenically cooled pin-tip surface.

A high voltage is then applied to the pin-tip surface and this generates helium ions. The helium, which is resting on the pin's radius surface, occupies sites like those in the metal itself. If a high voltage is applied and if the helium atoms that condensed on the solid surface are ionized, they can then be projected 10 centimeters from the pin tip to the phosphorus-coated plate or fiber optic window. In the process of moving 10 centimeters, the image of the pin tip is magnified from one to three million times. This high degree of magnification of the solid surface allows one to see on the phosphor-coated plate or fiber optic window the individual atom sites on the solid surface from which the helium ions came. Because of the high magnification, each individual atom site on the solid surface is readily seen in the FIM. A typical image seen when a pin tip of iridium is magnified from one to three million times is seen in figure 2-58.

Each individual white spot in figure 2-58 represents an individual atom site on the solid surface of the iridium. The rings seen are associated with the various atomic planes in the solid surface. Since the pin tip has a radius, there are a number of atomic planes exposed in the process of generating the image, and the various rings on the solid surface represent those particular atomic planes.

When one particular orientation is known, the ideal surface can be generated by computer analysis using crystallographic considerations. For example, in figure 2-58, if the central plane is known in the center of the photomicrograph, all the adjacent planes can be defined by computer analysis and compared to the image actually obtained. Thus, any disregistries or deviations from the ideal structure that has been determined by computer analysis of the real surface, such as that shown in figure 2 58, can be readily ascertained.

Another distinct advantage in using the FIM by the tribologist is that the FIM is essentially a tool which provides an asperity-free surface. The pin tip being examined in figure 2-58 has a radius that is actually free of asperities. When imaging first takes place at the relatively high voltages, the high voltage concentrates at the asperities that are present on the solid surface and causes localized evaporation of those asperities because of the high concentration of energy at those locations. Prior to obtaining an image of the complete surface, little localized bright spot images are obtained; these are constantly changing as evaporation takes place until, ultimately, the asperities are completely evaporated from the solid surface and only the image of the entire surface is asperity free. Thus, for the tribologist who is concerned and interested in the contact of asperity-free surfaces with solids, he has such a surface in the field ion tip.

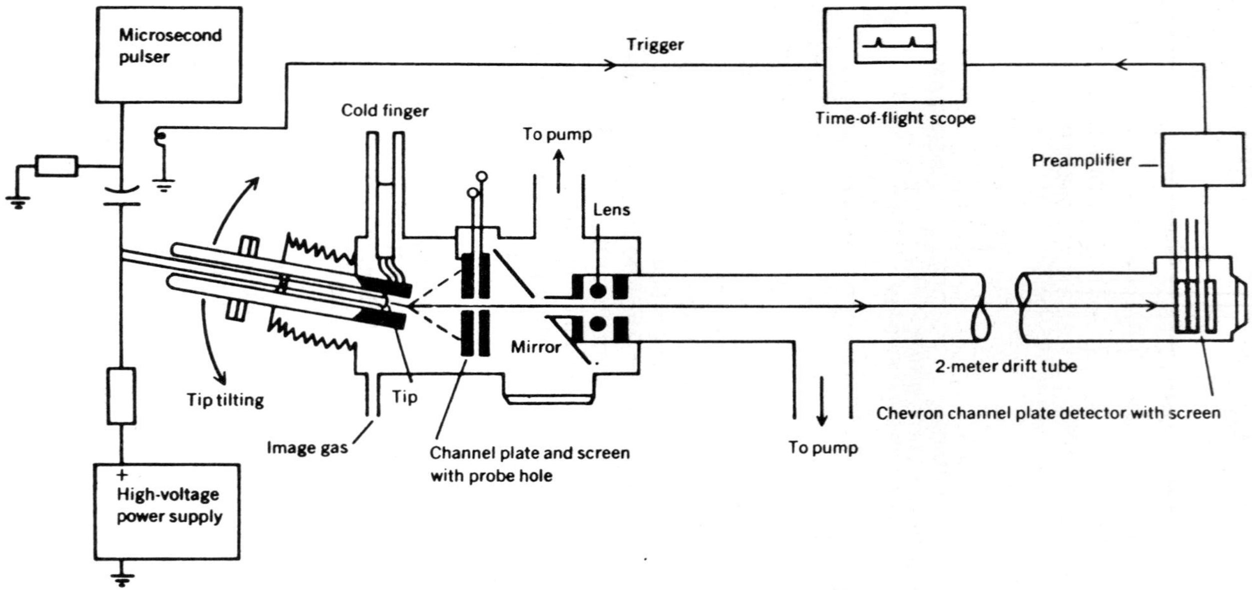

Figure 2-60. – Schematic drawing of atom-probe field ion microscope (ref. 54).

face layer. Since this is done in a vacuum system, the atoms in that outermost layer evaporate from the solid surface and travel in straight lines toward the channel plate. The particular atom of interest passes through the hole in the channel plates, through the hole in the mirror and the lenses, and down the time of flight mass spectrometer tube where the detector at the end senses the particular atom of interest and measures the m/E ratio, gaining some information about the atomic species present on the solid surface. It is, then, a conventional mass spectrographic analysis of the atoms using a time of flight mass spectrometer and a scope.

The combination of the atom probe with the FIM provides the tribologist with surface characterization tools to identify the structural arrangement of atoms on the solid surface (as well as the chemistry of clusters of atoms) or even single atoms on the solid surface. A typical atom probe spectra from a region of a solid surface of a pearlitic steel is presented in figure 2-61. The mass spectrum is for the cementite phase of the pearlitic steel (ref. 55). The FIM was used to locate the cementite phase on the surface; that particular phase was analyzed with the assistance of the atom probe, and the spectrum of figure 2-61 gives the carbon and iron peaks as well as the minor peaks of silicon, chromium, and copper, which are probably present as impurities.

Neutral and Ion Impact Radiation

Another technique has recently been developed to study the composition of solid surfaces by employing neutral and ion impact of a solid surface and then measuring the visible, ultraviolet, and infrared radiation that produced when the beam of low-energy ions or neutral particles impacts on a solid surface. The solid surface constituents are determined by an identification analysis of the optical lines that are produced in the collision process. The ions and neutrals involved attain low energy; this condition makes the technique very sensitive to the surface regions since the range into the solid is limited to at most a few monolayers because of the energy of the incident projectiles. The technique is fairly nondestructive; the damage to the sample surface can be even further minimized by using low bombardment energies as well as low current densities. Since surface atoms are being removed continuously by the incoming beam of ions or neutrals, the measurements (as a function of bombardment time) can be used to provide depth profile information on the solid. The measurement of optical radiation induced by the heavy particle impact on the solid surface requires monoenergetic beams of ions and neutrals in the range of 10 to 10 000 electron volts.

An apparatus used to study solid surfaces with this technique —a neutral beam and ion beam source of low energy—is shown schematically in figure 2-62. Using a neutral beam has a distinct advantage over some of the previously discussed systems in that it does not provide for the buildup of surface charges on nonconducting surfaces that may be examined. The first step in using this device (shown schematically in fig. 2-62) is to generate a source of ions or neutrals as the bombarding species. Electron bombardment ionization, usually of nitrogen molecules or one of the rare gases, is

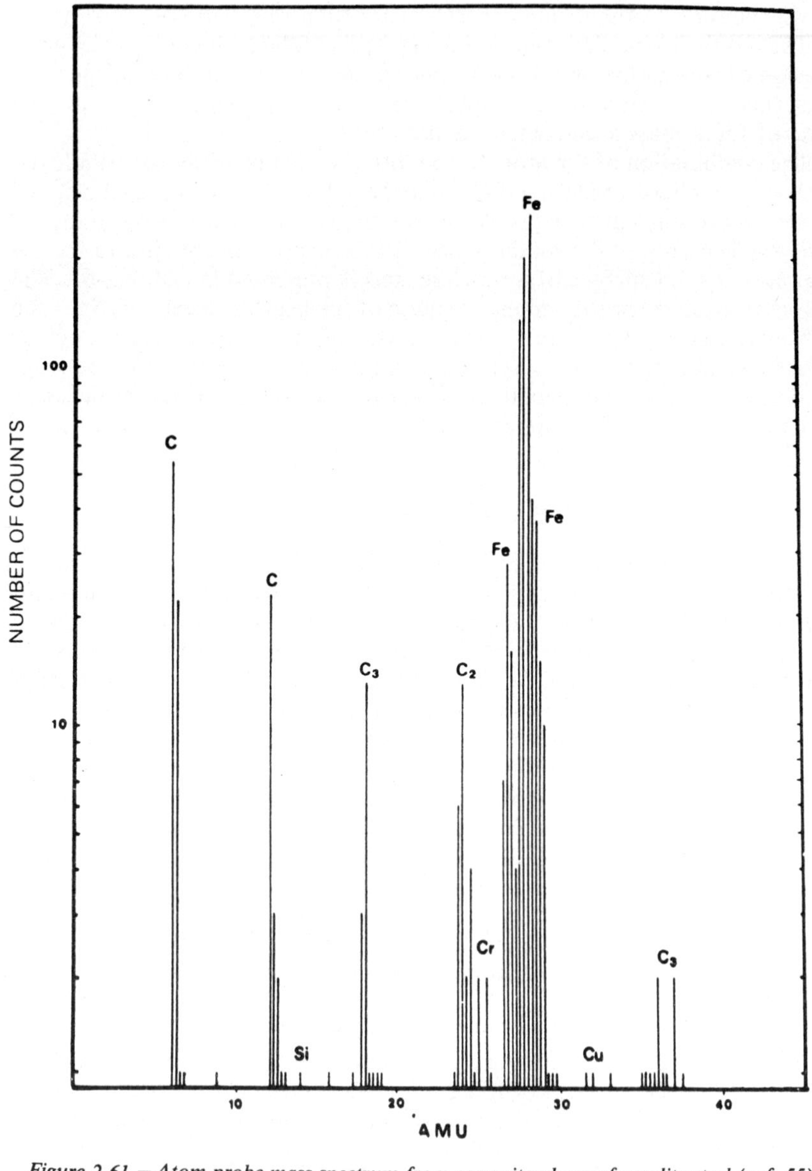

Figure 2-61.—Atom probe mass spectrum from cemenite phase of pearlite steel (ref. 55).

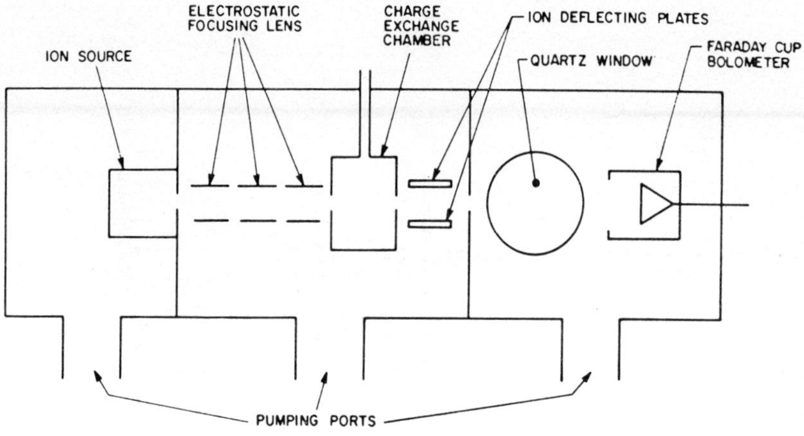

Figure 2-62.—Schematic diagram of apparatus to produce low-energy ion and neutral beams.

generally used, and the location of this is indicated in figure 2-62 as the ion source. Ions are then withdrawn from the source region, accelerated to the desired energy, and focused into a charge exchange chamber where fast neutrals can be created from ions by means of resonant charge transfer. Since the process involves very little momentum transfer, the resulting neutrals have essentially the same energy and direction as the initial ions. In general, the gas pressure in the charge exchange chamber is adjusted to produce approximately 30-percent neutralization of the incident ion beam. After passing through the charge exchange chamber the remaining ions can be deflected out of the beam, leaving only neutrals to enter the target chamber.

Alternately, to use ions as the bombarding species, gas is removed from the charge exchange chamber and the ion deflecting voltage is reduced to zero. The solid to be bombarded is oriented in the target chamber with the surface at an angle of approximately 45° to 60° with respect to the direction of the incident beam. The target chamber is equipped with a quartz window for viewing the interaction region, and photons, which are produced as a result of the particle solid collisions, are focused by a lens, usually quartz, into a monochrometer. The monchrometer and a cooled photomultiplier are used to record the spectral distribution of radiation produced in the collision process. Single photon counting techniques are used to enhance the sensitivity of the system for detecting radiation. The flux of the neutral beams in the target chamber is measured by a bolometer, which can also be operated as a Faraday cup to measure the ion current. The neutral beam intensity varies as a function of beam energy from 1×10^{-8} ampere at 30 electron volts to 1×10^{-6} ampere at 4000 electron volts. The actual beam diameter which impinges on the target surface, or the sample surface, ranges from 2 to 3 millimeters. In those experiments where depth profile is desired, higher beam current densities may be used.

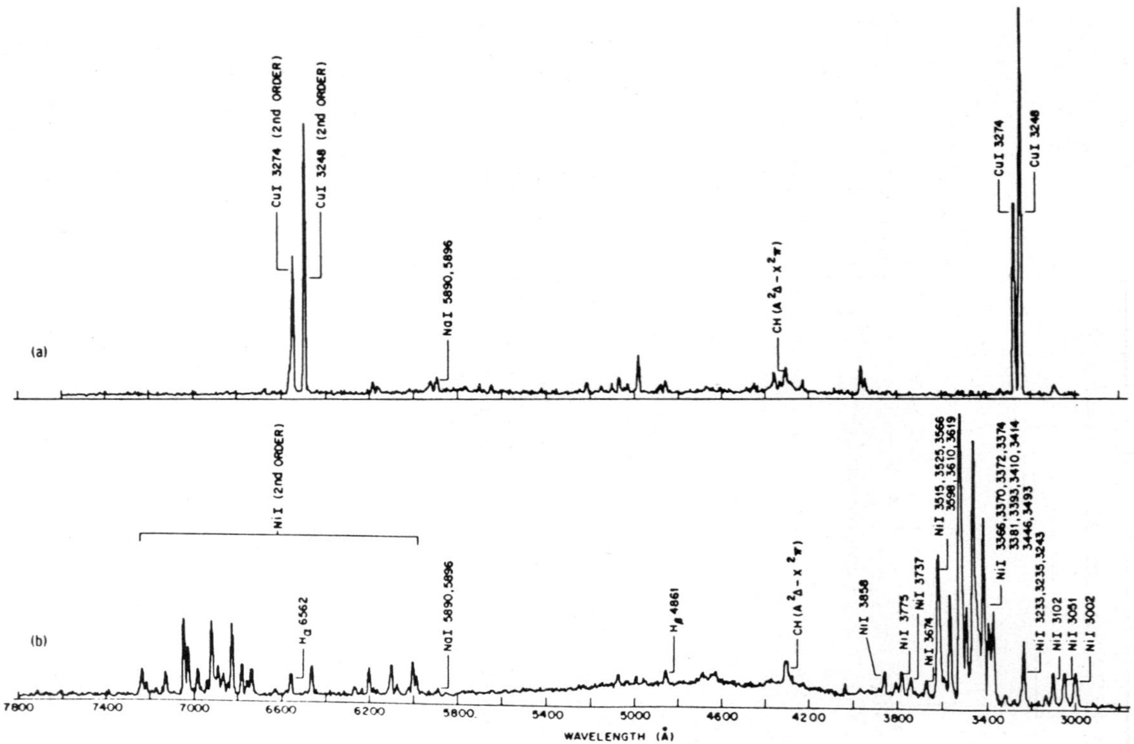

(*a*) *Copper.* (*b*) *Nickel*

Figure 2-63. – Spectra of radiation produced in impact of Ar^+ (4 keV) on copper and nickel. Lines arising from excited states of sputtered copper, nickel, and various contaminants are shown (ref. 56).

There are basically two kinds of measurements made from these experiments. The first of these is the spectral analysis of the emitted radiation obtained by scanning the monochromator through the wavelength range, typically 2000 to 8500 angstroms. The second measurement involves observing the intensity of a single line over a limited segment of the spectrum as a function of beam energy or as a function of bombardment time at fixed energy using either a monochromator or an interference filter. A wide variety of beam species and target materials can be used to make measurements.

The intensities in the individual spectral lines are proportional to the absolute concentrations of the specific surface constituents within the penetration depth of the incident beam of ions or neutrals. An example of the type of radiation emitted from excited states of surface species in low energy ion neutral impact on the solid surface is illustrated in figure 2-63 (ref. 56). The spectral distribution of radiation produced in the impact of argon ions at 4000 electron volts on copper and nickel surfaces is shown in figure 2-63. Most of the prominent lines in these spectral scans have been identified as coming from low-lying energy levels of neutral nickel and copper sputtered off the surface in excited states by the incident ion beam.

In addition to spectral lines from the surface target material—namely, copper or nickel—radiation is also observed, which is characteristic of surface contaminants; the molecular radiation occurs at 4300 and 3900 angstroms as electron transitions arising from the CH molecule. The origin of this radiation is believed to be due to the collisional excitation of adsorbed hydrocarbon contaminants present on the surface of the copper. If the copper surface is cleaned with the argon ion beam, the contaminants can be completely removed from the solid surface and the peaks occurring at 3900 and 4300 angstroms disappear.

Radiation Energy Sources

There are a number of surface analytical tools that employ radiation sources to analyze or energize the solid surface for analysis. One such tool employs the Mössbauer effect or, as it was earlier known, gamma ray resonance spectroscopy. The Mössbauer effect has been intensely studied since its discovery in 1958, but it has not been widely used in surface analysis until recently (ref. 57). Most of the studies employing the Mössbauer effect have been with bulk solids and, until very recently, surface effects were generally disregarded. The presence of a surface, however, does affect Mössbauer measurements; this effect can, therefore, be used to gain some information on the solid surface. The Mössbauer analytical spectroscopy is relatively simple in operation and in concept. And the measurements of nuclear effects—that is, the detection of recoiless gamma ray adsorption—is relatively easy.

The basic principle involved in Mössbauer measurements is shown schematically in figure 2-64. The sample to be analyzed and the gamma ray source are moved relative to each other. The absorption of gamma rays by

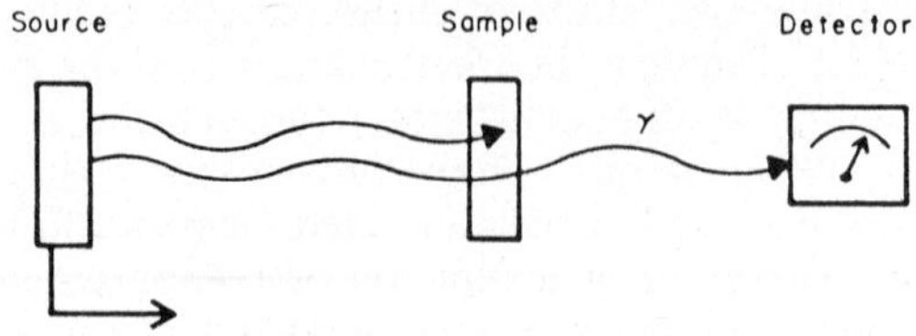

Figure 2-64. – Schematic of Mössbauer experiment. Sample and source are moved relative to each other, and absorption of gamma rays by sample at specific relative velocities is measured.

the sample at specific relative velocities (of sample and source) is then measured by the detector. In practice, a variety of methods are employed with wide variations in the hardware used; the results given by the various Mössbauer spectrometers differ, however, only in the precision achieved.

There are generally two groups of instruments, those which use a constant velocity drive of some sort and those which use drives having continuously variable velocity. With a constant velocity drive, a measurement of absorption is made at a particular fixed velocity of the emitter or absorber. The velocity is then brought to a new constant value, another absorption measurement is made, and so on. A variable velocity drive sweeps continuously and repetitively over a cycle, and the absorption measurements over a small increment are made. The various absorption measurements are then coherently added and stored by a multichannel analyzer.

An example of a simple Mössbauer spectrum is shown in figure 2-65. A radiation source consisting of cobalt 57 diffuses into a stainless steel surface (ref. 58). The gamma rays were directed toward the stationary adsorber, and the transmitted counts were stored in a 100-channel analyzer used as a velocity sorter. The Doppler velocity is positive when the source moves toward the adsorber, negative when the source recedes. The spectrum consists of a plot of absorption in arbitrary units versus the Doppler velocity. Each point in obtaining the absorption is measured over a small velocity segment. The parameters that can be obtained include the width and depth of the absorption line. The half-width is related to L. The depth of the absorption line indicates the magnitude of the effect, but the absolute measurement of this quantity is difficult. Of great interest is the fact that the resonance absorption band is not centered at zero velocity but is displaced to a positive velocity. This displacement is referred to as the isomer shift, and this is analogous to the chemical shift which is observed in AES. The shift can give much information about the composition of the species present on the solid surface.

Electron Paramagnetic Resonance (EPR) and Nuclear Magnetic Resonance (NMR)

Electron paramagnetic resonance (EPR) and nuclear magnetic resonance (NMR) are techniques that have been used for some time for bulk studies.

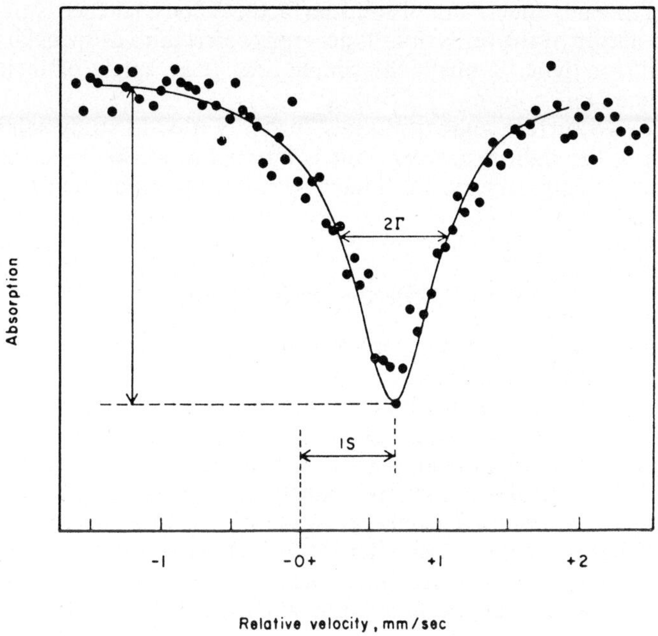

Figure 2-65.—Mössbauer spectrum of ferricinium picrate. Sample was at liquid-nitrogen temperature; IS denotes isomer shift (ref. 58).

They do, however, have capabilities for use in surface analysis. EPR detects the presence of unpaired electrons through their magnetic moments, and NMR detects the nuclei with net spin magnetic moments. The detection sensitivity of EPR is about 10^6 times that of NMR. This, therefore, tends to make the EPR system a more useful surface tool than the NMR.

The measurement of magnetic resonance is achieved by detecting entities with net magnetic moments. The behavior of the moments can be followed as various perturbations are applied—for example, (1) heating and cooling the sample, (2) exposing sample to various ambient environments, (3) exposing sample to radiation, (4) in the case of tribology, applying mechanical stresses on the substrate material, or (5) applying any combination of these various perturbations on the solid surface. The data that have been obtained using these techniques include positive identification of adsorbed species and the particular adsorption sites, wave functions of electrons on adsorbed species and on the surface itself, and even electric field strengths at the surface. Most of these details could not be obtained by using other surface analytical techniques. Detailed descriptions of these devices and their usefulness and application are presented in the literature (refs. 59 to 62).

Light Source Analytical Tools

When light impinges on a surface, the light reflects from that solid surface, and this reflection process is a surface phenomenon. Therefore, light can be used to study the nature of solid surfaces. When a surface is flat and smooth the nature of the reflections is generally referred to as specular—for example, mirror light. It obeys the simple laws (e.g., angle of incidence equals angle of reflection).

There are two surface analytical tools currently used to study solid surfaces which utilize light principles. One is internal reflection spectroscopy and the other is ellipsometry. Both have been used to study thin films on solid surfaces.

Internal Reflection Spectroscopy

When light approaches an interface from a medium which is optically less dense to one which is optically more dense, such as an oil film present on a solid surface, the reflection is called external reflection; its dependence on polarization and angle of incident light is shown schematically in figure 2-66. The data in figure 2-66 are for an interface of water on germanium where the refractive index N for water is 1.33 and for germanium is 4. The reflectivity can be modified by the presence of the film on the surface, especially with the use of parallel polarization near Brewster's angle. An example of this is shown in figure 2-67, where a silicon dioxide film of 500 angstroms is present on an aluminum surface. Although the film is

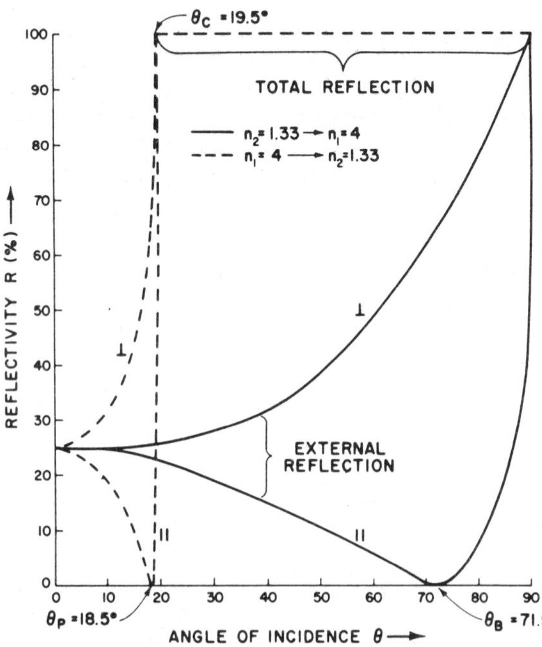

Figure 2-66.—Reflectivity as function of angle of incidence for interface between media. (Indices, $n_1 = 4$ and $n_2 = 1.33$, for light polarized perpendicular, $R_\perp$, and parallel, $R_{\parallel}$, to plane of incidence for external reflection (solid lines) and internal reflection (dashed lines). Angles θ_c, θ_B, and θ_p are critical, Brewster's and principal angles, respectively.)

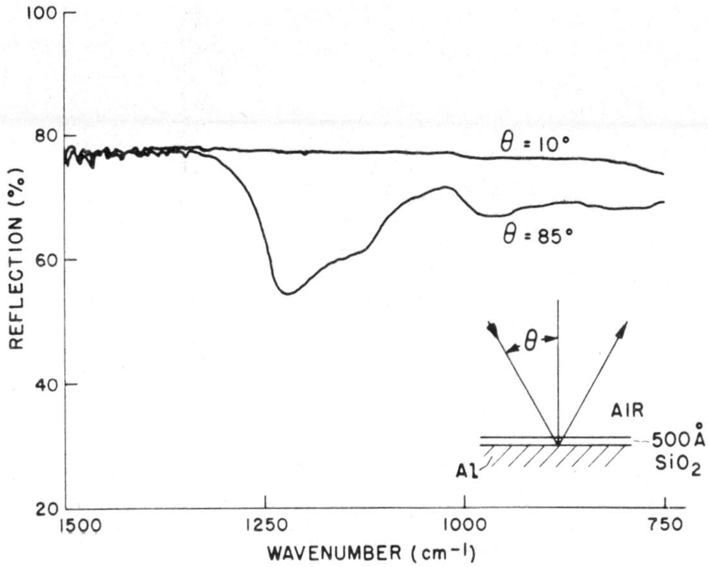

Figure 2-67. – Spectra showing power of specular external reflection near Brewster's angle for study of thin films on metal surfaces. SiO_2 film which is undetectable at 10° gives signal at 85° of 30%.

undetectable with near normal incident impingement of the light beam, it exhibits about 30-percent reflection losses at large angles of incidence. The reason the film is undetectable at normal incidence is that the incoming and reflected light waves interact to set up a standing wave with a node, zero electric field, at the reflecting surface for internal reflection.

External reflection has been highly neglected as a method of studying surfaces, but it is now being more widely used. It has been used to detect the presence of monolayer films adsorbed on surfaces. For a clean surface, reflection measurments can also yield information about the change in optical constants of the material (resulting from a change in atomic bonding) as one moves from the surface to the bulk. The reader is referred to summary papers on the subject in references 63 to 66.

When light propagates toward the interface from a medium which is more dense to one which is less dense, the reflection is called internal reflection; this reflection behaves much the same as external reflection from normal incidence until such time as a critical angle is reached (see fig. 2-66). Beyond the critical angle there is total reflection and the interface acts as a perfect mirror. There are many common examples of total internal reflection. The hypotenuse of a prism on the side of a glass of water suddenly appears silvered if viewed at certain angles. Perhaps the most common example of total reflection that occurs in nature is the mirage (fig. 2-68). In the mirage, a layer of warm air adjacent to a hot road or a sand surface is optically less dense (i.e., it has a lower refractive index) than the cooler air above it; for large angles of incidence, the condition for total reflection is

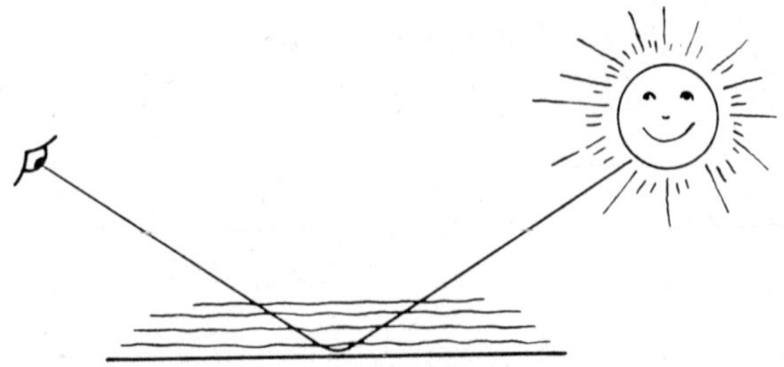

Figure 2-68.—Mirage, an example of total internal reflection found in nature. Light at grazing incidence is totally reflected by less-dense warm air near road surface.

satisfied. Thus, when looking at the road one sees the sky, which gives the impression of the presence of water in the distance.

When using light techniques it is desirable to have atomically flat, smooth surfaces. However, as is well known, surfaces are not well behaved and do not have this particular type of topography. As a result, in reflecting light off the surface, one obtains the kind of information presented in figure 2-66. The reflectance is not perfect, but it is diffuse, and it is commonly called just that, diffuse reflectance. Practical examples of diffuse reflectance include the reflection of light from terrain, fabrics, papers, and tribological surfaces. There is no rigorous solution to a general diffuse reflectance case, since the reflectance is a combination of external specular reflection, internal reflection, transmission, and scattering; reflectance is, therefore, dependent on the surface size, shape, refractive index, adsorption coefficient, polarization, etc. In spite of this difficulty, empirical methods have been developed to treat certain of these cases. For examples, the reader is referred to the reference work of Wendlandt and Hecht (refs. 64 and 66).

Ellipsometry

Another analytical surface tool for studying thin films on solid surfaces using light source is ellipsometry. Ellipsometry is a study of the state of polarization of light reflected at abnormal incidence from the surface under investigation. It is an extremely sensitive tool that can detect and measure the films on the surface even if the films are only a few atomic layers thick. A detailed analysis of the theory involved in ellipsometry is available in the literature. Only the basic essential elements are discussed herein.

In ellipsometry the incident beam of light is polarized. When the plane of vibration is arranged at 45° to the plane of incidence, resolved components from the plane of incidence and at right angles to it are equal in both intensity and phase on reaching the specimen's surface. In the majority of cases the light parallel to the plane of incidence is the more strongly attenuated as a result of reflection, and also a phase change is introduced between the two components. At one particular angle of incidence, the principal angle, the light reflected from the plane of incidence is of minimum intensity.

When the surface is covered with transparent film, such as an oil film, there is a further angle of incidence, the Brewster angle, at which reflection at the environment-film interface falls to zero independently of the film thickness; this means that the entire component of light is transmitted through the film. As the tangent of the Brewster angle of incidence is equal to the refractive index of the film, this affords an in situ method of experimental determination of the refractive index of thin films of lubricating materials. Reflection at the film-covered metal surface is strongly influenced by the interference between light reflected at the two interfaces, the environment-film interface, and the film-metal interface. It is this influence that is responsible for the development of interference colors or tarnishes appearing on surfaces.

If the solid substrate being examined is a dielectric material and there is no surface film present on it, then, if a plane of polarized light at 45° to the plane of incidence is incident on it, the reflected light also is plane polarized. Then the ratio of the reflected amplitude R_p polarized parallel to the plane of incidence R_s gives the tangent of the elliptometric parameter ψ.

If there is a surface film, or if the substrate itself is absorbing, then R_p and R_s are no longer in phase and the reflected light is elliptically polarized. The phase difference Δ $(\sigma - \sigma_s)$ is the second elliptometric parameter where σ_p and σ_s are the phases of R_p and R_s, respectively. Exact equations of ellipsometry, relating the measured quantities Δ and ψ to the optical parameters of the substrate and the film, were derived at the turn of the century. Since the exact equations of ellipsometry are rather involved, they cannot be converted analytically; it is only with the recent advances of the computer technology that the equations could be solved by numerical methods. Thus, the computer has made the ellipsometer a very practical surface analytical tool, particularly for examining the presence of thin films such as liquids on solid surfaces.

The experimental apparatus (fig. 2-69) employed to study surface films with ellipsometry is relatively simple (ref. 17). A beam of monochromatic columnated light is sent through a polarizer and a quarter wave plate before it is reflected from the sample and analyzed with a nicol. The fast access of the quarter wave plate is set at either 45° or -45° with respect to the plane of incidence. With this arrangement the positions of the analyzer and polarizer which yield minimum intensity of the transmitted light can be used to directly determine Δ and ψ. It can be shown with the help of the exact ellipsometric equations that, in general, at very small film thicknesses the ellipsometric parameter Δ is a function of the film thickness and that ψ is essentially a constant. Many experimentalists have used this property to study oxidation or corrosion during the early stages of the growth of films on solid surfaces by following variations of Δ. Vedam (ref. 17) has shown that the variations in ψ can be used to characterize, quantitatively, the surface layer of the substrate itself. This has been extremely useful in identifying where damage, such as strain, has occurred in the outermost layers of the solid surface.

The ellipsometric technique is sufficiently sensitive that small changes in the amount of damage in surface layers as well as changes in thickness of

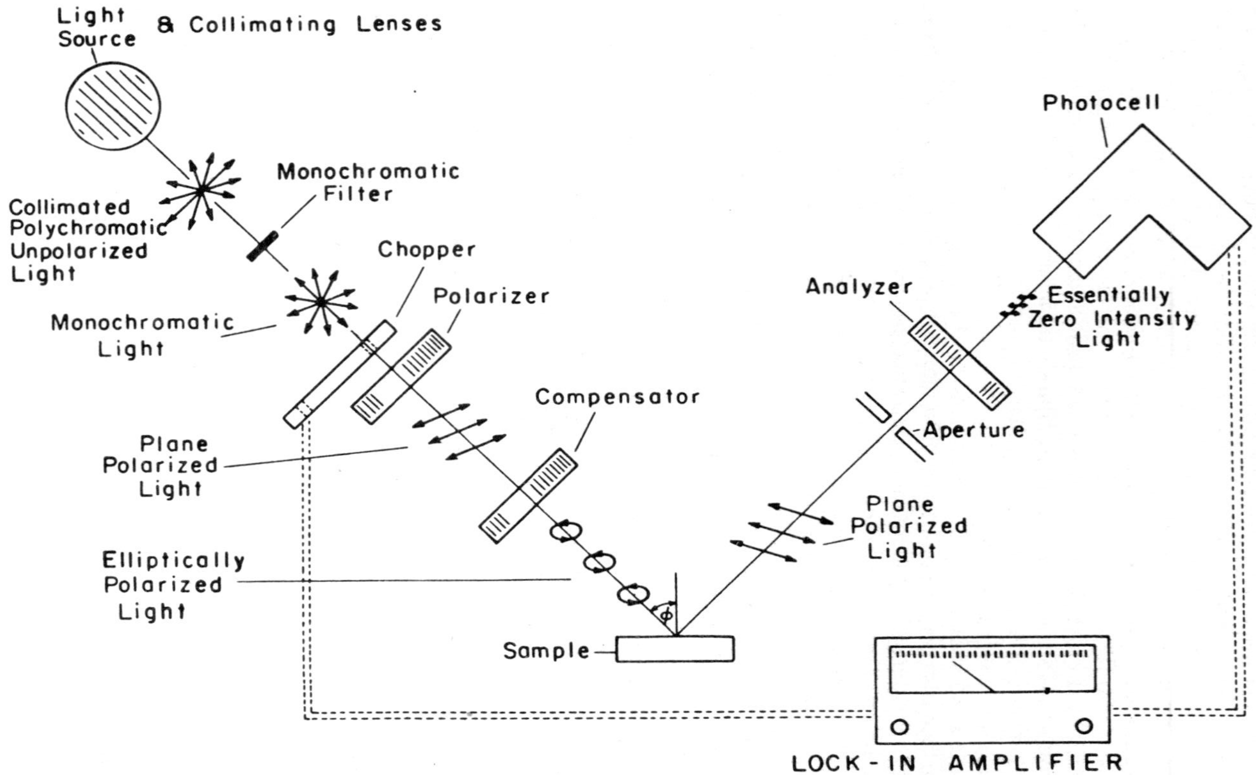

Figure 2-69. – Schematic diagram of experimental arrangement for ellipsometry (*ref. 17*).

films present on the solid surfaces can readily be detected. An example of this is shown in the data of table 2-I where a silicon single crystal surface has been treated by various techniques—mechanical polishing, sputtered by argon ion bombardment, cleaved, cleaved and annealed, and chemically polished. All of these different surface treatments resulted in obtaining different Δ and ψ values, as indicated in the table. As has already been explained, the variation in Δ is attributed to differences in film thickness (in this case, silicon oxide which has formed on the single crystal silicon surface after or during preparation) and the ψ variation in table 2-I is attributed to variations in damages in the surface layer.

Another interesting example of the information that can be obtained with ellipsometry is presented in table 2-II. Table 2-II lists the various Δ and ψ values observed on two samples of vitreous silica which were, respectively, polished with diamond paste and cerium oxide. The table indicates that the value of ψ on the specimen mechanically polished with diamond paste is 0.52° instead of 3.79° corresponding to the ideal strain-free surface. Computations with the ellipsometric equation reveal that the surface layers of the diamond-polished specimen have been permanently densified during mechanical polishing. The optical parameters of the densified layer has a refractive index of 1.530 and a thickness of 950 angstroms. With the specimen polished using cerium oxide, however, the surface appeared to be free of densified layer, as can be seen from the value of ψ measured in the specimen. The hardness of vitreous silica and cerium oxide are 4.9 mohs and 5.0 mohs, respectively. This may be why polishing with cerium oxide does not permanently damage the surface layers of vitreous silica.

The results of removal of the densified surface layers by etching the vitreous silica specimen polished with diamond paste are also presented in table 2-II. From these data it can be seen that both ψ and Δ increase rapidly toward the values corresponding to the ideal case with moderate etching. On continued etching, however, they overshoot the ideal value. With diamond paste polishing, this may be a result of the surface roughness on the microlevel which develops on a vitreous silica specimen.

TABLE 2-II. – VITREOUS SILICA[a]

Surface history	Polished with Diamond Paste ψ (in degrees)	Δ	Polished with Cerium Oxide ψ (in degrees)	Δ
Ideal case	3.79	360.00	3.79	360.00
Mech. polish	0.52	290.00	3.79	359.64
2 min etch in 10% HF	2.55	334.00	3.77	359.93
4 min etch in 10% HF	3.93	4.00	3.79	0.11
6 min etch in 10% HF	4.16	3.04	3.80	0.29

[a]Reference 17.

Infrared Spectroscopy (IR)

Infrared spectroscopy (IR) is a powerful tool that is becoming increasingly important in the study of solid surfaces. The technique is sufficiently sensitive to permit identification and measurement of fractions of monolayers of chemisorbed species. The sensitivity of the infrared measurements can be increased by multiple internal reflection (MIR) techniques; with such techniques, even monolayers can be measured very accurately. The utilization of polarized light for infrared MIR studies an indicates the molecular orientation of deposited material on the crystalline surface.

An example of the information that can be generated from IR is indicated in the data in figure 2-70. Figure 2-70 presents four metals to which carbon monoxide has been adsorbed: copper, platinum, nickel, and palladium (ref. 67). From the infrared spectra it is possible to gain insight, as already mentioned, into the molecular orientation of the adsorbed carbon monoxide on the surface of the various metals. This was done for the data of figure 2-70 by Eischens (ref. 67). It can be seen that, for copper and platinum, the carbon of the carbon monoxide forms a single bond with the metal surface with three bonds remaining to the oxygen; with nickel and palladium there are two carbon to metal bonds and two carbon to oxygen bonds. Hence, the mechanism of bonding is different for carbon monoxide to the four metals copper, platinum, nickel, and palladium. The ability to distinguish the form of bonding to the solid surface is extremely important in such studies as tribology. The reader is referred to the work of Eishens for an excellent

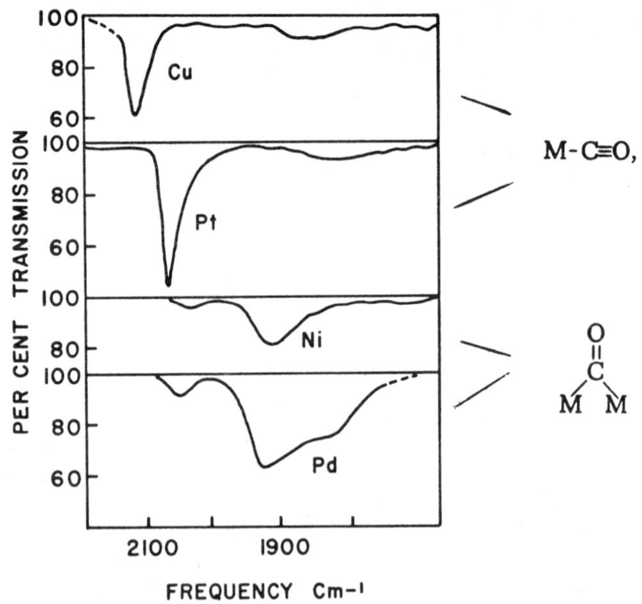

Figure 2-70. – Infrared spectra for chemisorption of carbon monoxide to various metals and resulting surface structures (ref. 67).

review on the subject of IR (ref. 67). The infrared technique has been very effective in the analysis of various films deposited on tribological surfaces, including such things as lacquers or varnishes that normally develop on pistons of reciprocating engines.

Surface Cleaning

Through the years various techniques have been employed to clean surfaces in tribology. These have included (1) conventional solvent cleaning with various organic solvents, (2) using mineral acids, (3) using abrasives (including abrasive papers and powders), (4) heating surfaces to high temperatures, and (5) chemical etching. Most of these techniques, however, while they may remove bulk surface contaminants, they do not generate a clean surface in a strict sense. In a vacuum environment, heating surfaces to very high temperatures may remove contaminants that are adsorbed on the solid surface, but it does not necessarily remove chemical compounds that are formed on the solid surface. Even in those situations where the heating is sufficient to cause dissociation of the chemical compounds (e.g., oxides), diffusion in the heating process causes the migration of bulk contaminants to the surface and thus can produce surface and solid state contamination.

In the course of studies using the analytical surface tools described thus far, it has very frequently been found that, when one had prepared a relatively clean surface, modest amounts of heating resulted in the segregation of bulk contaminants to the solid surface. Currently, probably the single most effective technique used to prepare atomically clean surfaces in a system for analytical surface analysis is the use of ion bombardment, where rare gas ions bombard or strike a specimen surface. In the process of striking the solid surface they knock off or liberate surface layers. The process is shown schematically in figure 2-71. The rare gas ions can be argon, xenon, or another noble gas which is ionized and then projected or propelled to a negatively charged surface. This process causes the knock off or removal of the surface layers. In the case of metals where there are oxides and adsorbates present on the solid surface, the argon or inert gas ion bombardment causes removal of these species by etching away the surface layers. The process can be continued so that, once the surface contaminants are removed, the material can be examined in bulk by selectively etching away the bulk material layers one at a time.

This process of examining the bulk structure or composition by sequentially removing layers with inert ion gas bombardment is frequently referred to as depth profile analysis, and it is used very effectively in XPS (X-ray photoelectron spectroscopy) as well as in AES analysis for analyzing bulk chemistry. The two surface analytical tools, XPS and Auger, are used in conjunction with the ion bombardment process or depth profiling to study the profile or the composition from the outermost surface layers into the bulk material. The process of argon ion or inert gas ion bombardment of a solid surface is a very widely used tool today for cleaning surfaces in surface physics work. Care must be taken when using this approach, however,

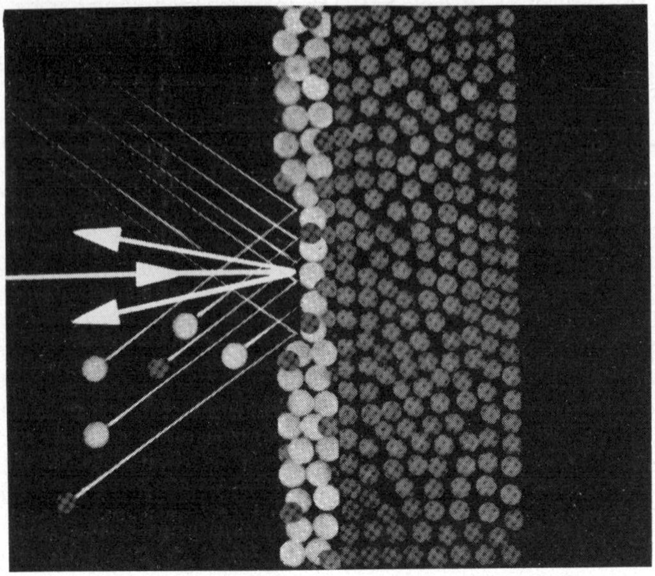

Figure 2-71. – Rare gas ion sputter etching process (ref. 68).

because if the incident ions are of relatively high energy they can produce strain in the lattice of the outermost atomic layers of the solid surfaces. This has been verified by LEED patterns obtained on clean metal surfaces such as iron that have been argon or inert gas ion bombarded for cleaning purposes. Residual strains exist in the metal surface after bombardment. These strains, however, are not very deep and are in the first few atomic layers of the solid surface; these strains can readily be removed by relatively mild heating of the surface.

Summary

There are a wide variety of analytical surface tools that can be used to study solid surfaces. These tools use various excitation sources and mechanisms to record the chemistry and structure of solid surfaces. Most of the devices described use one of the three or four energy sources depicted in figure 2-72. Figure 2-72 shows the excitation sources described thus far: (1) electron bombardment of the solid surface or the impingement of a beam of electrons in various energy ranges at the solid surface, (2) ion bombardment using such techniques as ion-scattering spectroscopy and secondary and mass spectroscopy, (3) X-ray radiation such as that utilized in X-ray photoelectron spectroscopy, and (4) ultraviolet light.

From the interaction of the excitation sources with the solid surfaces of materials, various forms are emitted from the solid surfaces and detected by various techniques which include X-ray wavelength analyzers or energy

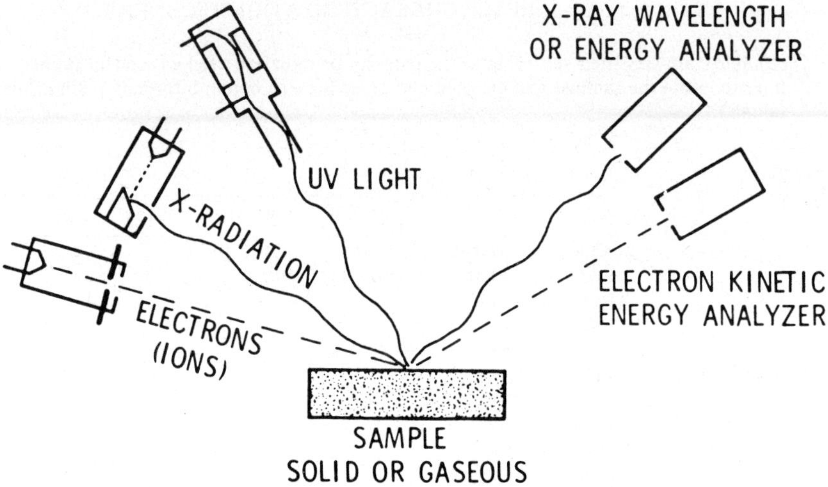

Figure 2-72. – Excitation of sample and recording of consequential events.

analyzers and electron energy analyzers. In addition to those shown schematically in figure 2-72, there are ion detectors for detecting ions liberated from solid surfaces—for example, mass spectrometry.

During the last 15 years surface analytical tools and techniques have been developing very rapidly. In fact, it is probably safe to say that nearly one new device per month is appearing to assist the scientist interested in characterizing solid surfaces. Because of the proliferation of surface analytical tools, acronyms have developed for describing these particular devices. For example, in this chapter we referred to X-ray photoelectron spectroscopy as XPS, low energy electron diffraction as LEED, and so on. The techniques described and the tools detailed in this chapter are by no means all inclusive. A number of these devices are displayed by their acronyms in table 2-III (ref. 68) to give the reader some indication of the wide variety of tools available to the surface physicist, chemist, or metallurgist.

In table 2-III the various techniques are classified in terms of the incident probe (as shown in fig. 2-72) where the incident probe may be photon, electron, neutral ion, phonon, or an E/H field. Then the the various techniques are classified (by acronym) under detection techniques whether photon, electron, neutral, ion, phonon, or E/H field. The greatest number of detection devices use either photon, electron, or ion detection systems. This list is by no means all inclusive; in all probability, while this chapter is being written, additional devices are appearing on the horizon and will be incorporated into the arsenal of tools used by the surface and analytical scientists in future years.

Table 2-IV (ref. 69) is a key to the acronyms appearing in table 2-III. Some of the more familiar devices, those that appear to be used most readily today are Auger electron spectroscopy (AES), appearance potential

TABLE 2-III. – SURFACE CHARACTERIZATION TECHNIQUES[a]

[Techniques are classified according to the particles or radiation used as incident probes (i.e., to excite the sample) and detected quantities (i.e., to obtain information about the sample).]

Incident probe	Detected particle					
	photon	electron	neutral	ion	phonon	E/H field
photon	ATR	AEAPS	LMP	LMP		
	COL	AEM	PD	PD		
	ELL	AES				
	ESR	PEM				
	EXAFS	PES				
	IRS	SEE				
	LS	SEXAFS				
	MOSS	UPS				
	NMR	XEM				
	SRS	XES				
	XRD	XPS				
electron	APS	AEAPS	ESDN	ESDI		
	BIS	AEM	SDMM			
	CL	AES				
	CIS	DAPS				
	EM	EELS				
	SXAPS	HEED				
	SXES	IS				
		LEED				
		SEE				
		RHEED				
		SEM				
		SLEEP				
		STEM				
		TEM				
neutral	NIRS	SEE	AIM		HA	
			MBRS			
			MBSS			
ion	GDOS	IMXA	ISD	GDMS		
	IIRS	INS	SDMM	IMMA		
	IIXS	SEE		ISD		
	NRS			ISS		
				NRS		
				RBS		
				SIIMS		
				SIMS		
phonon	ES	TE	FD	SI	ASW	
	TL					
E/H field	EL	FEES	FDM	FIM		CPD
		FEM	FDS	FIM-APS		SC
		ITS		FIS		

[a]Reference 68.

TABLE 2-IV. – ACRONYMS AND ENTRIES USED IN TABLE 2-III[a]

Acronym	Entry
AEAPS	Auger-electron appearance-potential spectroscopy
AEM	Auger-electron microscopy
AES	Auger-electron spectroscopy
AIM	Adsorption isotherm measurements
APS	Appearance-potential spectroscopy
ASW	Acoustic surface-wave measurements
ATR	Attenuated total reflectance
BIS	Bremsstrahlung isochromat spectroscopy
CIS	Characteristic isochromat spectroscopy
CL	Cathodoluminescence
COL	Colorimetry: IR, visible, UV, X-ray, and γ-ray absorption spectroscopy
CPD	Contact potential difference (work-function measurements)
DAPS	Disappearance-potential spectroscopy
EL	Electroluminescence
ELL	Ellipsometry
EELS	Electron energy-loss spectroscopy
EM	Electron microprobe
ES	Emission spectroscopy
ESDI	Electron-stimulated desorption of ions
ESDN	Electron-stimulated desorption of neutrals
ESR	Electron-spin resonance
EXAFS	Extended X-ray absorption fine structure
FD	Flash desorption
FDM	Field-desorption microscopy
FDS	Field-desorption spectroscopy
FEM	Field-emission microscopy
FEES	Field-electron energy spectroscopy
FIM	Field-ion microscopy
FIM–APS	Field-ion microscope – atom probe spectroscopy
FIS	Field-ion spectroscopy
GDMS	Glow-discharge mass spectroscopy
GDOS	Glow-discharge optical spectroscopy
HA	Heat of adsorption
HEED	High-energy electron diffraction
IIRS	Ion-impact radiation spectroscopy
IIXS	Ion-induced X-ray spectroscopy
IMMA	Ion microprobe mass analysis
IMXA	Ion microprobe X-ray analysis
INS	Ion-neutralization spectroscopy
IRS	Internal reflectance spectroscopy
IS	Ionization spectroscopy
ISD	Ion-stimulated desorption
ISS	Ion-scattering spectroscopy
ITS	Inelastic tunneling spectroscopy
LEED	Low-energy electron diffraction
LMP	Laser microprobe
LS	Light scattering
MBRS	Molecular-beam reactive scattering
MBSS	Molecular-beam surface scattering
MOSS	Mössbauer spectroscopy
NIRS	Neutral impact radiation spectroscopy
NMR	Nuclear magnetic resonance
NRS	Nuclear reaction spectroscopy
PD	Photodesorption
PEM	Photoelectrom microscopy
PES	Photoelectron spectroscopy
RBS	Rutherford backscattering spectroscopy
RHEED	Reflection high-energy electron diffraction
SC	Surface capacitance
SDMM	Scanning desorption molecule microscopy
SEE	Secondary-electron emission
SEM	Scanning electron microscopy
SEXAFS	Surface extended X-ray absorption fine structure
SI	Surface ionization
SIIMS	Secondary-ion imaging mass spectroscopy
SIMS	Secondary-ion mass spectroscopy

[a]Reference 68.

spectroscopy (APS), ellipsometry (ELL), electron microprobe (EM), field ion microscopy (FIM), field ion spectroscopy (FIS), low energy electron diffraction (LEED), and, to a lesser extent, reflection high energy electron diffraction (RHEED). Probably the most useful device for the tribologist is the scanning electron microscope (SEM), which, when used in conjunction with X-ray dispersive analysis, is an extremely useful analytical tool. To a lesser extent, the ion techniques (ISS) and secondary ion mass spectroscopy (SIMS) are also used.

A subject which has not been discussed thus far is the nature of the analytical surface tool and the interaction of the excitation source with the solid surface. Is it, for example, destructive or nondestructive? In some situations when used in tribological systems it is important that the surface not be destroyed. For example, in a quality control analysis of a gyro-bearing surface during operation, one would not want to use a destructive technique which would prohibit the further use of the bearing. Some of the techniques described in this chapter are destructive and some are not. This is also true of the techniques not described in the chapter but appearing in tables 2-III and 2-IV.

Figure 2-73 gives an indication of some of the devices available for obtaining compositional analysis of solid surfaces as a function of depth. Such an analysis uses the various tools that can provide compositional information and is restricted to those devices. Among the nondestructive techniques are nuclear back scattering (NBS) and electron microprobe analysis. When one wishes to obtain information as a function of depth, there are a host of techniques that can be utilized; these techniques analyze the surface with sputter etching or ion bombardment to provide indepth information. These would all be destructive if a sputter etch or ion bombardment is used for

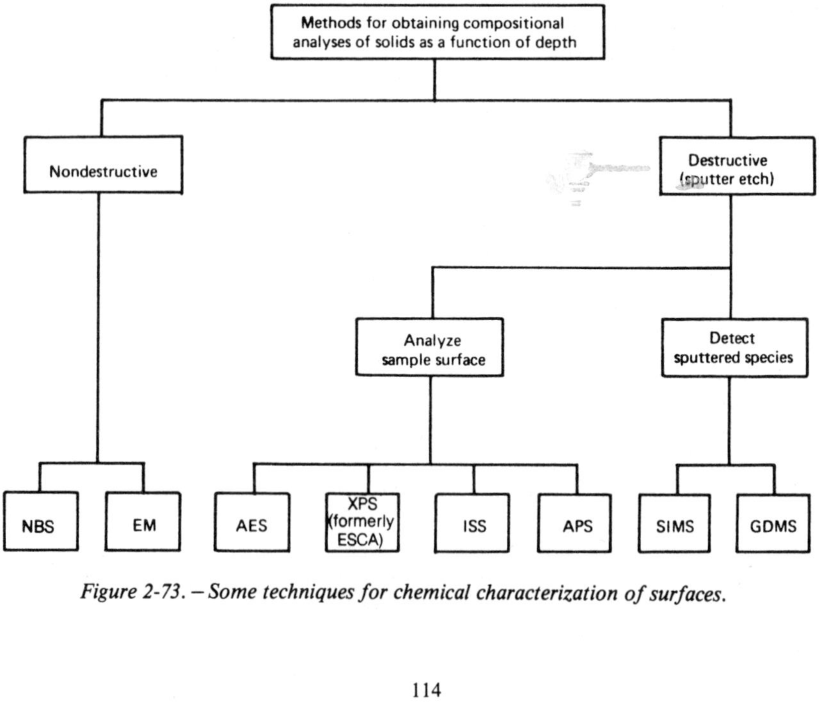

Figure 2-73. – Some techniques for chemical characterization of surfaces.

depth profiling. However, the destructive techniques can be subdivided into those techniques that analyze the sample surface as well as those that detect the sputter or ion removed surface species. Among those techniques that analyze the sample surface are Auger emission spectroscopy (AES), or X-ray photoelectron spectroscopy (XPS), ion scattering spectroscopy (ISS), and appearance potential spectroscopy (APS).

Those techniques that detect the species which are actually removed from the solid surface by sputter bombardment include secondary ion mass spectrometry (SIMS) and glow discharge mass spectrometry (GDMS). These two devices require the removal of the surface species in ionic form for detection purposes. Inherent in these devices is destruction of the solid surface. Even if depth profile information were not required and if it was simply a requirement that the surface composition be analyzed, both SIMS and GDMS would involve the destruction of the surface in the sense that solid surface material would be removed in order to accomplish the analysis of the surface composition. With AES and XPS, however, it is not necessary to actually remove surface material; it is only necessary to excite the solid surface and to liberate electrons from the surface to produce an analysis. The liberation of electrons from the surface is not surface destructive as is the liberation of ions when SIMS or GDMS is used.

Some of the capabilities and limitations of the devices detailed in figure 2-73 are presented in table 2-V. In addition to a separation of the devices which are destructive to the sample from those than are nondestructive, we can see that the limitations in the elements can be detected. For example, with AES, all the elements except hydrogen and helium can be detected on a

TABLE 2-V. – COMPARATIVE TABLE FOR THE VARIOUS TECHNIQUES USED FOR THE CHEMICAL CHARACTERIZATION OF SURFACES

	NBS	EM	AES	ESCA[f]	ISS	SIMS	GDMS	APS
Destructive to sample (in general)	No	No	No	No	No	Yes	Yes	No
Elements that can be detected	heavy	$Z \geq 4$ [e]	$Z \geq 3$	$Z \geq 3$	$Z \geq 3$	All	All except He, Ne	$Z \geq 3$
Elemental identification [a]	F	G	E	E	E	G	G	E
Sensitivity (typical, in monolayers)	50	5	~ 0.01	< 0.01	~ 0.01	< 1	~ 1	≤ 0.1
Detectability (*i.e.*, ppm) [b]	NA	100	< 1	NA	NA	1	100	NA
Results are (in principle) [c]	Abs	Abs	Abs	Abs	Abs	Abs	Abs	Abs
Depth probed (in Å)	10^4	10^4–10^5	15–20	15–75	3	$\sim 5 \times 10^4$	10–10^4	~ 10
Depth distribution of elements [d]	Yes	Yes	Y/d	No	Yes	Yes	Yes	Y/d
Chemical (*i.e.*, binding) information	No	Yes	Yes	Yes	No	No	No	Yes

a—E, Excellent; G, Good; F, Fair
b—NA, Not applicable
c—Rel, Relative; Abs, Absolute
d—Y/d, Yes, if destructive
e—Z, atomic number
f—also XPS

solid surface. The same is true for XPS (ESCA) and ISS. Some devices such SIMS can detect all the elements present in the solid surface. Note that with respect to sensitivity, AES, XPS (ESCA), and ISS have very good sensitivities; they are able to detect as little as 0.01 of a monolayer on a solid surface. This kind of sensitivity is extremely important in tribology where fractions of monolayers can influence adhesion and friction results. Note that, with the depth of sample analyzed, some techniques are very deep probes while others are strictly surface. For example, with the NBS, the probe depth is 10^4 angstroms; with AES, it is only 15 to 20 angstroms; and in ISS, it is only 3.

Extremely important information that can be obtained by these analytical surface tools is chemical information—that is, information about compound formation. How do elements exist on a solid surface? Are they combined in compounds and if they are, what is the structure of that particular compound? Some of the analytical tools can provide this information and others can provide no more than a simple elemental analysis. Table 2-V shows that a number of the devices do provide information about binding or chemical states; these include the electron microprobe, AES, and XPS or ESCA. But notice that while the electron microprobe can provide this kind of information the depth of analysis (depth of the probe) is much deeper than that of AES analysis and XPS or ESCA. AES and XPS are more truly surface analytical tools and even here distinctions can be drawn. While AES can provide information on chemical binding states from chemical shifts of the elements in the Auger spectra, it is nowhere as effective as XPS in identifying the particular compound in which an element finds itself. The real tool for determining chemical composition beyond the elements is XPS (or ESCA).

The data presented in table 2-V simply compare those techniques that provide chemical characterization of the solid surfaces. In addition to these, we have discussed other devices (such as FIM, LEED, and HEED) which can provide information about the structural arrangement of materials on the solid surface. The number of tools that are available to provide structural information about a solid surface are much less than those that can provide chemical analysis.

No attempt has been made in this text to describe in detail each of the devices presented in chapter 2. The intent was merely to provide the tribologist with some background information on the basic rudimentary features of some of the tools that are available for use in surface analysis. References are provided for the individual who is interested in getting more specific details about a particular technique.

Appendix – Etchants[a]

Crystals	Etchant	Remarks	References
Aluminum	(1) 47% HNO_3, 50% HCl 3% HF temp: room ref.: (*1*)	With Fe impurity	(*1*) S. Amelinckx, *Phil. Mag.* [7] **44**, 1048 (1953).
	(2) HF temp: room ref.: (*2*)	After decoration in vacuum at 350°C	(*2*) P.B. Hirsch, R.W. Horne, and M.J. Whelan, *Phil. Mag.* [8] **1**, 677 (1956).
Antimony	(1) CP 4 temp: room; time: 2-3 sec plane: (111) ref.: (*3*)	Cleaved surface	(*3*) J.H. Wernick *et al.*, *J. Appl. Phys.* **29**, 1013 (1958).
	(2) 1 pt HF, 1 pt Superoxal, 4 pt H_2O temp: room; time: 1 sec plane (111) ref.: (*3*)		
Bismuth	1% I_2 in methanol temp: room; time: 15 sec plane: (11) ref.: (*4*)	After prepolishing	(*4*) I.C. Lovell and J.H. Wernick, *J. Appl. Phys.* **30**, 234 (1959).
Brass	0.2% $Na_2S_2O_4$ temp: room; time: 60 sec plane: (111) ref.: (*5*)	Electrolytic: 0.1 A cm^{-2}	(*5*) P.A. Jacquet, *Compt. Rend.* **237**, 1248 (1953).

[a]Reference 7.

Crystals	Etchant	Remarks	References
Cadmium	(1) 2 pt orthophosphoric acid, 2 pt glycerine, 1 pt water time: 20-40 sec ref.: (*6*)	Electrolytic: 0.9-10 v after polishing in same bath at 2.1-2.2 v for 9-12 min	(*6*) A.A. Predvohitiv and N.A. Tiapunia, *Phys. Metals Metallog* (*U.S.S.R.*) (*English Transl.*) **7**, No. 6, 55 (1959).
	(2) Sat. sol. of picric acid in acetone time: 2 min ref.: (*7*)		(*7*) J. George, *Nature* **189**, 743 (1961).
Copper	(1) Impure: 60% H_3PO_4 in H_2O temp: room; time: 10 sec ref.: (*8*)	Electrolytic	(*8*) F.W. Young and N. Cabrera, *J. Appl. Phys.* **28**, 339 (1957).
	(2) Pure: 4 pt sat. aqueous $FeCl_3$, 4 pt HCl, 1 pt acetic acid plus few drops bromine temp: room; time: 15-30 sec ref.: (*9*)	Rinse in NH_4OH, prepolish necessary	(*9*) L.C. Lovell and J.H. Wernick, *J. Appl. Phys.* **30**, 590 (1959).
	(3) Several etchants planes: (111) (110) (100) ref.: (*10*)	Distinguish fresh dislocations and dislocations with Cottrell atmosphere	(*10*) F.W. Young, *J. Appl. Phys.* **32**, 192 (1961).
Germanium	(1) CP4 time: 1 min; planes: (111) (100) ref.: (*11*)		(*11*) W.G. Johnston, *Gen. Elec. Rept.* 61-RL-2649M (1961); *Progr. Ceramic Sci.* **2**, 1 (1962); R.D. Heidenreich, U.S. Patent 2, 619, 414 (1952); F.L. Vogel *et al.*, *Phys. Rev.* **90**, 489 (1953).
	(2) 40 ml HF, 20 ml HNO_3, 40 ml H_2O, 2 gm $AgnO_3$ time: 1 min; planes: (111) (110) ref.: (*11*)		

Iron	(1) 2% Nital containing 2% saturated picral time: 15 min ref.: (*12*) (2) Saturated picral time: 4 min ref.: (*12*)	 Anneal at 1750°F to decorate dislocations.	(*12*) L.C. Lovell, F.L. Vogel, and J.H. Wernick, *Metal Progr.* **75**, No. 5, 96 (1959).
Silicon iron	(1) 133 ml acetic acid 25 gm CrO_3, 7 ml H_2O temp: room time: 5-20 min ref.: (*13*) (2) Dilute phosphoric acid (200 gm/liter H_3PO_4) temp: 20-25°C; time: 4-12 min ref.: (*14*)	Electrolytic: 30 mA/cm^2 after predecoration Electrolytic: 0.15 A/cm^2	(*13*) W.R. Hibbard and C.G. Dunn, *Acta Met.* **4**, 306 (1956). (*14*) S. Felin and G. Castro, *Acta Met.* **10**, 543 (1962).
Nickel-manganese	100 cc H_3PO_4, 100 cc ethanol temp: 40°C time: 2 min ref.: (*15*)	Electrolytic: 2 A/cm^2 Cu-cathode	(*15*) T. Taoko and S. Aoyagi, *J. Phys. Soc. Japan* 11, 522 (1956).
Niobium	(1) 10 cc H_2O, 10 cc H_2SO_4, 10 cc HF, few drops superoxil temp: room ref.: (*16*) (2) H_2SO_4(95%), HNO_3(70%), HF(48%) in ratio 5 : 2 : 2 ref.: (*17*)	Agitate specimen in solution Strongly orientation-dependent	(*16*) A.B. Michael and F.J. Huegel, *Acta Met.* **5**, 339 (1957). (*17*) R. Bakish, *Trans. A.I.M.E.* **212**, 818 (1958).

Crystals	Etchant	Remarks	References
Silicon	(1) 1 HF, 3 HNO_3, 10 CH_3COOH time: 10 min to "overnight" ref.: (*18*)		(*18*) W.C. Dash, *J. Appl. Phys.* **27**, 1193 (1956).
	(2) 10 cc hydrofluoric acid (37-38 %), 10 cc HNO_3, 10 cc glacial acetic acid temp: 30-35°C; time: 2-3 min ref.: (*19*)	Adding 15 cc double distilled water reduces time to 1.5-2 min	(*19*) N.N. Sirota and A.A. Tonoyan, *Proc. Acad. Sci. U.S.S.R., Phys. Chem. Sect.* (*English Transl.*) **134**, 987 (1960).
Tantalum	5 pt H_2SO_4, 2 pt HNO_3, 2 pt HF temp: room; plane: (112) ref.: (*20*)	After predecoration	(*20*) R. Backish, *Acta Met.* **6**, 120 (1958).
Tellurium	(1) 3 pt HF, 5 pt HNO_3, 6 pt acetic acid time: 1 min; temp.: room plane: $(10\bar{1}0)$ ref.: (*21*)	Cleaved surface	(*21*) L.C. Lovell, J.H. Wernick, and K.E. Benson, *Acta Met.* **6**, 716 (1958).
	(2) Conc. HNO_3 ref.: (*22*)		(*22*) A.I. Blum, *Soviet Phys.—Solid State* (*English Transl.*) **2**, 1509 (1961).
	(3) 43 gm H_3PO_4(d = 1.55), 1 cc conc. H_2SO_4, 5 gm CrO_3 time: 1-3 min temp: 90-100°C ref.: (*22*)	Another etchant with nearly the same composition (84.4 gm H_3PO_4, 2 cc H_2SO_4, 4.4 gm CrO_3) works at 150-160°C	
Tungsten	(1) 2 pt $CuSO_4$(25 %), 1 pt conc. NH_4OH ref.: (*23*)		(*23*) U.E. Wolff, *Acta Met.* **6**, 559 (1958).
	(2) 32.7 gm $K_3Fe(CN)_6$, 4.78 gm NaOH, 107 ml H_2O plane: (110) ref.: (*24*)	No guarantee that all pits correspond to dislocations	(*24*) I. Berlec, *J. Appl. Phys.* **33**, 197 (1962)

Uranium	170 cc H_2SO_4(98 %), 100 cc H_2O(dist.), 5 gm CrO_3, 25 cc dehydrated glycerol time: 60 sec ref.: (*25*)	Electrolytic: 2 A/cm² similar etchants with slightly different composition are given in same reference	(*25*) A. Bassi, S. Granata, and G. Imarisio, *Energie Nucl.* **8**, 744 (1961).
Zinc	(1) 160 gm chromic acid, 50 gm hydrated sodium sulfate, 500 ml H_2O ref: (*26*)		(*26*) J.J. Gilman, *Gen. Elec. Rept.* 56-RL-1575 (1956); *J. Metals* **8**, 998 (1958).
	(2) 2 gm $NH_4(NO_3)_2$, 10 cc NH_4OH, 50 cc deionized water time: 10 sec; temp: room ref.: (*27*) for other etchants: see ref.: (*28*)	Rinsed in water in methyl alcohol and dried in a stream of air	(*27*) P.P. Sinka and P.A. Beck, *J. Appl. Phys.* **33**, 625 (1962). (*28*) V.M. Kosevich and V.P. Soldatov, *Soviet Phys.—Cryst.* (*English Transl.*) **6**, 347 (1961).
		Nonmetals	
$CaCO_3$	HCl (10 %) time: 10-60 sec; plane: (010) cleavage ref.: (*29*) other etchants: see ref.: (*30*) (*31*) (*32*)		(*29*) R.C. Stanley, *Nature* **183**, 1548 (1959). (*30*) G.B. Rais, *Dokl. Akad. Nauk S.S.S.R.* **117**, 419 (1957). (*31*) H. Watts, *Nature* **183**, 314 (1959). (*32*) R.E. Keith and J.J. Gilman, *Acta Met* **8**, 1 (1960).
CaF_2	Conc. H_2SO_4 time: 10-30 min; plane: (111) ref.: (*33*) other etchants: see ref.: (*34*)		(*33*) W. Bontinck, *Phil. Mag.* [8] **2**, 561 (1957). (*34*) W.G. Johnston, unpublished data, cited in ref. (*26*).

Crystals	Etchant	Remarks	References
CdS	Vapor of conc. HCl time: 5-10 sec; plane: (0001) ref.: (*35*) other etchants: see ref.: (*36*) (*37*) (*38*) (*39*)		(*35*) D.C. Reynolds and L.C. Greene, *J. Appl. Phys.* **29**, 559 (1958). (*36*) J. Nishimura, *J. Phys. Soc. Japan* **15**, 232 (1960). (*37*) J. Woods, *Brit. J. Appl. Phys.* **11**, 296 (1960). (*38*) A.J. Eland, *Philips Tech. Rev.* **22**, 288 (1961). (*39*) Zh. G. Pizarenko and M.K. Sheinkman, *Soviet Phys.—Solid State* (*English Transl.*) **3**, 838 (1961).
GaAs	2 HCl, 1 HNO_3, $2H_2O$ time: 10 min; plane $(\bar{1}\bar{1}\bar{1})$ ref.: (*40*) other etchants: see ref.: (*41*) (*42*) (*43*)		(*40*) J.G. White and W.C. Roth, *J. Appl. Phys. 30*, 946 (1959). (*41*) H.A. Shell, *Z. Metallk.* **48**, 158 (1957). (*42*) J.L. Richards, *J. Appl. Phys.* **31**, 600 (1960). (*43*) J.L. Richards and A.J. Crocker, *J. Appl. Phys.* **31**, 611 (1960).
InAs	0.4 *N* Fe^{++} in conc. HCl time: 3 min; temp: 25°C; plane: $(\bar{1}\bar{1}\bar{1})$ ref.: (*44*) other etchants: see ref.: (*45*) (*46*)		(*44*) H.C. Gatos and M.C. Lavine, *J. Electrochem. Soc.* **107**, 427 (1960). (*45*) J.W. Faust and A. Sagar, *J. Appl. Phys.* **31**, 331 (1960). (*46*) E.P. Warekois and P.H. Metzger, *J. Appl. Phys.* **30**, 960 (1959).
InSb	25 HNO_3, 20 CH_3COOH, 10 HF, 1 Br_2 time: 5 sec; plane: $(\bar{1}\bar{1}\bar{1})$	Stop etching by adding H_2O	(*47*) W. Bardsley and R.L. Bell, *J. Electron. Control* **3**, 103 (1957).

	ref.: (*47*) other etchants: see ref.: (*44*) (*45*) (*48*) (*49*) (*50*) (*51*) (*52*)		(*48*) J.W. Allen, *Phil. Mag.* [8] **2**, 1475 (1957). (*49*) R.E. Maringer, *J. Appl. Phys.* **29**, 1261 (1958). (*50*) J.D. Venables and R.M. Broudy, *J. Appl. Phys.* **29**, 1025 (1958). (*51*) H.C. Gatos and M. Lavine, *J. Appl. Phys.* **31**, 743 (1960). (*52*) H.C. Gatos and M. Lavine, *J. Electrochem. Soc.* **107**, 433 (1960).
KBr	Glacial acetic acid time: 3 sec; plane: (100) ref.: (*53*) other etchants: see ref : (*54*)	Rinse in CCl_4	(*53*) P.R. Moran, *J. Appl. Phys.* **29**, 1768 (1958). (*54*) J.S. Cooks, *J. Appl. Phys.* **32**, 2492 (1961).
KCl	Conc. sol. polyvinyl butyrol in buryl or ethyl time: 5-7 sec; plane: (100) ref.: (*55*) other etchants: see ref : (*53*) (*54*) (*56*) (*57*) (*58*) (*59*)		(*55*) M.P. Shaskolskaya, W. Yang-Wen and K. Shu-Chao, *Soviet Phys.—Cryst.* (*English Transl.*) **6**, 220 (1961). (*56*) S.A. Slack, *Phys. Rev.* **105**, 832 (1957). (*57*) M. Sakamoto and S. Kobayashi, *J. Phys. Soc. Japan* **13**, 800 (1958). (*58*) L.W. Barr *et al.*, *Trans. Faraday Soc.* **56**, 697 (1960). (*59*) J.J. Gilman, W.G. Johnston, and R. Sears, *J. Appl. Phys.* **29**, 747 (1958).
KI	Isopropyl alcohol time: 25 sec; plane (100) ref.: (*53*) other etchants: see ref.: (*53*)		

Crystals	Etchant	Remarks	References
LiF	$H_2O + FeF_3$ ($\sim 2 \times 10^{-6}$ molar) time: 1 min; plane: (100) ref.: *(60)* other etchants: see ref.: *(60)* *(61)* *(62)* *(63)* *(64)* *(65)* *(66)*	Distinguishable between aged and fresh dislocations	*(60)* J.J. Gilman and W.G. Johnston, "Dislocations and Mechanical Properties of Solids," p. 116. Wiley, New York, 1957. *(61)* J.J. Gilman and W.G. Johnston, *J. Appl. Phys.* **27**, 1018 (1956). *(62)* L.S. Birks and R.T. Seal, *J. Appl. Phys.* **28**, 541 (1957). *(63)* A.A. Urusovshaia, *Soviet Phys.—Cryst.* (*English Transl.*) **3**, 731 (1958). *(64)* A.P. Kapustin, *Soviet Phys.—Cryst.* (*English Transl.*) **4**, 247 (1959). *(65)* M.P. Ives and J.P. Hirth, *J. Chem. Phys.* **33**, 517 (1960). *(66)* A.R.C. Westwood, H. Opperhauser, and D.L. Goldheim, *Phil. Mag.* [8] **6**, 1475 (1961).
MgO	1 H_2SO_4, 1 H_2O, 5 NH_4Cl sat. solution time: 15 min; plane: (100) ref.: *(67)* other etchants: see ref.: *(34)* *(68)* *(69)* *(70)*	Distinguishes between fresh and aged dislocations	*(67)* R.J. Stokes, T.L. Johnston, and C.H. Li, *Phil. Mag.* [8] **3**, 718 (1958). *(68)* R.J. Stokes, T.L. Johnston, and C.H. Li, *Phil. Mag.* [8] **4**, 137 (1959). *(69)* J. Washburn, A.E. Gorum, and E.R. Parker, *Trans. A.I.M.E.* **215**, 718 (1958). *(70)* T.K. Ghash and F.J.P. Clarke, *Brit. J. Appl. Phys.* **12**, **44** (1961); A.S. Keh, *J. Appl. Phys.* **31**, 1538 (1960).

NaCl	Glacial acetic acid time: 0.1 sec; plane: (100) ref.: *(71)* other etchants: see ref.: *(34)* *(53)* *(54)* *(71)* *(72)* *(73)* *(74)*		*(71)* B. Jessensky, *Nature* **181**, 559 (1958). *(72)* S. Amelinckx, *Acta Met.* **2**, 848 (1954). *(73)* S. Mendelson, *J. Appl. Phys.* **32**, 1579 (1961). *(74)* M.P. Shaskolskaya and S. Jui-Fang, *Soviet Phys.—Cryst.* (*English Transl.*) **4**, 74 (1960).
NaF	Same as the one mentioned for KCl ref.: *(54)*		
NiO	Hot nitric acid ref.: *(75)* other etchant: see ref.: *(76)*		*(75)* T. Takeda and H. Kondoh, *J. Phys Soc. Japan* **17**, 1315 (1962). *(76)* T. Takeda and H. Kondoh, *J. Phys. Soc. Japan* **17**, 1317 (1962).
PbTe	10 ml H_2O + 5 gm NaOH + 0.2 gm I_2 time: 5 min.; temp.: 94-98°C plane: (100) cleavage ref.: *(77)* other etchants: see ref.: *(77)* *(78)*	Rinse in water	*(77)* B.B. Houston and M.K. Norr, *J. Appl. Phys.* **31**, 615 (1960). *(78)* G.P. Tilly, *Brit. J. Appl. Phys.* **12**, 524 (1961).

Crystals	Etchant	Remarks	References
SiC	Fused borax temp: 800-1000°C; plane: (0001) ref.: (*79*) other etchants: see ref.: (*80*)		(*79*) R. Gevers, S. Amelinckx, and W. Dekeyser, *Naturwissenschaften* **39**, 448 (1952). (*80*) F.H. Horn, *Phil. Mag.* [7] **43**, 1210 (1952).
UO_2	H_2O_2, H_2O, and conc. H_2SO_4 in proportion 6 : 3 : 1 by volume time: 3 min; temp: 20°C; plane (111) (100) ref.: (*81*) other etchant: see ref.: (*81*)		(*81*) A. Briggs, Harwell Report AERE M-859 (1961).
ZnS	Aqueous H_2O_2 solutions of 7.5% to 30% time: 10 min-1 hr temp: 60-80°C ref.: (*82*)	Wash with KCN solution, water, acetic acid, water, dry at 100°C	(*82*) P. Goldberg, *J. Appl. Phys.* **32**, 1520 (1961).

References

1. Bowden, F. P.; and Tabor, D.: The Friction and Lubrication of Solids. Part I. Oxford Clarendon Press (London), 1950.
2. Samuels, L. E.: Damaged Surface Layers: Metals. The Surface Chemistry of Metals and Semiconductors, Harry C. Gatos, ed., John Wiley & Sons, Inc., 1960, pp. 82-103.
3. Sliney, L. E.: Dynamics of Solid Lubrication as Observed by Optical Microscopy. ASLE Trans., vol. 21, no. 2, 1978, pp. 109-117.
4. Smithell, C. J.: Metals Reference Book. Vol. I. Fourth ed., Plenum Press, 1967.
5. Metals Handbook. Vol. I: Properties and Selection: Irons and Steels. Ninth ed., American Society for Metals, 1978.
6. Gilman, J. J.: Effect of Imperfections on Dissolution. The Surface Chemistry of Metals and Semiconductors, H. C. Gatos, ed., John Wiley & Sons, Inc., 1960, pp. 136-150.
7. Amelinckx, S.: The Direct Observation of Dislocations. Academic Press, Inc., 1964.
8. Kubaschewski, O.; and Hopkins, B. E.: Oxidation of Metals and Alloys. Second ed., Butterworth and Co. (Publishers) Ltd. (London), 1962.
9. Gwathmey, A. T.; and Lawless, K. R.: The Influence of Crystal Orientation on the Oxidation of Metals. The Surface Chemistry of Metals and Semiconductors, H. C. Gatos, ed., John Wiley & Sons, Inc., 1960, pp. 483-521.
10. Taylor-Hobson Model 3 Talysurf Instruction Manual. Taylor-Hobson, Ltd., Leicester, England.
11. Williamson, J. B. P.: Topography of Solid Surfaces—An Interdisciplinary Approach to Friction and Wear. NASA SP-181, 1978, pp. 85-142.
12. Shah, G. N.; Bell, A. C.; and Malkin, S.: Quantitative Characterization of Abrasive Surfaces Using a New Profile Measuring System. Wear, vol. 41, 1977, pp. 315-325.
13. Young, R. D.: Field Emission Ultramicrometer. Rev. Sci. Instrum., vol. 37, no. 3, Mar. 1966, pp. 275-278.
14. Jones, R. V.; and Richards, J. C.: Recording Optical Lever. J. Sci. Instrum., vol. 36, Feb. 1959, p. 90-94.
15. Thompson, A. M.: The Precise Measurement of Small Capacitances. IRE Trans., Instrum., vol. I-7, no. 3-4, Dec. 1958, pp. 245-253.
16. Zinke, Otto H.; and Jacovelli, Paul B.: Magnetic Flux Sensor. Rev. Sci. Instrum., vol. 36, no. 7, July 1965, pp. 916-920.
17. Vedam, K.: Characterization of Surfaces. The Characterization of Materials in Research: Ceramics and Polymers, J. J. Burke and V. Weiss, eds., Syracuse University Press, 1975, pp. 503-537.
18. Rabinowicz, E.; and Tabor, D.: Metallic Transfer Between Sliding Metals: An Autoradiographic Study. Proc. Roy. Soc. (London), ser. A, vol. 208, no. 1095, Sept. 24, 1951, pp. 455-475.
19. Friedlander, G.; Kennedy, J. W.; and Miller, J. M.: Nuclear and Radiochemistry. Second ed., John Wiley & Sons, Inc., 1964.
20. Overman, R. T.; and Clark, H. M.: Radioisotope Techniques. McGraw-Hill Book Co., Inc., 1960.
21. Cohen, A.: Use of Radioisotopes in Metallurgical Research. Metals Rev., vol. 2, no. 10, 1958, p. 143.
22. Burwell, J. T.; and Strang, C. D.: On the Empirical Law of Adhesive Wear. J. Appl. Phys., vol. 23, no. 1, Jan. 1952, pp. 18-28.
23. Merchant, M. E.; Ernst, M.; and Krabacher, E. J.: Radioactive Cutting Tools for Rapid Tool-Life Testing. Trans. ASME, vol. 75, no. 4, May 1953, pp. 549-559.
24. Swink, Laurence N.; and Brau, Maurice J.: Rapid, Nondestructive Evaluation of Macroscopic Defects in Crystalline Materials: The Laue Topography of (Hg, Cd)Te. Metall. Trans., vol. 1, Mar. 1970, pp. 629-634.
25. Walker, G. A.: Structural Investigation of Thin Films. J. Vac. Sci. Technol., vol. 7, no. 4, 1970, pp. 465-473.
26. Berg, Wolfgang: Uber Ein Rontgenographische Methode zur Untersuchung von Gitterstorungen an Kristallen (An X-Ray Method For Study of Lattice Disturbances of Crystals). Naturwissensckoften, vol. 19, 1931, pp. 391-396.

27. Barrett, Charles S.: A New Microscopy and Its Potentialities. Trans. AIME, vol. 161, 1945, pp. 15-64.
28. Newkirk, J. B.: Subgrain Structure in an Iron Silicon Crystal as Seen by X-Ray Extinction Contrast. J. Appl. Phys., vol. 29, no. 6, June 1958, pp. 995-998. (See also The Observations of Dislocations and Other Imperfections by X-Ray Extinction Contrast. Trans. AIME, vol. 215, June 1959, pp. 483-497.)
29. Siegbahn, K.; et al.: ESCA Atomic Molecular and Solid State Structure Studies by Means of Electron Spectroscopy. Almqvist & Wiksells (Uppsala), 1967.
30. Siegbahn, K.; et al.: ESCA Applied to Free Molecules. North-Holland Publishing Co. (Amsterdam), 1969.
31. Baldwin, B. A.: Relationship Between Surface Composition and Wear: An X-Ray Photoelectron Spectroscopic Study of Surfaces Tested with Organo-Sulfur Compounds. ASLE Preprint No. 75-LC-2D-4, Oct. 1975.
32. Laird, Campbell: Electron Microscopy. Characterization of Solid Surfaces, P. F. Keane and G. D. Larrabee, eds., Plenum Press, 1974, pp. 75-106.
33. Booker, G. R.; et al.: Some Comments on the Interpretation of the 'Kikuchi-Like Reflection Patterns' Observed by Scanning Electron Microscopy. Phil. Mag., vol. 16, no. 144, Dec. 1967, pp. 1185-1191.
34. Weiss, Brigette; Hughes, C. Wesley; and Stickler, Roland: SEM Techniques for the Microcharacterization of Metals and Alloys—I. Prakt. Metallogr., vol. 8, no. 8, Aug. 1971, pp. 477-491.
35. Ruff, A. W.: Studies of Deformation at Sliding Wear Tracks in Iron. NBSIR-76-992, National Bureau of Standards, 1976 (AD-A021295).
36. Stewart, Ian M.: Microstructural Studies Using the Electron Microprobe Analyzer. Microstructural Analysis: Tools and Techniques, J. L. McCall and W. Mueller, eds., Plenum Press, 1973, pp. 281-285.
37. White, E. W.: Application of Soft X-Ray Spectroscopy to Chemical Binding Studies with Electron Microprobe. Microprobe Analysis, C. A. Anderson, ed., John Wiley & Sons, Inc., 1973, pp. 349-369.
38. Chang, Chuan Chung: Analytical Auger Electron Spectroscopy. Characterization of Solid Surfaces, P. F. Kane and G. B. Larrabee, eds., Plenum Press, 1974, pp. 509-575.
39. Miyoshi, K.; and Buckley, D. H.: Anisotropic Friction and Wear of Single Crystal Manganese-Zinc Ferrite in Contact with Itself. NASA TP-1339, 1978.
40. Grant, J. T.; and Haas, T. W.: Auger Electron Spectroscopy Studies of Carbon Overlayers on Metal Surfaces. Surface Sci., vol. 24, 1971, pp. 332-334.
41. Chang, C. C.; and Boulin, D. M.: Oxide Thickness Measurements up to 120 Angstroms on Silicon and Aluminum Using the Chemically Shifted Auger Spectra. Surface Sci., vol. 69, 1977, pp. 385-402.
42. Ferrante, John: Auger Electron Spectroscopy Study of Initial Stages of Oxidation in a Copper-19.6-Atomic-Percent-Aluminum Alloy. NASA TN D-7479, 1973.
43. Davisson, C.; and Germer, L. H.: Diffraction of Electrons by a Crystal of Nickel. Phys. Rev., vol. 30, no. 6, Dec. 1927, pp. 705-740.
44. Farnsworth, H. E.: Penetration of Low Speed Diffracted Electrons. Phys. Rev., vol. 49, Apr. 15, 1936, pp. 605-609.
45. Bradshaw, A. M.; Menzel, D.; and Steinkilberg, M.: The Adsorption of Oxygen on W(100): A LEED, XPS and UPS Study. Proceedings of the Second International Conference on Solid Surfaces, Vacuum Society of Japan and International Union for Vacuum Science, Technique and Applications, 1974, pp. 841-845.
46. Rubin, Sylvan; Passell, Thomas O.; and Bailey, L. Evan: Chemical Analysis of Surfaces by Nuclear Methods. Anal. Chem., vol. 29, no. 5, May 1957, pp. 736-743.
47. Ball, D. J.; et al.: Investigation of Low-Energy Ion Scattering as a Surface Analytical Technique. Surface Sci., vol. 30, 1972, pp. 69-90.
48. Nicolet, M. A.; Mayer, J. W.; Mitchell, I. V.: Microanalysis of Materials by Backscattering Spectroscopy. Science, vol. 177, no. 4052, 1972, pp. 841-849.
49. Brown, F.; and Mackintosh, W. D.: The Use of Rutherford Backscattering to Study the Behavior of Ion-Implanted Atoms During Anodic Oxidation of Aluminum, Argon, Krypton, Xenon, Potassium, Rubidium, Chromium, Chlorine, Bromine and Iodine. J. Electrochem. Soc., vol. 120 , no. 8, Aug. 1973, pp. 1096-1102.
50. Peisach, Max; and Poole, D. O.: Analysis of Surfaces by Scattering of Accelerated Alpha Particles. Anal. Chem., vol. 38 , no. 10, Sept. 1966, pp. 1345-1350.

51. Carbonara, Robert S.: Ion-Scattering Spectroscopy for Microstructural Analysis, Microstructural Analysis. Tools and Techniques, J. L. McCall and W. M. Mueller, eds., Plenum Press, 1973, pp. 315-329.
52. Fujiwara, K.: The Effect of Silicon on the Adhesion of Noble Metals. Wear, vol. 51, 1978, pp. 127-136.
53. Vasile, M. J.; and Malm, D. L.: Study of Electroplated Gold by Spark Source Mass Spectrometry. Anal. Chem., vol. 44, no. 4, Apr. 1972, pp. 650-655.
54. Muller, E. W.; and Tsong, T. T.: Field Ion Microscopy. Elsevier Publishing Co., 1969.
55. Brenner, S. S.: Application of Field Ion Microscopy Techniques to Metallurgical Problems. Surface Sci., vol. 70, 1978, pp. 427-451.
56. White, C. W.; Simms, D. L.; and Tolk, N. H.: Surface Composition by Analysis of Neutral and Ion Impact Radiation. Characterization of Solid Surfaces, P. F. Kane and G. B. Larrabee, eds., Plenum Press, 1974, pp. 641-662.
57. Frauenfelder, H.: The Mössbauer Effect. W. A. Benjamin, 1962.
58. Goldanskii, V. I.: The Mössbauer Effect and Its Application in Chemistry. Consultants Bureau, 1964.
59. Pake, G. E.: Paramagnetic Resonance. W. A. Benjamin, 1962.
60. Slichter, C. P.: Principles of Magnetic Resonance. Harper & Row, 1963.
61. Abragam, A.; and Bleaney, B.: Electron Paramagnetic Resonance of Transition Ions. Oxford Clarendon Press, 1970.
62. Ingram, D. J. E.: Free Radicals as Studied by Electron Spin Resonance. Academic Press, Inc., 1958.
63. Harrick, N. J.: Internal Reflection Spectroscopy. Interscience Publishers, 1967.
64. Wendlandt, W. W.; and Hecht, H. G.: Reflectance Spectroscopy. Interscience Publishers, 1966.
65. Kortüm, G. (J. E. Lohr, transl.): Reflectance Spectroscopy. Springer-Verlag, 1969.
66. Wendlandt, W. W.: Modern Aspects of Reflectance Spectroscopy. Plenum Press, 1968.
67. Eischens, R. P.: Chemisorption and Catalysis. The Surface Chemistry of Metals and Semiconductors, H. C. Gatos, ed., John Wiley & Sons, Inc., 1960, pp. 421-438.
68. Weber, Roland E.: Auger Electron Spectroscopy for Thin Films Analysis. Res. Dev. Magazine, vol. 23, no. 10, Oct. 1972, pp. 22-28.
69. Powell, C. J.: Surface Characterization: Present Status and the Need for Standards. Appl. Surface Sci., vol. 1 , no. 2, Jan. 1978, pp. 143-169.

CHAPTER 3

Solid Surfaces in the Perfect State

All solids consist of atoms or molecules. Let us assume that we are examining a metal on an atomic scale and that we can get in and see each individual atom in the solid surface. If we could, we might see a simplified diagram of that particular atom as shown schematically in figure 3-1. A single metal atom, completely black in figure 3-1, in a solid surface would be surrounded by adjacent atoms of like nature. For example, if the metal's bulk were iron, each individual atom in the bulk surrounding the individual black iron atom would be bonded to that atom. The bonding to like atoms would occur within the bulk of the material; at the surface, however, there would be no bonding of the atom to the side away from the bulk solid. That is, at a free surface there is not the same type of bonding to a surface atom as occurs to that same atom when it is immersed in the bulk of the solid. There is a greater degree of bonding of atoms in the bulk of the solid than there is at the surface. This is discussed in more detail later.

The atom that is held in a lattice by bonding to adjacent atoms of like

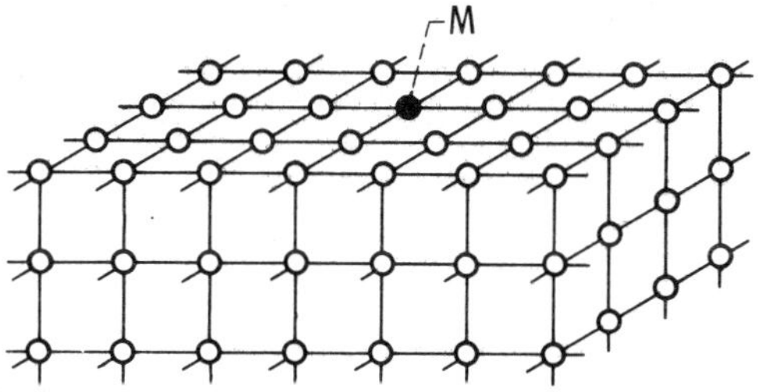

Figure 3-1. – Atom in solid surface.

nature is held with a certain specific amount of energy which keeps it bound to the solid. This energy is referred to frequently in the bulk state as the cohesive energy or the binding energy of the atom to like atoms in the bulk material. The cohesive or binding energy for various materials in the periodic table is presented in table 3-I at absolute zero. Values for the cohesive binding energies are presented both in electron volts per atom (the upper figure in each block) and in kilocalories per mole (the lower figure in each block) for each of the elements presented in the table. The information provided in table 3-I tells us how strongly a particular atom is bound in the bulk of the solid. Thus, for example, we might anticipate that tungsten with a binding energy of 8.66 electron volts per atom would be much more strongly bonded in the lattice than would an iron atom with a binding energy of only 4.29 electron volts.

Accompanying the bulk cohesive energies of solids are many of their physical and mechanical properties. The elastic moduli of metals are related to their cohesive energies. In the transition metals, for example, the higher the cohesive energy, the higher its elastic modulus and the more difficult it is to deform the metals. Likewise, for the transition metals, the melting points

TABLE 3-I. – COHESIVE ENERGIES OF THE ELEMENTS

[Energy required to form separated neutral atoms from the solid at 0 K; values in parentheses are at 298.15 K or at melting point, whichever temperature is lower. Upper value in eV/atom, lower in kcal/mole. Cohesive energy data furnished by Leo Brewer and reduced by S. Strassler.]

Li 1.65 38.0	Be 3.33 76.9											B 5.81 134.	C 7.36 170.	N (114)	O (60)	F (20)	Ne (0.50)
Na 1.13 26.0	Mg 1.53 35.3											Al 3.34 76.9	Si 4.64 107	P (79.2)	S 2.86 66.1	Cl (32.2)	Ar 0.080 1.85
K 0.941 21.7	Ca 1.825 42.1	Sc 3.93 90.6	Ti 4.855 112.0	V 5.30 122.	Cr 4.10 94.5	Mn 2.98 68.7	Fe 4.29 98.9	Co 4.387 101.2	Ni 4.435 102.3	Cu 3.50 80.8	Zn 1.35 31.1	Ga 2.78 64.2	Ge 3.87 89.3	As 3.0 69.	Se 2.13 49.2	Br 1.22 (28.2)	Kr 0.116 2.67
Rb 0.858 19.8	Sr (39.1)	Y 4.387 101.2	Zr 6.316 145.7	Nb 7.47 172.	Mo 6.810 157.1	Tc	Ru 6.615 152.6	Rh 5.752 132.7	Pd 3.936 90.8	Ag 2.96 68.3	Cd 1.160 26.76	In 2.6 59	Sn 3.12 71.9	Sb 2.7 62.	Te 2.0 46	I (25.6)	Xe (3.57)
Cs 0.827 19.1	Ba 1.86 (42.8)	La 4.491 103.6	Hf 6.35 146.	Ta 8.089 186.6	W 8.66 200.	Re 8.10 187.	Os (187)	Ir 6.93 160.	Pt 5.852 135.0	Au 3.78 87.3	Hg (0.694) (16.0)	Tl 1.87 43.2	Pb 2.04 47.0	Bi 2.15 49.6	Po (34.5)	At	Rn
Fr	Ra	Ac															

Ce 4.77 110	Pr 3.9 89	Nd 3.35 77.2	Pm	Sm 2.11 48.6	Eu 1.80 41.5	Gd 4.14 95.4	Tb 4.1 94	Dy 3.1 71	Ho 3.0 70	Er 3.3 77	Tm 2.6 59	Yb 1.6 36	Lu (4.4) (102)
Th 5.926 136.7	Pa 5.46 126	U 5.405 124.7	Np 4.55 105	Pu 4.0 92	Am 2.6 60	Cm	Bk	Cf	Es	Fm	Md		Lw

for the materials are related generally to the cohesive binding energy. Those metals with higher cohesive binding energies have higher melting points. Therefore, one would expect rhenium, which has a cohesive binding energy of 8.10 electron volts, to melt at a much higher temperature than, say, nickel, which has a binding energy of 4.435 electron volts. And this is, in fact, the case.

In addition to modulus of elasticity and melting point, many of the other properties of solid materials are related or can be related to cohesive binding energies. In accordance, knowing the cohesive binding energies is very useful to the tribologist. The binding energy tells how strongly a particular atom immersed in the bulk is bonded to its neighbors and what amount of energy is required to liberate that atom from its neighbors. The binding energy at the surface is somewhat less than that of the cohesive energy because the atom, as was mentioned earlier, is only bound on the sides within the bulk to like atoms. It is not bound at the free surface to like atoms. While the cohesive or binding energy of an atom in the surface is less than the cohesive or binding energy in the bulk solid, there is a direct relationship between the energy of an atom bound in the surface to the energy of an atom bound in the bulk. The energy associated with the atom bound in the surface is referred to frequently as surface energy. This is discussed in more detail later.

Bulk cohesive energies are important in tribology. When adhesion occurs at the interface with two dissimilar materials in contact and fracture occurs, the fracture does not usually occur in the adhesive junction but rather in the cohesively weaker of the two materials. In such situations, the cohesive binding energy can predict the amount of energy required to fracture at an adhesive junction.

Elemental metals and alloys are probably the most frequently encountered materials in the solid state of concern to tribologists. There are, however, a host of other solids which have entirely different binding than metals that frequently must be considered. Thus, for the metals in the transition series (in table 3-I), the binding is a metal to metal bond; therefore, in figure 3-1 the atom in the surface has a metallic bond to like atoms in the bulk. If, however, other possible solids are considered, other types of bonding can occur for solids, and some of these are shown schematically in figure 3-2.

Bonding

There are generally two types of solids, crystalline and amorphous solids. At this point, only crystalline solids are discussed. For crystalline solids, where there is a regular order or array to the atoms, a number of types of bonding are found (indicated schmatically in fig. 3-2). Figure 3-2(a) shows one of the weakest types of bonding that occurs for materials in a solid; it is associated with, for example, the attractive forces that develop between like atoms of materials (e.g., argon in a solid state). The bonding is relatively weak for the argon atoms in the bulk, and this relatively weak bonding of

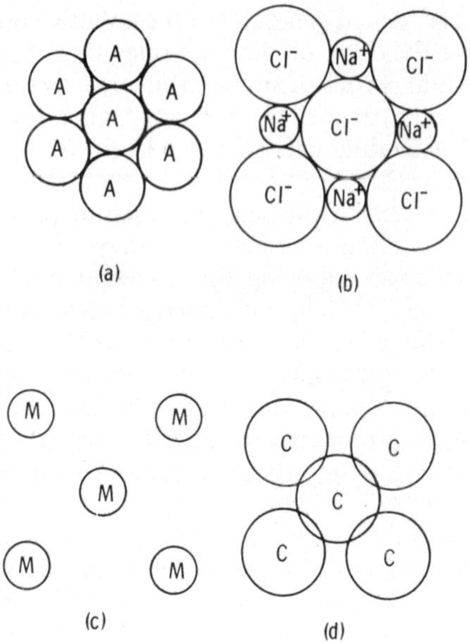

(*a*) *Van der Waals* (*crystalline argon*).
(*b*) *Ionic* (*sodium chloride*).
(*c*) *Metallic.*
(*d*) *Covalent* (*diamond*).

Figure 3-2. – Principal types of crystalline bonding forces.

neutral atoms is frequently referred to as van der Waal's forces or van der Waal's bonding. The relative weakness of this bond can be seen in table 3-I where the cohesive energies of the elements are presented. For argon the cohesive binding energy is 0.08 electron volt, which is very low compared to the metallic bond. Van der Waal's bonding is frequently encountered in tribological systems with the adsorption of materials such as liquid lubricants to a solid surface. This is not the bonding associated with the first monolayer of lubricants adsorbed on the surface because this may be a chemisorbed layer. Rather, it is the bonding of the liquid molecules to like molecules on the surface after the first monolayer has been developed. It also is the bonding that occurs in many conventional lubricants in the bulk.

Another type of bonding in crystalline solids that can occur is what is commonly referred to as the ionic bond (fig. 3-2(b)). This occurs in solids that are not elements but rather compounds of one element with one or more other elements. A classic example of this is sodium chloride (ordinary rock salt or table salt). Sodium chloride crystallizes in a form wherein the sodium ions and the chlorine ions bond to one another. We have a charge or electron transfer from the alkaline metal atoms (sodium) to the halogen atoms (chlorine); this results in ions being held together by electrostatic forces between the positive and the negative ions. Ionic bonding can be very strong because there is an electron charge transfer. The bonding between anions and cations (negatively charged ions with positively charged ions) results in very strong bonding forces in ionic solids. A classic example of this would be aluminum oxide, a tribological material which has inherently high strength.

A third type of bonding occurring in solid material is one that has already been discussed; it is the metallic bond found in all metals. In the metallic bond depicted in figure 3-2(c), the metallic ions are immersed in a sea of electrons. The electrons move freely between the cores of ions allowing for and contributing to many of the metallic materials properties. This freedom of movement in the sea of electrons imparts to metals their good thermal and electrical conduction characteristics. With metals, there is a wide range of strengths that can be found; the binding energies may be relatively weak, such as those associated with metals like indium and thallium, or they may be relatively strong, such as those seen with metals like osmium and tungsten.

A fourth type of bonding is that seen in crystalline solids such as the diamond (fig. 3-2(d)). It is referred to commonly as covalent bonding. In covalent bonding, neutral atoms appear to be bound together by overlapping their electron distributions. Covalent bonding is an extremely strong bond as evidenced by the high melting points and the high hardness of material such as the diamond.

The ionic bond (fig. 3-3(a)) is represented by an array of ions where we have alternating positive and negative charged ions making up the crystalline solid. In covalent bonding (fig. 3-3(b)), like atoms are bonded in an array to similar atoms, whereas in metallic bonding (fig. 3-3(c)), an array of cations (positively charged ions) is immersed in the sea of electrons which are free to move about. In addition to the ionic, covalent, metallic, and van der Waal's bonding shown in figure 3-2, there are other types of bonding that are experienced by the tribologist working, in particular, with liquid lubricants. In organic molecules, such as liquid lubricants, there are molecules which have nonpolar structures and there are others which have polar structures. A nonpolar molecule is shown schematically in figure 3-3(d). With a nonpolar molecule there is no preferred charge on the molecule.

In the polar molecule (fig. 3-3(e)), however, a dipole moment develops in the molecule such that one portion of the molecule is positively charged while another portion is negatively charged. This variation in charge of the molecule can be fundamentally very important where charge differences also exist in the solid surface. For example, if a solid surface is rich in electrons, making it attract the positive end of a polar molecule, that particular end of the molecule may then bond to the solid surface and thereby determine to a great extent the properties of the solid surface. Thus, understanding the dipole moment of molecules, particularly lubricating molecules, and the nature of charges on the surface can give considerable insight into the nature of the bonding that occurs at the solid surface and the type of properties that are exhibited by the surface layers as a result of that bonding. The dipole activity of various molecules can be obtained from conventional chemical texts (ref. 1). A variation does exist in many molecular structures.

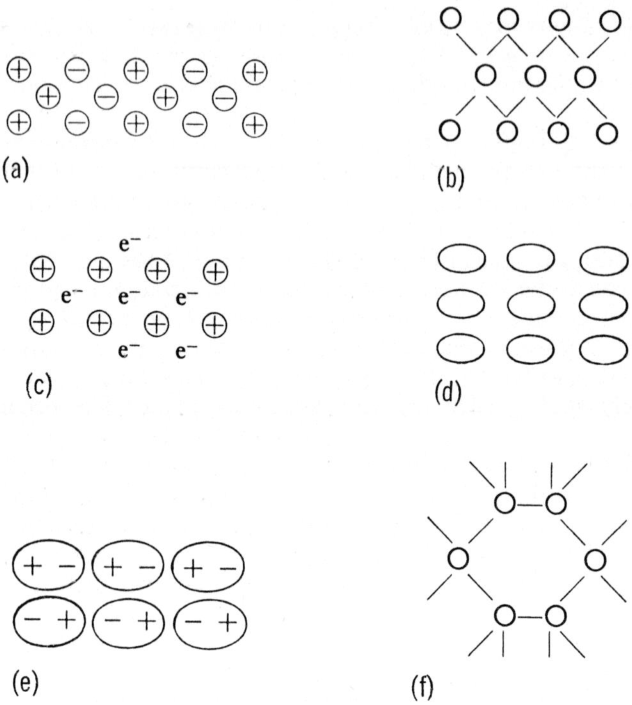

(*a*) *Ionic; array of ions.*

(*b*) *Covalent atomic; array of atoms.*

(*c*) *Metallic; array of cations and mobile electrons.*

(*d*) *Nonpolar covalent molecular; nonpolar molecules.*

(*e*) *Polar covalent molecular; array of dipolar molecules.*

(*f*) *Chelate; molecules in crystal vacancies.*

Figure 3-3. – Nature of bonding in solids.

Another form of bonding that can be found in solids is that associated with large molecules such as encountered in lubricants. When a molecule is very large, it may develop a ring structure or what is referred to as a chelate structure (fig. 3-3(f)). Such a structure has its own characteristic properties which differ from other bonds; these properties are detailed again in the chemical literature.

It was mentioned earlier that solids may exist in either crystalline or amorphous form. Amorphous solids lack any sort of regular registry or pattern. With crystalline solids, however, there is a regular pattern or array of atoms in the solid. In most instances, the tribologist is concerned with crystalline materials although, to some extent, occasions do arise where amorphous materials are of concern. Crystalline solids include most of the metals and alloys of interest to the tribologist as well as inorganic solids, ionic compounds, and even some lubricating species such as solid lubricants.

In figure 3-1, the presence of an atom on the surface of the solid was discussed. In that case, it was a simple schematic and we talked in terms of an atom somewhat in the abstract. Since the atom can not be seen with the naked eye, it was simply a diagrammatic sketch of an atom. We can see in figure 3-1 that there is some regular order in the bonding of one atom to like atoms in the solid surface. In crystalline solids, from an understanding of the basic physics of solids, one can computer simulate what solid surfaces should actually look like; this has been done for an iridium tip using the field ion microscope (FIM) (fig. 3-4(a)). Years ago one talked in terms of atoms on solid surfaces in the abstract, never having had the opportunity of actually seeing atoms. But today one can, in the laboratory, very routinely examine individual atom sites on a solid surface with the aid of the FIM.

In the computer simulation shown in figure 3-4(a), each individual black spot in the computer simulation represents an individual atom site on the solid surface. The rings represent atomic planes, and these atomic planes are called out with numbers in parenthesis; the numbers referring to the specific atomic plane that is being shown in the computer simulation. Figure 3-4(b) is an actual photograph of an iridium pin tip taken in the FIM (previously described in ch. 2) at a magnification of 1 million. It allows the researcher to see individual atoms on the solid surface. Each white spot in the photomicrograph (fig. 3-4(b)) is an individual atomic site.

In the photomicrograph, the atoms appear as white spots because the imaging voltage produces the brightest intensities at the atom sites of the layers that are being imaged on the solid surface. So instead of appearing black, as in the computer simulation, each individual atom sight appears as a white spot. Again, as in the computer simulation, the rings relate to the atomic

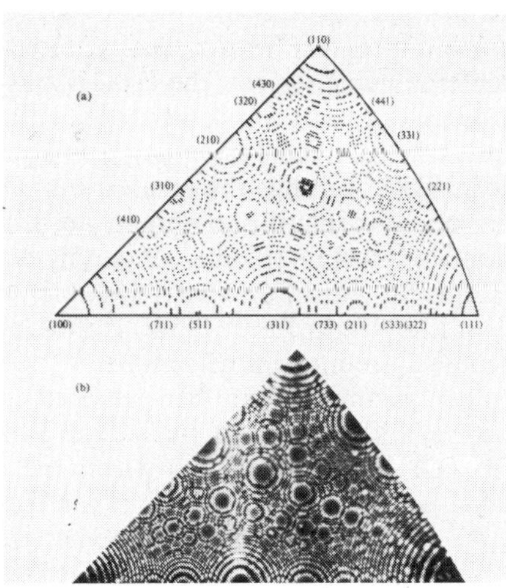

(*a*) *Computer simulated image.*

(*b*) *Ion image.*

Figure 3-4. – Computer simulated image of face-centered-cubic lattice compared with actual field ion micrograph of 900 Å iridium tip.

planes that are exposed. A direct comparison of figures 3-4(a) and (b) indicates that the real surface looks exactly like the computer simulation of figure 3-4(a). So we can, with the aid of computers, simulate what a real surface looks like without ever seeing that solid surface. The computer simulation in figure 3-4(a) is for a perfect solid surface. As we will see, real surfaces do not generally exist in the perfect state but contain defects and irregularities as is true of so many things in nature. This topic is discussed later in this chapter. The reason that a number of atomic planes can be seen in figures 3-4(a) and (b) is that the images produced in figures 3-4(a) and (b) are for an iridium tip radius of about 900 angstroms. We are, therefore, looking at a curved surface. As a consequence, a number of atomic planes are exposed in the process of generating the radius. If the surface were perfectly flat, we would tend to see one particular atomic plane.

Crystal Structure

What are atomic planes? The atoms in crystalline solids tend to bond to one another in a particular arrangement or array which is reproducible throughout the solid. And the atomic density varies in different directions through the crystal. If one had a cube (1 cm on a side) of a particular metal, for example copper, and if that cube were sawed at various angles, we would cut through various atomic densities of atoms. These variations in atomic density through the crystalline solid are referred to as atomic planes. They are the planes on which the atoms lie in the crystalline solid and bond to one another. The solid is made up of the individual atoms bonded within the crystalline planes and the planes are bonded to each other throughout the solid.

Figure 3-4 is an iridium surface with a face-centered-cubic crystal structure. In the face-centered-cubic system, the (111) plane shown in figure 3-4 in the lower right corner of both the computer simulation and the actual field ion micrograph is the particular crystallagraphic plane having the maximum atomic density. That is, it has the greatest number of atoms stacked along that particular plane so that if one were to cut through a crystal along the (111) orientation the greatest number of atoms would be intersected. This particular plane in the face-centered-cubic system is depicted schematically in figure 3-5(b). In figure 3-5(a) is the elementary unit cell of the face-centered-cubic system. A large number of elements, particularly the transition metals, crystallize in the face-centered-cubic form; this particular crystal structure is, therefore, of considerable interest. The elementary unit cell (fig. 3-5(a)) is the smallest unit within the crystalline solid that represents the atomic arrangement throughout the entire solid. The cube shows that atoms are at the four corners of the cubes, plus an additional atom is in the center of each of the six faces of the cube. This structure is distinguished from a body-centered-cubic structure which is normal for many metals. In the body-centered-cubic crystalline structure, instead of having an atom in the center of each of the six faces of the crystalline solid in the unit cell there is a single atom located in the center of the cube itself, and the atoms at the eight corners of the crystalline cube bond to that single

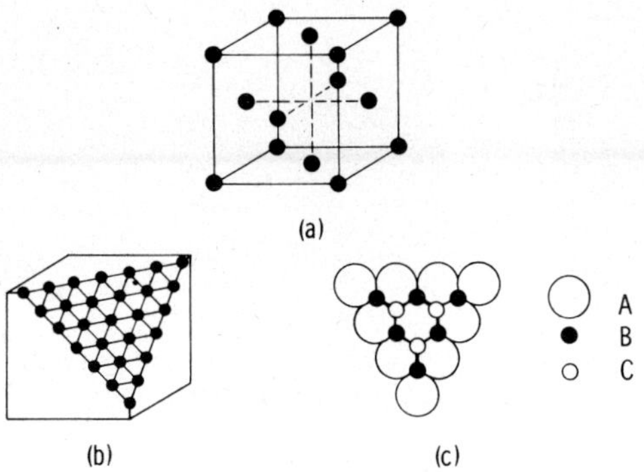

(*a*) *Elementary unit cell.*
(*b*) *Arrangement of atoms in* (*111*) *plane.*
(*c*) *Stacking sequence of* (*111*) *planes.*

Figure 3-5. —Face-centered-cubic structure.

central atom. For the face-centered-cubic structure (fig. 3-5(a)), the atoms on the four corners are bound to the central atom in addition to being bound to each other.

If we cut through the cube in figure 3-5(a) so as to pick up the atoms that make up the (111) surface referred to previously in reference to figure 3-4 (i.e., the high atomic density plane where we intersect the greatest number of atoms), we would get a slice through the crystal such as that shown in figure 3-5(b). In the (111) plane, the atoms are bonded to each other in a triangular array; one atom is bonded by a number of nearest neighbors—six to be exact.

The (111) orientation is, then, a diagonal slice through the crystalline solid. Now the (111) planes which cut through the crystalline solid repeat themselves in this particular structural arrangement of atoms on every fourth plane. Three planes have varying orientations, and the fourth repeats the orientation of the first plane. This is shown schematically in figure 3-5(c), which indicates a stacking sequence of the (111) planes. We have orientation A followed by orientations B and C, and then orientation A repeats: A, B, C, A, B, C, A, B, C, and so on throughout the crystal lattice.

In both the face-centered-cubic system and the body-centered-cubic system, there are three principle planes which are of most frequent interest and which are the most frequently referred to when discussing crystallography (the crystalline nature of solids). These particular planes are the (100), the (110), and the (111) surfaces which were already discussed in reference to figures 3-5 and 3-4. If we examine the cube surface, we see these particular planes (fig. 3-6). Figure 3-6(a) presents three (100) planes in a solid surface. The (100) planes are, essentially, the faces of the cubic structure shown in figure 3-5(a). In the face-centered-cubic unit cell there would be five atoms in this particular plane for the unit cell, whereas with the body-centered-cubic structure, there would be four in this particular plane. The (110) surface is depicted in figure 3-6(b). There are six of these planes that cut through the crystal at various orientations.

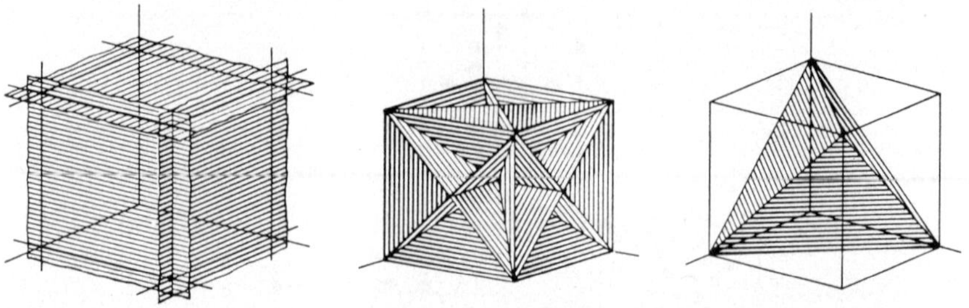

(a) *Three {100} planes.* (b) *Six {110} planes.* (c) *Four {111} planes.*

Figure 3-6.—Planes of possible slip in cubic crystal. Planes with indices higher than {111} are not shown.

Figure 3-6(c) presents the (111) planes in the face-centered-cubic crystal structure. As was mentioned earlier, the (111) planes are the high atomic density, low surface energy planes in the face-centered-cubic crystal system. These particular atomic planes are the ones that undergo relative motion one to another with deformation such as that encountered in tribological systems under loading and sliding conditions; this is discussed later (p. 157) in reference to certain crystalline defects.

In reference to figure 3-5 it was mentioned that the atoms on the (111) planes are packed in the close-packed array. This can be seen more effectively in the diagrammatic sketch of figure 3-7; three different atomic planes in the face-centered-cubic system are presented. The uppermost figure in figure 3-7 is for the (111) surface, and it can be seen that this particular surface has the atoms in a close-packed array. The close-packed array is a packing such that the atoms occupy the least amount of space possible per unit area; stated another way, the greatest number of atoms can be squeezed into a fixed space with this particular packing arrangement. This is the arrangement that would occur, for example, in the racking of billiard balls.

The use of the numerical designation (111) with parenthesis about it was taken from the crystallographic notations referred to as the Miller-Bravis indices. They are simply a numerical designation of various atomic planes in the bulk of the solid surface. These numerical indications are thoroughly described in standard texts on crystallography (e.g., ref. 2). Good reference sources for a description of the Miller-Bravis system would be Barrett's book on metallurgy (ref. 3) or Cullity's text on X-ray diffraction (ref. 4).

When the number is in parentheses, it refers to a specific atomic plane. When braces are used about the number, it refers to the entire family of planes. Thus, in figure 3-6, reference is to all of the (111) planes in the cubic solid, while in figure 3-7, the use of the parentheses refers to a specific plane—namely, the (111) plane.

In addition to a numerical connotation for the various planes in the crystalline solid, there are also directions in which planes of atoms move relative to each other. These crystallographic directions are also noted numerically (as indicated in fig. 3-7). In figure 3-7, the [110] direction is depicted for all three atomic planes: the (111), (100), and (110). When it is a

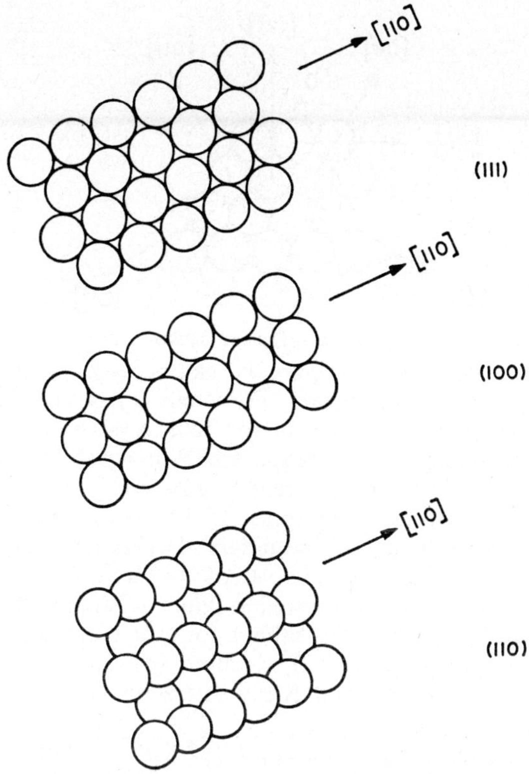

Figure 3-7. – Appearance of several crystal planes.

single direction, individual brackets are used about the number; when it refers to a family of directions, carets are used about the number.

In the middle schematic of figure 3-7 you see the packing arrangement for atoms in the (100) orientation. This would be the orientation on the face of the cubes in figure 3-6(a). The face planes in the face-centered-cubic system have atomic packing such as that depicted in figure 3-7 in the middle schematic.

The third arrangement of atoms in a crystal lattice is the (110) orientation depicted in figure 3-7 as the lower schematic. The atoms in this particular arrangement are really not all in the surface plane. There are three rows of atoms uppermost in the surface plane, with two intermediate rows slightly below the three upper rows of atoms. In contrast, all of the atoms in the (111) and (100) surfaces are in the same plane.

If we look at in particular the (111) crystal plane in figure 3-7, there are various crystallographic directions on that particular plane; these are shown in figure 3-8. The directions are depicted numerically in the brackets. And if one follows any particular crystallographic direction, there is a change in atomic packing from that direction to an adjacent direction, and the directions repeat themselves periodically where the packing is the same or identical. The symbol b in figure 3-8 is the spacing between adjacent atoms in the crystal lattice.

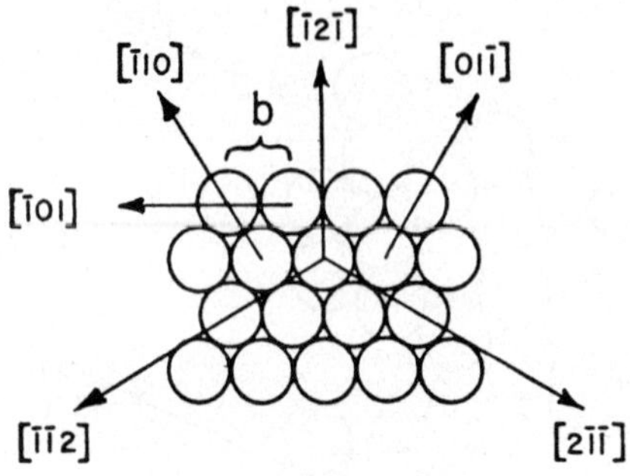

Figure 3-8. – The (111) plane.

The orientations presented in figure 3-7 can occur as planes in the bulk crystal or they can be surface planes. For single crystals of materials, any one of these orientations in a face-centered-cubic crystals or body-centered-cubic crystal might occupy the entire surface. With a polycrystalline surface where there are grain boundaries present, adjacent grains may have and generally do have different orientations. If the orientations were not different there would not be a grain boundary. Hence, a polycrystalline surface may be represented by a host of different orientations. The crystallographic connotations presented in figure 3-7 are but three representative orientations. There are many others; for example, one is referred to a standard crystallography text (ref. 2), metallurgical texts (ref. 3), solid-state physics texts (refs. 5 and 6), and X-ray diffraction techniques reference sources (ref. 4).

When discussing figure 3-1, it was indicated that the atoms in the surface plane of the solid were bound beneath it to atoms in the bulk as well as to atoms within the plane itself; there was, however, nothing above in the free space bonding to the atom. This is unlike the bulk solid where an atom immersed in the bulk solid is bound on all sides by like atoms. In the face-centered-cubic system, for example, an atom in the bulk is bound by the twelve nearest neighbors; consequently, there is bonding of a single atom, say of a metal, to twelve metallic atoms about that one single atom. On the surface, however, of a face-centered-cubic metal, if we look at a (111) surface plane (a high atomic density plane) the atom in the surface would not be bound by twelve atoms as it would be in the bulk, but rather it would be bound by nine atoms. There are three absent atoms that would normally be above the surface and, as a consequence, there is an absence of bonding to surface sites above the solid surface. When the surface is in the atomically clean state, a high degree of surface energy exists, and this may be measured on a solid surface.

You may recall, with reference to table 3-I, cohesive binding energies were discussed. These were the energies associated with bonding a single atom in the lattice of the bulk solid. This was the energy associated with pulling, or plucking out, that one atom from its twelve nearest neighbors (for the case of face-centered-cubic crystals). Of course, if the atom is in the surface, the binding energy is associated with nine atoms rather than twelve. The energy is still the binding energy of an atom to like atoms; the number involved has, however, been reduced. Now instead of talking in terms of cohesive energies we should talk in terms of surface energies (surface energy is the energy with which the atoms in the surface plane are bonded to like atoms in adjacent planes). It would be the energy required to separate two (111) planes from each other by splitting or pulling apart the atomic bonds across the interface and liberating two free, atomically clean surfaces.

An example of how the coordination number (number of nearest neighbors or atom has) can vary on the surface of a solid is given with the aid of the data in table 3-II. The coordination number for different surface sites is indicated. An atom in the corner site has a coordination number of bonding of 5, an edge site has a bonding of 7, and the (111) face has a coordination number of bonding to like atoms of 9. On a (100) face the coordination number of bonding is only 8.

TABLE 3-II. – COORDINATION NUMBER FOR SURFACE ATOMS

Surface site	Coordination number
Corner	5
Edge	7
(111) face	9
(100) face	8

In chapter 2 it was indicated that real solid surfaces are not smooth topographically but rather contain surface irregularities or asperities—hills and valleys in the solid surface. These are imperfections in the bulk solid surface on a macroscale. Even, however, on a microscale (at atomic and subatomic levels), variations in surface properties and characteristics do exist.

Electronic Structure of Surfaces

The variation in atomic packing indicated in figure 3-7 for the three different crystallographic planes results in differences in the electronic nature of the solid surface. If we go a step below the packing of the atoms and look at the electron density or electron distribution on the surface of the solid for the various atomic planes depicted in figure 3-7, it is the electrons involved in a solid surface that contribute to its binding to other solids, to adsorbed species, and to lubricants. Therefore, the electron distribution on the solid surface is of great importance.

If we look at the electronic arrangement of the solid surfaces we find that there are variations in the distributions of electron densities on the solid surface. For example, in the upper schematic of figure 3-9, we see the distribu-

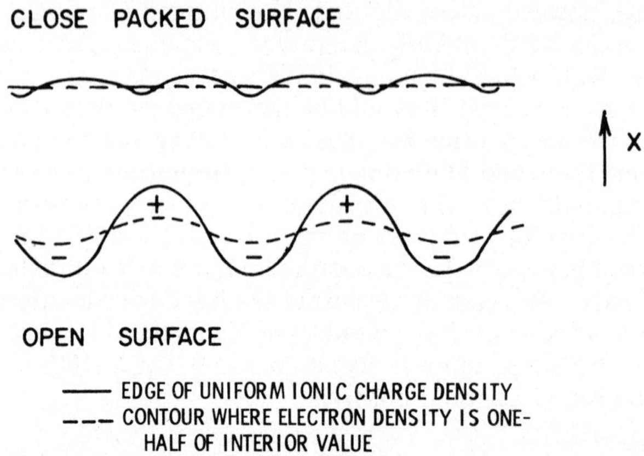

Figure 3-9.—Approximate electron density variation on jellium surfaces with periodic positive charge boundaries (ref. 7).

tion of electron density for the jellium surface (i.e., the electron surface) of the close-packed-atomic plane, such as the (111) plane depicted in figure 3-7. In this particular orientation of a close-packed surface, there are no sharp periodic variations in positive and negative charges; the positive charges are associated with the ion cores, and the negative charges are associated with the sea of electrons surrounding the ion cores, as was discussed in figure 3-3.

From an electron point of view, the close-packed surface is relatively smooth. If, however, we looked at a more open surface such as the (110) surface in the lower schematic of figure 3-9, we see that the distribution of positively charged ion cores and electrons is somewhat more perturbed. There is a distinct hill and valley configuration to the electron distribution in the solid surfaces. The positive ion cores lie somewhat above the surface, and the negatively charged electron density is somewhat to the interior. This variation of the electron density in the solid surface contributes to differences and variation in the reactivity or interaction of the solid surfaces with other materials. The energetics of the surface also vary. For example, the close-packed surface (the (111) surface in the face-centered-cubic system of fig. 3-7 and the upper schematic of fig. 3-9) has a relatively low surface energy (ref. 7). In fact, it has the lowest surface energy in the face-centered-cubic system. In contrast, however, the more open structure (the (110) surface in fig. 3-7 and also on the lower schematic of fig. 3-9) is a much more active surface and has a higher energy surface. Hence, it is much more reactive with other solids that come in contact with it and with species from its environment as well as with lubricants.

The structural differences seen between atomic planes in figure 3-7 can be detected with some of the analytical tools referred to earlier. For example, with ordinary X-ray diffraction techniques, it is possible using Bragg reflections to show the differences in the orientation of the various crystallographic planes of a solid. The X-ray differences among the three planes of figure 3-7 are shown in figure 3-10. A comparison of the patterns

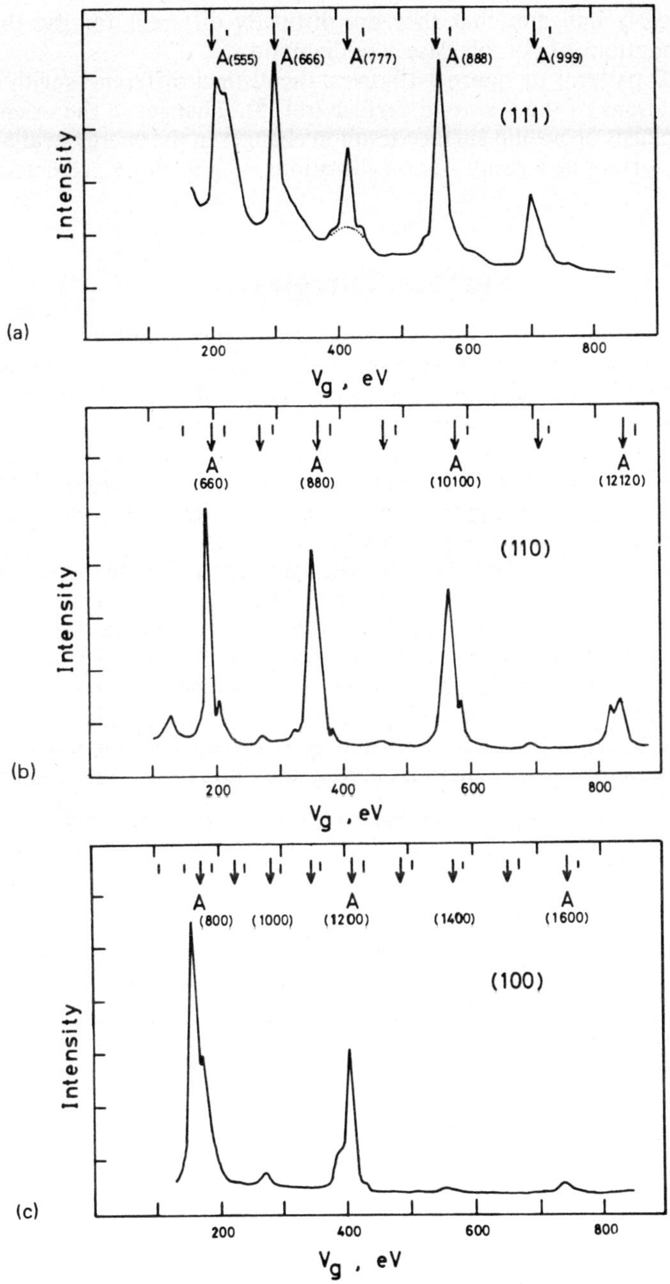

(*a*) (*111*) *Plane; 8°*. (*b*) (*110*) *Plane; 6°*. (*c*) (*100*) *Plane; 6°*.

Figure 3-10. – Intensity-voltage curves from (*00*) *beam at various angles of incidence to three planes of diamond. Vertical scale is in arbitrary units; Bragg positions are located by short bars; arrows indicate Bragg position corrected for constant inner potential; A denotes peaks that are allowed by structure factor* (*ref. 8*).

obtained clearly indicates that they are distinctly different for the three atomic orientations of, in this case, the diamond.

The LEED patterns of figure 3-10 reveal the distinct differences with the three orientations of the diamond surface (ref. 8). Changes in the orientation of the planes on a solid surface result in changes in the energy available at the solid surface as a result of considerations such as those presented in figure 3-9.

Surface Energies

The surface energy dependence on orientation is presented for the face-centered-cubic and body-centered-cubic metals (table 3-III). For the face-centered-cubic metals, all the orientation surface energies are referred to the (111) high atomic density, low surface energy plane as being the reference surface, and the variations are from that particular reference orientation. Likewise, for the body-centered-cubic system, the surface energies for the (110) surface are used as a point of reference, and the various deviations from that for other planes are presented in table 3-III.

All of the energies in table 3-III for the face-centered-cubic system are greater than the surface energy for the (111) surface, with the one anomalous situation of the nickel (100) as observed by Mykura (see refs. 9 to 12). This may be a deficiency in the experimental technique used for measuring the surface energies, because one would anticipate that the values would be greater for the (100) surface than the (111) surface; in fact, this was observed by Sundquist in his data and also by Winterbottom and Gjostein. Likewise, in the body-centered-cubic system, all planes other than the

TABLE 3-III. – STRUCTURAL DEPENDENCE OF SURFACE ENERGY

(a) Face centered cubic

Plane	Sundquist Au, Ag, Cu, Ni	Winterbottom & Gjostein Au	Mykura Ni
(111)	1·00	1·00	1·00
(100)	1·047	1·072	0·95
(311)	1·119	1·065	1·00
(110)	1·15	1·047	1·01
(210)	1·16	1·055	1·00

(b) Body centered cubic (ref. 12)

	W 1200°K	Ta 1400°K
(110)	1·00	1·00
(100)	1·022	1·050
(211)	1·019	1·068
(111)	1·025	1·083
(310)	1·025	1·065
(411)	1·038	1·070

(110) surface should have a higher surface energy than the (100) surface, and we see this is true for tungsten and tantalum.

Since the energy required to separate two atomic planes of the material varies with crystallographic orientation in the solid (that is, it takes less energy, for example, in the face-centered-cubic system to cleave or to separate (111) surfaces than it does to cleave or separate (110) surfaces), one might anticipate that when a crystalline solid is fractured, the fracture process takes place along the weakest crystallographic planes in the solid. Thus, for a face-centered-cubic crystal, one might anticipate that the fracture or cleavage would occur for a brittle solid (such as diamond or silicon) along the (111) planes, the high atomic density, low surface energy planes. The low surface energy means that these planes require the least amount of energy to separate one plane from another in the material. In fact, this is physically observed. The industrial cutting of diamonds for jewelry and industrial cutting tools relies very heavily on the fracture or cleavage of the natural diamond along the cleavage plane, which is the (111) surface in the diamond, a face-centered-cubic structure.

Cleavage Planes

The planes along which cleavage occurs in various materials are presented in table 3-IV for various structures: diamond (diamond, silicon, and germanium), zinc blend, metal oxides, layer lattice (molybdenum disulfide,

TABLE 3-IV. – PRIMARY CLEAVAGE AND SLIP PLANES FOR VARIOUS SOLIDS

Structure	Material	Cleavage	Slip
Diamond	C(diamond), Si, Ge	{111}	{111}
Zinc blende	ZnS, ZnO, InSb, GaAs, SiC, BN	{110}	{111}
Metal oxides	Al_2O_3 (alumina, sapphire), SiO_2 (quartz, fused silica), BeO, TiO_2	Crystallographic anisotropy not pronounced	
Layer structures	Mica, C(graphite), MoS_2	(0001)	(0001)
Sodium chloride	NaCl, NaBr, NaF, LiF, KCl, PbS, MgO, AgCl	{100}	{110}
Hexagonal close packed (h.c.p.)	Zn, Mg, Cd, Be, Sn, Ti, Co, Zr	(0001)	(0001), $\{10\bar{1}0\}$
Body centred cubic (b.c.c.)	α-Fe, Li, Na, K, V, Cr, Mn, Cd, Mo, W	{100}	{112}, {110}
Face centred cubic (f.c.c.)	Cu, Ag, Au, Ni, Al	None	{111}

graphite, and mica), sodium chloride, hexagonal close packed, body centered cubic, and face centered cubic.

In addition to the cleavage plane in table 3-IV there is also a slip plane. The slip plane is that plane along which, if cleavage does not take place, slip takes place in the solid. (This subject is discussed in more detail with reference to defects in solids and solid surfaces, particularly dislocations.) When a load or stress is applied to the crystal, the crystal moves, deforms, or changes its shape; this occurs along well-defined crystallographic planes. Again, these are the planes of weakest bonding—namely, the (111) planes in, for example, the diamond structure and the face-centered-cubic structure of the metals.

For some materials, cleavage does not occur because of the extreme ductility of the material. This occurs, for example, with the face-centered-cubic metals. It is noted in table 3-IV that there are no cleavage planes indicated for the face-centered-cubic metals. Hence, the deformation mechanism, or the mechanism of distortion of face-centered-cubic metals, is by slip, and the material does not fracture in a brittle fashion along cleavage planes as does a material like diamond.

With structures such as molybdenum disulfide, both slip and cleavage can be observed. For example, it is very easy to cleave molybdenum disulfide along the basal planes. These are the (0001) planes indicated in table 3-IV. The weakest bonding occurs between adjacent basal planes in graphite, mica, and molybdenum disulfide. As a consequence, it is very easy to separate the basal planes in the layer-lattice structure of these particular materials. In addition to cleavage, however, if a load is imposed on the surface of the basal plane of molybdenum disulfide single crystals and tangential motion is initiated, plastic deformation strain in the solid surface is observed. This strain takes place along well-defined planes (0001) so that we can, in some of these layer-lattice structures, see brittle fracture take place in both cleavage and slip.

Cleavage Strengths

With perfect crystalline solids, the strength of the solid is extremely high. When one analyzes the bond strengths based on the energies required to fracture atomic bonds along slip or cleavage planes in the crystalline solids indicated in table 3-IV, it becomes evident that the theoretical strengths for most crystalline solids are extremely high (ref. 13). Estimates of theoretical cleavage strengths for some of the materials are indicated in table 3-IV and are included in table 3-V; the estimated theoretical cleavage strengths are listed for iron, aluminum, beryllium metal, and magnesium oxide (which is representative of the metal oxides in table 3-IV). The estimated strengths are based on cleavage in the particular direction indicated on the cleavage plane indicated in table 3-IV.

The cohesive force between adjacent planes is listed as g. For example, with beryllium, it would be between adjacent basal (0001) planes and its value is 1900 dynes per square centimeter. This value is nothing more than twice the surface energy of the crystalline solid surface; that is, it is the sur-

TABLE 3-V. – ESTIMATES OF THEORETICAL CLEAVAGE STRENGTH[a]

Material	Direction	g (dynes/cm)	σ_f 10^{11} dynes/cm^2	σ_f 10^6 psi	σ_f/E
α-Fe	⟨100⟩	4000	4.3	6.2	0.23
Al	⟨111⟩	1350	1.9	2.8	0.20
Be	⟨0001⟩	1900	6.5	9.5	0.13
MgO	⟨100⟩	3200	6.1	8.8	0.15

[a] *Note.* g = planar cohesion $\approx 2\gamma$
γ = surface tension
σ_f = theoretical cleavage stress
E = tensile modulus in given direction

face energy of the solid times two because two surfaces are being generated when the solid along a cleavage plane is pulled apart.

The theoretical cleavage strengths are extremely high for all four crystalline solids presented in table 3-V. The reason the values are much higher than actually observed is discussed shortly in reference to the presence of defects in solids. The data in table 3-V are for perfect solids—that is, solids in the absence of defects such as dislocations, vacancies, grain boundaries, interstitials, and other imperfections in the crystalline solid.

Shear Strength Solids

Not only is theoretical cleavage strength of a crystalline solid extremely high in the absence of defects, but so also is the shear strength which is dependent on slip in the crystalline solid. This is indicated in the data of table 3-VI for the same four materials as indicated in table 3-V (ref. 13).

TABLE 3-VI. – ESTIMATES OF THEORETICAL SHEAR STRENGTH[a]

Material	Plane	τ_m 10^{11} dynes/cm^2	τ_m 10^6 psi	τ_m/G
α-Fe	{110}	0·7	1.0	0.10
Al	{111}	0.12	0.2	0.05
Be	{0001}	0.6	3.8	0.15
MgO	{110}	1.4	2.2	0.20

[a] *Note.* τ_m = theoretical shear stress
G = shear modulus on given plane

Estimates of theoretical shear strength are presented (table 3-VI) along the slip planes associated with the particular crystal systems involved. For alpha iron it would be the (110) plane, for aluminum (111), beryllium (0001), and magnesium oxide (110). Again, as was observed in table 3-V, for cleavage strengths, the shear strengths are extremely high. They are much above what is actually observed in practice in shearing these metals and magnesium oxide. It must be remembered, however, that these values are for perfect crystalline solids—that is, crystalline solids that are free of defects.

In the case of the theoretical cleavage strength, it is nothing more than the strength required to rupture or fracture the bonds across adjacent cleavage planes in a crystalline solid. On the other hand, the shear strength is the force required to initiate shear when one attempts to slide one crystallographic plane in the crystalline solid over another. Thus, in one case, a tensile force (cleavage direction) is required to pull the crystal apart, and in another, a shear force (tangential or shearing direction) is required.

Both of these properties of crystalline solids are important to the tribologist. The cleavage strength is important in the fracture of adhesive junctions, and the shear strength is important in adhesion and friction. If adhesion occurs at an interface between two solid surfaces in solid-state contact, the force to initiate tangential motion (friction force) is a shearing force. If two single crystals are brought into solid-state contact and attempts are made to move one over the other, the force required to initiate that movement is a shear force. If, however, adhesion occurs at the interface and one simply wishes to separate the solid surfaces by applying a normal force, the force required to fracture the adhesive junction may be a cleavage force (for brittle solids).

In figure 3-10 the LEED intensity patterns were for crystalline solids. With crystalline solids, distinct patterns can be seen for variations in orientation of the solid. In addition to crystalline solids, however, there are a host of solid materials (some of importance in tribology) that are noncrystalline (amorphous); that is, instead of the atoms being packed in specific arrangements, the atoms or the molecules within the solid are bonded in a disordered state. The well-defined crystallographic planes detailed in table 3-IV are, therefore, absent in amorphous materials. A classic example of an amorphous material is ordinary glass. With glass or other amorphous solids the crystallographic planes for cleavage and shear are not present. In general, the glasses are considered to be homogeneous isotropic solids. There are some microstructural characteristics of glasses, but, in general, it is fair to say that they are relatively homogeneous, randomly bonded isotropic solids.

Normally, crystalline solids have inherently better strength than amorphous materials. This is demonstrated in the data of figure 3-11 for silica glass in both crystalline and amorphous forms (ref. 14). For fused silica (the amorphous form), at low temperatures (200 K), the strength is somewhat lower than that for quartz (the crystalline form). As the temperature is increased, however, the intrinsic strength for the crystalline form of quartz decreases because of the presence of the preferred slip planes in the

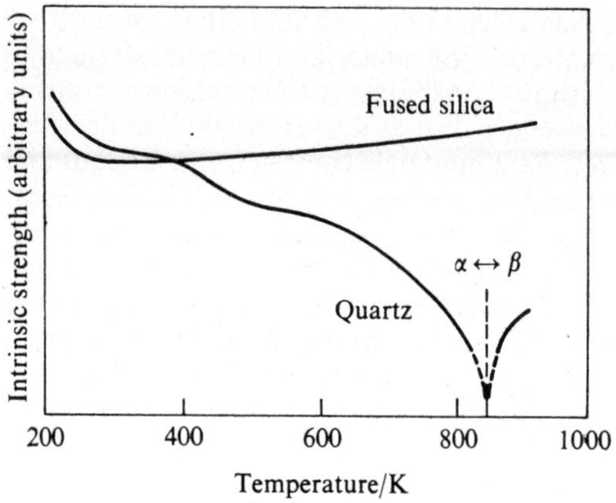

Figure 3-11.—Intrinsic strength of crystalline and amorphous forms of SiO_2. Data taken from Hertzian fracture tests in vacuum; quartz undergoes second-order, displacive $\alpha \leftrightarrow \beta$ transformation at $T = 846$ K, in which stacking SiO_4^{-4} tetrahedra pass into disordered state in thermal modes; fused silica is effectively disordered at all temperatures (ref. 14).

crystalline structure (fig. 3-11). These planes are absent in the fused silica and, as a consequence, with an increase in temperature there is very little change observed in the intrinsic strength of the fused (amorphous) silica.

There is a transformation in the crystalline form of quartz, which is commonly referred to as the alpha-beta transformation, at about 850 K. As that transformation is approached, there is a marked decrease in the intrinsic strength of the material with the increased mobility of atoms associated with the actual crystal transformation. After the quartz goes through the transformation there is again an increase in the intrinsic strength of the quartz in the beta form associated with the new crystal structure that is formed. The inherent weakness occurs only in the transformation region, where the change in structure is taking place. The structure change gives rise to a very large amount of mobility in the atoms and permits a much greater ease of shearing in the solid and a reduction in its ultimate intrinsic strength.

The data of figure 3-11 were obtained in a vacuum environment, which raises an interesting point in the study of solid materials for tribological applications. The environment in which the material finds itself is extremely important since it can have a marked influence on the mechanical behavior of solid surfaces. For example, the silicon dioxide, either in the fused silica form or in the quartz form, is extremely sensitive to moisture. The presence of adsorbed water on the surface of either quartz or fused silica can markedly alter its mechanical behavior; therefore, it is extremely important in measuring fundamental properties to control the environment, whether it be on silicon dioxide or other materials.

From a basic understanding of the nature of bonding in solids and properties such as cohesive energies as are presented in table 3-I, one can calculate the actual strengths and many other mechanical properties of solid materials based on atomic bonding; that is, the atomic theory of materials can be used to calculate the mechanical behavior and properties of solids as was done for theoretical cleavage and shear strengths (tables 3-V and 3-VI).

When making such calculations one finds that for most solid materials the calculated strength of the material (intrinsic strength) far exceeds the measured strength in the laboratory. The reason for this is the presence of defects in real materials. Just as the real world is not perfect, neither are real materials perfect. Real materials contain imperfections and defects in their structure in the bulk as well as in their surface and surface regions. As a consequence, one does not measure in the laboratory the calculated theoretical strengths of solids but rather strengths which are appreciably less than that. These measured strengths take into account the presence of all the defects in the solid. The materials which probably come closest to achieving theoretical strength are single crystal whiskers. These solids usually contain a minimum of defects, and on some occasions only a single defect—namely, a screw dislocation (to be described in more detail shortly). Because of the presence of even this single defect, however, the strengths are not what they would be for that material based on theoretical calculations.

While defects that are present in crystalline solids may be envisioned as contributing to a lesser degree of perfection (i.e., the material is no longer perfect with these defects present), it might be anticipated that they are always detrimental. But this is not the case. Defects can be a mixed blessing. In some circumstances they are definitely a disadvantage, and one would like to have materials approaching theoretical strengths. In other instances, however, the presence of the defects offers distinct advantages in materials, particularly in the tribological area. This is discussed in greater detail in a later section.

Real Surfaces

So far we have discussed solid surfaces, crystalline and amorphous, in the perfect state. As mentioned previously, real solid surfaces contain defects and irregularities as well as imperfections; these defects, irregularities, and imperfections have a marked influence on the surface chemistry, the surface physics, and surface metallurgical properties as well as the structural behavior of the solid surface. Some of the defects that are present in real surfaces are now discussed, and their influence on properties is indicated.

Surface Defects

Some typical defects that may be associated with a solid surface are shown schematically for a single crystal surface in figure 3-12. These imperfections include such things as a kink step in the solid surface and an actual step, which might be associated, for example, with cleavage of brittle solids where cleavage occurs along the cleavage plane and moves along the plane. This cleavage results in a step in the surface where the cleavage transfers from one crystallographic plane to another in the crystalline solid. It also occurs with movement from one crystalline plane to another crystalline plane of the same crystal set—for example, from one (111) plane in a face-centered-cubic material to another (111) plane in the face-centered-

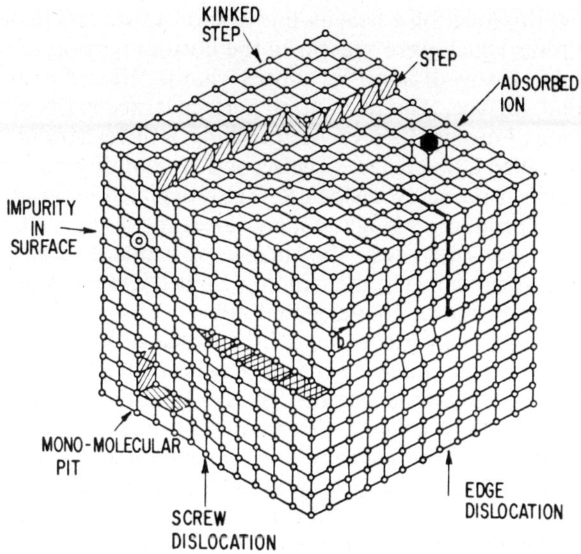

Figure 3-12. – Surface imperfections on crystal.

cubic metal. In this movement from one (111) plane to another (111) plane, a ledge or step develops in the crystalline solid surface. These steps are almost always seen in the cleavage of brittle materials. The cleavage of mica is a classic example. If a piece of mica is pulled apart by cleaving it along the cleavage plane and then a surface profilometer is run across the surface of the mica, the profilometer detects steps in the surface associated with an imperfection in the cleavage process. If there were perfect cleavage, it would take place along one single plane throughout the entire material. In practice this is not experienced. What does result is the formation of steps in the solid surface. Normally, these steps are only observed in brittle materials and not seen in ductile surfaces.

Another imperfection that exists in the surface is that of an adsorbed species (ion). When a surface is atomically clean, the atoms in the surface layer are set in perfect registry. That surface registry may differ from the registry of the bulk solid; for the surface, however, the registry is fairly uniform and there is no disturbance of that atomic lattice registry. Placing an adsorbed atom, however, on the solid surface in a particular location produces a disregistry of the atoms because the bonding process (the adsorbed atom to the atoms in the solid) causes a localized disturbance or imperfection in the otherwise perfect lattice.

Other defects that can occur in the surface of crystalline solids, and which play a very heavy role in tribological behavior, are dislocations. They come in two varieties or forms. The first is the edge dislocation depicted schematically in figure 3-12 at about 5 o'clock. Two rows of atoms on either side of the arrow at the bottom of figure 3-12 are for the edge dislocation. These two rows of atoms move up into the crystal surface and then start to spread apart; at this location a heavier line of atoms squeezes in between the two rows of atoms which were present in the bottom portion of the figure. This additional half-row of atoms produces what is referred to as a dislocation in the solid surface. The normal or perfect lattice exists without this center half-plane of atoms. Its presence produces a disregistry in the atomic lattice which is called a line defect. The second type of dislocation found in practical materials is the screw dislocation, also depicted in figure 3-12. The screw dislocation produces a ledge of material. It moves up in the solid in a spiral path, and as it emerges at the surface, it produces a step or ledge. Because of their significance in the tribological behavior of materials these two defects are discussed in greater detail later.

Another defect that can occur in a solid surface is the creation of a vacancy or the presence of a vacancy site. A vacancy is nothing more than an atom missing in a particular lattice site (fig. 3-13). If the missing atom is in an elemental material it is called a vacancy; if it is in a molecular solid it is called a molecular pit. The vacancy and the monomolecular pit are the same. The monomolecular pit simply applies to a crystalline solid other than an elemental material. The solid could be a compound or a crystalline organic material such as a polymer.

Still another type of defect that can occur in solid surfaces is interstitial atoms. Such an atom squeezes betweem adjacent atoms in the lattice (fig. 3-13). It is a foreign atom that does not belong in the crystal lattice. When the interstitial atom is not of the same chemistry as the host lattice it is generally an impurity. The impurity atom can occupy one of two sites. It can occupy a site or take up a position in the regular crystal lattice (as indicated in fig. 3-13), or it can actually substitute for one of the parent atoms on the crystalline solid surface (as indicated in fig. 3-12) where the impurity atom in the surface simply replaces one of the parent atoms in the basic crystal lattice.

In addition to the defects presented in figure 3-12 for a single crystal solid surface, there are other defects that exist in real surfaces. These defects include stacking faults and grain boundaries in polycrystalline materials. It must be remembered, however, that there are grain boundaries that surround the single crystal. If that single crystal were in a polycrystalline matrix, there would be bonding of that single crystal to adjacent single crystals or grains in the polycrystalline matrix. The grain boundary itself is a very severe defect in a crystalline solid.

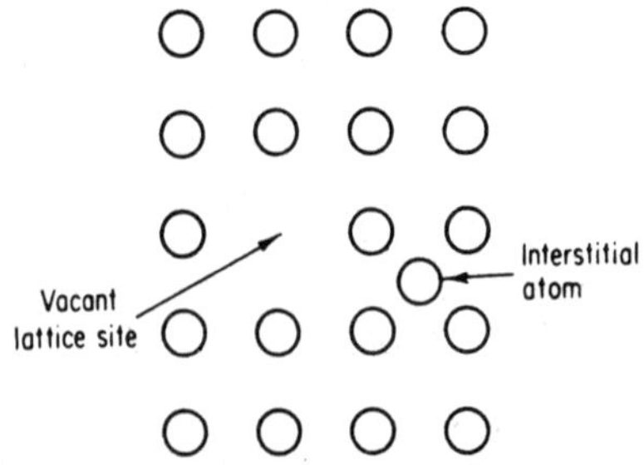

Figure 3-13. – Vacancy and interstitial crystal defects.

Before the mid-1950's one could talk in terms of vacancies and interstitials in solid surfaces and draw schematic diagrams such as those in figures 3-12 and 3-13 to depict schematically what these defects looked like, but no one had actually seen one of these defects and we only drew inferences to their existence from indirect experimental measurements. With the advent of the FIM, however, it became possible to see these defects on a solid surface. The presence of atoms in solid surfaces at a single site (such as interstitials or substitutional atoms or the presence of vacancies where a single atom is involved) are commonly referred to as point defects, because they are located at one point in the solid surface or the bulk of the material. Figure 3-14 shows a field ion micrograph with a vacancy in a solid surface. This photomicrograph was taken by Professor Muller and some of his colleagues (ref. 15). The vacancy in the solid surface is very evident; it is the black spot in the photomicrograph where we see the rows of atoms (the white spots). If the atom were present in that location all of the surface would consistent of rows of white spots.

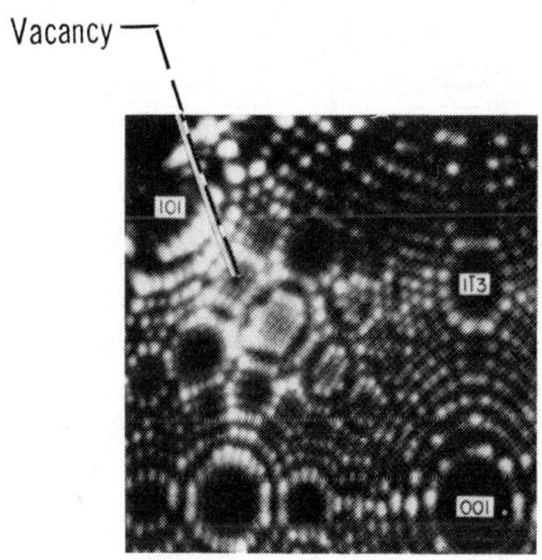

Figure 3-14.—Field ion photomicrograph of vacancy in the (203) plane of platinum (ref. 15).

Substitutional or Interstitial Atoms

In reference to figures 3-12 and 3-13, the presence of impurity atoms in the forms of substitutional and interstitial atoms was discussed. Where small concentrations of impurity may be present in elemental metal, that impurity on the surface can take up a position replacing one of the surface atoms of the metal; such an atom would be called a substitutional impurity atom (fig. 3-15(a)). The impurity can also fit into an interstitial site, as indicated in figure 3-15(b). The size of the particular impurity atom, whether in a substitutional or an interstitial position, can have a profound influence on the nature of the solid surface and the lattice registry in that solid surface. If the atom is relatively large (larger than the host atoms which surround it), it causes a bulging or strain in the crystal lattice. Since the lattice must accommodate it, it undergoes some change. With the interstitial atom, the degree of bonding to the interstitial atom (by the parent atoms in the

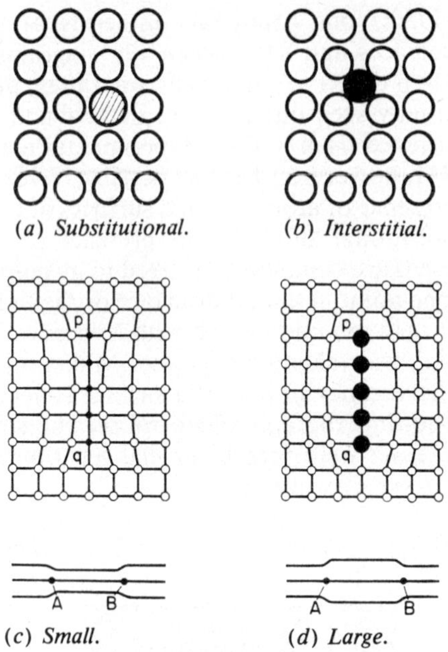

(*a*) *Substitutional.* (*b*) *Interstitial.*

(*c*) *Small.* (*d*) *Large.*

Figure 3-15. – Impurity atoms in crystal lattice.

solid surface) and its size can produce disregistry in the coherency of the crystal lattice in the solid surface. If the size is relatively large, it can produce displacement of the atoms in the solid surface, and this, in turn, produces a strain among the nearby atoms (as can a large substitutional atom). Where the size is very small, it also produces a strain; in addition, the parent atoms are being pulled toward the impurity atom, and this disturbs the lattice registry in and about the impurity atom. This effect is shown schematically in figures 3-15(c) and (d). In figure 3-15(c), an atom of the solute (impurity atom) is smaller than the parent atoms of the host material. The atoms around the impurity atom tend to be drawn toward it and produce the lattice strain. When the impurity atom is relatively large, it causes a bulging in the crystal lattice of the parent atoms as seen in figure 3-15(d). In figures 3-15(c) and (d) the effect of size is shown for a substitutional impurity atom (fig. 3-15(a)). The same type of relationship can exist for the interstitial atom; however, with the interstitial atom, the effects may be even more severe because this atom does not accommodate itself in the regular crystal lattice of the solid surface.

Dislocations

Dislocations are displayed in figure 3-16 with the anticipation that they are clearer than those in figure 3-12. The edge dislocation is seen as the half-row of atoms that does not belong in the lattice and thus causes a disruption of the lattice registry. The screw dislocation is shown in figure 3-16(b) as a

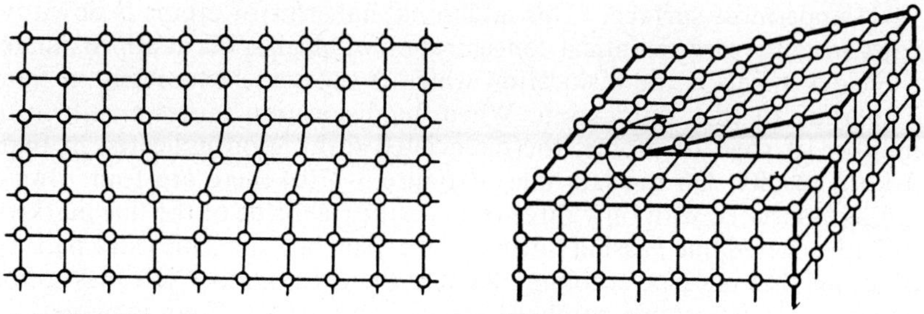

(*a*) *Edge dislocation.* (*b*) *Screw dislocation.*

Figure 3-16. – Dislocations.

step in the surface generated by the emergence of a screw dislocation along with the emergence of a line defect perpendicular to the solid surface. This pie-shaped wedge generated on the solid surface is similar to the one observed in figure 3-12.

Individual vacancies in solid surfaces and interstitials can be seen with an FIM. If the FIM can show individual point defects in solid surfaces, then it certainly can show a dislocation, which is a line defect (fig. 3-26). A line defect is associated with a row of atoms rather than with a single atom site on the solid surface. As might be anticipated, the FIM is a very useful tool for examining and studying dislocation structures.

With the aid of a computer the FIM can do much to explain dislocation behavior in solid surfaces (ref. 16). As an example, figure 3-17(a) shows a computer simulation of a perfect face-centered-cubic metal surface. (In contrast to field ion micrographs, the atoms are represented by black spots.) As was mentioned in chapter 2, in the FIM the surface is a radius; thus, one is looking at a number of atomic planes simultaneously. In one small, elliptically shaped area (upper region of fig. 3-17(a)) are the atoms associated with the (111) high atomic density, low surface energy plane for the face-centered-cubic metal system. This surface in figure 3-17(a) is nearly perfect except for the presence of an extra row of atoms between the rows below the

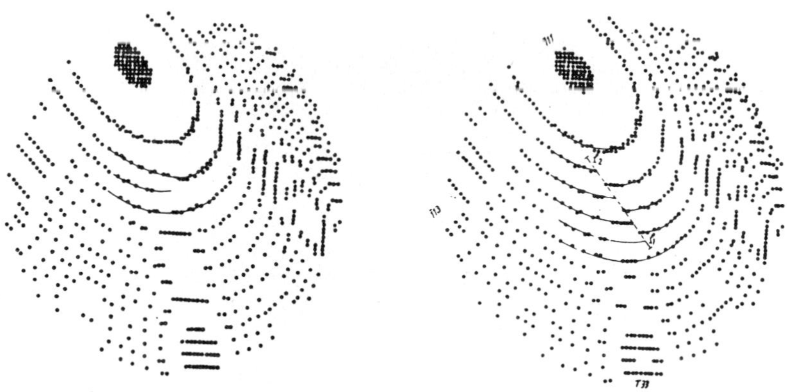

(*a*) *Perfect dislocation.* (*b*) *Dissociated dislocation.*

Figure 3-17. – Examples of simulated contrast for perfect and dissociated dislocations in face-centered-cubic metal (ref. 16).

(111) uppermost surface. (This additional half-row of atoms is down two rows from the large elliptical concentration of black spots at approximately 6 o'clock.) This is the dislocation which is shown schematically in figure 3-16(a) as an edge dislocation. When the dislocation causes a shift in the number of rows of atoms, the dissociated dislocation indicated in figure 3-17(b) results. On the left side of figure 3-17(b) there are four rows of atoms associated with only three rows on the right side of the line marked 1 to 2 (the dotted line passing through the crystal lattice). This extra half-row of atoms that must be accommodated in the crystal lattice is a dislocation. Because this dislocation produces a shift in a number of atmic planes, it is referred to as a dissociated dislocation.

Just as with the vacancies, so too with the presence of dislocations in solid surfaces. These line defects are readily discernible in a solid surface that has been examined in the FIM; figure 3-18 is a field ion micrograph of a platinum metal surface showing dislocations. Platinum is a face-centered-cubic metal and the field ion micrograph (fig. 3-18) shows a number of planes: the (113), the (113), the (113), and the (001). Some of the dislocations are indicated with small delta arrows in the photomicrograph. A typical line dislocation is depicted just above the (001) location at approximately 11 o'clock and six atomic rows out from the center of the (001) surface. Another line dislocation can be seen at approximately 1 o'clock, and a screw dislocation emerged on the solid surface at approximately 11 o'clock. A screw dislocation is better seen in the photomicrograph of figure 3-19 for the body-centered-cubic metal molybdenum. The central portion of this figure shows a (111) plane of the body-centered-cubic molybdenum imaged in the FIM. The micrograph shows the emegence of a screw dislocation at the solid surface. This dislocation causes a wedge shape of missing material in the outermost plane of the (111) surface. This is the point of emergence of the screw dislocation. The dislocation moves in the crystal normal to the (111) surface which is being imaged in the photomicrograph. Its emergence at the surface produces this step. In the pie-shaped wedge, which is dark, the gradation in coloı is due to the step nature of the wedge.

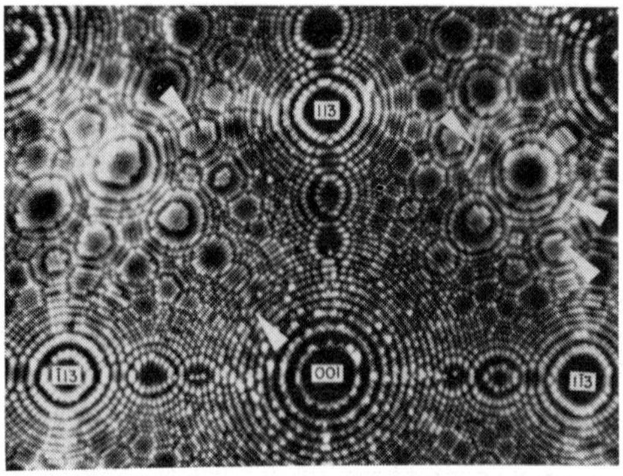

Figure 3-18. – Several dislocations revealed by field ion microscopy on platinum surface (ref. 15).

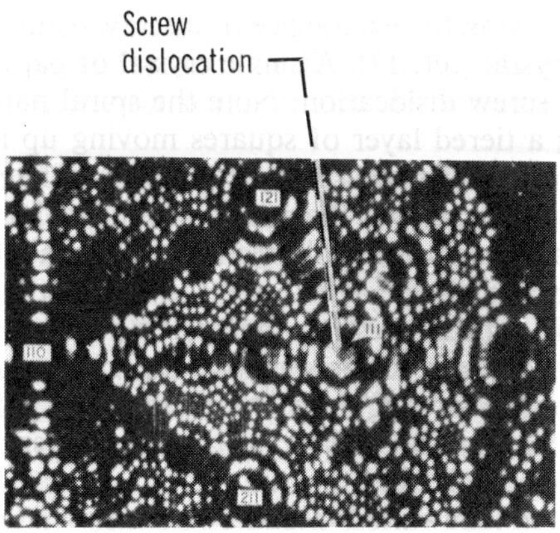

Figure 3-19.—Field ion micrograph of screw dislocation on (111) plane of molybdenum (ref. 15).

In tribological systems, the concern is with metals, alloys, and other solid materials; hence, dislocations of the line type and the screw type are important. Some may feel that beyond the solid surface itself, dislocations are of minimal importance. In tribological studies, however, dislocations can occur not only in metal or alloy surfaces but also in the lubricants themselves. Organic lubricating structures do have crystallinity to them and with that crystallinity can be associated the presence of dislocation defects. An example of this is presented in figure 3-20, which is an ordinary

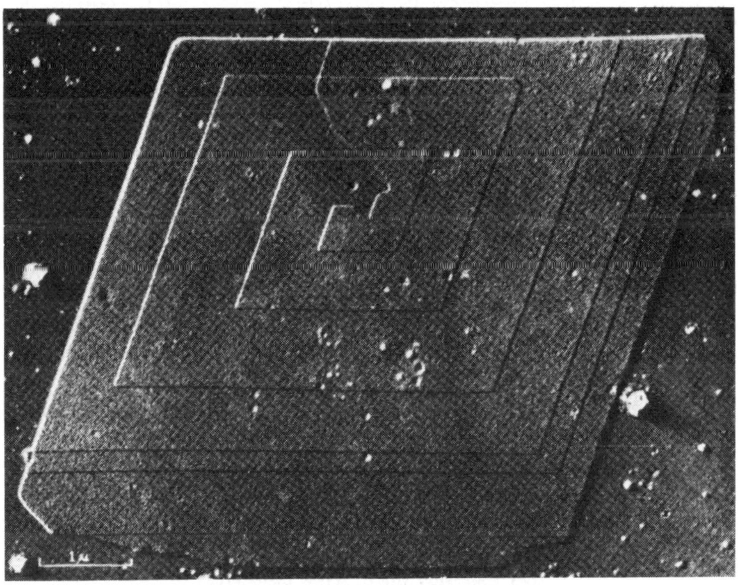

Figure 3-20.—Photomicrograph of paraffin crystal out of which grew screw dislocation (ref. 17).

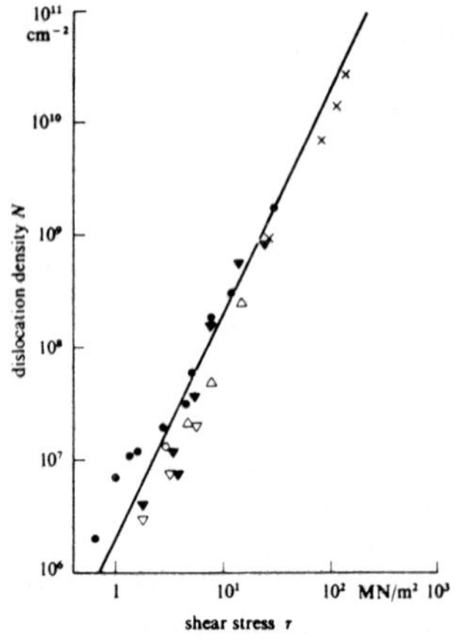

Figure 3-22. – Dislocation density N measured from etch pits on Cu single crystals deformed to shear stress τ (ref. 18).

face temperatures occur repeatedly, the process of strain hardening (producing an increase in dislocation concentrations) plus recrystallization (producing a reduction in dislocation concentration) repeats itself many times on the surface. The recrystallization process brings about a change in grain size. Some of the energy that is put into this system is dissipated in the formation of these new defects—namely, grain boundaries in the crystalline solid. The grain boundaries are defects but of a much larger size and scale than are the dislocations. They are of greater magnitude.

Both point defects (vancancies as well as dislocations) alter the mechanical properties of materials. One mechanical property of considerable interest to the tribologist and frequently used by him is the hardness of materials. Hardness is strongly affected by the concentration of vacancies as well as the concentration of dislocations. These defects have real practical significance in the mechanical behavior of solid surfaces. This is demonstrated schematically for high purity iron in figure 3-23, which shows a plot of dislocation density and vacancy concentration as functions of annealing temperture for ordinary high purity iron (ref. 19). In addition, there is a curve for the strength or hardness of the high purity iron as a function of the annealing temperature. As the temperature to anneal the high purity iron is increased, a reduction results in both concentration of vacancies and concentration of dislocations. Energy is supplied to this system to relieve the lattice disregistry that has been created by introducing the vacancies and dislocations. The atoms wish to rearrange themselves so as to reach a minimum energy state by eliminating vacancies and dislocations.

An interesting observation to be made, in figure 3-23, is that a very modest temperature is required to produce a marked reduction in concentration of vacancies. A simple increase in temperature to 200° C is sufficient to reduce vacancy concentration to a very low level. In practical tribological systems, these kinds of temperatures at surfaces and interfaces can be achieved very readily so that vacancies are easily removed from the sur-

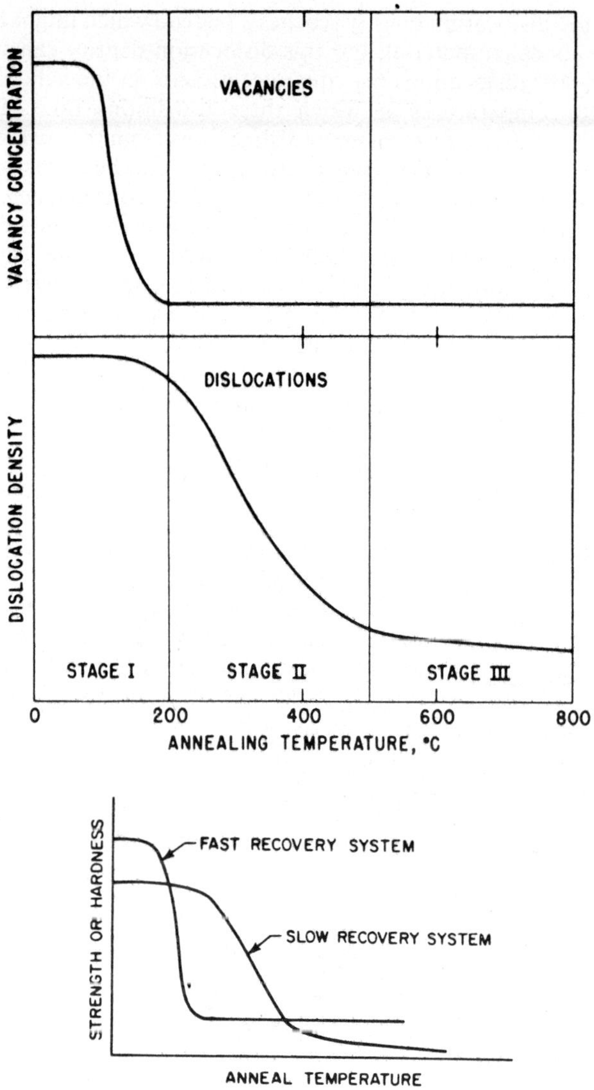

Figure 3-23. – Softening studies in high purity iron (ref. 19).

faces. The dislocation densities are, however, more difficult to remove and the temperatures required are somewhat higher.

Figure 3-23, stage 2, shows that the dislocation density is reduced in the 200° to 500° C region. In practical tribological systems, temperatures as high as 200° C are observed in lubricated systems, but the higher end of stage 2 (500° C) may occur only in solid lubricated systems or in systems where there is no lubrication (i.e., where the surfaces are running dry). In those situations, the dislocation concentration can be reduced considerably.

In stage 3, the dislocation density reaches a plateau which might be expected for a fully annealed material, but this dislocation density can still be very high, 10^4 to 10^6 dislocations per square centimeter in the solid surface.

The exact temperatures at which these transitions take place in the removal or liberation of vacancies or dislocations from the surface are functions of the rate at which the strain energy is put into the system and the rate at which the temperature is increased locally. As a consequence, if the process is a relatively slow one, the recovery is relatively slow and a lag exists in the reduction of both dislocations and vacancies as indicated by the mechanical behavior shown in figure 3-23 for the slow recovery system.

In most tribological systems, however, recoveries can be fairly rapid in that localized surface heating occurs very rapidly because of asperity interactions; also, since the cooling occurs very rapidly there is a quenching effect. In addition, the amount of strain energy that is put in occurs rapidly where there are repeated cycles of sliding or rolling over the surface. The energy is put in very rapidly and, as a consequence, a very fast recovery may be experienced; such a system may be classified as a fast recovery system (fig. 3-23).

The data of figure 3-23 should assist the mechanical engineer who is accustomed to thinking in terms of mechanical properties, such as tensile strength or hardness, because these basic properties used on a routine basis are definitely related to atomic defects in crystalline solid materials. Atomic defects on such a small scale as vacancies (i.e., point defects in the solids) as well as line defects of atoms in the solid (namely, dislocations) strongly influence hardness and tensile strength measurements.

The data of figure 3-23 for the tensile strength or hardness obviously are bulk mechanical measurements, and the dislocation and vacancy concentrations presented reflect the response of the material in the bulk. However, it must be remembered that the surface also has its own mechanical behavior and mechanical properties (in contrast to bulk properties). While it is very difficult to measure hardness or tensile strength of a surface layer of atoms, it is important to recognize that the same defects that exist in the surface as in the solid bulk will have a similar profound influence on surface strength as they do on bulk strength. Care and concern must be paid to their presence. With surfaces, even lattice strain can have a marked influence on the surface energy characteristics.

Stress Effects

Figure 3-24 shows the distortion that can be generated in a lattice of a surface as a result of applying a compressive load such as might be encountered in a practical tribological system where one solid surface is loaded against another (ref. 20). The load can produce a strain or distortion in the surface lattice as indicated. The application of the load and the removal of the load can cause an alternating application of lattice distortion by generating and relieving that distortion. This repeated process can introduce defects in the crystalline solid surface.

X-ray topography has been used to establish the presence of lattice distor-

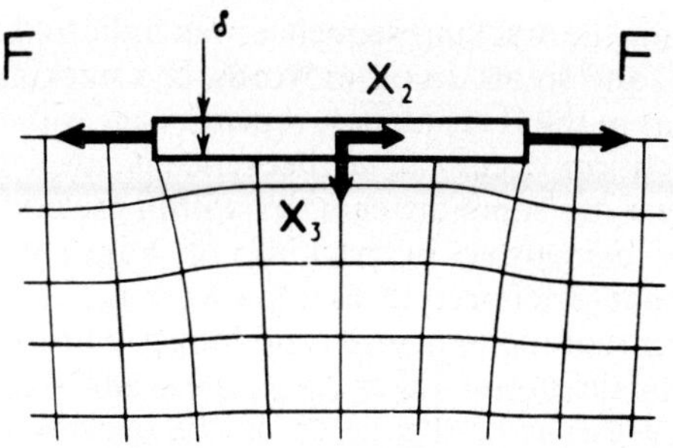

Figure 3-24.—Computed lattice distortion under compressed surface layer (ref. 20).

tion, or lattice strain, as a function of the application of load. The repeated application and removal of a load producing a lattice strain is a prime example of surface-initiated fatigue cracking that occurs in practical tribological devices. When stresses are sufficiently high, dislocations can be generated in the surface; the motion of these dislocations on the surface toward surface impurity atoms can cause a pileup of dislocations and ultimately a formation of surface cracks as a result of repeated stress applications. This basic concept is depicted schematically in figure 3-25; where S is the source or point of application of an applied stress.

Dislocations can move along the surface glide plane and be pinned by the impurities in the solid surface; dislocation coalescense can occur at points B and B′ (fig. 3-25). The energy which is put into the surface as a result of repeated loading must be absorbed in some fashion. The mechanism for absorbing the energy, over and above the accumulation and coalescense of dislocations at points B and B′, is the dissipation of energy in the formation of a surface microcrack which develops at point C. It is a surface-originated crack starting at the coalescence of dislocations (where they are pinned by a surface impurity) and then moving subsurface to the point C.

In the discussion of figure 3-5 a stacking sequence was mentioned for atomic planes in the crystalline solid. For the (111) surface in the face-

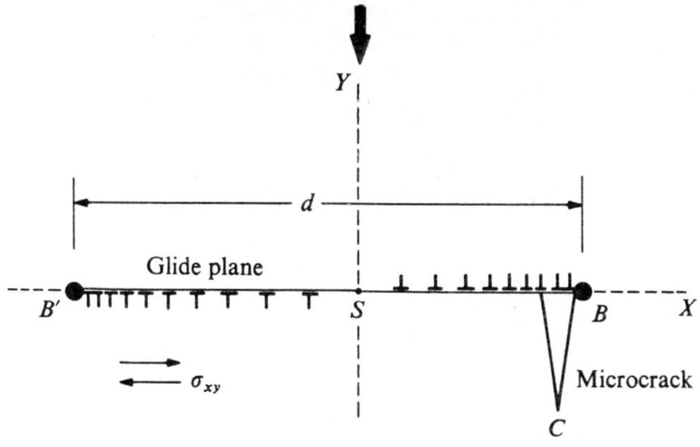

Figure 3-25.—Nucleation of microcrack by pileup mechanism. Source S generates dislocation loops which pile up at barriers B and B′ in glide plane; stress concentration at B (or B′) nucleates crack BC.

centered-cubic system, the stacking sequence was indicated as being A, B, C, A, B, C, A, B, C, and so on. In other words, the stacking sequence, the arrangement of atoms in the (111) planes, repeats itself with every third row of atoms. This repetition occurs in a theoretically ideal crystal system. Where, however, there are some disregistries within the crystalline solid of atomic planes, this disregistry in packing of adjacent atomic layers manifests itself in what is referred to as a stacking fault. The stacking sequence is disrupted so that the sequence is no longer uniformly A, B, C, A, B, C, A, B, C. When a disruption takes place, the atom layers or rows in the surface indicate the presence of this defect in the packing arrangement of atoms. This is shown schematically in figure 3-26 in the field ion micrograph for a cobalt pin tip with a large number of stacking faults running through the center region of the crystal in and about the (1010) prismatic plane located near and slightly below the center of the photomicrograph. The crystallographic planes manifesting the stacking faults are above and below the (1010) prismatic orientation. Hexagonal metals are particularly prone to a formation of stacking faults, as are face-centered-cubic metals.

Because of these planar defects in the crystalline solid, the energies associated with the stacking faults are different from those that would be associated with a normal stacking sequence. As a consequence, the properties of the crystalline solids can be markedly different, and these differences are a direct result of the stacking fault energy; various other properties have been and can be related to the energies associated with these stacking faults.

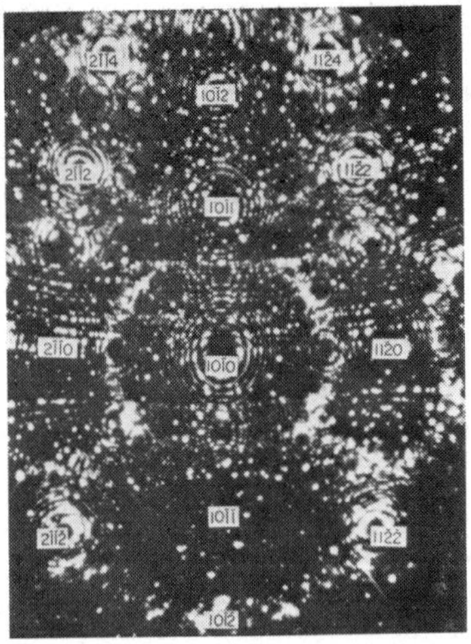

Figure 3-26. – Cobalt tip with many stacking faults (ref. 15).

Grain Boundaries and Their Energies

Another defect in real crystalline solids which was mentioned earlier is the grain boundary. The defects presented in figure 3-12 were only for a single crystal surface. If single crystals are entrained in a polycrystalline matrix,

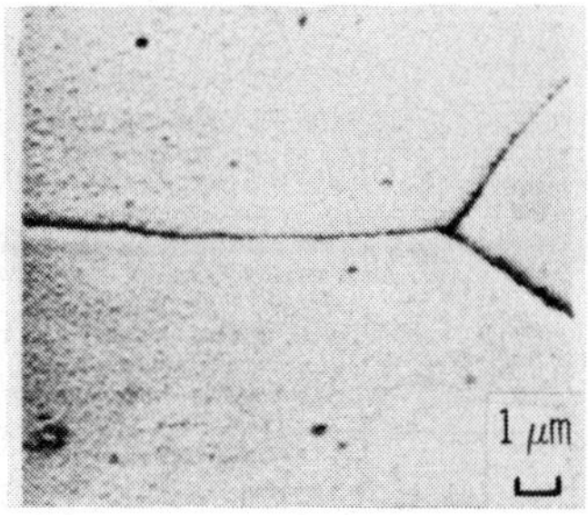

Figure 3-27. – Grain boundary in molybdenum metal.

they are adjacent to grain (or single crystal) boundaries which would separate them from adjacent grains in the polycrystalline solid.

Grain boundaries occur in nearly all practical tribological surfaces. Although they are extremely important, frequently very little attention is paid to them and to their real significance in the behavior of solid surfaces. A typical grain boundary on a molybdenum metal surface is shown in figure 3-27. There are basically three grains shown in the photomicrograph of figure 3-27 with a grain boundary separating the three individual grains. The very presence of the boundary indicates that the orientations of the adjacent grains at the surface are different. If this were not so, no grain boundary would be present. For example, if the grain in the upper portion of photomicrograph is a (111) surface, the two adjacent grains, the one to the right and the one in the lower portion of the photomicrograph, of necessity, must have different orientations because if they had identical orientations to the upper grain there would be no boundary.

Grain boundaries can develop with metals in the solidification process. They are nucleated by stresses, impurities, and all sorts of things which tend to destroy the uniform homogeneous behavior of the material. An atomic picture of the grain boundaries can be seen in the photomicrograph in figure 3-28, which is a field ion micrograph of a tungsten surface showing a grain boundary running through the tip. The grain boundary is located just below

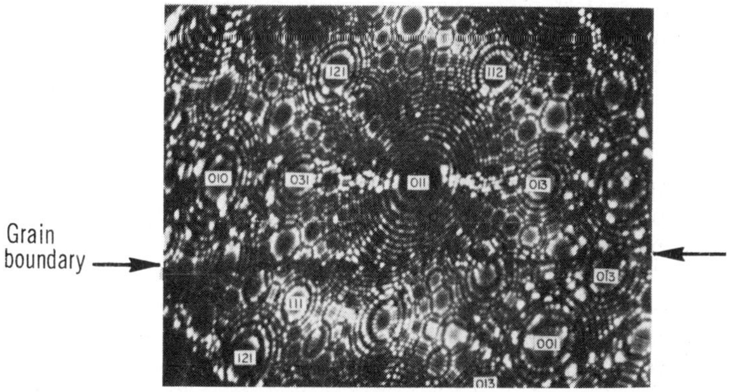

Figure 3-28. – Field ion micrograph of tungsten surface revealing grain boundary.

the (011) surface. It is readily apparent that the presence of the grain boundary produces atomic disregistry in the crystalline solid. Notice that (in fig. 3-28) the atomic arrangement on one side of the boundary is different from that on the other side; there is no matching of the orientations on the two sides of the boundary. The grain boundary itself is a region of high dislocation concentration because it is a defect in the crystalline solid and dislocations coalese in the grain boundary region. The greater the mismatch in orientation between adjacent grains, the greater the difference in energy associated with the boundaries (ref. 21).

In figure 3-29 an equilibrium diagram is presented for (1) two adjacent grains and (2) the free surfaces as well as the grain boundary between the two grains. Where the two grains are at different surface orientations, the boundary that connects the two adjacent crystalline grains has an energy of its own. That energy is reflected by the orientation differences of the two adjacent grains. The greater the mismatch in adjacent grains, the greater the grain boundary energy at the surface of the boundary.

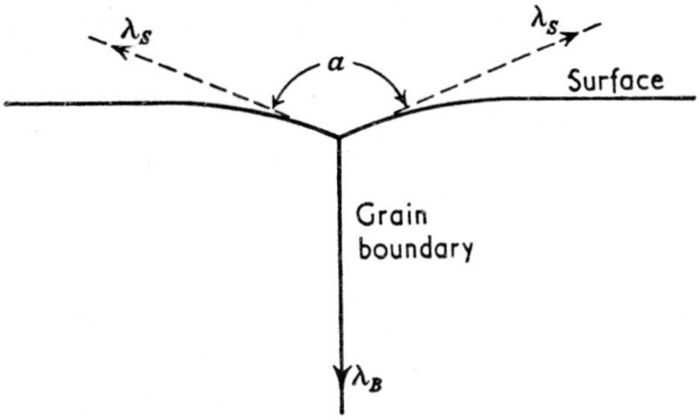

Figure 3-29. – Equilibrium between grain boundary and surface free energies (ref. 21).

This difference in grain boundary energy is presented in the experimental data of figure 3-30 where the relative grain boundary energy is plotted as a function of the difference in orientation between adjacent grains in a silicon-iron tricrystal with a common [100] axis perpendicular to the plane of the specimen (ref. 22). The solid line is a theoretical curve for energy based on differences in grain boundaries, and the data points are results obtained from experiment. Up to an orientation difference of about 30°, the grain boundary energy increases with differences in orientation. Beyond that, a decrease in energy takes place.

It might be assumed that the greater the degree of mismatch between adjacent orientations of grains and crystalline solids the wider the gap that forms the grain boundary, since the lattice disregistry in the grain boundary necessary to accommodate the mismatch between adjacent orientations of the grains would be greater. In fact, however, this is not observed. As McLean has indicated (ref. 21), the higher the degree of mismatch between adjacent orientations (the greater the difference in orientation θ, indicated in fig. 3-30)), the smaller the grain boundary angle (displacement between adjacent grains due to mismatch). For further discussion of this particular aspect of grain boundary behavior, the reader is referred to the works of

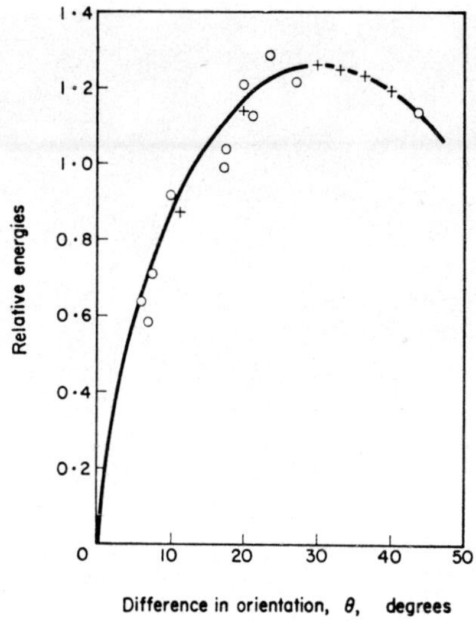

Figure 3-30. – Variation of relative grain boundary energy with orientation for silicon-iron tricrystals with common [100] axis perpendicular to plane of specimen. Full line is theoretical curve, and points are experimental values (ref. 22).

McLean on grain boundaries (ref. 21). For our purposes, it is important to recognize that the grain boundary serves as a link between the adjacent grains.

Sputter etching of the solid surface brings out the topography in the grain boundary region. Figure 3-31 is a photomicrograph of a copper surface which has been sputter etched. Figure 3-31(a) is a sputtered etched single grain surface; notice that the topography of the grain is very rough as a result of the sputter etching process. Figure 3-31(b) shows three grains, which all etch differently, adjacent to a grain boundary; the boundary itself has been etched as indicated by the heavy black markings in two directions for two of the boundaries—the boundary between the (310) and (100) planes and the (310) and (110) planes. Since the boundary between the (110) and (100) planes is much thinner and narrower than the boundaries referred to previously, a difference in angle or mismatch between the adjacent orientations is indicated. Actually, the degree of mismatch between the adjacent planes having a narrow grain boundary is greater than the degree of mismatch between the adjacent planes having the wider boundary as was discussed in reference to the energy considerations of figure 3-30.

Ordering

When a metal, for example, has more than one constituent in the bulk (i.e., two or more elements are present), a number of different phenomena can occur which can bring about a marked alteration in the nature of the solid surface. There are many changes in the surface that can take place as a result of the presence of a second element in the bulk. One of these changes is referred to commonly as order/disorder reactions in solid state. A

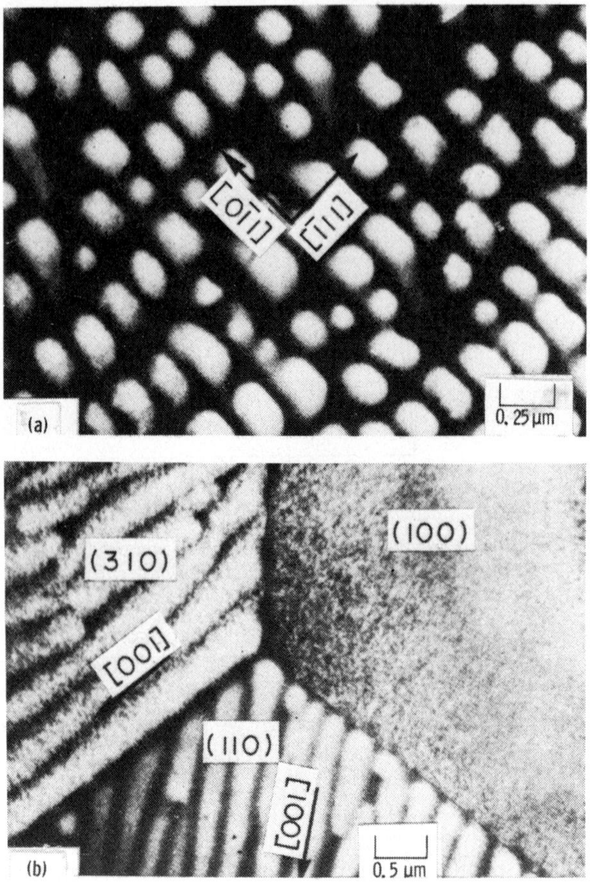

(a) (211) Surface.

(b) Three different and adjacent grains.

Figure 3-31. – Photomicrographs of copper after sputter etching with 2×10^{17} xenon ions per square centimeter at 60 keV (ref. 23).

number of elemental metals when alloyed with a second metal can develop in the bulk as well as at the surface. This is referred to as ordering—that is, where the second element or solute element developed and deposited in the solvent element (the parent) takes up a regular structure. When this takes place, a compound or structure results that is referred to as an ordered structure (fig. 3-32). This figure shows an arrangement wherein there are two different atomic species, the white circles represent one and the black circles represent a second. A number of binary alloy constituents exhibit order/disorder reactions. Iron-aluminum, iron-silicon, copper-aluminum, tin-copper, indium-copper, and the gold-copper alloy system exhibit ordering or order/disorder reactions.

With proper treatment of the material, the atoms in the surface layers as well as in the bulk can undergo an ordering such as that shown in figure

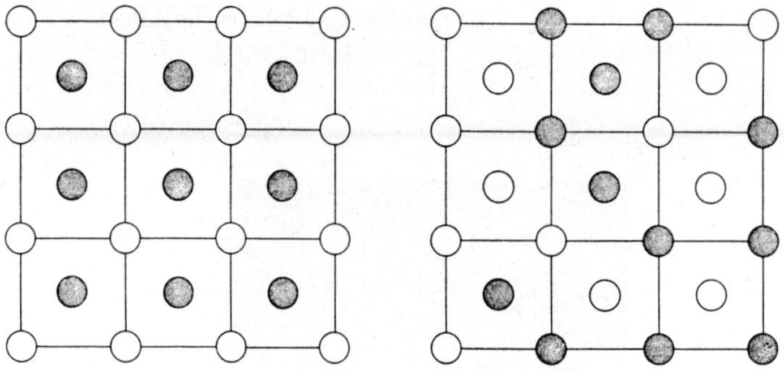

(*a*) *Ordered.* (*b*) *Disordered.*

Figure 3-32. – Ordered and disordered arrangements of AB ions in alloy AB.

3-32(a) where the solute atoms take up regular positions in the crystal lattice relative to those positions of the solvent atoms; an ordered crystallographic structure therefore results. The dark circles representing the solvent take up regular periodic positions throughout the crystal lattice in relation to the solvent atoms (open circles), and the entire surface of the solid is represented by that particular ordered structure. When the atoms are distorted in a surface layer and take up positions which are random rather than ordered, we have what is commonly referred to as a disordered structure (fig. 3-32(b)). The black circles of the solute atoms are in random sites relative to the positions of the solvent atoms; they do not take up a regular registry in the crystal lattice. The physical properties of these two particular materials are markedly different. For example, hardness in the copper-gold system is different for the ordered copper-gold alloys CuAu and Cu_3Au than it is for the disordered alloys of CuAu and Cu_3Au.

Elastic modulus as well as other mechanical properties are also influenced by ordering. In binary systems the ordering exists over a particular temperature range; transformation can occur as it does for crystallographic structures wherein one transforms from an ordered to a disordered state. Most binary metallic systems show transformation from ordered to disordered at some specific temperature as indicated in figures 3-32(a) and (b). Accompanying this transformation from an ordered to a disordered state is a change in the mechanical behavior of the materials; any mechanical change, in turn, affects tribological behavior. Such things as friction and wear are influenced by the order/disorder reaction; experimental data in the literature have established that this does exist.

To the tribologist, it may appear that the mechanical properties are causing the changes in the tribological behavior; in fact, however, a much more fundamental property, the atomic arrangement of the solute relative to the solvent atoms, is responsible for the mechanical property change, which in turn is reflected in tribological behavior.

The difference between an ordered and disordered crystalline state of material on the solid surface can be seen in the photomicrograph in figure 3-33. This is a field ion micrograph for a platinum-cobalt alloy. The platinum-cobalt (PtCo) alloy, which contains 50 atomic percent platinum and 50 atomic percent cobalt, undergoes an order/disorder transformation.

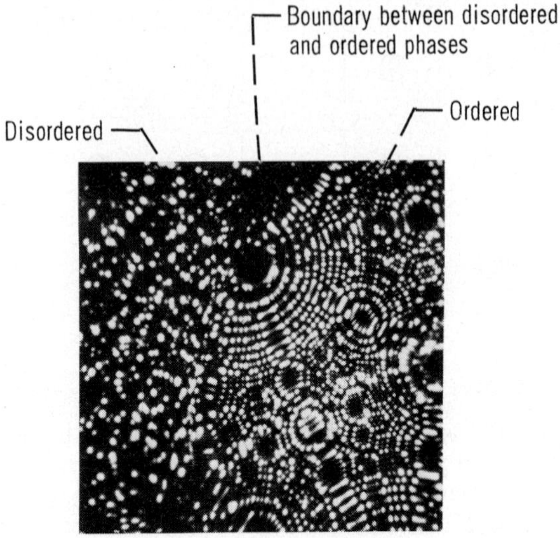

Figure 3-33.—Field ion micrograph indicating contrast between ordered and disordered PtCo (ref. 15).

The alloy can exist in the ordered or disordered state; actually, in a particular sample of material, one may have regions of both ordered and disordered states (fig. 3-33). Basically, the right side of the figure shows an ordered structure in the PtCo alloy, and the left side shows a structure in the disordered state. Comparing the two regions, which should be nearly identical in atomic arrangement, shows a marked difference in the arrangement of the atoms on the solid surface when moving from the disordered to the ordered structure. With the FIM we can see the atomic detail (shown only schematically in fig. 3-32) on a real surface (fig. 3-33).

Once an additional element is added to a simple elemental solid, a number of other changes can take place at the surface of the binary solid composition. These changes can include the order/disorder reactions just referred to as well as many other surface processes which can alter surface behavior and in turn influence tribological properties.

Surface Segregation

Figure 3-34 presents a schematic representation of some phenomena which can occur at a solid surface when a second element is present that can bring about changes in the surface character (ref. 23). Figure 3-34(a) shows what is referred to as segregation. When atom A is dissolved in solvent atom B and heating or straining of the bulk solid occurs, atom A frequently segregates at the surface; that is, it diffuses out of the bulk in the near surface regions and segregates out at the surface. This segregation process can occur in a number of alloy systems, and the segregation markedly alters surface properties such as surface energy, which in turn influences such tribological characteristics as adhesion.

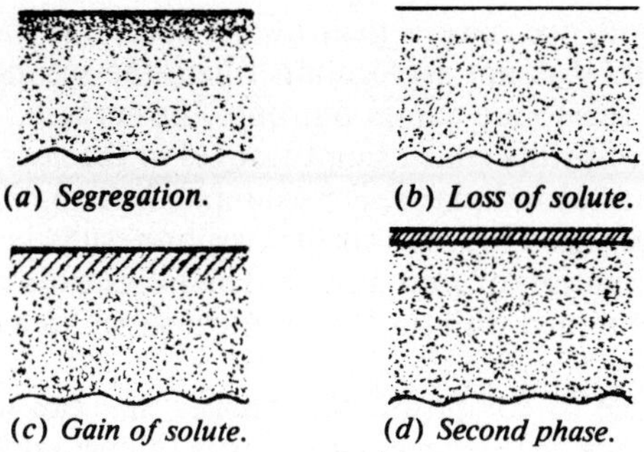

(*a*) *Segregation.* (*b*) *Loss of solute.*

(*c*) *Gain of solute.* (*d*) *Second phase.*

Figure 3-34. – Schematic illustrations of some ways in which compositional alteration can arise at or near surface (ref. 23).

Surface segregation has been observed in a number of practical alloy systems. It has been observed in copper-aluminum, iron-aluminum, and iron-silicon alloys. For example, in the copper-aluminum alloy (1 atomic percent aluminum in copper), a modest amount of heating to 200° C is sufficient to cause the aluminum to segregate on the solid surface such that the entire outer layer of the solid consists primarily of aluminum atoms. The properties of the surface are then really a measure of the properties of aluminum rather than properties of the copper-aluminum alloy. This segregation phenomenon is extremely important in tribology because the surface may not be what we think it is when we are considering, for example, the interaction or reaction of the solid surface with various environmental species such as a lubricant. Thus, when one has a steel containing a small concentration of an alloying element such as silicon, it may not be an interaction of the iron with some lubricating species in the environment (to form a surface compound that provides the lubricating film) but rather the interaction of the silicon (which has segregated on the surface) with the lubricant that forms the protective surface film. So understanding the segregation phenomena is very important in understanding the basic lubrication phenomenon.

Another phenomenon that can take place is just the opposite of surface segregation; this phenomenon is the loss of the solute from the near-surface region (fig. 3-34(b)). This can occur by more than one mechanism. One such mechanism is where the solute is somewhat volatile, and the more volatile species is lost from the solid surface because of heating. This might occur, for example, where a normal gaseous species is combined or dissolved in a solvent metal.

The opposite of volatilization is where the material is driven into the bulk—that is, where it is driven from the surface into the bulk and forms a structural phase or compound with the bulk material. This is a loss of solute and leaves the surface denuded or deficient in the alloy constituent (fig. 3-34(b)). In such a case, the surface behaves more like the solvent than the solute; in the segregation phenomenon, however, the reverse is true—that is, the surface behaves more like the solute than the solvent. Again, in both situations it is an elemental surface behavior rather than a binary alloy surface behavior.

In a third situation, we have a gain of the solute in the near surface regions (fig. 3-34(c)). It is not segregation where the second phase of the alloying element or the solute comes out onto the surface. Rather, in this case, there is a gain of solute which simply moves to the near surface region and enriches it with a second phase. For example, in iron nitriding, the impregnation of nitrogen into the near surface region results in a case hardening effect by the formation of iron nitride. The nitrogen is not actually on the surface but is combined with the iron in the near surface layers or in the surficial region.

The fourth example of compositional changes that can take place with materials and influence the surfaces of solids is the actual formation of a second phase at the surface (figure 3-34(d)). This occurs, for example, where element A (dissolved in element B, the solvent) actually forms a compound of some type with the solvent element B. If we keep increasing the concentration of the solute atom in the solvent such that we exceed the solubility limit, where element A can no longer be absorbed into solvent element B, then a compound or some second phase forms. That compound or phase may actually form at the surface. If it does, then the surface properties are markedly different from the other properties of the bulk alloy composition; the surface properties are those of the compound or second phase as opposed to the alloyed solvent properties. A classic example of this might be dissolving carbon in iron to form ordinary steel. When the concentration of the carbon and iron get sufficiently high, a second phase (iron carbide) begins to form. That second phase, if it forms at the solid surface, has markedly different properties than the bulk steel, and these properties are reflected in tribological measurements.

The segregation of an alloy constituent in a metal to its solid surface can markedly change the nature and composition of that solid surface, as previously mentioned. For example, small amounts of oxygen dissolved in bulk metals like titanium, zirconium, or indium can, upon heating, segregate onto the solid surface. Furthermore, when the metal is taken to its melting point, the segregation process can reverse itself, and the segregated species can redissolve in the molten metal and consequently leave the surface denuded of the segregated species. This is demonstrated with the aid of figure 3-35 for the diffusion of oxygen to and from an indium surface at temperatures below, at, and above the melting point (ref. 24). The data for figure 3-35 were obtained over a range of temperatures from 120° to 200° C. The melting point of indium is approximately 158° C.

Auger peak intensities for oxygen are plotted in arbitrary units as a function of temperature in figure 3-35. The Auger analytical surface tool analyzes the oxygen concentration to as little as a hundredth of a monolayer on a solid surface, and its depth of penetration is only four or five atomic layers. Thus, the oxygen plotted on the ordinate of figure 3-35 truly represents the oxygen concentration at the surface of the indium. It is readily apparent that the oxygen concentration on the surface of the solid indium (on the left of the figure) at temperatures below the melting point is much higher than the concentration of oxygen on the surface as the melting point is approached. As the melting point is approached there is a rapid or sharp decrease in the surface concentration of oxygen and, upon achieving

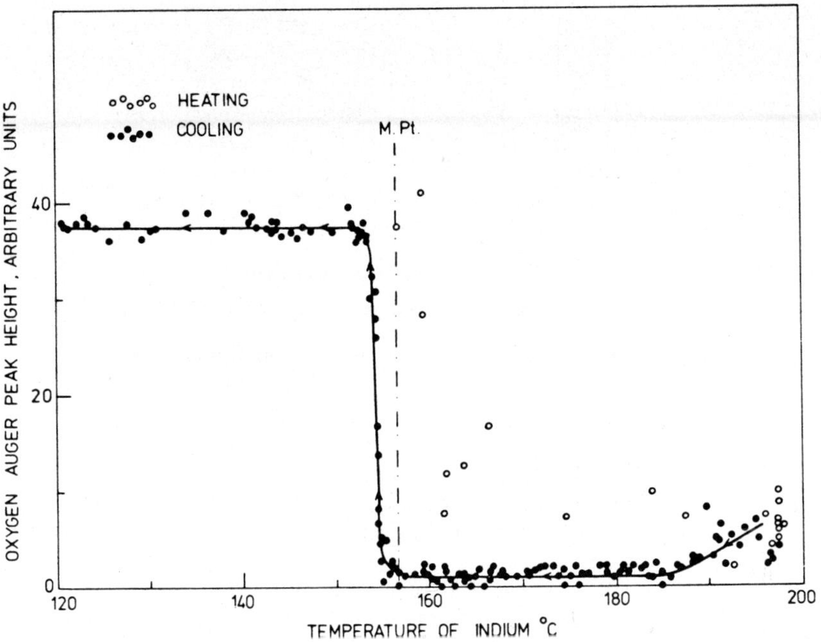

Figure 3-35.—Diffusion of oxygen to and from indium surface through melting point (ref. 24).

the melting point, the oxygen concentration is very low and remains low at temperatures to 200° C. The process is reversible and, if the indium is cooled to 120° C, the oxygen would again segregate on the solid surface and thus leave the indium surface rich in oxygen and the bulk of the indium devoid of the oxygen that is present above the melting point (fig. 3-35). Thus, dissolved elemental species present on a metal surface can become extremely mobile upon liquefication and dissolve in the bulk metal. When cooled, if the concentration of that particular species exceeds the solubility in bulk, segregation of the species that is dissolved in the molten metal occurs on the solid surface.

This is analogous to the situation involved in supersaturating a liquid. For example, if sugar is dissolved in water and the water is heated, a higher concentration of sugar can be dissolved at elevated temperatures than at room temperature. As the temperature is lowered, however, the sugar crystallizes out of the solution (if nuclei are available) because the solution is essentially supersaturated. In essence, the indium of figure 3-35 in the molten state is supersaturated with oxygen and, when cooling to the solid state, segregation takes place at the surface.

The data of figure 3-35 only indicate that the oxygen concentrates on the surface and that there is less on the surface in the molten state than there is when the indium is in the solid state. The question is how much is present. If there is an extremely small concentration of solute on the surface of the solvent its significance and importance in tribology may be relatively minimal.

If the data, which are isotherms for carbon segregation on an iron (100)

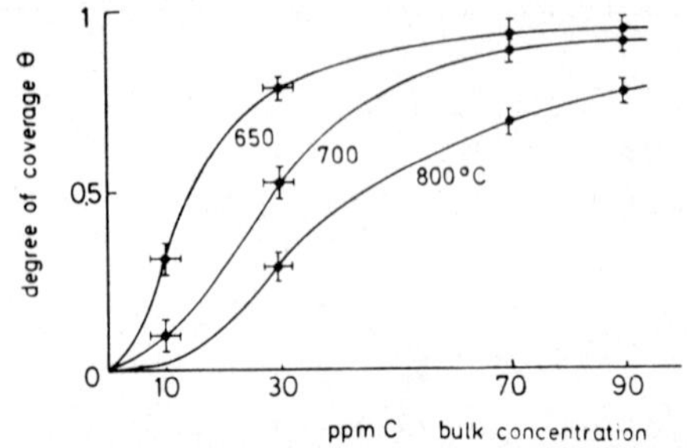

Figure 3-36. – Isotherms of carbon segregation on Fe(100) at 650°, 700°, and 800° C.

single crystal surface at 650°, 700°, and 800° C, in figure 3-36 are examined, one sees that the surface coverage can be fairly complete (ref. 25). The maximum concentration of carbon in the bulk examined was 90 ppm of carbon. There are basically four compositions of the iron single crystal incorporating carbon: 10, 30, 70, and 90 ppm of carbon in the iron single crystal. This is plotted on the abscissa of figure 3-36. The ordinate shows the degree of surface coverage, with the maximum being 1 (unity) for complete surface coverage of the iron by carbon.

Examining the data of figure 3-36 reveals that at 30 ppm of carbon (in the bulk) the surface concentration is maximum at 650° C. It is approximately three-quarters of a monolayer on the solid surface; that is, three-fourths of a monolayer of carbon covers the solid surface. This coverage is enough to have a very strong influence on tribological properties. As the temperature is increased from 650° to 700° C, the concentration of carbon on the surface is seen to decrease. The carbon begins to diffuse into the bulk iron and is lost from the surface. At 800° C there is still a further decrease in the concentration of carbon on the iron surface, and the surface coverage now is only about one-fourth of a monolayer. So we have gone from three-fourths of a monolayer at 650° C to one-fourth of a monolayer at 800° C. Simply increasing the temperature by 150° C is sufficient to bring about a loss of one-half of a monolayer of carbon from the solid surface; that carbon is dissolved in the bulk iron.

The difference (with change in temperature) in the surface coverage at a concentration of about 30 ppm of carbon is fairly great (one-half of a monolayer). As the concentration of carbon in the iron increases, however, the difference (with change in temperature) is less marked, as shown in figure 3-36 at a concentration of 90 ppm of carbon. At 90 ppm of carbon, increasing the temperature from 650° to 800° C results in a change of from approximately nine-tenths of a monolayer to about seven-tenths of a monolayer of surface coverage. Thus, the temperature effect is more pronounced at the lower concentrations of the dopant or solute in the solvent than it is at higher concentrations.

It is important to recognize from the data of figure 3-36 that extremely small concentrations of an impurity atom are required to play a very pronounced role on the surface of solids. As little as 10 ppm of carbon in the bulk iron single crystal are sufficient at 650° C to produce a quarter of a monolayer of surface coverage of the iron by carbon. This surface segregation is extremely important because it alters the interaction behavior of the iron with other metals or nonmetals in tribological systems. It also impedes or inhibits the interaction of the solid surface with lubricating species; it essentially poisons the solid surface. From a tribological point of view, this poisoning of the solid surface is good with respect to the reduction in adhesion of two solid surfaces in solid-state contact; it is, however, detrimental to effective lubrication when the presence of the carbon inhibits the interaction of the lubricant with the clean metal surface which is normally very active.

Other species beside oxygen and carbon segregate to the surface of metals and influence the surface behavior. For example, nitrogen is an element which is frequently found dissolved in elemental metals. Since it is in the environment, nitrogen frequently is found on the surface of metals; it might, therefore, be anticipated that nitrogen would play an important role in the surface physics and chemistry of metals.

Two iron single crystal (100) surfaces were doped with nitrogen. One single crystal was doped with 150 ppm of nitrogen, and the second was doped with 530 ppm of nitrogen. The surfaces were then examined with AES analysis to determine the concentration of nitrogen on the surface of iron (ref. 25). The results of those experiments are presented in figure 3-37. The degree of surface coverage θ is plotted as a function of temperature. The surface coverage of nitrogen on the iron for both compositions at temperatures to 500° C is essentially unity; that is, there is a complete monolayer of nitrogen adsorbed on the (100) surface of the iron due to segregation of the nitrogen from the iron to the surface. The fact that the iron contains 150 ppm in one case and 530 ppm of nitrogen in the other case does not alter the surface effect. In both cases the surface is covered with essentially a monolayer up to 500° C.

At temperatures above 500° C, however, the amount of surface coverage is reduced more rapidly with the sample of iron doped with the lower amount of nitrogen (150 ppm). At 700° and 800° C there is some spread in the data, with the iron sample containing the higher concentration of nitrogen (350 ppm) maintaining a higher degree of surface coverage at these

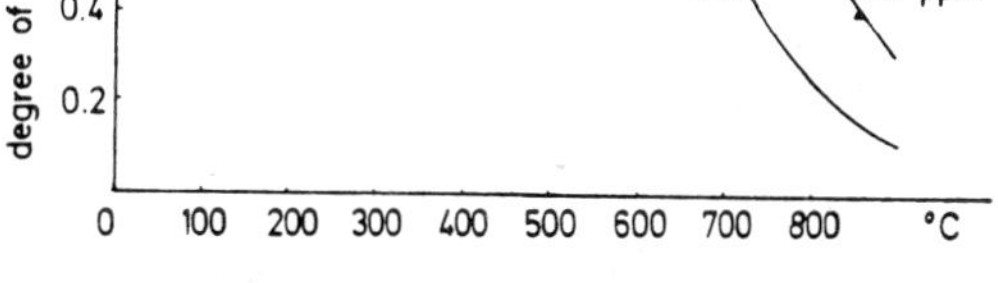

Figure 3-37. – Degree of nitrogen coverage on iron as function of temperature, calculated from kinetic results on nitrogen desorption from iron compared with data from AES measurements on nitrogen segregation.

higher temperatures than does the sample containing the lower concentration of nitrogen. The point to be made with figure 3-37 (as it was with fig. 3-36) is that extremely small concentrations of the nonmetal dissolved in the metal can play a very important role in the surface chemistry of elemental metals. This, in turn, can very strongly influence tribological behavior.

The segregation observed in figures 3-36 and 3-37 was for the segregation of carbon and nitrogen on the surface of single crystals of iron of the (100) orientation. Earlier the subjects of atomic density and surface energy of the surface planes of the various metals were discussed. It was indicated that certain atomic planes are of a higher atomic density than others and that they accordingly have lower surface energies. Surface orientation plays a role in the amount of segregation that occurs. One usually sees a lower concentration of the segregated species on the high atomic density, low surface energy planes than on the low atomic density, high surface energy planes; that is, the impurity atoms dissolved in the bulk segregate much more readily to the surface of low atomic density, high surface energy planes. Thus, if the single crystal surface were of an iron (110) surface rather than of a (100) surface, the surface coverage anticipated might be less.

In addition to orientation effects in the surface segregation of nonmetallic alloying elements, the presence of more than one alloying element—that is, more than one solute in the solvent metal—can affect the surface segregation process. Very frequently, for example, the presence of one nonmetallic element retards or impedes the segregation of a second alloying element to the surface of the metal. This has been observed when both carbon and sulfur are present in iron, and, frequently, iron contains both carbon and sulfur as bulk impurities. The presence of the carbon inhibits the segregation of sulfur to the iron surface until the carbon concentration in the bulk iron reaches a sufficiently low level where the sulfur begins to segregate on the solid surface. Again, the concentration of both nonmetallic species in the metallic element iron are of impurity level concentrations.

An earlier discussion indicated that grain boundaries are basically a defect in the perfect single crystal surface and that they are sites of high energy. They are higher energy sites than are the surface orientations of the grains adjacent to the boundary itself. In light of the foregoing discussion, one might anticipate that, in the segregation process, the impurity or solute atoms would segregate fairly readily at a grain boundary since it is a high-energy site. This fact has been observed for a number of impurity elements present in bulk metals such as iron.

One such example is presented in figure 3-38 for tellurium dissolved in iron (ref. 26). This is a polycrystalline sample of iron containing a very small concentration of tellurium (as indicated in the abscissa of the figure). The tellurium in the grain boundaries is seen to increase with an increase in the amount of tellurium in the iron. The grain boundary composition is shown as a percent of a monolayer that is concentrated in the grain boundary region; the plot reflects anywhere from 0 to 60 percent of a monolayer of tellurium. With as little as 10^{-4} percent of tellurium in the bulk of the iron, the concentration in the grain boundary can be as much as 60 percent

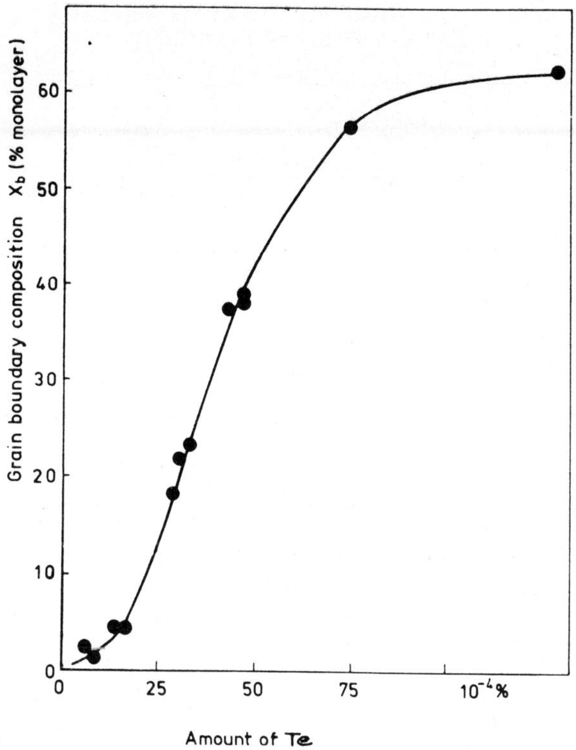

Figure 3-38.—Intergranular concentration of tellurium as function of bulk concentration in Fe-Te alloys.

of a monolayer. The data in figure 3-38 again serve to stress the importance of small concentrations of impurity atoms on the behavior of solid surfaces.

While the majority of the grain boundary itself is subsurface, the grain boundary does terminate at the surface. As a consequence, the concentrations reflected in figure 3-38 are on the surface in the region of the grain boundary. This increased or enhanced concentration of the impurity atom in the grain boundary region affects the tribological behavior. For example, rubbing the surface or sliding across the surface can cause the material concentrated in the grain boundaries to become smeared out or rubbed across the entire surface. They serve as reservoirs for the impurity which is present in the bulk. This can be desirable, as in the case of tellurium and iron: tellurium reduces the adhesion and friction of iron. However, it can be undesirable, as in the case of aluminum and iron: aluminum increases both adhesion and friction for iron (see discussions in chs. 5 and 6).

Using the atom probe in conjunction with the FIM (described in detail in ch. 2), Tsong was able to identify the segration to grain boundaries of chromium in a 410 stainless steel; he established that not only do nonmetallic impurities segregate to grain boundaries of metallic systems but that even metallic components segregate to grain boundaries in metal to metal systems (ref. 27). Some of the data obtained by Ng and Tsong are presented in table 3-VII.

TABLE 3-VII. – GRAIN BOUNDARY SEGREGATION DATA IN 410 STAINLESS STEEL[a]

Conditions[b]	F_{Cr}, percent
In grain boundary	17.07±3.22
Just outside grain boundary	7.76±3.38

[a]Reference 27.
[b]Tip annealed at 500° C for 3 min.

Table 3-VII shows that the concentration of chromium in the grain boundary is approximately twice that of the area outside the grain boundary (17.07 to 7.76 percent). The atom probe hole to detect the chromium atoms covered a surface area about 10 angstroms in diameter, while the grain boundary itself is no more than 3 angstroms wide. The real concentration of chromium in the grain boundary may, thus, actually be higher than is shown in table 3-VII. Furthermore, the evaporation rate in the grain boundary is higher than just outside the grain boundary. This may contribute to an accelerated loss of chromium from the boundary and thus promote a higher concentration in the grain boundary of chromium. By utilizing iron 54, whose isotopic abundance is known, as a calibrator, Ng and Tsong were able to realize different evaporation rates inside and outside a grain boundary and produce field evaporation of chromium in those regions.

There have been two general mechanisms proposed to account for the segregation of alloying elements (both metallic and nonmetallic) to the surfaces of solids from within the bulk. One concept involves reducing the surface energy to the lowest energy state possible; this can be accomplished frequently by segregating an alloy constituent to the solid surface. This mechanism can account for the segregation of alloy constituents to solid surfaces.

A second proposed mechanism accounting for the segregation of alloying elements to the surfaces of solids involves the concept of strain energy. If the foreign atom dissolved in the parent lattice causes strain in the parent lattice (as discussed earlier in this chapter), then the goal is to achieve the lowest energy state possible. This can be accomplished by rejecting the foreign species which is producing the strain in the crystal lattice. As a consequence, the lattice squeezes out the foreign species from the matrix of the bulk solid to the surface region. There are some who propose that the segregation process be achieved by a reduction of surface energy and, at the same time, a relief of strain in the crystal lattice.

There is no question that the presence of the impurity species does affect the surface energy of the solid. The effect can be very pronounced or it can be relatively minimal. Figure 3-39 shows the variation in surface energy for iron with phosphorus dissolved in the iron. Even though the concentrations of phosphorus are low, they are much higher than the concentrations of carbon and nitrogen in iron discussed previously (figs. 3-36 and 3-37).

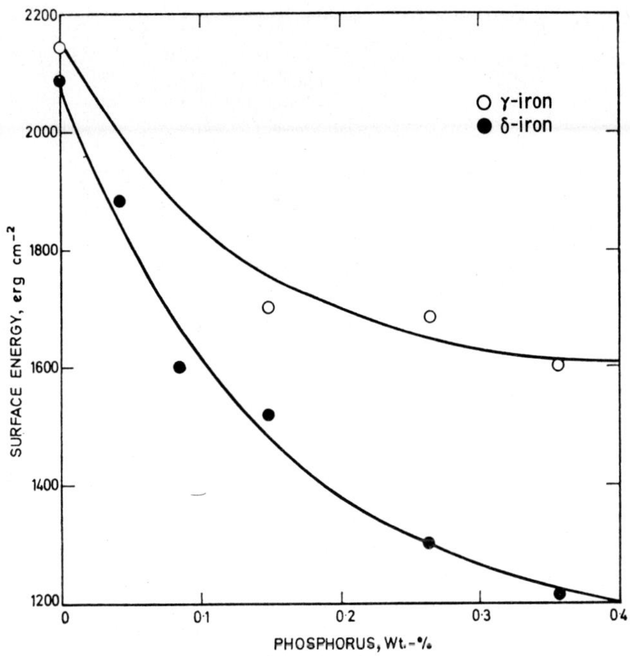

Figure 3-39. – Variation of surface energy of δ-iron (1450° C) and γ-iron (1350° C) with bulk phosphorus content (ref. 28).

The surface energy of the phosphorus-iron alloy is reduced markedly by the presence of phosphorus for two crystalline forms of the iron, delta and gamma (ref. 28). Both forms of iron (delta and gamma) have their individual surface energies of approximately 2100 ergs per square centimeter (fig. 3-39). With delta iron, the additions of small concentrations of phosphorus bring a rapid reduction in the surface energy. With approximately 0.36 percent by weight of phosphorus in the iron, the surface energy has been reduced from approximately 2100 ergs per square centimeter to just over 1200 ergs per square centimeter (nearly 100-percent reduction in the surface energy). Again, this small concentration has a marked surface effect because of its role in the segregation process. With gamma iron, the effect is not as great as it is with delta iron (fig. 3-39). The delta iron is a body-centered-cubic structure, while the gamma iron is a face-centered-cubic structure. The delta iron has a lower density than the gamma iron so the phosphorus may diffuse more freely and readily through the crystal lattice and thereby segregate on the surface, giving rise to a more marked reduction in surface energy than is observed with the gamma iron.

The basic interactions of a nonmetal with a metallic element relative to its presence on the surface can be summarized, from an energy standpoint, with the aid of the line drawing in figure 3-40. In this figure the interaction of carbon with a nickel (111) surface is depicted schematically (ref. 29). At zero energy level we have carbon in graphite (graphitic carbon). To isolate the carbon atom from a bulk graphite body by vaporization requires 7.5

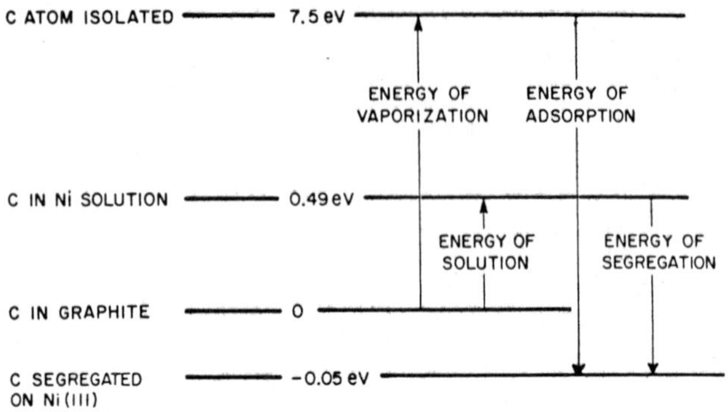

Figure 3-40.—Schematic diagram of atomic energy level for carbon-nickel (111) system (ref. 29).

volts, which is the energy of vaporization indicated in figure 3-40. Once a carbon atom is in the isolated free state, then it may interact with the clean nickel surface by adsorbing to the solid surface. The energy of adsorption would then be the energy associated with the carbon atom becoming bonded to the solid surface (7.5 eV, energy required for isolation of the carbon atom initially from a graphite body plus 0.05 eV associated with the attachment of carbon to the nickel (111) surface). If carbon is dissolved in nickel in the molten state, it requires 0.49 electron volt, and this is referred to in figure 3-40 as the energy of solution. This is the energy to dissolve carbon into nickel within the solubility range of carbon in nickel. If that carbon is once dissolved into nickel and then segregates to the surface, the energy associated with segregation is 0.49 electron volt plus an additional 0.05 electron volt, associated with the segregation of carbon to a nickel surface.

Environmental Interactions with Real Surfaces

Most real solid surfaces are not atomically clean when examined with surface analytical tools. The solid surfaces all contain adsorbed species (physically adsorbed material or chemically adsorbed), chemical reaction products such as oxides, or other films which have formed as a result of interaction with the environment. The environment may be a conventional, normal atmospheric condition or some artificially imposed environment such as a lubricating film. The simplest and most common occurrence recognized with real surfaces is that nearly all surfaces contain adsorbates, either physically adsorbed or chemisorbed material, and the adsorbate gets on the adsorbent surface by interacting with it. The interaction is very specific and has a characteristic energy which reflects both the surface energy of the solid which is interacting as the adsorbent and the energy of the particular species that is doing the adsorbing. As the adsorbate approaches the solid surface, an energy of interactions arises as a function of the distance that the adsorbent is away from the adsorbate (ref. 30). This energy to distance relationship is shown schematically in figure 3-41. The l on the abscissa refers to the distance with which an atom in the adsorbent

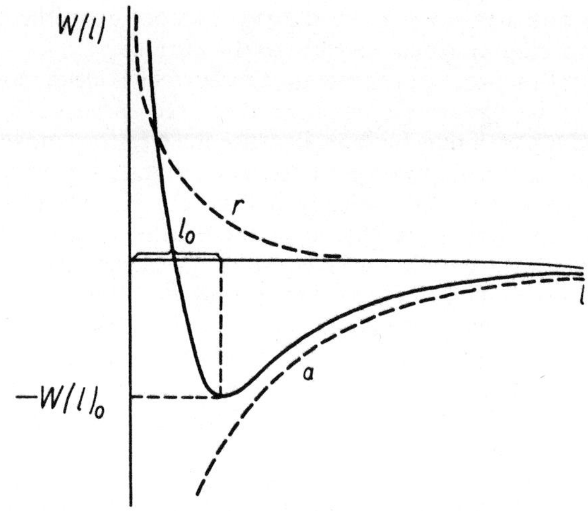

Figure 3-41.—Interaction energy as function of distance between adsorbate and adsorbent (ref. 30).

(the solid surface) and an atom associated with the adsorbate (the adsorbing species) are separated. There is, from the energy diagram, an optimum distance where the energy associated with bonding occurs. This is the inner section of l_0 with the line generated by $-W(l)_0$. That point of intersection is the equilibrium binding energy associated with the bonding of an adsorbed species with a solid surface. It is a function of both the adsorbent and the adsorbate species.

As the adsorbate is moved closer to the solid surface, the energy goes from a very strongly negative energy value—that is, a strong binding energy—to repulsive energy. This repulsive energy becomes strongly positive as the distance between the adsorbate and adsorbent is reduced. If the distance between adsorbent and adsorbate atoms is increased away from equilibrium binding energy force located at the intersection of l_0 and $-W(l)_0$, the binding force decreases and ultimately, when the distance is sufficiently large, drops to zero. The attractive force is essentially zero.

Physical and Chemical Adsorption

The fundamental energy diagram in figure 3-41 can be used to consider the binding of physically adsorbed species and chemisorbed species to solid surfaces of all types. This basic energy diagram and the adsorption process associated therewith is a fundamental step in the development of surface films of all kinds. In the formation of surface compounds (e.g., oxides due to oxidation of metal surfaces) the precursor step is the initial adsorption to the metal surface prior to the oxidation reaction. In the process of physical adsorption, the equilibrium position of energy for binding on the lowest energy state in figure 3-41 is a function of a balance between a negative charged layer and a positive charged layer; this occurs in the surfaces between the adsorbing molecule or atom and a surface.

In the adsorption process, the metal, if its surface is clean, contains a sea of electrons at the surface (as has been discussed earlier). The adsorbing molecules interact with that surface. Because the electrons move on the surface of a metal, the concentration of electrons slightly subsurface has been reduced as is indicated schematically in figure 3-42. There is an uneven distribution of electrons when the migration or the movement of electrons out of the metal surface to the surface results in a depletion of electrons in a near surface of the metal (ref. 31). This produces an imbalance in electron distribution and charging results. At the surface a negatively charged layer associated with the band resulting from the electrons that have spewed onto the surface for binding purposes has developed, and in the near surface region (subsurface) a band or region results where the layer is positively charged due to the depletion of electrons resulting from the migration of electrons to the solid surface. Consequently, dipole holes exist, as indicated in the schematic of figure 3-42. It is this kind of electron distribution of the surface layer of a metal that occurs with the adsorbing physically of molecules. The bonding associated with the negative and positive charged layers is relatively weak and is of van der Waal's nature as mentioned previously.

Not only do species physically adsorb to real surfaces, but real surfaces also contain chemisorbed species. In fact, most active metal surfaces do have chemisorbed species present on their solid surfaces, and it is very difficult to avoid chemisorption in active gaseous environments with clean surfaces such as metals. When a metal surface interacts with gases in the environment, the principal gaseous species has encountered are nitrogen, ox-

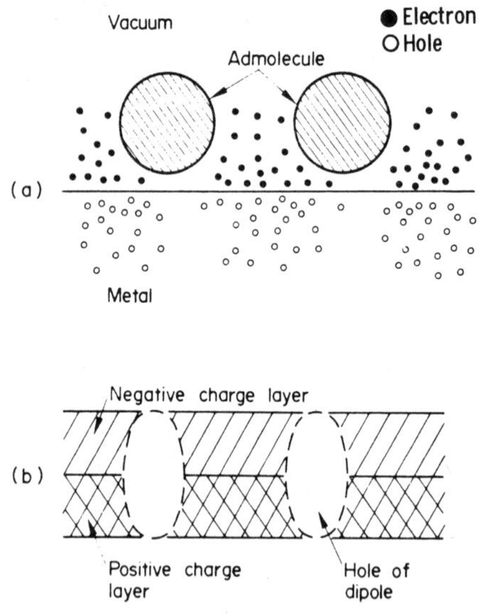

Figure 3-42. – Schematic illustrations of electron distribution in surface layer of metal with physisorbed molecules.

ygen, and hydrogen. All of these gaseous species exist as diatomic molecules rather than in atomic states. In order for chemisorption (i.e., chemical bonding) of the species to occur to the solid surface, the diatomic molecule must first become dissociated, and a certain quantum of energy is required to achieve this dissociation. This energy requirement is shown schematically in figure 3-43 with the gaseous molecules nitrogen, oxygen, and hydrogen. The energy indicated near the ordinate is that associated with generating the individual atoms from the diatomic molecule (the energy required to break the molecule into individual atomic constituents) (ref. 32).

When the individual gaseous atoms interact with a metal surface, chemisorption occurs and the atomic species become bonded to the solid surface. Chemisorption is a much stronger bonding than that associated with physical adsorption, and it is very specific. The energies associated with the chemisorption process indicated in figure 3-43 vary as a function of two variables: (1) the metal surface to which adsorption is taking place, and (2) the adsorbing species. Varying the adsorbing species with a fixed metal surface results in changes or differences in energies of chemisorption. Likewise, maintaining the same gaseous species but changing the metal surfaces results in variations in the chemisorption energy. For example, the energy of adsorption associated with the bonding of ethylene to a clean iron surface is much greater than is the energy associated with the bonding of ethane. Both gaseous molecules contain two carbon atoms and hydrogen in their structure. Ethylene, however, is a doubly bonded species, while ethane contains only a single bond between adjacent carbon atoms. This difference in degree of bond saturation and the amount of associated hydrogen with the molecular structure makes a difference in the binding energy of these two species to the iron surface.

If one goes a step further and takes a triple-bonded, two carbon molecule such as acetylene, one finds that the bonding of acetylene to a clean iron surface is stronger than the bonds associated with either ethylene or ethane. Again the triple bond and the higher degree of unsaturation in the acetylene

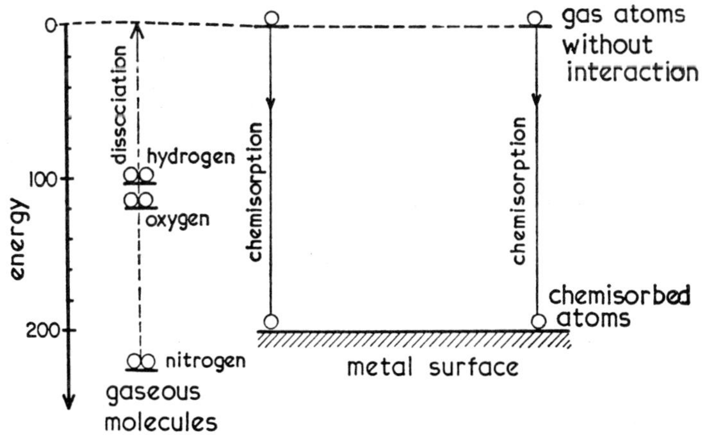

Figure 3-43.—Energies of two atoms as gaseous diatomic molecule, gas atoms, and atoms chemisorbed on metal surface (ref. 32).

accounts for stronger bonding of the acetylene to the iron surface than is experienced with ethylene or ethane. In fact, as the degree of unsaturation increases, the binding energy of the basic hydrocarbon structure, the two-carbon atom, to the iron surface increases. Conversely, if one maintains the same particular molecule, such as ethylene, and adsorbs it to different metal surfaces, the chemisorption energy varies. For example, the energy of adsorption of ethylene to gold is much less than it is to a titanium surface. Thus, titanium is much more reactive with ethylene than is iron, and iron, in turn, is much more reactive with ethylene than is gold. These differences are extremely important in tribology because they give the tribologist an indication of the strength of the chemisorbed bonds that are formed and an indication of their consequent durability or resistance to dissociation in practical lubrication devices.

A clean metal surface continues to adsorb a particular atom in the process of chemisorption until that surface becomes saturated with the adsorbing species. The adsorption from fractions of monolayers to full monolayers on solid surfaces of various gaseous species has been followed very effectively in recent years with analytical surface tools such as XPS, AES, and SIMS.

In figure 3-44 the relative oxygen intensities (detected by XPS, AES, and SIMS) are plotted as functions of exposure to oxygen (oxygen dosage in langmuirs); that is, the uptake of oxygen on a clean nickel surface (ref. 33) is a function of oxygen exposure. There is a greater oxygen signal from the solid surface with increased dosage.

As shown in figure 3-44, the oxygen dosage required to produce peak in-

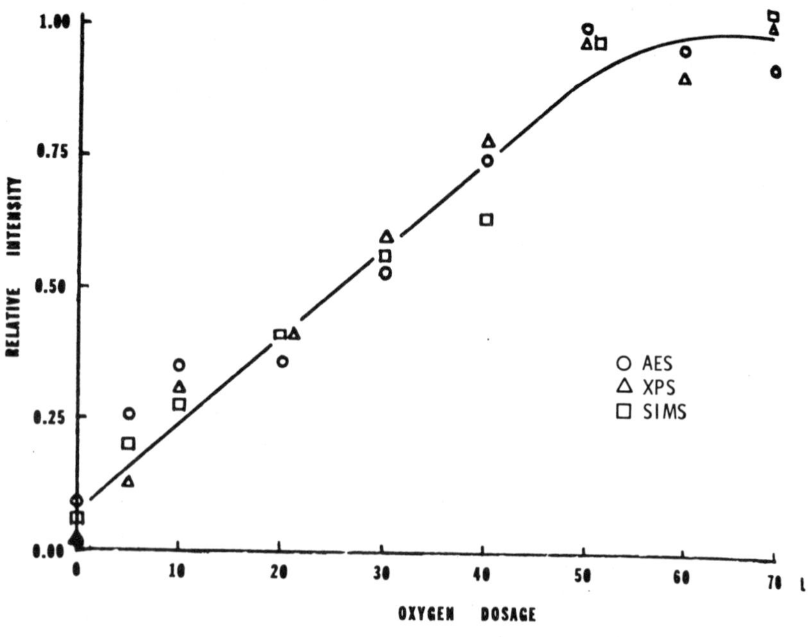

Figure 3-44. – Normalized intensities of XPS, AES, and SIMS oxygen signals as function of oxygen dosage on sputter-cleaned Ni foil (ref. 33).

tensities agrees very well with all three analytical tools (XPS, AWS, and SIMS). At approximately 60 langmuirs (fig. 3-44) the surface is becoming saturated with oxygen, and the chemisorption process ceases. Any further addition of oxygen to the surface can only be present on the surface as physically adsorbed oxygen or, if sufficient energy is available to form an oxide, by oxidation of the metal surface to form nickel oxide. But chemisorption is essentially a monolayer process, and once the surface becomes covered with a monolayer of oxygen, chemisorption ceases as a process. Before additional oxygen can be taken up by the solid surface, one of the other two mechanisms must begin to operate, either physical adsorption or oxidation.

All real metal surfaces contain at least chemisorbed gaseous species, and in an oxygen-containing environment such as air the surface chemisorbed species are usually principally oxygen. They may contain other adsorbed species mixed with the oxygen, but the principal species found on the solid surface is usually oxygen. Most often the presence of the oxygen has resulted in the oxidation; that is, the oxygen present on the solid surface has gone from simply chemisorbed oxygen to the formation of metal oxides. While it is chemically adsorbed to the solid surface, the oxygen is very strongly bonded and very highly specific in its bonding characteristics to metal surfaces. Another character of chemisorption, which sets it aside from chemical reaction or chemical bonding, is that the chemisorbed species can be recovered in its initial identity without having dissociation or reduction of the particular adsorbed species taking place.

If a molecule of a particular species (say hydrogen sulfide) adsorbs to the solid surface, the hydrogen sulfide would adsorb as hydrogen sulfide and the sulfur would bond to the surface with the hydrogen remaining attached to the sulfur. In fact, this does not happen; the hydrogen dissociates from the sulfur. If it does not and the sulfur bonds to the metal surface, then theoretically the hydrogen sulfide can be removed. Even though it is chemisorbed by the sulfur atom to the iron surface, it can be removed as hydrogen sulfide by a desorption process that is, by supplying sufficient surface energy to the solid surface to break the sulfur to iron bond. In true chemisorption this is effectively and easily accomplished as demonstrated by the data of figure 3-45.

Figure 3-45 presents the heat of desorption for sulfur from a platinum (100) surface as a function of the surface coverage, assuming that the surface coverage does not exceed one-half of a monolayer. An examination of the data of figure 3-45 indicates that the energy required to desorb a chemisorbed sulfur atom from a solid surface increases as the surface coverage decreases (ref. 34). Thus, the binding of an absorbed species to a clean metal surface is a function of surface coverage, the binding energy. The greater the surface coverage, the lower the binding energy. The energy varies with surface coverage. The less of the absorbed species that remains on the surface the higher the energy required to remove the remaining adsorbate. This energy can be quite appreciable. For example, in figure 3-45, with a half of a monolayer of surface coverage of sulfur on the platinum (100) surface, the energy of desorption or the heat of desorption is approximately 60 kilocalories per mole. When the surface coverage is reduced to a

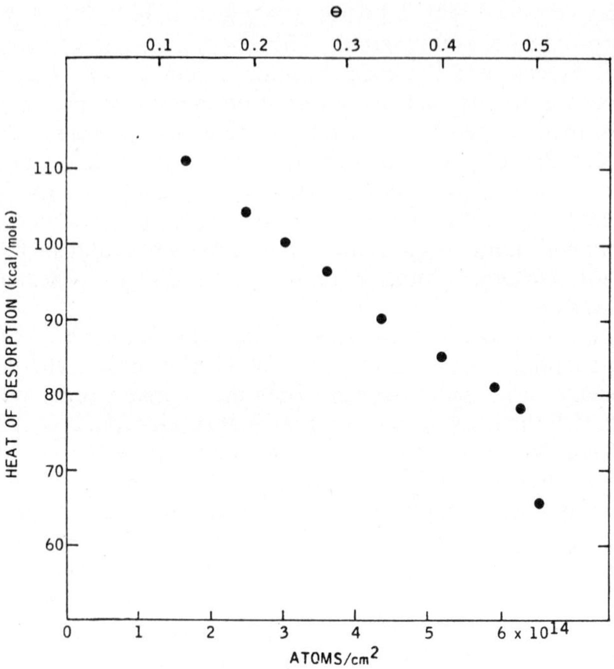

Figure 3-45. – Heat of desorption of sulfur from Pt(100) as function of coverage (ref. 34).

tenth of a monolayer on the solid surface, the heat of desorption increases to 110 kilocalories per mole. This is nearly a twofold increase in the energy required to remove a sulfur atom from a solid surface as a result of removing the concentration of sulfur from the solid surface.

Chemical Reaction

The curve presented in figure 3-45 is general in nature and applies to other adsorbates which chemisorb to metal surfaces. In other words, as the surface coverage decreases, the binding energy of the atoms on the solid surface adsorbed thereto increases. When the adsorbing species interacts with the solid surface to form a compound by a reaction mechanism, simple desorption can no longer take place and the heating of a surface must be such as to supply sufficient energy to cause dissociation of bonds formed in compounds as opposed to the heat of desorption associated with removing chemisorbed surface atoms or molecules.

When a chemical reaction takes place, both the adsorbing species and the adsorbent (the metal surface) undergo a change. This change in the character of both the metal and the oxygen is reflected in the data of figure 3-46 (ref. 35), which presents Auger spectra for clean and oxidized aluminum surfaces. The upper spectra indicates the pure aluminum film. The very large peak near the ordinate is the aluminum peak. A small oxygen peak is seen at approximately 500 electron volts, but not very much oxygen

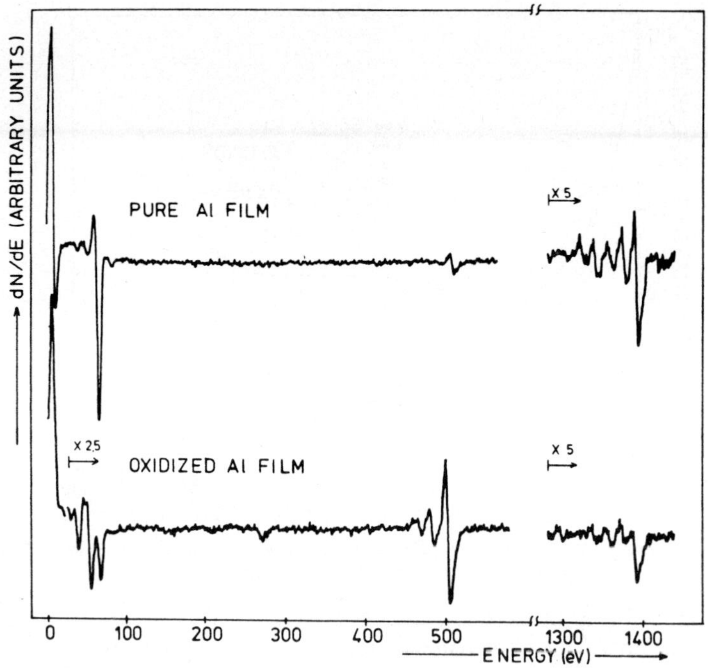

Figure 3-46.—Auger electron spectra from clean and oxidized Al surfaces (ref. 35).

is actually present on the solid surface. When the aluminum surface is oxidized—that is, oxygen interacts with the solid surface to form aluminum oxide Al_2O_3—a distinct change in the shape of the aluminum peaks takes place in the Auger spectra as indicated by the lower spectra in figure 3-46. The oxygen peak intensity is increased as reflected by the larger oxygen peak at approximately 500 electron volts, and the aluminum appears as a pair of peaks rather than the single peak observed with the pure aluminum. Thus, surface detection tools such as AES pick up the differences with the solid surface when chemical reaction has taken place on a solid.

With the chemisorption of oxygen on the pure aluminum, the aluminum peaks seen in the upper portion of figure 3-46 for pure aluminum film would still be reflected in the Auger spectra, but there would also be an oxygen peak as there is in the spectrum for oxidized aluminum. Only when the surface has been oxidized, however, does the aluminum peak undergo the change observed in figure 3-46. When the surface is oxidized, shifts in the aluminum peak take place as a result of the formation of the oxide; that is, the chemical reaction produces a change in the binding energy of the aluminum which is reflected in the Auger spectrum. A careful examination of figure 3-46 indicates this shift.

The presence of the various species or forms of aluminum on a solid surface and their relative intensities as functions of oxygen uptake are presented in figure 3-47. In this figure the aluminum peak of figure 3-46 (which occurs at 64 eV) is plotted as a function of uptake of oxygen. We can see that the relative intensity of the aluminum peak decreases as the

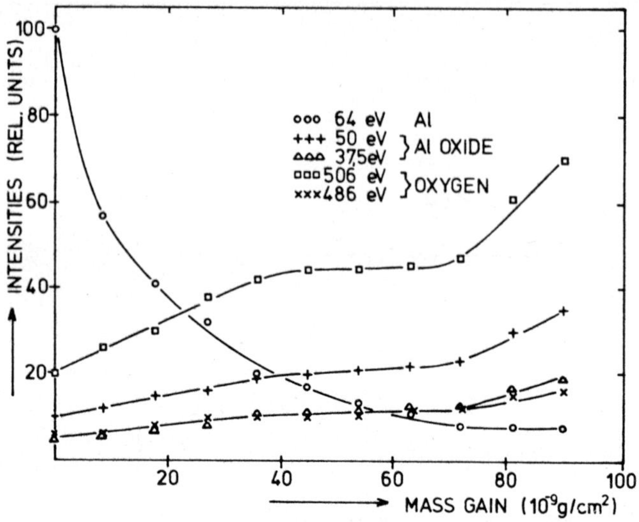

Figure 3-47.—Peak to peak heights of AES intensities of 64.50, and 37.5 eV Al and 506 and 486 eV oxygen lines as function of oxygen uptake (ref. 35).

aluminum is being covered by adsorbed oxygen. This occurs because the primary beam of electrons associated with the AES analysis is being screened by the layer of oxygen that is adsorbing on the aluminum surface and minimizing or limiting the number of primary electrons that get through to the aluminum. Also, the adsorbed oxygen layer limits the number of ejected secondary electrons or Auger electrons coming out of the surface from the aluminum layer. The oxygen produces a shield to keep out the secondary electrons. The overall effect is a reduction in the intensity of the aluminum peak with increased exposure to oxygen (fig. 3-47).

In the formation of aluminum oxide, the aluminum single peak becomes double peaks at 50 and 37.5 electron volts. The change in concentration of these peaks with increasing exposure to oxygen is also presented in figure 3-47. As the aluminum decreases in intensity, the uptake of oxygen associated with chemisorption of oxygen on the surface increases in intensity as indicated by the squares in figure 3-47. The reason for this is that there is no fixed ratio of oxygen to aluminum as there is in the compound. Instead, the oxygen continues to adsorb over the oxide until the surface becomes saturated or a monolayer forms (as was observed with the adsorption of oxygen on a nickel surface, fig. 3-44).

When oxides form on metal surfaces they can form in a random fashion on the surface of the metal or they can take on the structural characteristics of the surface being oxidized. With metals, for example, the oxide may develop in an epitaxial manner on a particular orientation of the metal. Thus, if a copper surface has a (110) orientation, the oxide developed on the surface may have that same orientation. This epitaxial development of oxide (where the oxide takes on the character of the surface metal) has been observed in a number of metallic systems. Table 3-VIII presents some results obtained for two such metals, copper and silver, where the oxide

TABLE 3-VIII. – EPITAXY OF OXIDE FILMS ON MAJOR FACES OF METALS[a]

Metal	Oxide	Surface preparations	Oxidation conditions	Parallel planes		Parallel planes		Remarks	Misfit percent
				Metal	Oxide	Metal	Oxide		
Cu	Cu_2O	Single	O_2 at	{001}	{111}	$[1\bar{1}0]$	$[1\bar{1}0]$	Electron	18
(fcc)	(cubic)	crystal	150°			$[1\bar{1}0]$	$[\bar{1}10]$	micro-	
		spheres	to			[110]	$[1\bar{1}0]$	scope	
		and flat	350° C,			[110]	$[\bar{1}10]$	obser-	
		surfaces	1 to 10	{110}	{110}	$[1\bar{1}0]$	$[1\bar{1}0]$	vations	
		electro-	mm,	{111}	{111}	$[1\bar{1}0]$	$[1\bar{1}0]$	indicate	
		polished	15 sec			$[1\bar{1}0]$	$[\bar{1}10]$	very	
		and an-	to	{113}	{110}	$[1\bar{1}0]$	$[\bar{1}10]$	smooth	
		nealed	2 hr	{012}	{012}	$[1\bar{1}0]$	$[1\bar{1}0]$	sur-	
		in H_2				$[0\bar{1}1]$	$[\bar{1}10]$	faces,	
						[011]	[110]	free of	
								facets or	
								terraces	
Ag	Ag_2O	Single	O_2 at	{001}	{111}	$[1\bar{1}0]$	$[1\bar{1}0]$	Parts of	16
(fcc)	(cubic)	crystal	200°	{110}	{110}	$[1\bar{1}0]$	$[1\bar{1}0]$	crystal	
		spheri-	to	{111}	{111}	$[1\bar{1}0]$	$[1\bar{1}0]$	show	
		cal sur-	370° C,					signs of	
		faces	pres-					terraces	
		grown	sures					and	
		on W	up to					facets	
		strip	100						
		heater	atm						

[a]Reference 36.

formed on copper is the lower oxide Cu_20 and the oxide formed on the silver is the oxide Ag_20. Cu_20 is the lower oxide of copper, Cu0 being the higher oxide; with silver the only oxide formed is Ag_20.

Single crystal spheres were oxidized in an oxygen environment. The results presented in table 3-VIII are from the classic work of Gwathmey and Lawless (ref. 36). In table 3-VIII is a column for parallel planes (i.e., where the plane of the oxide matches the plane of the metal) and a column for parallel axes. The parallel planes column includes both metal surface plane orientation and oxide orientation. For example, for the (110) metal surface plane, the same plane matches on the surface the (110) oxide plane. Likewise, with the (111) plane of the metal the surface oxide also has a (111) orientation. In addition to the matched parallel planes of the oxide in the metal, the axes of the metal and oxide can also be parallel as indicated in a

the surface energetic considerations. This is commonly referred to as reconstruction. The term reconstruction is used because the surface is not the same as the bulk lattice. It undergoes a change in stucture, or a reorientation or reconstruction of the atoms occurs in the lattice at the outermost layer. This is frequently observed in LEED studies of single crystal metal surfaces. Gold is a classic example of a material in which reconstruction at the solid surface has been observed. There is some controversy in some of the metal systems (such as gold) as to whether or not the LEED patterns truly represent reconstruction at the solid surface or rather, simply, are the result of bulk material impurities that surface. However, there are situations where reconstruction does distinctly occur on a metal solid surface. The reconstruction process as well as segregation, chemisorption, and compound formation are indicated schematically in summary form in figure 3-48, which presents some possible surface events that can take place. These have already been discussed.

The important thing to recognize with figure 3-48 is that all these events depicted schematically can take place at a surface and they do change the nature and character of the solid surface. Thus, when such a surface interacts with another solid in tribological systems, the interaction is influenced and affected by such things as reconstruction, segregation, chemisorption, and compound formation.

References

1. Adamson, Arthur W.: Physical Chemistry of Surfaces. Second ed., Inter Science Publishers, 1967.
2. Phillips, F. C.: An Introduction to Crystallograpy. Second ed., Longmans, Green & Co. (London), 1956.
3. Barrett, C. S.: Structure of Metals, Crystallographic Methods, Principles and Data. McGraw-Hill Book Co., Inc., 1943.
4. Cullity, Bernard D.: Elements of X-Ray Diffraction. Addison-Wesley Publishing Co., Inc., 1956.
5. Kittel, Charles: Introduction to Solid State Physics. Third ed., John Wiley & Sons, Inc., 1966.
6. Azaroff, Leonid V.: Introduction to Solids. McGraw-Hill Book Co., Inc., 1960.
7. Juretschke, H. J.: Electronic Properties of Metal Surfaces. The Surface Chemistry of Metals and Semiconductors, H. C. Gatos, ed., John Wiley & Sons, Inc., 1960, pp. 38-53.
8. Lurie, P. G.; and Wilson, J. M.: The Diamond Surface: I. The Structure of the Clean Surface and the Interaction with Gases in Metals. Surface Sci., vol. 65, 1977, pp. 453-475.
9. Mykura, H.: The Variation of the Surface Tension of Nickel with Crystallographic Orientation. Acta Metall., vol. 9, no. 6, June 1961, pp. 570-576.
10. Sundquist, B. E.: A Direct Determination of the Anisotrophy of the Surface Free Energy of Solid Gold, Silver, Copper, Nickel and Alpha and Gamma Iron. Acta Metall., vol. 12, no. 1, Jan. 1964, pp. 67-86.
11. Winterbottom, W. L.; and Gjostein, N. A.: Determination of the Anisotrophy of Surface Energy of Metals, Part II: An Experimental γ-Plot of Gold. Acta Metall., vol. 14, no. 9, Sept. 1966, pp. 1041-1052.
12. Müller, A.; and Drechsler, M.: Eine Messung Der Anisotropie Der Oberflächenenergie Von Reinem Wolfram Mit Dem Feldionenmikroskop. Surface Sci., vol. 13, 1969, pp. 471-490.
13. Averbach, B. L.: Mechanisms of Fracture. The Science of Materials Used in Advanced Technology, Earl R. Parker and Umberto Columbo, eds., John Wiley & Sons, Inc., 1973, p. 138.

14. Swain, M. V.; et al.: A Comparative Study of the Fracture of Various Silica Modifications Using the Hertzian Test. J. Mater. Sci., vol. 8, 1973, p. 1153.
15. Muller, Erwin W.; and Tsong, Tien T.: Field Ion Microscopy. Elsevier Publishing Co., Inc., 1969.
16. Bowkett, K. M.; and Smith, D. A.: Field Ion Microscopy. North-Holland Publishing Company (Amsterdam), 1970.
17. Anderson, N. G.; and Dawson, I. M.: The Study of Crystal Growth With the Electron Microscope. III—Growth Step Patterns and the Relationship of Growth Step Height to Molecular Structure in n-nonatria-contane and in Stearic Acid. Proc. Roy. Soc. (London), ser. 1133, vol. 218, 1953, pp. 255-268.
18. Haasen, P.: Physical Metallurgy. Cambridge University Press (London), 1978. Originally published in German by Springerverlag under the title Physilalische Metallkunde (Heidleberg), 1974.
19. Hays, C.: Electropolishing of Thin Metal Foils. Metallographic Specimen Preparation: Optical and Electron Microscopy, J. L. McCall and W. M. Mueller, eds., Plenum Press, 1974, p. 318.
20. Amelinckx, S.; et al., eds.: Modern Diffraction and Imaging Techniques in Materials Science. North-Holland Publishing Company (Amsterdam), 1970.
21. McLean, Donald: Grain Boundaries in Metals. Oxford Univ. (Clarendon) Press (London), 1957.
22. Read, W. T.; and Shockley, W.: Imperfections in Nearly Perfect Crystals. Wiley & Sons, New York; Chapman and Hall, London, 1952, p. 352.
23. Westbrook, J. H.: Surface Effects on the Mechanical Properties of Nonmetals. Surfaces and Interfaces II Physical and Mechanical Properties, J. J. Burke, N. L. Reed, and V. Weiss, eds., Syracuse University Press, 1968, ch. 3, pp. 95-138.
24. Gettings, M.; and Riviere, J. C.: Precipitation and Resolution of Impurities at the Surface of Indium on Traversing the Melting Point. Surface Sci., vol. 68, 1977, pp. 64-70.
25. Grabke, H. J.; et al. Equilibrium Surface Segregation of Dissolved Nonmetal Atoms on Iron (100) Faces. Surface Sci., vol. 63, 1977, pp. 377-389.
26. McMahon, C. J., Jr.; and Marchut, L.: Solute Segregation in Iron-Based Alloys. J. Vac. Sci. Technol., vol. 15, no. 2, Mar.-Apr. 1978, pp. 450-466.
27. Ng, Yee S.; and Tsong, T. T.: ToF Atom Probe FIM Investigation of Surface Segregation in Dilute Alloys. Surface Sci., vol. 78, 1978, pp. 419-438.
28. Hondros, E. D.; and McLean, D.: Surface Energies of Solid Metal Alloys. Surface Phenomena of Metals. S.C.I. Monograph no. 28, Society of Chemical Industry (London), 1968, p. 39.
29. Blakely, J. M.; and Shelton, J. C.: Equilibrium Adsorption and Segregation. Surface Physics of Materials, J. M. Blakely, ed., Vol. I, Academic Press, 1975, pp. 189-239.
30. Davison, S. G., ed.: Progress in Surface Science. Vol. 1, part 1, Pergamon Press, Ltd. (Oxford), 1971.

    Goodwin, T. A.; and Mark, P.: The Influence of Chemisorption on the Electrical Conductivity of Thin Semiconductors.

    Wojciechowski, K. F.: The Quantum Theory of Adsorption on Metal Surfaces.
31. Takaishi, T.: Interactions Between Physically Adsorbed Molecules. Prog. Surface Sci., vol. 6, no. 2, 1975, pp. 45-62.
32. Anderson, J. R.: Chemisorption and Reactions on Metallic Films. Vol. I. Academic Press, 1971.
33. Conner, G. R.: Combination Analysis of Metal Oxides Using ESCA, AES, and SIMS. J. Vac. Sci. Technol., vol. 15, no. 2, Mar.-Apr. 1978, p. 343.
34. Fischer, Traugott E.; and Kelemen, Simon R.: The Adsorption of Sulfur on the Platinum (100) Surface. Surface Sci., vol. 69, 1977, pp. 1-22.
35. Benndorf, C.; Seidel, H.; and Thieme, F.: Initial Oxidation of Aluminum Films Investigated by AES, Work Function and Gravimetric Measurements. Surface Sci., vol. 67, 1977, pp. 469-477.
36. Gwathmey, A. T.; and Lawless, K. R.: The Influence of Crystal Orientation on the Oxidation of Metals. The Surface Chemistry of Metals and Semiconductors, H. C. Gatos, ed., John Wiley & Sons, Inc., 1960, p. 495.
37. Holloway, P. H.; and Hudson, J. B.: Kinetics of the Reaction of Oxygen to Clean Nickel Single Crystal Surfaces. Part I: Nickel (100) Surface. Surface Sci., vol. 43, 1974, pp. 123-140.

CHAPTER 4

Tribological Surfaces

The surfaces used in lubrication, friction, and wear components vary in initial surface topography depending on the particular use involved. One very common surface is the polished surface; this is a metal surface machined or cut, abraded, and finally polished with a fine powder abrasive (such as aluminum oxide or silicon carbide).

If a metal surface that is very highly polished with something such as aluminum oxide (with a very fine particle size of the order of 1 μm) is examined macroscopically, the surface looks almost like a mirror. However, on a microscale, the surface topography or the profile of the surface is not at all atomically smooth but rather contains the surface irregularities or asperities referred to earlier. A typical polished surface for aluminum is shown in figure 4-1 together with the surface profile trace for that surface. The photomicrograph indicates that, at high magnification, the surface is really not atomically smooth and contains surface irregularities which become apparent at high magnifications in the SEM. Beneath the photomicrograph is a profile trace that was obtained with the diamond stylus surface profilometer moving across the metal surface. It indicates hills and valleys, or irregularities, in the solid surface. The surface irregularities are relatively minimal, and the surface profile does not show much detail. On some occasions, when preparing surfaces for tribological applications, the solid surfaces are electropolished or chemically polished with reagents. Generally, a chemical polish produces a finer surface structure than does mechanical polishing with abrasives such as aluminum oxide. With the chemical polish, the solution tends to concentrate its reaction at the high spots on the solid surface and to chemically dissolve those spots; this leaves the surface in a much smoother state than would be obtained with the mechanical polish. With mechanical polishing, there are still fine grooves and scratches of the size reflected by the particle size of the abrasive grit (fig. 4-1).

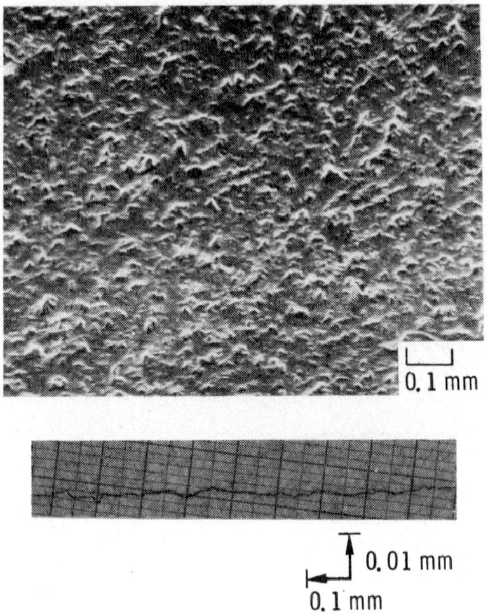

Figure 4-1.—Photomicrograph and surface profile of aluminum surface polished with 1-micrometer aluminum oxide powder.

The photomicrograph in figure 4-2 is for an electropolished aluminum surface. The photomicrograph indicates some holes or defects in the surface. Inclusions and other impurities are brought to the surface and exposed by the electropolishing process. The aluminum appears to be relatively dirty in the sense that there are defects in the solid surface. The inclusions in the surface of the aluminum in figure 4-2 can result from a number of sources. They can be in the bulk of the alloy as inclusions in the alloy material itself or they can come to the surface as a result of having been imbedded in the surface by polishing agents or abrasive materials. A typical tribological surface, for example, which is in rubbing contact with another harder surface, may accept the hard wear particles which become imbedded there as a result of the rubbing process. These particles then become exposed at the surface when that surface is electropolished. It is not uncommon to find wear surfaces that look like the photomicrograph of figure 4-2 after the worn surface has been electropolished. The bulk electropolished surface, however, is devoid of scratches and gross mechanical defects as is indicated in the surface profilometer trace beneath the photomicrograph in figure 4-2. This surface has a much finer topography than that seen in figure 4-1. The hills and valleys are minimal with very little undulation in the nature of the solid surface.

In some lubrication applications, very rough surfaces are used. For example, where solid film lubricants are burnished onto a surface, it is frequently advised that the surface be grit-blasted or bead-blasted prior to burnishing

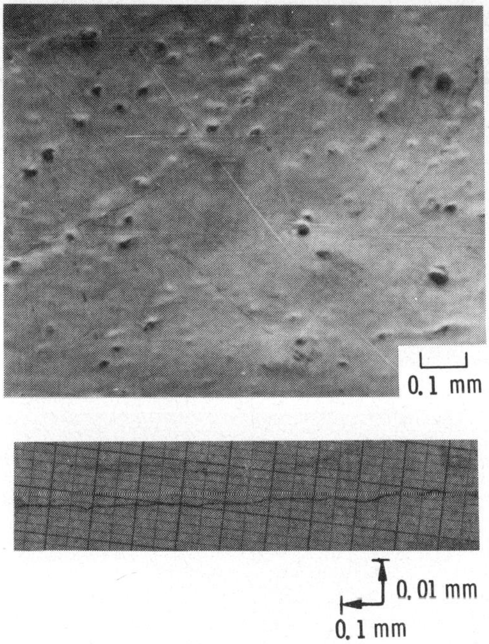

Figure 4-2. – Photomicrograph and surface profile of electropolished aluminum surface.

the molybdenum disulfide so that pockets develop on the solid surface to act as reservoirs to retain and hold a solid film lubricant; the solid lubricant, therefore, becomes available at the surface during the rubbing process and provides a continuous reservoir of lubricating material. Thus, very frequently in the use of solid film lubricants a rough rather than a highly polished surface is used. A typical grit-blasted aluminum surface that might be employed for such burnishing operations is shown in figure 4-3. The photomicrograph in figure 4-3 reveals readily the hills and valleys associated with grit-blasting the surface. The pockets in the surface are a function of the size of the grit that is being used to bombard the surface; the larger the grit, the larger the valleys generated.

Beneath the photomicrograph in figure 4-3 is the profile trace of the grit-blasted surface shown in the photomicrograph (fig. 4-3). The surface is extremely rough as indicated by the surface profile trace. When this trace is compared with the surface profiles of figures 4-1 and 4-2, it is apparent what grit-blasting can do to roughen a solid surface.

Probably the smoothest surface that the tribologist encounters in practical lubrication systems is the cleaved surface, which is the smoothest atomically. Generally, however, the cleaved surface is only seen in inorganic crystals and materials, such as mica, that can be cleaved along certain crystallographic planes. Although it does not normally exist in metals, some

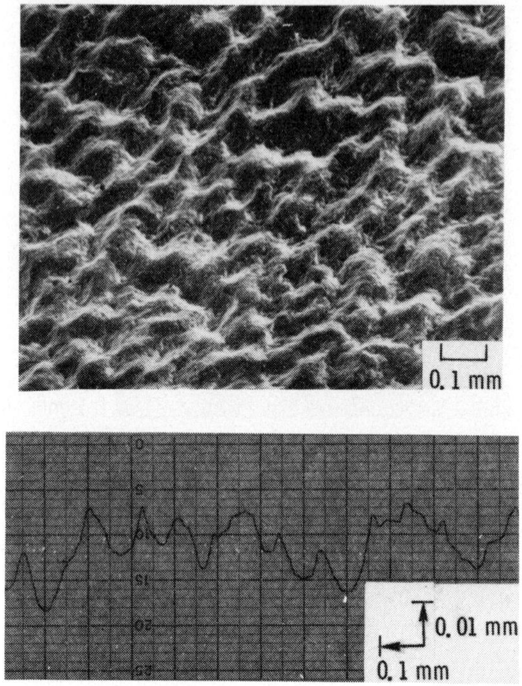

Figure 4-3. – Photomicrograph and surface profile of grit blasted aluminum surface (25 m grit).

metals, such as the hexagonal-close-packed ones, can be cleaved at cryogenic temperatures along their natural cleavage planes. If, however, one cleaves crystalline quartz, a surface profile such as that shown in figure 4-4(a) is obtained. The surface profile is extremely smooth, and it can be compared to a 1-micrometer standard surface roughness (fig. 4-4(d)). This quartz surface is very smooth and contains very few defects or undulations. Figure 4-4(b) presents a cleaved mica surface. Just as was observed with the cleaved quartz, a mica surface is extremely smooth when a surface profilometer is run across it. The irregularities or defects seen in the metal surfaces prepared by other techniques (figs. 4-1 to 4-3) are not seen in these surfaces.

An experimentalist may be interested in maintaining a very smooth surface and still desire a relatively active surface (say a metal) so that he can study lubricant interactions with a metal in the clean state. This result can be achieved by depositing (vapor deposition, sputtering, or ion plating) onto a cleaved inorganic surface (either quartz or mica) a thin film of the metal of interest. That was done for the surface of figure 4-4(c) where a thin film of iron was sputter deposited on a quartz substrate. From an examination of figure 4-4(c), it can be seen that the iron takes on the surface topography of the quartz.

(*a*) *Quartz.*

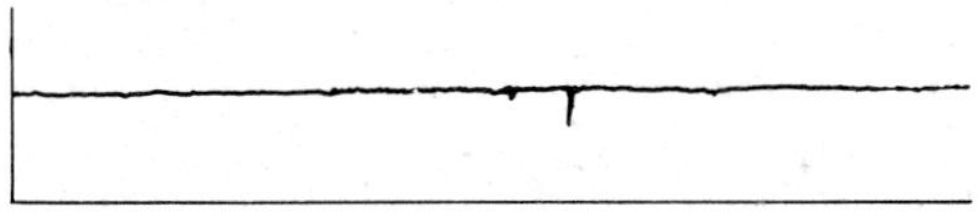

(*b*) *Mica.*

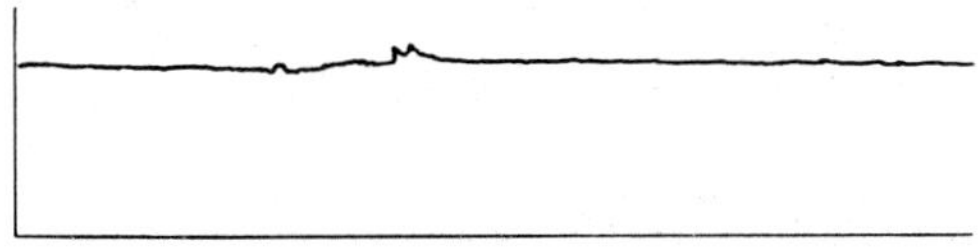

(*c*) *Iron on quartz.*

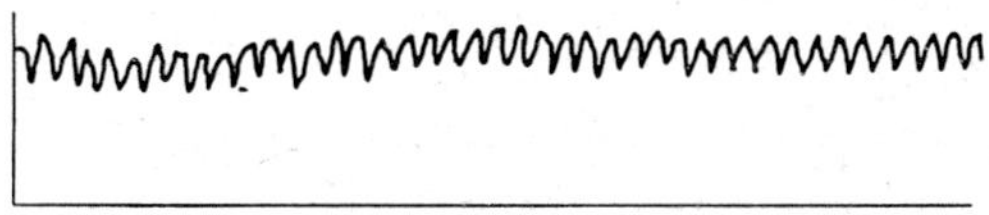

(*d*) *1.0-Micrometer standard steel surface roughness.*

Figure 4-4. – Surface profiles of various materials used in sliding friction studies.

Since the deposition was done in a vacuum environment, the surface layers in figure 4-4(c) consist of clean iron, which can then be studied for its interactions with lubricants. This type of a film preparation is extremely helpful when an analytical tool such as ellipsometry is used to study lubricant films on a metal surface. In ellipsiometry it is very desirable to have a relatively smooth surface such as that obtained by cleavage of inorganic crystals or semiconductor materials.

The most common technique for obtaining the surface profiles presented in figures 4-1 to 4-4 is the surface profilometer (discussed in ch. 2). Because the diamond stylus has a radius at its end of finite size, the tracing process is not an exact one and the true surface profile can not really be obtained. First, there is the limitation associated with the radius of the stylus that prevents an accurate profiling of the solid and its surface. Second, the nature of the instrumentation tends to distort the real nature or image of the solid surface because the vertical magnification is greater than the horizontal magnification.

Figure 4-5 is a profile modification that develops as a result of the stylus radius. The sharp peaks present on the real surface are traced over by the stylus. The radius of the stylus does not permit it to follow the exact contour of the solid surface. The dotted line in figure 4-5 reflects the surface profile obtained by the stylus as opposed to the true surface profile. Thus, the experimentalist who is working with surface profile techniques must

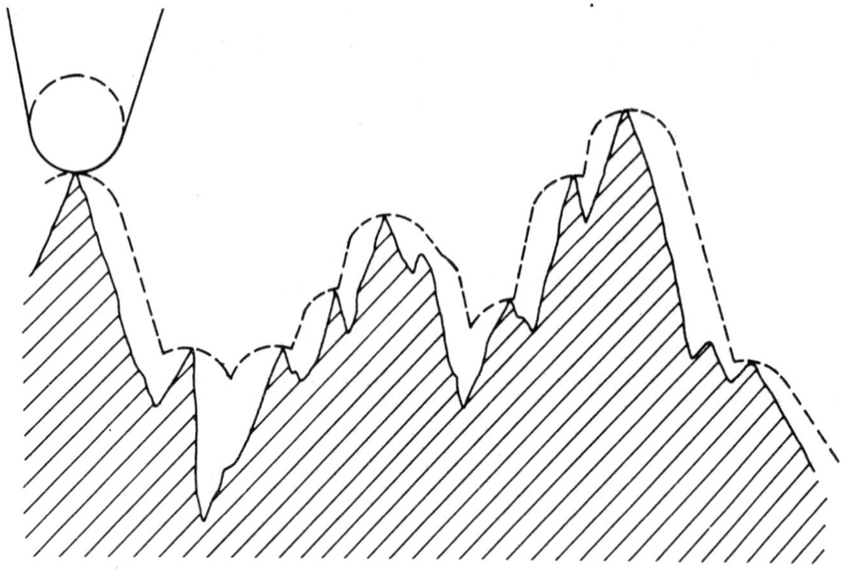

Figure 4-5.—Profile modification due to stylus radius.

remember this inherent limitation when using the physical device to determine a surface profile. With an extremely rough surface, the distortion resulting from the stylus radius is much greater.

Rough and Smooth Surfaces

The tracing on the rough surface is much more difficult to follow than the one on the smooth surface. As a consequence, the difference in the real topography from the traced topography would be greater for the rough surface.

A considerable amount of research effort has gone into studying the true topography of solid surfaces. Of all the investigators studying surface topography, Williamson has probably done more than any other investigator to define the true nature of the tribological surface (ref. 1). During the course of his studies, he has, with the aid of the computer, tried to determine from surface profile traces the actual nature of a solid surface by removing the inherent errors and defects incorporated in surface measuring techniques. For example, he has shown that, while real surfaces from a surface profilometer point of view appear to be very rough and jagged, when a detailed computer analysis is made of that surface the surface is not at all as it appears to be from the surface profile trace. Typically, asperity angles are no more than 15° as opposed to the very sharp angles heretofore thought to exist.

Figure 4-6 presents two profile traces. Figure 4-6(a) is a profile obtained on a surface with a surface profilometer. On surface profilometers the ver-

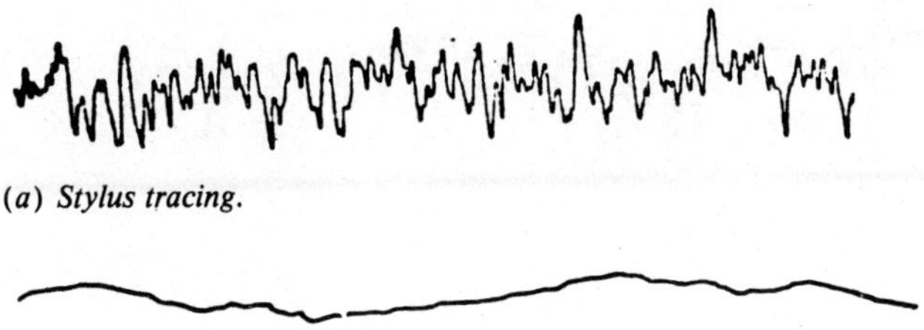

(*a*) *Stylus tracing.*

(*b*) *Actual topography.*

Figure 4-6. – Surface profile from stylus tracing and actual surface topography (*ref. 2*).

tical magnification is generally much greater than the horizontal magnification. As a consequence, the surface profile trace is distorted in the vertical direction. The amplitudes of the irregularities in the vertical direction are much greater than those in the horizontal direction. Figure 4-6(b) is the actual surface when all the errors associated with the measuring device or technique are eliminated from the surface profilometer tool and the computer is used in assisting to identify the nature of the actual solid surface. Figure 4-6(b), then, represents the true surface topography (ref. 2).

The techniques used by Williamson are very effective in resolving the complexities of the surface profile and giving a characteristic picture of the real solid surface. Unfortunately for the average tribologist in the laboratory, it is difficult to incorporate the techniques employed by Williamson. However, the surface profilometer is a very useful tool for identifying and comparing solid surfaces. The device can be used to compare one surface with another and to show the differences that have taken place in the surface as a result of, for example, the process of wear. In the sliding, rolling, or rubbing process, the nature of the topography of a solid surface can change markedly from that prior to the initiation of relative motion between two solid surfaces.

Steel is probably one of the most commonly used tribological materials. In bearings, for example, a steel surface may be in the ground state. An examination of the ground surface with the Talysurf, a surface profilometer instrument, reveals a trace such as that shown in figure 4-7(a). Note that the magnification in figure 4-7(a) is 5000 in the vertical direction and 25 in the horizontal direction. When rolling or sliding contact occurs between two solid surfaces over a period of time, the surfaces wear in the adhesive mode and the topography of the surfaces change from that characteristic of the ground surface shown in figure 4-7(a) to one similar to that shown in figure 4-7(b). (Note that the vertical magnification has changed from 5000 to 250.)

In figure 4-7(b) the surface becomes highly irregular with a back and forth transfer from one surface to another in the adhesive wear process. Since the vertical magnification has been changed from 5000 to only 250, an interpretation of figure 4-7(b) shows the surface profile is extremely rough where adhesive wear has taken place. In contrast, when the surfaces are rubbed in such a manner as to produce abrasive wear of the surfaces on a

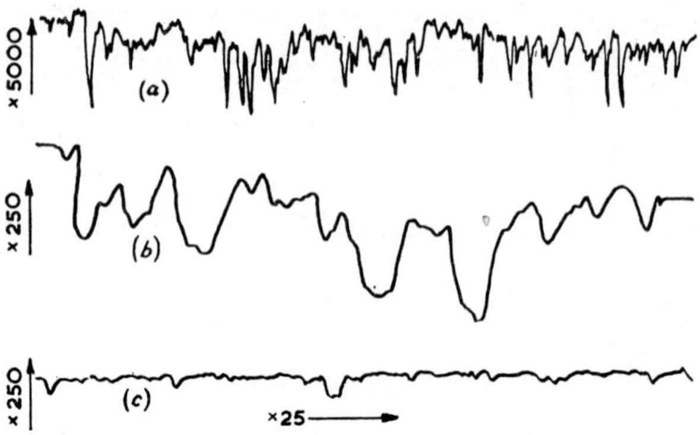

(*a*) *Normal ground surface.*
(*b*) *Normal seized surface.*
(*c*) *Slowly worn surface.*

Figure 4-7. – Talysurf traces of steel surfaces (ref. 3).

fine scale, a polishing action really occurs and the surface profile is somewhat like that seen in figure 4-7(c). The surface is relatively smooth compared to that shown in figure 4-7(b) where adhesive wear takes place. Typically, iron oxide (Fe_2O_3) is present on the solid surface of steel; Fe_2O_3 is essentially jewelers rouge and is a very fine abrasive. In dry sliding, Fe_2O_3 can produce an abrasive wear action. This is frequently observed in oxygen-rich environments also. The abrasive action of the FeO_3 removes the asperities so that the surface becomes fairly flat and regular (fig. 4-7(c)).

The data in figure 4-7 are extremely important for they show that surface topography in a dynamic sliding or rolling system is continuously changing (ref. 3). Thus, when a surface profile is measured at a certain time on the solid surface, it does not necessarily mean that that particular profile exists a few minutes later when sliding, rolling, or rubbing is again initiated. The surface topography is constantly changing in a dynamic system where two solid surfaces are in sliding, rubbing, or rolling contact under an imposed load.

Not only does the topography of the surface change continuously with the rubbing process, but the mode of wear can markedly alter the topography observed. Adhesive wear can produce a very rough surface topography such as that seen in figure 4-7(b); very smooth wear, which is associated with either abrasive or corrosive wear where the surface is actually polished, can give a surface such as that shown in figure 4-7(c). The corrosion process can also produce a smoother surface due to preferential chemical attack at the tips of the surface asperities; this chemical attack results in polishing. Frequently this is observed with chemical additives in oils, such as extreme pressure additives or antiwear additives. They initially react at the asperity tips to form protective surface films. The tips are gradually worn away with continued solid-state contact until the entire surface becomes fairly smooth as a result of the chemical action at these locations.

While the condition of the various surfaces prior to sliding, rolling, or rubbing contact between two solid surfaces may be markedly different, the rubbing process alone can produce marked changes in solid surface topographies and can generate a new surface that reflects the nature of the tribological process rather than the initial differences. For example, in figure 4-8 three surface conditions were examined initially: a brushed unworn surface, a ground unworn surface, and lapped unworn surfaces (ref. 4). These surface profiles are presented in the top, third, and fifth surface profile traces in figure 4-8 (reading top to bottom). Intermediate between these profiles are the profiles obtained for the worn surface on the three initially existing surface conditions. For the brushed unworn surface, after wearing, the tips of the asperities have been worn down and flats have been generated on the solid surface. The pockets, pits, or valleys in the surface still remain. The hills have essentially been worn down.

This wearing down of the hills reflects an abrasive or corrosive wear process as opposed to adhesion, which does not normally occur in this manner but rather as shown in figure 4-7(b). The flattening of the surface of the hills is reflective of a gradual abrasive or corrosive process. With the ground unworn surface, the topography of the worn surface after some sliding is not too different from that surface prior to running. A comparison between the lapped unworn surface and the worn surface shows even less difference in

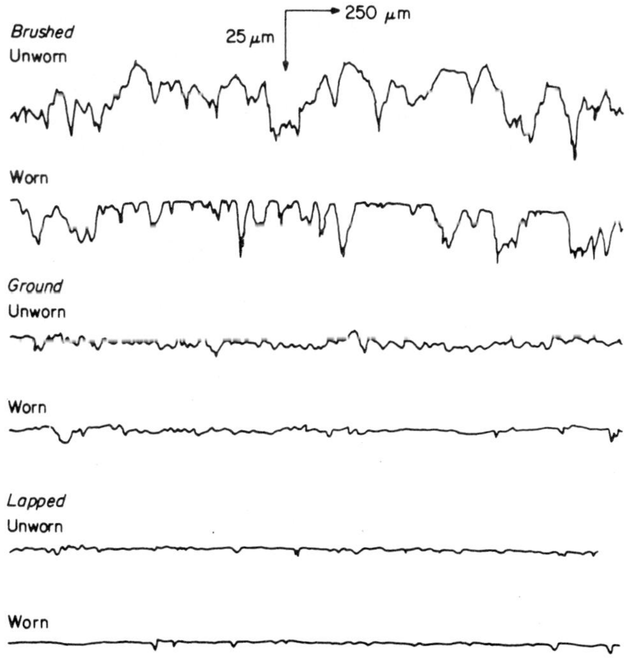

Figure 4-8. – Friction probe surfaces (ref. 4).

topography as a function of the wear process where the wear processes of all three systems are essentially the same. Thus, the initial surface topography has an influence on the final topography after some rubbing, sliding, or rolling contact has occurred across the surface. With a very rough surface, such as the brushed unworn surface, the tips of the apserities are worn away but valleys or pits are left in the surface. For the ground surface, very little difference between the unworn and the worn surfaces is observed; even less difference is observed for the lapped surface after wearing.

If the wear process were to continue for the brushed unworn surface, the wear process would eventually progress to the point where the valleys would be worn away and the surfaces for all three initial starting conditions would be approximately the same; this result reflects the nature of the actual wear process in generating its own surface. It is simply a function of the time, speed, and load in the mechanical system that produces the actual wear that is necessary to bring about a likeness of the surfaces when the initial surface topographies vary. This particular concept is extremely important because very frequently a designer asks what the most desirable initial surface topography of a machine element is. The answer obviously is that while the initial surface topography may vary from system to system after some period of wear of a piece of machinery the surface topography is that generated by the wear process. In other words, the machine element generates its own surface topography despite differences in the initial topography.

Adhesive wear is probably the most devastating type of wear encountered. It causes the greatest amount of disruption of solid surfaces from a surface profile point of view primarily because the wear can take place in a back and forth transfer mode; that is, material can transfer from one surface to another and then back again. This frequently results in the surface being extremely rough; material is plucked out of the surface as well as being transferred back to it, resulting in a profile that looks something like that shown in figure 4-9. For lack of a better term, positive and negative wear volumes are used to reflect the form of wear to the surface (ref. 5). The wear track width is indicated in figure 4-9 by W and the maximum amount

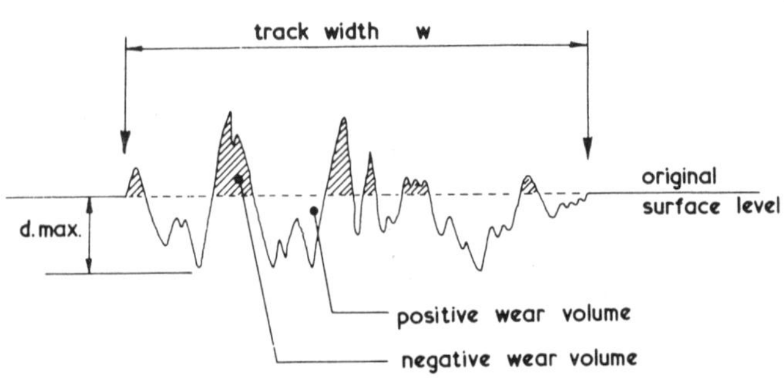

Figure 4-9. – Schematic drawing of wear scar. Maximum pit depth, d max.; accumulative track width, w; vertical scale magnified (ref. 5).

of wear occurring surface or subsurface by d. The original surface level is shown.

The positive wear volume reflects that amount of material that has been lost from the surface as a result of the adhesive plucking of material from the surface. It leaves pits, voids, cavities, or valleys beneath the original surface level. In addition to this, however, material can be transferred from the contacting surface back to the surface shown in figure 4-9. When that occurs, material builds up above the original surface level because of adhesive transfer as indicated by the shaded areas above the original surface level (fig. 4-9). This is referred to as negative wear material; it is the material that is transferred from an opposite surface. Thus, for example, if a copper surface were sliding on a steel surface, the shaded areas (if the original surface of fig. 4-9 were a steel surface) would reflect transfer of copper to the steel surface. This would represent negative wear to the steel but positive wear to the copper. If copper picked up steel from the steel surface and left voids, the surface would reflect a cavity of steel below the original surface level.

The data of figures 4-1 to 4-9 show that initial surface topographies can vary in practical systems and that these surface topographies change markedly in the sliding, rolling, or rubbing process associated with practical machinery. Also, a topography which may be initially present on the solid surface is changed very rapidly once solid-state contact occurs between two solids. A new topography is generated that is continuously changing with the sliding, rolling, or rubbing process. Depending on the wear mechanism, an initially smooth surface may become rougher if the wear mechanism is one associated with adhesive wear, or a very rough surface may become smoother if the wear process is one of abrasion or corrosion.

In addition to the normal wear processes which can bring about changes in surface topography, there are other mechanisms which tend to alter the profile of solid surfaces. These include massive adhesive transfer or welding of surfaces when clean surfaces are in sliding contact in an unlubricated state. Under such conditions, adhesion may occur on two different scales—minute or massive. When adhesion occurs on a minute scale, the transfer of material is as was observed in the surface profile traces of the earlier figures. In addition thereto, in the unlubricated state when the loads are sufficiently high and/or when the experiments are conducted in a vacuum environment where surface oxides cannot reform, massive adhesion that may occur at the interface between two solid surfaces can result in the plucking out of large amounts of material from one or another of the surfaces. This phenomena is very frequently observed under severe operating conditions (vacuum, inert, and reducing environments with metals in sliding, rolling, or rubbing contact in the absence of lubrication or when a lubrication system undergoes failure).

The adhesion process is initiated by either the wearing away of residual surface oxides so that metal to metal contact can occur in the metal to metal system or by the rupture of the surface oxides due to plastic deformation of the surfaces in solid-state contact under load. For example, an aluminum surface contains a thin film of aluminum oxide from 100 to 1000 angstroms thick depending on the condition of the aluminum. This oxide is very hard and brittle. The aluminum beneath the oxide is, however, relatively ductile, and should a specimen of another material (e.g., a steel ball) be pressed against the aluminum surface, the elastic limit of the aluminum is exceeded as load is increased and the aluminum undergoes plastic deformation. When it does this, if the aluminum and steel ball are enclosed in a non-oxidizing environment (vacuum, an inert atmosphere, or a reducing atmosphere), the relatively brittle aluminum surface oxide is broken up much like a thin film of ice on a lake. When the aluminum oxide film is broken up, it exposes the nascent metallic aluminum beneath the oxide. This nascent aluminum is very reactive and interacts with the steel surface to form strong adhesive bonds. When the steel ball is then removed from the aluminum surface or is moved tangentially across it, aluminum would be plucked out of the surface and transfered to the steel ball. An example of such a transfer is seen in figure 4-10. The adhesive particle is a relatively large, irregularly shaped particle which has adhered to the surface of the second body and projects a considerable distance above the solid surface.

The changes in surface topography by particles transferred from one surface to another, as indicated in figure 4-10, are extremely destructive to practical tribological systems. For example, if a close tolerance bearing would have adhesive transfer occurring such as that depicted in figure 4-10, even a single particle of such a transfer could be sufficient to destroy the operating clearances and cause premature failure of the bearing.

Development of Transfer Films

While adhesive transfer can take the form shown in figure 4-10 (where a large particle or piece of material is transferred from one surface to another), it can also take other forms which are not entirely detrimental to a tribological surface but may, in fact, be beneficial. One such transfer is the transfer of carbon to mating metal surfaces when carbon is in sliding contact with metals. Carbon is used in sliding contact with metals in mechanical

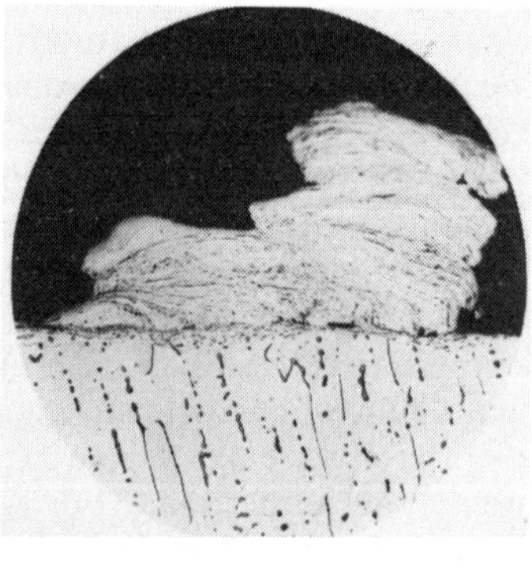

Figure 4-10. – Severe surface welding resulting from unlubricated sliding.

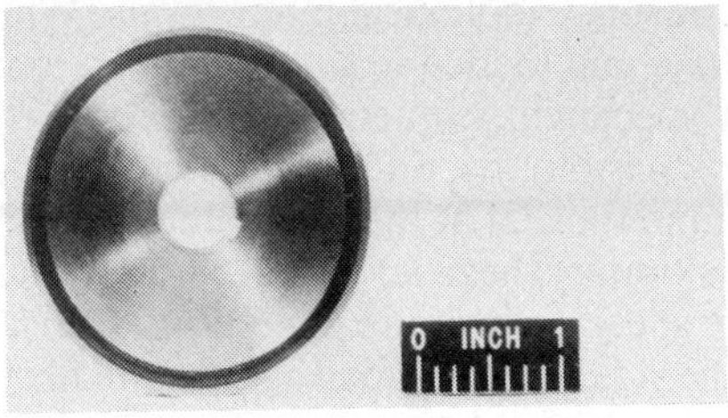

Figure 4-11. – Transfer film formed during sliding contact with carbon in dry air.

dynamic seals where one is sealing either a fluid or a gas. A ring-shaped seal slides against a metal surface and transfer of the carbon to the metal surface takes place. This type of transfer is shown in figure 4-11 where the carbon was slid against the metal disk surface in an air environment. When normal metal oxides are present on the metal surface, a thin film of carbon transfers to the metal surface. The transferred carbon film is very thin and essentially follows the surface topography of the metal. The valleys between asperities are filled by the carbon that is transferred to the metal surface. Since the film, in general, is fairly thin and uniform over the entire surface, after some period of sliding, carbon is essentially rubbing on carbon. Interestingly enough, the wear behavior of the carbon, as will be discussed in chapter 8, is extremely sensitive to the presence of the carbon transfer film. If the carbon transfer film does not develop, carbon wears at a very high rate. If the carbon transfer film does develop, the wear of carbon is appreciably less. Again, environment is extremely important in the formation of the carbon transfer film (fig. 4-11).

If the experiment of figure 4-11 were conducted in a vacuum environment and the metal surface were atomically clean prior to the carbon being brought into contact with the metal, a carbon transfer film would not develop. In fact, metal would be found in some cases transferred to the carbon surface; for example, with carbon sliding against copper in a vacuum environment, copper transfers to the carbon surface with no visible evidence of a carbon transfer film developing on the copper. This system, therefore, becomes one of copper sliding on copper as a result of the adhesion of the copper to the carbon; the friction coefficient becomes very high, and the surface of the copper is very rapidly disrupted by the adhesion (now) of copper to copper with radical transformation in the surface topography of the copper surface. If, however, the copper surface oxidizes before it is placed in a vacuum system, and the carbon slides against the copper oxide, a transfer film of carbon to the copper is observed just as is observed in figure 4-11 for the carbon sliding on the oxidized metal surface in dry air.

Surface and Subsurface Stresses

In addition to the adhesive transfer to solid surfaces that can alter the surface topography in a friction, wear, or lubricated system, the surface can undergo other changes during tribological interactions. When two solid surfaces are placed in contact and loaded (e.g., the conventionally used pin-on disk or rider on a flat specimen geometry frequently used by experimentalists in the field of adhesion, friction, wear, and lubrication), various stresses develop in the materials. If at the same time tangential motion is introduced, a zone of maximum shear stress develops subsurface. This particular region where the shear stresses are maximized (subsurface in the flat material as a result of the rider being loaded against the flat) is referred to as the zone of maximum subsurface shear stress. This was recognized by early rolling element bearing researchers, and Rabinowicz used a schematic (fig. 4-12, ref. 6) to show it. The figure depicts the point subsurface in the flat where the zone of maximum subsurface shear stress is located.

While a rider is sliding on the flat surface, in addition to the zone of maximum subsurface shear stress, a zone of maximum tensile stress develops behind the actual point of contact. As the rider moves tangentially in the direction indicated in figure 4-12, behind it a wake of tensile stress develops in the flat which can produce surface cracks particularly in brittle materials (such as glass). In addition to the cracks that can develop in brittle materials at the zone of maximum tensile strength, cracks can also develop near the zone of maximum subsurface shear stress. Fracture cracks in a surface such as glass in the compressive or tensile stress zones are shown in figure 4-13. Figure 4-13(a) reveals the presence of fracture cracks in the glass surface in the compressive zone. The crack is located in the interface between the rider and the glass disk surface. The fracture cracks (curved in shape) that are found behind the actual contact zone are the tensile fracture cracks that develop behind the contact zone between rider and glass disk. Figure 4-13(b) reveals the presence of these curve cracks.

Figure 4-13(c) reveals the surface stress components that developed in the sliding direction in a glass surface. The stresses are both compressive and tensile relative to the Hertzian radius r. It is apparent that if a surface profile trace were obtained parallel to the wear track on this surface the surface topography would reflect the presence of the cracks in the tribological surface. The profusion and concentration of cracks developed in a surface

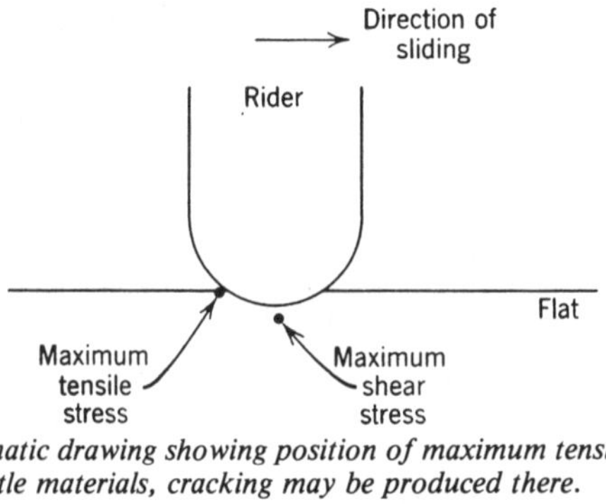

Figure 4-12. – Schematic drawing showing position of maximum tensile stress behind region of contact. In brittle materials, cracking may be produced there.

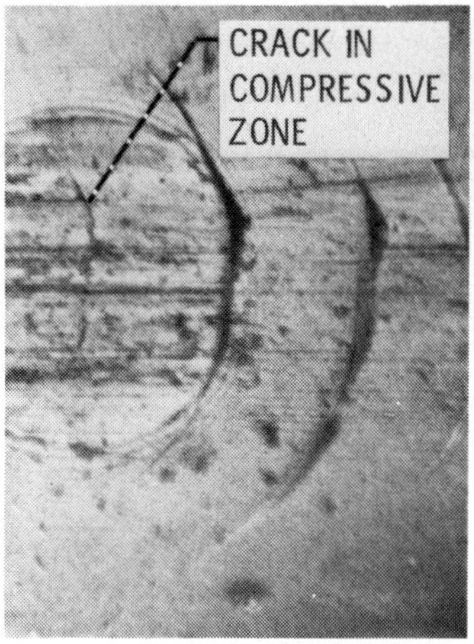

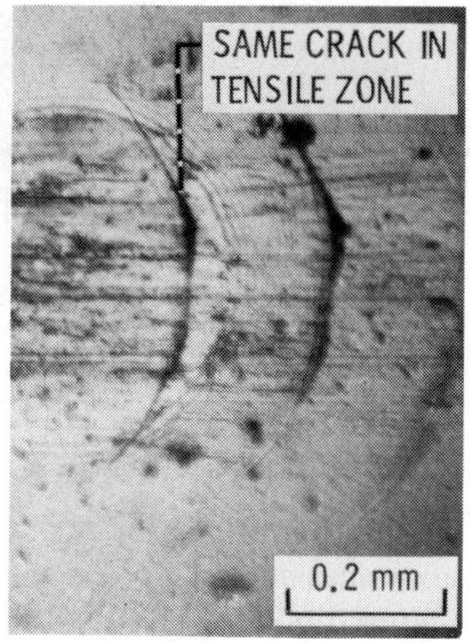

(*a*) *Crack in compression.* (*b*) *Crack in tension.*

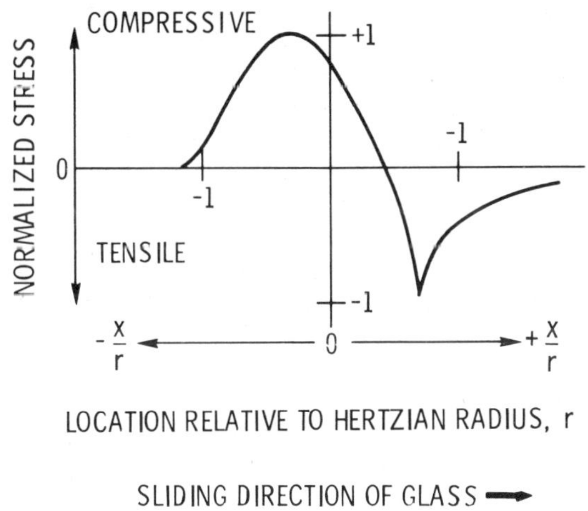

(*c*) *Surface stress component in sliding direction.*

Figure 4-13.—Crack formation and propagation in unlubricated contact. Load, 13.2 newtons (3 lb); original magnification, 150.

(particularly of a brittle material) can be very great and have a marked effect on the topography of a solid surface.

In figure 4-14 are two photomicrographs for a silicon (111) single crystal surface after sliding an iron single crystal across that solid surface. The adhesion of the iron to the silicon produces strong bonding at the interface. When tangential motion is initiated, the tensile force to the rear of the contact of the iron with the silicon produces sufficient force to produce the fracture cracks observed. Since the iron bonds fairly strongly to the silicon,

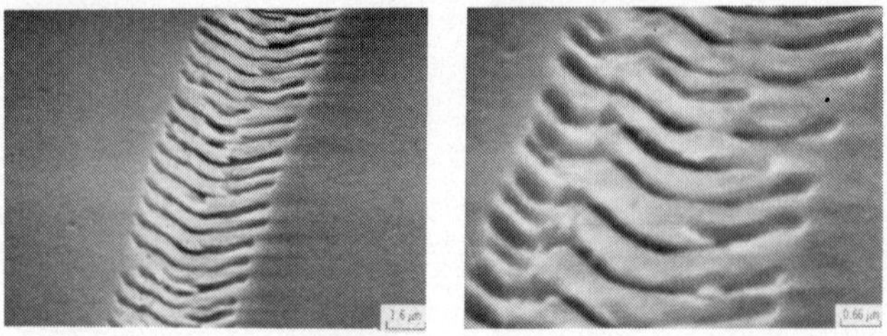

Figure 4-14. – Wear track made by single-crystal iron (110) sliding across silicon (111) surface. Sliding speed, 0.7 millimeter per minute; temperature, 23° C; pressure, 10^{-8} newton per square meter.

repeated adhesion and cracking occurs over the entire length of the contact zone during sliding, reflecting the high concentration of cracks present in the photomicrograph of figure 4-14. The development of cracks in the surface of a solid as a result of solid-state contact is frequently a precursor to the generation of wear particles. Once a crack forms in the surface, it is relatively easy for material to fracture out of the surface and leave it. A defect is already present for initiating the removal of a particle from the solid surface when the crack is formed.

In single crystal materials, these cracks have been observed to form along slip bands in the material (i.e., along the preferred cleavage or slip planes). For example, in copper sliding across a copper single crystal surface, cracks in the surface have been observed along (111) slip bands. Since surface initiated cracks may run subsurface in the material, wear particles are formed from these cracks. Another example of the liberation or the formation of wear particles from crack formation is demonstrated with the photomicrographs of figure 4-15 which shows wear tracks on an aluminum single crystal surface after having been in sliding contact with polytetrafluoroethylene (PTFE). The interesting observation that can be made from figure 4-15 is that the crack develops in the aluminum surface along a (111) plane as a result of a soft polymer (PTFE) sliding against the metal surface. The aluminum is in a clean state. With sliding or tangential motion, the crack is initiated as it is in brittle materials. Because the aluminum is a crystalline material, instead of the crack having a curved shape as was observed in amorphous glass in figure 4-13, the crack forms an irregular straight line reflecting the (111) orientation of the crystal at the surface. In these experiments, adhesion of the PTFE to the aluminum occurs and, with tangential motion, fracture cracks develop in the aluminum. They occur repeatedly across the surface as indicated in the low magnification photomicrograph in the upper part of figure 4-15.

An increased magnification of one of the cracks is shown in the lower photomicrograph of figure 4-15. The sharp edge of the crack is readily apparent. The interesting observation to be made from this photomicrograph is that, in addition to the crack, a piece of wedge-shaped material has been literally scooped out of the aluminum surface leaving a pit in the surface.

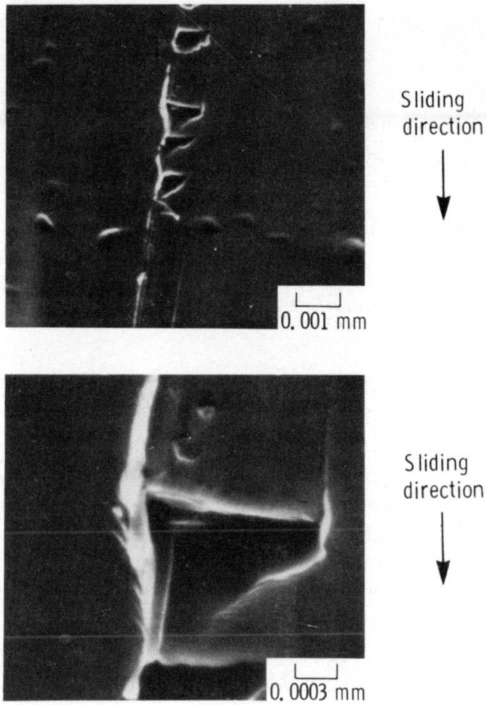

Figure 4-15. – Wear track on aluminum single crystal surface. Slider, PTFE; load, 200 grams; temperature, 23° C; single pass.

This scanning electron photomicrograph shows the depth of the pit and indicates that the surfaces are extremely smooth. Even the edge of the crack is extremely smooth as is the base of the crack. Where the material has been removed, the ledge left behind is extremely smooth in topography. Evidence for plastic deformation exists at the edges of the wedge away from that orientation where fracture occurred. The edge of the pit which runs parallel to the wear track shows evidence of plastic deformation as does the edge at an angle of approximately 45° to the pit.

Fatigue.—The photomicrographs of figures 4-13 to 4-15 that reveal the formation of cracks in surfaces are obtained with materials in sliding contact. Cracks can also develop in materials with rolling contact. The cracks develop basically in two forms as was observed by early rolling element bearing researchers and depicted by Rabinowicz in his text on friction and wear of materials (ref. 6). Figure 4-16 presents a schematic from his reference work on the possible mechanisms for the formation of cracks as a result of fatigue failures in rolling contact; Rabinowicz depicts a ball rolling across the surface with the generation of a surface crack as a result of repeated stress cycles across the surface (fig. 4-16(a)). In addition to the sur-

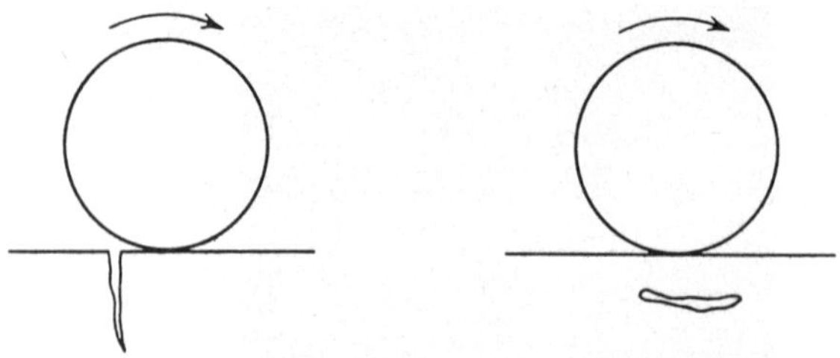

(*a*) *Surface crack.* (*b*) *Subsurface crack.*

Figure 4-16. – Typical surface fatigue failures in early stages (ref. 6).

face initiated crack, he also indicates the formation of subsurface cracks in the zone of maximum subsurface shear stress (fig. 4-16(b)).

The formation of surface cracks alters surface topography immediately. The subsurface cracks obviously are not reflected in the surface topography of a material until such time as the crack propagates and moves to the surface, as it does in many practical tribological systems such as ball and roller bearings. At that time, a wear particle is liberated from the solid surface and leaves a pit in the surface that is reflected in the topography of that surface.

MacPherson and Cameron in their studies of fatigue have shown the development of these surface cracks; one such crack is depicted in figure 4-17 (ref. 7). Figure 4-17(a) is a scanning electron micrograph of the surface in the actual contact zone showing a surface initiated microcrack. It is obvious from the crack that, were one to examine the topography with the surface profile device, this surface would be very irregular in the region of the crack. Figure 4-17(b), which is a cross section of the surface initiated crack, shows the crack moving to some depth. It can be seen from figure 4-17(b) that a crack in the surface produces a discontinuity in the surface and a defect in the tribological surface not present prior to the tribological interaction of two solid surfaces.

Plastic deformation.—The surface topography of tribological surfaces can be altered not only by adhesion and adhesive transfer wear particles from fracture cracking but also by ductile machanisms. The sliding, rolling, or rubbing contact of two materials in the solid state can produce a change in surface topography for ductile materials. With ductile materials (unlike brittle materials), when two surfaces are placed in contact and the load exceeds the elastic limit of one of the two materials, plastic deformation occurs. The material flows in a plastic manner and the surface topography is markedly changed. This frequently is observed when the two solid surfaces placed in contact have markedly different mechanical properties (e.g., when a steel ball is placed against a copper surface). The mechanical properties of the copper (its hardness, elastic limit, etc.) are much less than those of the steel. As a load is applied to the steel ball, the copper deforms plastically when the elastic limit of the copper has been exceeded and an indentation is left in the copper.

(*a*) *Surface.*

(*b*) *Subsurface.*

Figure 4-17.—Fatigue scoring with surface initiated crack formation (ref. 7).

This deformation is even more severe where the harder surface of the two surfaces in contact has other than a spherical shape. For example, in cutting and grinding operations, the basic mechanism involves the use of an abrasive grit material with sharp cutting edges; these sharp edges rake out or remove material from a softer workpiece because of the hard cutting edge of the abrasive. Even a single pass of one of these grits across the surface produces a marked change in surface topography results. A number of investigators through the years have examined the plastic deformation of ductile materials in the contact of relatively hard materials against softer materials. Cocks (ref. 8) has studied the process as well as Tsukizoe in Japan (ref. 9). Courtell and his colleagues in France have done a very detailed analysis of the topography of the softer surface as a result of single interactions with a hard, high strength material (ref. 10). More recently, Sakamoto and Tsukizoe have done similar experiments with steel in contact with copper (ref. 11).

Figure 4-18 shows the displacement of copper from a copper disk surface by a mild steel slider with a 100° cone end (ref. 11). This would be very analogous to a single grit in an abrasive grit material. A furrow is generated in the copper surface as a result of the mild steel having made a single pass along the surface. The large furrow has material buildup on either side of the furrow. At the end where the cone stops sliding against the copper there is a buildup of material (figs. 4-19(a) and (b)). Figure 4-19 reveals the buildup of material in front of the cone on the copper surface where sliding has stopped. The furrows generated in the soft copper are analogous to furrows generated while plowing a field. Material is moved to either side of the furrow and ahead of the sliding cone just as soil is moved with the plow. Hence, the term plowing is applied to this particular process where very hard material plastically deforms a softer material.

Simultaneous with the plowing of the surface and the disruption of the surface topography by the harder, higher strength material sliding against the softer, lower strength material, other processes take place which can

Figure 4-18. – Displacement of copper from copper disk by mild steel slider with 100° cone end (ref. 11).

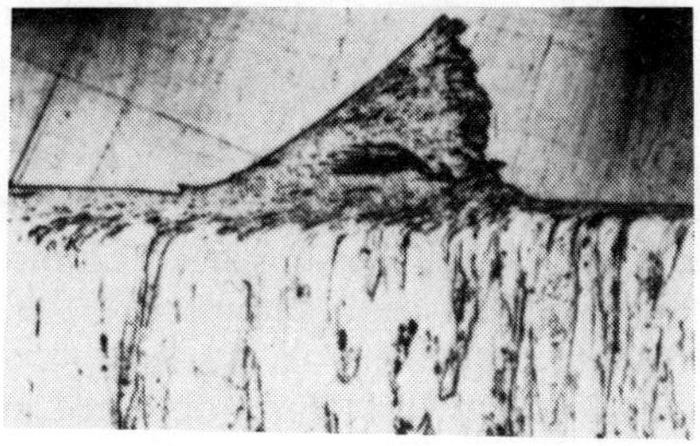

(*a*) *Sliding distance, 1.8 millimeters.*

(*b*) *Sliding distance, 4.8 millimeters.*

Figure 4-19.—Etched section of frontal bulge generated by mild steel 100° cone end rod sliding on copper (ref. 11).

further alter the surface topography. For example, in figure 4-18 the side of the furrow farthest from the legend indicates the presence of a crack in the material that has been swept to the side of the furrow. It also indicates, closer to the center of the photograph, removal of wedge-shaped material from the wall that has been plastically moved to the side of the furrow. In the photograph the crack indicates the initiation and the formation of a particle of debris that can be liberated from the solid surface with repeated passes. The location nearer the center of the photomicrograph, wedge-shaped area, reveals a place where a particle has already been removed from the surface as a result of the single pass sliding of the steel slider against the copper surface.

The fractures in the copper surface are ductile in nature as opposed to the brittle nature of the fractures that were observed earlier. The evidence for the ductile fracture is seen in the wedge-shaped area that reflects the removal of a piece of material along the edge of the track near the center of figure 4-18, as already discussed. The edges are very jagged or irregular and indicate a nonuniform tearing out of the wear particle from the surface. If the fracture were brittle in nature, one might anticipate a relatively smooth wall along the edges of the wedge-shaped area. Repeated passes over the same surface produce work hardening, in work hardenable materials such as metals, so that with repeated passes the depth of the furrow is altered by the prior condition of the surface. Topography, however, continues to change.

The plastic deformation of one surface when two surfaces are in solid-state contact can occur in the presence or absence of lubricants. In fact, in some instances, the presence of lubricants can increase the deformability of the solid surfaces by such mechanisms as the Rehbinder effect. Plastic deformation of the solid surface is, therefore, observed in the presence of lubricants. Those properties of solids which influence their mechanical behavior also influence the degree to which the surface topography is altered by sliding, rolling, or rubbing contact. For example, the presence of alloy constituents in metals can alter mechanical behavior, and this, in turn, alters the mechanism and the amount of plastic deformation that may occur on a solid surface. With metals, the presence of an alloying element in a simple binary system can create a situation where the alloy deforms to a lesser degree than does either of the elemental metals that go into making up the binary alloy. This effect is demonstrated in the surface profile traces of figure 4-20.

The surface profile traces in figure 4-20 are for iron-chromium alloys (ref. 12); included also, are surface profile traces for elemental iron and chromium. In the experiments conducted in mineral oil (fig. 4-20), a single crystal grit of silicon carbide slid across the surfaces of the elemental iron, elemental chromium, and the five iron-chromium alloys. The mineral oil was used to minimize adhesion effects of the metals to the silicon carbide and to maximize deformation effects (i.e., the effect of the alloying on deformation behavior). The surface profile traces show that the 9 and 14 weight percent chromium in iron alloys produce less surface deformation than either the elemental iron or elemental chromium surfaces. Hence, some alloying can be beneficial in reducing the amount of surface deformation.

The data of figure 4-20 are for two elemental metals alloyed in a simple binary system. The same effect can be produced to a marked degree with a carbon-iron binary system. The concentration of carbon necessary in iron to produce a marked change in surface deformation is considerably less than is necessary for metallic alloying elements such as chromium.

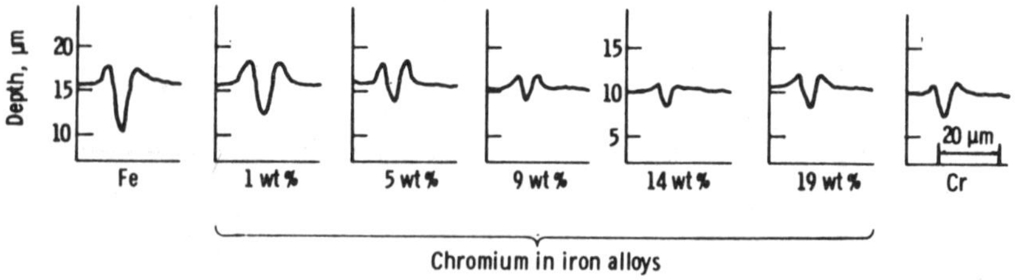

Figure 4-20. – Grooves on iron-chromium alloys, iron, and chromium. Single-pass sliding of 0.025-millimeter-radius silicon carbide rider in mineral oil; sliding velocity, 3 millimeters per minute; load, 0.1 newton (10 g); room temperature (ref. 12).

Thus far only the two-body interaction problem relative to the generation of the tribological surfaces has been discussed. There are, however, situations where a third solid element becomes involved in the generation of a tribological surface. One such example is the formation of wear debris in a mechanical device where the wear debris finds its way into the contact zone between two surfaces in solid-state contact. When the debris does so, it can abrade one of the two surfaces in the solid-state contact and produce a mild wear effect over a prolonged period of time. If, however, the particle size of the wear debris is relatively large or the particle is harder than either of the two surfaces in contact, severe surface damage can occur. That damage can take the form of surface indentations or deformation spots in the solid surface as, for example, where a ball and a race of a rolling element bearing make contact.

Formation of dents by hard particles.—If the hard particle of dirt or debris is trapped between the ball and race contact (e.g., a particle of silicon carbide) and the ball rolls over the particle, it could generate a dimple or dent in the solid surface. Such surface defects are observed quite frequently as a result of dirt in lubrication systems. An example of such a dent is shown in the photomicrograph in figure 4-21. The circular area in the center of the photomicrograph is a dent produced in the solid surface as a result of the presence of a trapped wear particle (ref. 13). The particle is fairly spherical. As a consequence, the dent also appears to be fairly spherical.

The presence of such surface defects are reflected in lubricated systems, and they do have an influence on lubricant behavior in a contact zone. The effect on lubrication can be seen in figure 4-22 where the surface of figure 4-21 is lubricated with an oil film and an elastro-hydrodynamic device is used to study the interface. A thin lubricant film is placed on both solid surfaces. The dent is on a steel ball, which is fixed, and a glass plate rotates under the steel ball. The interface is viewed and photographed through the glass plate. The rings in figure 4-22(a) around the contact zone are the

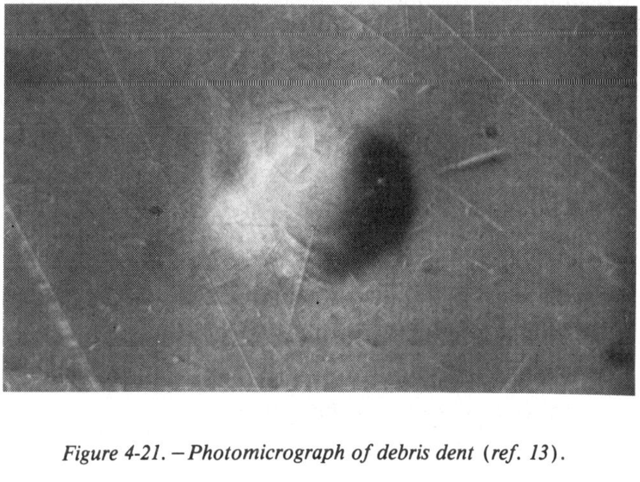

Figure 4-21. – Photomicrograph of debris dent (ref. 13).

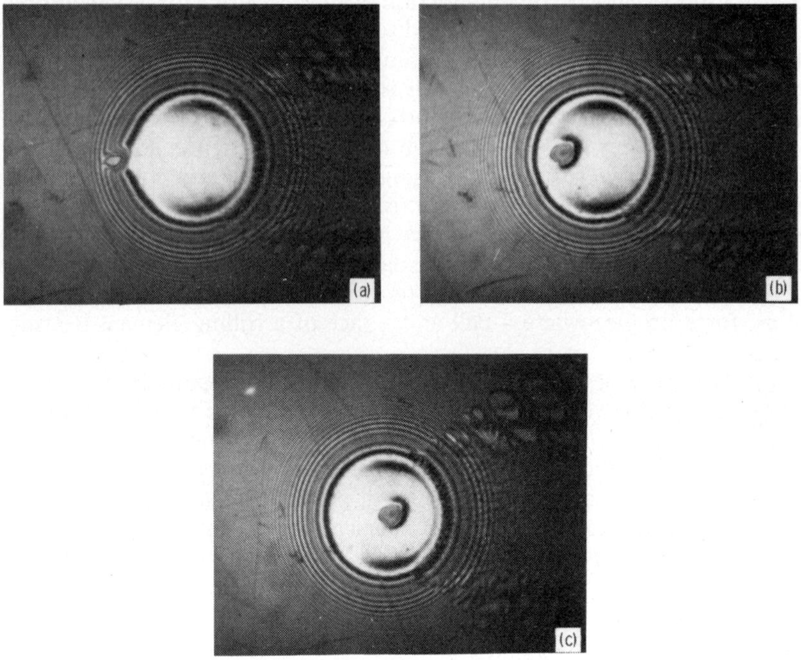

Figure 4-22. – Photomicrographs of dent in three different positions. $h_o = 0.14$ *micrometer;* $p_{max} = 1.2 \times 10^9$ *newtons per square meter;* $u = 0.0134$ *meter per second (ref. 13).*

Newton rings associated with the liquid lubricant film. The center white region is the actual contact region where the ball contacts the flat glass surface, and the small, circular or nearby circular object entering the central contact zone at about 9 o'clock is the dent in the steel ball as it approaches the contact region. This produces a profile of the oil film in and near the contact inlet such as that shown in figure 4-23(a) where the height (thickness) of the oil film is plotted (ordinate). The inlet is at the left of the profile curves.

The dashed line in figure 4-23(a) represents the oil film profile for a smooth surface in the absence of a dent. The hump in the curve reflects the influence of the dent in figure 4-22(a) as it approaches the contact zone. The dent produces a marked step in the lubricant film in the contact zone. Therefore, a change in the surface profile produces a change in the lubricant film profile in the contact region. In figure 4-22(b) the dent in the steel surface has moved from the entry zone of contact into the actual contact region itself. The presence of the dent is reflected again in the lubricant film profile (fig. 4-23(b)) where the hump in the curve again is associated with the presence of the dent in the actual contact zone. In figure 4-22(c) the dent in the surface has moved further into the contact zone and closer to the exit region. As it does so, the hump in the oil film profile moves also (fig. 4-23(c)).

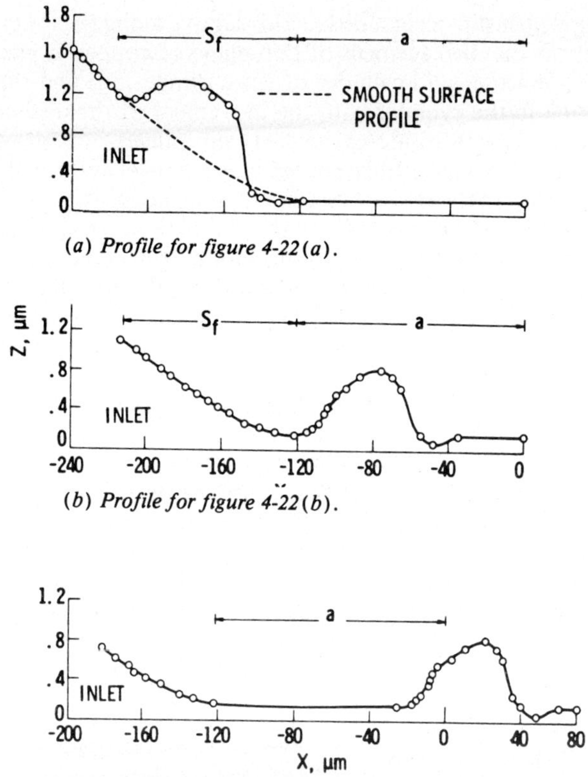

(*a*) *Profile for figure 4-22(a).*

(*b*) *Profile for figure 4-22(b).*

(*c*) *Profile for figure 4-22(c).*

Figure 4-23.—Dent profiles for figure 4-22. h_o = 0.14 micrometer; u = 0.0134 meter per second (ref. 13).

Thus, the presence of surface defects that are generated in tribological surfaces as a result of two solid surfaces in solid-state contact can alter behavior in lubricated devices.

Atomic Nature of Tribological Surfaces

When tribological surfaces are deformed, at a much finer scale than has been discussed so far in this chapter, much finer changes and more subtle changes take place in the solid surface as a result of the deformation process. For example, a close examination of a deformed tribological surface shows the presence of surface ledges or minute edges that develop in the deformation region as a result of the emergence of slip bands along the surface with the deformation process. Slip along slip bands is associated with the easy glide of dislocations along the preferred slip planes in a material. Hence, for face-centered-cubic metals which have the (111) crystallographic planes along which slip occurs, dislocations move along the (111) planes and cause or permit the displacement of one plane of copper atoms relative to another. If one has a solid cylinder of fixed dimensions and slip occurs, it produces a jog in the cylinder with the movement of the dislocations (and the slip associated with it) along the (111) slip planes.

On tribological surfaces, when the surface region is deformed plastically, slip bands develop in the surface region and the matrix about the slip region contains the undeformed material. In the slip region, however, the dislocations move in such a fashion so as to produce an emergence of slip bands at the surface, leaving small edges projecting above the surface like those found in figure 4-24. In figure 4-24, the lines that appear in a diagonal direction in two directions are slip bands; duplicate sets indicate cross slip in the material. A close examination of the solid surface reveals the emergence of small edges above the plane of the surface. These small edges are the slip bands (or slip material) that have emerged on the solid surface; these edges project above the solid surface and increase the total surface area. It is analogous to having a deck of cards and moving the cards such that various cards throughout the deck project above the edge of the deck.

In reference to an earlier figure, the point was made that there is a location subsurface where the zone of maximum subsurface shear stress exists. In this particular zone of shear stress the dislocation concentration is highest. Figure 4-24 shows the region where a void is subsurface in the solid. When the surface is stressed and plastic deformation occurs, dislocations

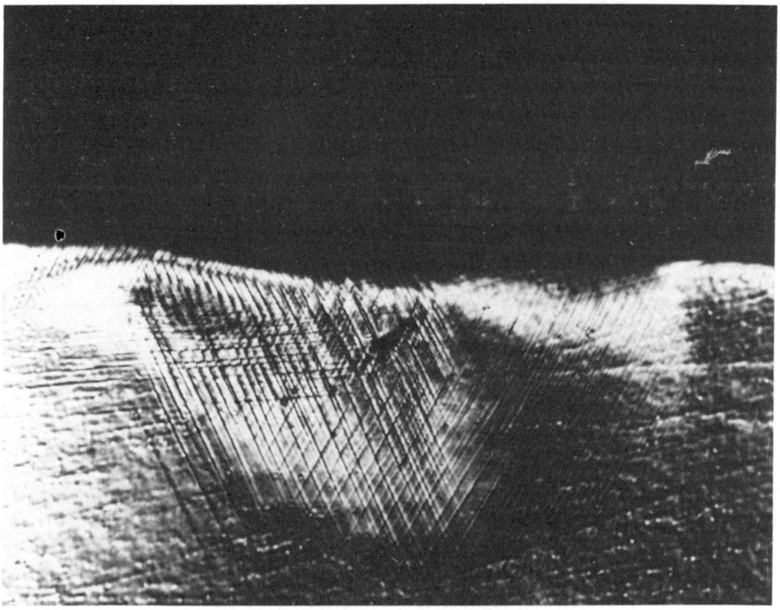

Figure 4-24. – Slip band emergence on tribological surface (ref. 10).

can coalesce along the slip bands in a particular region. With the dislocations coalescing there, a void develops slowly; this void builds with repeated stressing of the solid surface until the void begins to work its way to the surface. This is indicated in figure 4-24 where a crack or a fine line is seen emerging from the void at about 1 o'clock. The void is developing a crack that is working its way to the solid surface. Once the crack reaches the surface, the potential exists for the release of a wear particle when fracture occurs opposite the zone where the crack emerges on the solid surface. A wedge-shaped piece of wear debris can then be generated.

The slip bands emerging at the solid surface do change the surface topography. They produce a series of ledges or steps in the surface that did not exist before the deformation process. These ledges or steps can be thought of as microasperities that have developed in the deformation of the solid surface.

The photomicrograph in figure 4-24 indicates that very high degree of deformation with multiple slip occurs in the actual contact zone. Defects are not only generated in the surfaces of solids as a result of tribological interactions but they also develop subsurface as well. In figure 4-24 the dislocations are moving along slip bands which emerge at the solid surface. The processes of relative motion between two solids, however, can also initiate the formation of dislocations right at the surface. The mechanism for generating surface dislocations as a result of shear or relative tangential motion in a tribological system is shown schematically in figures 4-25(a), (b), and (c).

If two single crystals of the same material were brought into atomic registry across an interface, such that the atomic planes matched perfectly, and the crystal lattices were in complete registry, then the surface might look something like that depicted schematically in figure 4-25(a). There would be bonding of like atoms across the interface, and the perfect registry of the crystal lattice would be maintained. This is only theoretically possible because, in practice, it is not possible to achieve such registry across an interface.

The very presence of an interface means some discontinuity or disregistry between the two adjacent materials in solid-state contact. Thus, if there were two single crystals of copper brought into contact, an interface is always physically present. If we assume, however, for a moment that such a condition is possible, then the schematic of figure 4-25(a) would result where there would be perfect bonding across the interface and the lattice spacing (indicated as a) in figure 4-25(a) would be maintained in both surfaces and at the interface. If we assume that tangential motion was initiated between the two solids, then, as one surface was moved relative to another at the interface, a dislocation would develop in this surface. The dislocation might be a perfect dislocation such as that indicated in figure 4-25(b) where an additional half space of atoms is placed in the crystal lattice of one of the two solid surfaces in contact. Note that in figure 4-25(b) the additional half space of atoms in the crystal lattice produces a crowding of adjacent atoms in the crystal lattice and the lattice spacing shown in (a) for the solid surface is no longer maintained in the lower half of figure 4-25(b).

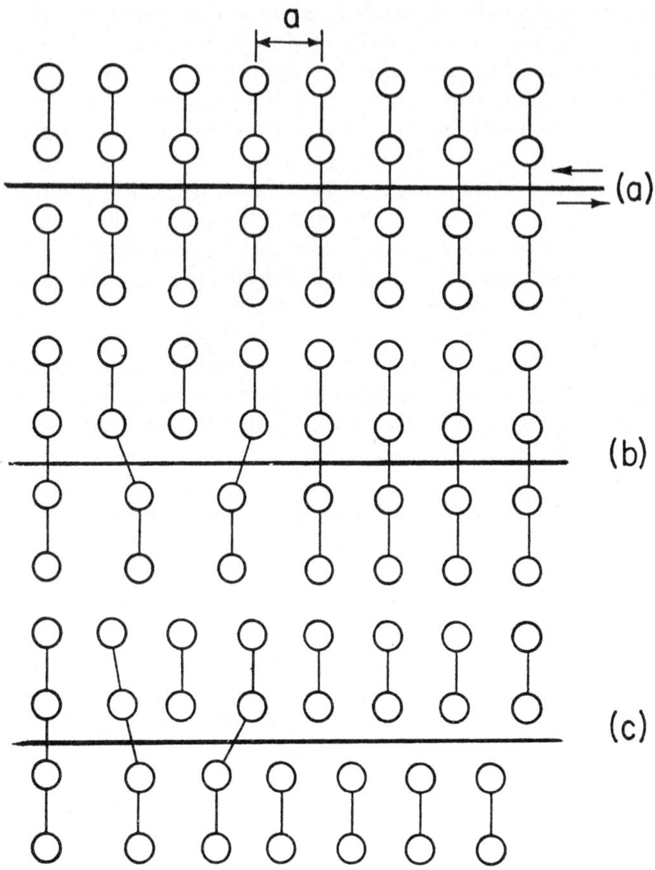

Figure 4-25. – Dislocations (ref. 14).

In figure 4-25(b), however, no tangential displacement has taken place of the upper solid surface relative to the lower solid at the interface. The presence of the dislocation is accommodated for in the crystal lattice of the two solids. In figure 4-25(c), however, we have what is referred to as an imperfect dislocation. When tangential motion is initiated and the surface of the lower solid is displaced a slight bit from that of the upper solid, a jog or step is produced (right side of fig. 4-25(c)). Multiplying this phenomena many times results in the steps at the solid surface (fig. 4-24). Repeated rubbing, rolling, or sliding over a surface usually results in an increase in the generation of dislocations. Most solid materials have a fixed dislocation concentration in the material prior to the initiation of sliding, rolling, or rubbing. With such contact, however, the concentration of dislocations can be increased by the mechanism shown in figure 4-25(c).

The generation of dislocations in tribological surfaces is demonstrated in some rolling experiments on magnesium oxide conducted by Dufrane and

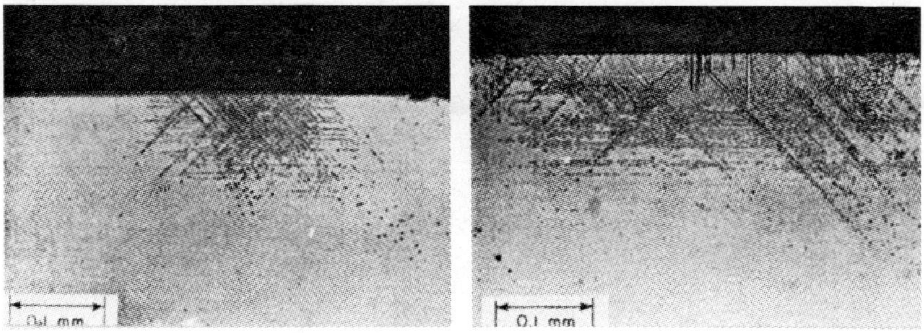

(*a*) *One cycle.* (*b*) *10^6 Cycles.*

Figure 4-26. – Effect of increasing rolling-contact stress cycles on track width. Load, 244 grams; rolling direction, [110].

Glaeser (ref. 15). In figure 4-26, the magnesium oxide surface had a steel ball rolled across it under a load of 244 grams in the [110] crystallographic direction. Figure 4-26(a) shows a cross section under the wear track with dislocation concentrations near the surface as a result of one rolling pass across the surface; etch pitting was used to bring out dislocation sites. The individual pit marks reflect the dislocation concentrations subsurface and pin point the actual location of the dislocations. The etch pits fall somewhat along lines similar to the slip lines seen in figure 4-24. The etch pits are actually etching out the dislocations along the slip lines, and it is for that reason that there is a similarity between figures 4-26 and 4-24. The etch pits are rather profuse in the actual contact region from subsurface to the surface. If the number of cycles in which a ball rolls across the surface is increased to a million cycles, the surface looks like that shown in figure 4-26(b) in cross section. The concentration of the dislocations has increased markedly. It is interesting to note that the depth to which the dislocations move subsurface has not increased markedly, but the density at the surface and near-surface region has become so profuse that the area encompassed by the dislocations becomes almost completely blackened by the high density of dislocation etch pits. Furthermore, the actual width of the deformation band at the surface (width of the wear scar) has increased considerably after one million stress cycles across the surface. There is no question that the properties of the surface in the actual wear contact zone are markedly different from the properties outside the contact zone, that area which has not undergone the deformation due to rolling contact.

Direct observation of dislocations with transmission electron microscopy can also reveal the dislocation concentration. The etch pitting technique is much simpler and quicker to use; the transmission technique, however, does show the actual dislocations themselves. Figures 4-27 and 4-28 reveal the change in dislocation concentration as a function of rolling passes on the magnesium oxide when the load is increased from 244 grams (used in fig. 4-26) to 570 grams. The load is identical in figures 4-27 and 4-28. The only

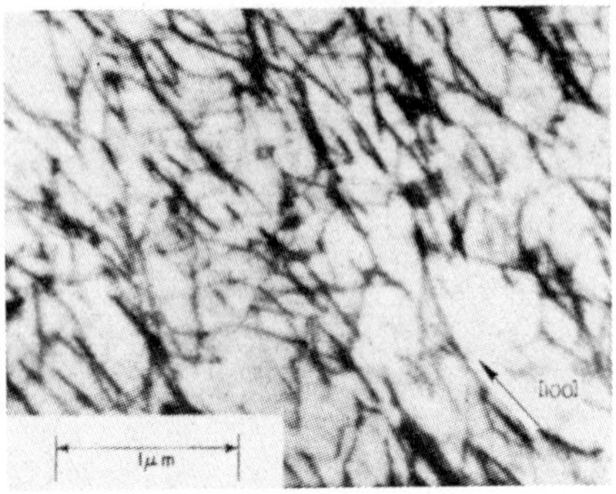

Figure 4-27. – Dislocation configuration after 10^3 rolling-contact cycles with 570 gram load.

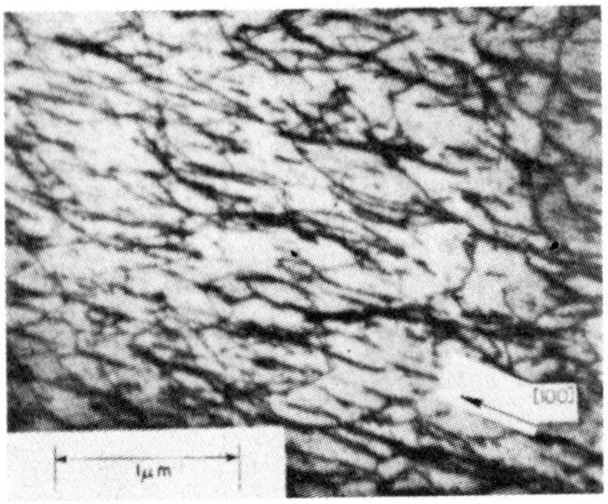

Figure 4-28. – Dislocation configuration after 10^5 rolling-contact cycles with 570 gram load.

thing that has been changed is the number of rolling cycles (contact cycles). Comparing figures 4-27 and 4-28 reveals a much higher concentration of dislocations in figure 4-28 than are present in figure 4-27; this indicates a higher degree of plastic deformation as a result of the increase in the number of stress cycles from 1000 to 10 000 cycles.

As mentioned earlier in the text, the presence of dislocations indicates the presence of defects in the surface of solids. The defect regions are regions of different energies. The preparation of tribological surfaces can have an influence on the final nature of the surface composition. For example, polishing with abrasive polish powders can produce changes in the surface of solids. Hard abrasive particles are very frequently used in the finishing operations for the preparation of tribological surfaces. Not only are they used in metals and alloys, but they are used for finishing inorganics and glasses as well.

As mentioned earlier, the environment can play a very strong role in the nature of the surface topography and the composition of a solid surface. For example, some ellipsometric studies of glasses polished with cerium oxide under oil or water reveal some striking differences in composition and density of the surface layers. Some ellipsometric results obtained from the polishing of glass surfaces with cerium oxide in oil or water are presented in table 4-I (ref. 16). In table 4-I there are four glasses which have been polished with cerium oxide using either oil or water as a lubricant. The wavelength of the incident-light source λ is in angstroms; the angle between the incident light beam and a normal to the solid surface φ is 59°; and the ratio of the reflected amplitude polarized parallel to the plane of incident light is given by ψ. The important thing to recognize with the data of table 4-I is that, when oil is used as a lubricant in every case, the surface layers are densified as shown by the decrease in the value of ψ from ψ_0. On the other hand, when water is used as a lubricant, some of the cations in the surface layers of the glass are leached out by the water with a consequent decrease in the refractive index. In other words, this leaching out of the cations has the opposite effect of densification, and what one observes is the overall effect; ψ still decreases, however.

In addition to the foregoing, changes in the experimental parameters can alter the densification of the solid surfaces. For example, the load that is applied while polishing the surface with cerium oxide influences the densification of the surface layers. The greater the load, the greater the densification; the lower the load, the less the densification of the solid surface of the four glasses indicated in table 4-I. Thus, the nature of a tribological surface is influenced by the environment in which the surface is polished.

The influence of environment (i.e., the lubricant) when polishing the four glasses was observed (table 4-I) to make a difference in the densification of the solid surface layers. The actual polishing material itself also makes a difference. If the material to be polished remains constant and the lubricant re-

TABLE 4-I. – GLASSES POLISHED WITH CeO_2[a]

[$\lambda = 5461$ Å; $\varphi = 59°$.]

Glass	ψ_0, deg	ψ_{oil}, deg	ψ_{water}, deg
Soda Lime (73:17) 0080	3.87	3.84	3.50
Borosilicate (66:24) 9741	5.06	4.80	4.95
Borosilicate (65:18) 7052	4.68	4.53	4.60
Aluminosilicate (62:17) 1720	3.21	3.13	3.16

[a]Reference 16.

TABLE 4-II. – AMORPHOUS SiO_2[a]

Surface history	Polished with diamond paste, ψ, deg	Polished with cerium oxide, ψ, deg
Ideal case	3.79	3.79
Mech. polish	.52	3.79
2 min etch in 10% HF	2.55	3.77
4 min etch in 10% HF	3.93	3.79
6 min etch in 10% HF	4.16	3.80

[a]Reference 17.

mains the same, changes in the surface condition are also observed with a change in the material used to polish the surface. These changes are reflected in the data of table 4-II for vitreous silica (ref. 17) where the vitreous silica surface has been polished by diamond paste or cerium oxide. Table 4-II also presents the surface history of the material.

Under *Surface history*, the ideal case is indicated as 3.79 for the ellipsometric parameter. It is interesting to note that the diamond is much harder than vitreous silica but that the cerium oxide and the vitreous silica have Mho hardnesses nearly the same (5.0 and 4.9, respectively). When the surfaces have prior mechanical polishing, there is very little change in ψ for the surface polished with cerium oxide; the value of the ellipsometric parameter remains essentially unchanged at 3.79. When the surfaces have prior mechanical polishing, the ellipsometric parameter drops to 0.52 after polishing with diamond paste. Computations with the ellipsometric equations that are used with the equipment reveal that the surface layers of the diamond polish specimen have been permanently densified; they are no longer strain free but are highly strained during mechanical polishing. The densified layer has a refractive index of 1.530 and a thickness of 950 angstroms. Thus, the diamond paste polishing produces a considerable amount of strain in the vitreous silica; the 3.79 ideal case reflects a strain-free surface, and the 0.52 ellipsometric parameter reflects the high degree of strain.

The specimen polished with the cerium oxide appears to be, from the ellipsometric parameter, free of the strained layer. The similarity of hardnesses between the vitreous silica and the cerium oxide may be why polishing with cerium oxide does not permanently damage or strain the surface layers of the vitreous silica.

If the surface, after straining (by polishing with diamond paste), is then chemically etched with hydrofluoric acid for various time periods, it can be seen that the ellipsometric parameter begins to return to the value for the ideal case. With 2 minutes of etching, the value increases to 2.55. With 4 minutes of etching, the value becomes 3.93 (overshooting the ideal case of 3.79). Likewise, after 6 minutes of etching the value becomes 4.16 (again overshooting the ideal value of 3.79).

The important thing to recognize in table 4-II is that the polishing activity produces a permanent change in the nature of the surface layers of the vitreous silica. This is true not only of glasses and silicates but of materials in general. The polishing activity generally produces a change when the polishing material is harder than the surface being polished. While the cerium oxide does not produce much surface straining of the vitreous silica, the diamond paste does produce strain and is a much better polishing agent for the vitreous silica. The differences in mechanical properties of the diamond and the vitreous silica appear to be necessary to accomplish the effective polishing action by abrasion. Vitreous silicas are relatively brittle materials and, therefore, abrasion removes material from the solid surface in the polishing action. In the mechanical polishing of metal surfaces extreme care must be taken because, in polishing, it is possible for the abrasive material to become embedded in the metal surface. Since most metals are ductile and the abrasive material is very hard and brittle, the abrasive can become buried in the surface layers. For example, aluminum oxide has been detected buried in gold and silver surfaces and silicon carbide has been found in copper surfaces when these materials were used as polishing agents for polishing the surfaces of the noble metals. The embedded abrasives in the ductile metals can then produce an abrading surface. A grinding wheel is nothing more than hard abrasive grits embedded in a soft resin matrix, and that is what results if the abrasives that are used to polish a soft metal surface become embedded in that metal. When that surface is brought in contact in lubrication systems with other solid surfaces, the embedded particles can act as abrasives toward the contacting surface. With extremely ductile surfaces, it is wise to use chemical techniques to polish the solid surface (either electropolishing or chemical polishing) in place of mechanical polishing as a final step in the preparation of tribological surfaces.

Metallurgical Effects

Straining

In the sliding, rolling, or rubbing contact of materials, the surfaces become strained as a result of the mechanical activity that takes place at the solid surface. The degree of strain is influenced by a number of mechanical factors (relative speed, load, geometric considerations) as well as other parameters (temperature and other factors involved in the actual system).

On polycrystalline metal surfaces or alloys, where differences in orientation exist at the solid surface, the amount of strain that occurs under a given set of mechanical parameters (such as load and speed) varies depending on the orientation. Thus, in a relatively strain-free surface, marked changes can exist in the surface strain pattern associated with each of the individual grains (depending on their orientation) after the system has been exposed to sliding, rubbing, or rolling contact.

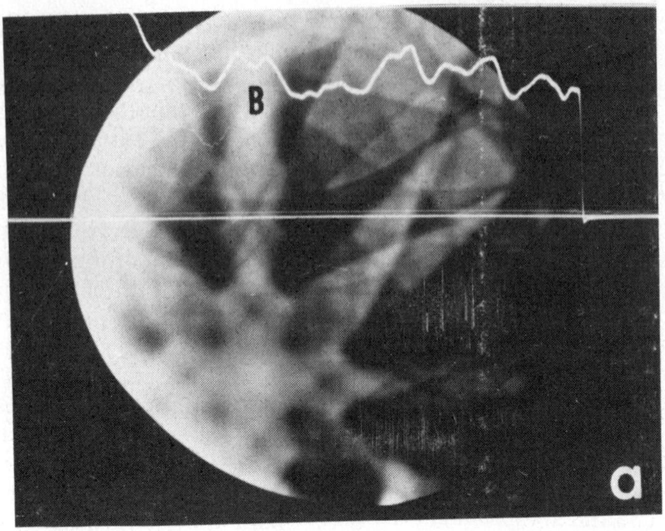

(*a*) *Grain at low strain* ($<1\%$) *location.*

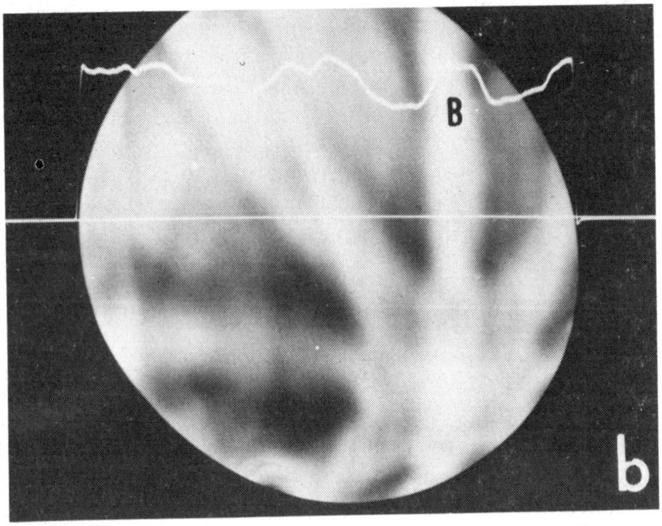

(*b*) *Grain at medium strain* ($\simeq 13\%$) *location.*

Figure 4-29. – SACP patterns from two grains of polycrystalline iron sample. Note contrast at 220 band marked B.

A number of surface tools are available to characterize the amount of strain that takes place in a metal surface. Probably one of the most effective is the electron channeling technique. It has been very effectively utilized by Ruff in following the strain in iron samples by examining the grains in the wear contact zone and measuring the amount of strain that occurs in these grains (ref. 18). Figure 4-29 presents two electron channeling patterns (SACP-selected area electron channeling pattern). Figure 4-29(a) is for an electron channeling pattern obtained in a relatively low strain region where there is a high degree of perfection in the pattern, and figure 4-29(b) is the pattern obtained from a grain where there was a much higher degree of strain (the pattern is not as sharp).

The variation in the strain energy introduced into a tribological surface is extremely important because most materials used in practical lubrication devices involve polycrystalline surfaces. When the surfaces are rubbed together or rolled one over the other, the amount of strain that is taken up by each grain varies as a function of the orientation. This produces a strain inhomogeneity across the surface. If one were to map a tribological surface, they would find that the strain energy varies from grain to grain and across the solid surface. This produces a very nonequilibrium state or condition. Such a condition can introduce other changes in the material such as recrystallization which is a function of both temperature and strain energy.

Reorientation (Ordering) and Recrystallization

The higher the degree of strain energy, the lower the temperature of recrystallization. Consequently, highly strained grains tend to promote recrystallization of the solid surface long before the surface may otherwise be ready for such recrystallization. The grains in a metal surface not only undergo a high degree of strain in a tribological surface, but the tribological surface may also contain grains which are highly oriented as a result of the sliding, rolling, or rubbing process. In other words, the grains tend to become reoriented at the surface so as to reflect the effects of the mechanical parameters imposed on the surface. This is demonstrated with the aid of figure 4-30, which shows what can happen to a surface during a sliding friction experiment (ref. 19).

Figure 4-30(a) shows the normal equiax grains (grains in the normal configuration in the annealed state) without preferred orientation or directionality to the individual grains. When sliding is initiated across the surface, however, the orientation tends to become more highly directional and progresses from (a) through (b), (c), and (d) developing a degree of directionality. At first, the grains simply change their direction as a result of the imposed directional movement (fig. 4-30(b)). With continuous sliding, and the strain associated therewith, the recrystallization temperature of the surface can be reduced appreciably. Recrystallization can occur in the surface, and this increases the number of grains from those seen in figure 4-30(a) to the number seen in figure 4-30(c). The first step in a friction experiment may be a preferred orientation of the grains otherwise lacking that orientation

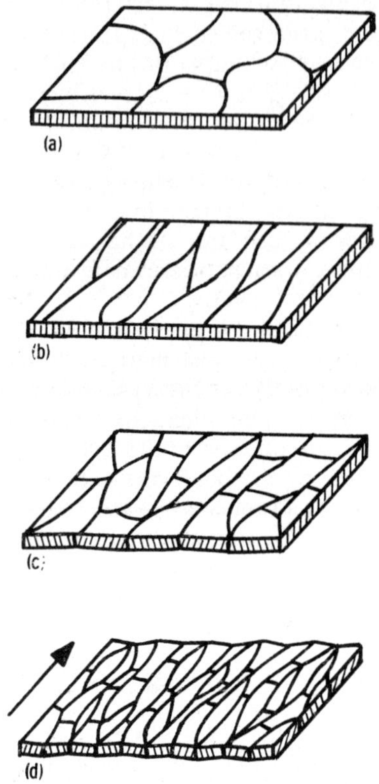

Figure 4-30.—Scheme of surface formation during friction (horizontal component of disorientation is shown).

(fig. 4-30(b)); subsequently, a recrystallization process takes place, as reflected in figure 4-30(c) where a larger number of grains are now present, still maintaining directionality. Because of imposed directional sliding, the recrystallization process can be repeated a number of times so that the grains continue to get smaller; this is reflected by the directional effects shown in figure 4-30(b). Such a result is seen in figure 4-30(d) where the grains maintain the orientation effects of figure 4-30(b); a much larger number of these grains exists following repeated recrystallization. Recrystallization is a common metallurgical phenomena that is extremely important in tribology because of the prominence of its occurrence. It probably occurs much more frequently than is recognized by the experimentalist in the laboratory. Many years ago, microscopic examinations of tribological surfaces indicated (near the surface that had experienced rubbing or rolling contact) a very fine grain structure almost lacking grain boundaries because of the extremely small size of the boundaries themselves. In fact, in many instances, many investigators considered that there was an absence of grain boundaries and that the surface layer in a rubbed, rolled, or sliding surface really was amorphous; that is, the general metallurgical structure was so changed by the mechanical activity at the solid surface that the crystallinity in the material was completely lost near the surface. This surface is markedly different from the bulk of the material, and it was referred to by early investigators as the Beilby layer.

Subsequent investigations, however, in more recent years have revealed that the Beilby layer is, in fact, a very fine grained structure that results from repeated recrystallization of the solid surface. The sliding, rolling, or rubbing contact at the surface coupled with high flash temperatures and bulk surface temperatures bring about a recrystallization of the solid surface. The recrystallization can occur repeatedly, and the grain size continues to be reduced by the recrystallization process because insufficient time is allowed for grain growth to occur. As a result, the grain size continues to get smaller until it appears that the surface lacks crystallinity. However, electron diffraction studies of the solid surface reveal that the surface is, in fact, crystalline in nature.

The recrystallization process is extremely important because the mechanical properties of the surfaces change drastically with recrystallization. Recrystallization brings about an annealing effect of the material which then alters the mechanical behavior. For example, the microhardness may drop drastically with recrystallization. Recrystallization is observed in a wide variety of materials at the solid surface. It is the tendency of tribological researchers to discount recrystallization as a process because of the high temperatures required for recrystallization to occur. It must be remembered, however, that recrystallization is aided by deformation; that is, the temperature needed to produce recrystallization is reduced by the addition of deformation to the solid surface. Deformation is an inherent part of the sliding, rolling, or rubbing of solid surfaces. As a consequence, the transformation temperature (the temperature associated with recrystallization) can be appreciably reduced by this process.

One of the most commonly used base materials in tribological devices is iron and its alloys. An examination of the recrystallization curves of iron shows that the recrystallization temperature can be appreciably reduced by plastic deformation as might occur in sliding, rolling, or rubbing contact. A recrystallization diagram for electrolytically refined iron annealed for 1 hour is presented in figure 4-31 as a three-dimensional plot of grain size and temperature for recrystallization as a function of deformation or strain that takes place in the metal (ref. 20). As the amount of deformation is increased, the temperature for recrystallization decreases. The recrystallization temperature can be dropped from 870° to 400° C with the addition of about 50 percent strain in the solid surface (fig. 4-31). Such strains are extremely common in friction and wear surfaces.

This reduction in the recrystallization temperature with deformation is very common among most metals and is not peculiar to iron. Titanium, for example, has a normal recrystallization temperature of approximately 900° C. Straining the surface about 60 to 70 percent can reduce the recrystallization temperature from 900° to about 400° C. Such a surface temperature (400° C) can readily be achieved in tribological surfaces where there are heavy loads, high speeds, or a combination thereof. It is very easy

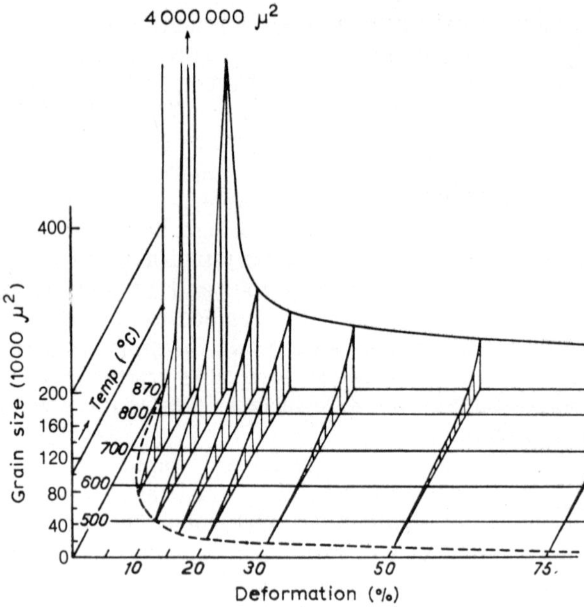

Figure 4-31.—Recrystallization diagram for electrolytically refined iron annealed 1 hour (ref. 20).

to introduce strain into a surface. For example, simply rubbing 600 grit silicon carbide paper across a metal surface can produce a marked amount of strain in the solid surface. The maximum amount of strain occurs near the surface; the degree of strain drops off subsurface, as might be anticipated, because the strain does not penetrate deeply into the system with such a relatively mild surface disturbance.

An example of the strain produced in a brass surface with abrasion of 600 grit silicon carbide paper is presented in figure 4-32; the amount of strain or percent of strain is plotted as a depth subsurface (ref. 21). It can be seen that a strain of 5 percent can develop in the region near the surface. The strain, however, drops rapidly with depth; at a depth of 10 microns, the amount of strain is less than 1 percent, and at 25 microns, it has been completely dissipated.

Rubbing a surface with 600 grade silicon carbide paper is a relatively mild surface disturbing process. Yet, 5-percent strain in the near-surface region can develop. Figure 4-32 shows the curve begins at approximately a 5-micron depth. The authors of figure 4-32 could not measure the strain at the solid surface. Had the curve been extrapolated to the surface, the amount of strain at the surface could conceivably be 20 to 25 percent. Thus, fine polishing, which is another term for mild abrasive wear of a solid surface, can produce marked strains in the metal surface. This strain influences the behavior of the solid when in contact with other solid surfaces.

Metallurgical changes in the solid surfaces can have a pronounced effect on other mechanical properties. For example, with strain, hardening fre-

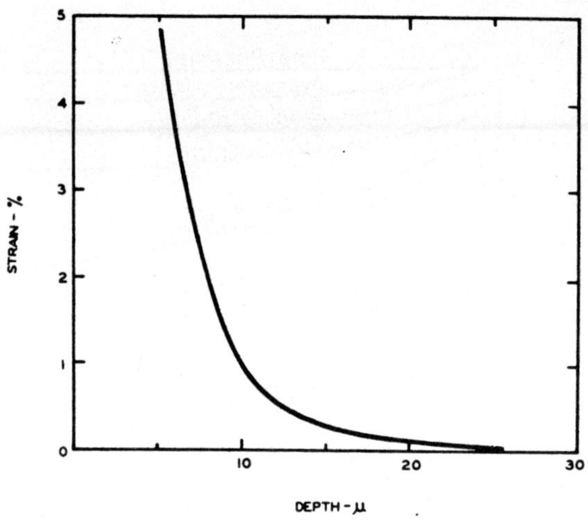

Figure 4-32.—Estimate of strain gradient in brass surface abraded on 600-grade silicon carbide paper (ref. 21).

quently occurs in metal systems. Some metals are more prone to strain harden than others. Aluminum, for example, strain hardens fairly readily with deformation. In addition to changes in hardness, a very common property of metals, which has been found to be affected by surface conditions, is the fatigue strength of metals. It is fairly well known that defects in solid surfaces and metallurgical changes in surficial regions can alter the fatigue strength of metals very markedly. One of the most commonly studied families of materials relative to the effect of surface conditions on fatigue strength has been steels. Normally of interest to the tribologist are wrought steel materials (cast steels are not used very widely in practical lubrication devices). However, if one examines cast steel surfaces that have been prepared in various fashions and relates the findings to fatigue strength, one discovers that fatigue strength is markedly altered by the surface condition. This effect is demonstrated in the data of figure 4-33.

In figure 4-33 the reduction in fatigue strength is plotted as a function of the ultimate tensile strength (ref. 22). The data are for surfaces that have been prepared in various ways: (l) fine polish, (2) average polishing, (3) good grinding, (4) fine turning, (5) rough turning, and (6) unmachined. The data in figure 4-33 indicate that the surface condition present on the solid surface can alter fatigue strength considerably for these various surface finishes. Fine polished and ground surfaces are commonly used in tribological systems. In some applications, even turned surfaces are found to be satisfactory for particular mechanical devices. While the data of figure 4-33 are somewhat prejudiced in the sense that one might expect cast structures to be much more sensitive in their fatigue strengths to surface conditions than wrought materials, the data do show the importance of surface condition on mechanical behavior of metals and alloys. The distinctions and differences among the various finishing techniques used in figure 4-33

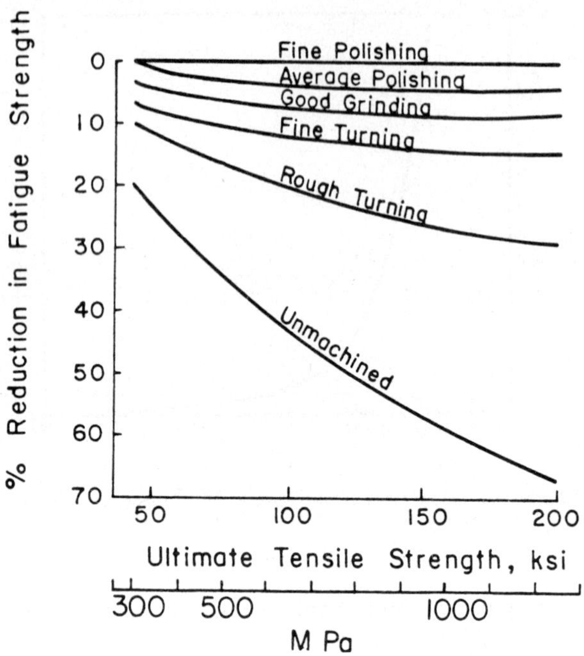

Figure 4-33.—Reduction in fatigue strength of steels with varying surface finishes and ultimate strengths (ref. 22).

tribological systems. In some applications, even turned surfaces are found to be satisfactory for particular mechanical devices. While the data of figure 4-33 are somewhat prejudiced in the sense that one might expect cast structures to be much more sensitive in their fatigue strengths to surface conditions than wrought materials, the data do show the importance of surface condition on mechanical behavior of metals and alloys. The distinctions and differences among the various finishing techniques used in figure 4-33 may not be as pronounced with the wrought surface, but, nonetheless, it is certain that they do exist.

In addition to recrystallization and straining of surfaces that occur in tribological devices, other metallurgical phenomena can occur that influence the friction, wear, and adhesion behavior of solids in contact. The fabrication process by which surfaces are frequently generated can, with certain materials, introduce preferred orientations. Materials which exhibit a high degree of anisotropic behavior are notorious with respect to being sensitive to alterations of orientation with various fabrication techniques. Hexagonal metals, for example, tend to orient fairly readily with a preferred orientation at the solid surface depending on the particular technique used to form that solid surface.

Rolling, drawing, sinking, and reducing are fabrication operations that can alter the surface texture of a metal. This is particularly true with the hexagonal metals zirconium, titanium, cobalt, and beryllium. Zirconium, for example, has a hexagonal crystal structure. That hexagonal structure orients in various forms on the solid surface depending on the fabrication operations that are conducted on the material in the process of generating the surface. Not only does the orientation vary with the particular operation used to generate the surface, but variations within the surface may also exist as a result of the fabrication operation. The effects of such preferred orientation resulting from fabrication are seen for zirconium in figure 4-34 where various fabrication or surface preparation techniques were used (ref. 23). In figure 4-34(a) the sheet is simply rolled with reduction in the sheet; this reduction operation produces a change in orientation. The inserted circles show the orientation and hexagonal unit cell of the zirconium before and after the rolling process. As the sheet approaches the rolls, the orientation of the hexagonal unit cell is in one direction. After the rolling operation has been completed, there is a modification in the orientation of the crystal in the surface layers.

Plug drawing of tubes produces a modification of orientation as indicated in the schematic of figure 4-34(b). Both figures 4-34(c) and (d) likewise show changes in crystallographic orientation at the surface with tube sinking and tube reducing operations. Adhesion and friction of metal surfaces are very sensitive to orientation at the solid surface, and the final surface that is used in the tribological device has a surface orientation that is influenced and determined by the process used to generate that surface.

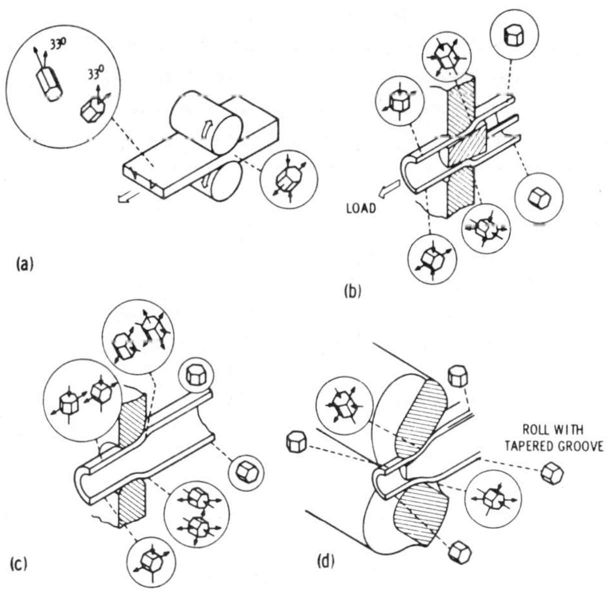

(*a*) *Sheet rolling.*

(*b*) *Plug drawing with wall thinning.*

(*c*) *Tube sinking with and without wall thickening.*

(*d*) *Tube reducing.*

Figure 4-34. – Preferred orientation for various types of fabrication of zirconium (ref. 23).

Ultimately, the orientation developed from the formation or fabrication process can dictate adhesion, friction, and even wear behavior. Thus, it is important to understand the influence of orientation on the behavior of surfaces in sliding, rolling, and rubbing contact.

Chemical Nature of Surfaces

In addition to the physical, atomic, and metallurgical properties of surfaces being important in the tribological behavior of materials, the chemistry of the surfaces is also extremely important. The chemistry of a tribological surface can be affected and determined by the environment in which the solid surface to be lubricated finds itself, or it can be influenced or determined by the solid bulk material. When a small amount of alloying element is present in a material, it can alter the surface behavior and change the nature of the solid surface. There are many alloying elements present in metals that can alter the tribological nature of the solid surface. Many of these elements are found as contaminants in conventional bearing, gear, and seal materials. When present in the proper concentrations they can, however, alter markedly surface effects. One such element is sulfur.

The presence of sulfur in iron has been known for some time to alter the machinability of steel and iron. Small concentrations of sulphur make the machinability of steel much more easily accomplished. Fundamental studies in examining various amounts of sulfur in iron reveal that very small concentrations alter the wear behavior of elemental iron markedly. An example of this is manifested in the photomicrographs of figure 4-35 where 0.45 percent of sulfur has been added to electrolytic iron. There are two photomicrographs in figure 4-35: (1) the microstructure of the electrolytic iron, and (2) electrolytic iron containing 0.45 percent sulfur. In the structure containing sulfur, the dark lines and the salt and pepper appearance are due to the formation of iron sulfide in the matrix. In the rubbing process, the iron sulfide becomes smeared out on the surface and reduces the wear behavior of the solid surfaces appreciably.

In the upper photomicrographs of figure 4-35, those to the left show the microtopography of the electrolytic iron disk surface as well as the rider after sliding contact. Those on the right show the electrolytic iron containing 0.45 percent sulfur. Identical mechanical conditions exist in both experiments. The natures of the surfaces, however, are markedly different. In the presence of the 0.45 percent by weight sulfur, the surface takes on a much smoother topography and is not severely disturbed as it is in the case of electrolytic iron. In the photomicrograph for the electrolytic iron rider, the rider material has actually undergone plastic flow. The rider geometry is no longer circular in the wear contact zone but shows evidence of trails or tails falling to the right of the rider specimen. With the 0.45 percent sulfur, the rider specimen wear scar is extremely smooth and its diameter is much smaller than that for the electrolytic iron. The sulfur containing specimen shows appreciably less wear.

Figure 4-35 represents an example of a small concentration of an alloying element markedly altering surface topography of a tribological surface. It

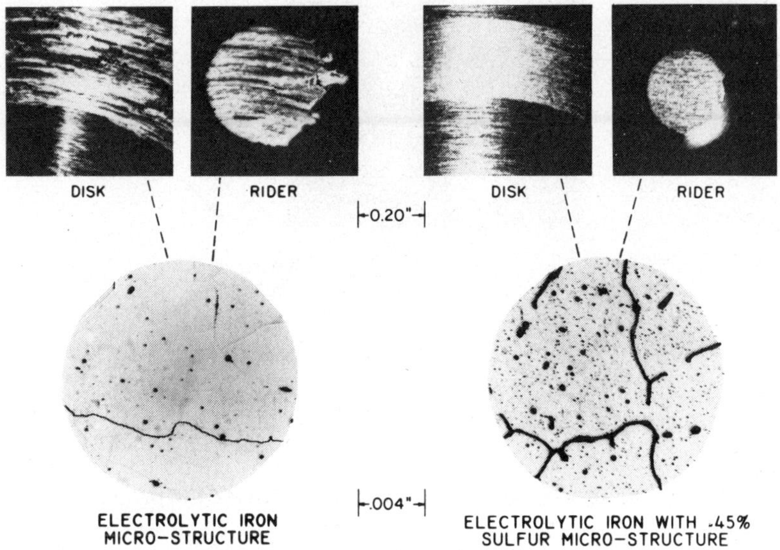

Figure 4-35. – Influence of sulfur addition to iron on sliding behavior in vacuum.

has been known for some time that certain materials have inherently good wear resistance and maintain relatively smooth surfaces in tribological systems. Materials that exhibit this property are the cast irons. For hundreds of years, cast irons have been used because they have inherently good wear resistance. The presence of carbon (in graphite form) in the cast iron is believed to be responsible for the good wear characteristics associated with cast irons. There are a number of cast irons, and they have varying compositions containing graphite (or amorphous carbon). Some have better wear resistance than others, but, in general, the cast irons have better wear properties than iron-base alloys, which do not contain graphite. An example of one such microstructure at the surface is presented in figure 4-36.

Figure 4-36 is a photomicrograph of a wear track on a gray cast iron composition with 3-percent carbon. Examining of the surface reveals black patches in the wear contact zone. A careful examination of these black patches with AES reveals them to be carbon. The graphite from the cast iron has become rubbed out over the surface and formed a protective surface film. In essence, the bulk structure of the cast iron carries in it a built-in solid lubricant. The graphite becomes smeared over the surface in the contact zone and provides a lubricating film; this minimizes the adhesive wear and transfers from one surface to another. An example of the effectiveness of the smearing out of the carbon across the surface is indicated in the photomicrograph in figure 4-36 by looking at the actual rubbing contact zone, both inside and outside. Outside the contact zone there is very little graphite; it is seen only as isolated little islands in the matrix. In the wear track (the wear contact zone proper), however, the concentration is much higher. That is because the small islands of graphite have become rubbed

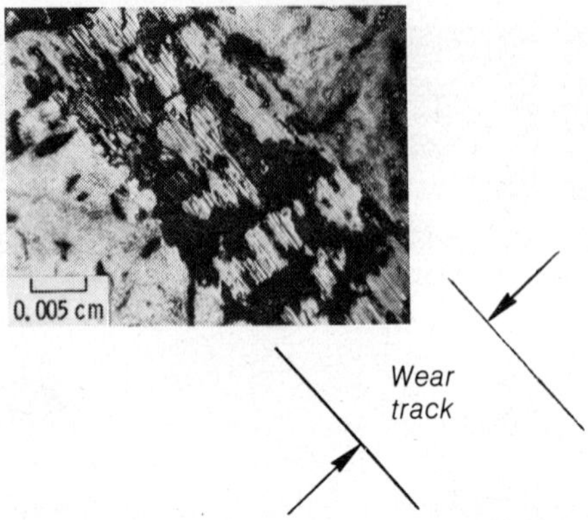

Figure 4-36.—Wear track on 3.02 percent carbon in gray cast iron. Sliding velocity, 5 centimeters per minute; load, 50 grams; temperature, 23° C.

out in the sliding process across the surface to provide the protective surface film.

One problem of using solid film lubricants effectively and providing protective surface films has been the ability to retain the solid film lubricant on a metal surface. When the solid film lubricant is inherently contained in the bulk of the metal or alloy, such as it is in the cast irons, this particular problem is avoided; the bulk of the composition serves as a reservoir for the liberation of the graphite and provides the protective surface film necessary to minimize adhesive wear.

The important point to be made with figures 4-35 and 4-36 is that the chemistry of a solid surface can be altered markedly by the presence of the alloying elements. They can appear at the surface and alter surface topography markedly, primarily by chemical effects. Chemical interaction with the surface or by the formation of surface films can prevent strong adhesion of one solid to another.

The tribological surface is the result of the surface interacting with the environment. This interaction can take the form, for example, of a metal interacting with the environment to form oxides, nitrides, or hydroxides; interaction with environments other than air can produce different surface films. Another mechanism for forming surface films which can alter the nature of tribological surfaces is that of other chemical interactions (lubricant, lubricant additive, or solid film lubricants with solid surfaces). These chemical reactions can also bring about changes in the nature of a surface. These changes take place by the simple interaction of the environmental species with the solid surface to form a surface compound. It generally reduces the surface energy markedly and changes the chemical composition of the solid surface. A further factor in practical tribological devices which even further alters the nature of the solid surface is the effect, on surface chemistry, of mechanical parameters (such as load and speed) of two surfaces in solid-state contact. Increasing the load or the sliding speed at an interface simply brings more energy to the interface, and the chemical nature of the surface, as is well known, is extremely dependent on energy. The greater the amount of energy available at the surface, the more likely chemical reactions will take place should they be favored.

An example of the way a tribological surface is affected by its chemical nature is given in figure 4-37 where an oxide-covered iron surface is examined in sliding friction experiments with an aluminum oxide rider in the presence of a vinyl chloride atmosphere. Vinyl chloride monomer was used as the environmental atmosphere. Vinyl chloride has a propensity to polymerize, and mechanical activity can assist in that polymerization process. This is demonstrated in the data of figure 4-37; as the load on the surfaces is changed, there is a change in the surface chemistry of the oxidized iron surface. At a load of 100 grams, in the vinyl chloride environment, iron and oxygen peaks associated with the iron oxide film are present on the iron surface. Also present are carbon and chlorine peaks associated with the adsorbed vinyl chloride. If the load is increased from 100 to 500 grams, the Auger spectra of figure 4-37(b) is obtained. With such an increase in load, there is a marked increase in the concentration of carbon and chlorine on the surface as is reflected in the heights of the Auger peaks for carbon and chlorine. With an increase in the load (and, hence, the quantum of energy available) at the solid surface, there is an increase in the concentration of vinyl chloride on the surface.

If the load is further increased to 800 grams, the film can no longer support the load and penetration of the film occurs by plastic deformation at asperity contacts. Additional nascent metal is seen in the Auger spectra (as reflected in greater Auger peaks for iron, fig. 4-37(c)). The load is sufficient to cause severe plastic deformation of the iron and additional exposure of iron. This is evidenced by the increase in the peak heights for the three iron Auger peaks and a decrease in the peak intensity of oxygen (compare the oxygen in fig. 4-37(c) with that seen in figs. 4-37(a) and (b)). It is readily apparent that the amount of oxygen on the surface has been appreciably reduced by the increased loading. Likewise, the amount of vinyl chloride in the surface has also been reduced, as indicated by the Auger peak heights for carbon and chlorine. The peak heights for carbon and chlorine in figure 4-37(c) are appreciably less than those in figure 4-37(b), and they are also less than those seen in figure 4-37(a).

Thus, at a load of 100 grams the vinyl chloride film that is present on the surface provides a protective surface film and keeps the rider from disrupting the iron oxide film and exposing the iron (see fig. 4-37). Increasing the load from 100 to 500 grams puts a greater demand on the system and requires the generation of a thicker vinyl chloride film on the surface. The system responds by more vinyl chloride interacting with the solid surface to form a thicker protective surface film. With a further increase in loading, however, the film can no longer support the load because the energy placed in the interface with the increased loading provides a severe amount of plastic deformation and the film cannot be replenished rapidly enough to prevent exposure of nascent metal. Figure 4-37(c) shows the exposure of metal as a result of breakdown of the lubricating film.

Figure 4-37 presents a classic example of two points: (1) analytical surface tools are effective in following the history of a lubricating species in its in-

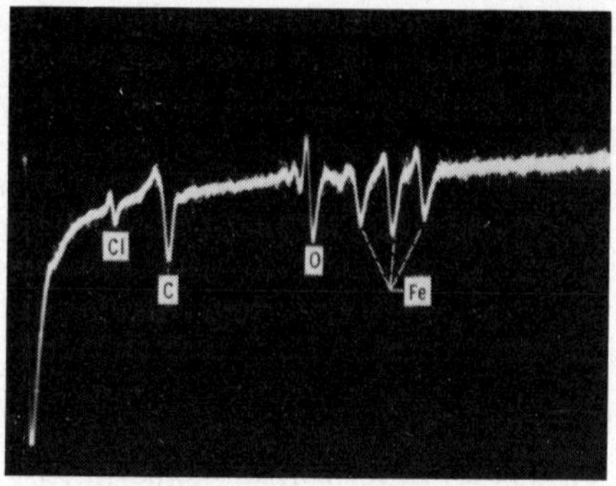

(*a*) *Load, 100 grams.*

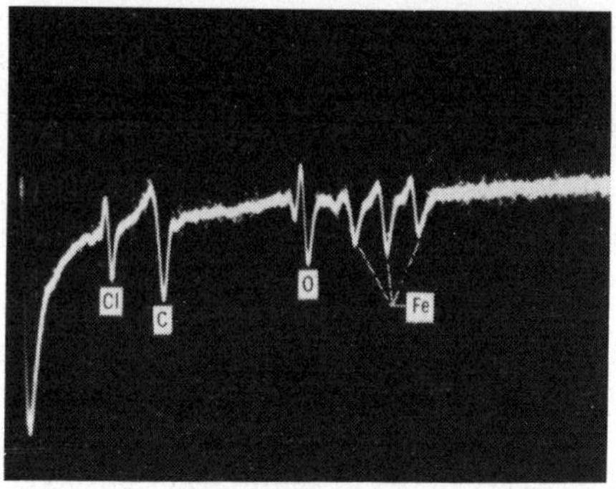

(*b*) *Load, 500 grams.*

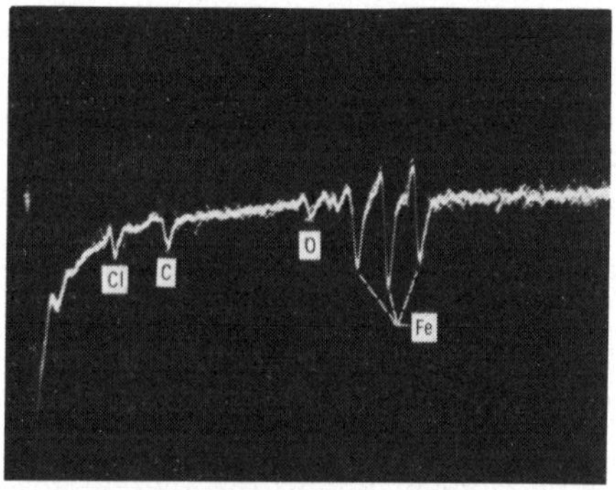

(*c*) *Load, 800 grams.*

CS-66320

Figure 4-37.—Auger spectra for oxide covered iron surface with vinyl chloride adsorbed during sliding at various loads. Ambient pressure, 10^6 torr of vinyl chloride; rider specimen, aluminum oxide.

teractions with a solid surface, and (2) the chemistry of the surface is dictated by mechanical interactions (to a degree). Chemical composition changes with a change in such mechanical parameters as load.

References

1. Williamson, J. B. P.: Interdisciplinary Approach to Friction and Wear. NASA SP-181, 1968, pp. 85-142.
2. Williamson, J. B. P.: The Theory and Practice of Tribology. Vol. 1, ch. 1, The Nature of Surfaces; sec. A, The Shape of the Surfaces. ASLE Lubrication Handbook (1980).
3. Barwell, F. T.; and Milne, A. A.: Lubrication of Materials in the Solid State. Physics of Lubrication, Brit. J. of Appl. Phys. Suppl. 1, 1951.
4. Wild, E.; and Mack, K. J.: Lubrication of Nuclear Reactor Components, Friction Systems and Liquid Sodium and Argon. Tribology, vol. 11, no. 6, Dec. 1978, pp. 321-324.
5. Hurricks, P. L.: The Friction and Wear Behavior of Amorphous Selenium Under Lightly Loaded Contact Conditions. Wear, vol. 47, 1978, pp. 335-358.
6. Rabinowicz, Ernest: Friction and Wear of Materials. John Wiley & Sons, Inc., 1965.
7. MacPherson, P. B.; and Cameron, A.: Fatigue Scoring: A New Form of Lubricant Failure. ASLE Trans., vol. 16, no. 1, 1973, pp. 68-72.
8. Cocks, M.: Shearing of Junctions Between Metal Surfaces. Wear, vol. 9, 1966, pp. 320-328.
9. Tsukizoe, T.; and Sakamoto, T.: Friction Between Hard Rough and Soft Smooth Surfaces. Jap. Soc. Mech. Eng. Bull., vol. 19, no. 127, Jan. 1976, pp. 54-60.
10. Barquins, M.; Kennel, M.; and Courtel, R.: Comportement De Monocristaux De Cuivre Sous L'Action De Contact D'Un Frotteur Hemispherique. Wear, vol. 11, 1968, pp. 87-110.
11. Sakamoto, T.; and Tsukizoe, T.: Deformation and Friction Behavior of the Junction in Quasi-Scratch Friction. Wear, vol. 48, 1978, pp. 93-102.
12. Miyoshi, K.; and Buckley, D. H.: Friction and Wear Characteristics of Iron-Chromium Alloys in Contact with Themselves and Silicon Carbide. NASA TP-1387, 1979.
13. Wedeven, L. D.: Influence of Debris Dent on EHD Lubrication. ASLE Trans., vol. 21, no. 1, 1978, pp. 41-52.
14. Weertman, Johannes; and Weertman, Julia R.: Elementary Dislocation Theory. MacMillan Co., 1964.
15. Dufrane, K. F.; and Glaeser, W. A.: Rolling Contact Deformation of MgO Single Crystals. Wear, vol. 37, 1976, pp. 21-32.
16. Yokota, H.; et al.: Ellipsiometric Study of Polished Glass Surfaces. Surface Sci., vol. 16, 1969, pp. 265-274.
17. Vedam, K.; and So, S. S.: Characterization of Real Surfaces by Ellipsometry. Surface Sci., vol. 29, 1972, pp. 379-395.
18. Ruff, A. W.: Studies of Deformation and Sliding Wear Tracks in Iron. NBSIR 76-992, National Bureau of Standards, 1976 (AD-A021295).
19. Garbar, I. I.; and Skorinin, J. V.: Metal Surface Layer Structure Formation Under Sliding Friction. Wear, vol. 51, 1978, pp. 327-336.
20. Burgers, W. G.: Rekristallisation Verformter Zustand Und Erholung. Handbuch der Metallphysik, Bd. III, Tl. 2, G. Masing, ed., Akademische Verlagsges ellschaft m.b.h., (Leipzeig), 1941, p. 356.
21. Samuels, L. E.: Damaged Surface Layers: Metals. The Surface Chemistry of Metals and Semiconductors, H. C. Gatos, ed., John Wiley & Sons, Inc., 1960, p. 92.
22. Mitchell, M. R.: Review of the Mechanical Properties of Cast Steels With Emphasis on Fatigue Behavior and the Influence of Microdiscontinuities. J. Eng. Mater. Technol., vol. 99, no. 4, Oct. 1977, pp. 329-343.
23. Campbell, I. E., ed.: High Temperature Technology. John Wiley & Sons, Inc., 1956.

CHAPTER 5

Adhesion

When two solid surfaces are brought into contact, adhesion or bonding across the interface can occur. With well-lubricated surfaces this is not generally observed. If two solid surfaces are clean and all of the adsorbates (oxides and lubricants) are removed, however, adhesion or bonding of one solid to another always occurs. The nature of the bond strength formed at the interface between the two solid surfaces is a function of the materials in contact. With metals, for example, it is a general observation that when two dissimilar metals are brought into contact the interfacial bonding between the two metals is stronger than the cohesive bonds in the weaker of the two metals. Upon application of separating forces to the bonded junction, fracture occurs in the cohesively weaker of the two materials, and this allows the cohesively weaker to be transferred to the cohesively stronger. Fracture generally does not occur at the interface.

A distinction must be made between adhesion and cohesion. In a strict sense, cohesion represents the atomic bonding forces associated within a material; that is, cohesion represents the forces that exist in the bulk of the material bonding one atom to another or one molecule to another. Thus, for example, if one cleaves a crystalline material in the bulk and generates two new surfaces, the bonds that are fractured are the cohesive bonds. When, however, two dissimilar (or even identical) materials are brought into solid-state contact with an interface, the bonding of the surface of one solid to the surface of another results in the formation of adhesive bonds. This is generally called adhesion as opposed to cohesion.

Both adhesion and cohesion are extremely important to the tribologist because they can dictate, to a very large degree, the adhesion and friction forces measured between two solid surfaces in contact. They are thus important in bearings, gears, seals, and electrical contacts. In addition, adhesive wear is one of the most severe types of wear encountered in practical lubrication devices, and adhesion and cohesion are of fundamental importance to this wear mechanism. In nearly all practical systems where two solid surfaces are in solid-state contact, some adhesion occurs, even under the most effectively lubricated situations. The function of a lubricant is simply to reduce and minimize the adhesive forces, not necessarily to eliminate them. The more effective the lubricant is, the better it does the job of minimizing the adhesive forces across an interface.

Solids can be divided into two classes: (1) the relatively brittle materials and (2) the plastic, or ductile, materials. Both types of materials are important since both are used in tribology.

Cleavage of Solids

Understanding cleavage of solids (both brittle and ductile) is important because it assists in the understanding of the surface energies associated with solid surfaces. Surface energy indicates not only the force with which surfaces can bond one to another but also the nature of interaction of clean solid surfaces with the environment. Probably the simplest way to demonstrate the cohesive bonding forces in a solid is to cleave it along its cleavage plane. This has been done by many investigators over the years. Probably the simplest, most direct, and straightforward method for cleaving surfaces is to use a simple wedge to separate the atomic planes in a relatively brittle material.

Obreimoff (in 1930) used a simple wedge to cleave the mica layers (ref. 1). The particular approach he used is shown schematically in figure 5-1 where a glass wedge of the thickness or height H is inserted between adjacent flakes of the mica. It is driven inward and develops a force in the upward direction F. A cleavage crack develops ahead of the glass wedge indicated in figure 5-1 as OC. The length of the crack is indicated by c, and the thickness of the wedge of cleaved mica or the sheet of cleaved mica is indicated by d. For a very simple experiment as indicated in figure 5-1, an indication of the cohesive bonding strength of the mica can be obtained.

Similar experiments to those of Obriemoff were conducted by J. J. Gilman and his associates. Instead of cleaving brittle solids with a glass

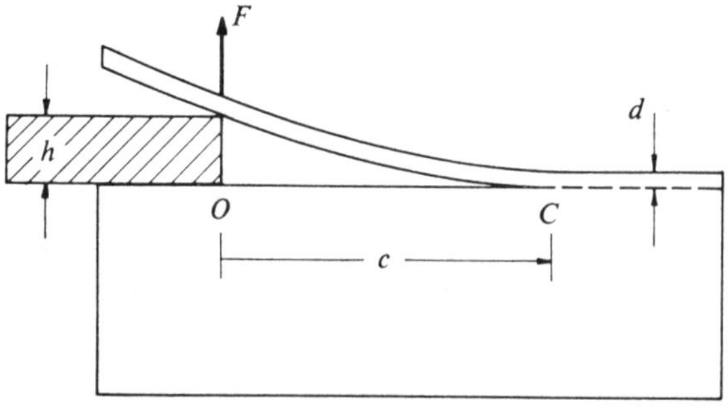

Figure 5-1.—Obreimoff's experiment on mica. Wedge inserted to peel off cleavage flake (ref. 1).

wedge, they used a hammer and chisel to cleave literally thousands of inorganic and other types of crystals (ref. 2). The cleavage experiment is important because with cleavage two new surfaces are generated. In the process, the energy required to cleave the crystal is a reflection of the energy of cohesive bonding.

Surface Energy Effects

The surface energy (energy associated with the freshly generated surface) is extremely important to adhesion, and this can be arrived at by cleavage experiments. Gilman expended considerable effort in studying the cleavage of solid surfaces as well as the generation of equations and theoretical approaches for calculating and determining the surface energy of solids. Other investigators have done likewise, and many equations have been prepared and put forth to predict surface energies of solid surfaces. The Gilman approach is a relatively simple one.

Gilman's simple equation for calculating surface energy is

$$\gamma = (E/y_o)(a_o/\pi)^2$$

where

a_o range or elastic distance of the attractive forces

E elastic modulus

y_o equilibrium lattice constant perpendicular to plane

γ surface energy

The desirable feature of this equation is that those parameters needed for making the calculation of surface energy are readily available.

When one attempts to cleave a crystalline solid, cleavage generally takes place along certain preferred planes in the crystalline solid. These are referred to as the cleavage planes. Various explanations have been given through the years to identify why certain planes cleave and others do not. Some have argued that it is the close-packed planes in the solid that preferentially cleave, but this does not explain the behavior of many crystalline solids with a zinc-blend structure (e.g., ZnS or InS) or a fluorite structure (e.g., CaF_2). Others have discussed the cleavage plane as being the plane with the minimum number of chemical bonds per unit area. This particular approach fails from the relatively vague meaning of chemical bond. Still another argument is that the anisotropic elastic constants determine the cleavage plane of a crystalline solid. And lastly, and probably the explanation which is most commonly subscribed to, the plane of minimum surface energy is the cleavage plane in the crystalline solid.

Even with this explanation there are exceptions. For example, both iron and tungsten have the lowest surface energy on the (110) planes yet these two particular metals appear to cleave along (100) planes. However, other body-centered-cubic metals, such as tantalum and vanadium, do cleave along (110) planes; that is, they follow an expected behavior based on the explanation of the lower surface energy plane being the cleavage plane.

Gilman, using the simple equation presented earlier, calculated the surface energies for the cleavage plane of various crystalline solids. The data generated by Gilman are presented in table 5-I for various brittle solids as well as for semiconductors and metals. Table 5-I lists various properties: (1) Young's modulus of elasticity, (2) lattice constant (parameter required in the equation in addition to Young's modulus), (3) lattice parameter a_o, (4) spacings of possible cleavage planes y, (5) surface energy for (100), (110), and (111) planes, and (6) observed and predicted cleavage planes. In addition, some experimental surface energy values are included.

These data are especially good in predicting the surface energies and cleavage planes of the brittle solid materials. With the metals, however, the prediction is not quite as good. For example, for copper, the calculations show that the lowest surface energy occurs on the (100) surface and the highest surface energy occurs on the (111) surface. This is contrary to measured observations which indicate that the (111) surface is the lowest energy surface. Even though it is not so indicated in Gilman's table, the data of other investigators reveal that cleavage does occur along the (111) surface in copper.

Another metallic system which does not calculate as has been experimentally observed is beryllium. The surface energy for beryllium in table 5-I indicates a higher energy on the basal plane than on the prismatic (1010) plane. This is contrary to experimental observations which show that the (0001) plane (the basal plane) has lower surface energy than does the prismatic (1010) surface.

The hexagonal metals zinc and cadmium do, however, indicate the surface energies in the proper directions; that is, they indicate higher surface energies on the prismatic (1010) surface than on the basal plane (the (0001) surfaces).

Graphite, which is known to cleave on the basal plane (and is very familiar to the tribologist), has a much higher surface energy at the edge sites of the crystallographic platelets—namely, on the (1010) surfaces rather than on the basal planes (the (0001) surfaces). For graphite, the difference is nearly 100 fold in the surface energy for the two orientations, indicating the high degree of anisotropic behavior of graphite.

While the data of table 5-I are not perfect in predicting the proper surface energy relationships and modulus of elasticity for the various solids, they do give an indication of trends; they also show that, in general, some relationship exists between the cleavage planes and the surface energy planes. In general, the cleavage planes are the low surface energy planes.

The real significance of table 5-I for the reader is that from basic, fundamental properties and the use of a simple equation calculations can be made to determine the surface energies of crystalline solids. The theoretical calculation can be a guide. The experimental values presented in table 5-I are to be used with extreme caution, because the best thing that can be said about experimental values of surface energy to date is that they are questionable; the reason is that it has been very difficult in the past to control the quality of the materials that have been cleaved experimentally to determine surface energies. Thus, it has been very difficult to get accurate surface

TABLE 5-I. – SURFACE ENERGIES[a]

Crystal	Structure	Young's Moduli (10^{11} dynes/cm²)			Lattice Constant (Å)	a_0	Spacings of Possible Cleavage Planes			Surface Energies $\gamma_{hkl} = \frac{E_{hkl}}{y_{hkl}}\left(\frac{a_0}{\pi}\right)^2$			Cleavage Plane		Surface Energy (ergs/cm²)		
		E_{100}	E_{110}	E_{111}			y_{100}	y_{110}	y_{111}	γ_{100}	γ_{110}	γ_{111}	Observed	Predicted	Experimental	Theoretical	Reference
MgO	rock salt	24.5	30.8	33.7	2.10	1.05	2.10	1.48	*	1310	2330	*	100	Yes	—	1362	17
LiF	rock salt	7.35	10.8	12.8	2.00	1.00	2.00	1.41	*	374	780	*	100	Yes	—	700, 169	18, 19
NaCl	rock salt	4.37	3.45	3.22	2.81	1.40	2.81	1.99	*	310	345	*	100	Yes	330	77–188	17, 20–24
NaBr	rock salt	3.48	2.72	2.53	2.97	1.48	2.97	2.10	*	248	288	*	100	Yes	—	76–177	17, 24
KCl	rock salt	4.80	2.30	2.00	3.14	1.57	3.14	2.22	*	318	271	*	100	No	—	56–163	17, 24
KBr	rock salt	3.70	2.15	1.88	3.29	1.64	3.29	2.32	*	306	253	*	100	No	—	54–151	17, 24
KI	rock salt	2.55	1.31	1.56	3.53	1.76	3.53	2.49	*	233	165	*	100	No	—	58–136	17, 24
PbS	rock salt	8.34	4.1	3.5	2.98	1.49	2.98	2.10	*	625	440	*	100	No	—	—	
AgCl	rock salt	3.21	2.27	2.06	2.87	1.43	2.87	2.03	*	230	230	—	—	—	—	—	
AgBr	rock salt	3.29	2.02	1.79	2.77	1.38	2.77	1.96	*	230	200	—	—	—	—	—	
CaF_2	fluorite	14.1	10.1	9.2	5.45	1.36	*	*	1.56	*	*	1100	111	Yes	—	—	
ZnS	zinc blende	4.88	7.4	8.9	5.43	0.96	*	1.92	*	*	360	*	110	Yes	—	—	
InS	zinc blende	4.14	6.2	7.5	6.45	1.14	*	2.28	*	*	359	*	110	Yes	—	—	
Glass		45.				0.8	> 1.6			< 1820					1214	—	25
Ge	diamond	10.2	13.7	15.5	5.66	1.23	1.41	2.00	3.26	1110	1050	726	111	Yes	—	—	
Si	diamond	13.1	17.5	19.9	5.42	1.17	1.35	1.91	3.12	1350	1270	887	111	Yes	—	1232	26
C	diamond	105.0	116.0	120.0	3.57	0.77	0.39	1.26	2.06	7050	5500	3500	111	Yes	—	5650	13, 27, 28
W	b.c.c.	38.9	38.9	38.9	2.87	1.24	1.58	2.23	0.91	4680	3320	8130	100	No	—	2680, 5510	13, 29
α-Fe	b.c.c.	13.2	22.2	28.4	3.16	1.37	1.43	2.02	0.83	1440	1710	5340	100	Yes	—	1450, 1600	29, 30
Na	b.c.c.	0.21	0.55	1.41	4.29	1.85	2.14	3.03	1.24	33	63	389			—	94–190	29, 31–33
Cu	f.c.c.	6.67	13.1	19.4	3.62	1.28	1.81	2.56	1.04	590	820	2980			1100, 1370	500–1210	29–35
		E_{0001}	$E_{10\bar{1}0}$				y_{0001}	$y_{10\bar{1}0}$		γ_{0001}	γ_{1010}						
Be	h.c.p.	29.7	26.6		2.29, 3.58	1.11	1.80	1.98		2060	1680		0001	No	—	—	
Zn	h.c.p.	12.5	10.9		2.66, 4.93	1.33	2.46	2.30		185	850		0001	Yes	—	600, 772	29, 35
Cd	h.c.p.	2.8	8.1		2.98, 5.62	1.49	2.81	2.58		226	730				—	546, 800	29, 35
C	graphite	1.8	113.0		—	0.71	3.4	2.46		27	2340		0001	Yes	—	—	
Te	selenium	2.06	4.28		4.5, 5.91	2.44	1.73	3.86		720	670		$10\bar{1}0$	Yes	—	—	

* Nonneutral planes

[a]Reference 2.

energies, because most of the surfaces that have been used to generate the surface energy measurements have not been well defined by surface analyses prior to making surface energy measurements.

All of the rock salt type crystalline structures in table 5-I are observed to cleave along (100) planes. It is also interesting to know that the diamondlike structures for materials such as germanium, silicon, and carbon all cleave along the (111) surfaces and that the hexagonal metals beryllium, zinc, and cadmium cleave along the (0001) or basal planes. This particular mechanism of cleavage along the basal planes does not occur for all hexagonal metals. The preferred cleavage plane varies with the particular metal just as it does with the face-centered-cubic metals. In table 5-I both tungsten and iron are observed to have cleavage planes of the (100) structure, yet the low surface energy planes for these metals are the (110) surfaces. Some investigators ascribe the difference between the actual observed cleavage plane and predicted cleavage plane (based on surface energy considerations) as being due to the impurities of the materials, since other body-centered-cubic metals, such as vanadium and tantalum do, in fact, cleave on the low surface energy or the (110) surfaces.

Ideally, one would anticipate, from a fundamental consideration of the atomic nature of solid materials, that the crystal planes in a solid that exhibit the highest modulus of elasticity normal to them, with the greatest atomic packing in the plane and the greatest amount of spacing between adjacent planes, would have the lowest surface energy. The reason is that, if the atomic packing is maximized within the plane, there would be fewer bonds available for bonding across the interface. As a consequence, the energy of binding to the next adjacent plane would be much lower.

It is relatively easy to experimentally cleave crystalline solids along the cleavage plane with brittle materials (such as the rock salt structures of table 5-I) using the approach (chisel and hammer) taken by Gilman and his coworkers. When ductile materials like metals are cleaved, however, an entirely different kind of behavior is observed. In ductile materials, before rupture of bonds occurs, there is the ability for the crystalline solid to undergo plastic deformation or strain. This plastic deformation or strain takes place before fracture sets ductile crystalline solids apart from the more brittle materials (where an atomic bond simply fractures on cleavage).

With the ductile materials, the attempt to separate or produce fracture along a plane in the crystalline solid results in coalescence of holes due to the generation of defects (such as dislocations) in the crystalline solid. The fracture then occurs by a stretching and rupturing of atomic bonds after the coalescence of defects to the bond site locations (fig. 5-2(a)) and/or the fracture of atomic bonds (fig. 5-2(b)). It is more like an unzipping process than like a simple fracture or rapid breaking of the bonds as observed with the more brittle materials.

The higher the surface energy of a solid surface, the stronger the bonds of cohesion (or adhesion of a metal to itself). A comparison of the metals platinum, copper, and zinc in table 5-II indicates that there is a relationship between the surface energies and the energy of self-adhesion or cohesion of metals (ref. 3). Platinum, copper, and zinc were examined in adhesion experiments, and the coefficients of adhesion are presented in table 5-II along with surface energy values for these three metals. The coefficient of adhesion is the ratio of the force to separate the solid surfaces once brought into solid-state contact to the load with which the surfaces are initially pressed into contact.

Platinum, which has the highest surface energy in table 5-II, also has the highest adhesion coefficient; zinc, which has the lowest surface energy, also has the lowest adhesion coefficient. These data are consistent with those one might anticipate from the energetics of the solid surfaces involved.

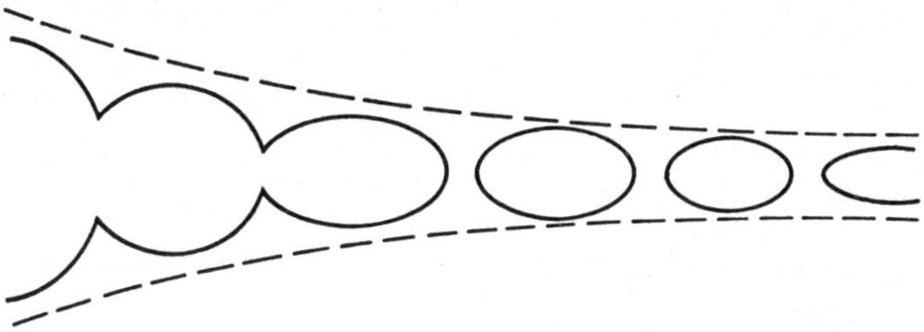

(*a*)

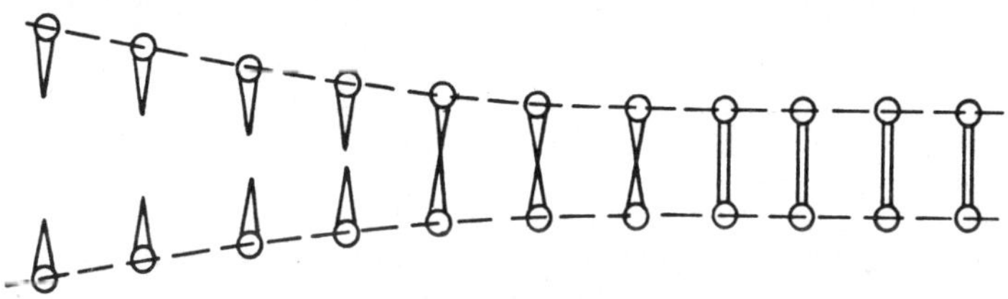

(*b*)

(*a*) *By growth and coalescence of holes through viscous or plastic flow.*
(*b*) *By stretching and rupture of atomic bonds.*

Figure 5-2. Cumulative dilutational fractures.

TABLE 5-II. – SURFACE ENERGIES AND SELF-ADHESION[a]

Metal	Surface energy[b] at 700° C, mJ/m^2	Median coefficient of adhesion
Pt	3200	0.8
Cu	2200	.7
Zn	740	.2

[a]Reference 3.
[b]All data extrapolated or interpolated except those for Al, Mg, Zn, and Sn.

Field Ion Microscope Adhesion Experiments

In practical tribological devices, metals are seldom in solid-state contact with themselves. More frequently, dissimilar metals are encountered in solid-state contact. Detailed studies have been conducted with dissimilar metals in contact using the FIM as a tool to study the nature of the contacts. Field ion microscopy was discussed briefly in chapter 2.

An FIM modified for the adhesion experiments is shown schematically in figure 5-3. The pin tip is mounted on two posts which connect directly with the cold finger for cryogenically cooling the one specimen or the pin tip. The FIM is inside a vacuum chamber. A bellows is mounted at the top of the chamber to allow deflection and movement of the pin tip. A high voltage lead is applied to the pin tip to bring voltage to the tip for imaging purposes. The system contains a phosphor-coated fiber optic window for imaging the pin tip. The phosphor-coated fiber optic window of figure 5-3 is approximately 10 centimeters from the pin tip. A vacuum system is provided for evacuating the system and a bleed-in valve for bleeding imaging gas into the chamber.

Adhesion experiments are conducted with a beam mounted inside the chamber that contains a flat specimen which is brought into touch contact with the pin tip and the load is applied by electromagnets. The forces of

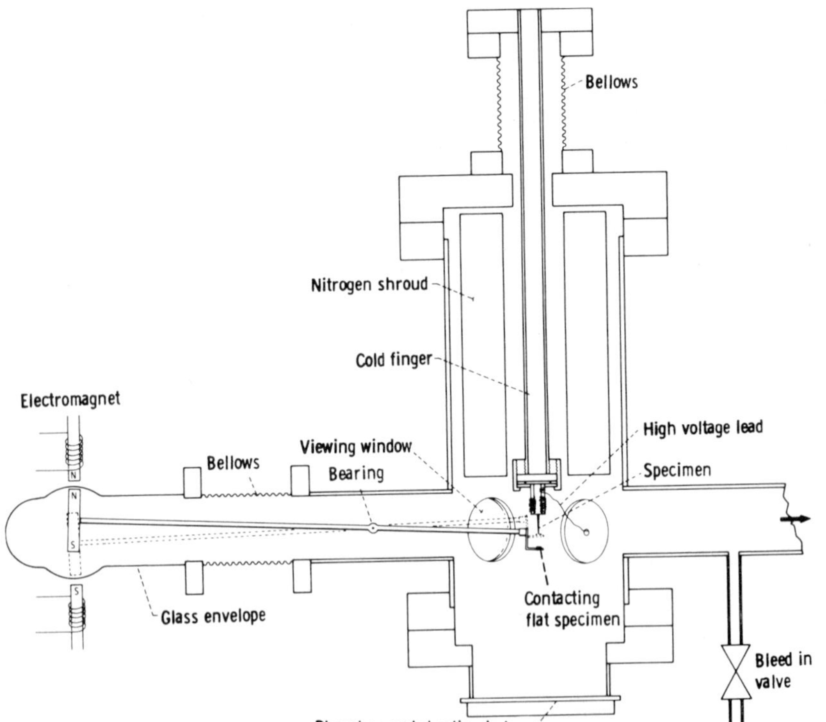

Figure 5-3.—Field ion emission microscope adhesion apparatus.

adhesion are measured by measuring the current on the electromagnets required to separate the two solid surfaces.

An enlarged view of the beam and its operation are shown schematically in figure 5-4. The beam is mounted on a taunt band rather than the bearing as indicated in figure 5-3, and a photocell is used to control the movement of the beam so the beam motion is relatively uniform. The electromagnets are outside the glass envelope, or vacuum tube, with a permanent magnet mounted inside the chamber. The permanent magnet is deflected by applying current to the electromagnets, and it can be moved up or down by the control of this current. The photocell is tied into the circuit for a feedback of the impulse to control the motion of the electromagnets. The emitter tip is shown making contact with the flat specimen at the end of the beam, opposite that containing the magnets.

The apparatus shown in figures 5-3 and 5-4 is extremely useful in studying the adhesive behavior of metals in solid-state contact because the FIM is still the only analytical surface tool that can give an atom by atom structural picture of a solid surface.

Figure 5-5(a) is a photomicrograph of a tungsten surface at an imaging voltage of 16 kilovolts prior to being contacted by a second solid surface. As was mentioned in chapter 2, each white spot indicates an atom site. Since the surface being viewed in figure 5-5(a) is a radius, a number of crystallographic planes are exposed; each ring represents a different atomic plane on the solid surface.

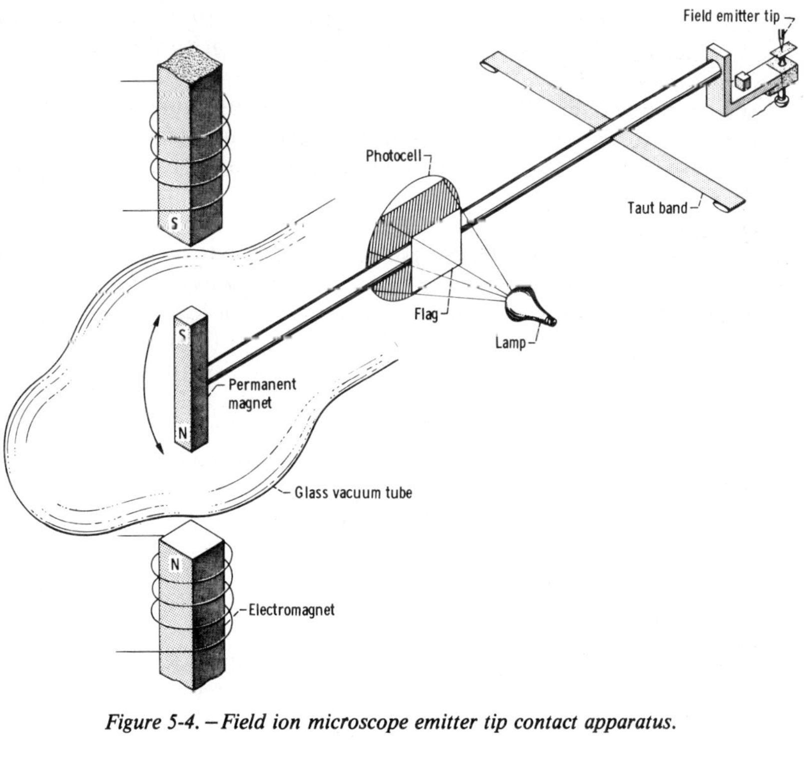

Figure 5-4. – Field ion microscope emitter tip contact apparatus.

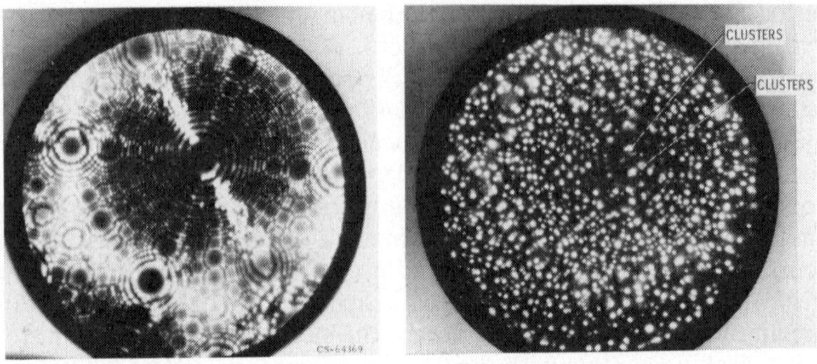

(*a*) *Before; at 16.0 kilovolts.* (*b*) *After; at increased image voltage.*

Figure 5-5. – Field ion micrographs of tungsten surface before and after being contacted by gold in adhesion experiment.

When a flat of gold is brought in contact with a tungsten pin tip, the image seen in figure 5-5(a) is markedly changed after that contact. This image is shown in figure 5-5(b). There are clusters of gold which have transferred to the tungsten surface. A very careful examination of the photomicrograph of figure 5-5(b) reveals the ring structure associated with the tungsten beneath the gold atoms which have adhered to the solid surface. For dissimilar metals in contact, the interfacial bonds are generally stronger than the cohesive bonds of the cohesively weaker of the two materials. The tungsten-gold contact supports that hypothesis in that the gold (the cohesively weaker of the two metals) is seen to transfer to the tungsten in clusters in figure 5-5(b).

A very careful and detailed analysis of the clusters of figure 5-5(b) reveals that there are three gold atoms bonded to each tungsten atom on the tungsten (110) surface; this indicates a possible compound formation on the solid surface with the structure WAu_3. A careful examination of the metallurgical literature fails to reveal any compounds between tungsten and gold in bulk alloy chemistry. These results, then, indicate the importance of characterizing surfaces very carefully because the behavior of metals and materials on the surface can be entirely different from the bulk.

The gold in figure 5-5(b) could be readily removed from the surface by field evaporation. Field evaporation consists of imposing a slightly higher voltage than the imaging voltage of 16 kilovolts shown in figure 5-5(a). The voltage is sufficiently high to cause the severing of the bonds between the gold atoms and the tungsten; this results in freeing the gold (gold comes off the surface and leaves the tungsten behind). Ultimately, one can reobtain the tungsten surface seen in figure 5-5(a).

Incorporating the atom probe permits the analysis of the species as they leave the surface. Since an atom probe allows an analysis of the chemistry of a single atom on the solid surface, it was used to identify the presence of gold on the tungsten in figure 5-5.

From a measure of the added voltage required to produce field evaporation of the gold from the tungsten surface, one can get an indication of the relative bond strength of the gold to the tungsten through some appropriate calculations. Such measurements were made for gold and a number of other metals in contact with tungsten, and the adhesive binding energies of these various metals to tungsten were calculated (ref. 4). The results of these measurements are presented in figure 5-6.

In figure 5-6 the adhesive energy is plotted as kilocalories per gram atom of the metal for various metals in adhesive contact with tungsten. The metal which exhibits the strongest binding force to tungsten is tungsten itself; that is, the energy of cohesion (self-adhesion) is stronger than the energies of dissimilar metals bonding to the tungsten surface. That is the first observation to be made from the data of figure 5-6.

Another observation to be made from figure 5-6 is that the members of the platinum metals family (osmium, iridium, and platinum) show a decreasing adhesive energy to the tungsten as the cohesive energies of the family members decrease. In other words, the energy of cohesion in osmium is greater than that in iridium, and that in iridium is greater than that in platinum. If the cohesive binding energies are compared with the adhesive energy values plotted in figure 5-6, a correlation of cohesive binding energies of the metals and the adhesive energies of the metals to tungsten is seen. The bonding at the interface between the metals and tungsten in each case of figure 5-6 is stronger than the bonding in the cohesively weaker of the two materials. In other words, in each case in figure 5-6, the metal in contact with tungsten is cohesively weaker than tungsten. Consequently, when the solid surfaces are separated after adhesive contact has been made, fracture occurs in the cohesive bonds in the metal brought into contact with tungsten. For that reason a correlation exists with the members of the platinum metals family and their cohesive binding energies.

Of all the metals brought into contact with tungsten in figure 5-6, the gold bonds to the tungsten in the weakest fashion. Despite this weak bonding, it was speculated that a gold-tungsten compound forms on the solid surface (fig. 5-5). Thus, despite compound formation, the bonding is still not as strong as it is for the other metals (platinum, iridium, osmium, rhenium,

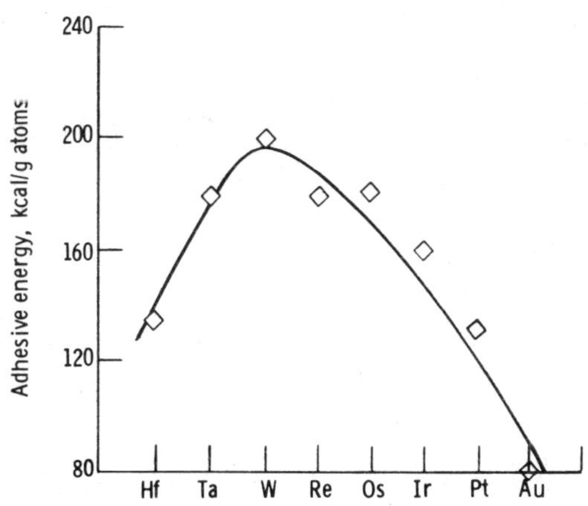

Figure 5-6. – Adhesive binding energy of 5d transition metals to tungsten.

tantalum, and hafnium) in contact with the tungsten. This is because the cohesive bonds in the gold are weaker than they are in any of these other metals.

The adhesion of all dissimilar metals is not characterized by the formation of possible surface compounds as is observed with gold and tungsten in figure 5-5. In the contacting of many metal couples, other phenomena are observed on the surface. In some instances, actual epitaxial transfer takes place; that is, the material that is bonded to the pin tip surface takes on the topography of the pin tip. The orientation of the surface film essentially mirrors the orientation of the substrate. The atoms are bound in the same lattice that is possessed by the contacted substrate (or pin tip).

In addition to epitaxial transfer (which occurs, e.g., for gold in contact with a pin tip of iridium), there are also situations where the transfer is in clusters on the solid surface. There are sites (where transfer occurs) which are preferred over other sites because of variations in surface energies on the solid surface of the two materials in contact.

Figure 5-7(a) shows a field ion micrograph of an iridium surface prior to being contacted by a platinum flat; the individual white spots again indicate individual atom sites, and the rings indicate atomic planes. The cross of white spots, in the center, indicates the zone decorations on the solid surface.

After the iridium surface has been contacted by a flat of platinum, the image obtained in figure 5-7(b) is observed; the iridium surface is almost completely covered by the transferred platinum. The very bright spots on the solid surface are clusters or clumps of platinum which have transferred to the iridium surface and have been plucked out of the platinum because of the fracture of cohesive bonds in the platinum. Again, the adhesive bonds at the interface are stronger than the cohesive bonds in the cohesively weaker of the two materials (platinum); hence, platinum transfers to the iridium surface.

Because the voltage which is used to image the surface is concentrated at these asperities, the pseudoasperities that are generated on the iridium surface by the transfer of the platinum are field evaporated very rapidly; after a short time, the parent iridium surface is obtained as it existed prior to the

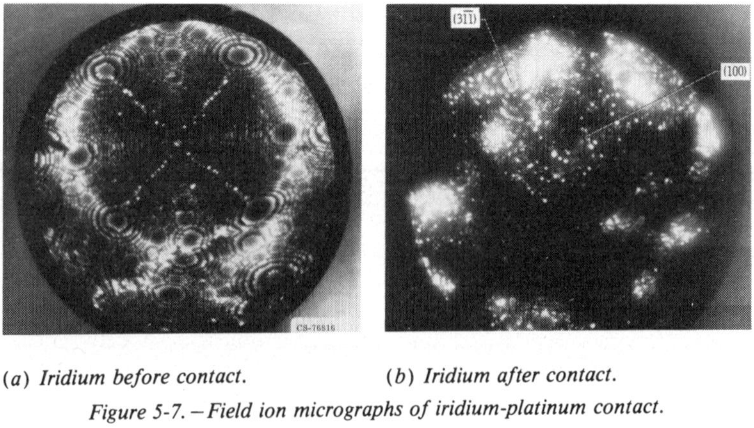

(a) Iridium before contact. *(b) Iridium after contact.*

Figure 5-7.—Field ion micrographs of iridium-platinum contact.

solid-state contact. The transfer of clusters of material is observed to occur most readily on high atomic energy surface planes.

It should also be noted that, in the experiments of figure 5-7, there was a slight bit of tangential motion or vibration imposed on the couple when in solid-state contact. In the absence of such vibration, there is an epitaxial transfer of platinum to the iridium. Simple touch contact and tensile fracture of the adhesive junction shows an epitaxial transfer of platinum to the iridium. If, however, one imposes a slight bit of tangential motion by allowing the vibration to exist in the system and to generate some tangential component of force, the results of figure 5-7(b) are obtained.

Thus, the manner of adhesion in the adhesive transfer from one surface to another is influenced by mechanical parameters as well as by the chemical nature of the solid surfaces and the surface energies involved. Simple touch contact with a tensile fracture of the adhesive junction yields epitaxial transfer of platinum to iridium; solid-state contact to the platinum-iridium under the same load conditions with a slight bit of tangential motion (due to vibration) markedly alters the transfer mechanism from epitaxial transfer of platinum to iridium to cluster transfer of platinum to iridium (fig. 5-5-(b)). The transfer shown in figure 5-7(b) generates the pseudoasperities, which can be harmful to machine elements when they occur.

Iron is probably the most widely used single element in tribologocal systems. It is used to make steel, and steels are, to a great extent, used in bearings, gears, and mechanical seals. The fundamental behavior of iron in adhesion is extremely important in tribology. Adhesion is important from the interaction of iron with itself, with other metals, and with environmental constituents.

LEED Adhesion Experiments

LEED is a very useful tool for studying the crystalline behavior of solid surfaces—that is, the structural arrangement of atoms in the outermost atomic layer of a solid surface and the changes that take place in that surface when an adhesion experiment in solid-state contact occurs.

In solid-state contact, however, it is always important to know the chemistry of the surface as well as the structure. Consequently, LEED and AES analysis are used in combination to characterize a solid surface in adhesive experiments. LEED gives the structural arrangement of atoms in the outermost atomic layer (as described in ch. 2), and AES analysis gives a chemical analysis of all the elements (except hydrogen and helium) present on the solid surface to a sensitivity of 1/100 of a monolayer.

The apparatus that is used for this type of study is shown schematically in figure 5-8. The apparatus consists of a small fiber specimen, which is a small cylinder that has a flat end on it, and a crystal specimen, which is about the size of an aspirin tablet. The crystal specimen can be a single crystal and the fiber a polycrystal, or both can be polycrystalline. However, if both are polycrystalline, the effectiveness of LEED is reduced because LEED is primarily for looking at single crystal surfaces. It can, however, be used to study individual grains on large grain polycrystalline samples. The electronics for the LEED system electron gun and detection screen are posi-

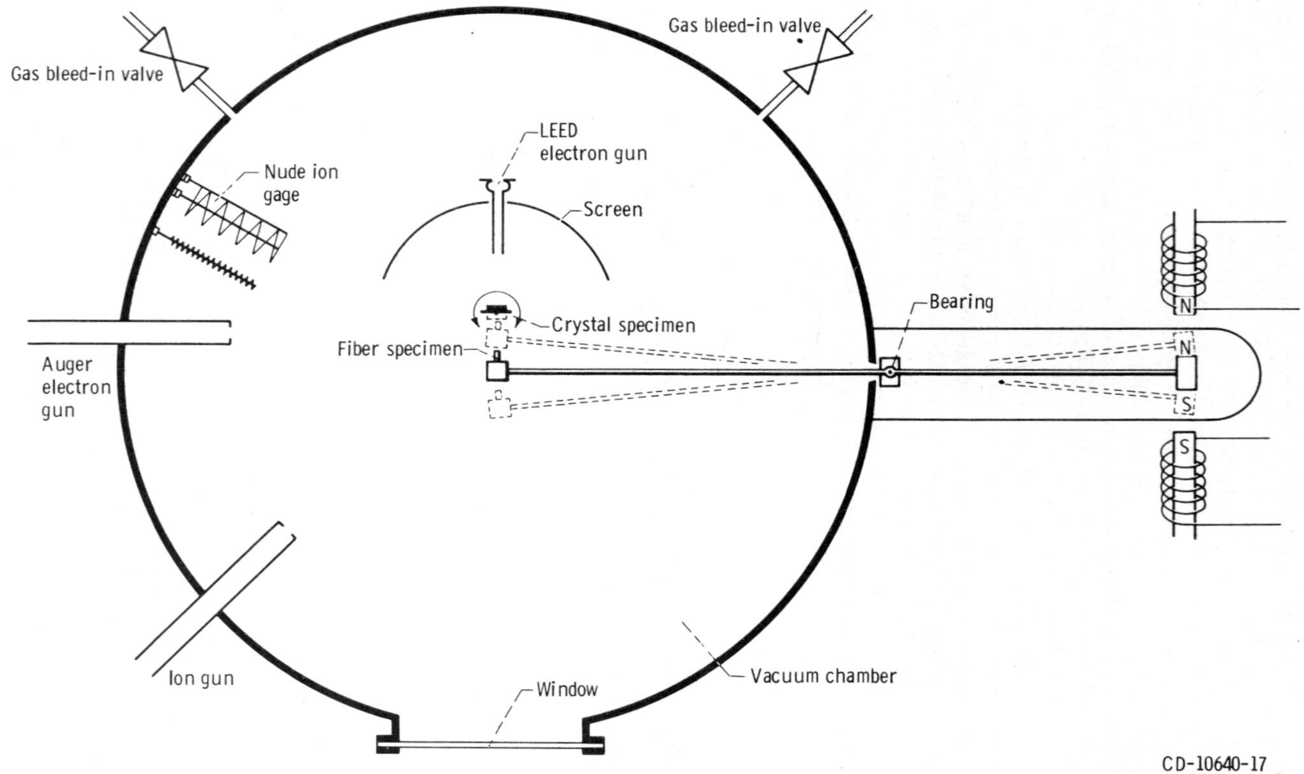

Figure 5-8. – Low-energy electron-diffraction (LEED) adhesion apparatus.

tioned such that a beam of electrons is directed at the crystal surface; the electrons are diffracted from the surface to the phosphor-coated screen where the individual diffraction spots are displayed. From the window a photograph can be taken of the diffraction pattern on the screen. The crystal can be rotated slightly (approximately 45°) and an Auger electron gun, the beam of which is directed at the crystal surface, can be used to detect the chemical species present on the solid surface. The screen, or grids, associated with the LEED system are used for Auger detection as well.

Once the crystal surface is very carefully characterized, the crystal can then be rotated 180° for adhesion experiments. If, however, as is generally the case, residual surface contaminants are present on the crystal surface, a specimen can be rotated so that it is facing the ion gun indicated in figure 5-8. The crystal surface can then be bombarded with a stream of argon ions in a voltage range of approximately 1000 volts to clean the specimen surface. The argon ions that impinge on the crystal surface remove adsorbates, oxides, and other surface contaminants to produce a clean metal surface. This surface can then be again characterized with LEED and AES analysis.

The system contains gas bleed-in valves for bleeding in environments to which the crystal is to be exposed. The adhesion forces between the fiber specimen and the crystal surface are measured by deflecting the beam inside the vacuum chamber. Two electromagnets are used to deflect the permanent magnet on the end of the beam opposite that containing the fiber specimen. Varying the current on the electromagnets controls the amount of deflection on the beam and the force with which the fiber specimen is pressed against the crystal surface. The entire system is evacuated by a vacuum pumping system consisting of absorption and ion pumps; a pressure of 10^{-10} torr is a typical operating pressure for the system during the course of experimentation.

The load can be applied to the specimens in contact by increasing the current to the electromagnet; that tends to force the fiber specimen more heavily against the crystal specimen. The adhesion force is then based on the current required to cause the electromagnet to pull the fiber specimen away from the crystal specimen surface.

If one were to construct a model of atoms on a solid surface from marble balls (i.e., to take the marble balls and place them in such an arrangement or array as to simulate a solid surface of one of the crystallographic faces of iron), say the (011) face, they would obtain an arrangement of atoms such as that indicated in the marble model of figure 5-9. In figure 5-9 the marbles are arranged just as atoms would be found on an (011) surface.

The black rectangle of figure 5-9 indicates the unit mesh which characterizes the atom arrangement on the solid surface that is detected by LEED, a_0 being the lattice parameter. The major azimuths are shown with the vectors indicated in the lower portion of the marble model. Thus, if a LEED pattern is observed for a single crystal iron (011) face, four white diffraction spots should be seen arranged in a rectangular array reflecting the arrangement of the atoms indicated in the rectangle (unit mesh) of figure 5-9; the four corners of the rectangle are the four atoms displayed and repeated in the pattern. The display is basically a display of the two-dimensional lattice.

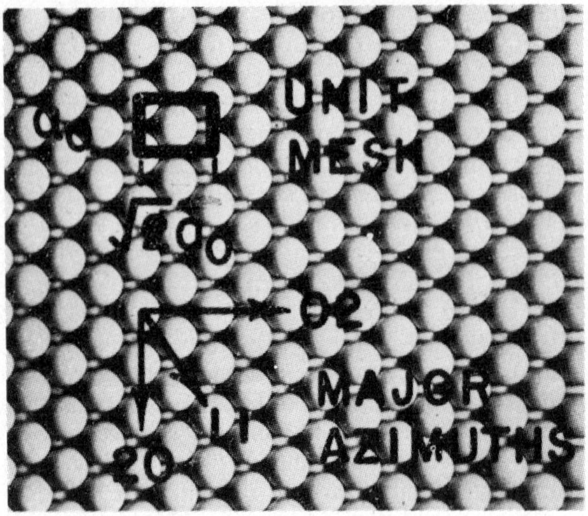

Figure 5-9. – Marble model of (011) face of iron.

Figure 5-10 shows three LEED patterns of an iron single crystal surface of the (011) orientation. A diffraction pattern for the iron (011) surface is shown in figure 5-10(a). The four very bright diffraction spots reflect the unit mesh shown in figure 5-9 from the ball model and reflect the iron atoms present in the solid surface; they also identify the orientation of those atoms.

This particular iron sample had been cleaned by bombardment and subsequent heating. However, the heating brought a contaminant to the surface which caused the additional diffraction spots to form a ringlike structure encompassing the iron diffraction spots. The additional small diffraction spots present in the LEED pattern of figure 5-10(a) are due to carbon. The carbon was identified with the use of AES analysis, and the Auger spectrum obtained for the surface presented in figure 5-10(a) is shown in figure 5-11. Auger peaks are detected for carbon as well as iron on the solid surface.

There are no other elemental peaks in the spectrum; this indicates that those diffraction spots not associated with iron must, therefore, be associated with carbon (fig. 5-10(a)). Other investigators have observed similar patterns with other metals. For example, carbon has been found to form a ring-type structure on a platinum surface.

If the surface of figure 5-10(a) is argon ion bombarded, the carbon contamination on the surface can be removed, as is revealed by the LEED pattern in figure 5-10(b). Removing the carbon, however, leaves the iron in a strained state because of the bombardment with the argon ions, and the diffraction spots are no longer circular and bright but become diffused and elongated.

AES analysis of the surface in figure 5-10(b) reveals that carbon was removed from the surface. A modest heating to 200° C for approximately 15 minutes is sufficient to produce an annealing of the strain in the outermost layer of the iron solid surface. Annealing and removing strain result in

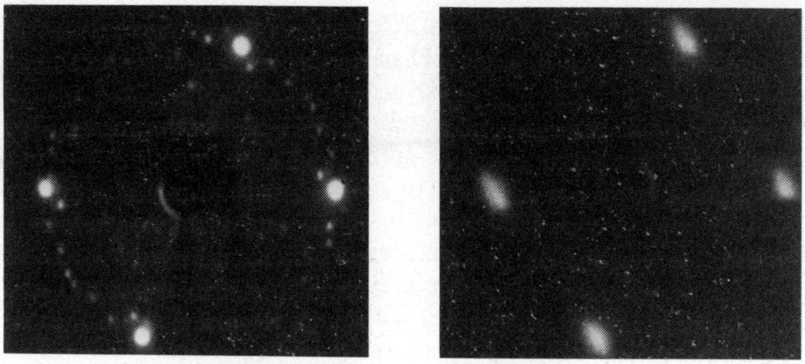

(*a*) *Carbon contaminants.* (*b*) *Argon bombarded.*

(*c*) *Clean surface* (*110 V*). CS-78-1970

Figure 5-10. – LEED patterns of iron (011) surface with carbon present and after argon ion bombardment.

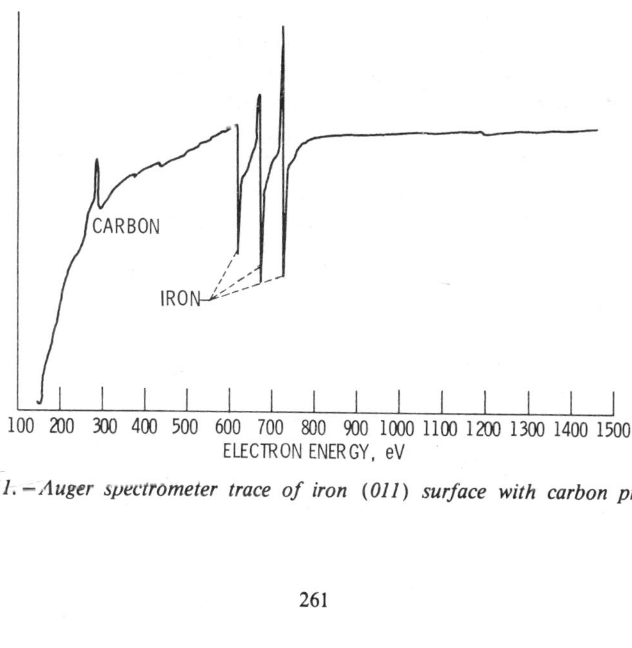

Figure 5-11. – Auger spectrometer trace of iron (011) surface with carbon present on surface.

the formation of sharp diffraction spots (fig. 5-10(c)) for the clean iron surface. Figure 5-10(c), then, is the LEED pattern for a clean iron surface; it reveals only four spots in a rectangular array such as was observed for the unit mesh in figure 5-9 of the ball model.

Work with LEED and Auger in the apparatus shown in figure 5-8 is instrumental in revealing the true nature of solid surfaces used in tribology. It became readily apparent in working with such surfaces that it is extremely difficult to obtain atomically clean surfaces. For example, a modest surface heating (fig. 5-10(c)) can result in the segregation of carbon to the solid surface with the surface becoming coated with carbon just as it was in figure 5-10(a). The bulk of the iron serves as a reservoir for carbon, even though the original iron material was triple zone refined (containing only parts per million of carbon); the carbon chooses to segregate on the solid surface of the iron with slight heating or straining of the metal. Even when all the carbon has been removed, additional heating brings about the segregation of other minor alloy constituents (which may be present in the iron in only impurity levels). For example, in some iron samples, the complete removal of the carbon resulted in freeing sulphur, and the surface became contaminated by sulphur as a result of the selective segregation of sulphur on the surface once the carbon was expended from the bulk iron.

It appears that carbon suppressed the segregation of sulphur. As long as carbon was present in the bulk metal at a fairly high concentration (parts per million), the presence of carbon was sufficient to prevent the segregation of the sulphur. Once the carbon concentration became sufficiently low, however, the sulphur segregated on the solid surface where it was observed with both LEED and AES analysis.

The difficulty in obtaining a clean iron surface suggests that much of the research work in adsorption and oxidation which took place in the 1940's and 1950's with surfaces that had supposedly been characterized may be relatively useless because of the absence of well-defined surfaces. The iron that many of the researchers may have been working with in adsorption and reaction studies may not have been clean iron surfaces at all. Rather, they may have been selectively contaminated surfaces, either from contaminants in the environment (from contaminanted vacuum systems) or from contaminants that may have come from within the bulk to segregate on the surface (as with carbon and sulphur). All the impurities, whether they come from the environment or from the bulk metal themselves, poison any surface reaction or surface chemistry. These interactions would be particularly detrimental to any studies of chemical kinetics because extremely small amounts on a surface can markedly poison such kinetics.

A typical iron surface prior to cleaning contains a number of surface contaminants. In figure 5-12(a) is an Auger spectrum (obtained in the apparatus of fig. 5-8) of an iron surface prior to cleaning; no heating or bombardment with argon ions was done to the surface. The Auger spectrum shows that a number of elements are present on the surface. There are Auger peaks (three of them) associated with iron, an oxygen peak associated with iron oxide and adsorbates, a carbon peak associated with segregated carbon and/or adsorbed carbon monoxide and carbon dioxide, and a sulphur peak.

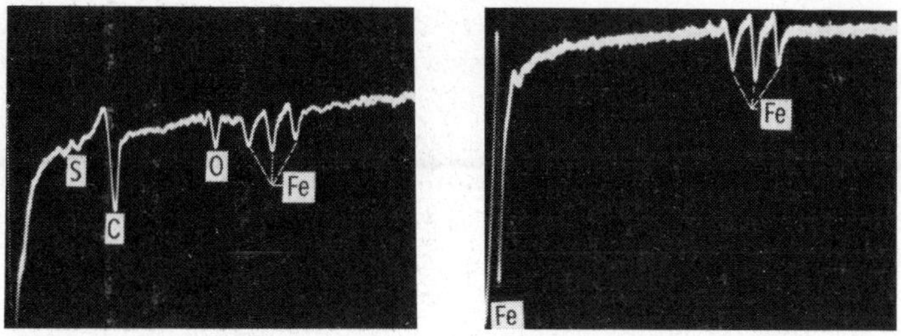

(*a*) *Before.* (*b*) *After.*

Figure 5-12. – Auger spectra for iron surface before and after sputter cleaning. High purity iron vacuum zone refined.

Normally the spectrum for clean iron contains four Auger peaks, three high energy electron Auger peaks and one at a much lower energy. The low energy electron peak for iron is extremely sensitive, however, to surface contaminants, and the presence of the slightest amount of surface contamination completely masks its presence as is done in figure 5-12(a). Only three high energy Auger peaks for iron are detected. The presence of the sulphur, carbon, and oxygen contaminants on the surface mask the low energy Auger peak.

If the surface is very carefully cleaned by argon ion bombardment and a subsequent anneal to remove strain, the Auger spectra of figure 5-12(b) is obtained. All the sulphur, carbon, and oxygen have been removed by the argon ion bombardment. The only peaks detected in the Auger spectrum are the four peaks for iron, the three high energy peaks which are shown in figure 5-12(a) plus the low energy Auger electron peak that is very prominent in figure 5-12(b). It was completely absent in figure 5-12(a) when surface contaminants were present.

A combination of the Auger spectrum of figure 5-12(b) plus the LEED pattern of figure 5-10(c) characterizes an iron (011) surface both structurally and chemically for adhesion studies. The surface, when it is characterized as clean with these two tools, is then in an atomically clean state.

The tribologist is interested in an atomically clean metal surface because, in even the most well-lubricated practical systems, solid-state contact can occur as a result of intimate interactions between two solid surfaces with deformations occurring at the interface and the resultant exposure of nascent metal. Thus, an atomically clean surface represents the worst possible condition with respect to destruction of surface topography, friction, adhesion, and adhesive wear.

Adhesion to Iron of the Noble Metals Copper, Silver, and Gold

The noble metals are used in frequent tribological couples to contact iron surfaces. Copper, in particular, dates back many years in its interactions with iron. For example, in automotive applications, conventional leaded bronze bearings are copper-base alloys that contact shafts of steel or other iron-base materials. In these contacts, even with effective lubrication, transfer of copper to the steel has been observed. This reflects adhesion of copper to iron (of the steel) with adhesive transfer taking place even with effective lubrication of the surfaces.

To gain a better fundamental understanding of the nature of the interaction between the noble metals and iron, some adhesion experiments were conducted in the apparatus of figure 5-8 with noble metals contacting the (011) surface of iron.

LEED patterns obtained after the three noble metals copper, silver, and gold were in adhesive contact with iron are presented in figure 5-13. The micrographs of figure 5-13 indicate that all three noble metals transfer in the same way to iron; that is, the basic diffraction patterns, obtained for the noble metals having contacted iron, are identical. This is shown schematically in the lower left of figure 5-13. While the diffraction spots are not readily apparent in all three photographs, the basic patterns are identical on careful examination.

Thus, the first observation to be made from the LEED patterns, and confirmed by AES analysis, is that there is transfer of all three noble metals; that is, all three noble metals adhere to the iron surface in an orderly fashion. The diffraction pattern obtained reflects an ordering on the solid surface; it is not a random transfer, but rather an orderly one.

The second observation to be made from the LEED pattern coupled with Auger analysis is that the interfacial bond formed between all three noble metals and the iron is stronger than the cohesive bond in the weaker of the two materials. In this case, the noble metals are all weaker cohesively than the iron, so the noble metals, therefore, transfer to the iron. This is similar to the observations that were made earlier with various metals in contact with tungsten. The cohesively weaker always transfers to the cohesively stronger and leaves the transferred material on the cohesively stronger surface.

The third observation to be made from the LEED data is that all three noble metals transfer in the same way to the solid surface of iron; that is, given a fixed iron orientation, all three noble metals brought into solid-state contact with that orientation of iron transfer and arrange themselves on the solid surface in the same way, despite differences in lattice registry among the various noble metals with the iron. These results suggest a certain degree of mobility in the adhesively transferred material because it is hard to conceive that, upon solid-state contact, the noble metals would all bond to the iron in the same manner. Since there are differences in lattice spacings in the surface planes of all three noble metals, it might be anticipated that this difference in spacing would be reflected in the way the noble metals transfer to the iron. The fact that it is not seems to indicate that some rearrangement or some movement takes place on the solid surface of the iron after the material has been transferred.

In the adhesion and bonding of an iron (011) surface, the possible sites where this bonding can occur must be considered. At this stage in the development of LEED, considerable uncertainty exists in the interpretation

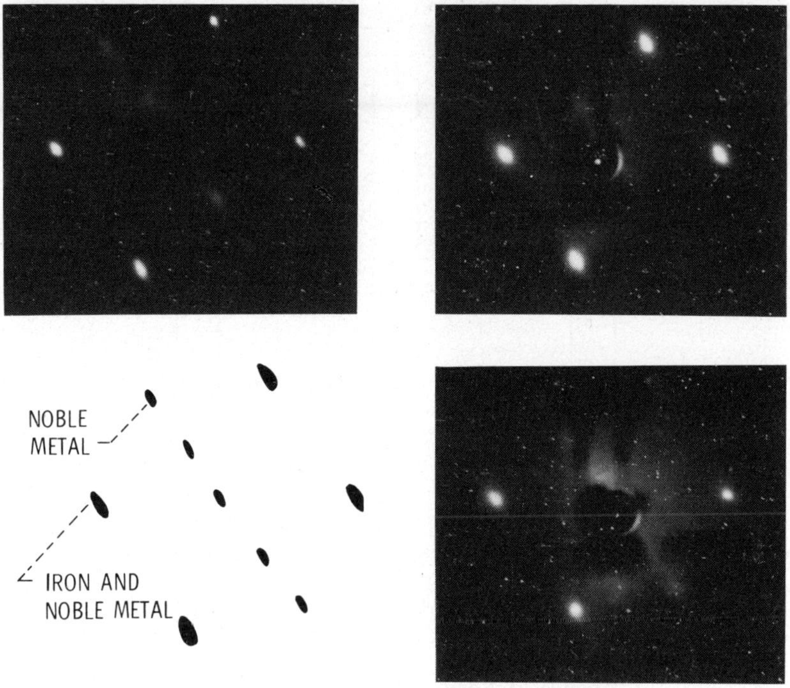

Figure 5-13. – LEED photographs of iron (011) surface after adhesion to noble metals.

of diffraction patterns. Nonetheless, LEED is very useful in establishing such relationships as that revealed in figure 5-13—namely, that all three noble metals do transfer in an orderly manner to an iron surface.

One can speculate from the possible sites for bonding on the iron (011) surface the possible locations of adhered atoms or adsorbates. Figure 5-14 is a schematic representation of the possible bonding locations of foreign atoms to an (011) surface of a body-centered-cubic face (be it iron or any of the other body-centered-cubic metal such as tungsten, vanadium, molybdenum, tantalum, etc.).

Three possible sites shown in figure 5-14 are (1) the "bridge" sites (where bonding occurs), (2) the "well" sites (deep part in the crystal lattice on the iron or other body-centered-cubic (011) face; these sites are formed when four atoms come together and a well exists between them), and (3) the maximum coordination sites (those locations where one has the maximum number of coincidence for coordination bonding to the solid surface).

One would generally anticipate that one of these particular sites is involved in the majority of bonding of various dissimilar species to an iron or other body-centered-cubic metal (011) surface or face.

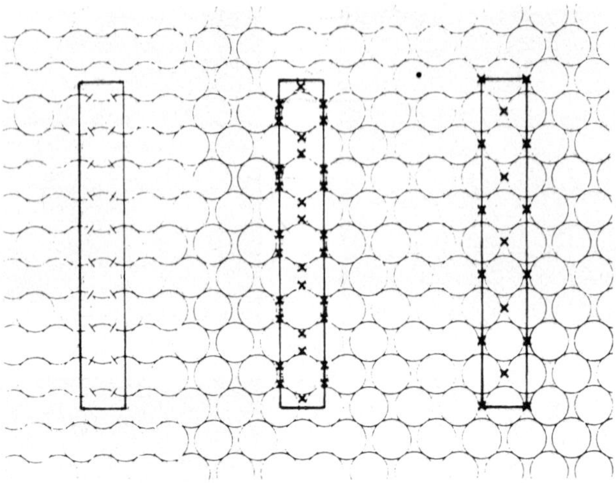

Figure 5-14. – Possible sites on (011) body-centered-cubic face.

Influence of Various Properties of Metals on Adhesive Bonding of these Metals with Iron

The adhesion force of various metals to an iron surface has been measured. Some of the results obtained together with other properties of these metals (cohesive energy, atomic size, valency states, and solubility in iron) are presented in table 5-III. In this table the adhesion force for iron to itself is shown to be in excess of 400 dynes.

Table 5-III shows that cohesion or self-adhesion gives much stronger adhesive forces than does the adhesion of any other metal to iron. Also, that solubility in iron seems to have very little influence on the adhesive behavior of the metals to iron. Lead, for example, which is insoluble in iron, has a relatively strong adhesive bonding force to iron when compared with metals such as silver and gold (which have limited solubility).

Furthermore, metals such as platinum and aluminum, which have similar solubilities (20 atomic percent for platinum and 22 atomic percent for aluminum in iron), have markedly different adhesive forces to iron. Platinum exhibits an adhesive force of 100 dynes (table 5-III), while aluminum exhibits an adhesive force in excess of 250 dynes, which is 2.5 times greater for aluminum in contact with iron than for platinum in contact with iron. Thus, the solid solubility itself does not have a pronounced influence on the adhesive behavior of dissimilar metals in solid-state contact.

TABLE 5-III. – SOME PROPERTIES OF VARIOUS METALS AND FORCE OF ADHESION OF THESE METALS TO IRON

Metal	Cohesive energy, kcal/g-atom	Atomic size, Å (10^{-10} m)	Valency states	Solubility in iron, at.%	Adhesion force to iron[a], dynes
Iron	99.4	2.86	2, 3	-----	>400
Cobalt	101.7	2.50	2, 3	35	120
Nickel	102.3	2.49	2, 3	9.5	160
Copper	80.8	2.551	1, 2	<.25	130
Silver	68.3	2.883	1	.13	60
Gold	87.6	2.877	1	<1.5	50
Platinum	134.8	2.769	2, 4	20	100
Aluminum	76.9	2.80	3	22	250
Lead	47.0	3.494	2, 4	Ins.	140
Tantalum	186.7	2.94	5	.20	230

[a]Applied load, 20 dynes; temperature, 20° C; ambient pressure, 10^{-10} torr.

The same load of 20 dynes is applied to all the metals in contact with iron in table 5-III. Thus, with some metals such as lead, for example, the deformation in the real area of contact is much greater than it is for a metal like tantalum, which has a relatively high modulus of elasticity. The real area of contact for tantalum is much smaller. However, the cohesive binding energy with lead is approximately one-fourth that of tantalum. As a result, when these two effects are combined and the adhered junction is pulled in tensile fracture, fracture in the lead results in transfer to the iron surface over a much greater area than it does for the tantalum-iron contact where iron transfers to the tantalum.

The reasons for the very strong adhesive bonding forces of metals such as aluminum to iron is discussed in chapter 6 relative to the nature of the *d* valence bond character of metals or the chemical activity, as it were, of metals. Those metals that are chemically more active seem to exhibit much stronger forces of bonding, adhesion, and friction than do the less active metals.

It is apparent from table 5-III that when dissimilar metals are brought into contact there are differences in the adhesion properties of these metals. When one of the two solid surfaces in contact is held constant and the other is varied, there is a variation in adhesive bonding forces with some metals exhibiting stronger bonding to iron than others.

Effect of Crystollographic Orientation on Adhesion

There are even more basic properties of metals which influence the adhesive behavior of metals in solid-state contact. For example, when the crystallographic orientation of the solid surface was discussed earlier, it was indicated that the surface energy as well as the coordination numbers of the different atomic planes of a crystalline solid vary with orientation as do other mechanical properties—for example, the modulus of elasticity. It might, therefore, be anticipated that differences in adhesion behavior would exist for different crystallographic orientations of solid surfaces in contact. Experimental evidence, in fact, bears this out.

Since copper is a commonly used tribological material much like iron, its basic properties are of considerable interest. Adhesion measurements were made, in the apparatus of figure 5-8, for gold in contact with various orientations of copper: the (111), (100), and (110) orientations. Surfaces were characterized with both LEED and Auger before and after adhesive contact.

The measured adhesion forces are presented for the three orientations together with some other properties of these surfaces in table 5-IV. Because of the differences in packing, there are different coordination numbers at the surface of these three planes of copper. The (111) surface has a coordination number of 9, which means that there are 9 atoms bonded to a single atom exposed on the copper (111) surface. With the (100) orientation there are only 8 atoms bonded to that one surface atom, and with the (110) surface there are only 7. The atomic arrangement for these surface atoms is shown also in table 5-IV.

Just as the coordination number varies, so does the number of surface atoms. The maximum number of surface atoms is on the (111) surface, which is the close-packed surface; that is, it is the orientation which gives the maximum packing of atoms in a particular plane. At the surface a very high concentration of surface atoms exists for this particular orientation.

TABLE 5-IV. – SOME PROPERTIES OF THREE PLANES OF COPPER TOGETHER WITH MEASURED ADHESIVE FORCES TO THOSE PLANES

Copper surface plane	Coordination number of surface	Atomic arrangement of surface unit mesh	Number of surface, atoms /cm^2	Elastic modulus, dynes/cm^2	Surface energy, ergs/cm^2	Force of adhesion to gold,[a] mg
(111)	9	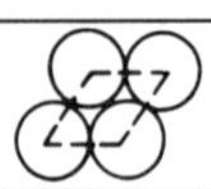	1.7×10^{15}	19.4×10^{11}	2499	80
(100)	8		1.5×10^{15}	6.67×10^{11}	2892	185
(110)	7	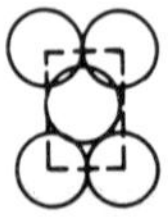	1.1×10^{15}	13.1×10^{11}	----	390

[a]Applied load, 20 mg; Au (100) surface; contact time, 10 seconds.

The least number of atoms exists for the (110) orientation, and an intermediate number of atoms exists for the (100) orientation.

Just as the coordination number (i.e., the number of bonding of like atoms to a particular atom of the surface) decreases and the number of surface atoms decreases, the modulus of elasticity also varies. The maximum modulus of elasticity is observed on the (111) surface and the lower moduli of elasticity are observed on the (100) and (110) surfaces. There appears to be a juxtaposition for the elastic modulus of the (100) and (110) surfaces. It is anticipated that the lowest modulus of elasticity would exist on the (110) surface.

Surface energy also varies with orientation but in an inverse manner to the modulus of elasticity, number of surface atoms, and coordination number. As the coordination number, the number of surface atoms, and the elastic modulus decrease, the surface energy increases on the solid surface. For the (111) surface, the surface energy is approximately 2500 ergs per square centimeter while for the (100) surface it is nearly 2900 ergs per square centimeter (or 400 erg/cm^2 difference), the (100) surface having a larger surface energy.

As discussed earlier, as the surface energy increases, the coordination number at the surface decreases, and the modulus of elasticity decreases, an increase in the adhesion force is anticipated; that is, the adhesion force would go in the same direction as the surface energy. And table 5-IV shows that this is, in fact, the case. The lowest adhesion force is found on the close-packed, high atomic density, low surface energy (111) planes; the maximum force of adhesion is found on the (100) surface (which is the surface that contains a lower coordination number, a lower number of surface atoms, and a lower modulus of elasticity than the (111) surface). Although the author could not find a value for the (110) surface, the surface energy is anticipated to be higher than that for the (100) surface.

Thus, there appears to be a relationship among various parameters: (1) adhesive bonding force of copper to gold, (2) cohesion of copper atomic planes to themselves, and (3) various basic properties. These properties include the coordination number, the number of surface atoms, the modulus of elasticity, and the surface energy. In general it appears that, as the coordination number decreases, the number of surface atoms decreases, the modulus of elasticity decreases, the surface energy increases, and the force of adhesion increases.

From the values presented in table 5-IV, it is anticipated that the researcher can predict adhesive behavior for different orientations based on such properties as coordination number, number of surface atoms, and surface energy, if reliable surface values can be obtained for these properties. The data of table 5-IV are for the three crystallographic planes of copper: (111), (100), and (110) orientations. The adhesion measurements were made for these copper orientations in contact with gold. With the copper in contact with itself, similar adhesive behavior is observed.

If two copper single crystals of the same orientation (e.g., a (111) orientation) were brought into solid-state contact and the plane and crystallographic direction between the two single crystals could be perfectly matched across the interface, atom to atom bonding would occur across the interface with a complete loss of the interface. This means that the two single crystals would become one continuous single crystal without any evidence of boundary between them because there would be a perfect match (lattice registry) with the two single crystal surfaces. This represents the ideal case and is not what is experienced.

The experimentalist who attempts to make adhesion measurements for like planes of like material in solid-state contact with matched crystallographic directions across the interface can expect the equivalent of a grain boundary at the interface since there is a high degree of mismatch in the crystal lattice. Defects (primarily interfacial dislocations) will occur to accommodate the misfit across the interface, and this region will become a highly defect-laden one. In practice, the interface exists and contains defects such as vacancies and dislocations. If perfect bonding occurs at the interface with the loss of the interface because of a perfect match in registry, the strength of the adhesive bond is essentially the tensile strength of a new single copper rod.

Even with imperfect bonding across the interface, however, the particular crystallographic plane involved exerts an influence on the measured adhesive forces, even though they are considerably less than the theoretical tensile strength of the material. This is reflected in the data of table 5-V for copper single crystals. The same three orientations that are reported in table 5-IV are included in table 5-V. The (100), (110), and (111) orientations of copper are brought into solid-state contact with themselves—that is, with another copper single crystal of the same crystallographic plane. Attempts were made to align crystallographic directions on the surface as closely as possible.

TABLE 5-V. – ADHESION COEFFICIENT FOR VARIOUS METAL SURFACES IN CONTACT

Material	Atomic planes		Adhesion coefficient
	Surface 1	Surface 2	
Copper (fcc)	(100)	(100)	1.02
	(110)	(110)	.61
	(111)	(111)	.30
	Poly-crystal	Poly-crystal	1.00
	(111)	(110)	.25
	(111)	(100)	.20
Cobalt (hcp)	(0001)	(0001)	<.05
Copper nickel (fcc/bcc)	(111)	(111)	.25
Copper cobalt (fcc/hcp)	(111)	(0001)	.10
Copper tungsten (fcc/bcc)	(111)	(110)	<.05

The data of table 5-V show that differences in adhesive forces do exist for the three crystallographic orientations. Just as was observed in table 5-IV, the lowest bond strength or adhesion force is found for the (111) surface, which is the high atomic density, low surface energy surface of copper. The other two orientations exhibited higher adhesive bonding forces.

Polycrystalline copper has many grain boundaries at the surface. It is interesting to compare the adhesive bond forces of polycrystalline copper to itself with the single crystal orientations presented in table 5-V. The data in table 5-V give a value of 1.00 for the adhesion coefficient for the polycrystalline copper in contact with polycrystalline copper. This value is near that of the maximum bond force ((100) vs. (100)) found for the copper single crystals. The polycrystalline surface represents an agglomerate or a mixture of various crystallographic orientations where each grain has a different orientation with its own characteristic surface energy. The surface represents an average of the energies of the various orientations of the surface grains. In addition, there is the energy associated with the grain boundary itself. As discussed earlier, the grain boundaries are high energy sites. Thus, since a greater degree of activity is expected at the grain boundaries, they certainly will play a role in the adhesive bond forces for metals in solid-state contact.

Polycrystalline copper shows adhesion of the same order of magnitude as the single crystal exhibiting the maximum bonding strength across the interface, as opposed to some average between the minima and maxima. This may result because of the influence of the higher energy grain boundary locations that tend to increase the binding force because of the increased energy at these locations.

If the atomic planes are deliberately mismatched, then an even greater degree of defect generation is anticipated at the interface. As one moves away from crystal perfection, the number of defects increases, and this results in an effect on the bond strength between the solid surfaces in contact. This is observed in table 5-V for the adhesion measurements of the (111) surface in contact with the (110) surface of copper and the (111) surface in contact with the (100) surface of copper. The data show that the adhesion forces of the dissimilar planes in contact are less in each case than the lowest value for like planes in solid-state contact (table 5-V).

Effect of Crystal Structure on Adhesion

Copper, a face-centered-cubic metal, exhibits relatively high adhesive bonding forces when in contact with itself. It has been reported that close-packed-hexagonal metals have inherently lower adhesion properties than do the face-centered-cubic metals. Adhesion measurements were therefore made with cobalt, which is a close-packed-hexagonal metal at temperatures below 417° C, in contact with itself; the contact was of single crystals of cobalt of the basal orientation (the (0001) plane in contact with the (0001) plane of cobalt) to gain an insight into the effect of crystallographic structure. The (0001) plane of cobalt is the high atomic density, low surface energy plane for the hexagonal metal. It is analogous to the (111) plane in the face-centered-cubic system and, therefore, to the copper (111) planes of table 5-V.

While some degree of anisotropy (with respect to crystallographic orientation and their properties) exists in the face-centered-cubic metals (table 5-IV), an even higher degree of such anisotropy exists for close-packed-hexagonal metals. The basal plane generally exhibits the highest atomic density, the lowest surface energy, and the highest modulus of elasticity normal to it.

From the foregoing it might be anticipated that the basal orientation of a hexagonal metal such as cobalt in contact with itself would exhibit the lowest adhesion forces. In order to make a comparison between the close-packed-hexagonal metal and the face-centered-cubic metal, adhesion measurements have been made with cobalt in contact with itself for the basal planes; the results are indicated in table 5-V. The adhesive bonding forces are extremely low for the basal orientation of cobalt in contact with cobalt, markedly less than that observed for the (111) planes of copper in contact with themselves. Thus, even though the (111) and the (0001) planes of copper and cobalt, respectively, are high atomic density, close-packed planes, differences exist in their adhesive behavior which reflect, in part, differences in the adhesive behavior of the different crystal structures. The close-packed-hexagonal metals exhibit markedly lower adhesion forces than the face-centered-cubic metals.

A logical question that can be asked is, "What happens when single crystals of, for example, copper and nickel (both face-centered-cubic metals) are placed into solid-state contact using the same crystallographic plane (the (111) plane)?" The adhesion measurement results in table 5-V show that the adhesive bond forces for copper in contact with nickel (where the planes at the interface are the same but the lattice parameters and the chemical nature of the two elements are different) are less than those forces observed for the matched crystallographic planes of copper in contact with copper. They are, however, equivalent to those forces for dissimilar planes of copper in contact with copper.

The adhesive bond forces are stronger for the dissimilar metals in contact than they are for cobalt in contact with itself (bearing in mind that copper and nickel are both face centered cubic and cobalt is close packed hexagonal). For many years it was believed that it was desirable to incorporate into practical tribological devices dissimilar metals in contact because this would tend to minimize or reduce adhesion and adhesive wear on the system. The data of table 5-V indicates that cobalt in contact with itself exhibits lower adhesive bond forces at the interface than does copper in contact with a dissimilar metal, namely, nickel.

Both copper and nickel are face centered cubic. For copper in contact with low energy cobalt (close packed hexagonal), the results in table 5-V show that the adhesive bond forces are appreciably less than those observed for copper in contact with nickel (face-centered-cubic metal). Thus, the influence of the hexagonal structure is reflected in the adhesion measurements of table 5-V for copper contacting cobalt. The adhesive bond forces are higher than those observed for cobalt in contact with itself. The face-centered-cubic metal copper in contact with the close-packed-hexagonal metal cobalt does not give as low adhesion measurements as does the cobalt in contact with itself. It is, however, appreciably less than for copper in contact with nickel (both face-centered-cubic metals).

Another common crystal system observed for metals and of interest to the tribologist are the body-centered-cubic metals. These include a wide range of transition metals. Copper was brought into contact in adhesion experiments with tungsten (body-centered-cubic metal) to determine the influence on interfacial bonds and their strengths of a body-centered-cubic metal in contact with a face-centered-cubic metal. The results of these experiments are presented in table 5-V. These data show that extremely low adhesion values were obtained for copper in contact with tungsten; the values were comparable to those obtained for cobalt in contact with itself. Tungsten is an extremely high modulus elasticity material with very high cohesive bonding energy, and the (110) plane in tungsten is the high atomic density, low surface energy plane. Therefore, minimum adhesion to that particular crystallographic orientation is expected.

It should also be indicated that copper in contact with other body-centered-cubic metals such as iron and molybdenum exhibits higher adhesive forces than copper in contact with tungsten. In fact, of all the systems examined by this author for copper contacting various body-centered-cubic metals, the lowest adhesive bond forces were measured for copper in contact with tungsten.

An examination of the bulk metallurgical literature fails to reveal any type of compound formation between copper and tungsten. They are not soluble, and no compounds have been reported in the literature. One might therefore expect a complete immiscibility of these two materials and, at the interface, the binding should be minimal because of the absence of any types of compound formation. Nonetheless, however, copper was observed to transfer to the tungsten surface in the experiments reported in table 5-V. The LEED measurements and AES analysis of the tungsten surface reveal the presence of copper. Not only was the presence of copper revealed, but the copper transferred to the tungsten surface in an orderly fashion and some surface work by an earlier investigator has indicated the formation of copper-tungsten compounds on the surface (ref. 5). This is extremely interesting because the metallurgical literature fails to reveal any type of compound between tungsten and copper. Again, as was discussed in reference to the field ion microscopy results in adhesion studies, extreme care must be taken in extrapolating bulk data to predict surface behavior.

Adhesion, friction, and wear are surface phenomena. These phenomena are strongly influenced by surface properties, and the surface characteristics of materials are frequently different from the bulk character of the materials. The coordination number at the surface is different, the types and nature of defects may be markedly different, and the metallurgy may be extremely different. Therefore, the particular characteristics of the surface must be studied to understand the role of the material in tribological behavior.

The metallurgist who is familiar with bulk metallurgy would not predict the compound formation observed for copper in contact with tungsten. Notwithstanding the formation of the compounds, the bond strengths are obviously relatively weak as indicated by the adhesive force measurements of table 5-V.

A number of conclusions can be drawn from the data presented in table 5-V. First, like planes in contact with like planes exhibited higher adhesive bonding forces than dissimilar crystallographic planes of the same metal in contact with themselves. Second, the orientation at the surface influences adhesive behavior. The high atomic density, low surface energy (111) surfaces of copper exhibit lower adhesive bond forces than the other crystallographic planes of copper (table 5-V). The polycrystalline form of a metal in contact with itself exhibits adhesive bond forces comparable to the highest values measured for single crystals in contact with themselves; this reflects, in all probability, the influence of grain boundary energies.

Cobalt, the close-packed-hexagonal metal, has markedly lower adhesion forces when brought in contact with itself than does copper when like or equivalent planes are compared. The adhesive bond forces of face-centered-cubic metals in contact with face-centered-cubic metals can be markedly higher, even though the metals are dissimilar, than for like metals in contact with themselves, such as cobalt. A face-centered-cubic metal in contact with hexagonal metal results in lower adhesive forces than the face-centered-cubic metal in contact with itself. The adhesive bond forces are, however, still greater than they are for the hexagonal metal in contact with itself; this indicates that similiar metal contacts may, in some instances, give lower adhesion results than dissimilar metals in contact. Lastly, adhesive bond forces can be weak where interfacial compounds form as a result of adhesive contact (as was observed for copper in contact with tungsten).

The data of table 5-V for copper in contact with nickel (dissimilar face-centered-cubic metals) raise the question of whether all dissimilar face-centered-cubic metals behave in a like manner with respect to the adhesion coefficient. In other words, if copper is brought in contact with face-centered-cubic metals other than nickel, are the adhesive forces comparable to those forces reported in table 5-V for copper in contact with nickel or do variations in adhesive forces exist if one of the face-centered-cubic metals is held constant? Are there any parameters or properties of face-centered-cubic metals beside orientation (reflected in table 5-V) that exert an influence on the adhesive behavior of face-centered-cubic metals in contact with other face-centered-cubic metals?

An examination of various properties of face-centered-cubic metals in contact with themselves and other metals of the same crystal system reveals that there are some fundamental properties that do influence the adhesive behavior of face-centered-cubic metals in contact with themselves and other face-centered-cubic metals. Experiments with the apparatus shown in figure 5-8 were conducted with the face-centered-cubic metal gold contacting other face-centered-cubic metals, and adhesive forces were measured for the (111) or high atomic density, low surface energy plane of gold compared to the analogous plane of copper reported in tables 5-IV and 5-V. These experiments were conducted with a number of face-centered-cubic metals to gain some insight into the influence of various properties on adhesion behavior.

Effect of Lattice Misfit (Disregistry) at the Interface on Adhesion

The results of these experiments reveal that the degree of lattice misfit (disregistry) at the interface can influence the adhesive forces measured for the metals in solid-state contact. The greater the degree of mismatch in the lattice parameters of the metals in solid-state contact, the more likely it is that defects will be generated at the interface; the greater the concentration of defects, the weaker the bonding and the lower the adhesive binding forces. Experimental results substantiate these hypotheses. The experimental evidence for the effect of lattice misfit on the adhesion behavior of face-centered-cubic metals is shown in table 5-VI. The adhesive bond forces are reported for a gold (111) surface in contact with (111) surfaces of other face-centered-cubic metals. The other face-centered-cubic metals examined included silver, aluminum, and copper. Silver and copper were incorporated or included because they are also noble metals and have many properties similar to those of gold.

Aluminum was considered for evaluation primarily because its properties differed from those of the noble metals—it being much more chemically active. Aluminum has lattice parameters fairly close to those of gold. The adhesion results presented in table 5-VI indicate that the adhesive bonding force of gold to itself is in excess of 400 dynes. Dividing the adhesive bonding force by the applied load would give an adhesion coefficient in excess of 20. Likewise, when the (111) plane of silver was matched with the (111) plane of gold, the adhesive bond forces (400 dynes) were comparable to that observed for gold in contact with itself. Table 5-VI shows that the lattice parameters for gold and silver are extremely close. There is only a 0.01-angstrom difference in the lattice match of silver and gold; the degree of misfit between the lattices of gold and silver is 0.19 percent (table 5-VI). When gold is brought into contact with a (111) surface of aluminum, so that the atomic planes are the same, the aluminum has a lattice parameter fairly close to that of gold but not quite as close as that of silver. There is a misfit of approximately 0.71 percent in the lattice parameters of aluminum and

TABLE 5-VI. – EFFECT OF LATTICE MISFIT ON ADHESION OF GOLD TO VARIOUS FCC METALS

Metal	Lattice parameter, Å	Percent misfit with gold (111)	Force of adhesion of gold (111) to a (111) metal surface[a]
Gold	4.078	0	>400 D
Silver	4.086	.19	>400
Aluminum	4.049	.71	>400
Copper	3.615	11.1	80

[a]Applied load, 20 dynes; contact time, 10 sec; temperature, 20° C; ambient pressure, 10^{-11} torr.

gold. The lattice spacings are, however, still fairly close and very high adhesive forces of aluminum to gold are seen (table 5-VI). Again, the bond forces exceed 400 dynes (i.e., an adhesion coefficient of 20).

Copper is a noble metal (as are gold and silver) having many properties similar to those of gold and silver except that copper has lattice parameters which are markedly different from those of gold and silver. The lattice spacing between adjacent copper atoms is 3.615 angstroms and that presents a lattice misfit or mismatch to a gold (111) surface of 11 percent (table 5-VI). When a (111) single crystal surface of gold is brought into adhesive contact with a (111) surface of copper with the resulting 11 percent mismatch in the crystallographic lattices, there is a very marked drop in the adhesion force. The adhesion force is only 80 dynes (an adhesion coefficient of 4, table 5-VI).

The results of table 5-VI, then, indicate that the degree of lattice misfit for face-centered-cubic metals can have an influence on the measured adhesion behavior. Where there is a large degree of misfit, the adhesive bond force appears to be lower than where the lattices are fairly closely matched (as it is for silver in contact with gold). These observations are entirely consistent with the fundamental behavior of materials at interfaces. Van Der Merwe has done a considerable amount of research on the generation of what is referred to as misfit dislocations at the interfaces of two solid surfaces in solid-state contact (refs. 6 and 7). For example, misfits that generate dislocations are observed at grain boundaries and for coatings developed in layers. The higher the degree of mismatch across the interface, the greater the concentration of misfit dislocations in the interfacial region. The higher the concentration of dislocations, the weaker the interfacial region because the defect concentration tends to inhibit or minimize the number of bonds that form across the interface.

In addition to the generation of defects, where there is a high degree of lattice mismatch, there is (where sufficient energy is absent for the formation of misfit dislocations) the creation of vacancies at the interface. Where bonding occurs at the interface, the bonds generally exist in a strained state.

When two solid surfaces of the same metal or dissimilar metals are brought in contact in the atomically clean state, adhesion always occurs at the interface. When attempts are made to separate those solid surfaces by applying a separating force (in tension) to produce tensile fracture of the junction, the process is essentially one of conducting a tensile test. In an attempt to separate or pull apart the solid surfaces, there is some critical force at which the surfaces begin to separate; below this force level, the atoms in the surface that are bonded to another behave in an elastic manner.

Rupture of Cohesive Bonds Elastic and Plastic Effects

It may be recalled that when two dissimilar metals are bonded to another the bond strength at the interface is stronger than the cohesive bonds in the cohesively weaker of the two materials. Thus, the tensile test or tensile experiment consists essentially of pulling until tensile fracture occurs in the cohesively weaker of the two materials near the interface region. With gold in contact with copper, adhesion occurs at the interface; then, when the specimens are pulled to fracture of the adhesive junction, fracture occurs in the gold and leaves it transferred to the copper surface. The tensile force required to produce fracture then is the force required to fracture the cohesive bonds in the gold.

The tensilelike behavior is reflected in the data of figure 5-15 for gold in contact with copper. These are two single crystals, and the measured adhesive forces are those required to fracture the cohesive bonds in the junction. This is plotted on the ordinate as a function of time to fracture. The force to fracture the cohesive bonds are reported in dynes and the time to fracture in seconds. At extremely low forces (less than 150 dynes), no matter how long the force is applied to the specimens bonded to another, fracture does not occur. This is because the forces applied are not sufficient to overcome the elastic limit of the bonds in the interfacial region of the gold and to bring about fracture. At some particular force (150 dynes in fig. 5-15), however, the elastic limit of the adhesive junction is exceeded and applied forces in excess of 150 dynes are sufficient to bring about plastic behavior at the interface; this ultimately produces fracture of the adhesive bonds. This necking or tensile behavior is in the data of figure 5-15. As the force to fracture the adhesive bonds is increased, the fracture occurs at lower times, reflecting some strain behavior at the interface prior to actual fracture of the cohesive bonds in the gold.

The data of figure 5-15 indicate that, in asperity contacts, necking down of the individual asperity junctions must occur before fracture (much as tensile specimens in a tensile test). One can envision each of the asperities in contact as being the equivalent of a mini tensile specimen with the gold bonded at one end to the copper surface. As the force exceeds the elastic limit in the gold, the asperity necks down until it produces a fracture. So, strain occurs in these asperities behind the actual points of real contact at the interface; the real area of contact represents the bonding of the gold to copper. The pulling to tensile fracture generates dislocations not only at the interface but in the interfacial region, subsurface from the actual interface.

Maugis et al. (ref. 8) have studied, in the electron microscope, the dislocation behavior of surfaces after adhesive contact of a fine tip on 4500-angstrom aluminum foils under loads of 1 to 200 micronewtons. Results of their studies reveal dislocations in the near surface regions as a

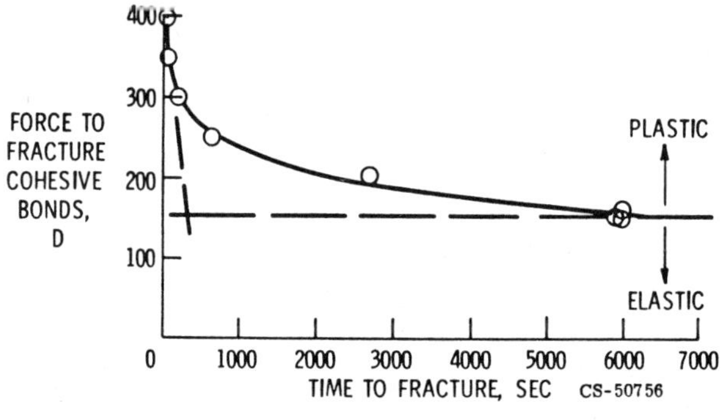

Figure 5-15. Time required to rupture cohesive bonds in gold. Adhesion of gold (110) to copper (110); load, 20 dynes; contact time, 10 seconds; pressure, 10^{-10} torr; temperature, 20° C.

result of adhesive contact. The dislocation concentration is maximum in the central zone of actual contact, and the dislocation density drops off as one moves away from the center of the actual contact zone. The dislocation densities observed by Maugis are plotted schematically in figure 5-16; the experimental results show a gradual dropping off of dislocation concentration outside the actual contact zone (a0a in fig. 5-16). The sharp dropoff indicated by the dashed curve b in figure 5-16 would be more consistent with the actual experimental observations of figure 5-15; that is, at some region outside the actual contact zone, the applied stress should be insufficient to produce plastic deformation, and there should be no dislocations generated. There is a critical stress or elastic limit which must be exceeded before dislocations become generated and plastic deformation occurs. This then should be a fixed distance along the surface away from the actual contact zone. The experimental results observed by Maugis et al. in figure 5-16, however, indicate that the dropoff in dislocations is gradual as opposed to being a sharp dropoff with no dislocations generated when a location on the surface is reached where the elastic limit is no longer exceeded.

One possible explanation for the observed experimental results of figure 5-16 is that of inhomogeneity in the strain behavior in the actual contact region. There is a nonuniformity in strain throughout the surficial area where contact has been made. As a consequence, there is not a sharp point of demarcation between the elastic and plastic regions in the material but rather it varies in different locations: (1) near the surface, (2) at the surface, and (3) subsurface. The consequence is that, when one measures the dislocation concentration in the general area, there is a gradual dropoff (an averaging out) of the observed dislocation concentrations rather than a sharp dropoff as would be predicted from a theoretical estimation (dashed curve of fig. 5-16(b)).

The size of the junctions formed at the contacting asperities (the real area of contact) is influenced by a number of properties, one of which is the load applied to the specimen surfaces in contact. When the load is sufficient to

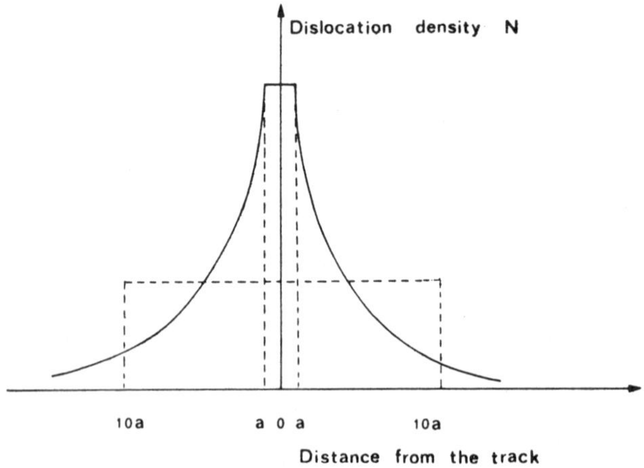

Figure 5-16.—Schematic variation of dislocation density from edge of track, as experimentally observed (solid line) and as taken for estimation (dashed line) (ref. 8).

exceed the elastic limit of the weaker of the two materials, plastic deformation takes place in the contacting asperities of the weaker of the two materials until the real area of contact increases to the point where the load can be supported by the area; growth of the junction then stops. As the load is increased, the real area of contact continues to increase. It might be anticipated that the adhesive force is a function of the applied load; that is, the greater the load, the greater the real area of contact and the greater the number of bonds that must be broken in the tensile fracture of the adhesive junction. This is, in fact, the case. If one metal is brought into solid-state contact with a second clean metal and the load is continuously increased, there is an increase in the adhesive force as is indicated by the data of figure 5-17 for gold in contact with a nickel surface.

With increasing load beyond 2 grams, there is a continuous linear increase in the adhesive force measured for the two solid surfaces when one pulls the specimens to fracture. The results of figure 5-17 have been observed for other similiar and dissimilar metals in contact in the clean state. In other words, as the applied load increases, the adhesive force (the force to fracture the adhesive junctions) increases.

The load was applied for 10 seconds before fracture of the adhesive bonds at the interface (fig. 5-15). The reason for the specific time of 10 seconds (also used in obtaining the data of fig. 5-17) is that the load contact time influences the adhesive forces where the load exceeds the elastic limit in the asperities of the cohesively weaker of the two materials in solid-state contact. Thus, in figures 5-15 and 5-17 the gold is shown deforming plastically in the asperity region at a force in excess of 20 grams.

If the elastic limit is exceeded (and it is in most cases), then sufficient time must be allowed for an equilibrium condition to be obtained where no fur-

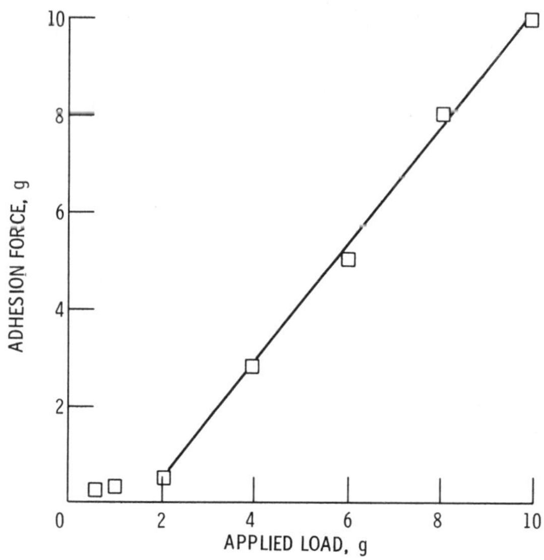

Figure 5-17.—Force of adhesion as function of applied load for gold (111) surface contacting clean nickel (011) surface. Ambient pressure, 1.33×10^{-8} newton per square meter (10^{-10} torr); contact time, 10 seconds; temperature, 23° C.

ther growth in the real area of contact occurs. On initial loading of one surface to another, the real area of contact grows as a function of the creep properties or the strain characteristics of the weaker of the two materials in solid-state contact. Thus, junction growth can occur with time after the load has been initially applied. The influence on the adhesive force of two materials in contact as a function of contact time prior to reaching the equilibrium condition (where there is no further growth in real area of contact as a function of strain or creep) is shown by the data of figure 5-18. In this figure a single crystal gold (111) surface is in contact with the copper-aluminum alloy (5 percent Al). The force of adhesion is plotted as a function of contact time. The applied load was 20 dynes (same as in fig. 5-15). It can be seen that with increase in contact time there is an increase in the force of adhesion. It should be indicated that beyond 10 seconds there is no further change in the force of adhesion; the curve flattens out and becomes parallel to the abscissa. The results in figure 5-18 indicate that, in making adhesion measurements, care must be taken to allow sufficient time for equilibrium conditions to be obtained.

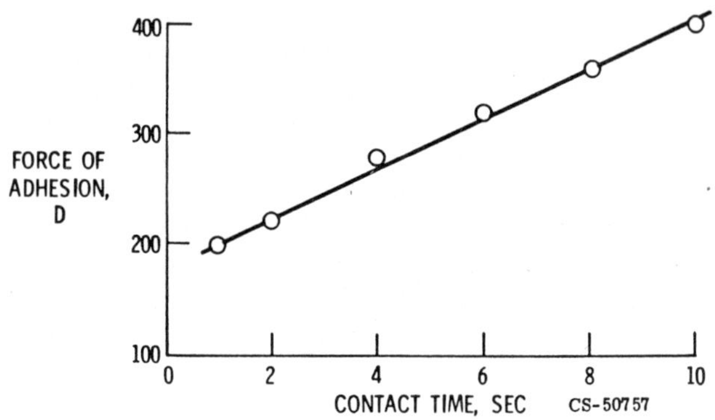

Figure 5-18.—Effect of contact time on adhesion of gold (111) surface to 5.0 percent aluminum in copper alloy. Load, 20 dynes.

Alloy Segregation Effects

Before the advent of analytical tools to characterize surfaces, most experimental researchers in the field of tribology (as well as in areas of corrosion, catalysis, and chemistry) assumed that a very small amount of impurity dissolved in the metal was relatively harmless as far as surface properties were concerned. Thus, when a researcher was working with 3 nines purity (99.9 percent) iron, he felt that was sufficient to characterize the surface as being one of iron. Careful examination of high purity metals (even triple zone refined metals containing only a very few parts per million of impurity), however, have shown that very small concentrations (parts per million) of alloying elements in most metals are sufficient to alter many of the bulk properties and even, to a greater degree, the surface properties. Such impurities can completely alter surface behavior. In recent years, a number of research papers have appeared in the literature showing the effects of various impurities in metals on the surface characteristics and surface behavior of these materials.

The effect of segregation of small concentrations of sulfur and carbon to the surface of iron has already been discussed; it has been indicated that even parts per million of these impurities dissolved in the bulk of the iron can alter surface adhesion behavior.

Copper-based alloys are widely used in tribology. Most of the bronzes and brasses have adhesion, friction, and wear properties that make them useful in one area or another of this field. A common class of bronzes that have been used in tribological applications are the aluminum bronzes.

Some adhesion experiments have been conducted in a vacuum environment with clean surfaces containing various amounts of aluminum (or tin) dissolved in copper. The results of these adhesion experiments are presented in figure 5-19; adhesive bonding force is plotted as a function of concentration of aluminum in the copper. Adhesion measurements were made with a gold (111) surface in contact with the copper alloys as well as with a copper (111) surface. In addition, a copper tin alloy (of 1 atomic percent tin) was also examined. The interesting observation to be made from the data of figure 5-19 is that with small additions of aluminum to copper (as little as 1 atomic percent) there is a marked increase in adhesive force (from 80 mg to in excess of 400 mg, a fivefold increase).

Additional increases in the amount of aluminum do not alter further the adhesive behavior of the copper-aluminum alloys. At compositions to 10 atomic percent of aluminum in copper, the adhesive bonding forces are in excess of 400 milligrams or five times the adhesive bonding force of copper to the gold. Now, if a single crystal aluminum of the same crystallographic orientation (the (111) surface, which is the low surface energy, high atomic density plane) is brought into solid-state contact with the gold (111) surface, obtain adhesive forces are obtained that are comparable to those obtained for the copper—1-atomic-percent-aluminum alloy; that is, the copper alloys containing as little as 1 atomic percent aluminum behave just as pure aluminum relative to adhesion.

With the addition of 1 atomic percent tin to copper, a decrease in adhesive bond force (from 80 to 40 mg) is observed. The copper-tin alloy shows this decrease only after the specimen has been heated to 200° C for a short period. If that surface is sputtered cleaned with argon, the adhesion force returns to the value for pure copper of 80 milligrams.

The data of figure 5-19 provoke the obvious questions: "Why the increase in the adhesive force of copper to gold with the addition of small quantities of aluminum, and why does the aluminum-copper alloy behave as if it were a pure aluminum surface?" In addition, "Why does the addition of a small amount of tin cause a reduction in adhesion on heating of the alloy to about 200° C, while sputtering that surface removes the effect of the heating and the adhesion coefficient for the sputtered alloy is the same as that of pure copper?"

The surface analytical tools (LEED and AES) now available to the tribologist are very useful in trying to answer these questions. LEED patterns were obtained for the pure copper and the copper containing various atomic percents of aluminum. A clean copper (111) surface gives a LEED pattern which consists of six diffraction spots in a hexagonal array. The LEED patterns obtained for the copper containing 5 and 10 atomic percent

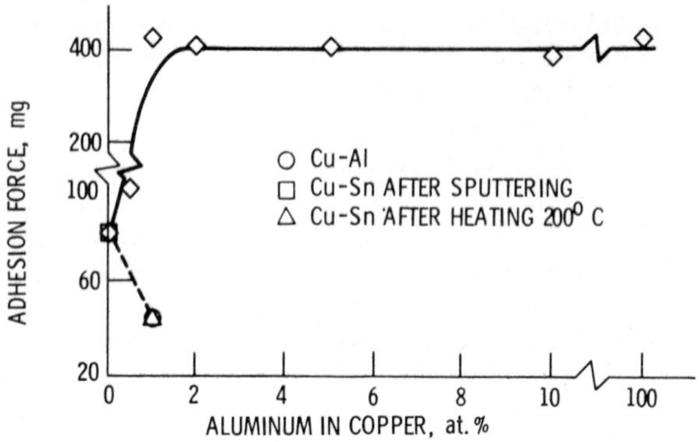

Figure 5-19. – Adhesive force of (111) gold to (111) surface of copper and copper alloys as function of bulk concentration.

aluminum are presented in figure 5-20. The LEED pattern for the 5 atomic percent aluminum and copper is shown in figure 5-20(a). The six bright diffraction spots on a hexagonal array out near the periphery of the pattern are those characteristic of a (111) face-centered-cubic surface, a close-packed surface. The rest of the background for a pure copper surface should be entirely black. A careful scrutiny of figure 5-20(a), however, reveals the beginnings of additional diffraction spots in the pattern near the center of the beam spot (i.e., between the location of the primary beam and the diffraction spots characteristic of the (111) surface).

The LEED pattern for the copper containing 10 atomic percent aluminum is presented in figure 5-20(b). This pattern shows six diffraction spots in addition to those expected for pure copper.

AES analysis of the copper-aluminum alloys along with the LEED patterns reveal the marked difference in the concentration of the aluminum on the surface with heating the surface. Figure 5-21 presents the aluminum surface concentrations where the aluminum peak intensity is normalized to a copper Auger peak and the intensity ratio is plotted as a function of

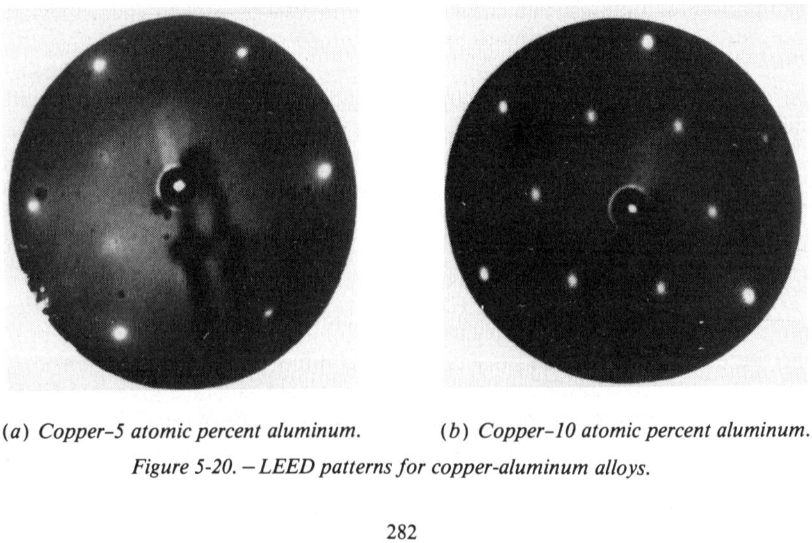

(a) Copper–5 atomic percent aluminum. *(b) Copper–10 atomic percent aluminum.*

Figure 5-20. – LEED patterns for copper-aluminum alloys.

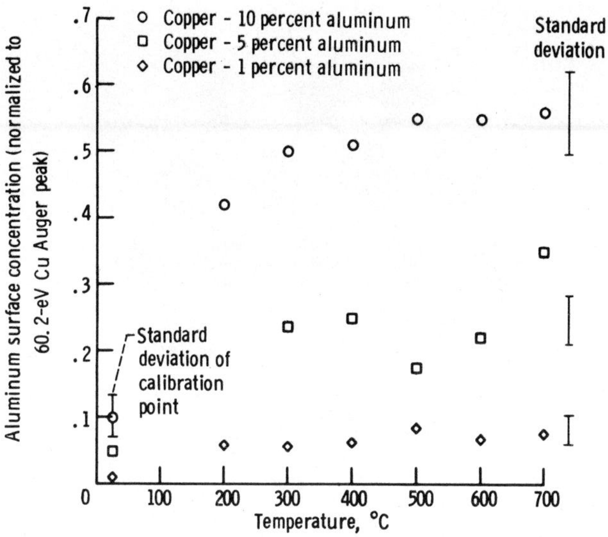

Figure 5-21. – Increase of aluminum surface concentration for copper – 1, 5, and 10 atomic percent aluminum alloys. For each point, crystal was sputtered, heated for 30 minutes, and then allowed to cool to room temperature.

temperature. As the alloy specimens are heated, the aluminum begins to segregate out onto the solid surface, and this segregation process occurs with as little as 1 percent aluminum in the copper. The segregation of aluminum to the surface is even greater with 5 percent aluminum in the alloy, and it is greater still with 10 percent aluminum in the copper alloy; that is, the surface composition is much richer in aluminum (as detected by AES analysis, fig. 5-21) than it is in the bulk.

The surface structure of the copper-aluminum alloys with segregated aluminum is shown schematically in figure 5-22. This figure is based on the Auger results of figure 5-21 and the LEED results of figure 5-20. An examination of the top view of figure 5-22 shows the substrate copper atoms (clear white circles) and the aluminum atoms (black dots). We see a fairly uniform distribution of aluminum relative to the substrate atoms. A side view reveals that the aluminum is actually segregated out onto the surface so that the outermost layer is a layer of aluminum atoms. While the atoms do not exist on the solid surface in a close-packed array as do the substrate atoms, they do completely form the outermost layer.

Given the model of the surface depicted schematically in figure 5-22, the explanation for the adhesion behavior of the copper aluminum alloys (fig. 5-19) becomes obvious. With mild heating, which occurs in baking out of vacuum systems, the aluminum segregates to the copper surface. When the copper-aluminum alloys are brought into contact with the gold, the only thing the gold sees are the aluminum atoms on the surface. Thus, as the gold approaches the surface, a layer of aluminum atoms is sandwiched between the substrate and the gold surface. As a consequence, the adhesion behavior is that of gold to aluminum rather than gold to a copper-aluminum alloy. The copper-aluminum alloy in contact with the gold, therefore, shows adhe-

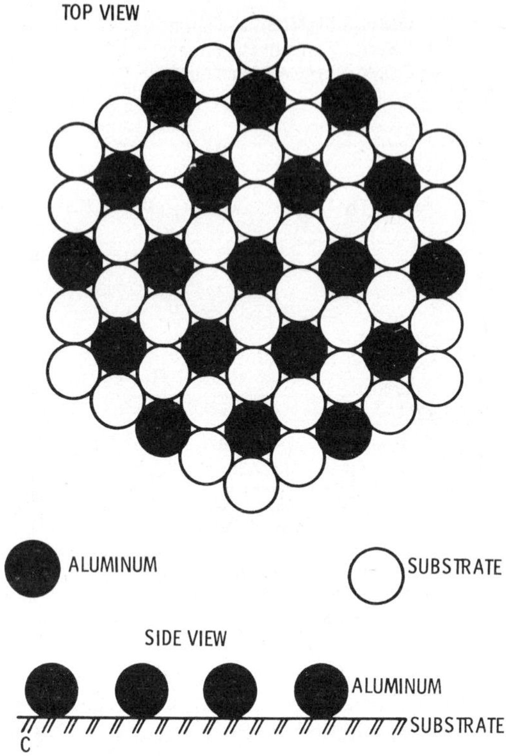

Figure 5-22. – Surface segregation of aluminum in copper-aluminum alloys.

sion behavior identical to that for a pure aluminum surface. It does not make any difference whether the surface is pure aluminum of (111) orientation or a copper-aluminum alloy containing in excess of 1 atomic percent aluminum because the aluminum in the alloy segregates to the surface and the mating gold surface sees only aluminum atoms. Therefore, the adhesion forces are the same for the alloys and the pure aluminum.

Alloying with tin was observed to cause a decrease in adhesion forces (fig. 5-19). LEED and Auger studies of the copper-tin alloy indicate that tin also segregates on the solid surface just as does the aluminum. Unlike aluminum, which causes an increase in the binding strength, the presence of tin causes a decrease in bonding strength and a consequent reduction of adhesion forces. In fact, with tin, the segregation occurs to even a greater degree than it does with the aluminum. This is indicated in the data of table 5-VII showing the maximum coverage of minor constituents on copper or iron alloy surfaces. The ratios of surface concentration to bulk concentration are presented in the table for copper-aluminum and copper-tin alloys as well as for the iron-aluminum alloy. With copper containing 1 atomic percent tin, there is a 15 to 1 ratio of tin atoms on the surface to those in the bulk, whereas with 1 atomic percent aluminum, the ratio is only 6.5 to 1. Thus, the tin segregates even more effectively at the surface of the copper than does the aluminum.

TABLE 5-VII. – MAXIMUM COVERAGE OF MINOR CONSTITUENT ON ALLOY SURFACES

Alloy	Ratio of Surface Concentration to Bulk Concentration	Atomic Size from Lattice Nearest Neighbor Distance	
Cu–1 a/o Al	6.5	Cu–2.556 Angstrom	–f.c.c.
Cu–5 a/o Al	4.5	Al–2.862	–f.c.c.
Cu–10 a/o Al	3.1	Sn–3.022	–Tetragonal
Cu–1 a/o Sn	15.0 ± 2	Fe–2.481	–b.c.c.
Fe–10 a/o Al	8.0		

Note: Atomic size gives a rough measure of the amount that the alloy atom strains the parent lattice.

The data in figures 5-19 to 5-22 are extremely important because they indicate that the adhesion behavior of elemental metals can be altered by selective alloying, particularly when segregation occurs. One can increase or decrease adhesion by proper selection of alloy constituents.

Iron-base alloys are widely used in tribology as has already been mentioned; table 5-VII presents segregation data for iron-aluminum alloys. The iron—10-atomic-percent-aluminum alloy was examined for comparison to the copper-aluminum alloys. It was found that the aluminum segregated to the surface of the iron just as it did to the surface of the copper; the ratio of aluminum atoms on the surface of the iron to those in the bulk was 8 to 1. It also produced an increase in adhesion to the iron surface.

The segregation observed for the copper-aluminum, copper-tin, and iron-aluminum alloys was irreversible; that is, once segregation occurred to the solid surface, the solute atom remained on the surface of the solvent, and temperature changes did not alter that behavior. The only way the enriched surface layer of solute could be removed was by sputter removal. Thus, the segregation process for these systems was a permanent change in nature of the solid surface from the chemical point of view.

There are, however, some alloying elements that segregate to the surface of metals and do not remain on the surface but rather return to the parent lattice; that is, the alloy constituent segregate on the surface with heating or straining of the bulk metal and, on cooling or relief of the strain, they return to the parent lattice. The segregation process, in this case, is a reversible one. Silicon is such an alloying element. When silicon is alloyed with iron, the silicon segregates to the solid surface of the iron on heating, and it returns to the parent lattice on cooling. Adhesion studies were conducted with iron-silicon alloys to determine the effect of the silicon on adhesion. Since the silicon returns to the parent lattice on cooling, the measurements were made at temperature (ambient to 700° C) with a gold single crystal (111) orientation in contact with a single crystal iron-silicon alloy of the (110) orientation (the high atomic density plane for the body-centered-cubic iron-silicon alloy).

The results of the adhesion experiments with the iron-silicon alloy are presented in figure 5-23. The adhesion coefficient (the force to fracture the adhesive junctions divided by the load) is presented as a function of

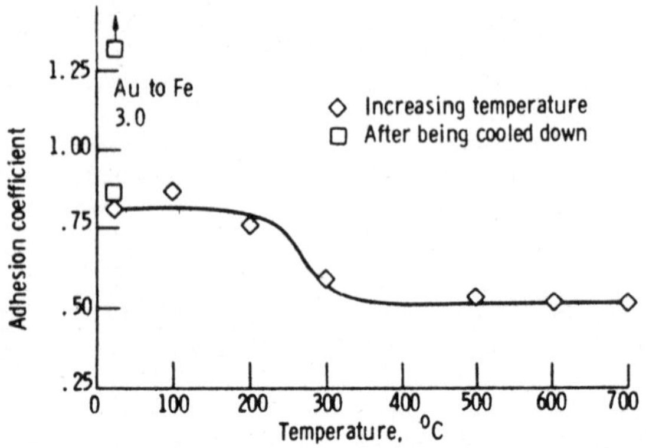

Figure 5-23. – Adhesion of gold (111) surface to iron – 6.5-percent-silicon alloy crystal (110) at various temperatures. Load, 1.0 gram; environmental pressure, 10^{-8} newton per square centimeter.

temperature. Also on the curve is the adhesion force for gold to elemental iron. The adhesion coefficient is approximately 3.0 for gold to iron, as indicated by the data point to the left side of the figure. With the iron—6.5-percent-silicon alloy, however, the adhesion coefficient is approximately 0.8. With heating, the silicon begins to segregate on the surface at approximately 300° C; that is, a sufficient amount of energy is supplied to the bulk alloy at 300° C to bring about equilibrium segregation to the surface of the iron. The adhesion data of figure 5-23 reflect a change in adhesive force with this segregation; at 300° C, a marked decrease is observed in the adhesive bonding force after which the adhesion coefficient remains constant to 700° C.

If the specimens are allowed to cool to room temperature, the adhesion coefficient which was obtained at room temperature is again obtained; this indicates the reversibility of the segregation process. The silicon segregates out at 300° C, and it returns to the parent lattice when the specimens are cooled to room temperature. Thus, the silicon moves in and out of the lattice with changes in temperature. Similar effects have been observed with the application of strain to the lattice. If oxygen is allowed to interact with the surface at temperatures in excess of 300° C so that the surface becomes oxidized, the silicon is frozen on the solid surface; this probably is due to the formation of silicon oxide, which ties up the silicon on the surface and prevents its return to the parent lattice in the bulk of the solid.

This discussion on segregation effects can be summarized as follows. In practical tribological systems, if alloying elements can segregate to the surface (in a manner such as the silicon in iron), it is then entirely conceivable that these alloy constituents will alter the adhesion behavior, particularly since temperatures can very readily exceed 300° C in tribological systems. Where the segregation is a reversible process, an alloying element can influence adhesion and friction of a tribological system and the experimentalist may never know that they have exerted such an influence because, at room temperature, the behavior is entirely different. A better understanding is needed as to why the silicon segregates in a reversible manner yet other alloying constituents in iron (such as aluminum, carbon, and sulfur) segregate on the surface in an irreversible manner.

Metal-Semiconductor Contacts

Thus far metals have been discussed in terms of contact with other metals or alloys. In practical lubrication devices, metals are in contact with nonmetals as well as metals. Nonmetallic surfaces with which metals are frequently brought into contact are semiconductor surfaces such as silicon and germanium. A question exists as to the adhesion behavior of metals in contact with semiconductors, very analogous to that for metals in contact with metals or alloys.

When the surfaces are clean, adhesion does occur and, as a general principle, fracture occurs in the interfacial region in the weaker of the two materials in solid-state contact. Fracture does not necessarily take place at the interface; the interfacial bonds are frequently stronger than the cohesive bonds in the cohesively weaker of the two materials. This is indicated in the scanning electron photomicrographs of figure 5-24 for a gold (111) surface contacting a silicon (111) surface. The white material in the photomicrograph represents the gold transferred to the silicon surface on simple touch contact with a 30-gram load after separation of the surfaces. The X-ray map, obtained with energy dispersive X-ray analysis in the SEM, is presented in the lower portion of figure 5-24. The map is for gold. The white appearing material in the photomicrograph is gold, which has transferred to the silicon surface.

Interfacial adhesive bonding has occurred, and the bond strength is stronger than the cohesive bonds in the gold so that, on separation, gold remains adhered to the silicon surface. It is rather interesting that silicon has a much higher cohesive strength than does gold. Gold and germanium (germanium is a semiconductor surface) have very comparable cohesive binding energies. If the experiment done with silicon in figure 5-24 is repeated with germanium, entirely different results are obtained; germanium is observed to transfer to the gold surface; that is, germanium is plucked out of the solid surface and found present on the gold as a result of adhesive contact. Again, as with silicon, the interfacial bonding is stronger than the cohesive bonding in the weaker of the two materials. The weaker of the two materials, however, in the gold-germanium contact is germanium rather than the gold as in the gold-silicon contact.

Iron has a stronger cohesive binding energy than does silicon. If a same type of experiment, as is presented in figure 5-24, is repeated with iron in contact with silicon, the silicon is seen to pluck out just as the germanium does when the germanium is in contact with gold.

The strong bonding and adhesive transfer of gold to silicon is extremely interesting also because (as was observed with the gold-tungsten and the copper-tungsten systems) the gold-silicon system forms no surface compounds yet strong adhesive bonding of gold to silicon is observed. In fact, the bond is sufficiently strong to cause the gold to remain on the surface with simple touch contact.

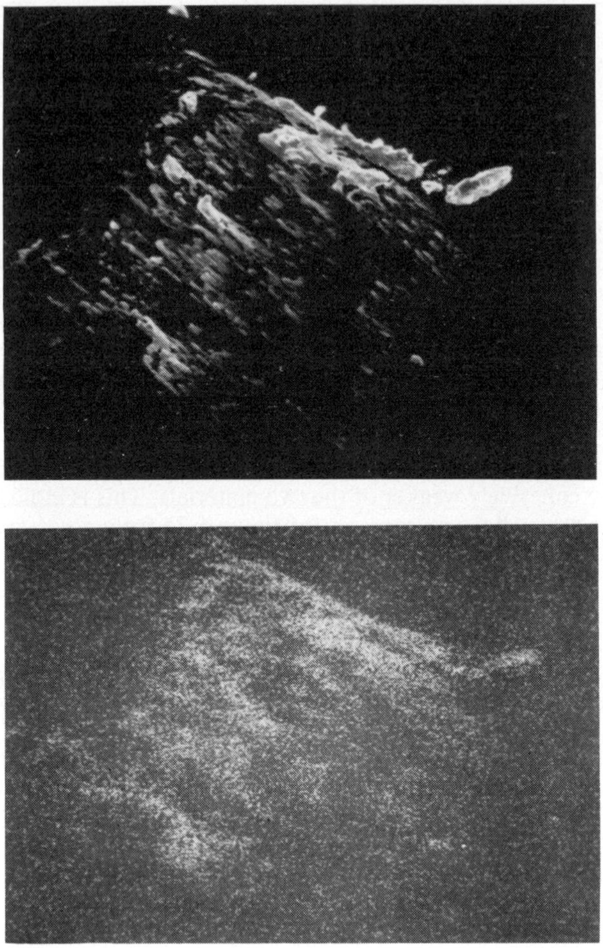

Figure 5-24.—Gold transferred to silicon (111) surface after adhesive contact. Load, 30 grams; sputter cleaned surfaces; temperature, 23° C; pressure, 10^{-8} newton per square meter.

Surface Contaminant Effects

Strong adhesion occurs with clean metals in contact with either clean metals or clean alloys and clean metals in contact with atomically cleaned semiconductor surfaces. A question exists as to what happens when one of the two solid surfaces is contaminated with, for example, a residual surface oxide in the case of metals. If only one of the surfaces contains an oxide, how does that influence or alter adhesion behavior?

Experiments with only one surface cleaned and the other containing residual surface oxides have been conducted and some interesting results are observed. In figure 5-25 two LEED patterns are presented for an iron (011) surface having been contacted by two different metals, an oxidized nickel

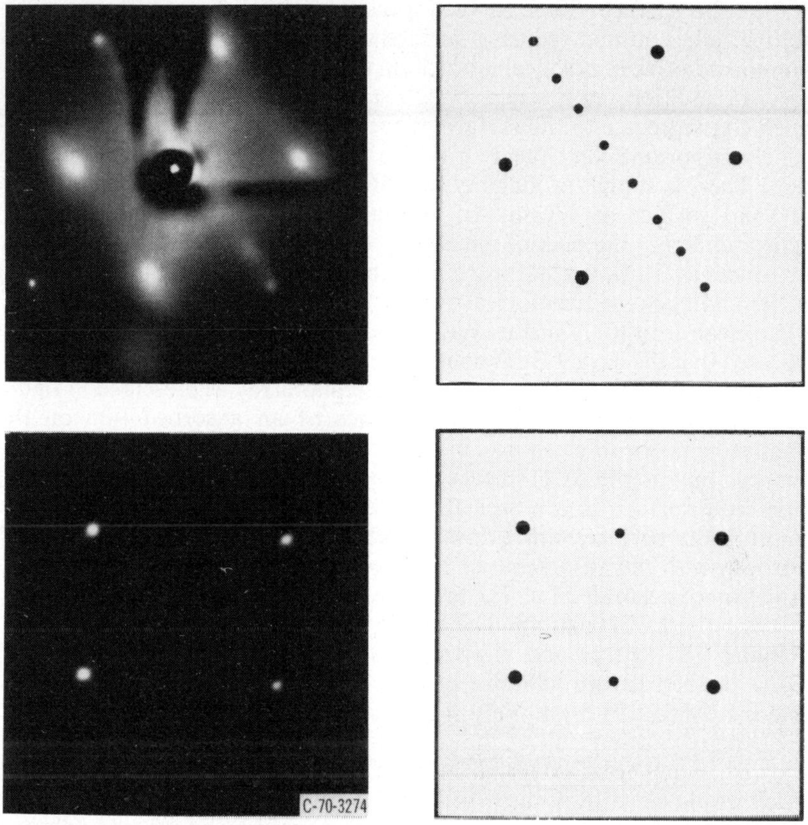

Figure 5-25. – LEED photographs of iron (011) surface after adhesive contact with nickel and tantalum.

surface and an oxidized tantalum surface. The annealed iron (011) surface had been sputter cleaned prior to solid-state contact. The basic diffraction pattern for the clean iron (011) surface is the rectangular array of four diffraction spots already discussed. In figure 5-25 an additional series of diffraction spots appears in the pattern after oxidized nickel has contacted the iron surface. Likewise, after oxidized tantalum has contact the iron surface, there are two additional diffraction spots seen in the diffraction pattern.

An AES analysis of the iron surfaces after such adhesive contact reveals the presence of oxygen on the iron surface. The iron has reduced both the nickel oxide and the tantalum oxide; that is, it has taken away some of the oxygen from these two oxides in the adhesion process. Both nickel oxide and tantalum oxide are fairly stable oxides. Despite this fact, however, these oxides are reduced by the nascent or clean iron. This observation is extremely important because it shows that the chemistry of the surfaces can be altered in the adhesion process, and there can be a reduction reaction when clean metals are brought into contact with other surfaces.

When a film is present on a solid surface, the amount of strain that occurs is considerably arrested; this depends to a great degree on the nature of film on the solid surface. Some surface species are much more effective in inhibiting adhesion and reducing surface strain than are others. Sulfur is a common additive in practical lubrication systems that interact with iron surfaces. In addition, most steels contain small amounts of sulfur in solution which can segregate to the surface of the solid steel. Thus, sulfur is an extremely important constituent. It is also frequently present in the environment. There is a high probability of sulfur interacting or being present on the solid surface as a result of (1) interactions with the lubricant, (2) segregation from the bulk of the alloy or metal, or (3) interactions with the environment. Thus, the sulfur-iron system is an extremely important one in understanding some tribological systems.

The clean iron (011) surface has already been discussed, and it has been indicated that the basic LEED structure for that surface is a rectangular pattern with four diffraction spots in a rectangular array as presented in figure 5-26. In order to determine the influence of an adsorbed film on the adhesive behavior of clean iron in contact with itself, hydrogen sulfide was adsorbed on the iron (011) surface. The adsorption of hydrogen sulfide to a clean iron surface is a dissociative adsorption process; that is, if one monitors the surface with a mass spectrometer, there is a liberation of hydrogen with the adsorption of hydrogen sulfide on the iron surface and formation of a sulfide film. The presence of a sulfide film is shown in figure 5-27 with the LEED pattern. The LEED structure is depicted in the pattern of figure 5-27 before and after adhesive contact. Adjacent to the basic LEED pattern before adhesive contact is the same surface region after adhesive contact by iron. Very little change has taken place in the basic LEED pattern as a result of the solid-state adhesive contact. If strain had occurred to any great extent in the solid surface as a result of contact, the lattice would have been disturbed and some marked change would have been observed in the basic structure observed for the iron prior to adhesive contact. The results in figure 5-27 indicate the effectiveness of sulfur on an

Figure 5-26. – LEED pattern of clean iron (011) surface (110 V).

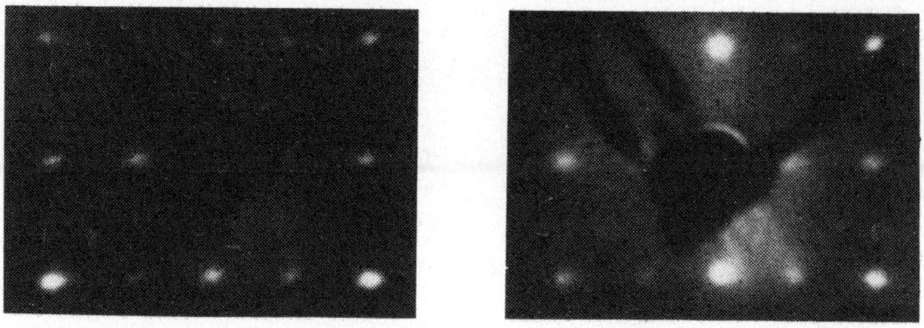

Before adhesive contact

After adhesive contact
3.6×10^{-3}-N force flat

Figure 5-27. – LEED patterns obtained from iron (011) c(2X4) surface after exposure to hydrogen sulfide.

iron surface in arresting adhesion and also in inhibiting deformation of the surface as a result of adhesive contact.

Sulfur is extremely effective in inhibiting adhesion; in fact, it is more effective than oxygen. Equivalent surface coverages of oxygen and sulfur were placed on an iron surface so as to produce the same basic LEED pattern that is observed in figure 5-27, and adhesion experiments were conducted on both surfaces to determine which films (oxygen or sulfide) were more effective in reducing adhesion. The results obtained in these experiments are presented in figure 5-28. The force of adhesion in dynes is plotted on the ordinate as a function of the applied normal load in dynes for an iron (011) surface covered with either an oxygen film or a sulfur film of equivalent surface coverage. For equivalent coverage of oxygen and sulfur on the surface, the force of adhesion is much lower for the sulfur-covered surface than it is for the oxygen-covered surface. With increasing load, the adhesive bond force increases for the surface with oxygen on the surface, as it would for clean metals in contact; that is, the oxygen film on the solid surface is penetrated by the contacting metals, and greater and greater metal contact occurs through the oxide film with increasing load. The sulfur film is, however, extremely effective in preventing metal to metal contact, even at the highest loads employed in figure 5-28. Surface analytical tools such as LEED and AES analysis can be used, then, to quantify the effectiveness of various surface films on the adhesion behavior for metals in contact. Comparisons can be made as to the effectiveness of these various films by obtaining LEED patterns and Auger spectra.

While the sulfur is extremely effective in inhibiting adhesion, it does not bond as strongly to the iron surfaces as does oxygen. If oxygen were bled into the system containing iron with sulfur on the surface, the oxygen completely displaces sulfur from the surface. Adhesive bond forces would, therefore, be comparable to those obtained for the iron surface covered with oxygen. While the sulfur film is present, however, it is much more effective in reducing adhesion than is the oxygen film.

There are many practical lubrication systems which involve metals in contact with ceramics or carbides (such as titanium carbide, iron carbide, or tungsten carbide). It is of interest to know the nature of the interactions of clean metals with these particular ceramic or nonmetallic systems. Hondros (ref. 9) has done considerable research on the adhesion of metals to

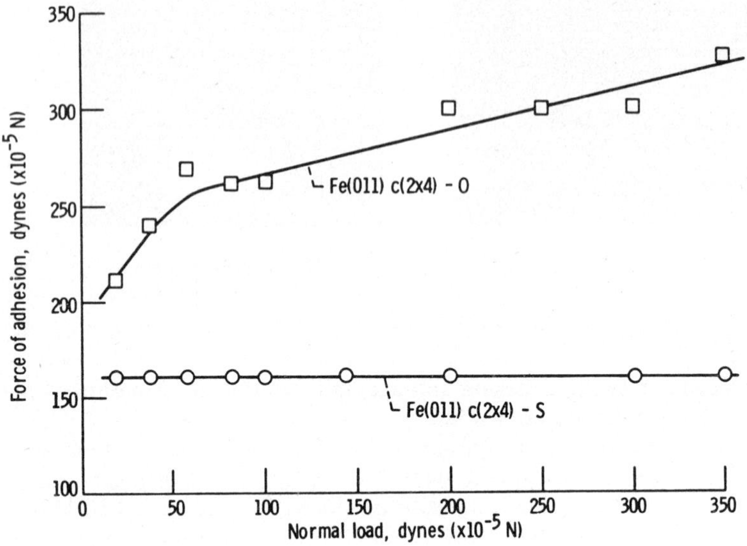

Figure 5-28. – Effect of oxygen and sulfur on adhesion of iron (011) to itself. Diameter of contacting flat, 3.0 millimeters; contact time, 10 seconds.

nonmetals (ceramics or carbides). He has attempted to relate the work of adhesion of metals and metal alloys to ceramic materials. He also has attempted to correlate this adhesion with interfacial free energy (the energy at the interface between the metal and the ceramic or carbide system). In the course of his studies, he found that some metals bond very strongly to ceramics or oxides (dielectric materials) while others do not, and he has ranked the adhesive behavior in relatively broad qualitative terms based on the nature of the metal and ceramic system. Some of his data are presented in figure 5-29 where there is plotted an inverse correlation between the work of adhesion and the interfacial energy (energy at the interface between the nonmetal and the metal). The nonmetallic components were aluminum oxide (Al_2O_3), silicon dioxide (SiO_2), iron oxide (FeO), and iron carbide (Fe_3C). From the data of Hondros, it appears that good adhesion occurs in the system having the lowest interfacial free energy.

When the noble metals copper, silver, and gold (which are clustered together) are in contact with aluminum oxide, the interfacial energy is relatively high and gives poor adhesion of these noble metals to aluminum oxide (fig. 5-29). This author has made similar observations for these metals in contact with aluminum oxide. In contrast, however, iron with a small amount of chromium dissolved in it gives very high adhesive bonding as is indicated in the figure; with 5 percent chromium in the iron, the work of adhesion indicates a good bond strength (fig. 5-29). If the concentration of chromium is increased from 5 to 15 percent, however, the bond strength becomes even greater. Both iron alloys containing chromium (5 or 15 percent) have bond strengths greater than those for elemental iron in contact with aluminum oxide.

Even the addition of 5 percent molybdenum to iron increases the adhesion of iron to aluminum oxide; the adhesive bond forces are still not as

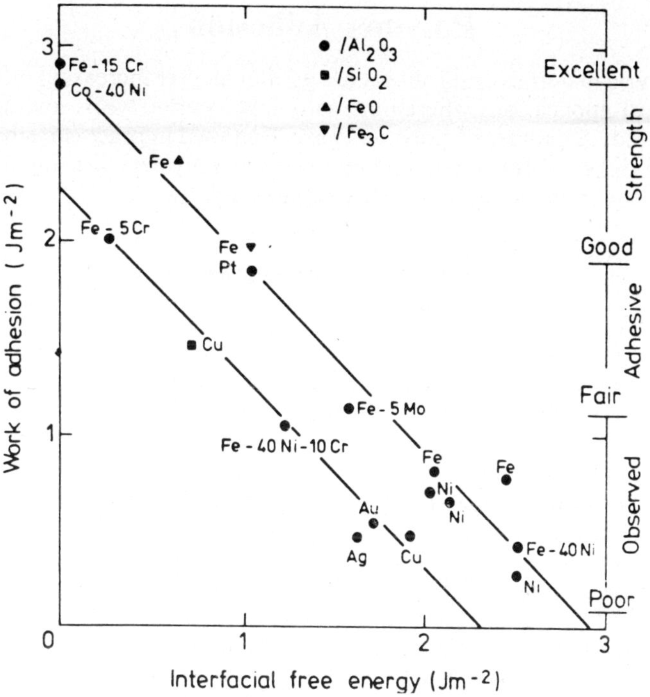

Figure 5-29.—Inverse correlation between thermodynamic work of adhesion and interfacial energy between metal and nonmetal. Solid lines indicate upper and lower quartiles of results (ref. 9).

great as they are for iron containing chromium. But, 5 percent molybdenum is not as effective as 5 percent chromium in increasing the adhesive bond strength. For certain applications, the bonding of metals to ceramic systems can be desirable. In most tribological applications, however, it is an undesirable phenomenon and one tries to minimize adhesion. Consequently, one should select materials for contact with ceramics (like aluminum oxide) that have relatively low works of adhesion, such as the noble metals. In systems where one intends to join ceramics to metals, strong adhesive bonding is important, and it becomes very significant to observe the effect of alloy constituents such as chromium in iron on the adhesive bond strength. The fact that adhesion can be altered by the addition of alloying elements (such as chromium to iron) is interesting. The equilibrium segregation concept which was discussed earlier in this chapter could be very effectively used to improve bond strength or to reduce it for metals in contact with ceramic systems as well.

The iron-aluminum alloy system, for example, might behave in a manner analogous to that for the iron-chromium system (fig. 5-29) because of the strong affinity of aluminum for aluminum oxide. It may have a tendency to improve bond strength of the aluminum oxide to the iron.

Polymer Adhesion

Polymeric materials and rubbers are used in an ever increasing number of mechanical applications where inherently low friction, wear, and adhesion are required. A number of polymers have been used in practical tribological systems; three of the most commonly used polymers are high density polyethylene, polyimide, and polytetrafluorethylene.

Adhesion Mechanisms

The adhesion behavior mechanism of polymers to themselves and to other surfaces has eluded researchers for many years. There are many explanations for the possible mechanisms for adhesion of polymers but none seem quite adequate to explain completely the behavior of polymeric materials. One of the most commonly held adhesion theories is that of Bowden and Tabor (ref. 10). They explain the behavior of the friction of polymers on the basis of simple adhesive bonding of the polymeric materials to another surface. For example, for polymers in contact with metals, the polymer bonds to the metal surface to produce strong adhesive bonds between the two surfaces. When any type of tangential motion is attempted, the shear that occurs in the polymer indicates that the interfacial bond is stronger than the cohesive bond in the polymeric material. This is analogous to the observations made earlier in this chapter for metals in contact with other metals. The bonds in the cohesively weaker of the two materials are fractured cohesively, the adhesive bond at the interface being stronger than the cohesively weaker of the two metals in contact.

When sliding is initiated between the surfaces for polymers in contact with glass or metals, shear takes place in the polymer and a polymer film is left on the solid surface. In other words, the interfacial bond formed between the polymer and the solid substrate (either glass or metal) is stronger than the cohesive bond in the polymer; shear, therefore, occurs in the polymer.

Another theory for the adhesion behavior of polymers is the mechanical theory. In this theory, the polymer interlocks around irregularities or pores in the surface of the substrate material. It is basically a mechanical type of interlocking. It is likely that this mechanism is only important in isolated cases such as roughened wood, cardboard, textiles, and certain polymeric forms. Mechanical adhesion should be enhanced by greater contact with the irregularities or pores of the substrates. The viscosity and certain other properties of the adhesive are, therefore, important on the basis of this particular mechanism.

Another mechanism for explaining the adhesive bonding of polymers is that the adhesive macromolecules are adsorbed onto the substrate surface and held there by various forces of attraction. The adsorption is usually physical (i.e., due to Van der Waals forces), but chemisorption may occasionally occur. This theory assumes a definite interface between the polymer and the substrate.

Yet another theory is the diffusion theory. In this theory the polymeric macromolecules diffuse into the substrate and eliminate the interface. Voyutskii, in particular, has been a strong advocate of the diffusion mechanism (ref. 11). His experimental evidence is based mainly on autoadhesion experiments (bonding experiments where the polymer and the substrate are identical as, e.g., the bonding of polyethylene to polyethylene). In particular, he studied the bonding of rubbers at elevated temperatures, and he found that the bond strength increased with various parameters: increased period of contact, increased temperature, increased pressure, decreasing molecular weight, and addition of plasticizers. Bond strength decreased with an increase in cross linking (ref. 11).

It is probable that diffusion occurs when two identical polymers are brought together at relatively high temperatures so that cohesion takes place. There is, however, very little direct evidence for diffusion in polymer-polymer contacts although various workers (including Bueche et al.) have provided good evidence for self-diffusion in the bulk phase (ref. 12). Therefore, there is no reason why diffusion should not take place if two strips of the same polymeric material are heated together. However, with chemically dissimilar polymers, such as polyethylene and an epoxy resin, diffusion is highly unlikely.

There are authors who have yet suggested another mechanism for polymeric adhesion. These authors include Deryagin and Smilga who have suggested a mechanism involving electrostatic forces arising from contact between the polymer and the substrate (ref. 13). The theory assumes that electrons are transferred from one material to another so that the surfaces end up being oppositely charged. The evidence put forth by Deryagin for this mechanism is that broken polymer junctions are observed to exhibit charging. This, however, does not seem to establish conclusively that the bonding or adhesion of the polymer to a substrate is due to electrostatic charges because the charge surface may have been produced during fracture of the adhesive junction.

While in any particular circumstance any one of the foregoing mechanisms may operate, it is believed that the adhesion theory of polymers, which is based on metal-metal behavior, is probably the most likely mechanism operating for polymers contacting polymers and other solid surfaces. Bickerman, who frequently argued a mechanical theory for the friction behavior of metals as opposed to the adhesion theory, has used the adhesion theory to explain the behavior of polymers. According to Bickerman, the attractive forces between two dissimilar molecules must always be greater than the attractive forces between two molecules of the same material (ref. 14). Thus, in a metal-polymer adhesive joint, the attractive forces between the metal and the polymer are greater than the attractive forces between the polymer molecules themselves; thus, cohesive failure in the polymer occurs. Bickerman consistently held this view, which is essentially the same as that for the adhesion theory of metals.

Tabor calculated, for an organic adhesive with a surface energy of 30 ergs per square centimeter, the force (6×10^8 dynes/cm^2) required to remove the polymeric material from the metal surface in direct tension (ref. 15). This calculation is consistent with the position taken by Bickerman; that is, the interfacial bond strength would be extremely strong. This strength would be beyond the cohesive bond strengths in the polymeric material itself. One would, therefore, predict fracture in the polymer rather than failure of the adhesive bond between the polymer and the metal surface.

If one subscribes to the theory that adhesion occurs to polymers, then bonding (both chemical and physical) must be important in the behavior of polymers in contact with different polymers. The difficulty of measuring the bond developed at the interface between polymers and other polymers is the low value of the adhesive forces for polymers in contact with various materials. For example, in table 5-VIII the adhesion of indium to various materials is presented (ref. 16). The adhesion coefficient is the force of adhesion divided by the applied load. The adhesion coefficient is of the order of unity for the indium in contact with diamond, glass, carbides, and various other metals. The adhesion coefficients for the plastics polyvinylchloride and PTFE are extremely low. In fact, for PTFE, the authors of table 5-VIII were unable to obtain a measurable adhesive force. These data, however, are for materials in contact in a normal air environment. The surfaces have not been atomically cleaned prior to making such contacts.

In a normal air environment, the evidence for chemical bonding playing a role in the adhesion of polymers to themselves and to other materials is, at best, indirect. For example, it is known that alcohols can adsorb very strongly to aluminum oxide surfaces. If hydroxyl groups are introduced into a polymeric material, the adhesion of the polymeric material to aluminum oxide increases. This is direct evidence for a chemical role played in the bonding or adhesion of polymers to other surfaces. Even in a normal environment, however, data do exist and indicate that there are differences in the bonding strengths of various polymers to substrates. For example, figure 5-30 presents data for specific adhesion (in kg/cm^2) of ice to three

TABLE 5-VIII. – ADHESION OF INDIUM TO VARIOUS MATERIALS[a]

Material	Coefficient of adhesion
Diamond	0.9 to 1
Glass	1
Tungsten carbide	1
Metals: Fe, Cd, Zn, Co, Ag, Pb, Cu, Au	1
Thick oxides of copper or silver	1
Rock salt	0.7
Plastics: polystyrene, Perspex	0.5 to 0.7
Plastics: P.V.C., polythene	0.02
Plastics: P.T.F.E.	0

[a]Reference 16.

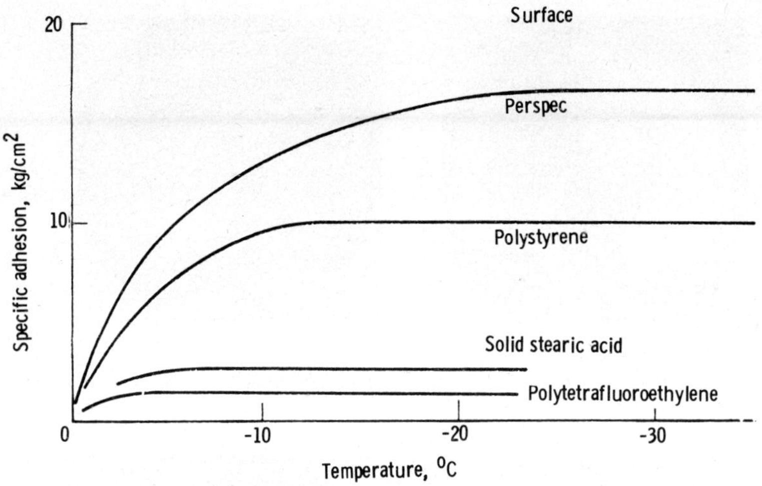

Figure 5-30. – Specific adhesion of ice to various organic surfaces (ref. 17).

polymeric materials as well as stearic acid (ref. 17). The data indicate specific adhesion at temperatures from 0° to -30° C.

Marked differences exist in the adhesion behavior of the various polymers to ice. The differences in the adhesive bond forces indicate that the nature of the interaction is somewhat specific. The ice in contact with all three polymers is the same, but the polymer structure and composition have changed, and distinct differences are observed in the adhesion behavior of the three polymers.

Attempting to establish the adhesive behavior of polymers to other substrates is extremely difficult in a normal air environment because surface films are present on both polymer and substrate. For example, the indium metal in table 5-VIII contains normal surface residual oxides which would inhibit adhesion. In the presence of the surface films, bonding across the interface of indium to the polymers may be of a physical nature and extremely weak; this could account for the relatively low adhesion values shown in table 5-VIII. If, however, the surfaces are very carefully cleaned in a vacuum environment and brought into contact, then the specific nature of the bonding should be directly observable. In studies like these the surface and the atom probe are extremely useful.

Polytetrafluoroethylene (PTFE)

Adhesion experiments conducted in the FIM with PTFE (polymer in table 5-VIII that showed essentially no adhesion to indium metal surface) in contact with a tungsten field ion tip (atomically clean) showed very strong adhesion.

Figure 5-31(a) shows the tungsten surface prior to solid-state contact with the polymer. The tungsten surface displays each individual atom site and

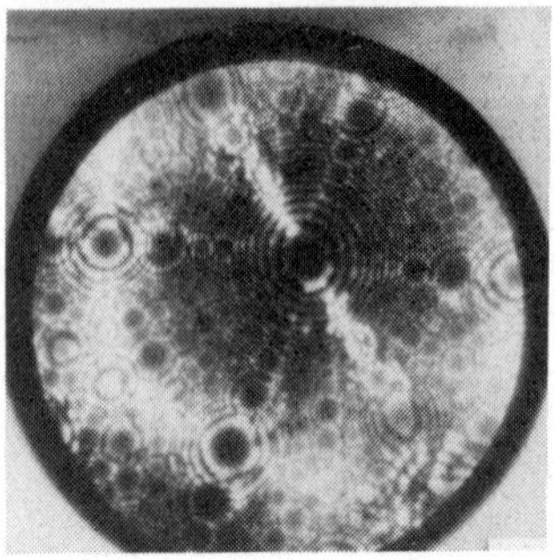

(*a*) *Tungsten.*

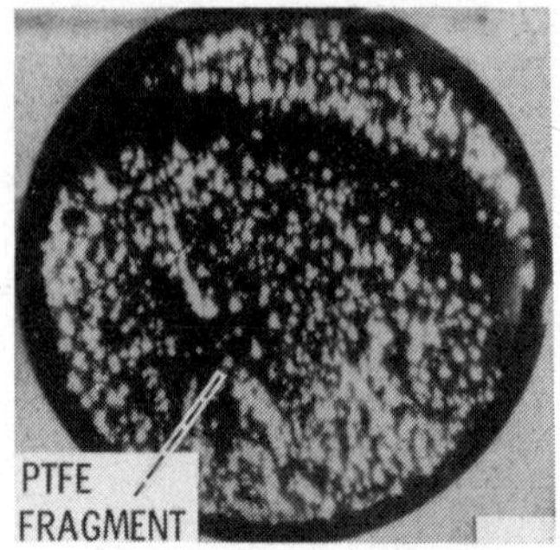

(*b*) *PTFE adhered to tungsten.*

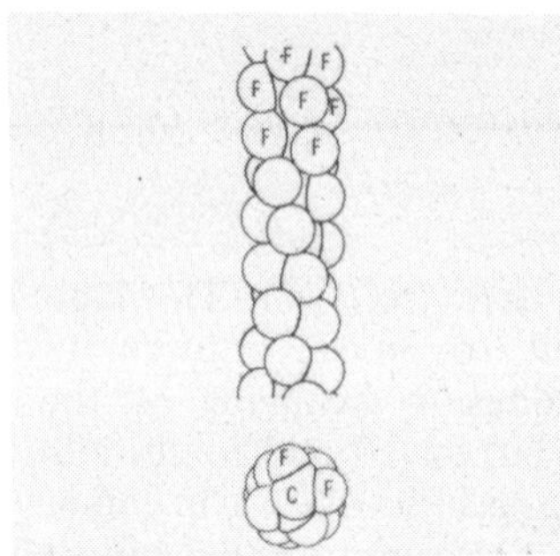

(*c*) *End and side views of PTFE chain.*

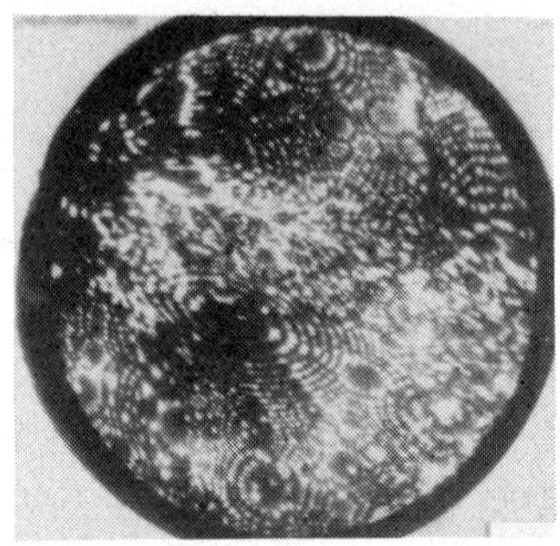

(*d*) *Tungsten after PTFE evaporation.*

Figure 5-31.—Field ion micrographs of adhesion of PTFE to tungsten.

the atomic planes involved. Figure 5-31(b) shows that same tungsten surface, after it was contacted by PTFE, with polymer fragments adhered to its surface. The tungsten tip in figure 5-31 is an atomically smooth surface (completely free of asperities). (Asperities or irregularities are burned off by the concentration of the imaging voltage at these high spots or surface irregularities until they are removed by field evaporation.) Since the tungsten surface is atomically smooth, the mechanical interlocking theory cannot be used to explain the bonding of the PTFE to the tungsten surface; however, the tungsten surface is completely covered with PTFE fragments.

The PTFE fragments on the solid surface exist as clusters of three atoms. It is believed that the polymer adheres to the solid surface by bonding the end caps of the polymer chain—that is, the carbon atom at the end of the polymer chain (fig. 5-31(c)). The CF_2 groups bond to the solid surface of the tungsten, and this accounts for the clusters of three atoms on the solid surface. Field evaporation and identification of the species in the atom probe revealed the presence of both carbon and fluorine on the tungsten surface. Independent studies with AES analysis for PTFE in contact with various metals have indicated that the bonding mechanism is carbon (from the end cap of the polymer chain) to the metal surface.

The bonding of the carbon in figure 5-31 would be of the nature of a tungsten carbide type of bonding. The bonding of the PTFE fragments (fig. 5-31) to the tungsten surface must be chemical in nature because the imaging voltage required to produce the photomicrographs of figure 5-31 is sufficiently high to cause desorption of any physically adsorbed species.

The chemical nature of the interaction of PTFE with the tungsten surface is such that when the polymer is completely removed by field evaporation from the tungsten the tungsten surface is as revealed in figure 5-31(d). A comparison of the tungsten surface after PTFE evaporation with that same surface prior to contact with the PTFE indicates a markedly different surface structure. A permanent change has taken place in the tungsten surface as a result of the PTFE contact. The chemical bonding of the PTFE by the carbon atom to the tungsten surface has produced a permanent strain in the tungsten surface.

The data of figure 5-31 then present direct evidence for chemical bonding of PTFE to a metal surface. Thus, with the aid of atomically cleaned surfaces and the FIM, it is possible to establish that chemical bonding does occur at the interface between very low surface energy polymeric materials (such as PTFE) and metal surfaces. The measured adhesion forces for tungsten to PTFE are extremely high. Since the polymer is seen to have transferred to the tungsten surface (fig. 5-31(b)), it can be assumed that the adhesion forces measured are attributable to the fracture of cohesive bonds in the PTFE. It is interesting to note that this is true only if the adhesive bonds at the interface between the polymer and the metal are stronger then the cohesive bonds in the PTFE. This is analogous to what was observed for dissimilar metals in contact where the cohesively weaker of the two materials fails in cohesion on tensile fracture of the adhesive junction; the interfacial bonds between the two dissimilar metals are being stronger than the cohesive bond in the cohesively weaker of the two metals.

The strong bond forces in the adhesive transfer of PTFE to the tungsten surface in figure 5-31(b) is not unique to PTFE in contact with tungsten. Other polymers also have been observed to transfer to metal surfaces and to exhibit very strong bonding. The bond strengths vary depending on the polymer in contact with the metal.

Polyimide

Polyimide, as mentioned earlier, is a widely used polymeric material in practical tribological systems and has a different structure than PTFE. It, however, also transfers to a tungsten surface and bonds chemically much like PTFE. Adhesion experiments have been conducted in the FIM with polyimide in contact with a tungsten surface; these experiments reveal, on separation of the solid surfaces, that polyimide transferred to the tungsten surface. The structure of the transferred polyimide is different from that observed with PTFE as indicated by the photomicrograph obtained in the FIM in figure 5-32. The polyimide is shown in a form of rodlike structures standing above the tungsten surface. Fractures again occurred in the polymer that is bonded to the tungsten surface. The tungsten surface lies beneath the rodlike structure seen in figure 5-32, and the tungsten orientation is the same as that seen in figure 5-31(a). Thus, the adhesive bonds at the interface between polyimide and tungsten are stronger than the cohesive bonds in the polyimide, and this causes fracture to occur in the polyimide and allows the polyimide to remain adhered to the tungsten surface.

It is interesting to note that the loading of the two surfaces in solid-state contact has an affect on the observed transfer characteristics. For PTFE in contact with the tungsten surface where the loads are extremely light (as observed in fig. 5-31(b)), small clusters of three atoms are found distributed over the surface with the carbon being bonded to the tungsten. When the load is increased, however, a structure analogous to that seen in figure 5-32 for the polyimide in contact with the tungsten is observed. Rodlike struc-

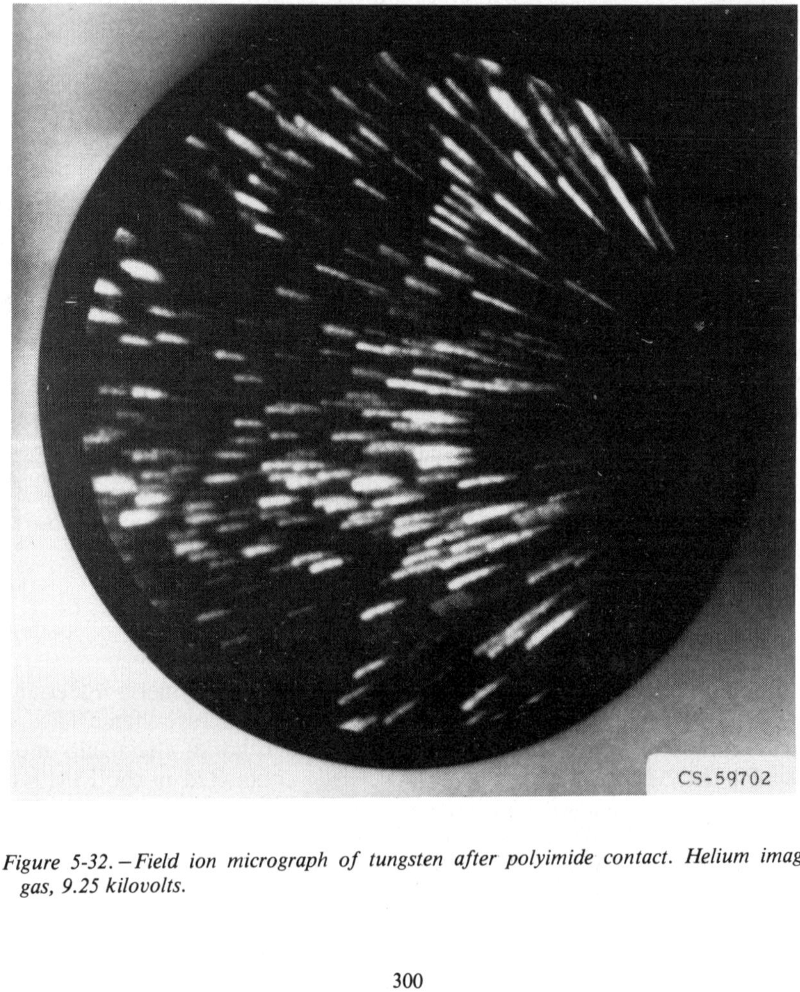

Figure 5-32.—Field ion micrograph of tungsten after polyimide contact. Helium image gas, 9.25 kilovolts.

tures appear on the surface. It is tempting to conclude from the rodlike nature of the transferred PTFE that a length of the polymer chain fragment has remained adhered to the surface. Instead of a cluster of three atoms being bonded at the surface, as was observed with light loads, a much larger or longer chain fragment remains adhered to the surface and the chain fragment stands above the surface like a bristle in a brush. However, this is not consistent with the observation made for the polyimide in contact with the tungsten. The polyimide has a ringlike structure (5 carbon atoms in a structure which would not give a linear type of orientation such as PTFE would). Thus, the rodlike structure of the polyimide on the surface is not related to the ring structure of the polyimide or the ring has been broken and a straight chain results.

Analogies Between Adhesion of Polymers and Adhesion of Metals

It appears from vacuum data and the use of analytical surface tools that both polymers and metals bond very strongly in a chemical manner to other substrate surfaces when the surfaces are clean. There is a degree of similarity between the adhesion of polymers and the adhesion of metals. The adhesion differences observed in air for metals and polymers may be related to the differences in mechanical behavior at the interface between polymers and metals. For metals in contact with other metals, and other substrates having high mechanical strength, elastic and plastic deformation can occur in the contact zone. In the asperity regions only, deformation occurs initially under applied loads, and nascent metal can be exposed so that clean metal surfaces can be generated even in an air environment; thus, adhesive bonding can occur.

Clean metal surfaces can be generated on both surfaces (so that there is metal to metal bonding), or it can occur in this surface where one surface is mechanically weaker than the other and thus experiences deformation. This is the case, for example, with indium in contact with some metals (table 5-III). The indium readily deforms plastically, and this disrupts the indium oxide (InO) and exposes clean indium to the other metal surfaces. With materials having extremely high elastic modulus (e.g., tungsten carbide or iron), the indium experiences most of the plastic deformation and the freshly generated (clean) indium metal can bond to either the oxide of iron or to the tungsten and the carbon of the tungsten carbide. Where the metals in contact with the indium are also soft (e.g., lead, cadmium, or zinc, table 5-VIII), deformation can occur in both metals at the interface and expose nascent metal and bring about contact of clean metals. Consequently, very high coefficients of adhesion can be measured in an air environment. Even with indium in contact with rock salt, clean surfaces can be readily generated for the indium.

With the indium in contact with the plastics (PTFE and polyethylene), however, it is very difficult to disrupt surface films on the polymer in solid-state contact because of the ease with which both the plastic and the indium deform plastically. There is a mutual flowing, as it were, without much disruption of surface films. Likewise, the adsorbed films (such as water vapor) that may be present on the plastics are not disturbed by the presence of indium on the surface. In fact, these films may be sandwiched between the plastics and the indium so that, with deformation, the increased deformation is like laying a blanket over a bed with sheets being sandwiched between. The sheets represent the oxides and the adsorbed layers of moisture. The sheets are not penetrated by the blanket. This is in sharp contrast with, for example, tungsten carbide, which is an extremely high elastic modulus material having a very high internal binding strength so that the asperities of the tungsten carbide can penetrate and puncture the oxides.

In a vacuum environment, the metal oxides can be stripped away. In the FIM, the pin tip, while it may have a very small radius of only 500 to 1000 angstroms, is still atomically smooth. There are no asperities on that surface; thus, the interaction is not one of an asperity of tungsten penetrating the PTFE but rather a relatively uniform contact of the PTFE with the smooth curvature of the tungsten emission tip. That is the reason that PTFE is found over the entire solid surface (fig. 5-31(b)). There are no clusters in certain areas of the solid surface, indicating that contact has only occurred in certain regions, but PTFE is fairly uniform and distributed over the entire surface, indicating contact over the entire radius with adhesive bonding. Thus, even though the tungsten surface does not consist of sharp asperities, chemical bonding to the PTFE does occur.

Even in air where the bonding of polymers to solid surfaces is relatively weak, the bond strengths can be increased by selectively doping the polymers with polar groups that bond fairly strongly to the substrate surface with which the compound is brought into contact. One can alter the adhesive behavior of a polymeric material to another surface by doping with materials that are highly polar and bond strongly to the substrate with which the polymer is brought into contact. Conversely, one can weaken the bond strength by introducing into the polymer bulk materials that are nonpolar and do not bond strongly to the surface with which the polymer is brought into contact. In the case where one is developing adhesives, it is desirable to have the former effect, and experiments have been conducted by various investigators to achieve stronger bonding by adding dopants to polymers which migrate to the interface and strengthen the interfacial bond.

Conversely, in the field of lubrication, one is interested in reducing the adhesion so that they can reduce the friction and the wear of the polymer in contact with other surfaces such as metals. In such cases, it is desirable to add materials that reduce the adhesive bond strength of the polymer to the metal surface. While nonpolar groups can do this in some instances, even polar groups (e.g., organic acids) may accomplish the same effect since the organic acid adsorbs to the surface. If the surface is a metal with which the polymer is in contact, the acid adsorbed to the metal surface then provides a lubricating film over which the polymer material can slide with a minimum amount of adhesion of the polymer to the metal; the acid acts as a lubricant and the polymer acts as a reservoir for the acid.

We can thus see the analogies between the adhesion behavior of polymers and the adhesion behavior of metals. It was also discussed in reference to the metals that other properties of metals (e.g., orientation and crystal structure) have an influence on the adhesion behavior of metals in contact with metals. Polymers are no different than metals. Again there is a close analogy. Various physical properties of the polymers exert an influence on its adhesion behavior, and it is well known that many of the properties of metals can be carried over into the field of polymers. For example, many of the structures that develop in metals can also be found in polymers. When one considers the material in the molten state and then selectively quenches or cools it, different mechanical properties are obtained. These differences in mechanical properties are directly ascribable to differences in structure.

In figure 5-33 are a schematic (showing various structures) and a stress strain curve. Phase I shows the molten state of a polymer; the particular molecules in the polymer have random orientation. As the specimen is quenched, a glassy (amorphous) state is obtained; this is indicated in figure 5-33 as phase II. By slow cooling rather than rapid quenching, one can generate crystallinity in the polymer (shown as phase III); ordering of the polymer structure occurs as a result of this slow cooling. This is analogous to the behavior of metals. If a molten metal is rapidly quenched (in cryogenic liquids), one can generate what is referred to as amorphous metals, metals lacking crystallinity. If, however, the metal is slowly cooled to room temperature (i.e., the normal type of cooling), the crystalline structure of the particular metal develops. Most metals exhibit crystalline structures because even rapid cooling rates are sufficiently long to allow some crystalline development.

Once the crystalline form of the polymer has been formed, the crystalline polymer can undergo differences in its characteristics. It can be cold worked to develop orientations or texturing, as metal surfaces can be textured. It can be annealed to regenerate a random orientation (fig. 5-33). If the glassy state is cold worked, it undergoes recrystallization and the recrystallized form can develop preferred orientation, just as do metals. The various

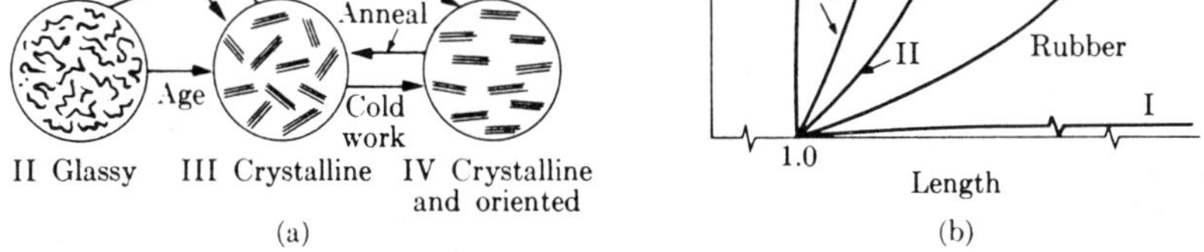

(a) Thermal and deformation treatments.

(b) Comparative stress-strain relations. (Roman numerals correspond to structures in (a).)

Figure 5-33. – Deformation of linear polymers.

structures that can be produced with polymers are very analogous to those for metals. The different structures depicted in figure 5-33 result in differences in mechanical behavior. This is seen in the stress-strain curves of figure 5-33(b) where stress is plotted as a function of length. The Roman numerals correspond to the structures of figure 5-33(a). The molten state is the weakest structure, as one might anticipate; it has the lowest mechanical strength. For reference purposes, both metal and rubber curves are included in the figure. It is found from a comparison of these stress-strain curves that the crystalline and glassy phases of the polymeric materials fall between the metals and rubber in their mechanical properties.

The structure influences the mechanical properties of the polymers; crystalline polymers exhibit greater mechanical strength than do the glassy polymers. Likewise, there are differences in adhesion for the various forms of the polymers. The adhesion of the crystalline polymer is less than the adhesion of the glassy or amorphous polymer. It would be interesting to see a detailed set of experiments conducted with a specific, highly characterized and well defined polymer system. The polymer should be studied in its various structural forms to determine the influence of these structures on the adhesion qualities of the polymeric material. To the knowledge of this author, this study has not been done.

Crystallization of polymers can occur to varying degrees. In some polymers the degree of crystallization can be controlled so that the structure consists of a mixture of amorphous and crystalline phases. Furthermore, even when the material is in a crystalline state the crystal structure or morphology of the crystals varies with the polymers; again, these can be controlled to some extent. For example, lamellar crystals can be formed in polymers by folding back and forth the polymer chain between the faces of the lamellae as indicated schematically in figure 5-34. This would be analogous to having two basal planes (in a material like graphite) sliding over each other. Shear would take place between adjacent lamellae where bond forces are anticipated to be relatively weak and the polymer would conceivably have low friction and adhesion characteristics. Bowden and Tabor have done considerable research with polymer structures and have indicated that the orientation of the molecular chains in polymers can influence the adhesive behavior of materials in solid-state contact (ref. 17).

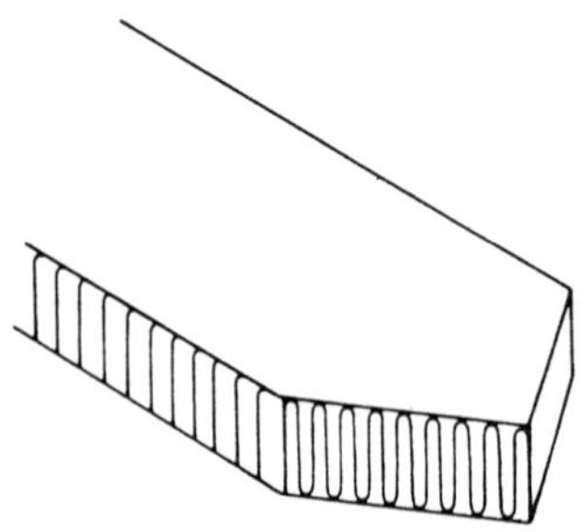

Figure 5-34. – Lamellar crystal and folding of polymer chain back and forth between faces.

The attempt to slide or move surfaces tangentially is affected by the amount of adhesive bonding that occurs at the interface. Bowden and Tabor have measured the adhesion component of friction for a steel hemisphere sliding on PTFE in two different directions relative to the polymer chains: (1) sliding along the polymer chains and (2) sliding across the polymer chains (ref. 17). They conducted these experiments with the two different orientations of the polymer and found the adhesion results presented in figure 5-35; the adhesive component of friction is plotted as a function of the load in grams. Curve I is for the steel ball sliding along the direction of the PTFE chains (molecular chains), and curve II is for the steel ball sliding across the direction of the molecular chains. There is a distinct difference in the friction properties for the two orientations of the polymer chain relative to sliding. Sliding along the direction of the molecular chain produces lower friction than sliding against it. Furthermore, the friction difference of about 20 percent is consistent over the entire load range. By analogy, the friction and adhesion properties of metals are influenced by the crystallographic orientation of the surface; adhesion properties of polymers apparently are also influenced by the crystallography of the polymer surface.

The adhesion properties of polymers can also be related to the changes or transitions that take place in the polymer. One particular transition that is very common in polymers and produces marked changes in its mechanical properties is a glass transition; it occurs where the polymer undergoes a transition from a glassy phase to a crystalline phase. With this transition, there is a marked change in the mechanical properties of the polymers. A

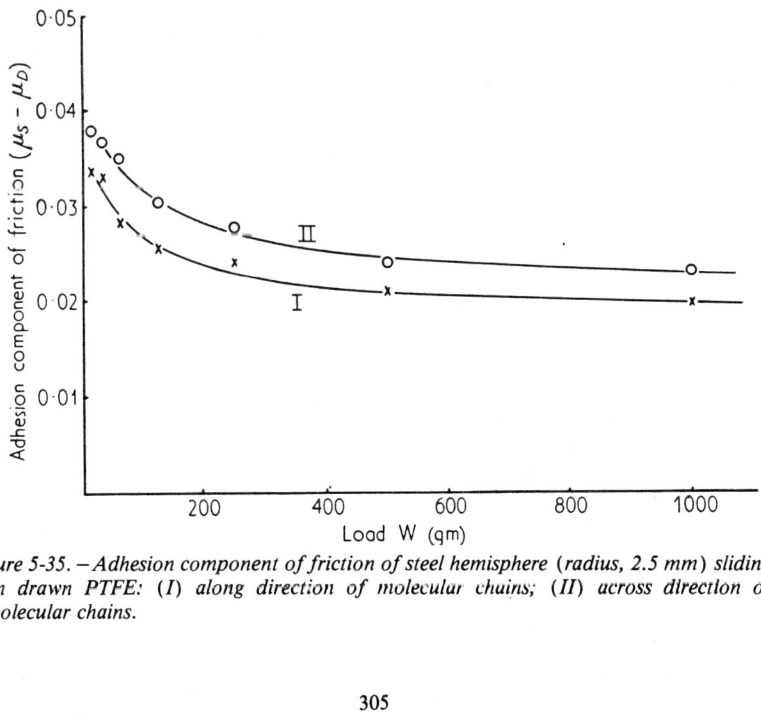

Figure 5-35. – Adhesion component of friction of steel hemisphere (radius, 2.5 mm) sliding on drawn PTFE: (I) along direction of molecular chains; (II) across direction of molecular chains.

degradation occurs in the mechanical properties in transforming from the glassy to the nonglassy phase as the melting point is approached.

Accompanying this transition is generally a change in the adhesion characteristics. The adhesive bonding of the polymers becomes greater after the transformation from the glassy to the crystalline state. This glass transition in polymers is related to other properties of the polymer—for example, melting point. The higher the melting point of the polymer, the higher the temperature at which the glass transition occurs. For most polymers, however, the glass transition occurs when the temperature is between -120° and 145° C. The glass transition temperatures for a host of polymers are presented in figure 5-36. The higher the melting point of the polymer, the higher the glass transition temperature (ref. 18).

1. poly(dimethyl siloxane)
2. *cis*-1,4-polybutadiene
3. *trans*-1,4-polybutadiene
4. *cis*-1,4-polyisoprene
5. *trans*-1,4-polyisoprene
6. polyethylene
7. polypropylene
8. polybutene-1
9. polypentene-1
10. poly-3-methyl butene-1
11. poly-4-methyl pentene-1
12. polyoxymethylene
13. polytetrafluoroethylene
14. polychlorotrifluoroethylene
15. isotactic polystyrene
16. poly(ethylene terephthalate)
17. isotactic poly(methyl methacrylate)
18. poly(*t*-butyl acrylate)
19. isotactic poly(isopropyl acrylate)
20. bisphenol polycarbonate
21. poly(vinyl carbazole)
22. nylon 6
23. nylon 66
24. poly(vinyl methyl ether)
25. poly(vinyl isobutyl ether)
26. poly(vinyl cyclohexanone)
27. poly(vinylidene chloride)
28. poly(ethylene oxide)
29. poly(propylene oxide)
30. cellulose triacetate
31. poly(methyl isopropenyl ketone)

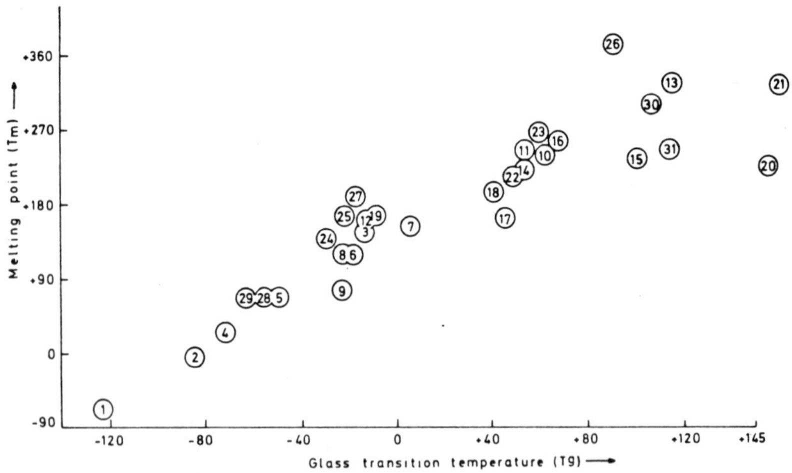

Figure 5-36.—Relation between glass-transition temperature and melting point.

Rubber Adhesion

The adhesion behavior of rubbers to varying substrates is extremely important in a number of practical application areas—for example, the rubber to glass contact in windshield wipers. Currently, an extremely interesting area for the application of rubber to metal adhesion is that of the radial tire with steel cords. It is important in the radial tire that the steel cord bond very strongly to the rubber. Unfortunately, experimentation has shown that the binding or adhesion of steel to rubber is very poor. There are, however, other metals to which rubber has been observed to bond strongly—for example, brass. As a consequence, when making wire for radial tires, the steel is generally coated with a film of brass by electroplating. Adhesion can be improved even further by adding vulcanizing agents to the rubber—namely, sulphur. If the concentration or composition of the sulphur at the interface is proper, very good adhesion and very good bonding of the rubber to the metal surface can be achieved. Because of the practical importance or significance of rubber bonding to metal, a considerable amount of research effort has been expended using analytical surface tools to characterize the adhesion process of rubbers to metals.

Data on the adhesion of selected materials to vulcanized rubber are presented in table 5-IX. The rubber, which contained sulphur as the vulcanizing agent, was vulcanized at 150° for 25 minutes (ref. 19). Table 5-IX includes the material surfaces with which the rubber was brought into contact, the adhesion levels, the composition of the interface as determined by X-ray photoelectron spectroscopy (XPS), and remarks on the nature of the adhesion. With iron and steel there is very poor adhesion of the rubber

TABLE 5-IX. – ADHESION OF SOME SELECTED MATERIALS TO RUBBER[a] (REF. 19)

Material	Adhesion level[b]	XPS of interface	Remarks
1. iron, steel	0	–	no adhesion
2. copper sheet	0	excess Cu_2S	some adhesion if undercured
3. copper-plated steel[c]	700–900	–	good adhesion if plating thickness <50 nm
4. steel + Cu_2S coating[d]	700–800	–	good adhesion for fresh Cu_2S layer (<50 nm)
5. zinc sheet	100–200	ZnS formation	poor adhesion
6. copper-plated zinc[c]	700–800	Cu_2S formation	good adhesion if plating thickness <50 nm
7. 70/30 brass sheet	700–1000	formation of Cu_2S and ZnS	good adhesion; level depends on surface preparation

[a] Vulcanized at 150°C for 25 min.
[b] In $N/64\ mm^2$.
[c] By electroless immersion plating.
[d] Prepared from sample 3 by reaction with sulfur in liquid paraffin at 180°C.

to the surfaces, and the bond forces for rubber to iron or steel could not be detected. With copper sheet, the adhesion level was O, but some adhesion occurred if undercured; XPS analysis of the interface reveals a reaction of the sulfur from the vulcanized rubber with the copper sheet surface to form excess copper sulfide (Cu_2S). Copper sulfide is a relatively brittle inorganic compound and, when formed at the interface in thick layers, provides a brittle region that can cause separation of the rubber from the copper surface. When copper is plated on steel in sufficient thickness, however, good adhesion can be obtained (table 5-IX) if there is not an excess of copper sulfide in the interface.

The data of the first three materials in table 5-IX (iron, copper, and copper-plated steel) produce several questions. What effect does the copper sulfide have since the plain steel and copper sheet exhibit no adhesion and yet there was adhesion with copper-plated steel? Why is it when the copper is plated on the steel the adhesion improves? An examination of the plated metal indicated that there was copper sulfide present, but not in excess, near the interface. It was anticipated that the copper sulfide in some way played a role in achieving the bonding since the bonding was very poor to the straight copper (with resulting excess of copper sulfide) and to the straight steel (without the benefit of the copper sulfide). The steel surface was, therefore, coated with a very thin coating (of the order of 15 nm) of copper sulfide; the results show that the adhesion of rubber to the surface was very good. Thus, copper sulfide in the proper concentration apparently promotes good adhesion of the rubber to the steel surface.

Brass consists of both copper and zinc, and if a zinc surface is examined, it is found that the adhesion for the rubber to zinc is poor. The zinc sulfide formed at the interface is detrimental to good adhesion. When the zinc was plated with copper and the copper sulfide was allowed to form on the surface of the copper plate, good adhesion was observed. Likewise, with a 70-30 brass sheet (where there is a mixture of zinc sulfide and copper sulfide) good adhesion was observed when the sulfides were mixed, provided the zinc sulfide was not present in excessive quantities. There had to be a proper ratio of copper sulfide to zinc sulfide. If the sulfides were present in excessive quantities, the interfacial region became brittle and bond fracture occurred between the rubber and the metal surface. With the 70-30 brass sheet and an optimum concentration of copper sulfide and zinc sulfide at the interface, very strong bonding and good adhesion occurred.

Use of XPS in Studies of Adhesion Interfaces

XPS was used to identify the rubber to metal interface in table 5-IX. It was of appreciable assistance in interpreting and understanding the mechanisms of interfacial chemistry in the rubber to metal system. To give the reader an indication of the usefulness of XPS, some discussion on the technique as it was used in relation to the data obtained in table 5-IV is now be presented. In essence, with XPS, one separates the various compounds or elemental materials by differences in their binding energies. XPS is an extremely sensitive tool, and the binding energies can be separated on the basis of differences to tenths of an electron volt.

The binding energies for some of the elemental metals in compounds involved in the adhesion of rubber to the metal surfaces shown in table 5-IX are presented in table 5-X. The binding energies are presented for copper, copper oxides, copper sulfide, zinc, zinc oxide, zinc sulfides, and sulfur (ref. 19); each material has its own characteristic binding energy. It should be noted that such compounds in the table may have more than one binding energy. The differences in these binding energies can be used to fingerprint the particular species that are present on a surface or at an interface between rubber and metals.

The type of analysis that can be generated from XPS in the peak intensities of the technique are presented in figure 5-37 for the rubber-brass system. The XPS peaks for copper and copper sulfide are presented on the left side of figure 5-37. The peak associated with the formation of copper sulfide is observed to shift to the right. The copper sulfide peak with the film grown on the copper surface at the interface in figure 5-37 shifts to the left, or a decrease in the binding energy with the formation of the Cu_2 as the film is grown on the copper surface. With zinc, the zinc sulfide at the interface appears to be analogous to the peaks obtained for zinc sulfide, since the zinc sulfide peak at the interface coincides in binding energy with the zinc sulfide generated from a powder.

From the peaks on the far right side of figure 5-37 the composition of the interface relative to sulfur can be seen. The interface reveals a peak for S_8 (unvulcanized rubber). This is for sulfur that is not consumed in cross linking in the rubber structure. There is a relatively large XPS peak associated

TABLE 5-X. – BINDING ENERGIES OF Cu $2_{3/2}$, Zn $2p_{3/2}$, AND S 2p IN SOME SELECTED COMPOUNDS AND AT THE RUBBER-TO-BRASS INTERFACE (REF. 19)

Sample	Binding energy (eV) [a]
Cu	932.2
Cu_2O	932.6
Cu_2S (powder)	932.4
Cu_2S (film grown on copper)	931.6
Interface rubber–brass [b]	931.6
Zn	1021.2
ZnO	1022.0
ZnS	1021.9
Interface rubber–brass [b]	1022.0
S_8 (unvulcanized rubber)	164.1
Vulcanized rubber	163.5, 164.1
Cu_2S	161.3, 164.0
ZnS	161.5
Interface rubber–brass [b]	161.3, 163.5, 164.0

[a] Using Au $4f_{7/2}$ = 83.6 eV as reference; estimated accuracy ±0.2 eV.
[b] Binding energies on both sides of the interface were identical.

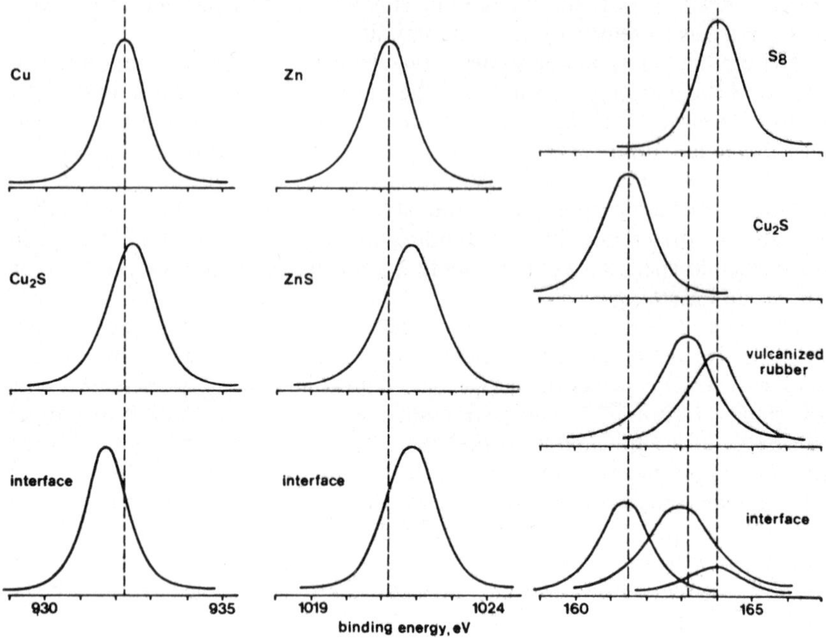

Figure 5-37.—2p photo lines of Cu, Zn, and S at rubber-to-brass or rubber-to-polyester interface following vulcanization at 150° C for 25 minutes (ref. 19).

with copper sulfide. Two additional large peaks at the interface are associated with the vulcanized rubber.

Depth profile analysis using argon ion bombardment or some other inert gas can be used with XPS to depth profile the interface region of rubber in contact with the brass. This was done in the brass-rubber adhesion experiments. Depth profile data for rubber in contact with brass where the adhesive junction was broken at liquid nitrogen temperatures following vulcanization of dry rubber to polished 70/30 brass at 150° C for 25 minutes are presented in figure 5-38. Concentration is plotted on the ordinate; on the abscissa is the depth profile into the brass and into the rubber, zero being the interface between the brass and the rubber. An examination of the elements near the interface shows that there is a rapid or sharp decrease in the concentration of copper and zinc as the interface is approached from the brass side. There is also an increase in the concentration of sulfur, oxygen, and carbon at that interface, or near the interface in the brass itself. From the rubber side, there is a decrease in the sulfur concentration as the interface is approached, the sulfur is apparently taken up in the formation of copper and zinc sulfide.

Copper and zinc are present in the rubber to a considerable depth, and the maximum concentrations of both the zinc and the copper in the rubber do not occur near the surface but rather slightly subsurface. For some unknown reason the adhesion of rubber to brass is very good in the dry

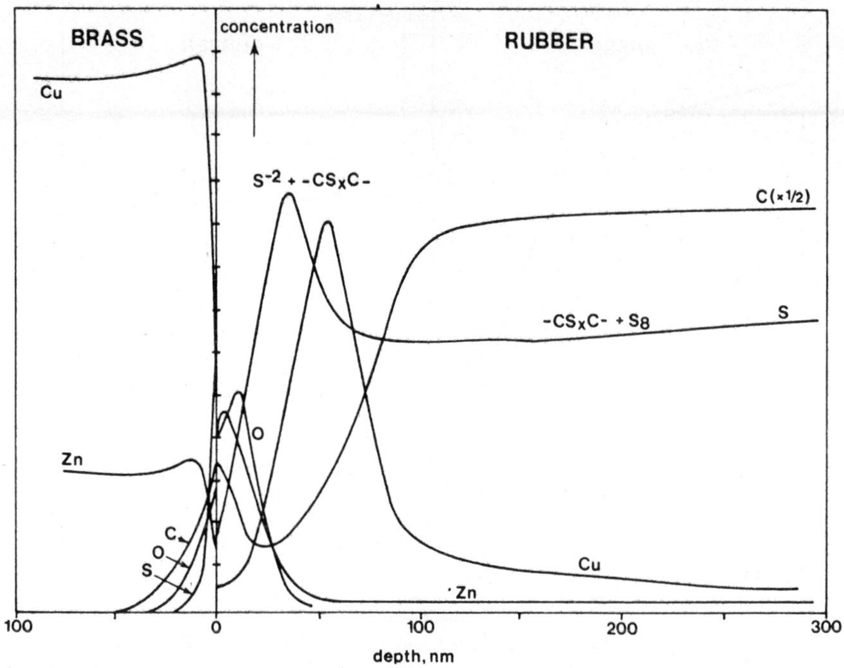

Figure 5-38.—In-depth concentration profiles of rubber-to-brass sample broken at liquid nitrogen temperature following vulcanization of polished 70/30 brass to dry rubber at 150° C for 25 minutes (ref. 19).

state; when the rubber surface is wet, however, the adhesion is extremely poor. In order to understand the reason for this difference in adhesion behavior, depth profile analyses were conducted on a rubber to brass joint that was formed with wet rubber; after adhesion, the rubber-brass junction was fractured at liquid nitrogen temperature. The results obtained in these depth profile analyses are presented in figure 5-39.

A comparison of the depth profile analysis of figure 5-38 to that of figure 5-39 indicates markedly different interfacial chemistry when the water vapor is present on the surface of the rubber. Copper and zinc behave entirely differently in the surficial region of the brass. The copper concentration drops off much more rapidly (as the interface is approached from the brass side) than it does in the absence of water vapor. The zinc increases in concentration as the interface is approached and then decreases at the actual interface itself. The amount of oxygen seen in the brass in the presence of the water film on the surface is much greater than was observed for the dry rubber in contact with the brass. Furthermore, the sulfur penetrates to a much greater depth in the brass in the presence of the water film than it did in its absence. Figure 5-39 shows that the sulfur has penetrated approximately 200 nanometers into the brass, whereas in figure 5-38 (the dry state) the sulfur concentration has terminated at approximately 25 nanometers beneath the surface; there is no sulfur observed beyond that depth.

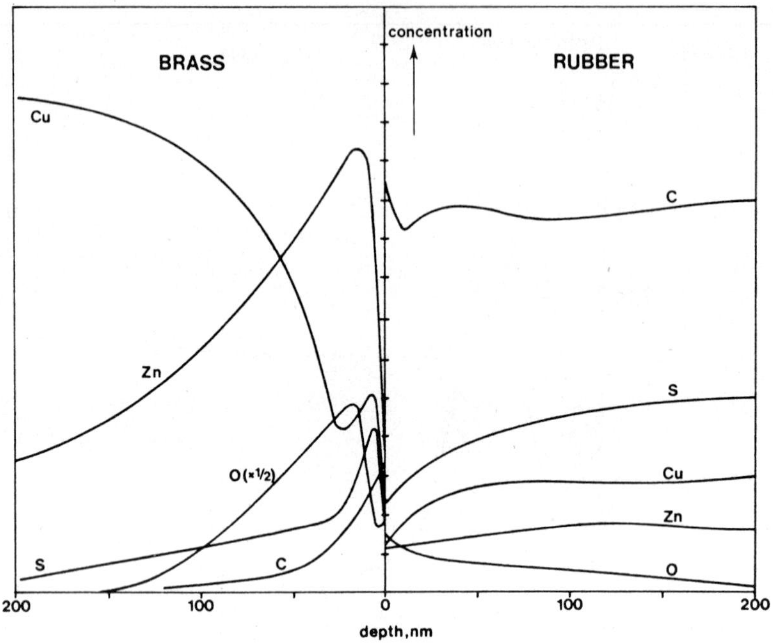

Figure 5-39.—In-depth concentration profiles of rubber-to-brass sample broken at liquid nitrogen temperature following vulcanization of polished 70/30 brass to wet rubber at 150° C for 25 minutes (ref. 19).

Likewise, the oxygen moves much deeper into the brass in the presence of the water film (approximately 150 nm compared to 40 nm in the absence of the water film). The presence of the water film on the surface of the rubber seems to promote the diffusion of oxygen, sulfur, and carbon into the brass.

Earlier in this chapter one of the mechanisms described for adhesion was the diffusion mechanism. In this particular instance, it appears that this type of a mechanism can and may be operating to promote adhesion to the surface because we see that oxygen, sulfur, and carbon have diffused considerably into the bulk of the brass. The diffusion, however, is detrimental to adhesion, contrary to the basic theory which indicates that diffusion promotes adhesion. Diffusion here tends to arrest adhesion because the rubber does not adhere as well to the brass in the presence of the water film as it does in its absence. Just as surficial chemistry in the brass is markedly altered by the presence of water film on the surface of the rubber, likewise the surficial chemistry in the rubber is also considerably different than it was in the dry state. There appears to be a greater concentration of zinc and copper in the rubber in the presence of the water film. The concentration of sulfur appears to have diminished considerably in the surficial regions of the rubber with no sharp dropoff at the interface as was observed in figure 5-38 for the dry rubber in contact with the brass. Furthermore, the carbon concentration does not decrease rapidly as the interface is approached, as was the case with the dry surfaces in solid-state contact.

The depth profile data of figures 5-38 and 5-39 are typical examples of how a surface analytical tool such as XPS can be used to assist in the understanding of the adhesion mechanism. The adhesion for the brass in contact with dry rubber is good; associated with that good adhesion is the interface chemistry revealed by the analysis of figure 5-38. Adhesion of the brass in contact with the wet rubber is relatively poor; associated with that poor adhesion is the chemistry of figure 5-39, an entirely different chemistry than that of figure 5-38.

References

1. Obreimoff, J. W.: The Splitting Strength of Mica. Proc. Roy. Soc., (London), ser. A., vol. 127, no. 804, Apr. 1930, pp. 290-297.
2. Gilman, John J.: Direct Measurements of the Surface Energy of Crystals. J. Appl. Phys., vol. 31, no. 12, Dec. 1960, pp. 2208-2218.
3. Sargent, Lowrie B., Jr.: On the Fundamental Nature of Metal-Metal Adhesion. ASLE Trans., vol. 21, no. 4, Oct. 1978, pp. 285-290.
4. Muller, Erwin W.; and Tsong, Tien T.: Field Ion Microscopy: Principles and Applications. American Elsevier Publishing Company, Inc., 1969.
5. Taylor, Norman J.: A LEED Study of the Epitaxial Growth of Copper on the (110) Surface of Tungsten. Surface Sci., vol. 4, 1966, pp. 161-194.
6. Van Der Merwe, Jan H.: Structure of Epitaxial Crystal Interfaces. Surface Sci., vol. 32, 1972, pp. 198-228.
7. Van Der Merwe, Jan H.; and Van Der Berg, N. G.: Misfit Dislocation Energy in Epitaxial Overgrowths of Finite Thickness. Surface Sci., vol. 32, 1972, pp. 1-15.
8. Maugis, D.; et al.: Adhesion and Friction on Aluminum Thin Foils Related to Observed Dislocation Density. ASLE Trans., vol. 21, no. 1, Jan. 1978, pp. 1-19.
9. Hondros, E. D.: Precipitation Processes in Solids. To be published in ASTM-STP. See also Seah, M. P.: Surface Science in Metallurgy. Surface Sci., vol. 80, 1979, pp. 8-23.
10. Bowden, Frank P.; and Tabor, D.: The Friction and Lubrication of Solids. Part 2, Oxford Clarendon Press (London), 1964.
11. Voyutskii, S. S. (S. Kaganoff, transl.): Autohesion and Adhesion of High Polyer. Interscience Publishers, 1963.
12. Bueche, F.; Chasin, W, M.; and Debye, P.: The Measurement of Self-Diffusion in Solid Polymers. J. Chem. Phys., vol. 20, no. 12, Dec. 1952, pp. 1956-1958.
13. Deryagin, B. V.; and Smilga, V. P.. Present State of Our Knowledge About Adhesion of Polymers and Semiconductors. Proceedings of the Third International Congress of Surface Activity (Vortraege Originalfassung Kongress Grenzflaechenaktive Stoffe, 3), vol. 2, 1960, pp. 349-367.
14. Bikerman, Jacob J.: Science of Adhesive Joints. Second ed., Academic Press, Inc., 1968.
15. Tabor, D.: Basic Principles of Adhesion. Reports on the Progress of Applied Chemistry, vol. 36, 1951, pp. 621-634.
16. Moore, A. C.; and Tabor, D.: Some Mechanical and Adhesive Properties of Indium. Brit. J. Appl. Phys., vol. 3, Sept. 1952, pp. 299-301.
17. Bowden, Frank P.; and Tabor, D.: Friction and Lubrication of Solids. Oxford Clarendon Press (London), 1950.
18. Brydson, J. A.: The Glass Transition, Melting Point and Structure. Polymer Science. A Materials Handbook, vol. 1, A. D. Jenkins, ed., North Holland Publishing Company (Amsterdam), 1972, pp. 193-249.
19. Van Ooij, W. J.: The Role of XPS in the Study and Understanding of Rubber to Metal Bonding. Surface Sci., vol. 68, 1977, pp. 1-9.

CHAPTER 6

Friction

The friction between two solid surfaces in solid-state contact is the resistance to tangential motion of one surface over the other, whether that motion be sliding, rolling, or rubbing contact. When two solid surfaces are brought into contact, adhesion can occur at the asperities between the two solid surfaces, and this adhesive force at the interface acts as a resistance to motion. Generally there are two types of friction, static and dynamic. Static friction is the force required to initiate motion between two solid surfaces, or the force necessary to break the junctions (the adhesive bonds) that form at the interface between two solid surfaces. Dynamic friction is the friction associated with one surface sliding, rolling, or rubbing over another. It is a dynamic measurement (average force measurement) during the rubbing, rolling, or sliding process.

In the field of tribology the term friction coefficient is frequently used to describe the resistance to tangential motion. The coefficient of friction is the frictional force divided by the load applied to the two surfaces in contact. The effective load is discussed in the next section.

Many years ago, Bowden and Tabor demonstrated that, when iron surfaces were heated in a vacuum system, the surfaces seized (i.e., one surface stuck to the other (ref. 1)). They interpreted this as the result of clean metal iron surfaces coming into solid-state contact. At the time the experiments were conducted, the vacuum systems available were of the order of 10^{-6} torr, and surface analytical tools were not available to define and characterize the surfaces and to establish that, in fact, the surfaces were really atomically clean. We know today that heating is not sufficient to clean iron surfaces. Bulk contaminants, present in such minor concentrations as parts per million in high purity (triple-zone refined) iron, can contaminate the solid surface of iron when cleaned by heating techniques. Such high-purity iron was not available at that time; thus, it is reasonable to assume that the surfaces of iron were not atomically clean in those early experiments. Notwithstanding this, the seizure of the solid surfaces was observed when they were heated in a vacuum system, indicating that the surfaces were clean enough so that strong adhesive bonding could develop between the two iron surfaces.

Bowden and Tabor found that when they admitted a small concentration of oxygen to the system, as indicated in figure 6-1, the friction coefficient decreased. Increasing the pressure by adding additional oxygen (from 10^{-4} to 10^{-3} mm mercury) resulted in a further decrease in friction coefficient. The friction coefficients are plotted (on the ordinate) in units of 1, 2 and 3, as opposed to tenths of a unit as the friction coefficient is normally reported. Thus, the friction coefficients in figure 6-1 are extemely high, and the presence of even small concentrations of oxygen is sufficient to bring about marked reductions in the friction coefficient. Ultimately, after exposure to oxygen at several millimeters of mercury for over 15 hours, the friction coefficient drops to about 0.5, which is the friction coefficient normally measured for iron with its normal oxides existing in an air environment at 760 torr (mm of mercury).

The data of figure 6-1 are not unique for iron in contact with itself, but they represent a curve that might be seen for most metals in contact with themselves; that is, in the clean state of a vacuum environment, metals will undergo seizure when brought into solid-state contact, and the admission of oxygen or other surface-active gases reduces the friction force for these metals. A transfer of material from one surface to another is generally associated with this seizure. This occurs when the specimens are separated. Even where like metals are in contact, the fracture frequently does not occur at the interface but rather in one of the two solid bodies. The effect of oxygen on reducing the friction is a function of the activity of the metal. The more active the metal chemically, the more likely a marked reduction in friction force will occur with small additions of oxygen to the system. For

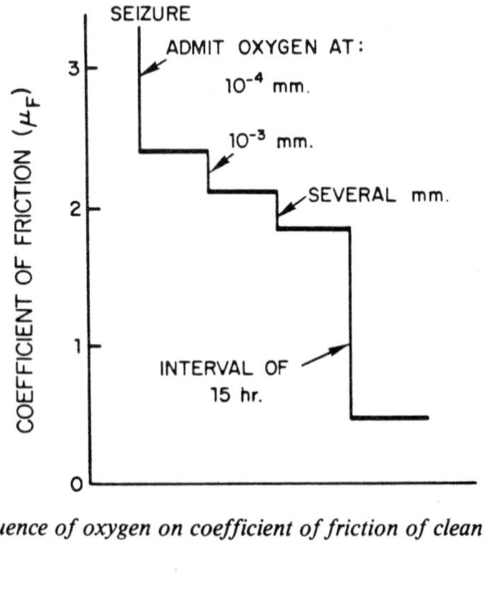

Figure 6-1. – Influence of oxygen on coefficient of friction of clean iron surfaces (ref. 1).

example, titanium scavenges oxygen fairly readily, and for equivalent concentrations of oxygen in a system, titanium is expected to undergo a much more marked reduction in friction than, say, would copper or iron.

Repeating the experiments of Bowden and Tabor with carefully defined, atomically clean surfaces (as determined by both LEED and AES analysis) indicates that the surfaces adhere to another with simple touch contact. No loading is required, and immediate seizure occurs for the iron in solid-state contact with itself. An Auger spectrum for an iron (001) single crystal surface after contact with an iron (001) single crystal surface is presented in figure 6-2. The Auger spectrum of figure 6-2 indicates iron peaks only. This absence of any other elements indicates that the surfaces are clean to within 0.01 of a monolayer.

Simple heating of triple-zone refined iron in a vacuum environment to 900° C does not produce the surface observed in figure 6-2. Generally, carbon, sulfur, or a combination thereof are present in the Auger spectrum for surfaces which have simply been heated. Despite this, however, seizure can occur as observed in figure 6-1 (ref. 1). Environment is an extremely important agent in controlling the friction coefficient of solid surfaces in solid-state contact. As has already been indicated with metals, one can go from complete seizure (which represents a coefficient of friction of infinity)

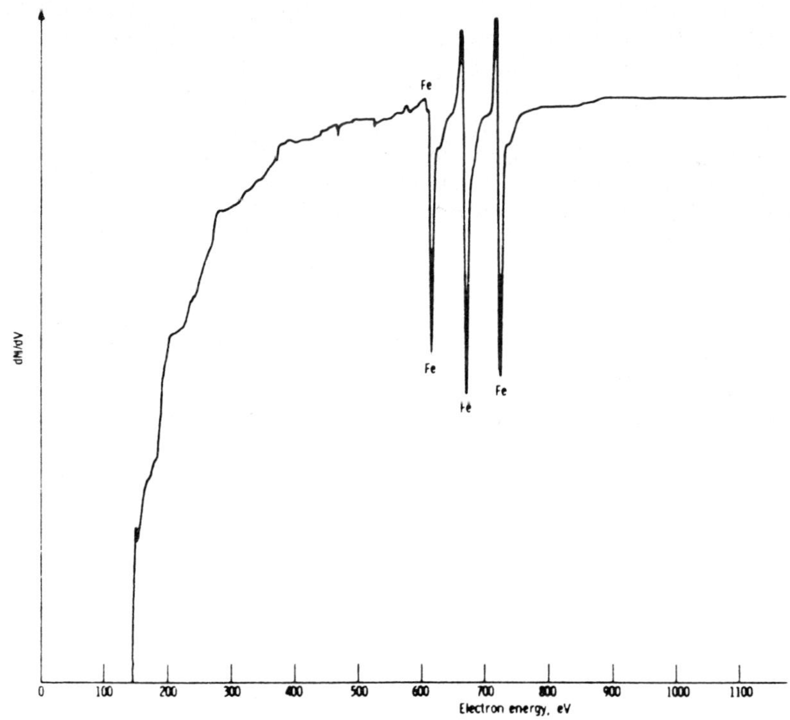

Figure 6-2.—Auger electron spectrometer analysis of iron (001) surface after adhesive contact with iron (001) surface.

to friction coefficients of 0.5 with the admission of oxygen to the surface. Ordinary air in the environment contains sufficient oxygen so that most metal surfaces are oxidized, and for most metals friction coefficients in a normal air environment in dry conditions of approximately 0.5 to 1.5 for sliding are obtained; that is, the friction coefficient for nearly all metal pairs does not vary greatly. Thus, oxygen in the environment is extremely important in providing a "lubricant" film to reduce the friction coefficient for metals in solid-state contact. It is a lubricant because it minimizes adhesion at the interface and thereby reduces the coefficient of friction. It also alters wear behavior (ch. 7).

It is easy to visualize that atomically clean metals, when brought into contact, might stick or adhere because of the extremely reactive nature of metal surfaces, particularly in the clean state. But what about metals in contact with other materials that are very resistant to influences by foreign substances—for example, metals in contact with a material such as diamond? Diamond is extremely hard and brittle, and metals are extremely plastic and deform readily. It might be anticipated that for a metal in contact with a metal junction growth can occur at the interface under an applied load that increases the adhesive bond forces between the metals. With a metal in contact with a material such as diamond, however, it might be anticipated that the brittle nature of the diamond would inhibit the formation of strong adhesive bonds at the interface. This is definitely not true. The adhesive bonds formed at the interface are chemical in nature and are a function of the chemical properties of the surface of the solid as opposed to the bulk mechanical properties of the solid. This is demonstrated in figure 6-3, where the coefficient of friction is plotted for diamond in contact with platinum and copper either in air or in vacuum. These data were taken from reference 2. In air, the diamond-platinum couple and the diamond-copper couple exhibit friction coefficients of the

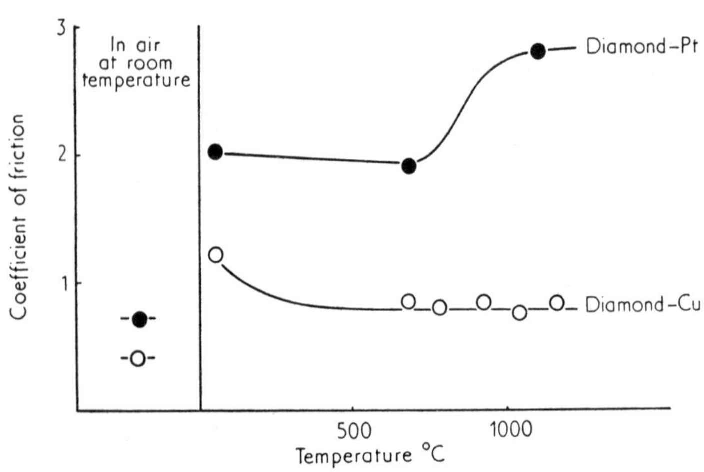

Figure 6-3.—Friction of diamond on metal after cleaning in vacuum. High friction, especially with platinum, shows that very strong interfacial adhesion can occur.

order of 0.4 to 0.7. In a vacuum environment, however, both couples exhibit higher friction coefficients than do the specimens in air. With heating in a vacuum environment, there is a further increase in the friction coefficient observed for the diamond-platinum couple, but a decrease is observed in the friction coefficient for the diamond-copper couple. The important conclusion to be drawn from figure 6-3 is that the presence of surface films on the metal influences the diamond to metal bonding (adhesion) and, consequently, the coefficient of friction. The oxidation of the metal surface is sufficient to reduce the friction because of reduced adhesion at the interface when the experiments are conducted in air. The data of figure 6-3 were actually obtained by Kenyon in his PhD thesis in 1956. Bowden and Tabor indicate, however, that the adhesive bond forces in Kenyon's data are so strong that they are equivalent to those in the metals platinum and copper; this indicates that strong bonding can occur not only with metals in contact with metals but for metals in contact with nonmetals such as diamond.

It can be argued (from the data of fig. 6-3) that if metal is in contact with a nonmetal surface and if the metal is atomically clean it is still a very active material. Therefore, it can interact with the less active diamond surface and be responsible for the difference in adhesive and frictional behavior observed in air and vacuum. This sounds plausible. Experiments conducted with diamond in sliding contact with itself show that the presence of surface films influences friction. Friction data obtained by Bowden and Hanwell are presented in figure 6-4 (ref. 3) where the friction coefficent is plotted as a function of the number of repeated cycles in a vacuum environment of 10^{-9} torr. The friction coefficient for diamond on diamond is initially relatively low (a little over 0.1), which is what is experienced in air. This friction coefficient may be associated with the presence of residual surface

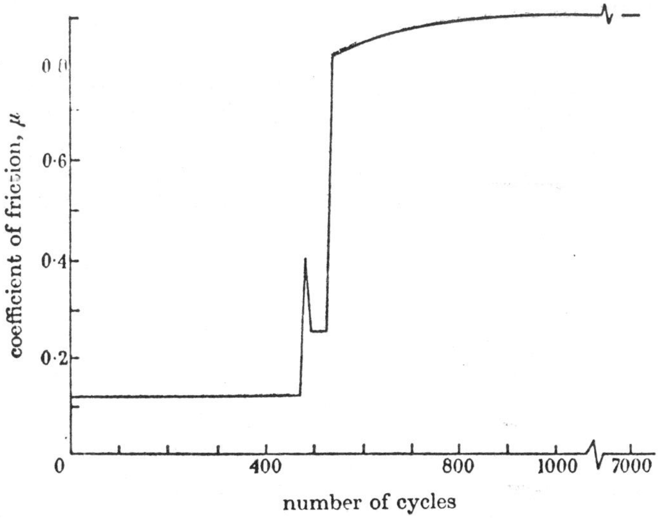

Figure 6-4. – Variation of friction of diamond on diamond with number of sliding traversals. One cycle represents one traversal in each direction; pressure, 5×10^{-9} torr.

contaminating films. Since the environment is a vacuum of 10^{-9} torr, it is presumed that the adsorbed films accounting for the friction coefficient are chemisorbed to the diamond, because physically adsorbed films would not be retained on the solid surface at these ambient pressures.

As the sliding process progresses with a repeated number of cycles the friction coefficient remains relatively low for almost 500 repeated cycles over the same surface. This indicates that the adsorbates present on the solid surface of the diamond are very tenacious and that they resist the mechanical activity and the energy generated at the interface from the rubbing process. They resist the tendency of that energy to cause desorption, which normally occurs when, for example, one heats a surface in a vacuum environment of this pressure magnitude. The heating brings about an induced desorption. The more stable the adsorbate, the higher the temperature necessary to activate desorption. In fact, an indication of the bond strength of adsorbates on surfaces can be obtained by simply measuring the temperature at which the adsorbate comes off the solid surface, as monitored by a mass spectrometer. This can be achieved by incrementally raising the temperature until desorption is observed in the mass spectrometer. The materials can be ranked based on these desorption characteristics. The data of figure 6-4 indicate that the adsorbates present on the diamond, which were not identified by the authors of figure 6-4, were relatively tenacious. After approximately 550 cycles the friction coefficient rose to a value in excess of 0.8, and, ultimately, after 1000 repeated cycles it reached approximately 0.9. So the friction coefficent increased nearly ninefold as a result of removing the adsorbed films present on the diamond surface. A friction coefficient of 0.9 is very high even for metals in contact with metals. Thus, the high friction coefficient for diamond in contact with itself indicates that diamond (like the metals in the earlier data presented) is sensitive to the environment.

Diamond consists basically of elemental carbon. It could be argued, from an academic point of view, that carbon certainly adsorbs oxygen, nitrogen, and other species that may be present in a normal air environment. These adsorbates have a strong influence on reducing the friction coefficient of metals in solid-state contact.

It might be further argued that, if oxygen bonds to the diamond to form a carbon monoxide or carbon dioxide like structure on the surface, these particular structures would inhibit adhesive bonding and, in turn, would account for the low friction coefficients measured for diamond in contact with itself. If, however, ceramic materials are examined that already contain oxygen or are, so to speak, saturated with oxygen, a similar friction behavior to that observed in figure 6-4 is obtained. Some sliding friction experiments were conducted, again in a vacuum environment, by Bowden and Hanwell with magnesium oxide. The results of these experiments are presented in figure 6-5 (ref. 3). The friction coefficient started out very low, slightly in excess of 0.3. With repeated cycles, however, the friction coefficient began to increase until at 1000 cycles it reached 0.8. It behaved like diamond in that, with repeated cycles, the friction coefficient increased in a vacuum environment. Because magnesium oxide contains ample amounts of oxygen, the thesis of oxygen desorption from the surface cannot account for the marked increase in friction coefficient of figure 6-5 like that for diamond on diamond (fig. 6-4). Oxygen plays an important role in reducing the friction coefficient; surface analytical tools such as LEED and AES have established that the presence of the oxygen on the surface does appreciably reduce friction for various materials. Also, the evidence presented in figure 6-5 indicates that, even with oxygen present in the molecular structure of magnesium oxide, strong bonds of adhesion can occur at the interface, and these strong bonds can cause very high friction coefficients for clean ceramic surfaces in solid-state contact.

In general, most classes of materials exhibit lower friction coefficients in air than they do in a vacuum environment with one notable exception. In soft glass on glass the friction coefficient is lower in a vacuum environment than in air. Removing adsorbates (or oxides in a case of metals) results in an increase in the coefficient of friction; that is, clean, solid material surfaces exhibit higher friction coefficients than do those containing contaminants. This is demonstrated in table 6-I where various classes of materials are presented together with friction coefficients measured in air, vacuum, or lubricated. Results are presented for soft glass on itself, sapphire (aluminium oxide, Al_2O_3) on itself, magnesium oxide (MgO) on itself, quartz on itself, sodium chloride on itself, lithium fluroide on itself, diamond on itself, and copper on itself. A comparison of the friction coefficients presented in table 6-I for the air environment with the friction coefficients presented in column 2 for a vacuum environment indicate that, in each case with the exception of soft glass, the friction coefficient is much higher in vacuum than it is in air.

The reason for the higher friction for glass in sliding contact with glass in moist air can be attributed to the chemisorbed water vapor on the glass

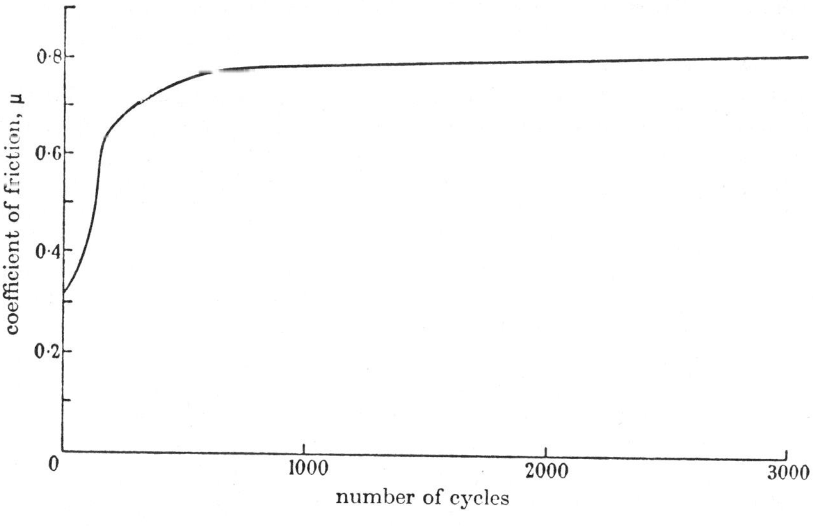

Figure 6-5.—Frictional behavior of magnesium oxide during large number of cycles. Pressure, 2×10^{-9} torr (ref. 3).

TABLE 6-I. – COEFFICIENT OF FRICTION FOR VARIOUS SOLIDS IN THREE DIFFERENT ENVIRONMENTS

Material combinations	Coefficient of friction		
	Air (moisture)	Vacuum (10^{-9} to 10^{-10} torr)	Lubricated (mineral oil)
Soft glass/ soft glass	1.0	0.5	0.28
Sapphire/sapphire	.2	.8	.20
Magnesium oxide/ magnesium oxide	.2	.8	.21
Quartz/quartz	.35	.7	.20
Sodium chloride/ sodium chloride	.70	1.3	.22
Lithium fluoride/ lithium fluoride	----	1.2	.22
Diamond/diamond	.1	.9	.05
Copper/copper	1.0	>100	.08

surfaces. If one very carefully controls the environment and admits dry air to a vacuum chamber where a glass surface has been thoroughly outgassed, the friction coefficient remains low. If, however, a small amount of moisture is admitted into the vacuum chamber, the friction coefficient immediately increases. This increase in friction coefficient is associated with chemisorption of the water vapor on the solid surface; the chemisorbed water vapor promotes bonding of the glass to itself and increases the friction coefficient.

The addition of the lubricant (a mineral oil) to the solid surface (table 6-I) produces a reduction in friction coefficient compared to results obtained in either air or vacuum. The mineral oil helps to reduce the adhesive bonding at the solid-state interface by reducing the clean solid to solid contact. In other words, the better a contaminant the lubricant is, the more effective it is in reducing friction coefficient.

It is important to measure the friction properties of various solids in the atomically clean state if at all possible because this gives a baseline measurement for what the friction behavior of the material is like based on its particular solid-state structure. Once the basic friction properties of the clean material are understood, the influence of adsorbates can better be understood. It might be concluded from the data of table 6-I and figures 6-1 to 6-5 that, for most solids in the solid-state contact, the clean surfaces exhibit higher friction coefficients than do the surfaces that contain adsorbates or chemical surface films (with the exception of glass in contact with glass). This is not, however, necessarily true. There are some metals whose oxides, when interacting with other ceramic materials (e.g., ceramic oxides), exhibit higher friction than do the clean metals in contact with those same ceramic oxides. For example, nickle oxide in contact with aluminum oxide (Al_2O_3) can exhibit a higher coefficient of friction than atomically clean nickel.

Metals exhibit high friction coefficients not only when they are in contact with other metals, ceramic oxides, and diamondlike materials but also when they are in contact with semiconductors. Germanium and silicon are two semiconductors that have relatively brittle natures. Relatively high friction coefficients are obtained for metals in contact with germanium or silicon. Sputter techniques can be effectively used to clean surfaces of semiconductors for friction experiments. In figure 6-6 two Auger spectra are presented. Figure 6-6(a) is an Auger spectra for a silicon (111) single crystal surface before sputter cleaning. The Auger peaks for the silicon are observed in the left side of the figure. In addition to silicon, carbon and oxygen are detected on the solid surface. If the surface is carefully sputter cleaned with argon iron bombardment, at -1000 volts dc, the silicon peak grows in intensity and the carbon and oxygen peaks decrease. Oxygen decreases until it is absent from the Auger spectrum. A small amount of carbon is still visible in the Auger spectrum, and this is believed to be due to a bulk contamination of the silicon with carbon. Repeated sputterings of the silicon surface always yield the same background concentration of carbon as that seen in figure 6-6(b).

The bulk of the silicon surface, however, has been denuded of surface contaminants such as oxides and carbon combined with oxygen in the form of carbon monoxide and carbon dioxide that would normally prevent adhesion of two solid surfaces in contact. Sliding friction experiments were conducted on the surface of figure 6-6(b) with an iron single crystal (110) surface sliding against the clean silicon single crystal (111) surface. The results of these friction experiments are presented in figure 6-7 where friction coefficient is plotted as a function of load for vacuum, oxygen at atmospheric pressure, and oxygen with a lubricating film of 0.2 percent oleic acid in a straight mineral oil.

It is apparent from figure 6-7 that extremely high friction coefficients are obtained for semiconductors in contact with clean metals just as was observed for metals in contact with clean metals and metals in contact with clean oxides. A friction coefficient in vacuum, at the very light loads, in excess of 3 was obtained; this decreased to 2 with increasing load. With the presence of oxygen on the surface, however, the formation of surface oxides appreciably reduced friction coefficient to less than half the value obtained in vacuum. If a thin film of oleic acid in mineral oil is placed on the surface, the friction coefficient drops to about 0.2 or one-fifth the value obtained with the oxygen containing surface film and only 10 percent of the value obtained for the clean surfaces. The environment makes a big difference in the observed friction behavior for materials in solid sliding contact. In general, the clean surfaces give the highest friction coefficients.

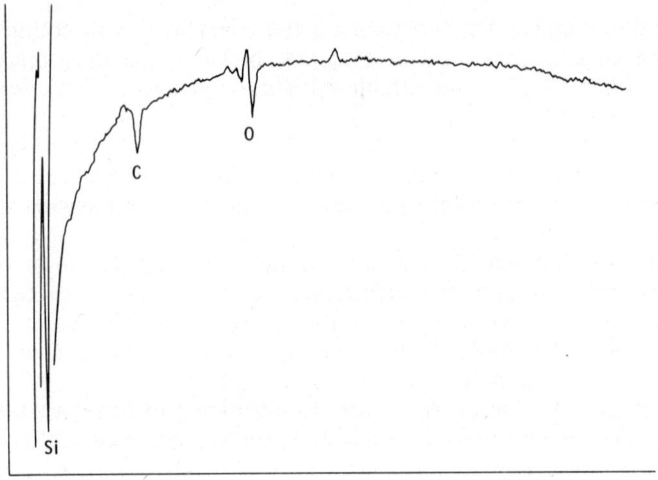

(*a*) *Before sputter cleaning.*

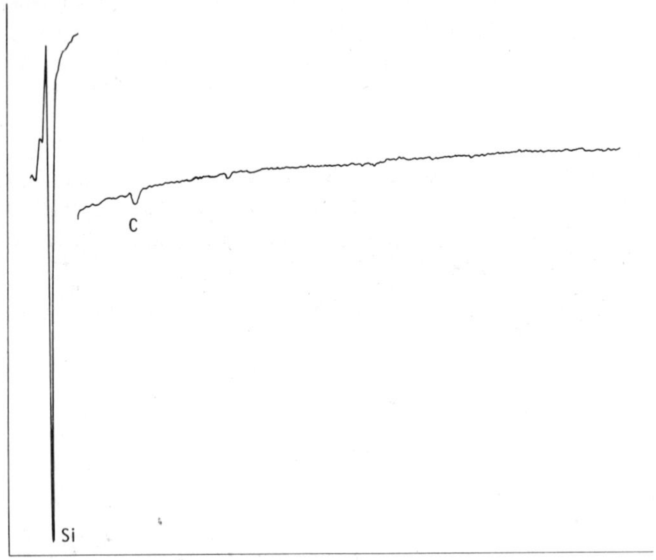

(*b*) *After sputter cleaning.*

Figure 6-6. – Auger emission spectroscopy spectra for silicon single-crystal surface.

Physical Character of Surfaces

The physical nature of a solid surface can influence the observed friction force for materials in sliding or rubbing contact. Many surfaces in nature are not smooth and uniform but rather are rough. For example, ordinary horsehair contains scales on the surface as indicated in the scanning electron micrograph in figure 6-8. The micrograph shows two horsehairs in contact;

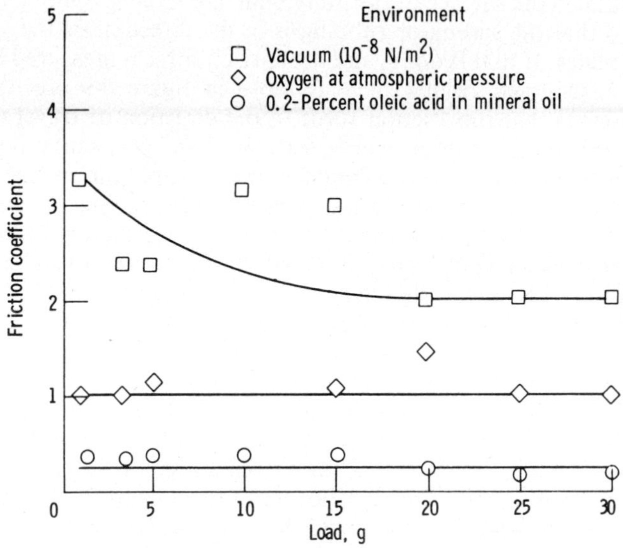

Figure 6-7. – Friction coefficient as function of load for single-crystal iron (110) sliding on single-crystal silicon (111) surface in vacuum, in oxygen, and lubricated with 0.2 percent oleic acid in mineral oil. Sliding velocity, 0.7 millimeter per minute; temperature, 23° C.

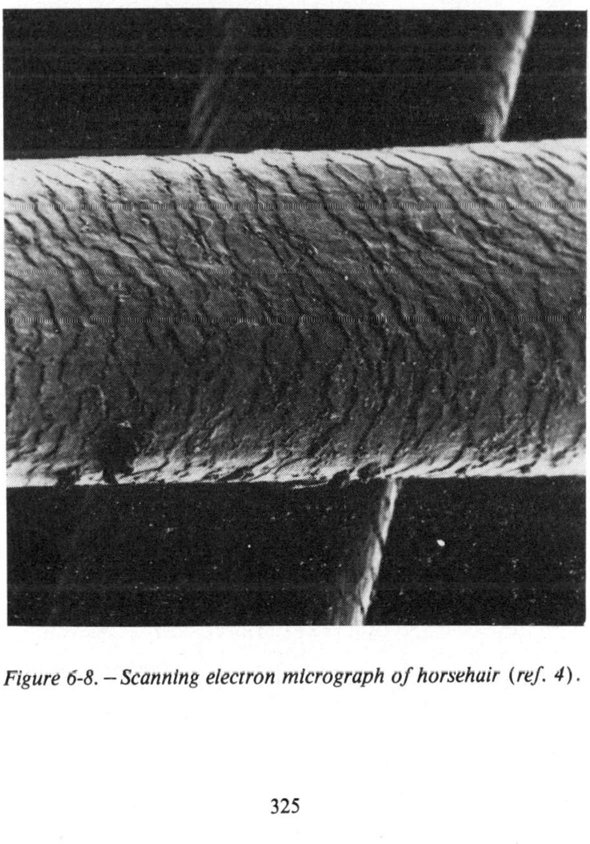

Figure 6-8. – Scanning electron micrograph of horsehair (ref. 4).

note the scales on the surface of the individual hair strands. One can rub the horsehair so that the sliding or rubbing is in the direction of the scales or against the scales. If that is done, and the friction force is measured between two scales in contact, results such as those in figure 6-9 are observed. Livesay observes that the friction force in the direction of travel with the scales is reflected in the upper curve (ref. 4). The lower curve represents sliding against the scales. The difference in the amplitude of the friction trace and the stick-slip behavior are greater where sliding is against the scales. The friction coefficients are different in the two sliding directions. A static friction coefficient of 1.56 is obtained for sliding with the scales, while the static value is 2.36 for sliding against the scales. The kinetic friction exhibits practically the same friction coefficient in both directions (with scales, 0.72; against scales, 0.73). Thus, the direction of sliding on the scales does not alter kinetic friction but it does influence static friction behavior.

The data of figure 6-9 were obtained in the very sensitive apparatus in which the horse hairs were loaded against each other with a 4-milligram load and with sliding conducted at the extremely small sliding velocity of 0.07 millimeter per second. The significance of the data of figure 6-9 is that, even though initial static friction coefficient differences might exist (based on differences in the direction of sliding), those differences are generally

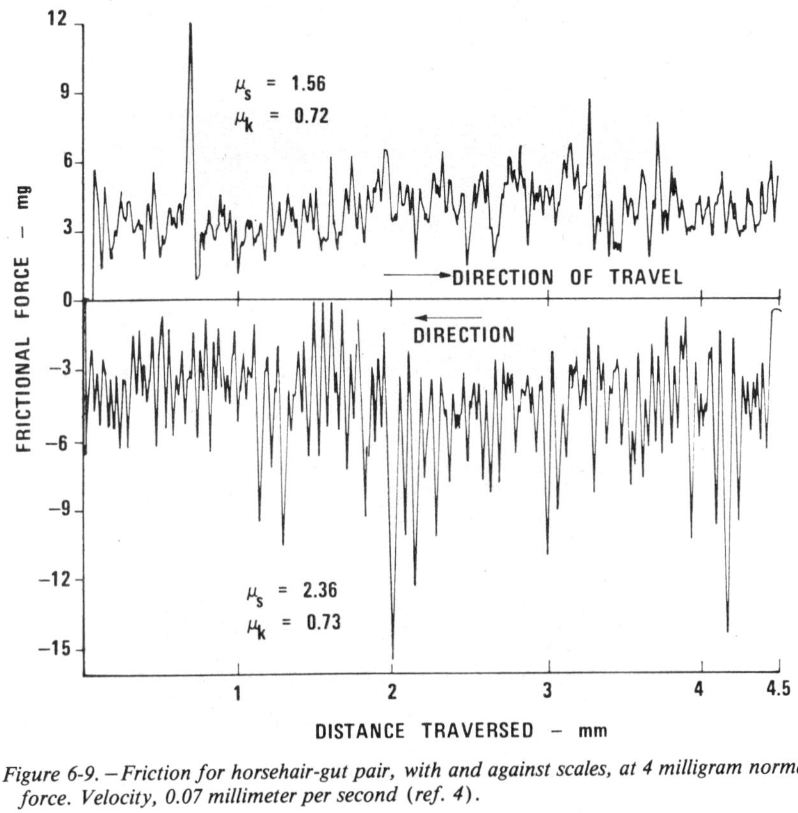

Figure 6-9. – Friction for horsehair-gut pair, with and against scales, at 4 milligram normal force. Velocity, 0.07 millimeter per second (ref. 4).

erased in kinetic measurements; that is, the surface topography does not seem to influence the friction behavior of these materials in solid-state contact.

While the amplitude of the friction traces might be markedly different in the two directions of sliding, the result (or average friction coeffcient) is essentially the same in the kinetic or sliding situation. This is as true for initial differences in friction behavior as for differences in surface topography. For example, Courtel and his coworkers have measured the friction force with repeated passes for aluminum surfaces that had been turned, polished, or electropolished, and they have determined the variation in friction coefficient with the differences in the surface topography of the three surfaces prepared by these operations (ref. 5). The results of their experiments are presented in figure 6-10.

In figure 6-10, friction coefficient is plotted as a function of the number of successive passes over the aluminum surface. The three curves are indicated by a, b, and c as the turned surface, surface polished with alumina, and electropolished surface, respectively. The electropolished surface is the smoothest and the turned surface is the roughest; the surface polished with alumina is an intermediate surface. The data indicate that, in the first pass, marked differences exist in the friction coefficients as a result of topography; the smoothest surface (electropolished) showed the lowest initial friction coefficient.

After approximately 100 passes across the surface, the differences among the friction coefficients of the various surfaces are markedly less than they were with the first pass across the surface. After 50 passes, the friction coefficients with all three initial surface conditions were essentially between 0.2 and 0.3. This is not unusual because in the sliding process with repeated

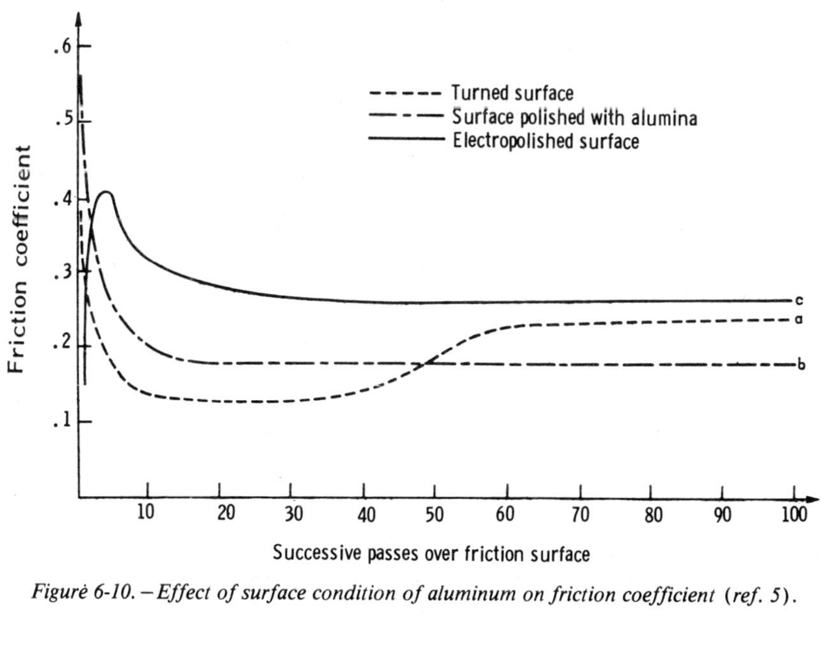

Figure 6-10. – Effect of surface condition of aluminum on friction coefficient (ref. 5).

passes over a solid surface the surface generates its own topography, which is characteristic of the particular system involved. It is a function of the materials in solid-state contact, the environment in which the materials find themselves, and the lubricant (or absence thereof).

The actual friction traces presented in figure 6-11 show the variation in the friction force with repeated passes over the same surface (ref. 6). The data in figure 6-11 indicate that the friction force does vary with repeated passes over the solid surface. Not only does the friction force vary, but the actual topography varies as indicated by figure 6-12. In this figure are seven surface profiles that match the seven friction profiles in figure 6-11.

A comparison of the seven friction force traces shows that differences exist; also, differences exist in surface profiles with the repeated passes. Thus, the sliding process generates its own characteristic surface, and the surface topography is probably changing continuously in most practical systems. These changes, then, influence the measured friction forces.

When two single crystals are placed in contact, the real contacts are the asperities. These contacts initially undergo elastic deformation, and then

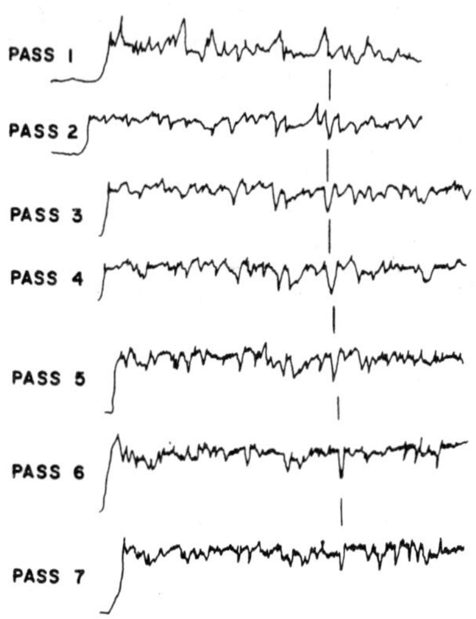

Figure 6-11. – Variation of friction force with number of passes over surface (ref. 6).

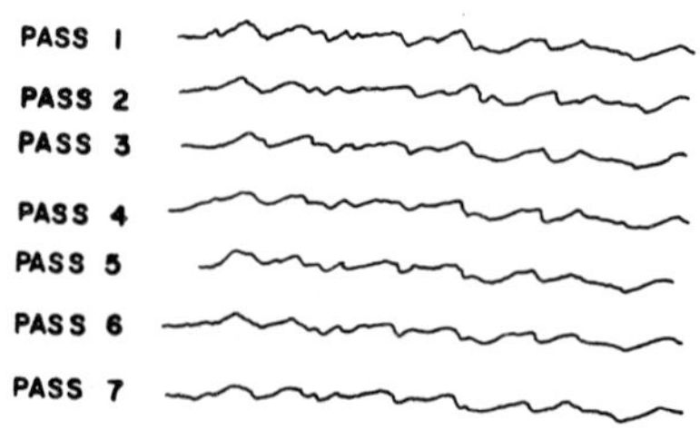

Figure 6-12. – Surface profile for seven passes (ref. 6).

plastic deformation as load is applied to the surfaces. Plastic deformation continues until the load can be properly supported. The interaction of the two surfaces (through asperities) is shown schematically in figure 6-13(a). Figure 6-13(a) shows crystal 1 sliding over crystal 2 with τ_a representing the shear force between the crystals at the interface.

When the surfaces are initially placed in contact at light loads without tangential motion, the deformation is elastic (fig. 6-13(b)). The surface deforms some and bonding of the surfaces may occur. When the load is removed, there is elastic recovery in the system and the surfaces return to their original states. This is observed in some systems with materials of high elastic modulus. For most practical metals, however, application of load generally results in some plastic deformation at the asperities, however small or localized. The plastic deformation is accommodated by the generation of slip lines for dislocation flow in the asperities on the solid surface (fig. 6-13(c)). Thus, in asperity contact with plastic deformation, dislocation generation occurs along slip lines, and this can initiate the formation of new surfaces in the actual contact zone. With elastic deformation, the initial surface topography can be recovered, whereas with plastic deformation, once it occurs, a permanent change in surface topography has taken place and this can alter interfacial friction behavior.

If the two crystal surfaces in contact in figure 6-13 have slipped on cleavage planes parallel to the sliding interface, shear occurs initially on a

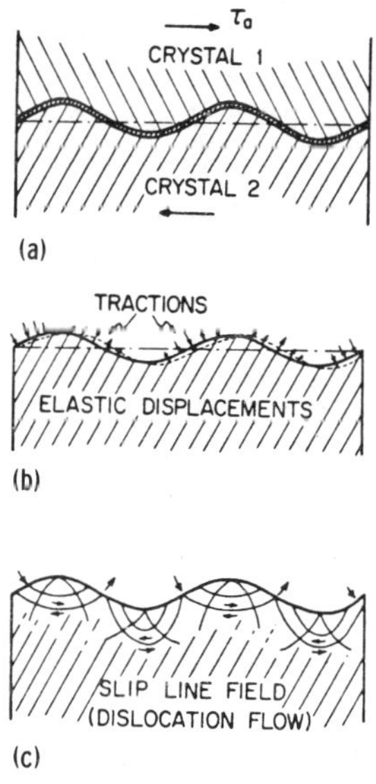

(*a*) *Two crystals sliding against each other.*

(*b*) *Elastic accommodation, by diffusion through bulk of crystal or in plane of boundary.*

(*c*) *Plastic accommodation, by dislocation motion of grains or crystals.*

Figure 6-13. – Sliding of crystals. There may be elastic or plastic accommodation, or holes may appear in boundary plane.

microscale along slip lines in the material. On a microscale this generates dislocations in the movement of the atoms in the surface layers over one another. This movement by dislocations is depicted schematically in figure 6-14. Figure 6-14(a) shows a close-packed array of atoms in the slip planes near the surface of the asperities (fig. 6-13(c)). With attempted shear tangential to the slip planes, the applied force causes displacements on the rows of atoms such as indicated in figures 6-14(b) and (c). These displacements are accommodated in the crystal lattice by a total displacement of the atomic rows relative to each other (fig. 6-14(d)). Thus, actual movement in the material can occur as a result of the generation and the dissipation of dislocations along slip planes. This slip process causes plastic deformation in the contact areas. When the slip planes are oriented so that they are parallel to the interface with the two crystals of figure 6-13(a) in solid-state contact, the shear between crystals (denoted by τ_a) in a mechanical sense causes shear in the interface region in rows of atoms analagous to that shown in figure 6-14. If it is on a much larger scale, the process terminates in fracture. The process starts as slip and is followed by the generation of dislocations and movements along slip planes; this ultimately leads to fracture and the separation of one slip plane from another.

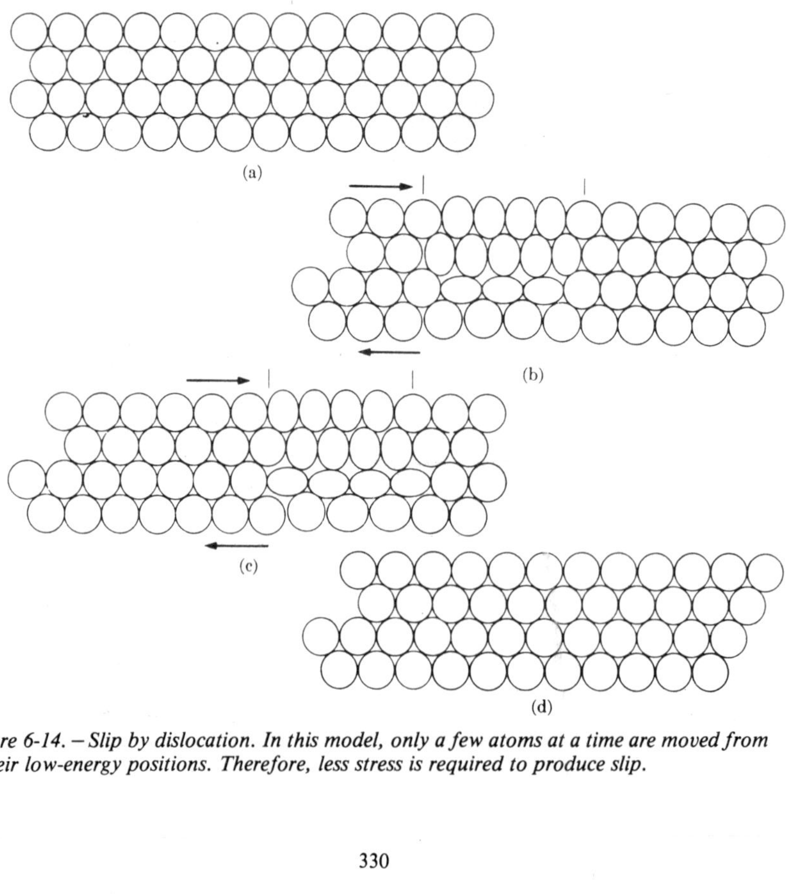

Figure 6-14. – Slip by dislocation. In this model, only a few atoms at a time are moved from their low-energy positions. Therefore, less stress is required to produce slip.

The energy dissipated in this process is reflected in the friction force measured between the two surfaces in solid-state contact. If adhesion occurs at the interface, and the adhesive bonds are stronger than the cohesive bonds (in the cohesively weaker of the two materials) as has been frequently mentioned throughout this text, then shear occurs in the cohesively weaker of the two metals.

A comparison of the slip and the shear process might, by analogy, be compared to the movement of cards in a deck. If a deck of cards is placed on the table and the finger is placed on the upper card and the cards are caused to be slightly displaced tangentially, one relative to another, there is an angle between the edges of the cards and the table top. These displacements are analogous to the generation of the dislocations in the solid; there is a small, minute displacement of one plane of atoms relative to another. Each card in the deck is analagous to a plane of atoms in the crystalline solid.

If one continues the movement of the cards over one another, until such time as the tangential motion brings about a complete separation of the cards one from another, the equivalent of complete separation of atomic planes in the crystal and solid exists. This might be visualized as occurring in the asperities if, in any one asperity, the individual crystallographic planes are all aligned parallel to the interface. In the shear process, initiation of tangential motion first causes generation and movement of dislocations along the slip lines or slip planes. This is micromovement or small displacements in the material. If the force is applied until additional dislocations are generated and if the process is continued, movement continues until total separation of two solid surfaces occurs.

As the load between two solid surfaces in contact is increased, the deformation in the asperity regions increases, first elastically and then plastically, until the load is supported. Even after the load has been fully applied and plastic deformation has occurred, additional creep can occur in the material that has undergone plastic deformation for the accomodation of the applied force. At some time, an equilibrium situation is reached where the applied load is supported by the real contact area generated as a result of the deformation in the contacting asperities. The friction force, then, is related to the amount of material that is in solid-state contact at the interface. The greater the real area of contact, the greater the force to achieve tangential motion of one surface over the other. In fact, this is what is observed experimentally. If one continues to increase the load for two surfaces in solid-state contact, the friction force increases with the applied load. This is demonstrated in figure 6-15 for aluminum sliding on a glass surface. The friction force between the aluminum and the glass is a direct function of the load applied to the surfaces in solid-state contact. The higher the load, the greater the friction force measured for the surfaces.

The friction force measured for the two surfaces in solid-state contact as a function of load is influenced by the nature of the surfaces and their surface chemistry. The friction force varies with the surfaces in various environments, but the films present on the surface may alter the amount of adhesion that occurs at the interface. If a film is present on the surface, the

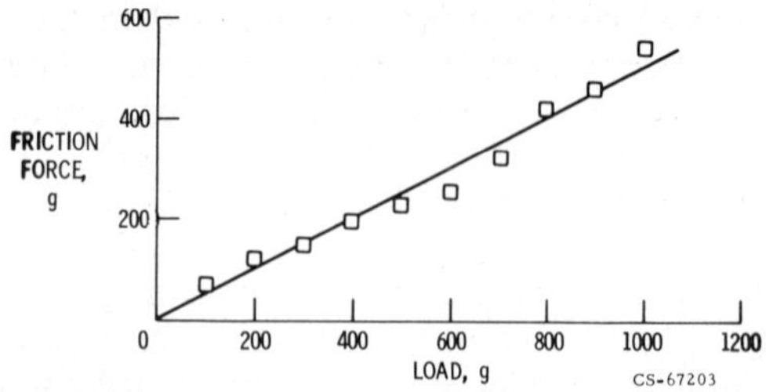

Figure 6-15.—Friction force as function of load for aluminum sliding on glass. Sliding velocity, 30 centimeters per minute; temperature, 23° C.

amount of solid-state contact (and, hence, bonding) that is expected to occur through the film is less than when the film is absent. As a consequence, the friction force measured is less where the adhesion is less. This influences the friction behavior with applied load as is indicated in figure 6-16 for glass sliding on glass. As indicated earlier in this chapter, glass is one of the few materials that exhibits a higher friction in moist air than it does in a vacuum environment in the clean state. In the clean state the bonding forces are weaker than they are when moisture is present on the solid surface (fig. 6-16).

In figure 6-16, the friction force is plotted as a function of load for glass sliding on glass in vacuum (10^{-10} torr) and in air saturated with water vapor. Figure 6-16 shows that the friction force is directly related to the applied load. The higher the load, the higher the friction force. Also, the environment makes an appreciable difference. In this instance, in the air

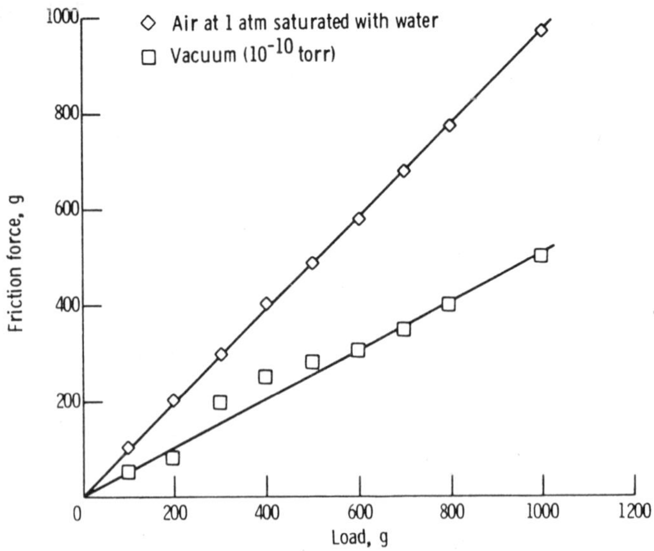

Figure 6-16.—Friction force as function of load for glass sliding on glass. Sliding velocity, 30 centimeters per minute; temperature, 23° C.

environment with water vapor the friction force is higher over the entire load range investigated than it is in vacuum. Normally, for most material combinations (metals in contact with metals, metals in contact with ceramics, or metals in contact with glasses), the reverse is anticipated; that is, the friction force would be higher with the clean metal surface than it would be with the surface containing the contaminating film. Glass is a somewhat unique case.

Where solid surfaces are atomically clean, and adhesion occurs at the interface, the strong bonding at the interface results in shear occurring in the cohesively weaker of the two materials. The average friction coefficient observed for these materials is, for all practical purposes, essentially a function of the shear strength of the cohesively weaker material. This generally is the case, although the opposing surface might be much different in mechanical properties and behavior when adhesion plays a dominant role. One can, however, selectively modify or alter one of the two surfaces in solid-state contact by changing its chemistry. Even this, however, in many instances does not markedly alter the observed friction behavior where adhesion is the dominant operating force in determining the friction between the surfaces in solid-state contact. For example, some experiments were conducted in our laboratory wherein simple binary alloys of iron and chromium were prepared and brought into contact with silicon carbide (ref. 7). Silicon carbide is an extremely hard, high elastic modulus matcrial. It was cleaned in vacuum and brought into contact with various materials: pure iron, pure chromium, and alloys containing various amounts of chromium and iron. The friction coefficients measured for these particular compositions are presented in figure 6-17. In figure 6-17 the friction coefficients for pure iron or pure chromium in contact with the silicon carbide surface were approximately the same, about 0.5. The addition of chromium to the iron, however, resulted in a marked increase in the fricton coefficient with stronger adhesion of the alloy to the silicon carbide surface

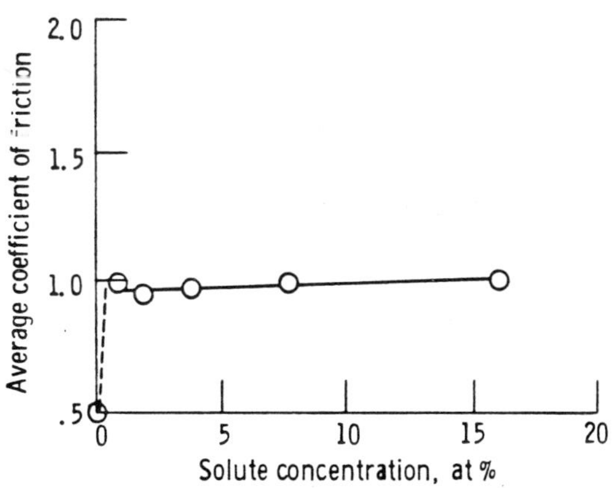

Figure 6-17. – Coefficient of friction for iron-chromium alloy as function of chromium concentration. Single-pass sliding on single-crystal silicon carbide (001) surface; sliding direction, ⟨10$\bar{1}$0⟩; sliding velocity, 3 millimeters per minute; load, 0.2 newton; room temperature; vacuum pressure, 10^{-8} pascal.

than to either of the elemental materials. With tangential motion, shear had to take place in the higher shear strength alloy. The variations in composition (from 5 to 20 weight percent chromium) did not result in marked differences in the friction properties of these alloys (fig. 6-17). Even though the concentration of chromium in the basic alloy varied, the friction coefficient measure was essentially the same, but higher than that of either pure iron or pure chromium.

The data of figure 6-17 reflect an adhesive behavior at the interface. The same alloy surfaces are again examined in contact with silicon carbide, but the surfaces are now lubricated with a mineral oil. The presence of the mineral oil on the surface completely masks the adhesion effect. Thus, the adhesion process is no longer the dominant process; rather, deformation becomes important. The alloys undergo plastic deformation with sliding of the very hard, strong, high elastic modulus silicon carbide across the surface.

Groove height measurements are presented in figure 6-18 for silicon carbide sliding across elemental iron and the various alloy compositions examined in figure 6-17. A variation in the depth of the groove formed exists with changes in the compositions; that is, the amount of plastic deformation that occurs with the silicon carbide in contact with the chromium alloys varies as a function of alloy content. The greatest amount of deformation was observed with the pure iron, and the least amount occurred for a 9-weight-percent chromium in iron alloy (fig. 6-18). The alloy compositions leading up to the 9 weight percent showed a progressive decrease in the amount of deformation that occurred. Thus, the data of figures 6-17 and 6-18 indicate that the behavior can be altered by the presence or absence of lubricating films. In figure 6-17, adhesion is maximized; in figure 6-18 with the lubricating film present, the frictional energy is dissipated in the deformation of the alloy surface.

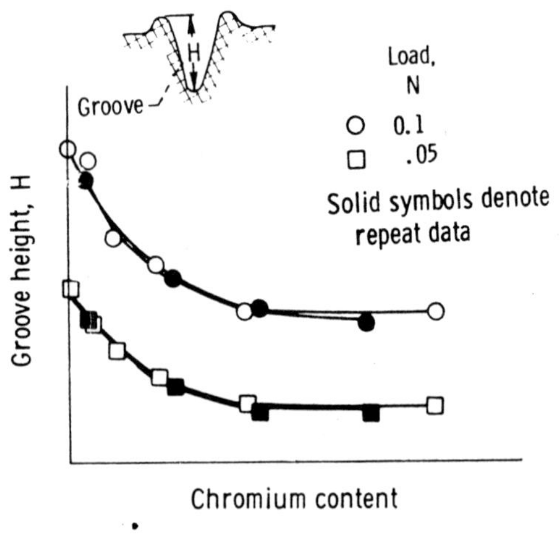

Figure 6-18. – Groove heights for iron-chromium alloys and pure iron as function of chromium content. Single-pass sliding of 0.025-millimeter-radius silicon carbide rider in mineral oil. Sliding velocity, 3 millimeters per minute; temperature, 25° C (ref. 7).

The friction coefficients, under lubricated conditions, for the composition examined in figure 6-18 are presented in figure 6-19. The curves reflect the changes in friction coefficients with the changes in the groove heights or depths (fig. 6-18). Iron had the greatest groove depth (amount of plastic deformation) as shown in figure 6-18, and in figure 6-19 it exhibits the highest coefficient of friction. As the amount of chromium is increased in the iron alloy, there is a decrease in the friction coefficient in the range of 9 to 14 percent where the amounts of plastic deformation observed in the groove height measurements of figure 6-18 are at a minimum. There is, correspondingly, a minimum friction coefficient (fig. 6-19).

These results indicate that the friction force in figure 6-19 in the presence of the mineral oil is a result of the deformation that occurs at the interface between the silicon carbide and the alloy with deformation taking place in the alloy compositions. The measured friction force, then, represents the resistance to motion caused by plastic deformation of the alloy with sliding. The deeper the silicon carbide penetrates the alloy surface the more resistence to tangential motion and the higher the friction force. The variation in friction force seen in figure 6-19 was not seen for the nonlubricated specimens with atomically clean surfaces where adhesion was strong (fig. 6-17). In figure 6-17 changing alloy composition had very little influence on adhesion; consequently, there was little influence on friction force.

These results indicate that the mechanism of friction has been completely changed from one involving adhesion to one involving plastic deformation. Thus, in understanding friction coefficients it is important to understand the underlying mechanism.

The situation of silicon carbide sliding on the iron chromium alloys in figures 6-18 and 6-19 where a mineral oil is present on the surface is schematically represented in figure 6-20(a) by a hard metal sliding on a soft metal. When one moves tangentially, the friction force is a function of the real area and the shear strength of the junction (fig. 6-20). With silicon

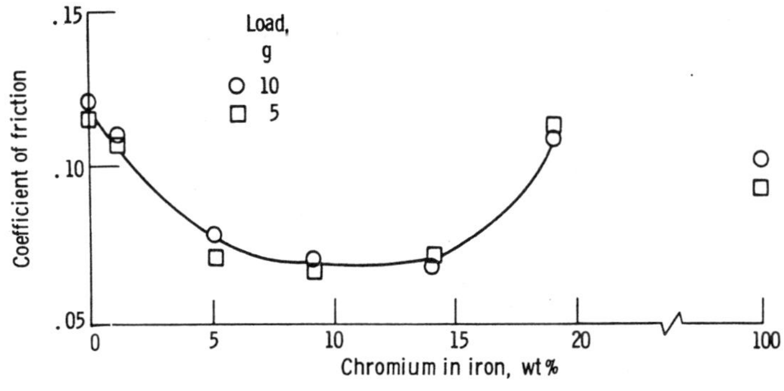

Figure 6-19. – Coefficient of friction for Fe-Cr alloys, iron, and chromium as result of single-pass sliding of 0.025-millimeter-radius silicon carbide rider in mineral oil. Sliding velocity, 3 millimeters per minute; room temperature (ref. 7).

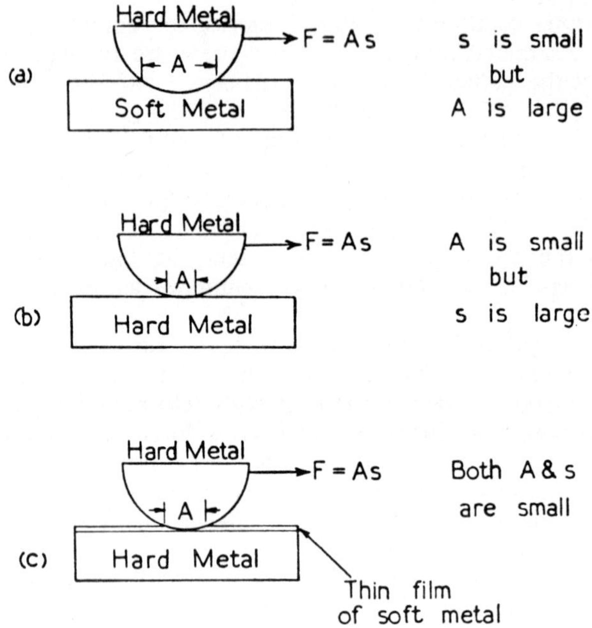

Figure 6-20.—Relation of friction to metal hardness. Low friction may be obtained by depositing a thin film of a soft metal on a hard metal substrate (ref. 1).

carbide in contact with a softer material (iron, chromium, or iron chromium alloys) the shear strength of the softer material is relatively low while the real area of contact is relatively large because deformation proceeds fairly readily.

Where two hard surfaces are in contact (e.g., silicon carbide in contact with silicon carbide or a hard metal with a hard metal) such as depicted in figure 6-20(b), the friction force again is a function of area in shear times the shear strength. With hard on hard surfaces, the area in real contact is relatively small but the shear strength is large because the hard or high elastic modulus materials have higher shear strengths and are more resistant to shear. As a consequence, the friction force for the two situations depicted in figures 6-20(a) and (b) may be the same even though the operating mechanisms at the interface may be different.

One can alter this situation, however, by imposing a low shear strength material at the interface. This is frequently accomplished when a lubricant film is placed at the interface. The shear is ideally to take place in the lubricant and not in either of the two solid surfaces in contact. This can be accomplished by applying a soft film to the hard surface of figure 6-20(b). When that is done, the results shown schematically in figure 6-20(c) are obtained where the hard metal is sliding on a hard metal and imposed at the interface area is a thin, soft metal film. In this situation, the area of contact is relatively small because the load is supported by the hard, high elastic moduli materials. And the shear strength is low because it is the shear strength of the thin, soft film imposed at the interface.

As mentioned earlier, the thin soft film could be a lubricant film or an organic material adsorbed on the solid surface. It can be a naturally occurring oxide (e.g., tin oxide on a tin surface) or it can be other substances present on the solid surface that have a lower shear strength than either of the solid surfaces in contact.

Adhesion plays an extremely important role in the friction forces measured for materials. The combination of adhesive bonding at the interface and the strength at the interfacial junction determines, to a large extent, the friction force that is measured. In addition, adhesive transfer can occur from one surface to another as fracture occurs in one of the two materials on either side of the interface.

This transfer is frequently observed in the sliding friction of carbons and polymers in contact with metal surfaces. Generally what is observed is that a polymer transfer film develops on a metal counterface in the process of sliding; ultimately, the polymer is sliding on a thin film of polymer. The polymer transfers by adhesion at the interface and shear in the polymeric material itself. Likewise, with carbons in an air environment, a similar mechanism generally operates; that is, adhesion of the carbon to a metal oxide occurs. With repeated passes, a carbon transfer film develops and the carbon is essentially sliding on itself. Changes in friction properties take place with these changes in interfacial behavior that have been brought about by an adhesive mechanism.

Some sliding friction experiments were conducted with the polymer polytetrafluoroethylene (PTFE) sliding on various metal surfaces in a vacuum environment so that the surfaces were essentially clean. Sliding was conducted at very slow speeds and at relatively modest loads. The friction coefficient for PTFE sliding on the metals was plotted as a function of angular distance or position around a circumferential wear path on the disk surface for a hemispherical pin sliding on a rotating disk. The friction results obtained in two such experiments are presented in figure 6-21, where

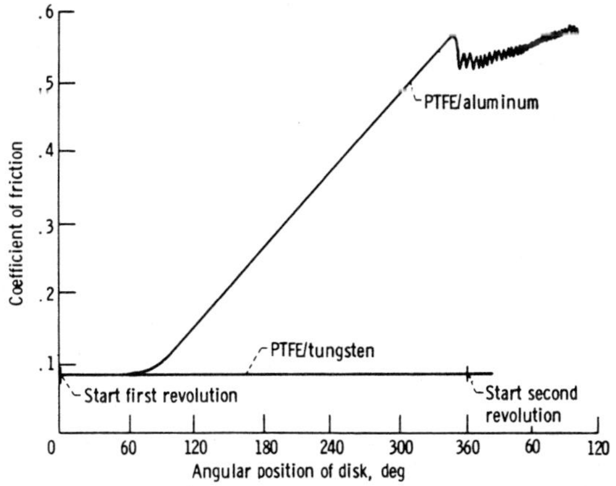

Figure 6-21. – Coefficient of friction of PTFE on aluminum disk and tungsten disk in vacuum. Sliding velocity, 0.07 centimeter per second; load, 250 grams.

the coefficient of friction for PTFE is plotted for that polymer in sliding contact with aluminum and with tungsten. Essentially, the entire friction traces are for one revolution; there is no repeated sliding over the same surface. In each step along the way, the polymer is essentially in sliding contact with metal; it is the first path of sliding. The data indicate that, with the polymer sliding on tungsten, the friction coefficient is relatively low, approximately 0.08, and relatively constant, unchanged over the entire one revolution along the disk surface. With the PTFE sliding on the aluminum, however, after approximately 60° of rotation on the disk surface, the friction coefficient begins to increase markedly. It is interesting to note that the friction initially is essentially the same whether the sliding is on aluminum or on tungsten. The reason for this is that adhesion occurs between the polymer and both metals (clean aluminum and clean tungsten).

As motion begins and the surfaces move tangentially, the adhesive bonds at the interface for both metals in contact with the polymer are stronger than the cohesive bonds in the polymer. Shear thus occurs in the polymer. As the motion continues, the friction coefficient remains essentially the same with both metals for 60° of rotation. The reason for this is that the property being measured, by friction, is the shear strength in the polymer. After 60° of rotation, however, the friction coefficient for PTFE sliding on aluminum begins to increase markedly; by the time one revolution is completed, the friction coefficient exceeds six times its original value. Why does the friction coefficient increase so markedly after 60° of rotation on the aluminum disk but not on the tungsten? The answer is found in a scanning electron microscopic examination of the PTFE rider surface after the one revolution. The results of such an examination are presented in figure 6-22, which shows the wear scar on the PTFE surface. In the upper photomicrograph is a bulge in the wear scar in the central region. If that particular region of the wear scar is magnified 10 times, the photomicrograph in the lower part of figure 6-22 is obtained. A careful examination of the bulge indicates a fragment of aluminum is embedded in the PTFE surface. Aluminum is actually transferred to the PTFE and rubs against the aluminum disk surface. Adhesion occurred at the interface with aluminum transferring to the PTFE. Aluminum became embedded in the surface and work hardened with sliding. Since the embedded aluminum fragment was much harder than the parent aluminum disk, it acted as a cutting tool for wear of the aluminum disk surface. Ultimately one has aluminum sliding on aluminum.

As sliding progresses on the aluminum disk, the friction coefficient begins to increase and to reflect the friction behavior for aluminum in contact with aluminum rather than PTFE. The shear properties of the PTFE reflect a friction coefficient of less than 0.1 as indicated by the friction data for PTFE and tungsten where there is no metal transfer to the PTFE rider. Thus, the difference between the PTFE sliding on aluminum and PTFE sliding on tungsten is that adhesive transfer of PTFE to tungsten occurs, and shear occurs in the PTFE giving a low friction coefficient of approximately 0.08 in vacuum. With sliding on aluminum, however, adhesion results in strong bonding at the interface and, rather than PTFE

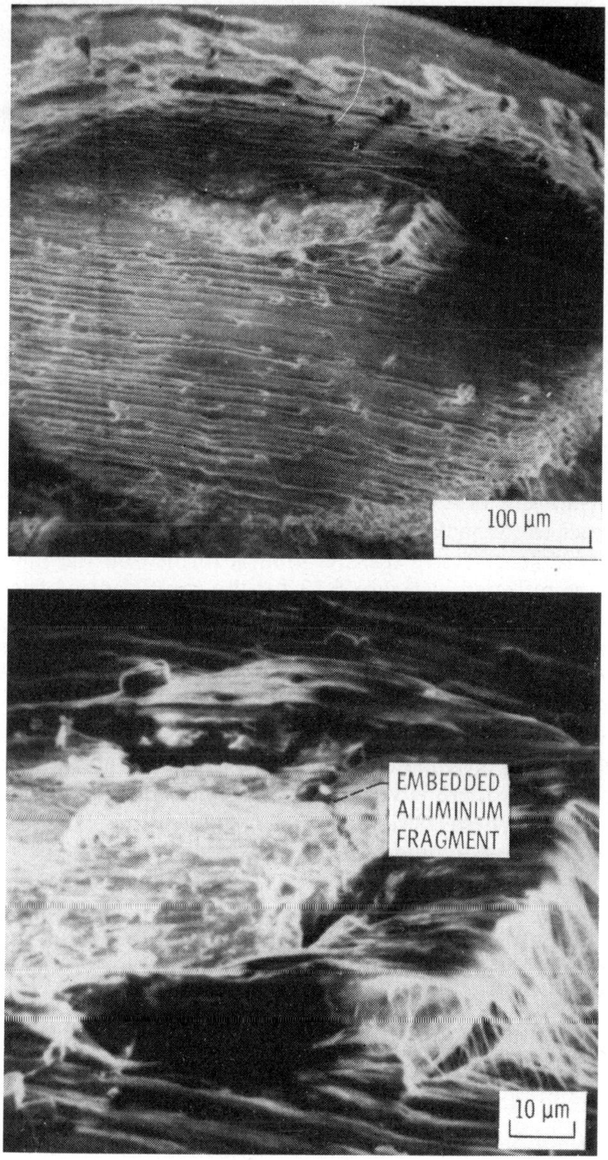

Figure 6-22. – PTFE-rider wear scar showing lodged metal fragment. Run on (110) surface; single pass; 250-gram load.

transferring to aluminum, aluminum is observed to transfer the PTFE. With continued sliding, the transferred aluminum begins to slide on the aluminum disk surface and the friction coefficient begins to represent that of aluminum sliding on aluminum rather than on PTFE.

The friction behavior of polymers in sliding contact with metals is influenced by the shear properties of the polymer. This is a common observation for most polymers in contact with clean metal surfaces. In other words, adhesion occurs at the interface and shear occurs in the polymer, and it is the shear strength of the polymer which is reflected in the measured friction force.

With PTFE, a transfer film develops during the first pass; that is, PTFE is observed on the metal surface with a single pass of a polymer across the surface. With some other polymeric materials, however, it is much more difficult to develop a transfer film, and it takes repeated passes before a transfer film develops on the mating surface. Polyimide is a polymeric material that is being used to a greater extent in practical tribological devices. It is a relatively hard, brittle material much like Bakelite.

When polyimide is brought into sliding contact with metals, a transfer film of polymer to the metal surface develops; however, it takes a number of passes before an effective transfer film is formed on the mating solid surface. This is reflected in figure 6-23 for the coefficient of friction of a filled polyimide composition (20 wt. % copper powder filled) sliding on stainless steel in a vacuum environment. The friction coefficient is plotted as a function of the number of passes over the same track. Initially, with the first pass across the surface, a very high friction coefficient of 0.6 is observed. With repeated passes, however, the friction coefficient begins to decrease, and after approximately 10 passes it reaches a value of about 0.15. With repeated passes there is very little additional change in the measured friction coefficient. These friction coefficients represent the development of a surface film of the polyimide on the stainless steel disk. The friction is very high before the transfer film develops; as the transfer film develops on the stainless steel, the friction decreases. After a transfer film is fully developed, so that the polymer is sliding on itself rather than on stainless

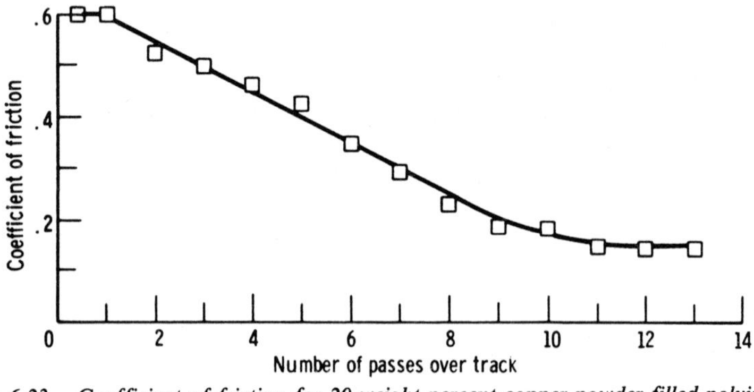

Figure 6-23. – Coefficient of friction for 20-weight-percent-copper-powder-filled polyimide sliding on 440C stainless steel in vacuum (10^{-10} mm Hg). Sliding velocity, 0.013 centimeter per second; load, 1000 grams; ambient temperature, 25° C.

steel, the friction forces are markedly less and reflect only the shear of the polymeric material itself.

The basic polymeric structure gives a brown color to the polymer. In the sliding process, the development of the polymer transfer film to the surface can be seen as it builds up on the stainless steel surface. After 13 passes, there is a visible film of polyimide on the stainless steel surface, and analytical tools are not needed to detect its presence.

The difference in friction coefficients between a single pass with a polymer in contact with stainless steel and 10 passes with a polymer in contact with stainless steel is that in the single pass adhesion of the polyimide to the metal surface does not result in transfer. As a consequence, the force measured does not reflect the shear strength of only the polymer. It is only after 10 passes where a complete transfer film is developed that the shear is almost exclusively in the polymer. Between 1 and 10 passes there are varying amounts of shear within the polymer and at the interface that account for the variation in friction coefficients.

In a host of practical devices, carbons are used in sliding contact with metals, usually steel surfaces or chrome-plated steel surfaces. A common application of carbon in contact with such metals as chromium or chrome-plated surfaces is in mechanical seals. Measurements made for the friction behavior of metals in contact with carbons indicate that, like polymers, a transfer film of the carbon must develop on the metal surface before an equilibrium friction force is measured for the carbon in contact with metals.

Generally the friction coefficient is higher for carbon in contact with the metal before a transfer film develops. Once a transfer film develops on the surface so that carbon is in contact with carbon, the friction coefficient is reduced appreciably. This is generally observed for the carbons in sliding contact with metals that are coated with surface oxide—that is, for metals that are present in a normal environment. If, however, carbon is brought into sliding contact with a clean metal surface, very frequently the adhesive binding at the interface between the carbon and the metal is sufficiently strong that transfer of metal to the carbon is observed. This, for example, is the case for carbon sliding on copper and silver in a vacuum system where the copper and silver surfaces have been cleaned by sputter bombardment. Sputter cleaning the surfaces results in strong adhesive bonds of the clean metal to the carbon body with the result that the shear takes place in the metal and the friction force recorded is for the metal in sliding contact with itself. Generally, when an oxide film is present on the metal surface (in an ordinary air environment), the carbon transfers to the metal surface by chemical bonding of the carbon to the oxygen and the oxide.

Studies have been conducted using analytical surface tools to follow the transfer of carbon to metal surfaces during the sliding process. AES analysis is a particularly useful tool for this study because it is very sensitive to carbon. Cylindrical mirror Auger analysis is suited to a sliding experiment because in situ studies of the transfer of carbon to the metal surface can be followed with repeated passes. Such studies have been done, and some results obtained for a carbon body sliding on a chromium surface are presented in figure 6-24.

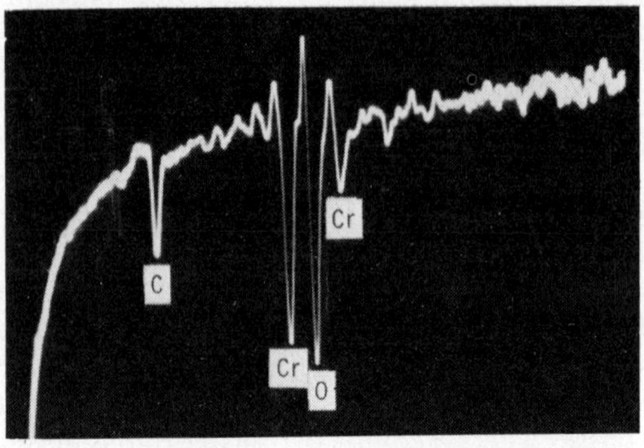

(*a*) *Before sliding.*

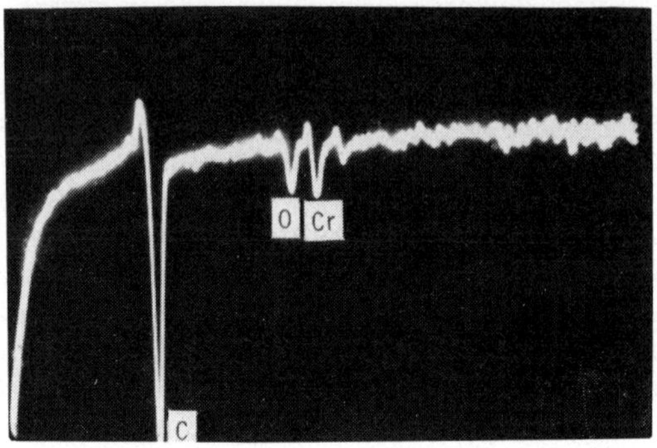

(*b*) *After 50 sliding passes.*

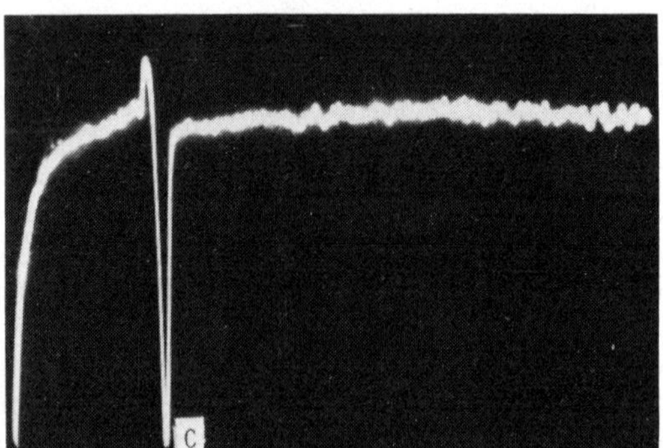

(*c*) *After 100 sliding passes.*

Figure 6-24. – Oscilloscope displays of oxide-covered chromium surface film and development of graphite transfer film.

In figure 6-24 are typical Auger spectra for an oxidized chromium surface prior to sliding, after 50 passes of sliding of a carbon body across the surface, and after 100 passes of a carbon body sliding across the surface. Before sliding (fig. 6-24(a)), the chromium surface reveals peaks for chromium, oxygen, and carbon. The chromium peaks, of course, are associated with the chromium metal, oxygen is associated with the oxide present on the surface (principally Cr_2O_3), and the carbon is associated with adsorbed carbon monoxide and carbon dioxide on the metal surface. When a carbon body is brought into sliding contact and after 50 passes over the same surface, the Auger spectrum of figure 6-24(b) is obtained. It is interesting to compare the Auger spectrum of figure 6-24(b) with that of figure 6-24(a). In figure 6-24(b) the chromium peaks have decreased markedly in intensity, indicating that the surface of the chromium is being covered. Also, the oxygen peak decreases very markedly in intensity, again indicating that the surface oxide is being covered. The Auger peak heights indicate the relative quantity of material on the surface, since the sensitivities in figures 6-24(a) and (b) are the same. All instrumentation settings were the same. Thus, the peak heights are a direct reflection of the amount of material present on the solid surface.

The carbon peak height increased markedly from figure 6-24(a) to figure 6-24(b). This marked increase in the carbon peak intensity is a direct result of the transfer of carbon to the chromium surface. After 100 passes of the carbon body across the surface we observe the Auger spectrum of figure 6-24(c); all the Auger peaks are completely absent from the spectrum except carbon. This means that the carbon transfer film is sufficiently thick so that all the metal and oxide are completely covered by the carbon. The film thickness is at least four or five atomic layers, the sensitivity (depth of penetration) of the Auger spectrometer. The carbon transfer film is, therefore, at least four or five atomic layers deep in figure 6-24(c). After some time the transfer of carbon to the chromium surfaces ceases, just as it does with the polymers, and the carbon is essentially sliding on a film of itself without further buildup of the carbon on the chromium surface.

For most metals in sliding contact with carbon, the development of a transfer film reduces the friction coefficient. This, however, is not the case for chromium. Table 6-II lists the friction coefficient for graphitic and amorphous forms of carbon sliding on copper, chromium, and aluminum. Friction coefficients are presented for the first sliding pass and after the stabilized friction has been achieved with the development of a full transfer film of carbon as determined by AES analysis. The surface conditions are also indicated as the normal oxide present and with a sputtered clean surface, where the oxide has been removed and AES analysis has been used to determine whether the surface is clean and free of oxides. After sputter cleaning, the surfaces of copper, chromium, and aluminum were essentially devoid of oxides before the first pass of sliding was initiated. The data of table 6-II indicate that with copper and aluminum reported passes over the surface with the normal oxide present and the development of a transfer film there is a reduction in the friction coefficient (stabilized average). With chromium, however, an increase in the friction coefficient is observed (table 6-II).

TABLE 6-II. – COEFFICIENT OF FRICTION AND CARBON WEAR FOR VARIOUS CARBON-METAL COMBINATIONS

Metal surface	Carbon	Surface condition	Friction coefficient		Wear volume per revolution, 10^{-6} cm^3
			First sliding pass	Stabilized average	
Copper	Graphitic	Normal oxide present	0.45	0.31	5.95
		Sputter cleaned	0.61	0.40	8.66
	Amorphous	Normal oxide present	1.2	0.7	2.95
		Sputter cleaned	1.5	0.83	3.80
Chromium	Graphitic	Normal oxide present	0.41	0.60	3.68
		Sputter cleaned	0.42	0.40	2.95
	Amorphous	Normal oxide present	0.80	0.86	2.44
		Sputter cleaned	0.80	0.80	1.47
Aluminum	Graphitic	Normal oxide present	0.80	0.50	8.90
		Sputter cleaned	1.0	0.45	7.95
	Amorphous	Normal oxide present	0.85	0.80	1.15
		Sputter cleaned	0.96	0.80	1.06

Temperature Effects

In 1935 Bowden and Ridler of Cambridge believed that the surface temperatures of metals in sliding contact must be extremely high because the asperity interactions dissipate a large amount of energy (ref. 8). In an attempt to establish a relationship between asperity surface temperatures and the rubbing process, they conducted some relatively simple experiments wherein two dissimilar metals were in sliding contact. They selected the materials, however, so that they would serve as elements of a thermocouple and determined the electromotive force generated with sliding.

They conducted sliding friction experiments in which they measured the surface temperatures both as a function of sliding velocity and as a function of load. They obtained some rather interesting results. Some of these data are presented in figure 6-25 where temperature is plotted as a function of the sliding velocity for constantan sliding on mild steel. In figure 6-25(a) there are three curves: curve 1 was for a load of 20 grams, curve 2 for 80 grams, and curve 3 for 102 grams. With increasing velocity at all three loads an increase in surface temperature was observed. The temperature at the 102-gram load at a sliding velocity in excess of 1000 centimeters per second exceeded 1000° C. This is for dry sliding of constantan on mild steel.

Just as Bowden and Ridler observed a speed or velocity effect (in fig. 6-25(a)), so also did they observe a load effect with changing velocity. The data are presented in figure 6-25(b) at three different sliding speeds: curve 1 is a speed of 20 centimeters per second, curve 2 is 485 centimeters per second, and curve 3 is 1100 centimeters per second. Even at a relatively modest load of 102 grams and a speed of only 485 centimeters per second, the temperature at the surface exceeded 400° C. This would be sufficient to cause recrystallization of a metal like iron.

The data of figure 6-25 are extremely significant because they help explain many of the changes in friction properties that had been observed with materials in sliding contact. Very frequently changes in sliding velocity produce changes in friction coefficients; changes in load also produce changes in friction coefficients. Understanding the relationship between surface temperature and sliding velocity or load can give insight into physical, metallurgical, and chemical changes that may be taking place at the surface and resulting in friction changes. It is easier to understand changes once it is understood that the temperatures at the sliding interface can be extremely high, easily in excess of 1000° C.

Bowden and Ridler recognized that these temperatures were occurring at the surface atoms in the asperities. They felt that there was a temperature limitation that would be achieved in the sliding process and that the limitation could be the melting point of the lower melting of the two metals in sliding contact. In the experiments they conducted with such metals as gallium, woods metal, and lead, they found that there was a limiting temperature that could be reached in the sliding process; this limiting temperature was the melting point of the low melting point metal. These high temperatures are localized at the asperity contacts while the mass of the metal was relatively cool (essentially at room temperature) with no evident signs of heating.

A point which is often overlooked by current researchers is that Bowden and Ridler also lubricated their surfaces with a conventional oil. Even with a well-lubricated surface running smoothly with a low friction coefficient, they found that surface temperatures over 600° C could be measured for the constantan-steel couple in sliding contact. Thus, even with effective lubrication, very high surface temperatures are readily achieved. Again, these temperatures can influence surface properties. The temperatures generated as a result of one asperity coming in contact with another and making solid-state contact is relatively short-lived (fractions of a second

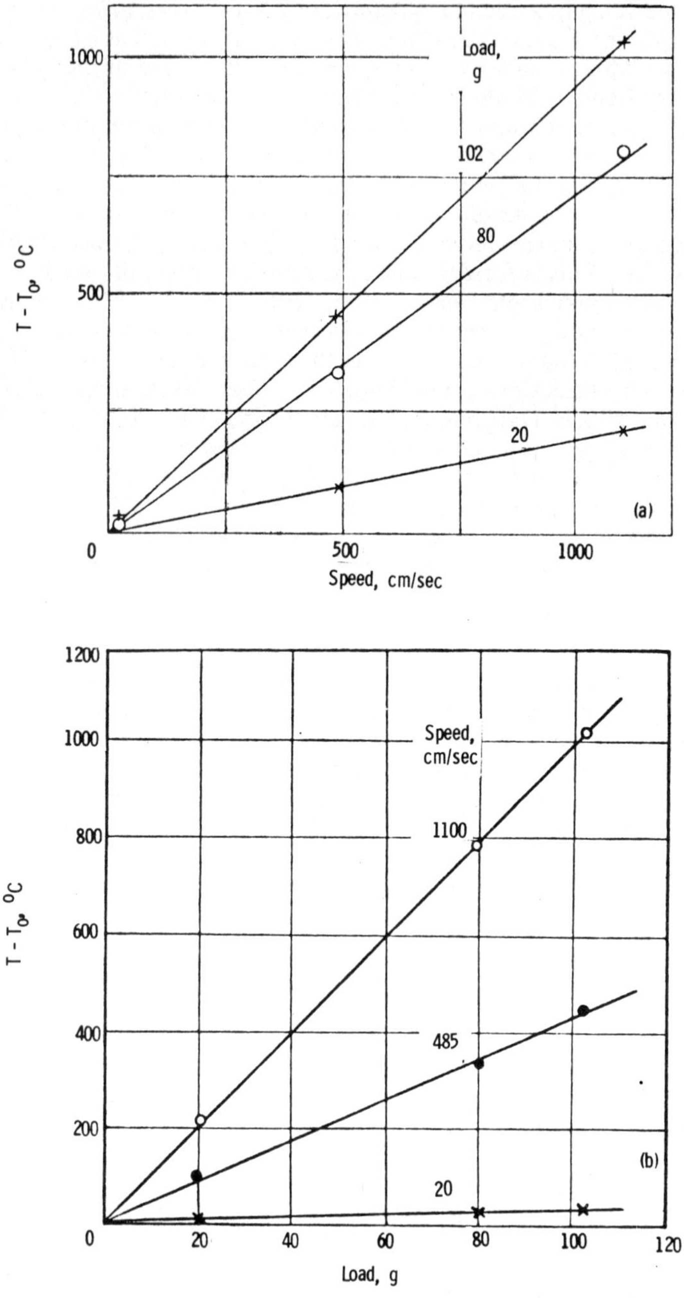

Figure 6-25. – Effect of sliding velocity and load on surface temperature of metals in sliding contact. Constantan on mild steel; initial temperature, 17° C (ref. 8).

only); these were referred to by Bowden as flash temperatures. With repeated sliding over the same surface, however, these localized flashes that are generated at the tips of the asperities in the surface atoms deposit a quantum of heat into the asperity surface region. That heat is dissipated since most metals that are used in tribological systems are good thermal conductors.

The heat is carried back into the asperity with the bulk of the solid acting as a good heat sink for the frictional heat generated at the interface. Thus, after a time, the near surface region becomes heated and the efficiency of dissipation of heat generated in the asperity is reduced because the difference between the asperity temperature and the bulk temperature of the solid ($T - T_0$) has been decreased. There is less gradient. The total surface temperature of a solid, after some period of running, thus reaches an equilibrium state where the surface temperature is a sum of two temperatures—the bulk surface temperature, which results from the repeated sliding across the surface, and the flash temperatures, which are continuously generated by asperity-asperity interactions. The two components of the total surface temperature are plotted in figure 6-26 as a function of sliding velocity for a brass specimen in sliding contact (ref. 9). An examination of figure 6-26 reveals total surface temperatures comparable to those predicted and measured by Bowden and Ridler in the 1930's. The total surface temperature has two components in figure 6-26; one is associated with the flash temperatures (asperity interaction temperatures), and the other is a result of bulk surface temperature heating. The bulk surface temperature keeps increasing with increasing sliding velocity; the higher the velocity, the higher the bulk surface temperature. With the flash temperatures, however, there is a relatively rapid increase in the flash temperature with an initial increase in sliding velocity and then a tapering off or an actual reduction (slight) in the flash temperatures with increased sliding velocity. The total temperature is the sum of these two: the flash and the bulk surface temperatures and the total surface temperature

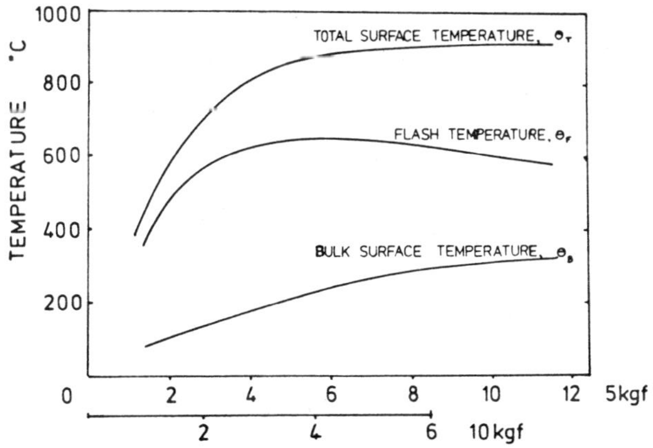

Figure 6-26. Variation of total surface temperature, flash temperature, and bulk surface temperature with speed for alpha brass specimens (ref. 9).

increases with increasing sliding velocity. In figure 6-26 there are two slopes to the curve for temperature as a function of sliding velocity. The very steep portion of the curve as the temperature rises from approximately 400° C to a value in excess of 800° C is associated with the increase in both surface flash temperature and bulk surface temperature with sliding velocity, the largest increase being in the flash temperature. When, however, the temperature exceeds 800° C and approaches the melting point of the brass, the rate of temperature rise drops off. The second portion of the total surface temperature curve in figure 6-26 reflects only slight changes, a very slight increase with further increase in sliding velocity. Some of the heat is absorbed in the process of converting the solid brass to liquid or molten brass; this is the heat of liquefaction.

Changes in the temperature of materials can also bring about changes in friction properties. It is important to understand what contributes to the surface temperature of materials so one can better appreciate those factors that contribute to frictional changes. For example, in figure 6-27 the effect of temperature on friction is plotted for PTFE (ref. 10). The friction coefficient is relatively low, approximately 0.1, at temperatures from 100° to -20° C. Suddenly, however, as the temperature is dropped further (from -20° to -60° C), there is an increase in the friction coefficient. The friction coefficient rises to approximately 0.2 and remains there at temperatures to -80° C. This change in friction properties with the change in temperature is reflected in other materials as well. For example, metals that undergo crystallographic transformations (i.e., that are polymorphic, changing from one crystal form to another) exhibit a marked change in friction properties.

Cobalt exhibits at room temperature a closed-packed-hexagonal structure and has low friction properties (approximately 0.35). If, however, the temperature is increased to approximately 417° C, a marked change in friction occurs for any further increase in temperature. The friction may increase many times as a result of further increases in temperature. This increase in friction is associated with the crystal transformation for cobalt from a close-packed-hexagonal structure below 417° C to a face-centered-cubic structure above 417° C.

If the coefficient of friction is plotted as a function of temperature for cobalt sliding on itself, an increase in friction coefficient is experienced at approximately 417° C. If, however, either the load or the speed are

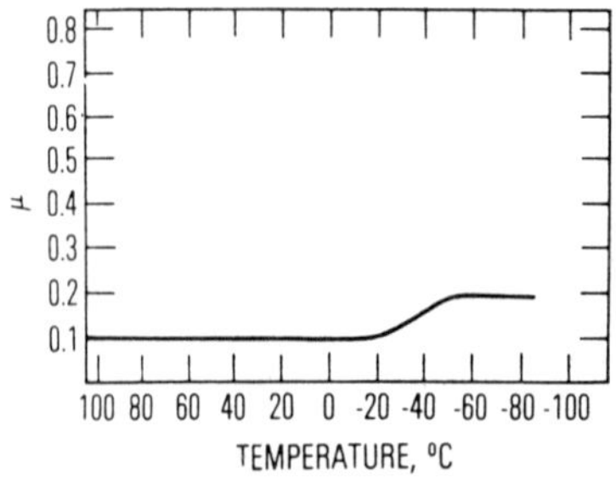

Figure 6-27. – Effect of temperature on friction of PTFE (ref. 10).

increased to a relatively high value—that is, increase the sliding velocity to a high sliding speed or apply sufficiently heavy load or combination thereof—the temperature at which an increase in friction coefficient is observed decreases; this author has observed the increase in friction coefficient at measured bulk temperatures as low as 350° C. Thus, the difference between 350° and 417° C is due to frictional heating under the applied load and at the speed involved. The total surface temperature is contributing over and above the actual ambient temperature to give rise to a structural transformation in the material. The transformation still occurs at 417° C, but that temperature is achieved much more readily when the sliding velocity and/or the load is increased.

Metallurgical Effects

As mentioned in the preceding section, the interfacial temperatures between two solid surfaces in solid-state contact can readily reach 1000° C in dry sliding; even under effective lubrication they can reach 600° C. Such temperatures can cause other marked changes in the behavior of material properties in the surface region. For example, the crystal structure of the surface can undergo recrystallization.

Recrystallization of metal surfaces in the pure state can occur at relatively modest temperatures. With recrystallization there is a change in the grain size of the material and also a relieving of the strain in the solid. These both produce changes in the mechanical behavior of the solid in contact. Table 6-III presents some approximate recrystallization temperatures for some metals and alloys. The data were taken from Guy (ref. 11). Examining the table reveals that relatively pure copper (99.999%) recrystallizes at a temperature as low as 120° C. We know from our previous discussion that these temperatures could easily be achieved in practical tribological systems. Thus, for copper surfaces in sliding contact, in many practical devices it might be anticipated that recrystallization occurs on a regular basis. Likewise, metals such as aluminum have recrystallization temperatures in the pure state as low as 80° C. Note that the addition of a small amount of impurity, however, to a metal can increase its recrystallization temperature markedly. For example, pure aluminum (99.999%) has a recrystallization temperature of 80° C while a less pure aluminum (99.0+%) has a recrystallization temperature of 288° C, a threefold increase in the recrystallization temperature (table 6-III). Note, however, that regular, conventional aluminum alloys have a recrystallization temperature of only 315° C. Even with deliberate alloying, the recrystallization temperature is not that much greater than is observed for normal, impure aluminum. Electrolytic iron has a recrystallization temperature of about 400° C and that the addition of carbon to produce carbon steels only increases the recrystallization temperature to 538° C. Of all the recrystallization temperatures in table 6-III, the recrystallization temperature for a low carbon steel is the highest reported therein. For the metals tin and lead the recrystallization temperature is below room temperature. Any mechanical

TABLE 6-III. – APPROXIMATE RECRYSTALLIZATION TEMPERATURES FOR SEVERAL METALS AND ALLOYS[a]

Material	Recrystallization temperature, ^{o}C
Copper (99.999%)	120
Copper, 5% zinc	315
Copper, 5% aluminum	288
Copper, 2% beryllium	370
Aluminum (99.999%)	80
Aluminum (99.0%+)	288
Aluminum alloys	315
Nickel (99.99%)	370
Monel metal	590
Iron (electrolytic)	400
Low-carbon steel	538
Magnesium (99.99%)	66
Magnesium alloys	230
Zinc	10
Tin	-4
Lead	-4

[a]Reference 11.

activity at the solid surface is almost certain to produce recrystallization at room temperature. Even with zinc, which has a recrystallization of 10° C, the slighest amount of mechanical activity of the surface produces recrystallization at the surface.

The mechanical properties can be markedly altered as a result of recrystallization as previously mentioned. Some of the effects are seen in figure 6-28, which shows an idealized curve for the behavior of the mechanical strength of materials as a function of the annealing temperature where T_{RX} is the recrystallization temperature and T_X the softening temperature of the material (ref. 12).

There are marked decreases in strength and grain size as the softening and recrystallization temperatures are approached (fig. 6-28). After the recrystallization temperature is reached, there is a notable increase in the grain size due to the recrystallization process and the relief of the strain associated with small grain structure.

With the recrystallization process, there are observed changes in the friction behavior for materials in sliding contact. This has been demonstrated in experiments conducted with single and polycrystalline materials sliding on an aluminum oxide surface in a vacuum environment

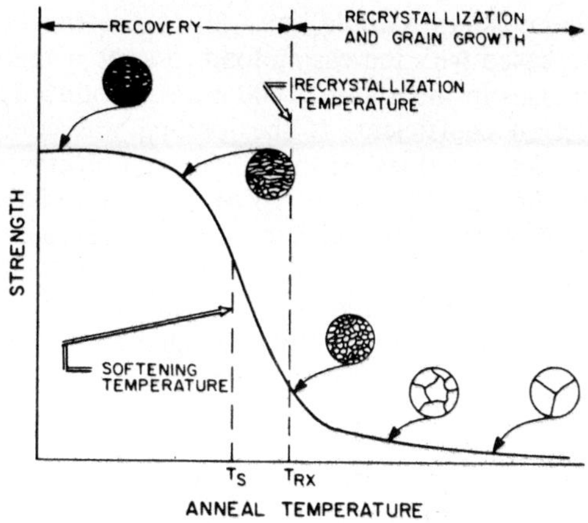

Figure 6-28. – Effect of temperature on mechanical strength (ref. 12).

where the load was used to cause recrystallization. This was the mechanism used to increase interface temperature.

Figure 6-29 shows friction coefficients plotted as functions of load for single and polycrystalline copper sliding in a vacuum environment against aluminum oxide. The single crystal was a (111) orientation sliding against the aluminum oxide surface, and the polycrystalline sample was a small grain size material. For the polycrystalline sample at a 100-gram load, the friction coefficient was 1.2. With increased loading, the friction coefficient decreased; at 400 grams the friction coefficient was approximately 0.9, and no further changes in the friction coefficient were observed at loads to 1000

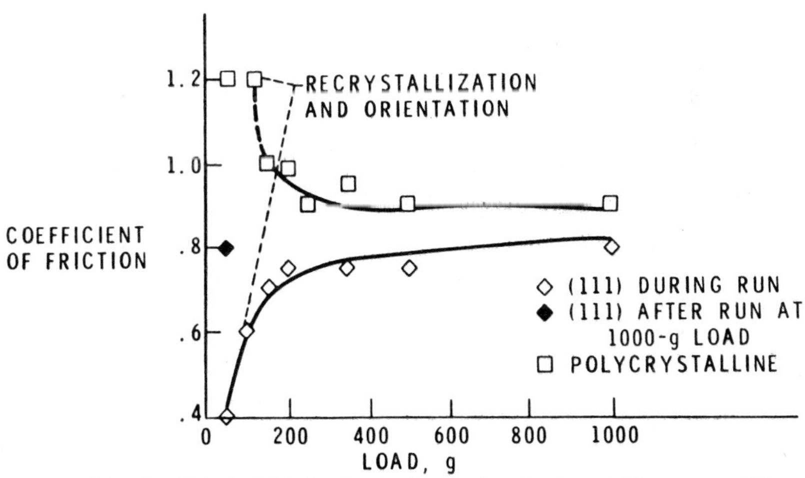

Figure 6-29. – Coefficient of friction for single-crystal and polycrystalline copper sliding on polycrystalline Al_2O_3 in vacuum. Pressure, 10^{-11} torr; copper oriented with (111) plane parallel to sliding interface; speed, 0.001 centimeter per second; no external specimen heating.

grams. With the single crystal sample, the friction coefficient at a 50-gram load was 0.4. It increased with increasing load to approximately 0.8 at 400 grams; further increases in the load to 1000 grams produced no additional increase in the friction coefficient. If the loading was continued on the specimens, the two friction values for the single and polycrystalline samples became essentially the same. Careful SEM analysis of the single and polycrystalline surfaces after the various loadings revealed that, in the polycrystalline sample, recrystallization had occurred at the 100-gram load with a texturing of the surface, and X-ray techniques indicated that that texturing was a (111) orientation. Further increases in load on the polycrystalline copper sample produced a further texturing of the surface; however, at 500 grams, no further change in the surface texture was observed. Basically, at the 500-gram load for the polycrystalline sample there was a recrystallized textured interface having a (111) orientation.

The identical changes occurred with the single crystal sample. Recrystallization took place at a 100-gram load with a polycrystalline interface being generated because of the strain placed in the crystal and the interface. Rubbing nucleated the formation of grain boundaries so that the single crystal surface became essentially a polycrystalline surface with increasing load. Thus, both single and polycrystalline samples of copper were identical structurally at the 1000-gram load. Relatively small differences in friction existed at the 1000-gram load, but at higher loads the friction became essentially the same because the interfaces or surface layers for the two specimens were identical. Thus, though the single crystal of the (111) orientation exhibited initially lower friction coefficients, with sufficiently high loading, the friction force reached that obtained for the polycrystalline material because of recrystallization. With the polycrystalline sample the orientation was also random.

The (111), which is the low friction plane of copper, was present at the interface but there was also a mixture of orientations at the interface that caused a higher friction value. The grain boundaries increase the shear strength of the interface and thereby increase the friction force measured. With recrystallization, a texturing developed on the surface, and this recrystallized texturing reduced the friction force because the textured orientations were all (111) orientations, which have the minimal friction force for the face-centered-cubic metal copper.

This author has through the years conducted analogous experiments for a number of other pure metals sliding against aluminum oxide. All the metals behave similarly if separate single crystal and polycrystalline experiments are conducted. If the single crystal is of the high atomic density, low surface energy plane (the preferred slip plane in the metal), the friction behavior is comparable to that observed for the single crystal of copper in figure 6-29. Likewise, the polycrystalline sample, if it is a random orientation, behaves analogously to the polycrystalline copper (fig. 6-29). As the load increases, the recrystallization process takes place and the friction properties for the single and polycrystalline metal become essentially the same (fig. 6-29). One very notable difference, however, is the load at which the friction coefficients become essentially the same. That varies with the particular metal involved. The higher the recrystallization temperature of the metal, the higher the load required to bring about a sameness in the surface friction properties and the surface morpholoogy of the two forms of the metal. Thus, tungsten, for example, which has a very high recrystallization temperature, requires a very heavy loading before sameness in friction values are observed for a polycrystalline sample of tungsten in random orientation and a single crystal of tungsten with a (110) orientation at the sliding interface.

Many other metals, such as the transition metals, fall between the copper and tungsten and the loads required to bring about the recrystallization, surface texturing, and sameness of friction values. A similar effect is observed for titanium in figure 6-30 where the friction coefficient in bargraph form is plotted for single and polycrystalline titanium sliding on a polycrystalline titanium surface. In figure 6-30 the coefficient of friction is plotted for 250- and 500-gram loads. The sliding velocity in figure 6-30 is slightly higher than that of figure 6-29. In addition, titanium is sliding on polycrystalline titanium rather than on aluminum oxide. Aluminum oxide is a good insulator; polycrystalline titanium is a good conductor. As a consequence, the frictional heating generated at the sliding interface may not be as great for the specimen configuration of figure 6-30 as for those of figure 6-29. A greater amount of loading may be necessary to produce the equivalent amount of interfacial temperature because both materials in contact in figure 6-30 are metals, while in figure 6-29 one is a metal and one is a ceramic material. Nonetheless (notwithstanding the foregoing differences), the results of figure 6-30 indicate that with increased loading from 250 to 500 grams there is a change in the friction properties of both the single and polycrystalline samples of titanium. With the single crystal of titanium, an increase in friction coefficient is observed with an increase in

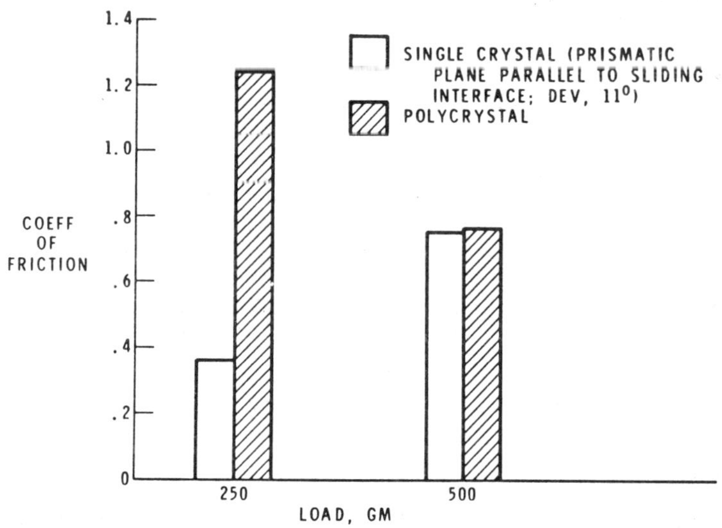

Figure 6-30. – Coefficient of friction for single-crystal and polycrystalline titanium sliding on polycrystalline titanium in vacuum. Pressure, 10^{-9} torr; speed, 2.28 centimeters per second.

the load from 250 to 500 grams, just as increased friction was observed with an increase in load for the single crystal of copper (fig. 29). Likewise, with the polycrystalline titanium, there is a decrease in friction coefficient when increasing the load from 250 to 500 grams; this is analogous to the decrease in the friction coefficient that is observed with polycrystalline copper in figure 6-29.

Again, an analysis with electron diffraction and SEM of the surfaces of both the single and polycrystalline samples reveals that the interface has basically the same structure for the two surfaces. A recrystallized preferred orientation is on the surface of both samples. Thus, the interface or the surficial region of the two forms of the material titanium are identical at the 500-gram load. As a consequence, the friction coefficient is identical for the two forms of titanium.

In figures 6-29 and 6-30 the loads required to bring about such changes are relatively modest and the sliding speeds are extremely low. If recrystallization can occur under these conditions, extreme care must be taken in interpreting data on the behavior of metals in practical tribological systems because almost certainly in many of these systems recrystallization is an integral part of the surface phenomena taking place in the sliding, rolling, or rubbing process. These changes bring about changes in friction behavior, mechanical behavior, and surface chemical behavior.

At sufficiently heavy load, plastic deformation and recrystallization can occur for materials in sliding contact. In the absence of loads sufficiently high to produce recrystallization, deformation and strain hardening can still occur for materials. Barquins, Kennel, and Courtel have done some detailed studies on the deformation behavior under light loads of metals in sliding contact and have looked at the behavior of the surfaces in friction measurements for various single crystal surfaces (ref. 13). They have observed the deformation patterns in the surfaces with the sliding friction process. Figure 6-31 is a photomicrograph from their work showing a wear track generated on a copper single crystal surface from a hemispherical rider having contacted that surface and moved tangentially a slight distance so as to produce an elliptical wear scar.

A close examination of the photomicrograh reveals a considerable number of slip lines along slip planes on the crystalline surface. In addition, in the upper right corner of the photomicrograph appear distinct grain boundaries. With tangential motion, a considerable amount of plastic deformation has occurred by slip along the crystallographic slip planes in the copper single crystal surface. The amount of strain is sufficient to produce localized recrystallization, and this causes the formation of grain boundaries in the upper and lower portions of the micrograph. In addition, in the upper portion of the micrograph twins appear as a result of the deformation process. The deformation in figure 6-31 is quite severe, and the slip lines on the surface indicate the degree of deformation. The pattern for slip lines generated on the surface as a result of the deformation process occur along well-defined crystallographic planes. These are shown in the photomicrograph of figure 6-32 (also from the work of Barquins et al.). In figure 6-32(a), the slip lines appear ahead of the slider with the sliding

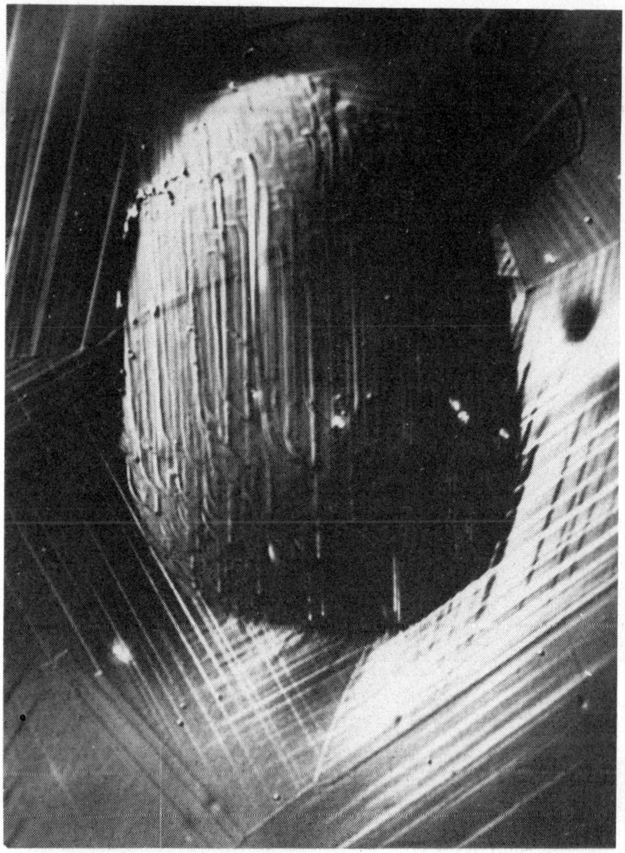

Figure 6-31. – Slip lines formed on friction surface (ref. 13).

process. In addition, slip lines appear on the right side as well as at the rear of the slider specimen. The highest concentration of slip lines appears ahead and behind the rider specimen; this indicates deformation or slip along the planes normal to the direction of sliding of the hemispherical rider on the copper surface. If the direction of the copper single crystal surface is changed from that presented in figure 6-32(a) to that presented in figure 6-32(b) where the orientation of the slider is changed, the slip lines still appear in the same crystallographic orientations. Changing the orientation of the rider does not alter the basic deformation mechanism in the single crystal of copper. The deformations still occur along the same well-defined crystallographic planes. The amount of deformation may vary with the orientation of the rider, but the concentrations occur in the same locations as were observed in figure 6-32(a). At sufficiently light loads, surfaces deform initially elastically. When the loads become sufficiently high, deformation occurs by a plastic mode (fig. 6-32), slip line formation being

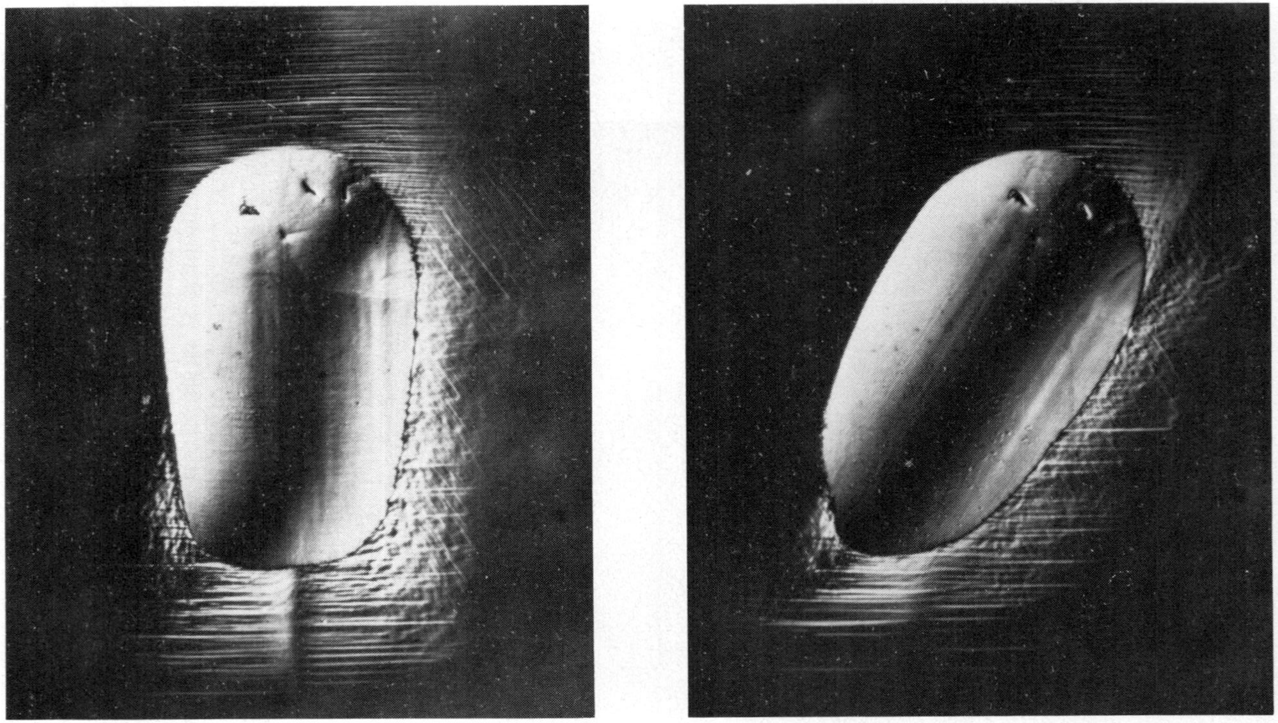

(*a*) *Orientation A.* (*b*) *Orientation B.*

Figure 6-32. – Slip with sliding.

the initial stage of plastic deformation of the solid surface. If the deformation process is continued with sliding by increasing the load, the deformation continues until bulk shear takes place in the metal near the interface period. At sufficiently high loads, recrystallation of the surface occurs (fig. 6-31).

Orientation Effects

Since most materials used in tribological systems are crystalline, there are anisotrophic properties to contend with. Mechanical properties, as well as the adhesion, friction, and wear behavior of these materials, are influenced by orientation of the crystalline solids. In studying the influence of orientation of crystalline solids on their behavior, most researchers use single crystal surfaces. In tribological studies, generally two solid surfaces are in contact. Thus, it requires a consideration of the orientation of both surfaces as opposed to a single surface.

A common specimen configuration used in tribological experiments is the rider on disk specimen combination wherein the disk rotates and the rider slides against the disk surface during the rotating process. One can selectively conduct experiments wherein both the rider and disk surface have preferred crystallographic orientations. It is, however, extremely difficult to get single crystals large enough to accommodate the fabrication of disk specimens. Therefore, a Bowden-Leben type apparatus can be used wherein the sliding is unidirectional on a flat so that a hemispherical rider specimen of the single crystal is made to slide unidirectionally on a flat single crystal surface. This has another distinct advantage in that the crystallographic directions are not continuously changed with the sliding process as occurs when the disk rotates relative to the rider disk contact. The crystallographic direction is continuously changing. With the Bowden-Leben type of sliding motion, this change in crystallographic direction is avoided where the flat surface is a single crystal. Using this technique, one can measure the effect of changes in crystallographic direction by selectively orienting the surface of the flat relative to the rider during the sliding process so that sliding is in the crystallographic direction desired.

If two perfectly matched single crystals of a metal are brought together, and both surfaces have the same crystallographic planes exposed on the solid surface, and if the crystallographic directions are perfectly matched across the interface as the two surfaces are brought into contact, the individual identity of the two solid surfaces would be lost and one continuous solid would be generated from the two individual single crystals. The reason is that, as the two surfaces approach one another, atomic bonding would occur across the interface and a continuous solid would exist through the interface with the interface being lost completely. In real situations, however, any attempts to match crystallographic planes and directions across an interface are not feasible; consequently, the best one can hope for is to achieve a situation analogous to a grain boundary where a large number of defect dislocations exist at the interface.

Attempts have been made by this author, for example, to match atomic planes and directions across an interface in sliding friction experiments; the technique used is shown schematically in figure 6-33. The rider orientations (both plane and direction) were matched across the interface for the riders and flat surfaces. The adhesion forces were measured with touch contact, and the friction force was measured with attempted tangential motion for the two surfaces in solid-state contact. Some experiments were conducted with copper (matched planes and directions) using three crystallographic matched sets of planes. The purpose of matching the directions was to see which of the three atomic planes matched across the interface would give the highest, or lowest, values of adhesion and friction. Results obtained in some experiments with matched planes and directions, such as shown schematically in figure 6-33, are presented in table 6-IV. The adhesion and friction results in table 6-IV are for matched planes and directions.

In table 6-IV are the properties of both single and polycrystalline copper. Studies were made for three orientations of the single crystal of copper: the (100), the (110), and the (111). In addition, a polycrystalline sample of copper is included for reference purposes, bearing in mind that the

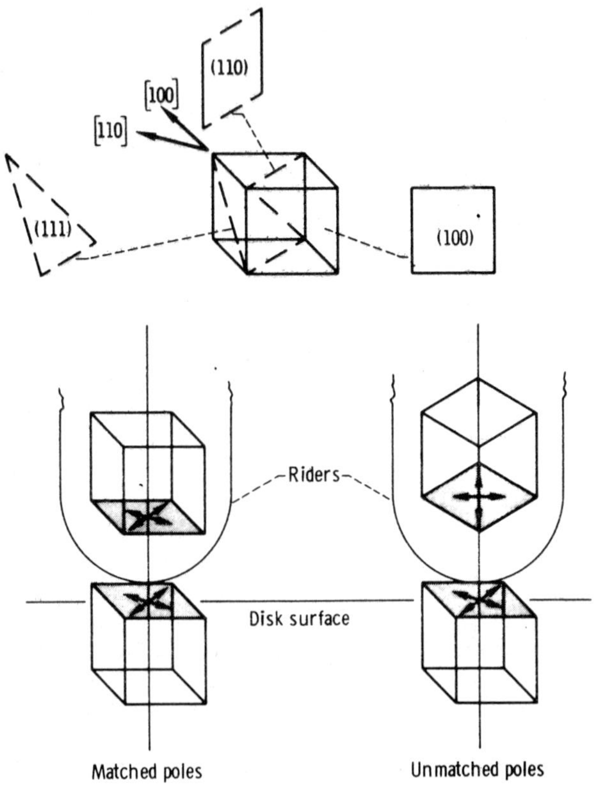

Figure 6-33. – Orientations; planes and crystallographic directions.

TABLE 6-IV. – VARIOUS PROPERTIES OF SINGLE AND POLYCRYSTALLINE COPPER (99.999 PERCENT)

Copper form and orientation	Young's modulus (10^{11} dynes $/cm^2$)	Adhesion coefficient before sliding[a] (matched planes)	Coefficient of friction during sliding[b]	Adhesion coefficient after sliding[c]
Single crystal (100) matched planes and directions	6.67	1.02	>40.0	>130
Single crystal (110) matched planes and directions	13.1	.61	>40.0	50.0
Single crystal (111) matched planes and directions	19.1	.30	21.0	10.5
Polycrystal	12.0	1.00	>40.0	100

[a]Load, 50 g; 10^{-11} torr.
[b]Load, 50 g; sliding velocity, 0.001 cm/sec; 10^{-11} torr.
[c]Load, 50 g; distance slid in preferred slip directions, 0.735 cm; 10^{-11} torr.

polycrystalline surface is a mixture of many different orientations with each adjacent grain having a different orientation than its nearest neighbor. The modulus of elasticity is also presented for these orientations in table 6-IV as are surface energy, friction, and adhesion (both before and after sliding).

The load applied to the crystals in contact was 50 grams, and both copper surfaces were cleaned in a vacuum environment at 10^{-11} torr. The sliding velocity for the sliding friction experiments was extremely low, 0.001 centimeter per second. The total distance slid was 0.7 centimeter. Sliding was conducted in a preferred crystallographic slip direction for the matched planes. An examination of table 6-IV indicates that the greatest adhesion forces were observed on the (100) planes, intermediate forces on the (110), and the lowest forces on the (111) planes. The (111) planes are the highest atomic density, lowest surface energy crystallographic planes in the face-centered-cubic system (to which copper belongs). Consequently, the minimum in adhesion is experienced for those particular crystallographic planes.

The adhesion coefficient (i.e., the force required to separate the solid surfaces divided by the applied load) is greatest on the (100) planes after sliding as well as before sliding. The adhesion is least on the (111) planes, again after sliding as well as before. The friction force is extremely high for the matched (100) and (110) planes (in excess of 40, the limit of measurement of the device). Severe cold welding or interfacial welding occurs at the interface. The lowest friction coefficient (21) was obtained on the (111) surface. This value of 21 is an extremely high coefficient of friction when compared to those for the same single crystal surfaces in an air environment. Severe adhesion at the interface gives rise to the very high friction forces observed for the matched planes and crystallographic directions.

It is interesting to compare the modulus of elasticity (table 6-IV) with the adhesion and friction coefficients. The (111) planes have the highest modulus of elasticity. They also have the lowest friction and adhesion coefficients.

With a polycrystalline sample, the adhesion coefficients approach those of the single crystal surface having the highest adhesion and friction characteristics. The polycrystalline sample consists of a number of orientations at the surface. One might anticipate, therefore, that the adhesion and friction properties would be an average of the various single crystal orientations. In the polycrystalline case, however, there are grain boundaries present. These grain boundaries are higher energy sites than even the highest energy grain orientations. Consequently, the measured adhesion and friction forces tend toward the higher values. This is true despite the fact that the modulus of elasticity is intermediate between the extremes of the modulus of elasticity for the single crystal surfaces. The surface energy, however, is higher. The surface energy for the polycrystalline case approaches that of the single crystal (100) orientation. This would then reflect on the various factors influencing adhesion and friction. On the basis of these results the surface energy would seem to influence, to a great extent, the measured adhesion and friction, since the surface energy for the polycrystalline surface is close to that of the high energy single crystal orientation rather than being intermediate as is the modulus of elasticity. Thus, in the case of the copper single crystals, the mechanical property of elasticity does not seem to be as critical as the surface energy in contributing to adhesion and friction.

In table 6-IV all the crystallographic orientations across the interface are identical. The atomic planes and directions are as close as it is possible to match them. The obvious question is what sort of friction behavior might be anticipated when the crystallographic orientations are different across the interface. If one deliberately mismatches the crystallographic planes, for example, does that alter the adhesion and friction behavior of materials in sliding contact? The evidence that it does is presented in table 6-V. In this table, the coefficients of adhesion and friction for various single crystal orientations of copper are presented—first for the matched (100) planes, then for the (110) plane in contact with the (100) plane, and lastly for the (111) plane in contact with the (100) plane. Minimum adhesion is observed before sliding with the (111) plane in contact with the (100) plane, intermediate adhesion with the (110) plane in contact with the (100), and the

TABLE 6-V. – COEFFICIENTS OF ADHESION AND FRICTION FOR VARIOUS SINGLE CRYSTAL ORIENTATIONS OF COPPER(10^{-11} torr, 20° C, 50 g)

Matched planes	Adhesion coefficient before sliding	Coefficient friction during sliding[a]	Adhesion coefficient after sliding
(100)/(100)	1.02	>40.0	>130
(110)/(100)	.25	>40.0	32.5
(111)/(100)	.20	>40.0	40.0

[a]Sliding velocity, 0.001 cm/sec; [110] direction; sliding distance, 0.735 cm.

highest adhesion before sliding for the (100) in contact with itself. The friction coefficients in all cases exceed 40 and are beyond the limit of measurement of the instrument. The adhesion coefficient for the matched planes was the greatest after sliding, with that for the mismatched planes being markedly less. Thus, as might be anticipated, mismatched orientations across an interface yield lower adhesion and friction than matched planes and crystallographic directions.

TABLE 6-VI. – COEFFICIENT OF ADHESION AND FRICTION FOR VARIOUS SINGLE CRYSTAL METAL COUPLES IN VACUUM (10^{-11} torr, 50 g)

Metal couples and orientations	Adhesion before sliding	Friction coefficient during sliding[a]	Adhesion coefficient after sliding	Soluable	Crystal structures
Cu(111)[b] / Cu(111)	0.30	21.0	10.5	S	fcc / fcc
Cu(111)[110][6] / Ni(111)[110]	.25	4.0	2.0	S	fcc / fcc
Cu(111)[110] / Co(0001)[11$\bar{2}$0]	.10	2.00	.5	S	fcc / hcp
Cu(111)[110] / W(110)[111]	<.05	1.40	.5	INS	fcc / bcc

[a]Sliding velocity, 0.001 cm/sec; sliding distance, 0.735 cm.
[b]Planes, ( ); directions, [ ].

Another question is what the influence of changing the materials at the interface is (i.e., bringing copper in contact with other metals where the crystallographic orientation is controlled and known). Changing the materials influences adhesion and friction behavior (table 6-VI). Table 6-VI gives the adhesion and friction properties for copper in contact with copper in the (111) orientation. The same orientation of copper is brought into sliding contact with nickel (face-centered-cubic metal like copper), cobalt (close-packed-hexagonal metal), and tungsten (body-centered-cubic metal). The general observation that can be made from the data of table 6-VI is that, when one brings dissimilar materials into sliding contact, the adhesion and friction forces measured with those dissimilar materials are less than the adhesion and friction forces for like materials of the same crystallographic orientation in contact. The cobalt and tungsten (in table 6-VI) are not face-centered-cubic metals, but the particular crystallographic planes selected (for cobalt the (0001) plane and for tungsten the (110) plane) are the highest atomic density, lowest surface energy planes. They are the equivalent planes in the hexagonal and body-centered-cubic systems, respectively, to the (111) plane in the face-centered-cubic system associated with nickel and copper. The adhesive forces for copper in contact with cobalt and tungsten are relatively low compared with copper in contact with itself and copper in contact with nickel. Likewise, the friction coefficient for copper in contact with cobalt is half that for copper in contact with nickel.

The solubility of the various metal systems is shown in table 6-VI; all the systems are soluble, except for copper and tungsten. However, caution should be exercised when using bulk solubility parameters. Even though copper is insoluble in tungsten in the bulk, copper does react chemically at the surface with tungsten to form surface compounds. Thus, in a strict sense, there is a solubility or affinity of copper for tungsten at the surface despite the fact that in the bulk copper is not normally considered to be soluble in tungsten. This is a caveat against using bulk properties to predict surface behavior. Adhesion and friction are very strongly dependent on surface properties and, in the case of copper and tungsten, adhesion occurs. Also, the friction coefficient is still relatively high, 1.4, compared with these same materials in an air environment. The reason for this high friction is that copper does adhere to the tungsten; as mentioned earlier, compound formation does exist for copper with tungsten at the surface (such formation has been identified with LEED analysis). Copper is observed to transfer to the tungsten surface in adhesion experiments.

In summary, the results of table 6-IV indicate that with matched crystallograhic planes across the interface the particular plane involved has an influence on adhesion and friction. The lowest adhesion and friction forces for matched planes and directions are observed with the highest atomic density, lowest surface energy planes—namely, the (111) planes with a face-centered-cubic metal such as copper. Table 6-VI indicates that, if the planes across the interface are mismatched, adhesion and friction are less than they are for matched planes. The results of table 6-VI indicate that, when dissimilar metals are brought into solid-state contact with equivalent planes from an atomic packing and surface energy point of view, adhesion and friction are less for the dissimilar metals in contact than they are for the metal in contact with itself.

The data of tables 6-IV to 6-VI are for copper, which in the atomically clean state is an extremely ductile material and very prone to adhesion. Consequently, copper is expected to adhere readily and exhibit high friction coefficients. With other metals, however, which are much more resistant to deformation (more brittle), it might be argued that these particular metals do not exhibit the strong adhesive forces that are experienced with materials like copper. This, however, is not true. Metals (e.g., tungsten) that have a very high modulus of elasticity and relative hardness (for the metal in the pure state) still exhibit strong adhesion at the interface. The real area of contact for an equivalent load, of course, is much less than it would be for a material like copper. Consequently, fracture occurs in the adhesive bonds over a much smaller area. Nonetheless, adhesion still occurs, and this is reflected in the measure of friction forces for very high elastic moduli, relatively brittle metals like tungsten (fig. 6-34).

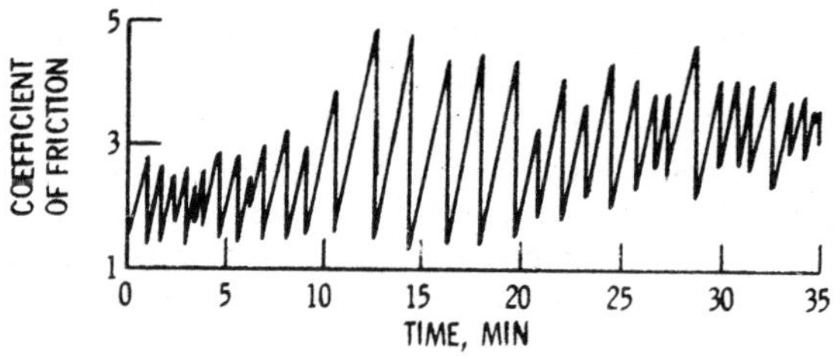

Figure 6-34. – Friction trace for clean (100) plane of tungsten sliding on like plane in vacuum. Sliding velocity, 0.001 centimeter per second; load, 50 grams; temperature, 20° C; ambient pressure, 10^{-10} torr.

Figure 6-34 shows a plot of the friction coefficient for a single crystal of tungsten with a (100) orientation sliding on a like plane of tungsten in a vacuum environment where the surfaces have been cleaned by bombardment. The sliding velocity is extremely slow (0.001 cm/sec) with a very light load (50 g) at ambient temperatures and with a vacuum of 10^{-10} torr. There is marked stick-slip behavior in the friction trace for tungsten sliding on itself. The friction coefficient is plotted as a function of time, and adhesion occurs and causes stick to take place. The friction force and, consequently, the friction coefficient rise sharply; then, at some point the tangential force applied by the mechanism turning the specimen is sufficient to overcome the adhesive bonds at the interface. Fracture occurs, the friction force drops sharply, and the slip associated with the stick-slip motion occurs. The process repeats itself, adhesion again occurs after the slip, the force builds up as the drive tries to move the surfaces to the point where the force applied to move one surface relative to another exceeds the interfacial adhesive forces. The interfacial adhesive bonds fracture, slip again occurs until another adhesive contact is made, which occurs very rapidly, and the process then repeats. Since the coefficient of friction (fig.

6-34) is relatively high (from 2 to 5), very strong bonds of adhesion can occur for clean tungsten in contact despite the fact that the material has a very high modulus and is relatively brittle. Nonetheless, when clean metal surfaces are brought in contact strong adhesion occurs, either from a metal to a metal or metals to nonmetals, as long as the metal surface is clean. Similar behavior has been observed for other high elastic moduli, relatively brittle materials such as iridium, tantalum, and rhodium.

The anisotropic friction characteristic of metals is not unique to metals in contact with metals; it also exists for other materials in sliding contact. For example, semiconductors behave like metals; that is, orientation influences the friction behavior. Likewise, inorganic crystals such as lithium fluoride, calcium fluoride, and sodium chloride exhibit an orientation influence on friction with sliding. Steijn has done a detailed analysis of the influence of orientation on the friction behavior of halide crystals (ref. 14).

Similar Elements

In the periodic table of elements there are elements that are grouped into various classifications based on similarities in their fundamental properties. For example, the platinum metals family (platinum, rhodium, osmium, and ruthenium) are grouped together because of the similarity of many of their properties. Likewise, the noble metals (copper, silver and gold) are classified together because, again, of the similarity of many of their properties. And similar comments apply to other materials in the periodic table (iron, cobalt, and nickel).

In the group 4 period of elements in the table, carbon, germanium, silicon, tin, and lead are present. These elements also have many similar properties; however, they have some distinctly different ones. For example, carbon and lead are conductors, tin is a conductor as a metal, and germanium and silicon are semiconductors. Despite these differences, however, they have certain properties which are similar. Of the elements in this particular group of the periodic table, those that are most closely related are germanium and silicon. Germanium and silicon have strictly semiconducting properties. In addition to being relatively brittle, they have other properties that are closely related. Despite this, if very careful control is taken to orient single crystals of germanium and silicon and conduct friction experiments with a known orientation of a metal surface in contact with these two surfaces, differences in friction behavior do exist.

In figure 6-35 friction coefficients are presented as functions of load for a gold single crystal (111) surface in contact with a germanium (111) surface and a silicon (111) surface. The data indicate that friction is markedly less with load over the entire load range investigated for the gold in contact with silicon than it is for gold in contact with germanium.

There is a good reason for the marked differences in the friction behavior observed in figure 6-35 for gold contacting the two elements. Adhesion occurs for the gold in simple touch contact to both silicon and germanium. If a simple adhesion experiment is conducted where silicon is contacted by

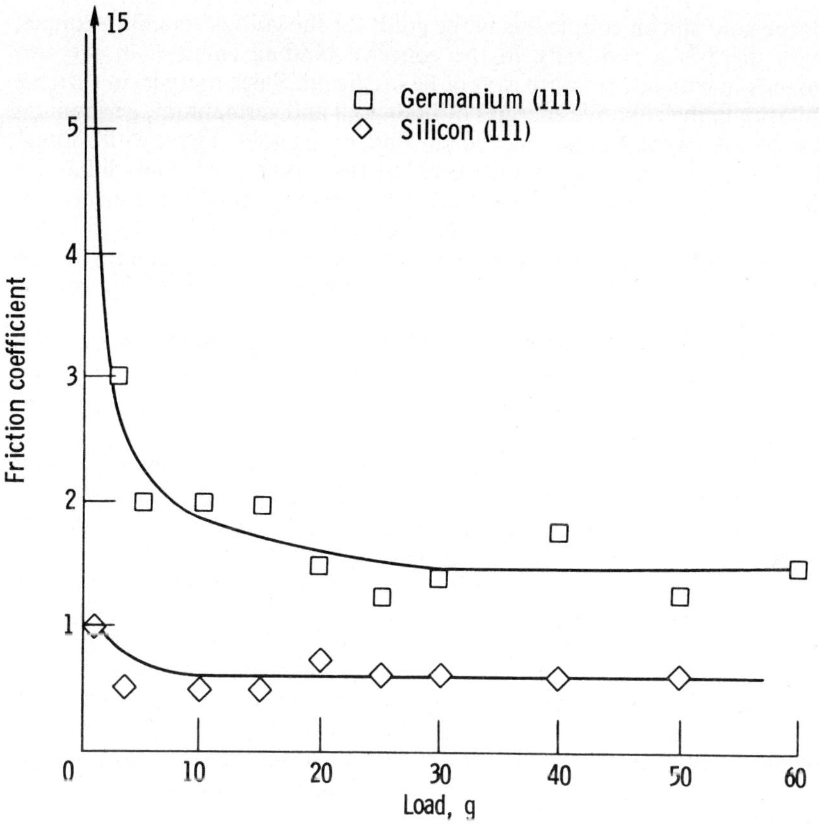

Figure 6-35. – Friction coefficient as function of load for single-crystal gold (111) sliding on single-crystal germanium and silicon (111) surfaces in vacuum (10^{-8} N/m²). Sliding velocity, 0.7 millimeter per minute; temperature, 23° C.

clean gold and the surfaces are separated, the silicon surface contains gold after the surfaces are pulled to fracture; that is, gold has transferred to the silicon surface, the adhesive bond at the interface being stronger than the cohesive bonds in the gold. Gold and silicon do not form any compounds, and such an interaction is not anticipated. Nonetheless, gold is observed to transfer to the silicon surface. Likewise, strong bonds of adhesion occur to the germanium surface when gold is brought into contact with a clean germanium surface; however, instead of gold transferring to the germanium surface as it did with silicon, germanium transfers to the gold surface and fracture is observed in the germanium.

An examination of the cohesive bond energies from an earlier portion of this text reveals that germanium and gold have nearly similar cohesive binding energies, while the cohesive binding energies of gold and silicon are markedly different. Silicon is much stronger cohesively than gold so, if the interfacial bond is very strong, fracture occurs in gold (the weakest region).

For the gold-silicon couple it is in the gold; for the gold-germanium couple, where there is a similarity in the cohesive binding energies in the two elements in contact, fracture cannot be predicted. Since there is an extreme similarity in the cohesive energies of both gold and germanium, germanium must be somewhat weaker since it transfers to the gold. Thus, with sliding, adhesion contributes to the measured friction force. For gold sliding on silicon, the shear strength of gold is reflected in the friction force measured, while for gold sliding on germanium, the shear strength (fracture strength, cleavage strength) of germanium is measured. This, then, accounts for the differences in the friction coefficients observed in figure 6-35 for gold in contact with germanium and silicon.

Similar differences have been observed in the friction behavior for other metals of the same family of the periodic table when in contact with a particular surface. Differences in friction coefficients are observed for the three noble metals copper, silver, and gold of the (111) orientation when sliding against the basal orientation of aluminum oxide. The highest friction forces are measured for the chemically most active (copper), and the least friction force is measured with the metal gold. With copper it is evident that adhesion of the copper to the oxygen in aluminum oxide occurs with the sliding process. It also occurs with the silver, but to a lesser degree. When the gold is in contact with aluminum oxide there is no evidence of transfer.

Adsorption studies have been conducted in the field ion and field emission microscopes to determine the binding of dissimilar materials in solid-state contact. Jones and Roberts (ref. 15) have shown, for example, that with the deposition of lead on the surface of tungsten there is relatively no lead present on the (110) surface (the low atomic density, low surface energy surface) of the tungsten. They conclude that the poor adhesion, or poor bonding, of lead to tungsten at this particular orientation accounts for the absence of lead on the (110) surface of tungsten. The field-emission microscope results indicate, in general, the poorest bonding or poorest adhesion to the high atomic density, low surface energy planes of various metals.

Grain Boundary Effects

For polycrystalline materials, the presence of grain boundaries in the material influences friction behavior. In the sliding process the grain boundary impedes this motion of dislocations and resists shear. Just as the bulk stress-strain properties are influenced or altered by the presence of grain boundaries (with a marked increase in the stress observed with strain for the polycrystalline material over the single crystal), similar behavior is seen in the near surface regions; that is, since the near surface dislocations in the sliding process are blocked in their movement by a grain boundary, they accumulate at the grain boundary and produce strain hardening in the surficial layers. This strain hardening makes it more difficult for sliding and increases the friction force for materials in sliding contact.

The grain boundary consists of a wall or layer of dislocations which accommodate the mismatch in the orientations of the grains adjacent to the

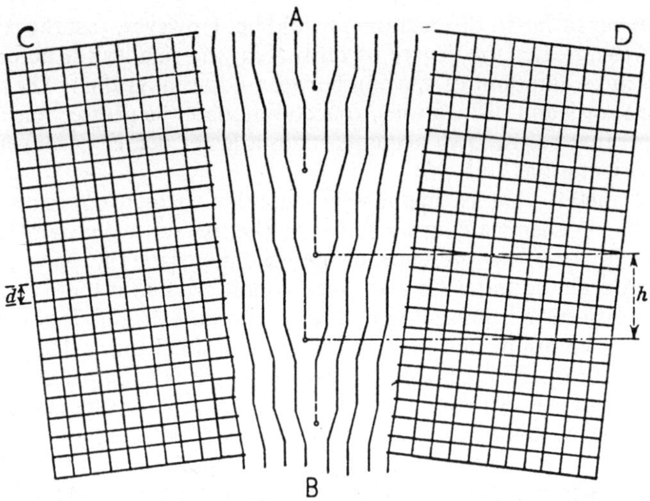

Figure 6-36. – Wall of dislocations causing sharp change of orientation.

boundary itself. McLean has aptly depicted this schematically (fig. 6-36, ref. 16). Figure 6-36 shows two grains, C and D, with their particular crystal lattice represented by the square patchwork, and a grain boundary A that is between the two crystallographic orientations of the individual grains C and D. The boundary at A tends to link, connect, or act as a bridge between the two adjacent grains. The dislocations fill in or act as a cement to make the transition from one orientation to another. The dislocations that develop near the interface are represented by the line AB through the crystal in a vertical direction. The dislocations that develop and accommodate the mismatch in the lattice are referred to as misfit dislocations. The lines to the left and right of the dislocations are crystal lattice lines; they accommodate the mismatch and the differences in orientation of the two adjacent grains. The mismatch in this region is not sufficient to produce the generation of individual dislocations, but it is sufficient to produce the strained lattice. These lines (three sets or three pairs of lines on either side of the center row of dislocations) are representations of the strain in the crystal lattice of grains C and D which helps to accommodate the mismatch between the dislocation network at AB and the normal crystal lattice for grain C and the normal crystal lattice for grain D. It is a further bridge linking the grains. Thus, there is a gradation in the mismatch in orientations. One goes from a relatively perfect lattice in either crystal C or crystal D to a region where the crystal lattice is extremely strained (that would be the region between the normal crystal lattice and the dislocations). Ultimately, the row of dislocations where the greatest amount of mismatch in the adjacent orientations occurs is accommodated.

One might normally think that the greater the degree of mismatch between adjacent grain lattices the wider the grain boundary (the amount and the number of rows of atoms necessary to take up the mismatch as well as the number of misfit dislocations) would be. However, just the reverse is observed; the greater the degree of mismatch, the smaller the actual grain boundary angle (amount of spacing between adjacent grains). The number of strained rows and dislocations to accommodate the greater mismatch is less than is that required to accommodate a lesser degree of mismatch in adjacent orientations (ref. 16).

Strained metal—that is, metal which contains a high concentration of dislocations—is chemically more active on the surface, as has already been discussed, because the presence of defects increases the energy in the material. A grain boundary is a strained condition in that there are a large number of dislocations present to help accommodate the misfit or mismatch in adjacent orientations plus there are rows of strained atoms that must help in accommodating the mismatch. Consequently, it is anticipated that these regions are high energy regions at the surface (i.e., the surface energy of the grain boundary would be higher than the energy in adjacent grains); this, in fact, is the case. The energy is greater at the boundary, and the boundary has its own characteristic energy which is separate and distinct from the energy of the grain on either side of the boundary.

Sliding friction experiments have been conducted across the surface of grain boundaries to measure the influence of the grain boundary on friction. Some experiments were conducted in this author's laboratory with single crystals and bicrystals of copper. The bicrystals were approximately the size of a dime and had a grain boundary running approximately through the center of the material. The one grain had a (111) orientation, and the other a (210) orientation. A polycrystalline copper slider slid across the surface, and the friction coefficient was measured during sliding in two directions—once across the grain boundary from the (210) grain to the (111) grain and then across the boundary from the (111) surface to the (210) surface.

The friction coefficients for these two conditions are presented in figure 6-37. The friction coefficient is plotted as a function of sliding direction (the direction being from the (210) to the (111) plane (fig. 6-37(a)) and from the (111) plane to the (210) plane (fig. 6-37(b)). The (210) surface is a lower atomic energy surface than the (111) surface. Figure 6-37(a) shows that the friction coefficient is much higher on the (210) grain than it is on the (111) grain. The grain boundary is depicted with the arrow in figure 6-37(a). There is a marked decrease in friction coefficient when moving from the (210) orientation to the (111) orientation. A peak appears in the grain boundary region in the friction coefficient. In sliding from the (111) surface to the (210) grain and moving across the grain boundary (fig. 6-37(b)), a distinct high friction region is associated with the grain boundary. On the (111) surface the friction is relatively low. Relatively high friction is observed when moving through the grain boundary region. Even though the friction drops off after passing over the grain boundary, it is still higher than that observed in the (111) surface. In general, the friction on the (210) surface is higher than that observed on the (111) surface; the grain boundary indicates its own particular characteristic friction. The results of figure 6-37 have been observed with other metals in sliding contact. The

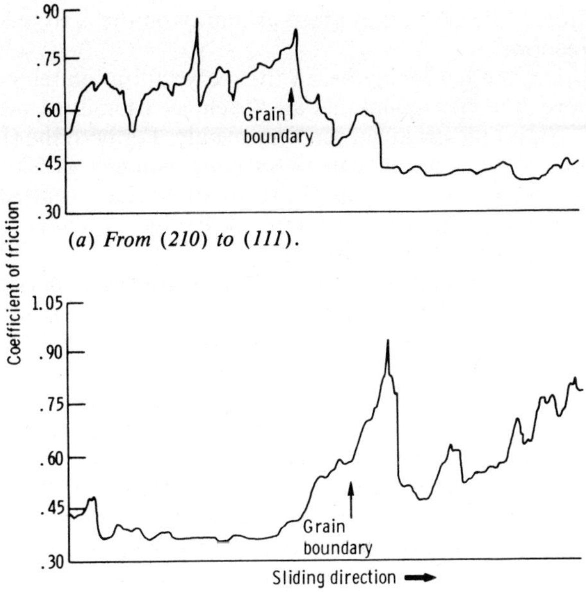

(b) From (111) to (210).

Figure 6-37.—Recorder tracings of friction force for copper slider sliding across grain boundary on copper bicrystal. Load, 100 grams; sliding speed, 1.4 millimeters per minute.

grain boundaries exhibit different friction properties than the bulk grains of the material do.

Crystal Transformations

Many metals in the periodic table exhibit more than one crystal structure; this is also true for nonmetals. For example, in quartz, the transformation from alpha to beta quartz is a change in the crystalline form. Some metals in the periodic table (such as magnesium, zirconium, titanium, cobalt, thallium, and tin) undergo crystal transformations. These materials are referred to as allotropic; they exist in more than one crystalline form. Tin, for example, transforms from a gray tin with a diamond-type crystal structure at 13° C to a white tin with a body-centered-tetragonal structure above that temperature. The gray tin has each tin atom tetrahedrally coordinated by four other atoms in the body-centered-tetragonal structure; the structure appears as a distorted diamond structure. Gray tin has a more symmetrical structure than white tin. Gray tin atoms can be pictured as stacking sheets composed of continuously linked, puckered hexagonal rings of tin atoms parallel to the (111) planes of the crystal. Shear takes place along these planes, typical of a diamond-type structure. With white tin, slip occurs on the (110) planes in the [001] direction at lower temperatures. At higher temperatures, slip also takes place primarily on the (110) planes but in the [111] direction.

With respect to strain hardening below the recrystallization temperature, tin behaves more like hexagonal metals (such as cadmium) and face-centered-cubic metals that strain harden very readily. For both the (100) and (110) orientations of tin, crystals can be strained as much as 500 percent with only about a factor of 2 increase in shear stress. Face-centered-cubic metals (such as copper) experience a factor of 500 increase in shear stress with as little as 50-percent increase in strain.

Friction coefficients are plotted as functions of temperature in figure 6-38 for tin and some tin alloys. The tin transformation indicated on the figure is approximately 17° C. Alloying alters the transformation temperature. The uppermost curve is for elemental tin. The coefficient of friction below the crystal transformation is relatively high, approximately 1.5. As the crystal transformation is approached, a marked decrease in friction coefficient is observed with crystal transformation from gray tin to white tin (from the cubic to the tetragonal form of the tin). With crystal transformation then, a change in friction properties is observed. The resulting transformation is reversible. One can run the experiment back in the opposite direction and retrace the curves; this indicates that the transformation is truly responsible for the changes in the friction properties observed.

Similar changes in friction behavior have been observed for other hexagonal metals in sliding contact. For example, as temperature is increased, cobalt undergoes a transformation at approximately 417° C from a close-packed-hexagonal to a face-centered-cubic structure. With the crystal transformation in cobalt, a marked increase in friction coefficient is

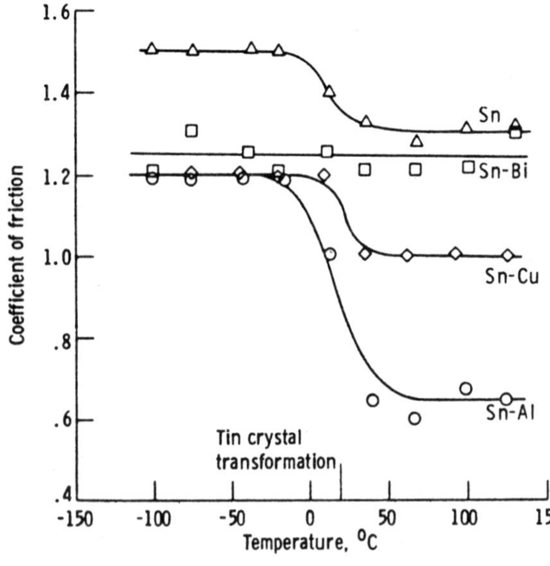

Figure 6-38. – Coefficient of friction for polycrystalline tin and tin alloys at various temperatures. Sliding velocity, 0.7 millimeter per minute; load, 10 grams; pressure, 1.33×10^{-8} newton per square meter (10^{-10} torr).

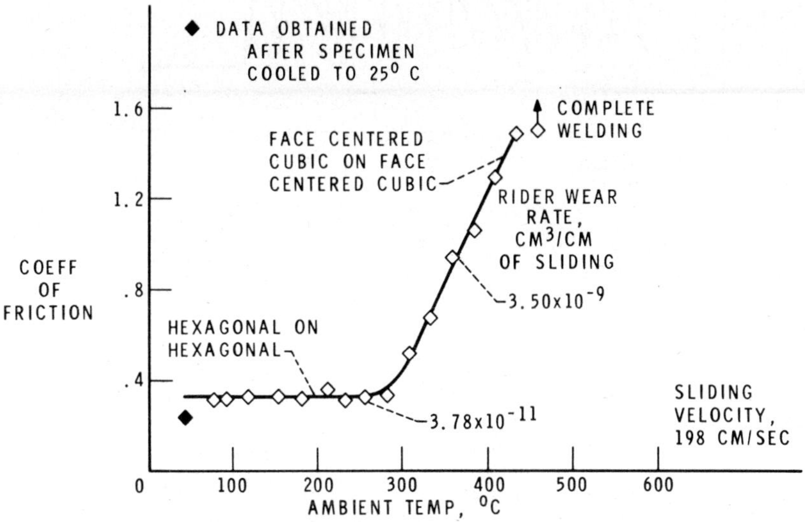

Figure 6-39. – Coefficient of friction at various ambient temperatures for cobalt sliding on cobalt in vacuum (10^{-9} to 10^{-7} torr). Sliding velocity, 197 centimeters per second; load, 1000 grams (9.8 N).

observed. Complete welding was observed with the face-centered-cubic form of cobalt in sliding contact with the face-centered-cubic form of cobalt (fig. 6-39). In sliding experiments below the crystal transformation temperature for cobalt, with the close-packed-hexagonal form sliding on itself, friction coefficients never exceeded 0.36. Thallium undergoes a crystal transformation from hexagonal to a cubic structure. It likewise undergoes a marked increase in friction behavior with a change in crystal structure.

Not only is there a change in the friction, but even the nature of the friction trace itself is influenced by the particular crystallographic form in which the metal exists. For example, gray tin is a relatively brittle material with very little ductility as opposed to the white tin with its tetragonal structure. Gray tin behaves similarly to the close-packed-hexagonal metal cobalt.

Friction traces obtained for the tin in the two crystallographic forms are presented in figure 6-40. Figure 6-40(a) is a friction trace obtained for tin in the white form (tetragonal) at 24° C. The trace shows a marked stick-slip behavior, which indicates adhesion and the breaking of adhesive junctions at the interface. In contrast, figure 6-40(b) shows for grey tin (where the crystalline form of the tin is diamond) at -46° C the friction trace is extremely smooth, even though the actual level is relatively high; this indicates an absence of the stick-slip behavior observed in figure 6-40(a) with white tin. Thus, not only the friction level but also the nature of the friction process is affected by the crystallographic structure of the material.

Other mechanical properties are also influenced by crystal transformations in allomorphic or polymorphic metals. A review of the

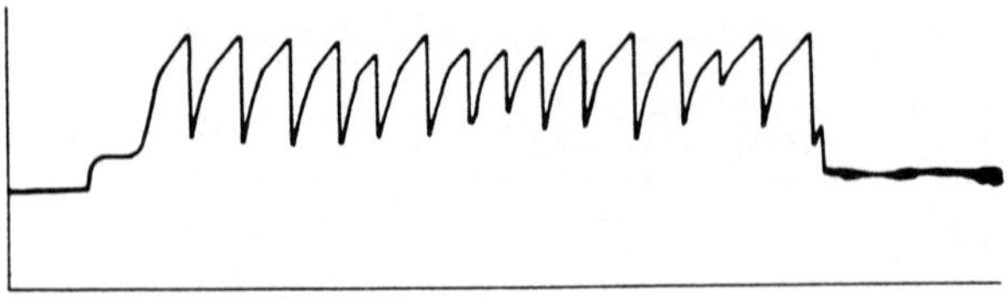

(*a*) *Temperature, 24° C; white tin.*

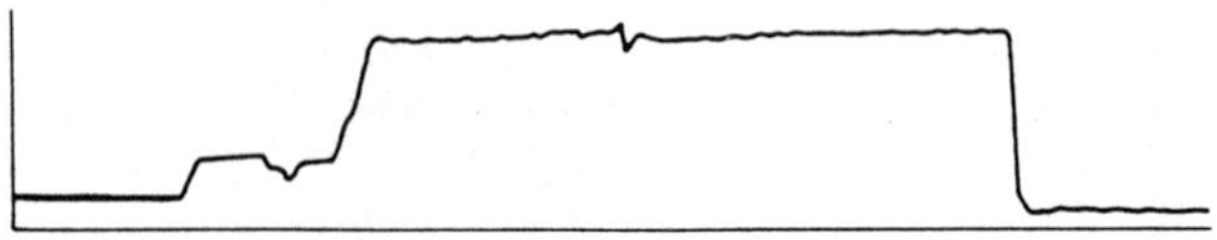

(*b*) *Temperature, -46° C; gray tin.*

Figure 6-40. – Friction traces for iron (110) sliding on tin (110) single-crystal surface at 24° and -46° C. Sliding velocity, 0.7 millimeter per minute; load, 10 grams; pressure, 10^{-8} newton per square meter (10^{-10} torr).

metallurgical literature indicates that there are a number of properties which are altered or influenced by crystallographic transformations.

The explanation for the difference in friction behavior with changes in crystal structure are associated with the different mechanisms of deformation that take place with the different crystallographic structures. For example, with cobalt (close packed hexagonal) below the crystal transformation temperature, the principal slip plane is the basal plane, the (0001) plane. Slip occurs in the ⟨1120⟩ crystallographic directions in that plane. Since there are three equivalent directions, there are essentially three slip systems that can operate with cobalt. As a consequence, with a limited amount of slip, the deformation is restricted in a metallike cobalt. If two pieces of cobalt are brought into solid-state contact under an applied load, the number of operable slip systems to accommodate the applied load with the deformation process is relatively limited. The true ultimate contact area is affected by the number of operable slip systems.

When tangential motion is initiated between two surfaces with a metal such as cobalt (in the hexagonal crystalline form), preferred orientation or texturing of the surface occurs, and the basal planes align nearly parallel to the surface when the material is polycrystalline; sliding, therefore, is essentially along the basal planes with shear occurring between basal planes. An examination of the metallurgical literature reveals that a very high degree of strain can be obtained in close-packed-hexagonal metals along the basal orientation before there is much increase in the shear stress. Thus, there is very little work hardening in the close-packed-hexagonal metals when shear takes place along or between basal planes. The real area of contact is minimized because of the limited number of slip systems when applying the initial load. When moving tangentially, the actual shear strength is kept to a minimum because of the limited number of operable slip systems and the preferred orientation or texturing of the basal planes. There is, consequently, low shear between adjacent planes. The friction force remains relatively low because both a and s (area in shear and shear strength) are relatively minimal.

In contrast, however, if the temperature increases above the crystal transformation temperature (above 417° C) for cobalt, one experiences a transformation from the close-packed-hexagonal to the face-centered-cubic structure. Associated with the face-centered-cubic structure are the (111) slip planes, which slip in the [110] direction. With this particular system there are 12 operable slip systems, as opposed to 3 in the close-packed-hexagonal system. Therefore, when a load is applied, there are many more crystallographic planes and directions available in the face-centered-cubic metal to accommodate deformation than there are in the close-packed-hexagonal metal. As a consequence, when the equivalent load is applied to the face-centered-cubic form of the metal cobalt, a greater real area of contact is realized at a fixed load because deformation can occur more readily.

Furthermore, when tangential motion is initiated, since there are a large number of operable slip systems available, and because of the crystal symmetry in the face-centered-cubic system, preferred orientation does occur, but not as readily as in the close-packed-hexagonal materials. And even when it does, there are intersecting slip planes which cause the development and the accumulation of dislocations at the intersection of these slip planes. What are referred to as Lomer-Cottrell dislocation locks form. They build up the stress and residual strain in the material, and these increase the shear strength near the interface. As a result, larger forces are required to shear the material. This is in contrast to the close-packed-hexagonal form of cobalt where shear can occur readily along basal planes. Consequently, both a and s are greater in the face-centered-cubic form of cobalt and friction forces are much higher with this crystalline form.

Elements can be added to metals that alter the crystal transformation. Some elements can arrest it, and some can promote its occurrence more rapidly. Adding, for example, bizmuth to tin arrests the crystal transformation, while adding copper and aluminum to tin has very little effect on the crystal transformation temperature.

Sliding friction experiments were conducted with alloys of tin and bizmuth, tin and copper, and tin and aluminum. The friction coefficients measured for these alloys are presented in figure 6-39 along with elemental tin, which has already been discussed.

An examination of figure 6-39 reveals that, while elemental tin undergoes a marked change in friction behavior with transformation from the gray to the white form of tin, the addition of bizmuth to the tin completely arrests the transformation so that no change in friction coefficient is observed. The tin continues to remain in one crystalline form. Consequently, there is a straight line relationship with friction over the entire temperature range examined in figure 6-39. The general level of the friction has been reduced for the alloys over that observed with the elemental metal.

The tin-copper alloy shows a reduction in friction coefficient analogous to that observed with elemental tin. In fact, the amount of reduction is almost identical to that observed with elemental tin; however, the curve for the tin-copper is displaced downward from that of elemental tin.

In contrast, when one adds aluminum to tin there is a marked decrease in the friction coefficient. While the aluminum does not arrest the transformation, it alters the nature of the crystalline material such that the amount of reduction that occurs is markedly greater (fig. 6-39) where the friction coefficient for the tin-aluminum alloy is decreased from 1.2 to about 0.6. Thus, the addition of alloying elements to the metal can alter its transformation behavior and, correspondingly, the friction coefficients observed for the metals in sliding contact.

Degree of Metallic Nature

With nearly all metals in the periodic table in the clean state a stick-slip type of behavior occurs; that is, adhesion occurs at the interface producing a stick-slip behavior if the surface is moved tangentially. This was observed in previous figures in this chapter, and it is also seen for other metals in the periodic table.

Group 4 of the periodic table was discussed earlier concerning the similarity in properties between the germanium and silicon semiconducting surfaces. Tin was also mentioned as a member of group 4. Tin is very interesting in that, if it is compared to germanium, many of its properties (for gray tin) are very comparable to the semiconductor germanium. Above the transformation, however, tin takes on the metallic properties of the metallic elements of the periodic table and lead, its sister element in group 4. Lead is beneath tin in the periodic table and is strictly metallic in nature; tin, however, because of its polymorphic nature, transforming from one form to another, has characteristics of both the semiconductor germanium and the very metallic element lead.

Above the transformation temperature, tin behaves very much like other metals in exhibiting stick-slip behavior. If the friction forces measured on a film of tin at 23° C (which is above the transformation temperature) are compared to friction for germanium, the friction traces are found to be entirely different (fig. 6-41). Figure 6-41(a) shows the friction forces plotted for germanium; a relatively smooth trace is observed. The peaks indicated on the right side of figure 6-41(a) are associated with the second startup of the germanium in a sliding friction experiment. It is a second static friction force measurement. The data of figure 6-41 indicate the static as well as dynamic friction for germanium. Under identical conditions, with a tin film, the friction coefficient is considerably greater and is characterized by stick-slip behavior (fig. 6-41(b)). The raggedy sawtooth friction for the tin film is characteristic of a metallic material. The load of 1 gram is extremely light.

Effective Shear Strength

The adhesion theory of friction states that friction is equal to the product of the real area of contact in shear and the shear strength at the interface.

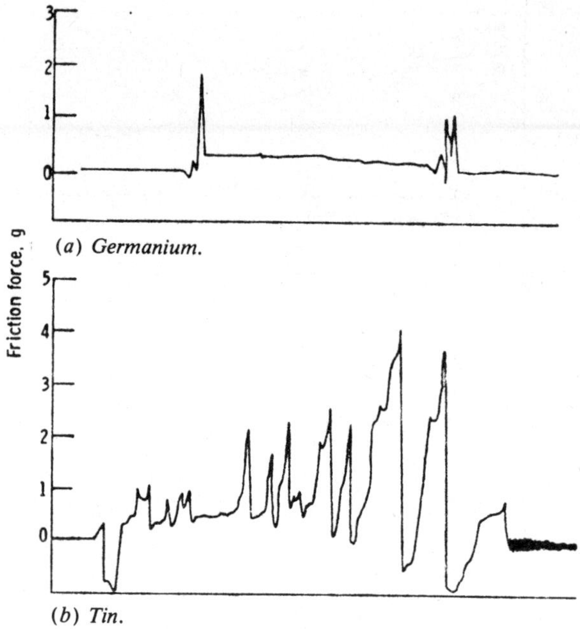

Figure 6-41. – Friction traces for 8×10^{-8} meter (800 Å) films of germanium and tin ion plated onto nickel (011) surface. Load, 1 gram; sliding velocity, 7.0 meters per minute; temperature, 23° C; pressure, 1.33×10^{-8} newton per square meter (10^{-10} torr).

This gives a measure of the friction force, and that divided by the applied load results in a coefficient of friction.

It might be reasonable to assume, therefore, that the shear strengths of metals in contact would have some influence on their friction coefficients. Some relationship should exist for shear strength with coefficient of friction, at least in a lubricated case. In the unlubricated case it is sometimes difficult to use the relationship of friction force as equal to the product of area in shear and shear strength because the area continues to increase as a result of junction growth with tangential motion, causing a rapid increase in friction to very high values and making friction coefficients somewhat meaningless in the sense that complete adhesion, or seizure, of surfaces can occur.

Miyoshi and Buckley have measured the coefficient of friction for various metals as a function of the shear strength of those metals. The surfaces here are lubricated with a mineral oil. Single crystals of silicon carbide grit were slid across the various metal surfaces. The friction coefficients were recorded and a measure was made of the groove depth or height formed on the surface as a result of the sliding process (ref. 17).

The results are presented in figure 6-42. In figure 6-42(a) the coefficient of friction for various metals is plotted as a function of the shear strength of the metals. Those metals with low shear strengths have the highest coefficients of friction, while those with high shear strengths have lower

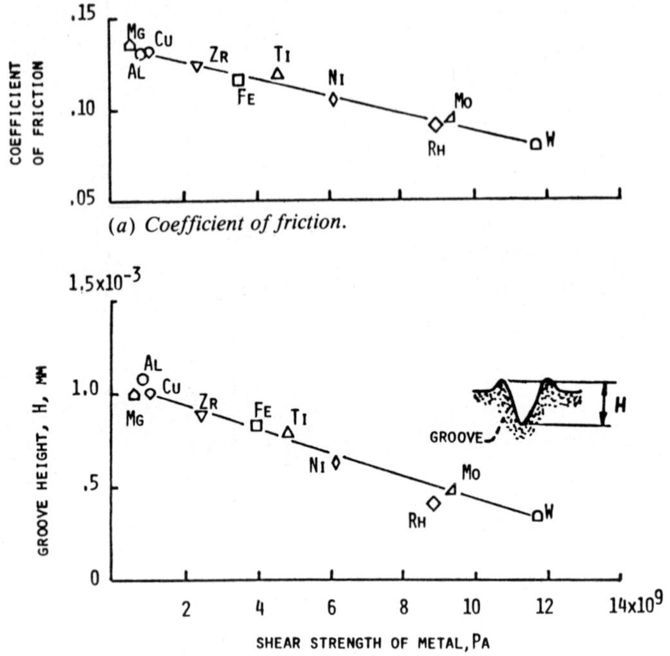

(*a*) *Coefficient of friction.*

(*b*) *Groove height.*

Figure 6-42. – Coefficient of friction and groove height as functions of shear strength for various metals (ref. 17).

coefficients of friction. Magnesium, aluminum, and copper have high coefficients of friction. With increasing shear strength there is a decrease in coefficient of friction; the friction coefficient is lowest with tungsten.

With the exceptions of magnesium and titanium, the metals in figure 6-42(a) are cubic in crystal structure. Magnesium is a hexagonal metal as is titanium, but these two metals exhibit multiple slip and, as a consequence, behave in many instances as if they were cubic rather than hexagonal.

In figure 6-42(b), the groove height (the furrow height), generated as a result of the silicon carbide single crystal sliding across the surface, is also plotted as a function of shear strength. With lubricated surfaces, the lower the shear strength, the deeper the groove, as might be anticipated. A greater degree of plowing occurs for low shear strength materials than for high shear strength materials such as tungsten. The interesting thing is that a direct relationship exists among the shear strength, groove height, and coefficient of friction. For these metals, the greater the shear strength, the lower the coefficient of friction and the less the depth of groove formed on the surface as a result of sliding contact. The shear is taking place by plowing the metal ahead of the silicon carbide slider. This shear process requires a certain quantum of energy, and that energy is reflected in the friction force or resistance to tangential motion of the silicon carbide as it moves across the metal surface.

With metals, such properties as crystal structure, orientation, and recrystallization influence the coefficient of friction at the surface for two solid surfaces in sliding contact. With nonmetals, there are many surface properties that also influence the friction behavior for materials in solid-state contact.

With polymers, for example, there are a number of surface properties which influence the friction coefficient for materials in sliding, rolling, or rubbing contact. One particular property of polymeric materials that has a marked influence on mechanical behavior at the surface of polymers is molecular weight. A particular polymer can exist in a wide range of molecular weights and, with a change in molecular weight, there is a marked change in mechanical behavior.

It may, therefore, be anticipated that the friction properties for sliding on polymers would also be influenced by molecular weight. Some experiments using various molecular weights of polyethylene oxide of high purity have been conducted and the friction coefficients have been measured over a range of molecular weights from 100 000 to 5 000 000. Sliding experiments were conducted for (1) polymer in sliding contact with itself, (2) polymer in sliding contact with a metal surface (iron), and (3) iron in sliding contact with the polymer. The reason for examining the various specimen configurations is that, in practice, polymers do not generally slide on themselves but are frequently in sliding contact with metals.

The results of the friction experiments are presented in figure 6-43 where the coefficient of friction is plotted as a function of molecular weight. The upper curve is for the polymer (polyethylene oxide) sliding on itself, and the lower curve is for the polymer sliding on metal.

There is an obvious difference in friction coefficient for the polymer sliding on itself from that of the polymer sliding in contact with the metal. There is basically no difference in friction behavior whether the polymer slides on iron or the iron slides on the polymer; essentially the same curve is obtained. With the polymer sliding in contact with itself, however, the friction force is much higher. The cohesion of the polymer produces stronger interfacial bonding forces than does the polymer bonded to the metal where shear occurs in the polymer.

A transfer film of polyethylene oxide was observed on the iron surface after sliding so that ultimately the sliding experiment is one of the polymer sliding on itself and the friction force reflects the forces required to shear the polymer (polyethylene oxide). The real area of contact for the polymer sliding on polymer is greater than it is for the polymer in contact with metal. This accounts for the higher friction coefficient measured for the polymer-polymer results.

With a load of 25 grams applied to the surface, the polymer is expected to deform. The area of contact would be much higher for the low molecular weight species of the polyethylene oxide because at the lower molecular weights the mechanical properties are much weaker for polyethylene oxide. Such properties as elastic limit, modulus of elasticity, and hardness are lower at the lower molecular weights. Therefore, for the same load, it might be anticipated that polyethylene oxide of 100 000 molecular weight would

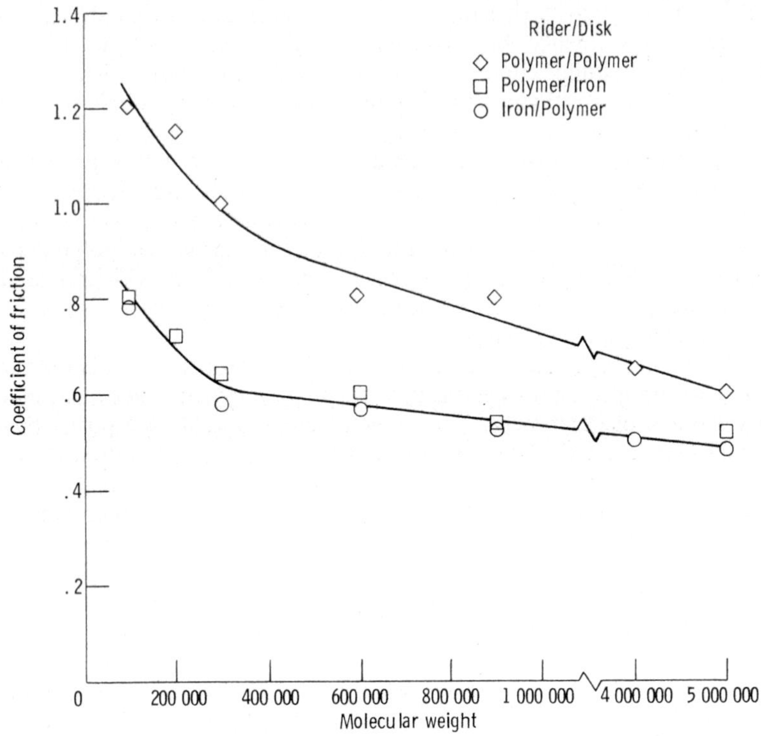

Figure 6-43. – Average coefficient of friction for polyethylene oxide polymer sliding on itself and iron as function of molecular weight. Sliding velocity, 0.1 centimeter per minute; load, 25 grams; temperature, 23° C; argon atmosphere.

have a much greater real area of contact than would one of 5 000 000 molecular weight. The shear strength, however, at 100 000 is much less than it is at 5 000 000. The important observation to be made from figure 6-43 is that the molecular weight of polymers has an effect on their friction characteristics. Just as there are metal properties at the surface which influence their function behavior, there are properties for polymers which influence friction characteristics.

Alloy Effects

In the discussion of recrystallization and crystal transformation, the influences of alloying elements on these properties and the relationship to friction behavior for metals in sliding contact have been covered. In general, however, alloying elements are observed to alter the friction behavior for most elemental metals. Another concept which has an influence on friction behavior is surface segregation. Certain alloying elements when dissolved in a solvent metal segregate to a solid surface on heating; that is, the concentration at the surface is greater for the solute than it is in the bulk of the solvent. The solute undergoes a migration to the surface and concentrates there; this phenomenon is frequently referred to as equilibrium segregation. The exact mechanism for segregation of alloying metals on the surface is not really known. Two hypotheses have been suggested. One is the lattice strain theory. If the solute atom dissolved in the solvent lattice is either larger or smaller than the parent solvent atom, it produces a lattice strain. Since that lattice strain brings about a higher energy situation in the bulk lattice, the system wishes to go to a lower energy state. In attempting to do so, the alloying element, the solute, is squeezed out of the solvent to the surface where it segregates. It is much like squeezing a grape out of its peel.

The second theory deals with the concept of surface energy. The clean metal surfaces are considered to be in a highly energetic state, and most solute elements present in the bulk when they segregate on the surface reduce the surface energy. By this process some researchers believe that the driving force to segregate the solute element to the surface of the solvent is the difference in energy (bulk lattice energy versus surface energy), and attempting to reduce the surface energy by segregating the alloy constituent to the surface reduces energy. The surface segregation of alloy constituents (equilibrium segregation) has a notable influence in a number of fields outside of tribology. Catalysis and corrosion, which are heavily concerned with the influence of alloy constituents, segregate on the surface of solids. Consequently, a considerable amount of research has been devoted to equilibrium segregation of various alloying elements so that the tribologist who is interested can find information on a particular system by reviewing the surface literature.

In the course of tribological experiments in our laboratory, a number of alloying elements were found which segregate to the surface of solvent metals and produce marked changes in adhesion and friction behavior. Aluminum segregates in the surface of copper as do indium and tin. Likewise, aluminium and silicon segregate on the surface of iron and alter its adhesion and friction behavior. When aluminum segregates on the surface of iron, increases in adhesion and friction are generally observed. With the copper alloys, the addition of aluminum increases adhesion because the aluminum segregates on the surface and produces an increase in the adhesive bonding forces to the solid surface; aluminum has a stronger adhesive bonding force than copper does to other solid surfaces. Indium and tin reduce the surface energy in the adhesive bonding to other solids; consequently, with the copper-indium and copper-tin alloys, reductions in both adhesion and friction are observed with these particular alloy systems.

A relatively unique system is the iron-silicon system. With silicon additions to iron, the silicon, on heating or straining, segregates on the surface of the iron just as does the aluminum. But unlike aluminum, which comes out on the surface and remains there with either strain or heating, the silicon appears on the surface and, if the surface is clean (not oxidized) and the specimen is cooled back toward room temperature, the silicon retracts into the bulk of the alloy; that is, the silicon segregates on the surface with heating or straining and then retracts (returns) to the bulk alloy on cooling or relief of the strain.

Despite this behavior, however, in the actual sliding friction process the silicon does exert an influence on the friction of iron. This is indicated in the data of figure 6-44 where the friction coefficient is plotted as a function of oxygen exposure for both iron and iron containing 3.5 percent silicon. In the clean state the iron underwent complete seizure. The silicon-iron alloy exhibited a relatively high friction coefficient as well. When oxygen was admitted into the system, the friction for iron began to decrease with increasing oxygen exposure. This result is probably explained by the formation of the iron oxides: first, the lower oxide of iron (FeO), then Fe_3O_4, and ultimately Fe_2O_3. With the 3.5-percent silicon-iron alloy, the friction coefficient also decreases with increasing exposure to oxygen. With silicon segregated on the iron surface, the uptake of oxygen is much greater than formation because silicon has a much stronger affinity for oxygen than iron does. Therefore, the friction decreased more rapidly for the silicon-iron alloy than it did for the pure iron. AES analysis of the iron and the silicon-iron alloy surfaces reveals a higher concentration of oxygen present on the surface of the silicon-iron alloy than on the iron at equivalent exposure levels. This indicates that the silicon promoted oxidation of the iron and thereby brought about a more marked reduction in the friction coefficient.

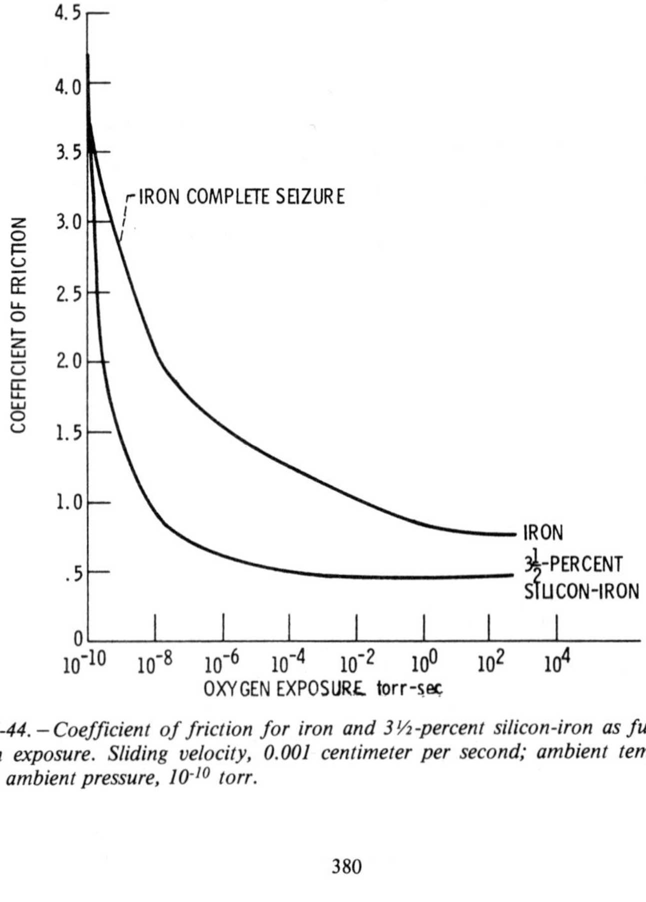

Figure 6-44.—Coefficient of friction for iron and 3½-percent silicon-iron as function of oxygen exposure. Sliding velocity, 0.001 centimeter per second; ambient temperature, 20° C; ambient pressure, 10^{-10} torr.

Other alloying elements can produce the same type of effect. For example, aluminum in iron increases adhesion and friction for iron in contact with other solids. If the surface is oxidized, such as was done in figure 6-44 for the iron-silicon alloy, the aluminum oxidizes very rapidly and produces an aluminum oxide layer with lower friction properties than those of the iron-aluminum alloy. This promotes a reduction in friction. The equilibrium segregation concept is extremely useful because it can be used in tribology; small concentrations of alloying elements can be added to produce marked changes in the friction properties of surfaces. For example, a small concentration of sulfur, phosphorus, or oxygen added to the bulk alloy can segregate to the surface and reduce adhesion and friction. Also, selective elements can be brought to the surface to interact with lubricating species that may be present in the environment. Thus, for example, where a particular lubricant of itself does not interact favorably with, say, an iron surface or an ordinary steel surface, the addition of a small concentration of alloying elements which segregate on the surface can help bring about an interaction if the alloying element or solute element interacts fairly strongly with the lubricant. Thus, lubricant interaction with a solid surface (enhanced lubrication) can be achieved by adding alloying elements to the base alloy that is to be lubricated. This is an area which has been barely examined by researchers in the field of tribology. The additions can also be used to reduce the corrosivity of lubricants with solid surfaces by adding alloying elements which, upon segregation, are less active at the surface than is the base material (solvent metal).

As was mentioned with the segregation concept, some alloying elements increase adhesion and friction and others reduce adhesion and friction. This is especially true where friction is influenced by the adhesion process when the surfaces are brought into contact. Not all alloying elements, however, bring about a reduction in the adhesion and friction of metals in sliding, rolling, or rubbing contact. Some alloying elements produce an increase in the adhesion and friction behavior of materials in contact despite the fact that the parent elements have lower friction and adhesion properties. This is demonstrated by the data of figure 6-45 where the average coefficient of friction is plotted as a function of chromium content for chromium-iron alloys. Friction data are also presented in figure 6-45 for pure iron and pure chromium. The friction coefficient for the pure elements (iron and chromium) is approximately 0.5. The addition of even 5-weight-percent chromium to iron, however, increases the friction force 100 percent; that is, there is an increase in friction coefficient from 0.5 to nearly 1 with the addition of chromium to iron. All the chromium alloys investigated in figure 6-45 have comparable friction coefficients of 1, or twice the value of either pure iron or pure chromium. In figure 6-45, silicon carbide was the slider specimen; a single crystal silicon carbide basal orientation was in contact with the iron-chromium alloys.

The surfaces were cleaned in a vacuum environment, and the friction coefficients reported in figure 6-45 reflect the influence of adhesion on friction force. Elemental metals (iron and chromium as well as the alloys) were found adhered to the surface of the silicon carbide after sliding; this

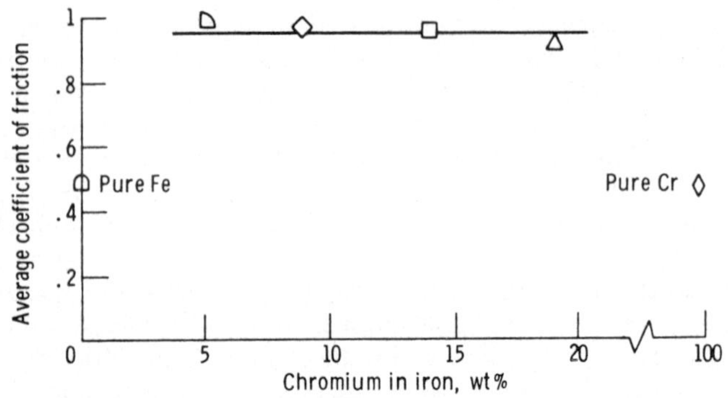

Figure 6-45. – Average coefficients of friction for iron-chromium alloys, iron, and chromium sliding on single-crystal silicon carbide (0001) surface in $[10\bar{1}0]$ direction. Single pass; sliding velocity, 3 millimeters per minute; load, 0.05 to 0.3 newton (5 to 30 g); room temperature; vacuum pressure, 10^{-8} pascal.

indicated that a strong influence of adhesion was present at the interface and that shear took place in the elemental metals and the alloys. The presence of the alloying elements, however, increases the shear strength of the elemental metals markedly (nearly twofold increase in friction). It is much easier to shear pure iron or pure chromium than it is to shear the iron-chromium alloys. Consequently, a much higher friction force is measured for the alloys than for the elemental metals in sliding contact with silicon carbide.

The surfaces are extremely important in the observed behavior because the adhesive bond at the interfaces is stronger than the cohesive bonds in the elemental metals. It was mentioned earlier, in reference to equilibrium surface segregation of alloy constituents, that when a difference in size exists between the solute and solvent atoms there is a tendency to squeeze the solute atoms out of the parent lattice of the solvent to reduce the energy. There are many situations, however, where the solute atom cannot be squeezed out of the parent lattice of the solvent so it remains locked (caged) in the lattice of the solvent metal. In such situations when there is a difference in atomic size, these differences in size can produce strain in the lattice just as they did in the equilibrium segregation situation. This straining of the bulk lattice of the material can then influence the friction behavior of the material because it influences the shear properties and the resistance to shear of the metal. The greater the amount of strain, the greater the resistance to shearing the metal.

Miyoshi and Buckley have studied the influence of alloying elements in iron on the friction behavior of elemental iron (ref. 18). Various elemental metals were added to iron to form simple binary systems where the atomic radius ratio of the alloying element to the iron was compared. The friction coefficients were then measured, and the friction properties of the binary alloys were related to differences in atomic radius ratio. Results obtained from some of the experiments are presented in figure 6-46. In this figure the coefficient of friction is plotted as a function of solute to iron atomic radius

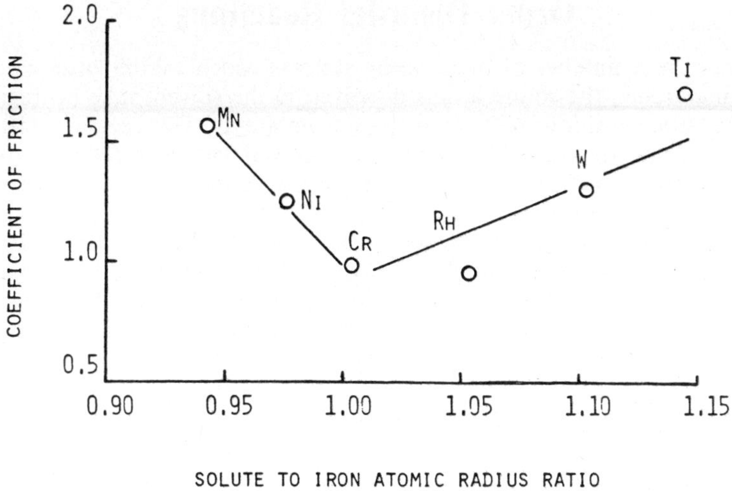

Figure 6-46. – Coefficients of friction for iron-base binary alloys as function of solute to iron atomic radius ratio (ref. 18).

ratio for the iron containing manganese, nickel, chromium, rhodium, tungsten, and titanium. The data in figure 6-46 show that a minimum coefficient is obtained when the atomic radius ratio is essentially 1; that is, the chromium element in figure 6-46 has an atomic radius the same as iron (atomic radius ratio of 1). Thus, the minimum amount of lattice strain would be produced by the presence of chromium in the iron lattice. If nickel is added to the iron lattice, the radius ratio decreases because the nickel is smaller than the iron. A strain results in the crystal lattice because of the smaller size of the nickel. Likewise, with manganese there is an even a greater disparity between the manganese and the iron and a greater strain exists in the crystal lattice; as a result, higher friction was observed than was observed with nickel.

This is where one deviates from an ideal radius ratio of 1. As the atomic size or atomic radius of the alloying element gets smaller, the friction coefficient increases because there is an increase in lattice strain. Conversely, if alloying elements are added which are larger than the parent iron solvent atom, a lattice strain is produced because of the difference or mismatch in atomic size of the solute atom and the solvent atom (iron). Data in figure 6-46 show the effect of adding atoms, such as rhodium, tungsten, and titanium, that are larger than iron to the parent lattice. As the atomic size of the solute atom increases, there is an observed increase in the friction coefficient because again there is an increase in the lattice strain as a result of the mismatch in atomic sizes of the solute atom with the solvent atom (iron). Thus, minimums in friction and strain are observed when the radius ratio is essentially 1; this is the case with chromium dissolved in iron (fig. 6-46).

Order-Disorder Reactions

There are a number of metal alloy systems which exhibit ordering. In ordering atoms, the solute atoms dissolved in the solvent take up regular atomic lattice positions relative to the solvent atoms. This regular ordering or positioning of atoms throughout the crystal lattice is referred to as ordering domains. It is analogous to a grain in a material. In the ordering domain, the structure is in the ordered state; that is, the solute A atoms are in regular specific positions relative to the solvent B atoms. Where A atoms are of comparable size to B atoms, the ordering is referred to as substitutional; where there is disparity in the sizes of the A and B atoms, the ordering is referred to as interstitial. The ordering in material systems is shown schematically in figure 6-47. Figure 6-47(a) detects the arrangement of atoms in the substitutionally ordered state, and figure 6-47(b) represents the atoms in the interstitially ordered state. There are a number of alloy systems which possess the order-disorder structures; iron-cobalt and gold-copper are two such systems. The mechanical properties of the alloys vary with ordering in an alloy.

There are transformation temperatures for order-disorder reactions in metal systems just as there are for crystal transformations. In addition,

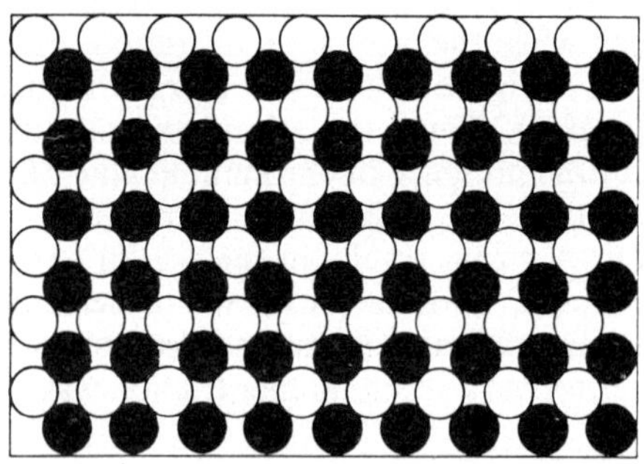

(*a*) *Ordered substitutional solid solution.*

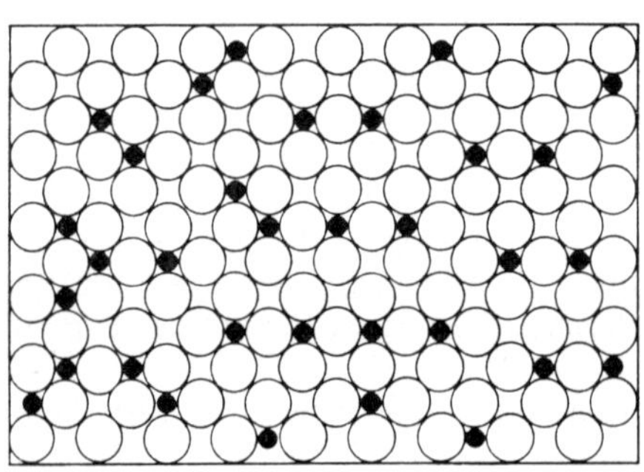

(*b*) *Interstitial solid solution.*

Figure 6-47. – Ordering in crystal lattice.

there are what are referred to as "long range" and "short range" ordering in alloy systems. The change in properties is not as marked with the short range ordering as it is with the long range ordering. Friction properties like other properties of alloy systems are influenced by order-disorder reactions.

Friction coefficients measured for particular systems (copper-gold alloy system and, more specifically, copper-gold alloy composition, Cu_3Au) are presented in figure 6-48. In figure 6-48, hardness, Young's modulus of elasticity, and the coefficient of friction for Cu_3Au are plotted as functions of temperature. The mechanical properties of this particular system are drastically affected by the order-disorder transformation (fig. 6-48).

X-ray analysis can be used to identify the ordered-disordered states for the material Cu_3Au. Selective heat treating can bring about a maximization of one or the other forms (either the ordered or disordered states for a particular material—e.g., Cu_3Au) or any other ordered structure. The data of figure 6-48 indicate that the hardness of the alloy Cu_3Au drops off with

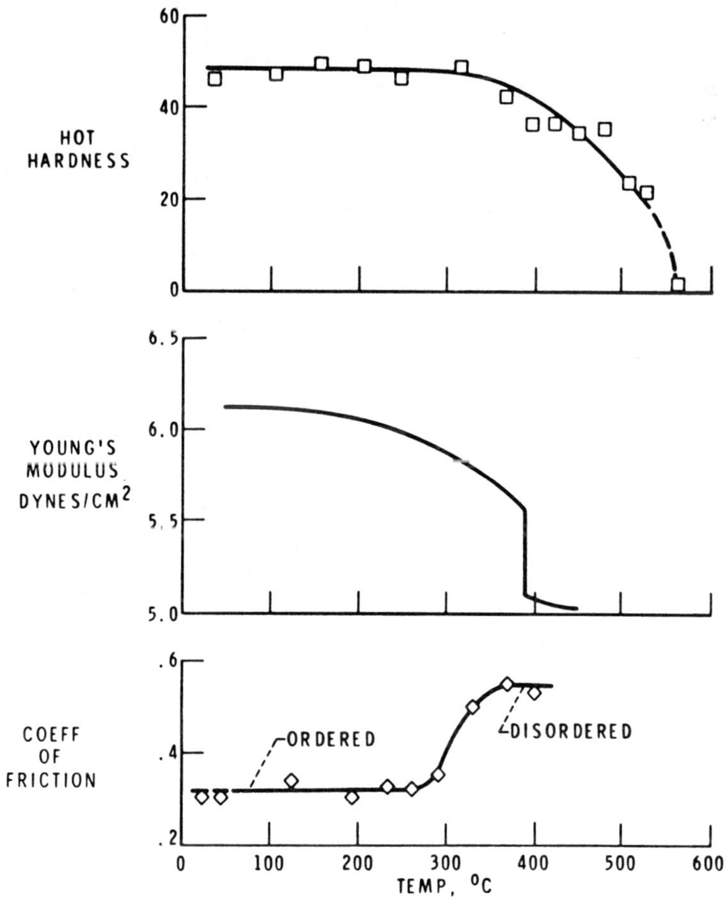

Figure 6-48. – Hardness, Young's modulus, and coefficient of friction for Cu_3Au sliding on 440C at various temperatures. Pressure, 10^{-9} torr; speed, 198 centimeters per second; load, 1000 grams.

temperature as does the modulus of elasticity. The friction coefficient, however, for the ordered state is relatively low and increases with the transformation from the ordered to disordered state. The area or region of the sharp decrease in modulus of elasticity (380° C) reflects the transformation from the ordered to the disordered state for Cu_3Au. Near that same region we see a marked increase in friction coefficient. In figure 6-48 there is a lag in the hardness; that is, the hardness remains high beyond the actual transformation temperature before the effects of the transformation are felt in the hardness measurements. In contrast, however, with friction coefficient measurements, the actual sensing of the transformation from ordered to disordered state occurs before the actual transformation temperature at 380° C, as indicated in the friction data. The friction begins to increase about 300° C, and this reflects the influence of the transformation from the ordered to the disordered state. The sliding velocity and the load applied to the surfaces put sufficient frictional energy into the interface so as to increase the surface temperature and bring about an earlier transformation than might normally be experienced. At very low speeds and very light loads, the curve for the coefficient of friction and the transformation from ordered to disordered state at 300° C could be shifted to the right; that is, it would require a higher temperature to bring about the change in friction coefficient where the sliding velocity and the loads are lower than those used in figure 6-48. The metallurgical literature reports many systems such as copper-gold. In the same binary system of copper and gold, another composition, namely, CuAu (50 at. % gold, 50 at. % copper) also exhibits order-disorder reactions just as does the Cu_3Au (75 at. % copper, 25 at. % gold). Thus, in a phase diagram for any one particular binary system, there may be more than one composition that is capable of being ordered.

Chemistry of Friction

In addition to the physical and metallurgical properties of surfaces influencing adhesion, friction, wear, and lubrication of solid surfaces, the chemistry of the solid surface is extremely important. The chemistry of both solid surfaces in contact, their reactivity, and their state are as important to the friction process as is the interaction of those surfaces with either the environment or the lubricating species. It must be remembered that, in addition to the normal surface chemistry that can exist between metals and either environmental constituents or the lubricant, there is the enhanced surface activity that is brought about by the sliding, rolling, or rubbing process itself; that is, the mechanical action or mechanical activity at solid surfaces tends to promote chemical reactions and produce surface chemistry that may be entirely different than that observed in static studies. For example, static corrosion of surfaces may be enhanced tremendously by the rubbing of two surfaces in solid-state contact. The straining of metal associated with the rubbing or rolling process also tends to promote surface chemical activity. Strained metal reacts much more rapidly with environmental constituents and lubricants than does nonstrained metal. Thus, the mechanical activity plus the enhanced surface temperatures on a solid surface tend to promote surface chemical reactions. These chemical reactions produce reaction products on the surface that can alter adhesion, friction, wear, and the ability of various species to provide lubricating films for solid surfaces.

A lubricant interacting with a metal oxide can produce entirely different reaction products than a lubricant interacting with a nascent metal. Thus, the condition of the surface from a chemical point of view is extremely important in surface interactions where lubricants are involved. One way of determining surface temperatures (since it is extremely difficult to measure the actual flash temperatures) is by chemical reaction. Recently, Hsu and Klaus at Penn State have used the reaction rates, as determined by the reaction products formed, to determine the actual junction or surface temperatures achieved in sliding systems (ref. 19). They calculated the surface temperatures and flash temperatures that would be generated using the Blok-Archard equation; they then actually measured from reaction rates the temperatures required to achieve such reactivity and found the temperatures in the actual rubbing experiments achieved by the sliding process.

The surface temperatures determined from the reactivity of the surfaces were much higher than those calculated by the Blok-Archard equation (fig. 6-49). In figure 6-49, junction temperature (asperity contact temperature) is plotted as a function of load. The surface temperature, as calculated from the Blok-Archard equation, increases with increasing load as might be anticipated because a greater amount of energy is being put into the surface with increasing load. The surface temperature achieved by using the reaction rates for the determination are also presented in figure 6-49. The temperatures, either calculated or determined from reaction rates, are markedly different. For example, for very light loads, a difference of 250° C exists between the temperature calculated from the Blok-Archard equation and that determined from reaction rates.

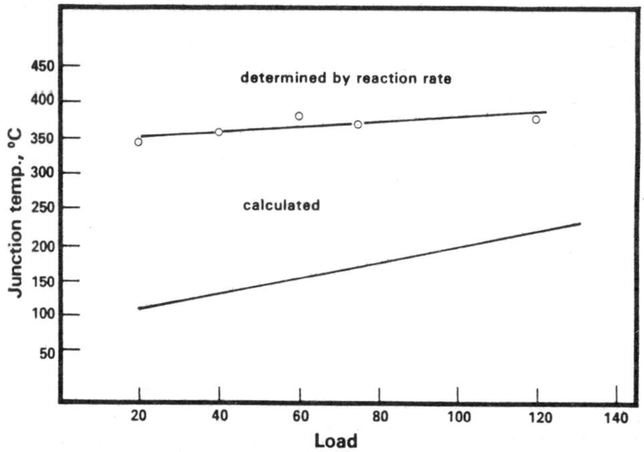

Figure 6-49. – Junction temperatures as function of load.

Since reaction rates represent the real solid surface and are a friction determinant, one is inclined to place more weight on the actual reaction rate determination of the junction temperatures. It would appear to be a much better indicator of what is actually being achieved at the surface, and it certainly is going to be the factor that influences friction behavior as opposed to using a calculated value that may or may not represent the real condition of the solid surface.

In addition to the physical and metallurgical properties of surfaces influencing adhesion, friction, wear, and lubrication of solid surfaces, the chemistry of the solid surface is extremely important. The chemistry of both solid surfaces in contact, their reactivity, and their state are as important to the friction process as is the interaction of those surfaces with either the environment or the lubricating species. It must be remembered that, in addition to the normal surface chemistry that can exist between metals and either environmental constituents or the lubricant, there is the enhanced surface activity that is brought about by the sliding, rolling, or rubbing process itself; that is, the mechanical action or mechanical activity at solid surfaces tends to promote chemical reactions and produce surface chemistry that may be entirely different than that observed in static studies. For example, static corrosion of surfaces may be enhanced tremendously by the rubbing of two surfaces in solid-state contact. The straining of metal associated with the rubbing or rolling process also tends to promote surface chemical activity. Strained metal reacts much more rapidly with environmental constituents and lubricants than does nonstrained metal. Thus, the mechanical activity plus the enhanced surface temperatures on a solid surface tend to promote surface chemical reactions. These chemical reactions produce reaction products on the surface that can alter adhesion, friction, wear, and the ability of various species to provide lubricating films for solid surfaces.

A lubricant interacting with a metal oxide can produce entirely different reaction products than a lubricant interacting with a nascent metal. Thus, the condition of the surface from a chemical point of view is extremely important in surface interactions where lubricants are involved. One way of determining surface temperatures (since it is extremely difficult to measure the actual flash temperatures) is by chemical reaction. Recently, Hsu and Klaus at Penn State have used the reaction rates, as determined by the reaction products formed, to determine the actual junction or surface temperatures achieved in sliding systems (ref. 19). They calculated the surface temperatures and flash temperatures that would be generated using the Blok-Archard equation, and then they actually measured from reaction rates the temperatures required to achieve such reactivity and found the temperatures in the actual rubbing experiments achieved by the sliding process.

The surface temperatures determined from the reactivity of the surfaces were much higher than those calculated by the Blok-Archard equation (fig. 6-49). In figure 6-49, junction temperature (asperity contact temperature) is plotted as a function of load. The surface temperature, as calculated from the Blok-Archard equation, increases with increasing load as might be anticipated because a greater amount of energy is being put into the surface with increasing load. The surface temperature achieved by using the reaction rates for the determination are also presented in figure 6-49. The temperatures, either calculated or determined from reaction rates, are markedly different. For example, for very light loads, a difference of 250° C exists between the temperature calculated from the Blok-Archard equation and that determined from reaction rates.

Since reaction rates represent the real solid surface and are a friction determinant, one is inclined to place more weight on the actual reaction rate determination of the junction temperatures. It would appear to be a much better indicator of what is actually being achieved at the surface, and it certainly is going to be the factor that influences friction behavior as opposed to using a calculated value that may or may not represent the real condition of the solid surface.

This author has taken some of Linus Pauling's valence bond data (ref. 20) for the transition metals to plot the coefficient of friction measured for various elemental metals in contact with themselves as a function of percent *d* valence bond character to see if any relationship exists between the basic chemical properties (namely, the *d* valance bond character of the transition metals) and the friction coefficients of metal surfaces. The results are presented in figure 6-50. The data in figure 6-50 were obtained at the extremely low sliding speed of 0.7 millimeter per minute with a load of 1

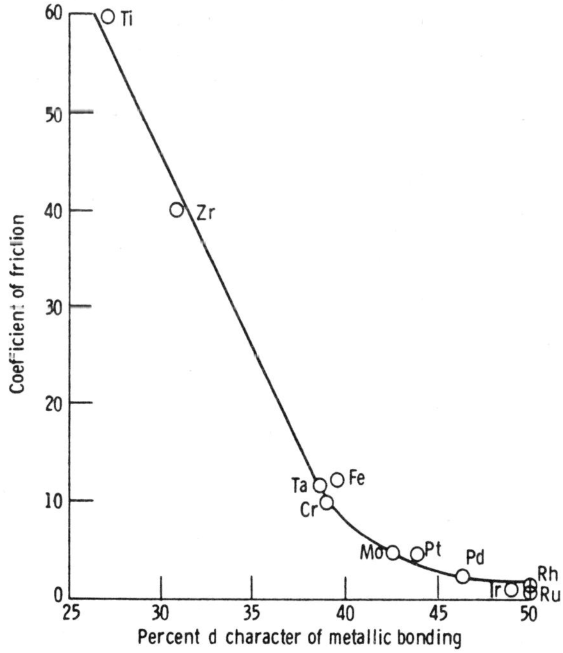

Figure 6-50. – Coefficient of friction as function of percent d bond character for various metals in contact with themselves in atomically clean state. Sliding velocity, 0.7 millimeter per minute; load, 1 gram; temperature, 23° C; pressure, 10^{-8} newton per square meter.

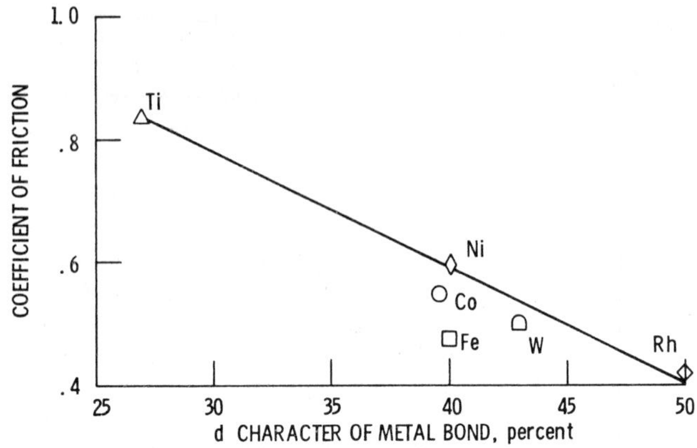

Figure 6-52. – Coefficient of friction as function of percent of d bond character of various metals in sliding contact with single-crystal manganese-zinc ferrite (110) surface in vacuum (10^{-8} N/m^{-2}). Single pass; sliding velocity, 3 millimeters per minute; load, 30 grams; temperature, 25° C.

surprising that titanium would exhibit very high coefficients of friction. In contrast, however, the interactions of rhodium with silicon and carbon are relatively weak.

The behavior for the transition metals in contact with silicon carbide is not unique to silicon carbide. An analogous type of behavior is seen for the transition metals in contact with other nonmetallic materials. For example, in figure 6-52, the friction coefficient is plotted as a function of percent *d* valence bond character for the transition metals in contact with magnesium zinc ferrite, a single crystal surface of the (110) crystallographic orientation (ref. 21). A comparison of the data in figure 6-52 with that in figure 6-51 indicates that the friction coefficient for rhodium is approximately the same (0.4) in sliding contact with silicon carbide or magnesium zinc ferrite. Thus, the relatively weak bonding of rhodium does not seem to be influenced much by the difference in the substrate material with which the rhodium is in contact. This may arise principally because of the relative weak bonding that does occur, and differences therein, therefore, would not make a significant difference in friction. In contrast, however, titanium, which exhibits high friction, has strong bonding to silicon carbide, and it exhibits even stronger bonding to magnesium zinc ferrite, as indicated by the friction coefficient of 0.8 as compared to 0.6 for silicon carbide. Despite this disparity, in both cases a direct relationship appears to exist between the friction coefficients measured for the transition metals and their d valence bond character in contact with the nonmetallic surfaces, magnesium zinc ferrite and silicon carbide.

With silicon carbide, transfer of metal to the silicon carbide surface was observed in sliding. Likewise, with the magnesium zinc ferrite, the adhesion at the interface was of such a nature as to result in the transfer of metals to the magnesium zinc ferrite. This occurred for both single and polycrystalline magnesium zinc ferrite. The transition metals depicted in figure 6-52 were all transferred to the magnesium zinc ferrite.

A question is suggested by the data of figures 6-50 to 6-52: Would the relationship between friction coefficient and *d* valence bond character still exist for a metal in contact with a nonmetal where the nonmetal is so weak that it transfers to the metal? To answer this question, some sliding friction experiments were conducted with a number of transition metals in sliding contact with pyrolytic boron nitride. Boron nitride has a hexagonal structure and should cleave fairly readily along basal planes in that structure. Friction coefficients for these experiments are presented as a function of percent *d* valence bond character in figure 6-53. In this figure the coefficient of friction is plotted as a function of percent *d* bond character for the metals sliding against pyrolitic boron nitride. The metals having the least percent d bond character had the highest friction coefficient, and those with the highest percent *d* bond character had the lowest coefficients of friction. (It is interesting to note that the friction level in figure 6-53 is higher than that observed in either figure 6-52 or 6-53 with the transition metals in sliding contact with silicon carbide or magnesium zinc ferrite.) There is a linear relationship just as there was in the other two figures with the metal titanium exhibiting the highest coefficient of friction and the metal rhodium exhibiting the lowest coefficient of friction. In contrast, however, to the data of figures 6-51 and 6-52 with the metals in sliding contact with pyrolytic boron nitride, instead of metal transferring to the surface of the nonmetallic material the nonmetallic material was observed to have transferred to the metal surface.

AES analysis was conducted on the metal surfaces, and boron nitride was found to be present on the surface as evidenced by the presence of both boron and nitrogen. An Auger spectrum obtained for the iron surface after sliding with pyrolytic boron nitride is presented in figure 6-54; the Auger emission spectra were obtained in the sliding wear track. In addition to the iron peaks, which are associated with the clean iron surface, peaks for boron and nitrogen are present in the spectra. No other peaks are detected. Thus, adhesion of the boron nitride occurs to the clean iron surface with the result that transfer of the boron nitride to the iron takes place. For all the

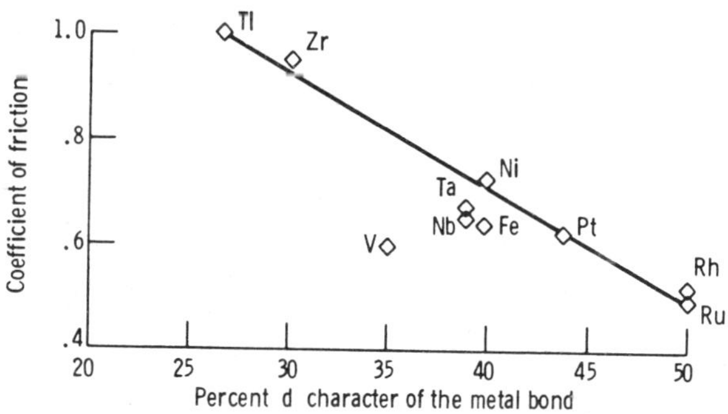

Figure 6-53. – Coefficient of friction as function of percent of metal d bond character for various metals in sliding contact with pyrolytic boron nitride. Sliding velocity, 0.7 millimeter per minute; load, 30 grams; temperature, 23° C; pressure, 10^{-8} newton per square meter.

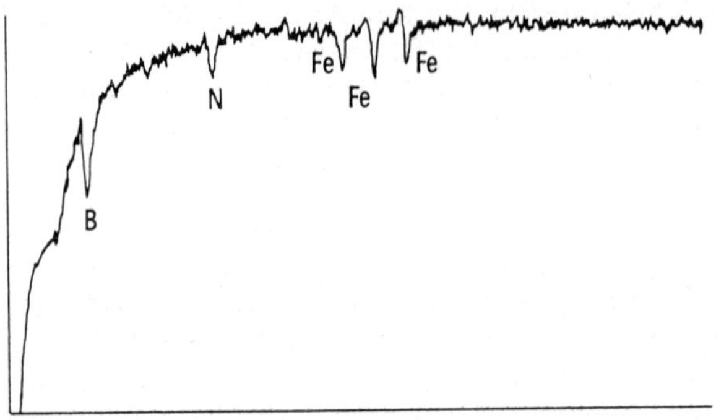

Figure 6-54. – Auger emission spectroscopy spectrum for iron surface after sliding pyrolytic boron nitride across that surface.

metals, shear took place in the boron nitride because boron nitride was detected on all of the metal surfaces. However, the chemical activity of the metal surface does still influence the friction coefficient observed because of the linear relationship of the friction coefficient to *d* valence bond character. Thus, there must be a greater degree of bonding at the interface with the lower percent *d* bond character metals than those that have a higher percent *d* bond character. This greater bonding would account for a greater area which must be sheared, resulting in the differences in friction, because the shear strength is basically that for boron nitride in experiments with each of the metals.

In figures 6-50 to 6-53, with the transition metals in sliding contact with nonmetals, where the metal transfers to the nonmetal or the nonmetal transfers to the metal, the *d* valence bond character has a direct influence on friction. The greater the percent *d* bond character, the lower the friction for the metals in either contact with themselves (fig. 6-50) or with nonmetals (figs. 6-52 and 6-53). With the metals in contact with magnesium zinc ferrite (fig. 6-52), the transfer of metal was observed to both the single and polycrystalline form of the ferrite material. With some systems, however, this is not the case. The friction behavior of the material is influenced by the crystal form in which the nonmetallic material exists. For example, with metals in sliding contact with aluminum oxide, when the metals are in contact with single crystal sapphire of the basal orientation, strong adhesion or adhesive bonds formed at the interface between the metals that are oxide formers and the oxygen of the aluminum oxide. With tangential motion, the interfacial bond (the metal to oxygen of the aluminum oxide bond) is much stronger than is the cohesive bond along the cleavage planes in the single crystal aluminum oxide. As a consequence, fracture occurs in the aluminum oxide along basal orientations, and this liberates free particles of aluminum oxide which have cleaved along the basal orientation.

The friction coefficients measured for a host of different materials in sliding contact with sapphire is approximately 0.2. The 0.2 friction coefficient represents the shear strength of aluminum oxide. It is essentially the same for all metals because the friction force is essentially a measure of the cleavage strength of aluminum oxide along the basal planes. These are exceptions, however, for the metals that do not form stable oxides—for example, gold and silver. When they are brought into contact with the single crystal aluminum oxide sapphire sliding on the basal orientation, the friction coefficients are approximately half those obtained with other metals—about 0.1 as opposed to 0.2. This difference in friction coefficient is due to the difference in bonding at the interface and where shear takes place. With the metals that are strong oxide formers, metal to oxygen bonding occurs at the interface and gives rise to a very strong interfacial bond and fracture in the aluminum oxide. With gold and silver in contact with single crystal aluminum oxide, however, there is very poor bonding at the interface because neither gold nor silver form very stable oxides. Consequently, instead of fracture taking place along cleavage planes in the aluminum oxide, shear actually takes place at the interface between the silver and gold and the surface of the aluminum oxide, the oxygen-rich surface of the aluminum oxide. Thus, this is one of the rare instances where the interfacial bonds are weaker than the bonds in one of the two materials on either side of the interface.

Essentially very little wear results. The mechanism for the change in character of the couple in contact is for the metals gold and silver to undergo plastic deformation with no evidence of transfer to the aluminum oxide surface. This behavior is different than that observed for both single crystal (fig. 6-52) and polycrystalline magnesium zinc ferrite, the friction coefficient is influenced by the *d* valence bond character of the metal. For the aluminum oxide case, the *d* valence bond character does not seem to exert much of an influence where the orientation is such that the weakest bonds near the interfacial region are the cohesive (cleavage) bonds in the cleavage plane of aluminum oxide. The magnesium zinc ferrite was not favorably oriented for cleavage.

When metals are brought into contact with polycrystalline aluminum oxide, however, grain boundaries are present with their variation in orientations. Thus, the behavior for the metals is analogous to that observed with metals in sliding contact with silicon carbide and magnesium zinc ferrite; namely, the metals are found to transfer to the surface of the aluminum oxide. The polycrystalline form of the aluminum oxide is much more resistant to fracture or shear in the interfacial region; consequently, the strong adhesion at the interface is analogous to that observed with the single crystals aluminum oxide. This results in shear occurring in the weakest region (the metal) producing transfer of metal to the aluminum oxide surface for all metals, except silver and gold, in contact with the polycrystalline aluminum oxide. Shear in the metal produces a marked increase in friction; the friction coefficients are four to five times greater when the metals are in sliding contact with the polycrystalline form than they are with the single crystal; this indicates that the shear strengths of the metals are greater than the fracture or cleavage strength along the basal orientation in the aluminum oxide. For some materials, whether or not transfer occurs is a function also of the development of a transfer film of material from one surface. This is a function of such things as orientation of the solid surface.

Where there are differences in orientation, there may be differences in observed friction behavior as a result of differences in adhesion characteristics at the interface. A classic example of this type of behavior is found with graphite. Single crystal graphite has a closed-packed-hexagonal lamellar structure with the basal orientation being the flat surface of the lamellae and the edges being the prismatic sites. The basal orientation is relatively inactive chemically. The edge sites are, however, the highly active region for pyrolytic graphite.

When graphite is in sliding contact with itself and with other materials, the influence of orientation (because of differences in chemical activity between the basal and the prismatic orientations) can be detected. It is relatively easy to remove the surface contaminants from the basal orientation of pyrolytic graphite. With argon ion bombardment, the contaminants present on the prismatic orientations can also be removed. The surfaces of pyrolytic graphite, both for basal and prismatic orientations, have been examined with AES analysis; the basal orientation has also been examined with LEED. In figure 6-55(a) is an Auger spectrum for the basal orientation of pyrolytic graphite showing the single (very high) Auger peak associated with carbon of the basal plane.

In figure 6-55(b) is an Auger spectrum for the prismatic orientation of pyrolytic graphite; again, the only peak seen in the spectrum is the carbon peak. In this case, however, the carbon peak is of much lower intensity than it was in figure 6-55(a). In examining the prismatic orientation (the edge sites) of the pryolytic graphite, it is much more difficult to pick up the carbon atoms. The concentration of carbon atoms within the beam spot size would be less than it would be for the basal orientation. Also for the basal orientation, there is maximum density of packing (or close packing) of atoms. Consequently, the carbon peak intensity is less on the prismatic orientation than it is on the basal orientation.

LEED analysis of the pyrolytic graphite surface reveals a ring structure (fig. 6-55(c)). If it were a perfectly ordered pyrolytic graphite surface of basal orientation, six diffraction spots would be seen in a normal hexagonal array. Repeated cleanings were performed on the pyrolytic graphite surface (including multiple heating and annealing) to bring about a perfect pattern for a high atomic density, close packing, which is characteristic of the basal orientation. The LEED pattern of figure 6-55(c) is characteristic of the carbon pattern seen on the surface of some metals (platinum and iron) where the surface was contaminated by carbon. Ring-type structures analogous to those in figure 6-55(c) have been observed on the surfaces of metals and are associated with carbon. Some researchers believe that the ring structure is a characteristic structure or pattern for graphite.

Once a clean pryolytic graphite surface is obtained in the vacuum environment, emission of contaminants in the system contaminate the prismatic orientation and are detected in the Auger spectra. For example, oxygen adsorbs on the prismatic sites and is detected by Auger analysis. On the basal orientation, however, even exposing the surface to oxygen at

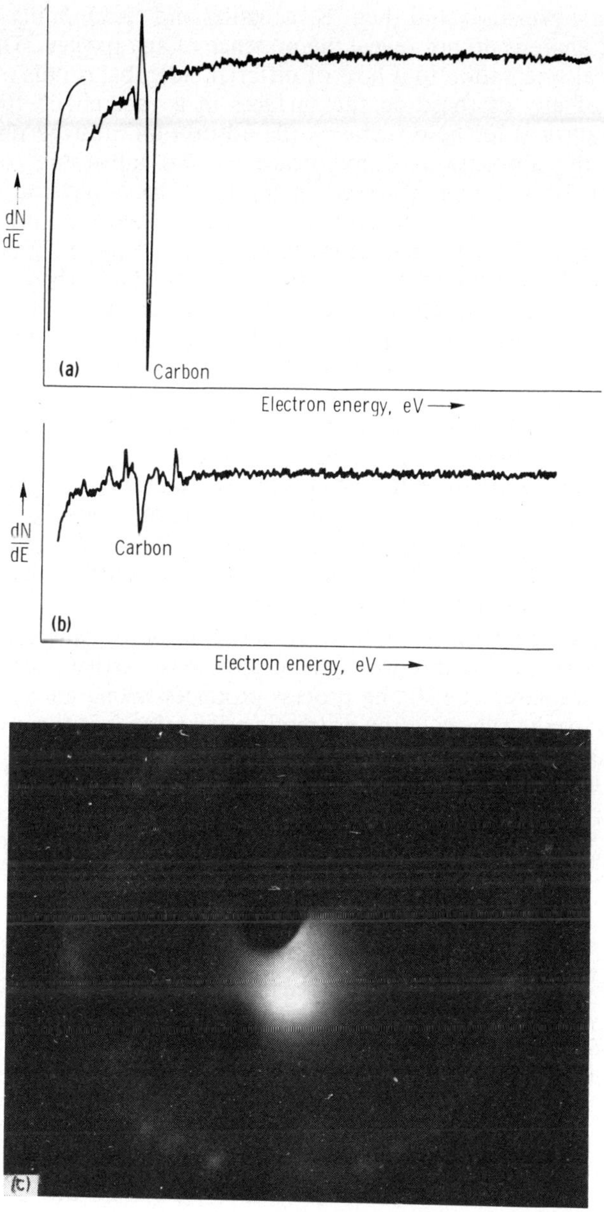

(*a*) *Auger spectrum for basal orientation.*
(*b*) *Auger spectrum for prismatic orientation.*
(*c*) *LEED pattern for basal orientation.*

Figure 6-55. – Surface analysis of pyrolytic graphite.

atmospheric pressures and then reevacuating and reexamining the surface with AES analysis do not reveal the presence of any oxygen. The exposure of the basal orientation to a host of different adsorbates fails to reveal the presence of any of these on the surfaces in a vacuum of 10^{-10} torr. If anything adsorbs to these surfaces, the adsorption must be physical. For example, the adsorption of hydrocarbons and lubricants to the basal orientation of graphite observed in air must be a physical adsorption phenomena since, in a vacuum environment, there are no absorbates detected on the basal orientation of pyrolytic graphite. Evacuating the system is sufficient to remove anything in the way of adsorbed material that may be physically present on the basal plane of the pyrolytic graphite. Adsorbates, however, are much more difficult to remove from the prismatic orientations, and techniques such as ion bombardment are necessary to achieve their removal.

Static friction experiments have been conducted with various metals in contact with the basal and prismatic orientations of pyrolytic graphite with both metal and graphite in the clean state. When clean metal surfaces are in sliding contact with pyrolytic graphite on the basal orientation, little or no transfer of metal is observed. For all practical purposes, it could be said that there is an absence of both adhesion and adhesive transfer of the metals to the basal orientation of pyrolytic graphite.

An entirely different result is encountered, however, when sliding occurs on the prismatic orientation (chemically most active surface) of the pyrolytic graphite. The sliding process produces strong adhesive bonds of the metals to the edge site (the prismatic orientations) of the graphite. This strong adhesion produces transfer of metal to the pyrolytic graphite surface in the process of sliding (i.e., shear takes place in the metal). An example of this adhesion process in shear is presented in the photomicrograph in figure 6-56. In this figure are a scanning electron micrograph (fig. 6-56(a)) and an energy dispersive X-ray analysis (fig. 6-56(b)). Figure 6-56(a) shows a particle of iron transferred to the prismatic orientation of the pyrolytic graphite. The transfer took place during the sliding of iron against the pyrolytic graphite prismatic orientation with shearing taking place in the metal. The metal particle that adhered to the surface appears as a flake, and the energy dispersive X-ray map for iron in figure 6-56(b) demonstrates that the transfer material is, in fact, iron.

In practical systems, the transfer of metal to the surfaces of nonmetallic materials is generally detrimental because, with repeated contact over the same surface, more and more metal transfers to the surface (such as iron to the pryolytic graphite prismatic orientation). Ultimately the metal is sliding on itself; this is a very destructive process to the system and to the surfaces in contact. It is an undesirable situation and one to be avoided from a practical point of view. This transfer to the prismatic orientation of pyrolytic graphite has been observed for metals other than iron. Cohesively stronger metals, such as tantalum, also have been observed to transfer to the prismatic orientation of the pyrolytic graphite.

Quantitatively, the difference in the friction behavior for metals in sliding contact with the two orientations of pyrolytic graphite are presented for two

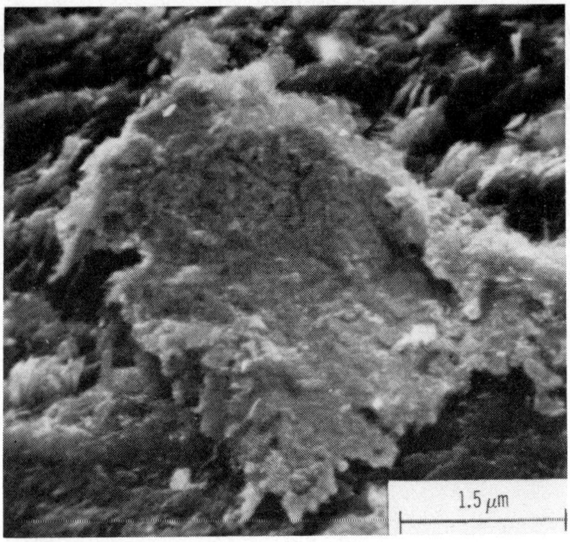

(*a*) *Scanning electron micrograph.*

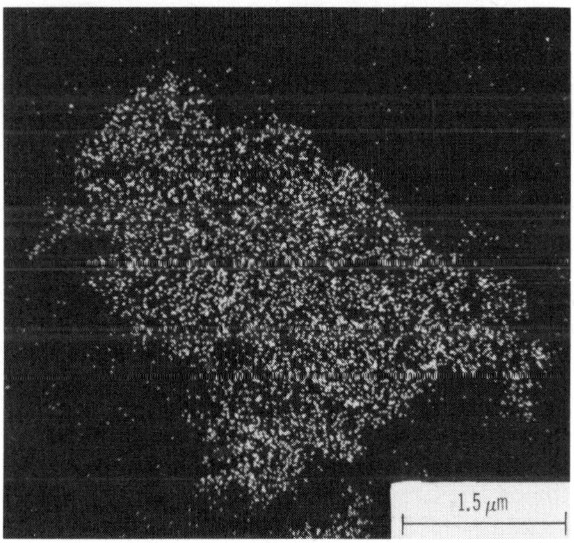

(*b*) *Energy dispersive X-ray analysis.*

Figure 6-56. – Scanning electron micrograph and energy dispersive X-ray analysis of iron particle transferred to prismatic orientation of pyrolytic graphite during sliding. Sliding velocity, 0.7 millimeter per minute; load, 10 grams; pressure, 10^{-10} torr; temperature, 23° C; 6000 counts EDAX.

temperatures in figures 6-57(a) and (b). Friction coefficient is plotted as a function of load for 23° C (fig. 6-57(a)) and 700° C (fig. 6-57(b)). At both temperatures, the prismatic orientation of the pyrolytic graphite exhibits a higher friction coefficient than is observed with the basal orientation. Bonding at the interface is relatively weak for the basal orientation of the pyrolytic graphite to the metal surfaces even though many of the metals (e.g., the transition series) form very stable carbides and one might anticipate strong metal to carbon bonding. With the prismatic orientation, however, the friction coefficient is higher at both temperatures because shear takes place in the metal. In general, the friction force level is higher at room temperature than it is at 700° C.

The friction behavior of solid surfaces in contact is influenced by various chemical properties of the solid surfaces. In addition to the *d* valence bond

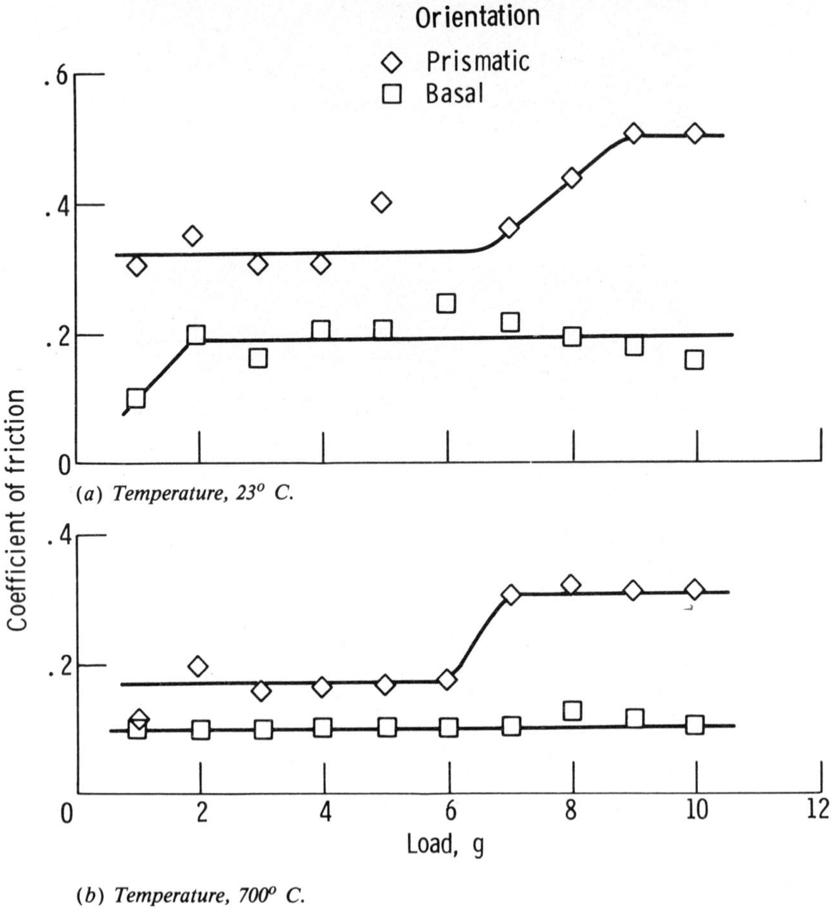

Figure 6-57. – Coefficient of friction as function of load for iron (110) single crystal sliding on basal or prismatic orientation of pyrolytic graphite. Sliding velocity, 0.7 millimeter per minute; pressure, 10^{-10} torr.

character just discussed, there are the interactions of the solid surfaces with the environment which can contribute to the chemistry and the friction behavior of solid surfaces. The interaction of environmental species with the solid surface introduces a change in the surface chemistry which can bring about or alter surface reactions. Such interactions of the environmental species with the solid surfaces alter the influence of various chemical properties of the surface on adhesion and friction behavior. Probably the most common environmental species found to interact with solid surfaces is oxygen. Oxygen is an integral part of the environment and, as a consequence, interacts with or is present on most solid surfaces. It is present as oxides on metals and as carbon monoxide and carbon dioxide in adsorbates on other surfaces; for ceramics, it may be present as water vapor condensed on the surfaces. Water is chemisorbed to the surface of glass (oxygen is present in the water). Carbon contains both adsorbed moisture and the gases carbon monoxide and carbon dioxide (which again contain oxygen). Even polymers are not immune to the influence of oxygen on their surfaces.

The presence of oxygen on all these surfaces has an effect on the chemical behavior of the solid surface whether the oxygen interacts with the solid surface by physical adsorption, chemisorption, or chemical reaction. It can be said with a fair degree of certainty that for all clean metal surfaces exposed to oxygen, with the exception of gold, oxides are formed or oxygen is chemibsorbed to the clean metal surface. Physical adsorption of oxygen to clean metals does not exist because of the strong interaction (activity) of oxygen with metals. Thus, exposure of clean metals to oxygen produces an immediate interaction of the oxygen with the metal surface. This interaction of oxygen with clean metals has an influence on friction, even for metals that have relatively slight activity with oxygen—for example, gold, which essentially does not interact with oxygen, and platinum, which interacts only weakly with oxygen. Despite these weak interactions with oxygen, the presence of oxygen on the surfaces of these metals still influences their friction behavior.

One can obtain a relatively smooth metallic surface by sputter depositing a film of metal on a quartz crystalline flat. This, then, presents a relatively smooth surface for friction studies. These films are useful where one is primarly concerned with the chemical nature of the metal and not so much its mechanical behavior or mechanical properties. The deposition of thin films on quartz substrates is ideal for this type of an experiment because the mechanical properties of the metal are not felt. At the interface, it is primarily the mechanical behavior of the quartz and the thin metal film (in its interaction with the environmental constituent). Surface profilometer traces across quartz crystals that have been coated with metallic films by sputter deposition show the surface to be extremely smooth. This topic was discussed earlier in chapter 2.

In general, for most active metals, the presence of oxygen on the surface, or the interaction of oxygen with the metallic film on a quartz substrate, reduces the friction force. This means that for metal to metal contact (or metal to metal film contact) the presence of an oxygen film on the metal surface reduces the adhesion and the friction.

Where the metals, however, are relatively inactive or react relatively weakly with oxygen, somewhat different results are observed. Some data obtained for a gold pin sliding across a platinum film (sputter deposited on a quartz substrate) are presented in figures 6-58(a) and (b). The friction force traces in figures 6-58(a) and (b) as functions of oxygen exposure and time were obtained in a vacuum environment for both clean and oxygen-exposed platinum films. The gold does not interact with the oxygen; only platinum interacts with the oxygen. When attempting to achieve tangential motion between the surfaces, there is a very high initial resistance to

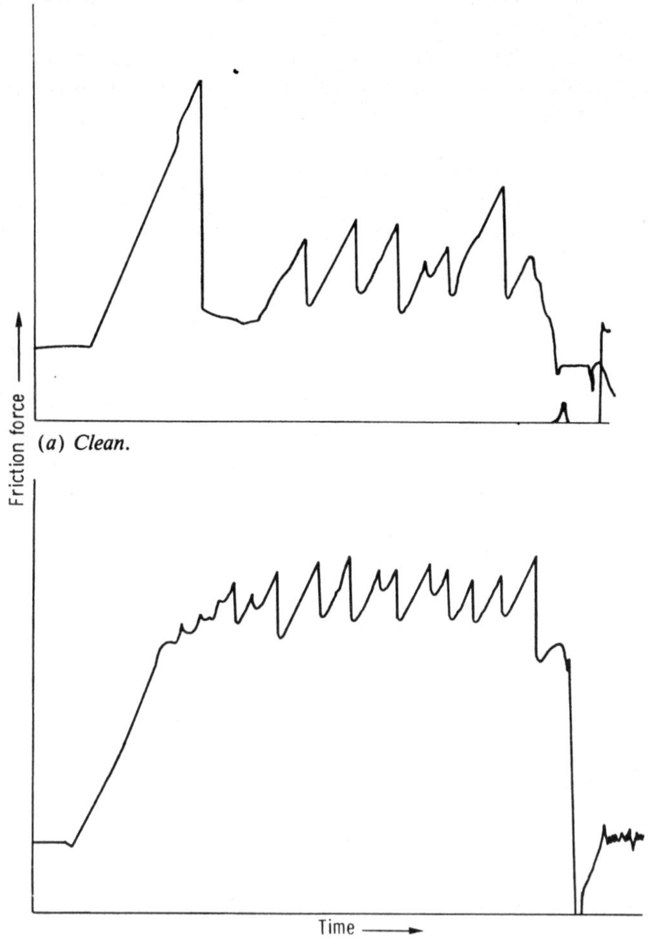

(b) *After oxygen exposure.*

Figure 6-58.—Friction traces for gold pin sliding on platinum film sputter deposited on quartz substrate. Sliding velocity, 0.7 millimeter per minute; temperature, 23° C; pressure, 10^{-8} newton per square meter; load, 30 grams.

tangential motion, and a stick-slip trace is obtained (fig. 6-58(a)); that is, strong adhesion of the gold to the platinum film has occurred, and the surface resists tangential motion. At some time, however, the adhesive bonding force at the interface is overcome by the tangential force, and slip occurs with a marked decrease in friction. With continued sliding, the stick-slip characteristic of the friction trace for clean metals in solid-state contact is obtained. With oxygen present in the film, however, the friction coefficient is as indicated in figure 6-58(b). There is no sharp static friction peak but rather a constant, high friction level that is somewhat different from that obtained for the clean surfaces in contact. There is a distinct difference, then, between a friction force trace obtained in figure 6-58(a) for the clean metals in contact and figure 6-58(b) for the metals exposed to oxygen. Both surfaces were examined with AES analysis to confirm the presence or absence of oxygen on these surfaces.

Another chemically active species that is frequently found in the environment of tribological systems is the element chlorine. Chlorine can be present in the environment as a gas or it can be present as part of an organic molecular structure in a liquid lubricant where extreme pressure or antiwear additives are used that contain chlorine in their structures. The interaction of chlorine with metal surfaces also has a relatively pronounced influence on the friction behavior of metals. It alters the friction characteristics rather markedly. Chlorine interacts with most metal surfaces of interest in practical tribological systems: copper, silver, iron, nickel, cobalt, and chromium. Just as with oxygen, the interaction of chlorine with the clean metal surface is not one of physical adsorption but rather one of chemisorption or chemical reaction. Its presence on metal surfaces is readily detected with surface analytical tools such as AES analysis. For example, in figure 6-59(a) is an Auger spectra for a copper disk specimen surface which has been exposed to chlorine (to saturation). In the Auger spectrum are the peaks associated with copper and the peak associated with chlorine. The peak associated with chlorine indicates a fairly high concentration of chlorine on the solid surface of the copper specimen.

The presence of chlorine on the copper affects the friction force or the force to initiate tangential motion between copper surfaces in sliding contact. This is indicated in the force measurements of figure 6-59(b) where the static friction coefficient is plotted as a function of the external motion actually achieved for the surfaces in contact. The data of figure 6-59(b) are essentially static friction measurements. For clean copper in contact with itself, which is indicated in figure 6-59(b) as the bare metal surface, the force continues to increase as a tangential force is applied to achieve relative motion; that is, the force associated with the resistance to tangential motion continues to increase first linearly and then with a decreasing slope. At no time is there a breakaway or separation of the surfaces. The material at the interface simply undergoes plastic flow. Adhesion or cold-welding has occurred at the interface, and when tangential motion is attempted, the surfaces simply flow plastically one relative to the other. This is the equivalent to a tangential tensile strength measurement or a shear strength measurement where the specimen has not taken to shear separation. There

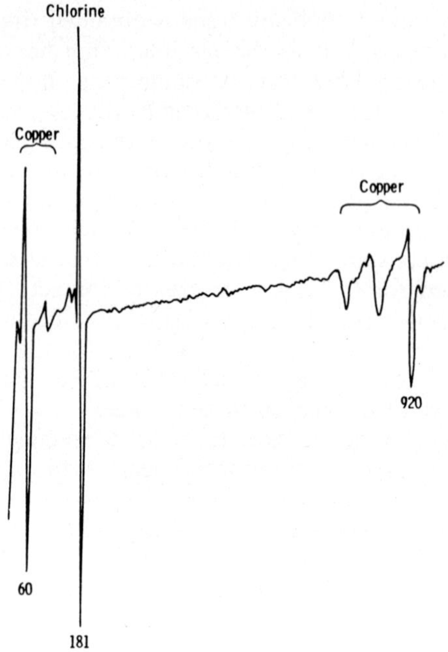

(*a*) *Auger spectrum of copper disk with saturation coverage of chlorine.*

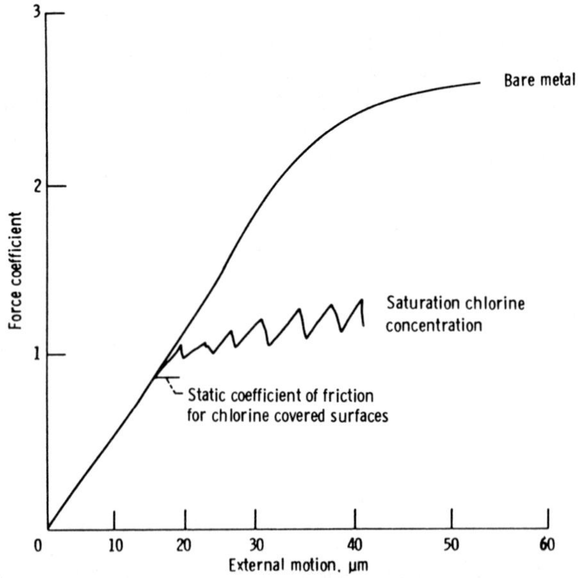

(*b*) *Force coefficient for copper on copper with and without absorbed chlorine as function of external motion of rider holder arm. Load, 100 grams.*

Figure 6-59. – Effect of chlorine on static friction of copper.

is first a relatively rapid increase in the force resistant to tangential motion, and, then, as plastic deformation (strain) begins to occur, the surfaces begin to move one relative to another but only plastically at the welded interface. Consequently, one observes the curve that is shown in figure 6-59(b) and indicated as the bare metal curve.

When the surfaces are coated with chlorine as shown by the Auger spectrum of figure 6-59(a), the presence of adsorbed chlorine on the surface alters that behavior markedly and the surfaces separate at the point indicated by an approximate force coefficient of 1. This, then, represents the true static coefficient of friction, and the sawtooth trace is a trace for dynamic friction. The one point where the saturated surface separates or digresses from the bare surface is that point in the curve where the static friction is measured. Beyond that point in the sawtooth trace one is experiencing dynamic friction force measurements, the static friction force being the force just necessary to initiate tangential motion.

It is of interest to note in figure 6-59(b) that even though the surface is coated with a considerable amount of chlorine, as indicated by the Auger spectra of figure 6-59(a), when the two solid surfaces are brought into contact deformation occurs at the interface and the asperities cannot completely penetrate the adsorbed chlorine layer to allow clean copper to contact clean copper. The interface, then, is represented really as a mixed interface wherein a portion of the actual interfacial real area of contact consists of copper bonding to copper and a portion wherein chlorine is sandwiched between the copper to copper surfaces. With chlorine sandwiched between the copper to copper surfaces the adhesive bonding is relatively weak; where the bonding is copper to copper, however, the bonding is strong. Since we have the metal to metal bonding over a portion of the real area of contact, the curve in figure 6-59(b) for the coated surface follows essentially the same trace as that for the bare metal. One must overcome the resistance to shearing the adhered junctions in the metal to metal interface. At some point, however, the force applied is sufficient to fracture those junctions and cause fracture. This then allows relative motion of the solid surfaces over one another and accounts for the dynamic friction tracing in figure 6-59(b). The stick-slip nature is still seen in the friction trace with the surface saturated with chlorine because, with sliding, some metal to metal contact can still occur through the adsorbed chlorine film. Consequently, one observes the stick-slip nature of the friction trace with tangential motion.

Even the presence of hydrogen on a metal surface can influence the friction behavior. Studies with tungsten sliding contacts have indicated that the presence of adsorbed hydrogen on a tungsten surface alters friction behavior. Thus, the atomic species that adsorbs on the surface or reacts with a solid surface is not critical. Almost any adsorbate present on the solid surface, particularly metal surfaces, influences the friction behavior, usually by reducing friction. There are, however, some instances where adsorbates on a solid surface actually increase friction. A classic example of this is water vapor on glass; here, an increase in the friction force is actually produced. For clean glass in a vacuum environment, the friction coefficient is approximately half of what it is in a moist air environment. So, in some instances, adsorbates produce an increase in friction forces at the interface. As a general rule, however, they bring about a reduction in friction forces. Likewise, as a general rule, for metals in sliding contact with metals, oxygen in the form of oxides on metal surfaces reduces the adhesion and friction at the surface. In some instances for metals in contact with ceramics where the metals are oxidized, however, the oxygen interactions of the metal oxide and the ceramic can produce stronger bonding than the clean metal to the ceramic. In such instances, higher friction forces are measured for the oxidized surfaces than are measured for the clean metal surface in contact with the ceramic (which may be an oxide such as silicon dioxide, aluminum oxide, or titanium dioxide (ref. 22)).

When two dissimilar materials or metals are in contact with one another at an interface and an environmental interaction takes place with one or both surfaces, the chemistry becomes much more involved. In such situations, one has the chemistry associated with the interactions of the dissimilar materials based on their chemical affinities for one another. There are the basic properties of the chemical nature of the solid surfaces, such as the concept of *d* valence bond character discussed earlier. There is also the interaction chemistry of the environmental constituent with the two solid surfaces.

It would appear that the best way to study such chemistry (because of its involved nature) involves (1) studying the basic interactions of elemental materials with themselves in the clean state, (2) studying the influence of an adsorbed environmental constituent on one of the two surfaces in contact (the second being clean), and (3) examining both surfaces after they interact with the environmental constituents.

The effectiveness of a lubricant is measured by its ability to prevent wear and to reduce friction. The only gage for the effectiveness of a lubricant is its ability to provide a protective film or coating on the surface to minimize friction and wear. The more effective it is in minimizing wear and reducing friction, the better the lubricant is. Even with effective surface lubrication, however, metal can transfer from one surface to another. This is seen frequently in bearings where dissimilar materials are in contact. A classic example of this is the copper-lead bearing alloy in contact with steel shafts where the copper is seen to transfer to the steel surface. This transfer is accomplished by an adhesion mechanism (adhesive bonding of the copper to the steel surface) even though the steel and the copper bearing are relatively well lubricated with a liquid lubricant on the surface.

This transfer effect can be seen on a much less complex scale by examining two elemental metals in solid-state contact and allowing a film of some material with good lubricating properties to develop on the surfaces. Experiments have been conducted over the years with various metals brought into contact with elemental iron in the presence of various halogenated hydrocarbons to determine the effectiveness of the halogenated hydrocarbons in preventing transfer of metal from one surface to another.

An Auger spectrum for an iron surface that has been exposed to methyl chloride is presented in figure 6-60(a); three high energy iron peaks are

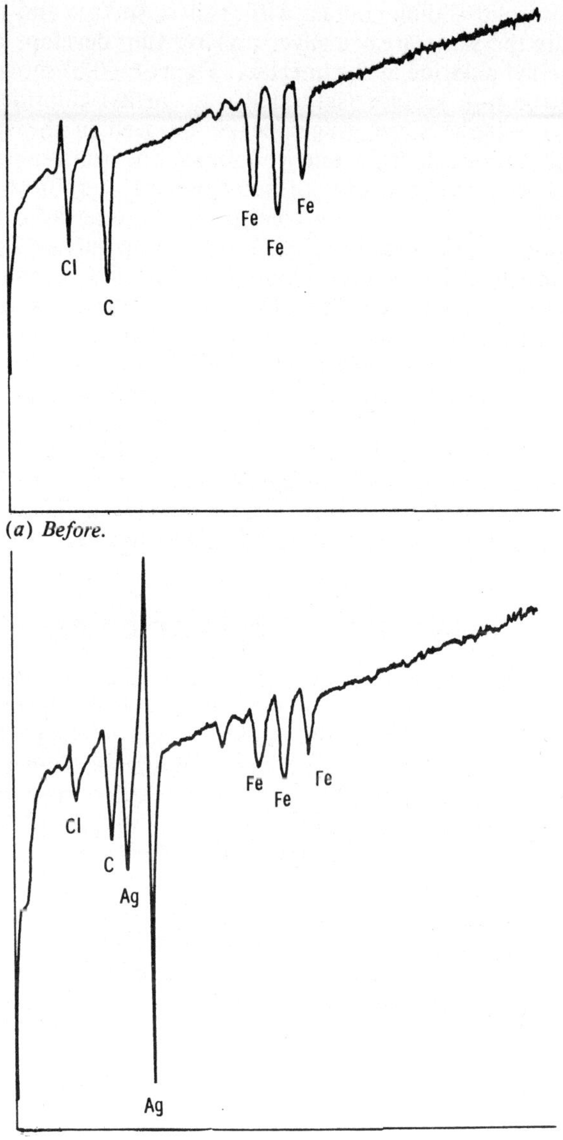

Figure 6-60. – Auger emission spectra for iron surface saturated with methyl chloride before and after 250 passes of silver rider over surface.

detected. The low energy iron peak is absent because the surface has become coated with the methyl chloride. Evidence of both the methyl group and chlorine on the surface is given by both chlorine and carbon in the Auger spectrum. Thus, it is reasonable to assume that the methyl chloride adsorbs on the solid surface as methyl chloride. If the iron surface of figure 6-60(a) is brought into solid sliding contact with a silver surface and if the silver rubs repeatedly the same area, a silver transfer film develops on the iron despite the methyl chloride at the interface. Figure 6-60(b) shows an Auger spectrum for the iron surface (saturated with methyl chloride) after 250 passes of silver. Large silver Auger peaks are seen in the spectrum in addition to the carbon, chlorine, and iron peaks. The iron peaks, of course, are associated with the iron disk or substrate; the carbon and chlorine reflect the presence of the methyl chloride at the interface. The methyl chloride has not been displaced or desorbed from the surface as a result of the rubbing process. The methyl chloride is sufficiently bonded to the surface so as to be maintained there. Despite its presence, however, silver apparently is able to contact iron and transfer to the iron forming a continuous transfer film of silver with the methyl chloride sandwiched between the silver and the iron substrate. Silver alters the friction behavior of the silver to iron contact. Ultimately, after 250 passes, silver is essentially sliding on a thin film of silver and the friction coefficients measured are relatively low and characteristic of silver sliding on silver. They do not reflect, however, the friction measurements made in the first passes where silver is sliding on the iron covered with methyl chloride.

Composition of Surface Films

The films present on solid surfaces vary widely in their chemistry. Even when the film is of a particular type (e.g., an oxide or chloride) different compounds can be formed (e.g., by the interaction of oxygen or chlorine with the surface, where the surface is a metal). With metals such as iron, the interaction with oxygen can yield three different oxides: the lower oxide of iron (FeO), an intermediate oxide of iron (Fe_3O_4), and the higher oxide of iron (Fe_2O_3). Any one or all of these oxides can be present on an iron surface. They can form as a result of the amount of oxygen available for surface film formation. With a sufficiently large concentration of oxygen (such as at atmospheric pressure and moist air), the higher oxides of iron are frequently being formed, namely, Fe_2O_3. This is the jeweler's rouge or rust oxide of iron, and it is abrasive. In contrast to the reddish Fe_3O_3, Fe_3O_4 and FeO are black, much softer, and not abrasive. Electron diffraction studies of films formed on surfaces indicate that these oxides vary in their friction properties, as do other metal oxides where more than one oxide form exists. An example of differences in friction behavior with different forms of a compound formed on the solid surface is indicated in the data of figure 6-61 for the oxides of Fe_2O_3 and Fe_3O_4. The Fe_2O_3 friction curve indicates a higher friction coefficient over the entire range of sliding velocities (ref. 23). Another example of metals where the friction properties of the oxides vary is copper. Copper has two forms of oxides that can form on its surface: Cu_2O is the lower oxide of copper and is rose, and CuO is higher oxide of copper and is black. Generally Cu_2O is found beneath CuO on the surfaces of copper where the surface has been heavily oxidized. Cobalt has two oxide forms. Friction experiments at various temperatures with these two oxides

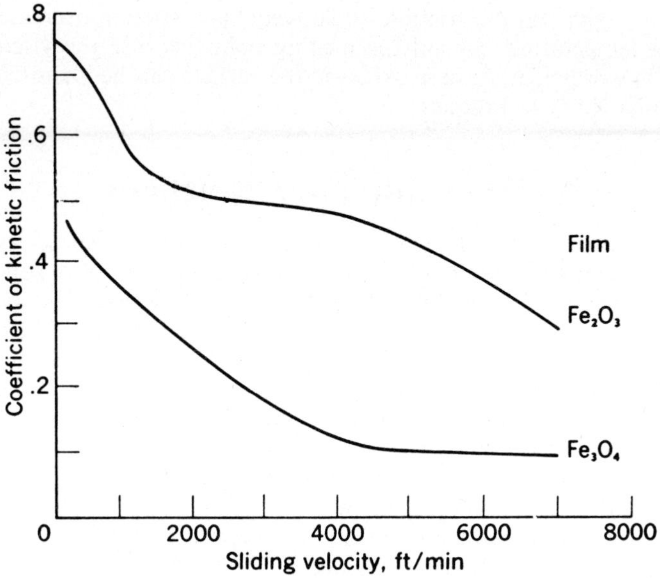

Figure 6-61. – Friction of preformed Fe_2O_3 and Fe_3O_4 films 1200 Å thick.

of cobalt indicate differences in friction behavior. The lower oxide of cobalt (CoO) exhibits a higher friction coefficient than the higher oxide of cobalt (Co_2O_3). This is the reverse of the behavior of the iron oxides.

Differences in friction behavior have also been observed for surface films of the halides of various metals. For example, iron chloride exists in two forms ($FeCl_2$ and $FeCl_3$), and the friction properties of these two chlorides are different. With inorganic compounds, such as the halides that have relatively low melting points, the form of the compound can also have an influence on friction behavior. If a specific chloride of iron such as $FeCl_2$ or $FeCl_3$ is examined as a function of temperature, it becomes readily apparent that as the melting point of these particular halides is approached, the friction properties decrease markedly with a change in form from the solid to the liquid state at the surface. The molten chlorides have much lower friction properties but they also are extremely corrosive to solid surfaces.

Even the presence of waters of hydration on an inorganic compound can influence the friction behavior of that particular compound. A classic example of this is boric oxide. Boric oxide (B_2O_3) exists at room temperature and in moist air as boric acid (H_3BO_3). If the friction properties of boric acid (i.e., the hyrdated form of boric oxide) are measured, the friction coefficients are less than those for the solid boric oxide. After the temperature of boric acid is raised beyond the melting point of the boric oxide, another change in friction properties occurs. This is the same change that occurred with the halides when the material changes from a solid to a liquid. After the melting temperature for boric oxide is exceeded, the friction associated with boric acid is low but the boric oxide becomes extremely corrosive. At the melting point, the viscosity of the boric oxide is very high and the friction is relatively high; friction decreases with increasing temperature beyond the melting point because the viscosity is reduced. The presence of these oxides in the surface can be identified with conventional X-ray techniques.

Surface Substitution Reactions

Many compounds are added to conventional oils to act as extreme pressure or antiwear additives in lubricants. There are also solid film lubricants that are applied to surfaces to reduce adhesion, friction, and wear. Many of these compounds rely on a reaction with the solid surface (particularly a metal surface) to provide a protective surface film. The reaction product used in inorganic compounds is a material which has inherently good resistance to adhesion and friction and affords surface protection from adhesive contact and transfer. Sulphur, chlorine, and phosphorus are common surface elements that are used in organic structures added to oils to provide extreme pressure or antiwear surface films by interacting with the metal surface to form metal sulfides, phosphides, or chlorides. The presence of these films on the surface is believed to give the surfaces better wear, friction, and adhesion characteristics than is observed in their absence. When the surfaces are in contact under heavy loads and the normal surfaces oxides are disrupted, the presence of the additives in the oils allows these active elements to interact with the solid surface to form inorganic compounds. The problem, however, is that most practical tribological systems contain many active elements aside from those deliberately introduced into the lubricant to provide a protective surface film. Thus, there may be competitive reactions going on at the surface for specific surface sites. This is true even among the additives themselves. For example, zinc dialkyldithophosphate, which is probably the most common antiwear and extreme pressure additive, contains a number of surface active elements (zinc, phosphorus, oxygen, and sulfur).

Any of these elements can be found on the solid surface depending on the metals involved in the system. One element may be present in one system and a different element in another system. If the mechanical properties are changed, a shift from the presence of one element to the presence of another element can be seen. The surface chemistry is influenced to a great degree by these competitive reactions. Also, the concentrations of the active species that can interact with a solid surface influence the final reaction products that are present on the surface to provide a protective surface film for that surface and thereby reduce friction. In nearly all liquid lubricants there is dissolved oxygen. Even if oxygen is not dissolved in the lubricant, oxygen is always present in the normal environment. Thus, there is an ample supply of oxygen available at solid surfaces in tribological mechanisms. Where surface reaction products form, they are relatively unstable in comparison with the oxides; it is therefore conceivable for a substitution type of reaction or displacement reaction to take place; that is, one type of compound may form initially on the solid surface and that compound may be displaced by another material.

The formation of sulfur or sulfide films may be displaced by other surface active elements such as oxygen where, for example, the oxide of the metal may be thermodynamically more stable than the sulfide. The sulfide may form initially simply because there is sufficient concentration of sulfur readily available at the surface for interaction with the solid surface. Even though oxygen may be present at lower concentrations in the environment, over a prolonged period of time, oxygen may displace sulfur, phosphorus, or chlorine. On a very fundamental level, this can be seen in surface analysis of ordinary iron surfaces that have been covered with sulfur films. The oxygen readily displaces the sulfide films.

Some experiments using AES analysis have demonstrated this displacement of one material from a surface by another. In figure 6-62, the Auger peak intensities are plotted for both sulfur and oxygen in arbitrary units as a function of oxygen exposure in langmuirs. The surface initially was covered to saturation with a sulfide film; that is, the iron accommodated as much sulfur as it possibly could on the solid surface. That would be indicated as the ordinate at 10^{-1} langmuir (between 45 and 50 arbitrary units). With exposure to oxygen, however, the sulfur concentration begins to decrease in intensity, and the oxygen concentration increases. This process continues with continued exposure to oxygen until after an exposure of 10^4 langmuirs the sulfur is completely displaced from the solid surface and all that is on the surface is oxygen; that is, an oxide has completely replaced a sulfide film at room temperature.

The sulfide film was generated by exposing the iron surface to hydrogen sulfide. That surface received 10^4 langmuirs of hydrogen sulfide, the same exposure that the surface with the sulfide film saw with oxygen. Yet, an equivalent exposure of oxygen was sufficient to completely wipe away the sulfide film. This is "static chemistry." These surfaces were not rubbed in sliding friction experiments. They were simply statically exposed. The dynamic process associated with sliding experiments and with rubbing machinery can bring about an alteration in this chemistry. It can either promote the oxide formation more rapidly or it can inhibit its formation under certain conditions. The important thing to recognize from figure 6-62 is that, in considering the chemistry of a solid surface and how it influences friction behavior, the entire system and all elements that may be present in the system which can interact with the solid surface to form surface films must be considered. The possibility of substituting or displacing reactions exists among the surface active species.

The friction coefficient associated with the data in figure 6-62 was not markedly altered by the presence of the oxide film. In fact, iron sulfide yields a friction coefficient of approximately 0.5 and iron oxide gives about an equivalent friction coefficient when the surfaces are completely saturated. Thus, the sliding friction experiments reveal very little change in the friction coefficient. But an examination of the surface with AES analysis reveals a complete substitution (an oxide film for a sulfide film) of surface films. In such situations, analytical surface tools are very useful,

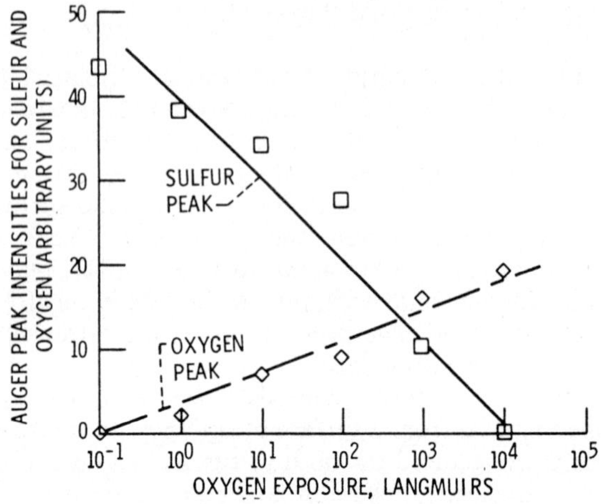

Figure 6-62. – Auger spectroscopy evidence for displacement of sulfur from iron surface by oxygen. Initial sulfide film formed by exposure of iron surface to 10 000 langmuirs of hydrogen sulfide at 23° C.

because friction coefficients do not necessarily reveal exactly what is happening on the surface. There is a marked difference in the surface chemistry between sulfides and oxides. For other systems, there are marked differences in friction behavior. For example, iron chloride yields a friction coefficient between 0.1 and 0.2. When the displacement of iron chloride by oxygen is studied, the friction coefficient is found to increase with increasing exposure to oxygen simply because the oxide has a much higher friction coefficient than does the chloride.

The films analyzed with AES analysis in figure 6-62 are extremely thin. In fact, they are too thin to be detected with, for example, the more conventional techniques such as electron or X-ray diffraction. The oxidation of surfaces which have compounds already formed is another mechanism for producing alterations or changes in friction characteristics (e.g., where solid film lubricants such as molybdenum disulfide are applied to surfaces). The oxidation of the molybdenum disulfide, which can be followed with AES analysis, yields results which indicate that the oxidation process causes a change in friction behavior. Again, as with the chlorides, the friction for molybdenum disulfide is lower than that of molybdenum oxide. Consequently, as the oxidation process takes place, the friction coefficient increases with increasing oxidation. The surface analysis then correlates the observed friction behavior with the oxidation of such compounds as molybdenum disulfide. Oxygen is an extremely reactive material and forms very stable compounds with most material surfaces. There are relatively few materials which interact with the solid surface that cannot be displaced by oxygen at some time in an actual piece of machinery where mechanical forces are interposed between the two surfaces in contact.

Role of Mechanical Surface Activity on Surface Chemistry

It is one thing to measure the static interactions of gaseous or liquid species with a solid surface as in conventional chemistry—that is, to study the interactions of various materials with a metal surface under quiescent or static conditions. In most tribological systems, however, there is mechanical activity at the surfaces. There is rubbing, rolling, or sliding contact, there are loads tending to push the surfaces together, and there is also some velocity (relative tangential speed) with which one surface moves across the other. There may be mechanical alignments involved that bring about localized higher energy situations that tend to produce irregular surface energy distributions. All of these factors tend to change the surface chemistry or the rate and manner in which materials interact with the solid metal surface.

Chemical reaction depends on the initial availability of a quantum of energy (activation energy) for reacting with a reaction that is exothermic, after which it proceeds without application of external energy. Where, however, the reaction is endothermic, some external forces are required to activate and maintain the reaction process. The presence of external forces such as load, speed, and variations in surface energetics play a role or part in the surface chemistry. For example, if one exposes a clean metal surface to almost any relatively active material, an interaction takes place. The strength of the bond between the metal surface and the species with which the metal surface interacts is a function of the chemistry of the particular metal and the interacting species. In general, most systems undergo some type of surface interaction. The only situation where a complete lack of activity might exist is the case where the metal surface would be exposed to inert gases such as argon, krypton, or neon.

With most metals and materials such as oxygen, chlorine, bromine, iodine, fluorine, and sulfur there is an interaction. Generally, the resulting interaction is either the formation of chemisorbed films or chemical compounds. There is no activation energy necessary to achieve the reaction of the species with the metal surface to form surface compounds. The reaction itself is exothermic, and energy is liberated when the surface active species is exposed to a very active clean metal surface. When, however, metal surfaces are already covered with surface films and those films are very stable (such as, e.g., metal oxides which are ever present), then the interaction of less active species with the metal surface is much more difficult to achieve. For example, in figure 6-62, oxygen formed a more stable compound when interacting with iron than sulfur did; that is, since iron oxide is much more thermodynamically stable than iron sulfide, it is easy for oxygen to displace the sulfur from the surface and form the thermodynamically more stable iron oxide. However, replacing oxygen on the surface with sulfur is much more difficult because thermodynamics are against such a substitution reaction taking place. Since the oxide is thermodynamically more stable, there is little chance for the sulfur to interact and displace the oxygen from the metal surface. It is here that the mechanical activity imposed at a solid surface begins to play an important role in the observed surface chemistry (and, accordingly, the friction behavior of materials in solid-state contact).

For example, if a clean iron surface is exposed to the relatively active gas vinyl chloride (unsaturated hydrocarbon containing chlorine), there is a strong propensity of iron to interact, and generally (with AES analysis of the iron surface) both carbon and chlorine are found on the surface; this indicates that the vinyl chloride has been chemisorbed. When, however, the solid surface contains an oxide layer and the iron surface is exposed to vinyl chloride, the interaction is not straightforward. There is less tendency for the vinyl chloride to interact with the metal surface, and any bonding to the oxide that occurs is relatively weak. If, however, mechanical energy is imposed at the surface by some mechanism, such as in altering load or speed, the surface chemistry of the oxidized iron surface can be altered, and this causes an increase in the concentration of the vinyl chloride at the surface. The activation process consists primarily of deformation (induced by the friction process), which exposes nascent iron for interaction with the vinyl chloride. It may be accomplished by altering the load or by simply increasing the sliding velocity. The temperature increases with increasing sliding velocity, and both concentration and temperature can supply the necessary activation energy to change the surface chemistry and, consequently, the friction behavior of solid surfaces.

Varying the load on an iron surface exposed to vinyl chloride causes marked differences in the friction behavior that is associated with changes in the surface chemistry. The data of figure 6-63 indicate friction coefficient and Auger chlorine peak intensity plotted as functions of load for vinyl chloride on an oxidized iron surface. At relatively low loads of 100 grams, the friction coefficient is extremely high. The surface concentration of chlorine is very low. As the load is increased, however, the concentration of chlorine in the surface also increases. Accompanying the increase in concentration of chlorine on the surface is a decrease in the coefficient of friction. It was discussed earlier that the friction coefficient for iron chloride was considerably less than that for iron oxide. Thus, once deformation results in exposure of nascent iron so that interaction with the vinyl chloride can occur, the presence of the chloride film produces a reduction in the friction coefficient. When a load of 500 grams is achieved on the surfaces, the friction coefficient is at the minimum. It has decreased appreciably from that observed at the 100-gram load. Corresponding with that decrease in friction is a maximization in the concentration of chlorine on the solid surface. If the load is increased beyond 500 grams, the friction coefficient begins to increase. This increase in friction coefficient again can be related to a decrease in the concentration of the chlorine on the solid surface.

At the higher loads, the mechanical activity or mechanical energy applied at the interface is greater than is necessary for achieving an optimum formation of a chloride film. The energy actually is of sufficient quantity to produce dissociation of the chloride film from the solid surface; that is, the chloride film is no longer stable on the solid surface because the energy is

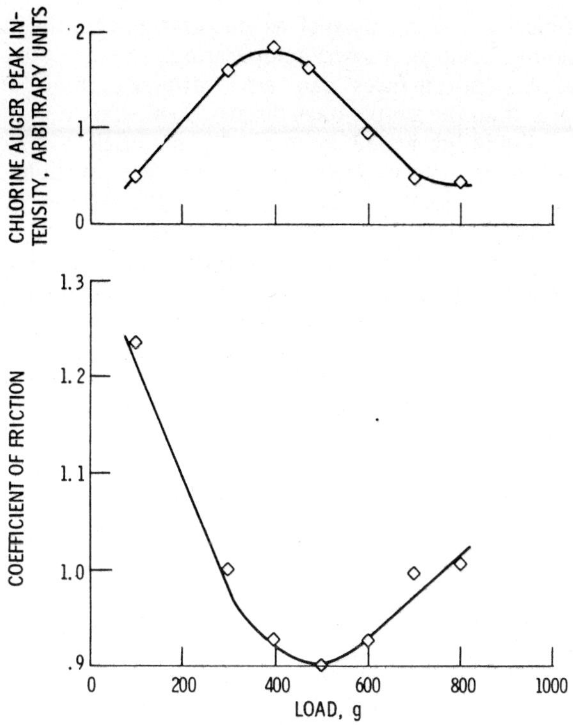

Figure 6-63. – Coefficient of friction and Auger chloride peak intensity as function of load for vinyl chloride on iron surface. Ambient pressure, 10^{-6} torr of vinyl chloride; rider specimen, aluminum oxide; sliding velocity, 30 centimeters per minute; temperature, 23° C; normal oxides present on iron surface.

sufficient to cause it to dissociate. Consequently, nascent metal is beginning to control friction behavior, and the friction increases as the concentration of surface chlorine decreases. If loading were continued, it would ultimately lead to catastrophic failure of the surfaces, because complete penetration of the surface film would occur and very severe seizure would take place.

By simply changing load or sliding speed, one can bring about changes in the surface chemistry. The compound that exists on the surface can be altered, for example, from one form of an oxide to another, and oxidation of compounds can occur at much lower ambient temperatures. For example, there are inorganic compounds on the solid surface that can provide effective lubrication. The effectiveness of these films can be changed by increasing either load or sliding speed. The increase in load or sliding speed produces an analogous effect to increasing ambient temperature. It reduces the ambient temperature at which a change is observed. For example, where a crystallographic change occurs at some specific temperature, an increase in sliding velocity or in load causes that change at a lower temperature. The same is true for surface chemistry where a chemical reaction may not occur until a certain temperature is reached;

the imposition of mechanical activity such as load and speed at the interface allows that reaction to occur at a lower temperature.

Mechanical activity at surfaces can also influence more complex molecular structures than the simple ones discussed thus far. For example, in most practical liquid-lubricated systems the materials are complex hydrocarbon molecules. Very frequently they are straight chain hydrocarbons, either principally or as additives in the system. Many of these organic molecules can have complex structures or, alternately, can be large with relatively high molecular weights. Organic species are extremely sensitive to the mechanical activity that takes place at solid surfaces. The rubbing activity on the surface of asperities produces a considerable amount of energy that can cause chemical changes in these organic materials. Thus, in addition to allowing or altering the reaction of surface-active species with the surface, where the structure of the active species is relatively large, the mechanical or rubbing actions at surfaces cause an alteration in the molecule itself.

A classic example of this is the formation of a friction polymer on metal surfaces as a result of the rubbing process. In many practical lubrication systems using conventional hydrocarbon lubricants, rubbing continuously on the solid surface produces a polymericlike deposit (black debris on the surface), which is a result of a polymerization of the hydrocarbon. The initial hydrocarbon that is placed on the solid surface is relatively stable, and, normally, in static contact with the metal surface, it never generates a polymeric film. However, the energy associated with rubbing can generate radicals; that is, the bonds in the complex molecular structure of the organic molecule can be ruptured and thus leave loose, very active end sites. In addition, when the rubbing process or the mechanical activity at the solid surface generates clean metal, the clean metal surface can also catalyze the generation of these radical species. The more active the metal, the more likely it is to contribute to the formation of radicals from organic structures. For example, one might expect titanium to be much more active in generating radicals where hydrocarbons are exposed to rubbing surfaces than where the hydrocarbons are in the presence of metals such as gold.

The ability of metal surfaces to generate radicals is highly specific, and some metals are much more likely to generate these films than others. There is no doubt that the mechanical action at the interface does contribute to the formation of these surface radicals which then combine to form larger molecular weight species. This process is depicted diagramatically in figure 6-64, where energy is applied to a simple hydrocarbon structure. In this case the indication is that the hydrocarbon structure loses a hydrogen ion and that there are other hydrocarbon species present in the environment (unsaturated hydrocarbon species). Those species then can interact with the radical that has been generated by the hydrocarbon losing a hydrogen ion. The carbon site that the hydrogen ion left becomes very active. In this particular form it is referred to as a radical, since there is an unsatisfied carbon bond in the surface. The carbon wants to form four additional bonds, but at the radical site there are only three bonds, one hydrogen and two carbons.

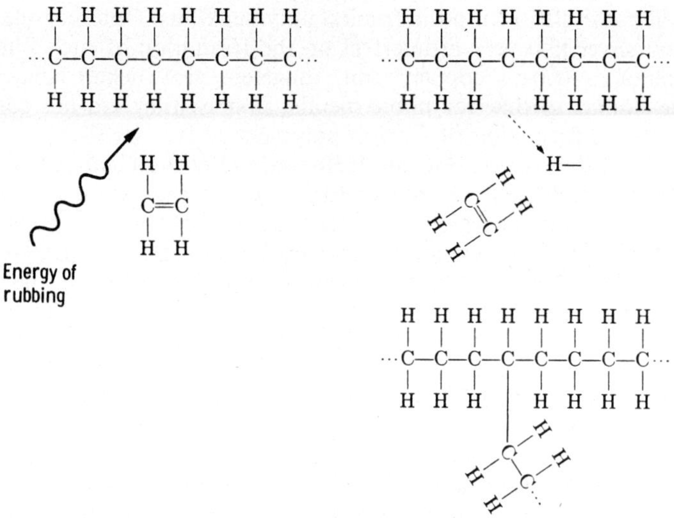

Figure 6-64. – Radicals and polymeric films formed from simple hydrocarbons as result of rubbing process and clean metal surfaces generated.

The mechanism in figure 6-64 is just one wherein radical formation can occur. The formation of the radicals need not operate by freeing a hydrogen ion. Where the molecular structure is a long chain hydrocarbon, carbon-to-carbon bond scission can occur in the molecule, and this liberates a very active carbon atom at an end site. It, then, would tie up or interact with other radicals of varying molecular weights to form larger molecular weight species than the parent molecule from which the radicals came. These larger molecular weight species continue to grow until ultimately one has a molecular weight of sufficient size to produce a solid in the liquid. This is essentially what happens when the molecular structure of the hydrocarbon is increased sufficiently.

In addition, bridges or cross-links can be formed between adjacent molecules by links of unsaturated species. This, then, would also contribute to forming very complex molecular structures that may be solid. Solids of this type are the polymeric films that have been observed frequently in tribological systems as a result of the sliding, rolling, or rubbing process (ref. 24).

Hermance and Egan were among the first to observe the formation of these polymericlike films on solid surfaces and to study them in any detail (ref. 25). They were the ones who first coined the term friction polymer. Since they did not observe the formation of these polymericlike films in the absence of rubbing, they used friction polymer to mean that it was associated with rubbing or mechanical activity of the surfaces. They found, with regard to metal specificity, that some metals form the polymers very readily. For example, when in contact with hydrocarbons, the metals palladium, platinum, ruthenium, molybdenum, tantalum, and chromium served as catalysts to form the friction polymer films. Other metals such as gold and silver had very little effect on the formation of such films. And some metals (such as copper, iron, tungsten, and nickel) behaved in a manner similar to that for noble metals; that is, they did not contribute heavily to the formation of friction polymers.

In general, the friction polymer forms with almost all hydrocarbons. The only case that is difficult to observe is the one with the low molecular weight members, for example, methane. Methane does not form the polymer very readily (ref. 25). The presence of unsaturation (double- and triple-bond) in the molecule, however, tends to promote the formation of the friction polymer; that is, the instability in the molecular structure tends to accelerate the formation of polymer film. Likewise, aromatic structures in the molecular species also promote the formation of the polymer film.

As was mentioned earlier, the polymer film formation can be catalyzed by metals; however, the polymer film can also form in the absence of metals. Polymer films have been observed on ceramic surfaces such as aluminum oxide. These films do not form as readily. With the metals there are two systems or parameters that cause the polymerization, the surface activity of the metal plus the mechanical activity imposed at the interface. With the ceramic materials there is only the mechanical activity. Although data are lacking in the literature, it would be interesting to make a direct comparison of a specific metal in both the oxidized and unoxidized states in reference to a specific hydrocarbon. Furthermore, the amount of polymerization that occurs under a given set of mechanical conditions of load and speed could be used to determine the respective roles of the metal and simple mechanical activity in the polymerization process.

Conventionally lubricated systems do not operate in a vacuum. Figure 6-64 shows the ability of the lubricant to interact with itself and generate larger species from the parent molecular structures. In addition to these types of interactions (i.e., the interaction of the lubricant with itself), interactions can also take place with other constituents present in the environment. As has already been discussed, oxygen is always present in mechanical systems, and it interacts with hydrocarbons and with the radicals that are generated in the rubbing and sliding process. It has been known for many years that, in reciprocating engines, a lacquer or varnish forms on the engine components. For example, pistons and piston rings develop a very hard varnishlike coating. It is a result of hydrocarbon degradation, which operates much as is shown schematically in figure 6-64. Instead of the degradation being simply the interaction of the hydrocarbon with itself, however, oxygen becomes a part of the reaction interacting with the hydrocarbon; consequently, unsaturated species can form hydrogen peroxide. This leads to the formation of ether groups or radicals that can polymerize to form polyethers or that can interact with more hydrogen peroxide and water to form alcohol and acidlike structures. The alcohol and acids can then interact with more hydrogen peroxide to lead to complete acid formation.

The organic acids interact with hydrocarbons to produce esters (lactones or lacquers). The polyesters or lactones are relatively high molecular weight

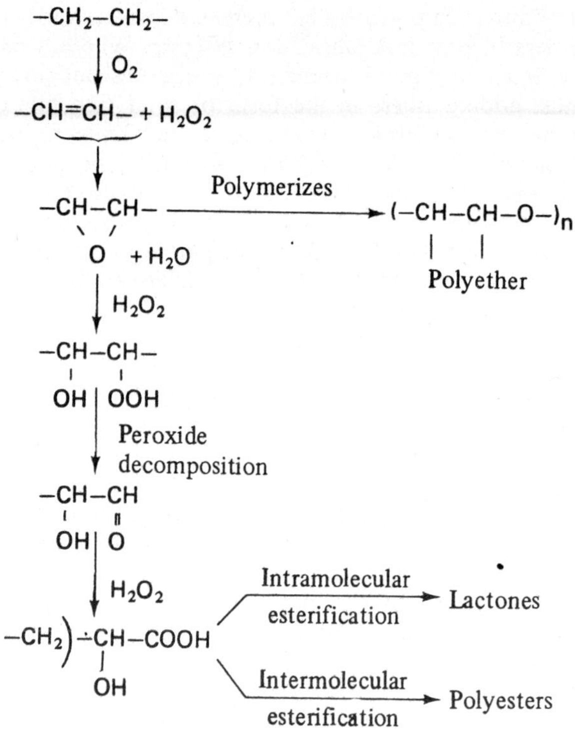

Figure 6-65. – Chemistry of lacquer formation (ref. 26).

species and are varnishlike in nature as shown schematically in figure 6-65. Analytical surface tools have been successfully used in following this type of surface chemistry by Burgess, Morris, and Vickars (ref. 26).

Polymeric materials, which are generated as a result of decomposition associated with the mechanical activity of hydrocarbons on surfaces, are also generated as a result of sliding, rolling, or rubbing contact with metal surfaces. Over the years polymeric compositions have been increasingly used in practical lubrication devices. The polymer compositions that are most frequently used include PTFE, nylons, polyethylenes, and high density polyethylenes; even epoxy compositions have been used as lubrication systems in gears, bearing retainers, seals, and other mechanical components. These polymers are not films but rather bulk, solid materials. Frequently they are used as self-lubricating components in a tribological system. For example, in a roller or ball bearing a cage of a plastic material is used to lubricate by transfer; that is, a transfer film of polymer develops on, for example, the balls in the ball bearing, and it is carried to the ball race contact zone. This, then, provides lubrication for solid surfaces in solid-state contact, prevents adhesion, and minimizes the friction force between two solid surfaces in contact.

The use of solid polymers has expanded to an even greater degree in tribology because of the advent of composite polymer compositions. The inclusion of additives in polymers has increased tremendously the potential use of polymers in practical lubrication systems. Where a basic polymer structure has inherently good mechanical properties but poor lubricating characteristics, adding fillers in the form of solid film lubricants to the polymer can improve its lubricating characteristics. Thus, for example, with many base polymers (such as PTFE, polyimides, or polyethylene) the addition of solid film lubricants (such as graphite and molybdenum disulfide) has been made to these materials to improve lubrication.

Polymers inherently are poor heat dissipators; when polymers are in contact with metal surfaces, therefore, the relatively low decomposition temperatures for most polymers make it difficult to use such compositions at relatively high speeds or under high load conditions. The heat generated at the interface due to the energy associated with the friction process (sliding or rubbing or rolling) is sufficient to cause decomposition of many of these polymeric materials. The addition of fillers such as metals, which improve the heat-carrying characteristics of these polymers, has increased their usefulness considerably.

Metals are probably the most common additive that is placed in the polymers to improve heat dissipation, where heat is a problem in a mechanical component using a polymer. In addition to solid film lubricants and metals being added to polymeric structures to improve lubricating characteristics and to improve heat-carrying capacity, respectively, other additives are put into polymers to improve mechanical strength and resistance to deformation. For example, in PTFE, it is common to add glass fibers. Glass fibers are added to give dimensional stability to the polymer in a mechanical device employing the polymer in a component such as a ball bearing cage. With glass, the basic polymer (PTFE) resists plastic deformation, which tends to distort the mechanical component and reduce its reliability. There are many polymer compositions that can be used in combination with a variety of fillers for practical lubrication devices, and each day the number being used increases.

A family of polymer compositions which has seen considerable exposure in this field is the epoxy compositions. These epoxy compositions have various properties which make them attractive for consideration in tribological devices. Generally polymer compositions such as epoxies do not slide on themselves in practical devices but most frequently are in contact with a metal. For example, in rolling element bearings, the balls and races would be metals (hardened steels) and the ball cage or retainer might be a polymer composition such as an epoxy. Generally during a run-in procedure a transfer film of the solid polymeric material is allowed to develop on the mating or contacting metal surface. In the ball bearing, the ball would be in contact with the retainer, and the retainer would be in contact with one of the two races. There is sliding, rolling, and rubbing contact in the bearing which helps to develop a transfer film of the polymer to the metal surfaces. The polymer transfer film that develops on the metal surfaces provides lubrication for the component; thus, because a film is interposed between the two metal surfaces, friction is reduced.

In theory, the transfer should be from the polymer to the metal surface, and this generally is the behavior observed. There are instances, however, where the polymer is sufficiently strong so that the interfacial bonds formed are stronger than the bonds in the cohesively weaker of the two materials. Since in the case of the very strong polymer this could be the metal, fracture might occur in the metal. Hence, metal would be transferred to the polymeric surface. Studies with various epoxies in contact with iron surfaces have demonstrated that, in some instances, the metal can actually transfer to the polymer rather than the polymer transferring to the metal. Experiments with iron in contact with epoxy using XPS (formerly ESCA) have demonstrated that this, in fact, does occur. The data in support of this are presented in figures 6-66 and 6-67.

Figure 6-66 is an XPS (ESCA) spectrum for an epoxy pin before sliding. There are no iron peaks present on the surface of the epoxy pin. Oxygen and carbon appear in the XPS spectrum as well as some silicon, which was transferred from the polishing paper to the epoxy surface during preparation. There are iron peaks in the XPS (ESCA) spectrum (fig. 6-67) after sliding the epoxy pin on an iron surface and achieving a steady friction coefficient. These peaks were absent in figure 6-66 before sliding was initiated. Thus, it becomes apparent, from the data of figures 6-66 and 6-67, that not only can polymers transfer to metal surfaces but also metals can transfer to polymer surfaces. In this case, the iron has transferred to the epoxy surface as evidenced by the XPS (ESCA) spectrum of figure 6-67.

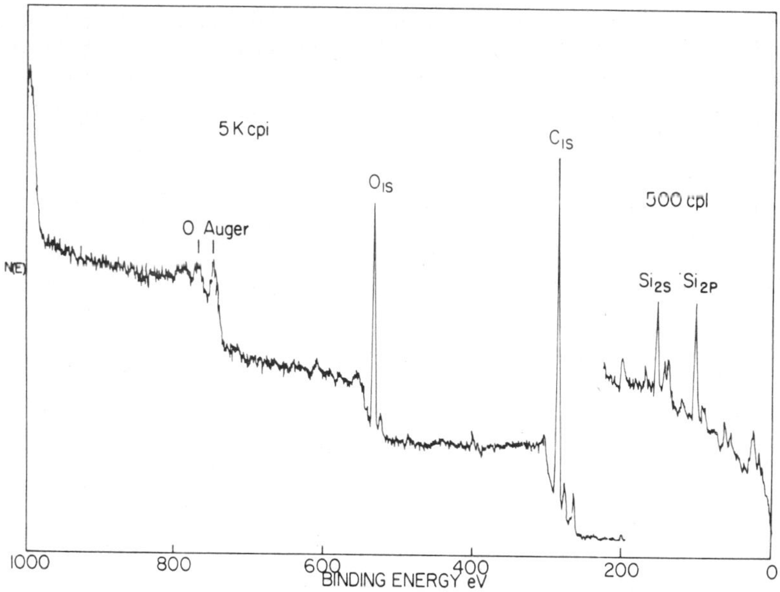

Figure 6-66. – XPS spectrum of epoxy pin before sliding. Note absence of iron peaks. Silicon peaks indicate that some silicon was transferred from polishing paper to epoxy surface during preparation.

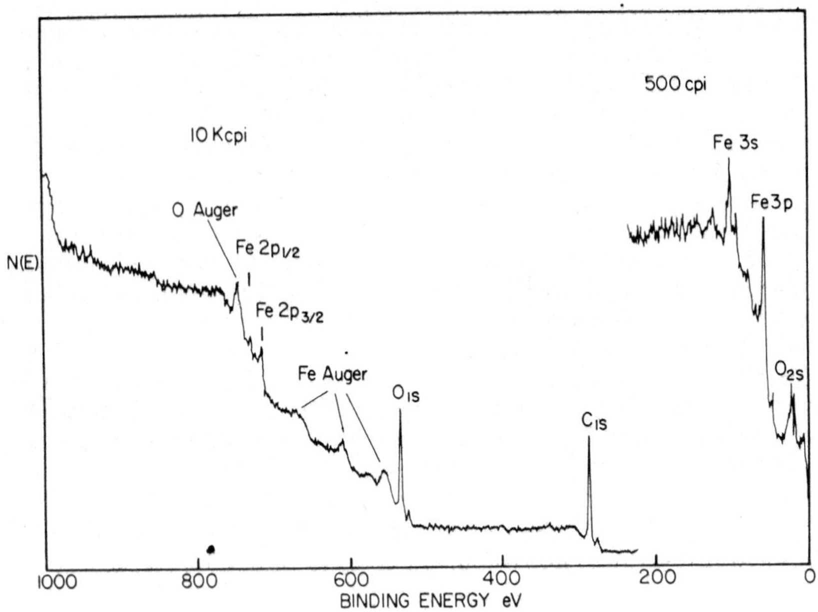

Figure 6-67.—XPS spectrum of epoxy pin after steady-state friction coefficient was achieved. Note presence of iron peaks.

Interaction of polymers in sliding contact with metals and their influence on the friction coefficient were discussed earlier in reference to PTFE in sliding contact with aluminum. In those studies aluminum transferred to PTFE and the friction became ultimately the friction of aluminum sliding on aluminum.

How does a polymer transfer to a metal? Is it by chemical bonding or simply mechanical transfer, where the asperities in the metal surface are covered with a thin film of the polymer? It is well known, for example, that if one grinds a steel flat surface in one direction, the grinding marks or grooves are oriented in one direction in the finishing operation. If, then, a polymer pin slides across the surface, first with the direction of grinding and then against the direction of grinding, the amount of wear or transfer of polymer to the metal surface is markedly greater when the polymer pin slides against the grinding marks rather than moving with the grinding marks. Thus, the mechanical action of the metal, acting as a shear or cutting edge to cut the polymer, is one mechanism of transfer of polymer to surfaces. But even in the absence of these surface irregularities such as grinding marks or other surface defects, the polymers have been observed to transfer to metal surfaces, even electrolytically polished surfaces and cleaved metal surfaces, at cryogenic temperatures where the metal surface is atomically flat or smooth. Furthermore, transfer has been observed in the FIM to atomically smooth pin-tips such as the tungsten-polymer transfer films discussed earlier; that is, PTFE transferred to tungsten on simple touch contact. Thus, strong bonds must form by other than simply mechanical transfer, since the pin tip in the FIM is atomically smooth and is not moved tangentially relative to the contacting surface. Bonding of the polymer to the metal surface must be more than simply van der Waals bonding because, if they were simply van der Waals interactions, the polymer film would not be present on the tungsten surface at the high voltages required for imaging the tip of the specimen in the field ion microscope. The polymer would field evaporate at much lower voltages. The fact that the polymer remains present on the solid surface to very high imaging voltages indicates that the adhesion of the polymer to the metal surface is chemical in nature.

Brainard and Buckley (ref. 27) and Pepper and Buckley (ref. 28) have postulated that the interaction of PTFE with clean metal surfaces is by the carbon bond; that is, it is carbon to metal bonding. The primary reason for concluding this is that in Auger studies of polymers in contact with clean metal surfaces the fluorine is susceptible to electron-beam-induced desorption. In other words, as the primary beam of electrons of the Auger analyzer is directed at the solid surface for analyzing the polymer film present on the metal surface, fluorine desorbs readily from the surface. One can literally watch the fluorine peak disappear in the Auger spectrum as one particular site on the solid surface is examined with the Auger analyzer until all that is left on the solid surface is carbon. Since the metal fluoride bond is much stronger than the carbon to fluorine bond, the bond would not be susceptible to electron-beam-induced desorption if the bonding were fluorine to metal. The fact, however, that the fluorine completely leaves the surface led Pepper and Buckley to conclude that the carbon to metal bonding was the one that was responsible for the transfer of the polymer film to very clean metal surfaces rather than a fluorine to metal bond that might intuitively be anticipated (ref. 28).

Other authors have concluded otherwise in different types of experiments. For example, Cadman and Gossedge have examined metal surfaces after rubbing with PTFE under atmospheric conditions and found that a film of PTFE transferred to the metal surfaces in all cases (ref. 29). Cadman and Gossedge, however, analyzed the polymer film on the surface using XPS with an argon depth profile analysis (argon etching). During those studies they found that metallic fluorides were present on the metal surface at the metal-polymer interface in two systems: (1) with the polymer sliding on a stainless steel surface, and (2) with the polymer rubbing a nickel surface. After the polymer rubbed across the metal surface, a razor blade was used to scrape away the excess polymer transfer film. This process was followed with an XPS analysis of the surface. Doing this repeatedly, Cadman and Gossedge found two different spectra for the nickel surface rubbed with PTFE (fig. 6-68). The curve, in figure 6-68(a), for PTFE rubbed on the nickel surface shows the presence of the fluorine 1s peaks. Figure 6-68(b) shows the same surface after it had been scraped with a razor blade. The fluoride peak for nickel fluoride is extremely weak. Nonetheless, one might conclude that some fluoride is present on the metal surface.

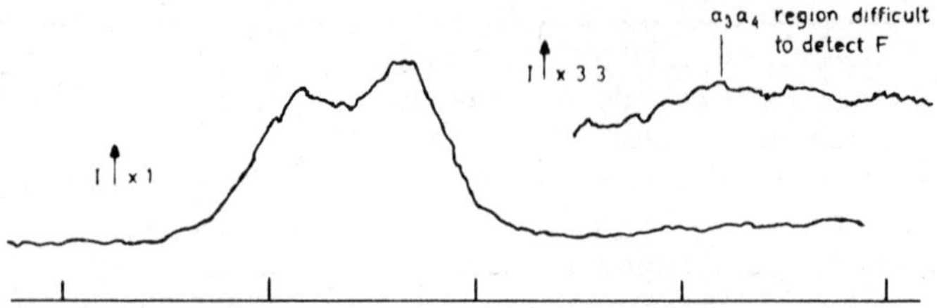

(*a*) *Rubbed with PTFE.*

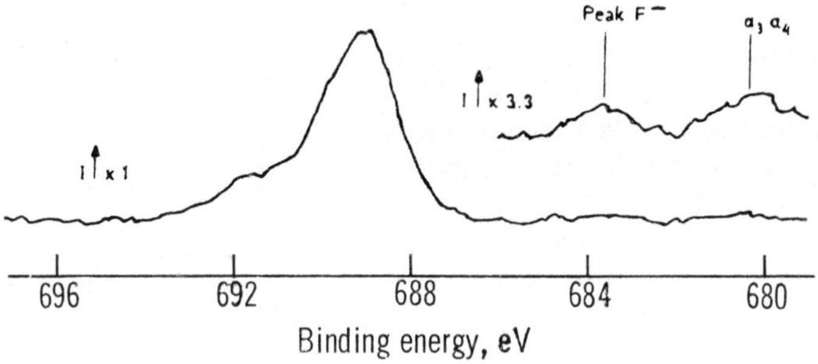

(*b*) *Rubbed with PTFE and scraped with razor blade.*

Figure 6-68. – F(1s) spectra for nickel rubbed with PTFE.

There are, however, instances where polymers do transfer to surfaces that do not react chemically with metals—for example, ceramic surfaces such as aluminum oxide. Although there is no question but that transfer films are observed on these surfaces, one cannot argue the formation of fluorides with the presence of the polymer transfer film. In such instances the mechanism must be other than chemical bonding or simple mechanical transfer.

Not only are the friction properties of metals, carbons, and polymers affected by the presence of surface films, but also the friction properties of ceramic materials in sliding contact with themselves are affected. When a ceramic material is in contact with a metal one expects surface films on the metals to alter friction behavior. But even ceramics (relatively hard and brittle materials) in contact with themselves seem to have a sensitivity to surface films in their friction behavior. Most ceramic materials adsorb such surface species as water vapor, carbon monoxide, carbon dioxide, and some other gases. Even the mechanical properties of some ceramic materials are altered or influenced by these adsorbed species. It is therefore logical to conclude that friction properties might also be influenced by adsorbed films on the surfaces of ceramic materials.

Probably one of the most widely used ceramic materials in the field of lubrication is aluminum oxide. It has been used in both single and polycrystalline forms as a high wear resistant solid surface. It also has inherently good friction properties. There are, however, many factors which influence the friction characterisitics for ceramics (such as aluminum oxide) even when they are in contact with themselves. Most ceramic materials, like metals, are highly anisotropic in their friction behavior; that is, they are very orientation sensitive. This is true whether they are in sliding or rubbing contact with themselves or with metals.

Friction data were obtained for aluminum oxide in a vacuum environment where the aluminum oxide surface was heated to high temperature in a vacuum of 10^{-10} torr to remove adsorbed water vapor. The friction coefficient was measured for the aluminum oxide in contact with itself in vacuum (10^{-10} torr) as well as in the presence of ordinary atmospheric air. Marked differences in friction behavior existed, and these differences can be attributed to the adsorbed films on the aluminum oxide in an ordinary air environment. Single crystal aluminum oxide (sapphire) in sliding contact with sapphire showed low friction coefficients in air (fig. 6-69); in vacuum, however, friction coefficients as high as 1.5 were obtained. With an increasing load, a decrease was observed in the friction coefficient.

Etch-pitting the surface of the sapphire with orthophosphoric acid revealed plastic deformation in the aluminum oxide in sliding contact. The

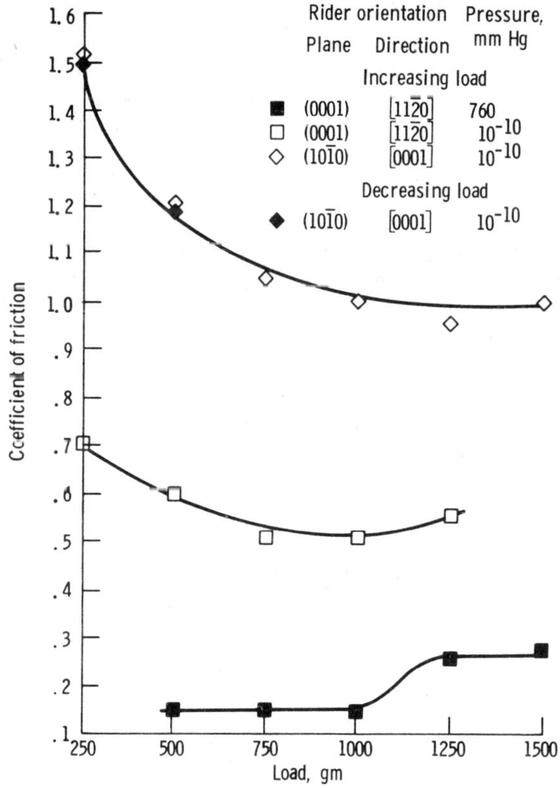

Figure 6-69. – Coefficient of friction as function of load for sapphire sliding on sapphire in vacuum (10^{-10} mm Hg) and air (760 mm Hg). Sliding velocity, 0.013 centimeter per second; temperature, 25° C; specimen outgassing with electron gun at 300° C; disk specimen was (0001) plane parallel to sliding interface.

prismatic orientation (1010) exhibited markedly higher friction coefficients than the basal orientation (0001) over the range of loads investigated in figure 6-69. Aluminum oxide (single crystal sapphire) has a rhombohedral crystal structure, a kind of hybrid hexagonal structure, and, as a consequence, its highest atomic density, lowest surface energy plane is its basal plane. As was observed with metals (also with ceramic materials), the lowest friction coefficients are on the highest atomic density, lowest surface energy planes. For aluminum oxide, this is the basal plane. Consequently, the friction coefficient for the basal orientation (0001) is approximately half that of the prismatic (1010) orientation in a vacuum environment (fig. 6-69). Both friction coefficients, however, are relatively high compared to the friction coefficients measured for the basal orientation of aluminum oxide at 760 torr in an ordinary air environment. At 760 torr the friction coefficient is between 0.15 and 0.25 for sapphire sliding on itself. Thus, the difference in friction behavior for sapphire in the presence and absence of adsorbates is relatively great. Adsorbates have a relatively profound influence on the friction behavior, even on materials that are relatively inactive (fig. 6-69).

Even at relatively high loads of 1000 to 1500 grams, the friction coefficient with the adsorbate present is still markedly less than it is in the absence of the adsorbate. The increase in friction coefficient observed at 1250 grams (in fig. 6-69) in the presence of the adsorbates may be attributed to an increase in the plastic deformation of the aluminum oxide and a greater amount of aluminum oxide to aluminum oxide contact through the adsorbed film. In these experiments it was not necessary to use any sophisticated analytical surface tool to determine the effects of adsorbates on friction properties. The aluminum oxide or single crystal sapphire specimens were cleaned in a vacuum environment by heating before the friction properties were examined. Using that as standard and then allowing the atmospheric air to enter the system to bring the system to atmospheric pressure are sufficient to reveal marked differences in friction behavior; these differences can be related directly to adsorbates that condense on the surface with a simple change in pressure. Thus, nothing more than a change of atmosphere and pressure caused a marked change in the friction behavior of aluminum oxide (fig. 6-69). What is normally considered to be a low friction material and a very wear-resistant substance (viz., aluminum oxide) can, under the proper conditions of surface cleanliness, exhibit very high friction characteristics. At the interface during these sliding friction experiments in both vacuum and air, plastic deformation of the aluminum oxide occurred. This was revealed on etch-pitting. Thus, a very brittle, high-strength material such as aluminum oxide in the sliding process undergoes sufficient loading at the interface or at the point of contact to deform plastically.

Steijn and others have measured the friction properties of a host of different ceramic materials and inorganic crystals, including metallic halides, and have found that there generally are anisotropic friction properties with these substances (refs. 30 to 32). Friction coefficient is very highly dependent on not only the crystallographic plane of sliding but also on the crystallographic sliding direction, as was observed in the investigation with metals. The friction properties of all these materials are also sensitive to the presence of adsorbed films. In some cases, adsorbates such as water vapor interact chemically with the surface. For example, the adsorption of water vapor on halide crystals such as sodium chloride alter the nature of the surface chemistry and certainly influence the friction behavior. Water vapor is probably one of the most influential adsorbates relative to its influence on friction characteristics. The presence of water vapor on metals influences friction properties. In carbons, a very marked influence on the friction behavior of carbon is observed in the presence of water vapor. Carbon materials and carbon bodies can be useless in the absence of water vapor, but they become very effective materials for good wear resistance and low friction characteristics when moisture is present. Likewise, the data of figure 6-69 show that the presence of adsorbates such as water vapor on ceramics such as aluminum oxide are very beneficial to their friction characteristics. Pumping out vacuum systems containing aluminum oxide samples (such as those used in fig. 6-69) reveals that, on bakeout, water vapor is liberated and is detected by a mass spectrometer in the system, indicating its prior presence on the surface of aluminum oxide. The tenacity of the water vapor to such substances as aluminum oxide indicates that the bonding of the water vapor to the aluminum oxide is not simply physical in nature, because one would anticipate that relatively modest vacuums would cause its removal, as is observed with carbon. Thus, in general, water vapor seems to be a very useful substance in the environment for it tends to control or reduce friction forces between two surfaces in rubbing, sliding, or rolling contact.

References

1. Bowden, F. P.; and Tabor, D.: Friction and Lubrication of Solids. Oxford Clarendon Press (London), 1950.
2. Bowden, F. P.; and Tabor, D.: Friction and Lubrication of Solids. Part II. Oxford Clarendon Press (London), 1964.
3. Bowden, F. P.; and Hanwell, A. E.: The Friction of Clean Crystal Surfaces. Proc. Roy. Soc. (London), series A, vol. 295, no. 1442, Dec. 1966, pp. 233-243.
4. Livesay, B. R.; and Belser, R. B.: Instrument for Measuring Small Friction Forces. ASLE Trans., vol. 23, no. 4, Oct. 1969, pp. 257-265.
5. Blouet, J.; and Courtel, R.: Sur L'Observation de Trois Phases Distinctes dans le Frottement Lubrifie de L'Acier sur L'Aluminum (On the Observation of Three Distinct Phases in the Lubricated Friction of Steel on Aluminum). C. R. H. Acad. Sci. (Paris), series C, vol. 273, no. 3, July 1971, pp. 220-223.
6. Elder, John A., Jr.; and Eiss, Norman S., Jr.: A Study of the Effect of Normal Stiffness on Kinematic Friction Forces Between Two Bodies in Sliding Contct. ASLE Trans., vol. 23, no. 4, Oct. 1969, pp. 234-241.
7. Miyoshi, K.; and Buckley, D. H.: The Friction and Wear Characteristics of Iron Chromium Alloys in Contact with Themselves and Silicon Carbide. NASA TP-1387, 1979.
8. Bowden, F. P.; and Ridler, K. E. W.: The Physical Properties of Surfaces, III—The Surface Temperature of Sliding Metals, The Temperature of Lubricated Surfaces. Proc. Roy. Soc. (London), series A, vol. 154, no. 883, May 1936, pp. 640-656.
9. Barker, K.; et al.: Physic Wear Behavior Produced by Thermal, Metallurgical, and Chemical Action. ASLE Trans., vol. 21, no. 4, Oct. 1978, pp. 323-328.

10. King, R. F.; and Tabor, D.: The Effect of Temperature on the Mechanical Properties and the Friction of Plastics. Proc. Phys. Soc. (London), vol. 66, pt. 9, no. 405B, Sept. 1953, pp. 728-736.
11. Guy, A. G.: Elements of Physical Metallurgy. Second ed., Addison Wesley Publishing Co., Inc., 1959.
12. Hays, C.: Electropolishing of Thin Metal Foils, Metallographics Specimen Preparations: Optical and Electron Microscopy. J. L. McCall and W. M. Mueller, eds., Plenum Press, 1974, p. 318.
13. Barquins, M.; Kennell, M.; and Courtel, R.: Comportement de Monocrist aux de Cuivre sous L'Action de Contact d'un Frotteur Hemispherique. Wear, vol. 11, 1968, pp. 87-110.
14. Steijn, R. P.: Friction and Wear of Single Crystals. Wear, vol. 7, 1964, pp. 48-66.
15. Jones, J. P.; and Roberts, E. W.: Adsorption of Lead on Low Index Plane of Tungsten. Surface Sci., vol. 62, 1977, pp. 415-430.
16. McLean, Donald: Grain Boundaries and Metals. Oxford Clarendon Press (London), 1957.
17. Miyoshi, K.; and Buckley, D. H.: Friction and Wear of Metals with a Single-Crystal Abrasive Grit of Silicon Carbide—Effect of Shear Strength of Metal. NASA TP-1293, 1978.
18. Mihoshi, K.; and Buckley, D. H.: Friction and Wear with a Single-Crystal Abrasive Grit of Silicon Carbide in Contact with Iron Based Binary Alloys and Oil—Effects of Alloying Elements and Its Contact. NASA TP-1394, 1979.
19. Hsu, Stephen M.; and Klaus, E. Erwin: Estimation of the Molecular Junction Temperatures in Four-Ball Contacts by Chemical Reaction Rate Studies., ASLE Trans., vol. 21, no. 3, July 1978.
20. Pauling, L.: A Resonating-Valence-Bond Theory of Metals and Intermetallic Compounds. Proc. Roy. Soc., series A, vol. 196, no. 1046, Apr. 1949, pp. 343-362.
21. Miyoshi, K.; and Buckley, D. H.: Friction and Wear of Single-Crystal and Polycrystalline Manganese-Zinc Ferrite in Contact with Various Metals. NASA TP-1059, 1977.
22. Pepper, S. V.: Shear Strength of Metal-Sapphire Contacts. J. Appl. Phys., vol. 47, no. 3, Mar. 1976, pp. 801-808.
23. Johnson, R. L.; Godfrey, D.; and Bisson, E. E.: Friction of Solid Films on Steel at High Sliding Velocities. NACA TN-1528, 1948.
24. Holm, Ragnar: Electric Contacts: Theory and Application. Fourth ed. Springer–Verlag (Berlin), 1967.
25. Hermance, H. W.; and Egan, T. F.: The Examination of Electric Contacts by the Plastic Replica Method. Trans. AIEE, Commun. Electron., vol. 76, pt. 1, 1957, pp. 756-762.
26. Burgess, J. E.; Morris, A L.; and Vickars, M. A.: The Chemistry of Lacquer Formation. American Chemical Society, Division of Petroleum Chemistry, vol. 13, 1968, pp. B91-103.
27. Brainard, W. A.; and Buckley, D. H.: Adhesion and Transfer of Polytetrafluoroethylene to Tungsten Studied by Field Iambicroscopy. NASA TN D-6887, 1972.
28. Pepper, S. V.; and Buckley, D. H.: Adhesion and Transfer of PTFE to Metals Studied by Auger Emission Spectroscopy. NASA TM X-68076, 1972.
29. Cadman, P.; and Gossedge, G. M.: The Chemical Nature of Metal-Polytetrafluoroethylene Tribological Interactions as Studied by X-Ray Photoelectron Spectroscopy. Wear, vol. 54, 1979, pp. 211-215.
30. Steijn, R. P.: Sliding and Wear in Ionic Crystals. J. Appl. Phys., vol. 34, no. 2, Feb. 1963, pp. 419-428.
31. Duwell, E. J.: Friction and Wear of Single Crystal Sapphire Sliding on Steel. J. Appl. Phys., vol. 33, no. 9, Sept. 1962, pp. 2691-2698.
32. Riesz, C. H.: Mechanism of Wear of Non-Metallic Materials. WADC TR-59-316, pt. 2, Wright Air Development Center, May 1960.

CHAPTER 7

Wear

When two solid surfaces are placed in solid-state contact, it is difficult to envision the absence of some wear even in the most efficiently lubricated systems because of asperity contact. There are, however, some theories which hold that, when the loads between two solid surfaces in contact are extremely light, there is an absence of wear between the solid surfaces. Wear may occur, but it is on such a small scale so as to be difficult to detect by most instruments and techniques conventionally used for analyzing wear. Wear can occur, for example, on an atomic scale where transfer occurs (at the atomic level) from one surface to another. The net effect of this scale of wear over long periods of time is that measurements do not yield noticeable wear. It must be remembered, however, that many of the techniques used for detecting wear are relatively crude when one is considering wear at an atomic level. It is believed that the case where wear is completely absent for two surfaces in solid-state contact is the exceptional. The more effective the lubricant, the greater its tendency to minimize or eliminate wear. Complete elimination, however, is very doubtful in most practical applications.

Stress State in Materials

When two solid surfaces are brought into contact in almost any tribological system, there is some force applied to one of the surfaces pressing it against the other in contact (whether it is in a bearing, gear, seal, or other mechanical component). The force applied to the surfaces in solid-state contact produces elastic deformation initially, and, if the load is sufficiently high, plastic deformation or strain is produced in the material. The application of a stress producing force has its effect in the material itself. For example, there is a region in the material near the surface where stress is maximized. For a long time that maximum shear stress was believed to have occurred on the surface. However, Davies in 1949 established that the maximum stress produced with elastic deformation of materials in solid-state contact does not really occur at the surface but rather slightly subsurface (ref. 1). This is depicted schematically in figure 7-1 where the position of the maximum shear stress for elastic contact of a sphere on flat is demonstrated. The d in figure 7-1 is the actual area of contact for the sphere with the flat. The subsurface eliptical rings indicate the various contours for maximum shear stress levels in the material flat by applying the load to the sphere. The zone of maximum shear stress lies considerably below the surface itself. As one goes deeper into the material from this zone, the stress in the material is dissipated.

Imagine the flat extending out of the paper (toward the reader) and the sphere contacting the flat; then, if the sphere were to slide tangentially across the flat (toward the reader), the zone of maximum subsurface shear stress would move toward the surface with tangential motion. The higher the friction force between the sphere and the flat, the closer to the surface the zone of maximum shear stress is. At some point when the friction force is sufficiently high, the zone of maximum shear stress occurs at the surface. The location of the maximum stress in the material is extremely important for two solids in solid-state contact, because the initiation of the formation of wear particles is frequently influenced by this location. For example, wear has been observed to be generated subsurface as well as in the surface regions of materials. The zone of maximum shear stress, therefore, has a very significant role in determining the origin of the wear particle.

The stress determination established by figure 7-1 is for an isotropic material, for example, an amorphous material like glass. With crystalline solids, however, which are most frequently encountered in tribological applications, this zone is affected by the crystalline nature of the material (its crystal structure, orientation at the surface, and other factors). The important concept to recognize, however, is that for solids there is a region where the stress is maximized; this region is subsurface or at the surface depending on the conditions of contact. Where the contact is simply static,

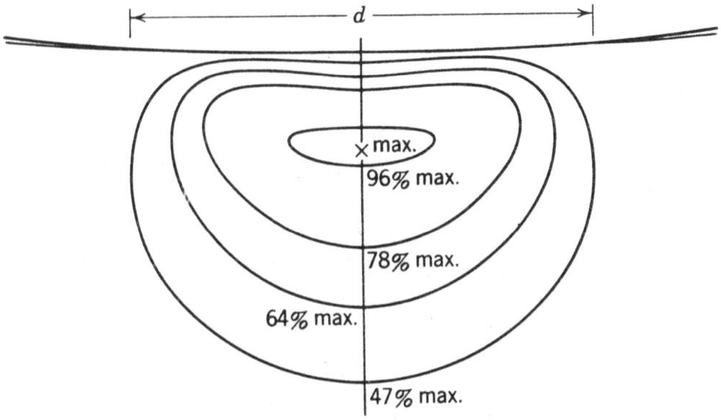

Figure 7-1. – Position of stress maximum for elastic contact of sphere and flat (ref. 1).

it is subsurface; when the materials are in sliding, rubbing, or rolling contact, the zone may occur close to the surface or even at the surface.

With most solids in contact, the contact occurs through asperities. It is difficult to imagine asperity contact without some plastic deformation; that is, some strain occurs in the asperities themselves, however slight. For most practical materials, the asperity contact (even under the lightest of loads) experiences very high localized stresses in the real area of contact of the asperities. This is generally sufficient to produce strain in the asperities near the tips. For example, in experiments with ceramic materials such as aluminum oxide (very high density, high strength materials), plastic deformation has been observed under relatively modest loading conditions. If plastic deformation can occur in a high strength – high elastic modulus material such as aluminum oxide, it is difficult to envision most metal systems not undergoing some plastic deformation when brought into solid-state contact under load. The deformation may occur only in the tips of the asperities. In the near future, it should be possible to identify the presence of dislocations in these asperity tips and to determine exactly under what conditions of mechanical loading the onset of plastic deformation or strain actually occurs.

In considering the contact of asperities of one surface with another, one must consider the properties of the two materials in solid-state contact. In some situations, the materials have similar mechanical properties, but in many practical devices, there are marked differences for the two surfaces in contact. For example, in the common journal bearing there is a considerable difference between the mechanical properties of the copper bearing and the mechanical properties of the steel shaft.

Where differences exist in mechanical properties, these differences are felt in the deformation at the interface. Material properties as well as mechanical properties influence whether, and to what extent, deformation occurs. If the asperity is thought of as a cone coming in contact with a flat surface, then asperity deformation can be schematically represented as in figure 7-2. We assume that the asperities are made of relatively soft

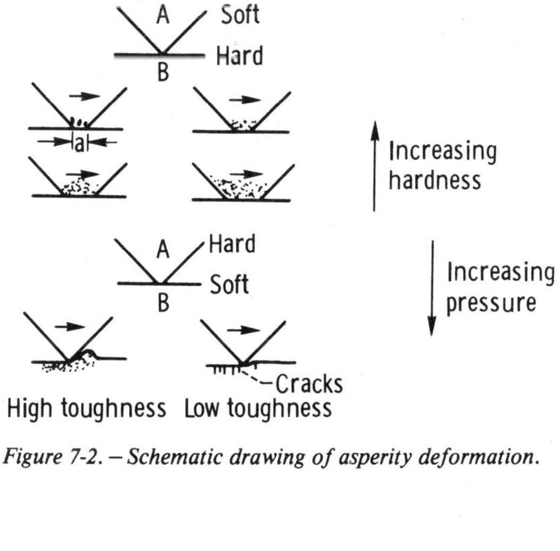

Figure 7-2. – Schematic drawing of asperity deformation.

materials (say indium, tin, or lead) in contact with a very hard substrate (e.g., tungsten). When the surfaces are brought into contact under an applied force, deformation occurs in the cohesively weaker of the two materials. In this case, the lead, indium, or tin deforms (first elastically and then plastically) and the conical nature of the asperity becomes flat to some extent and the real area of contact is represented somewhat by the area of deformation. In the surface and near-surface regions of the relatively cohesively weak material, a considerable amount of dislocation development is observed. The concentration of dislocations increases with the straining of the material occurring in the asperity.

Dislocation Concentration

If the load is increased further, the area grows as does the concentration of dislocations at the surface and in the near-surface regions. Progressive increases in the loading produce progressive increases in the amount of deformation at the surface in the cohesively weaker of the two materials. Not only are dislocations generated in the surface regions, but the dislocations also move further into the cohesively weaker of the two solids. The concept of the generation of dislocations associated with strain is extremely important relative to wear because the dislocations are line defects in solids. These line deflects can serve as sites for generating weak regions in the material where cracks and voids can develop along with wear particles. For example, the intersection of dislocations along slip lines can produce voids, these voids can grow with progressive deformation until in weakened subsurface regions a crack can be initiated. The crack can then subsequently work its way to the surface and generate a free wear particle. Likewise, in the surface regions, localized high concentrations of dislocations can serve as a source for initiating cracks that can progress subsurface if the load is increased or if cyclic loading conditions are imposed.

In many practical tribological systems, with deformation and strain, dislocation concentrations develop to a very high degree as the material at the interface becomes strained. However, where sufficient energy is added (due to, e.g., increased loading, sliding velocity, or ambient temperature), the material can recrystallize, and a sharp decrease in the concentration of dislocations at the surface or near-surface regions can result. Dislocation concentrations can grow from 10^6 to 10^{12} per square centimeter with straining. A rapid dropoff in dislocation concentration occurs when recrystallization of the material takes place. If the quantum of energy put in is reduced below that required for recrystallization, dislocation concentration again increases with increasing mechanical surface activity. This could occur practically, for example, where a variable sliding velocity condition exists.

Where operation occurs at relatively low sliding velocities under fixed load and fixed ambient temperature, the amount of frictional heat generated at the interface (under a given condition of surface lubrication) is relatively constant; with increased mechanical activity (e.g., sliding of the surfaces), an increase in the concentration of subsurface and near-surface dislocations occurs to some particular level, which remains relatively constant. Where, however, the sliding velocity in the system is increased for some reason, the maximum concentration of dislocations which the particular material can accommodate can be exceeded. When that occurs, recrystallization takes place, and the concentration of dislocations or line defects in the crystalline solid decreases and an annealing occurs in the material. Some data in this chapter will show that this occurs fairly often in tribological systems.

Generating these defects in high concentrations and with rapid quenching out produces rapid changes in the material, and it accounts for the initiation of certain wear processes. For example, where the asperity in contact with a solid flat surface is a cohesively strong material and the surface being contacted is a relatively weak one. Let the asperity be tungsten and the flat be indium, tin, or lead. On contact by the tungsten, when load is applied to it, plastic deformation occurs in the cohesively weaker material (indium, tin, or lead). At that time, an equilibrium condition is established where the area of contact is sufficient to support the load applied to the tungsten asperity. If any type of tangential motion is attempted, shear generally occurs in the cohesively weaker indium, tin, or lead. The surface of the weaker material is then plowed much like a farmer's plow plows a field.

Materials have varying degrees of toughness. Some are much tougher than others. Those materials with a high degree of toughness and a high modulus of elasticity do not deform readily, and there is less plowing of the surface by the tungsten asperity. For example, with tungsten in contact with rhenium there is very little plowing of the rhenium surface by the tungsten. For some solids, there is a complete absence of ductility; that is, the material is completely brittle. Glass is a very brittle material. There are many inorganic compounds that are also very brittle. When asperities contact these types of materials, the energy at the interface associated with loading instead of being absorbed in strain is dissipated in the generation of surface cracks. With tangential motion, the tangential forces impart energy at the surface, and this energy is again dissipated in the generation of surface cracks in the material as is indicated in figure 7-2.

The importance of various factors (deformation and its associated stresses, plastic flow, and fracture or crack formation in materials) and their contributions to wear have been observed experimentally. Studies with inorganic crystals such as magnesium oxide, lithium fluoride, and sodium chloride are very useful because these particular materials can be readily cleaved. These surfaces can be exposed to reveal the formation of fracture cracks, and they also can be etch pitted very readily to display dislocation generation and plastic behavior at the surface of the solids. These materials are generally thought of as not being ductile from most mechanical viewpoints; that is, they are considered to be relatively brittle materials that fracture easily. In sliding, rolling, and rubbing contact, however, these materials behave analogously to metals in that they also undergo plastic deformation, exhibit fracture behavior, and are sensitive to the presence of surface films.

One material which has been fairly thoroughly used to investigate the influence of stresses and the deformation behavior of materials in solid state contact is lithium fluoride. Lithium fluoride can be readily cleaved along natural cleavage planes normal to a rubbing surface, and it etch pits readily to reveal dislocations. Sliding friction experiments have been conducted with various materials in sliding contact with lithium fluoride to determine the influence of such factors as applied surface stresses on deformation and fracture behavior.

Effects of Surface-Active Films

Experiments have been conducted both in the presence of and the absence of surface-active films. The results of some of these experiments are presented in the photomicrographs in figure 7-3.

In figure 7-3 there are six photomicrographs. The upper three are for a lithium fluoride wear track that has been cleaved normal to the wear track so that it can be examined in cross section. The three photomicrographs represent three different loading conditions; the lightest load is to the left, the highest load to the right, and the intermediate load in the middle. A sphere was slid across the surface, which was lubricated with a paraffinic mineral oil.

In the left photomicrograph at a relatively light load, the concentration of etch pits subsurface indicates that plastic deformation has occurred in the lithium fluoride as a result of sliding across the surface. The etch pits follow the natural slip planes in the lithium fluoride and move from the surface subsurface. This indicates that the maximum concentration of dislocations is at the surface region, and it moves into the bulk of the solid. Just subsurface to the actual point of contact of the sphere as it moved across the surface is a void, a slight line under the surface which is a linear cavity in the

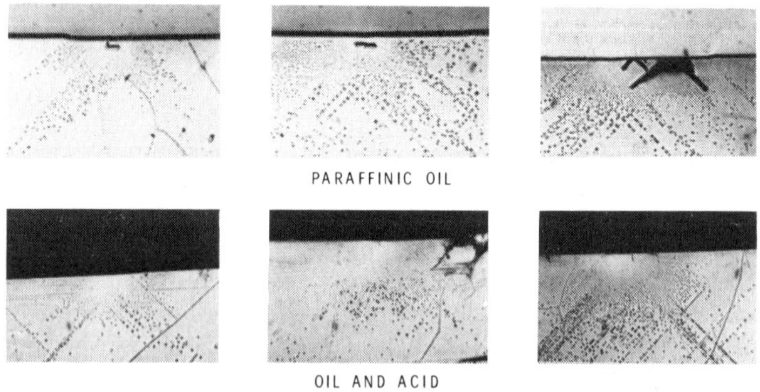

Figure 7-3. – Photomicrographs showing development of surface and subsurface fracture cracks in lithium fluoride in absence of surface active agent at various loads and absence of cracks in presence of surfactants. Speed, 0.005 millimeter per second; temperature, 20° C.

surface. This is the subsurface zone of maximum shear stress. The dislocations which have coalesced at this region because of the intersection along slip planes cause the void seen in the photomicrograph. Lithium fluoride is extremely brittle normally. If one were to take a hammer and rap the surface of lithium fluoride, it would break into pieces or shatter like glass. Nonetheless, in the sliding experiments contained herein, the lithium fluoride deformed plastically when a sphere slid across the surface. The plastic deformation was sufficient to produce a relatively high concentration of dislocations at the surface and subsurface as a result of that deformation. Furthermore, the coaslescence of dislocations in the zone of maximum subsurface shear stress produced a cavity or void beneath the surface.

In the middle upper photograph the lithium fluoride surface has been examined after sliding at a somewhat higher load. As the load increases, the amount of the plastic deformation also increases with the paraffinic mineral oil, and the cavity subsurface is now somewhat larger as well. Both the cavity and the amount of plastic deformation have increased under the higher load, indicating that there is an increase in strain in the material. This is the behavior of metals in contact. Lithium fluoride, a relatively brittle solid, therefore, behaves in much the same way in sliding conditions as does a ductile metal.

When the load is increased even further, however, the results seen in the cross-sectional photomicrograph of the wear track in the upper right of figure 7-3 are obtained. The surface again shows a considerable amount of deformation. The concentration of dislocations is so great that the region at the surface and just below the surface is black with their high density. With an increase in load, there was an increase in the friction force observed because of the greater intimate contact through the lubricating film. The lubricating film broke down at the interface, and contact of the sphere to the lithium fluoride crystal surface occurred. As a result, the friction force increased. Evidence for the zone of maximum shear stress (as shown in fig. 7-1) is found in the photomicrographs in figure 7-3. This zone moved from subsurface to the surface when the friction force increased. Instead of the voids developing subsurface as a result of dislocation coaslescence, the maximum stress now occurred at the surface. The response of the system to the stresses on the surface was the generation of fracture cracks in the surface moving from the surface to a subsurface zone. Two sets of those fracture cracks developed along the natural cleavage planes in the lithium fluoride, the cleavage planes being essentially the same as the slip planes in this particular crystal system. Thus, two chevron cracks are seen developing from the surface and moving subsurface.

The mineral oil present on the surface of the lithium fluoride does not lubricate the surface effectively. Since it acts more as a blanketing medium or inerting environment for the surface, the actual interaction effects of the sphere with the lithium fluoride result in changes to the lithium fluoride. There are, however, substances which are surface active; for example, organic acids interact with inorganic crystal surfaces like lithium fluoride. Oleic acid is one such acid.

A small concentration of oleic acid was therefore placed in the paraffinic mineral oil and the same experiments under the same load conditions as those in figure 7-3 were repeated. The only difference being that those experiments in figure 7-3 were conducted with the straight paraffinic mineral oil. The sliding velocity in all these experiments is extremely low (0.005 mm/sec). The photomicrographs obtained from these three sets of experiments are presented as the lower set of photomicrographs in figure 7-3. It is interesting to compare and contrast the photomicrographs for the lithium fluoride surface cross section when the oil contained the oleic acid. At all loads there is a complete absence of any subsurface cracking or void formation, and, in addition, no surface cracks are initiated in the sliding process. The damaged area in the center lower photograph on the right was a preexisting defect in the crystal. Under an identical set of experiments, simply adding oleic acid to the mineral oil influenced the deformation behavior of the material. All the energy put into the surfaces was dissipated in strain of the lithium fluoride, with no evidence for the formation of subsurface or surface-initiated cracks. The presence of a lubricant can markedly alter the changes that take place in the material as a result of solid contact.

The presence of the cracks in the upper photomicrographs is extremely important in wear because those cracks that are subsurface in the two photomicrographs on the left eventually work their way to the surface and liberate a piece of material, which fractures out of the solid surface. Likewise, those cracks in the upper right photomicrograph that originate on the surface and move subsurface ultimately result in removing a piece of material. This is evident in the photomicrograph by the very blackened area near the surface for the two larger chevron-shaped cracks moving subsurface. A piece of material has been plucked out of the surface at this particular site. This happens because of the weakening of the surface of the material with the cleavage process. Thus, these cracks, both subsurface and surface, serve as a site for the formation of wear particles. They are the precursors for the formation of wear particles which are ultimately lost from the solid surface. In the wear process, the environment plays an extremely important role in whether wear in a particular system occurs and, when it does, to what extent. Environment is one very important factor, as shown by the photomicrographs in figure 7-3.

The results of figure 7-3 are related to the influence of surface-active media on the mechanical properties of solid surfaces. In figure 7-3 the presence of the oleic acid improves the ductility or plasticity of the lithium fluoride surface. It is more accommodating or forgiving when loads are applied to the surface, and it complies by behaving in a plastic manner. In the absence of the acid, however, the surface is much more likely to be restricted in the amount of deformation or strain it accommodates. It undergoes some degree of plastic deformation as evidenced by the etch pits, which reveal dislocation sites in the material subsurface. However, the amount that it takes up is relatively limited, and if the quantum of energy put into the surface as a result of the sliding process is greater than can be accommodated by the material in plastic deformation, the additional energy must be dissipated in some fashion. The manner in which the lithium fluoride accommodates this additional energy is by undergoing fracture or by the formation of subsurface voids or cracks that are observed in the upper photomicrographs of figure 7-3. The presence of oleic acid on the surface, however, increases the ability of the material to deform in a plastic manner. This basic observation, which has been made for a number of materials, was first recorded by Rehbinder in Russia in 1928. Since that time many other investigators have observed the same phenomena in the mechanical testing of materials in the presence and absence of surface-active media. This subject is discussed in greater detail in chapter 9.

Generation of Defects

If brittle materials such as lithium fluoride undergo plastic deformation in tribological systems, it is then needless to say that metals in sliding, rubbing, and rolling contact deform in a plastic manner. With deformation the generation of defects is commonplace in tribological surfaces.

One of the principal forms of defects is the generation of dislocations in metals in sliding, rolling, and rubbing contact. Generally, with sliding, rolling, and rubbing contact, an increase results in the concentration of dislocations in the material. Evidence for this is shown, for example, in the electronmicrograph of figure 7-4, which is obtained from a copper rider rubbing against 440C stainless steel in liquid methane. Despite the fact that the copper was cooled by the liquid methane, a considerable amount of strain developed in the copper with sliding (ref. 2). The photomicrograph in figure 7-4 is a section in the rider approximately 200 micrometers below the wear scar surface itself. The high concentration of dislocations is evident; dislocation density is so great in some areas that the areas appear to be completely black from dislocation tangles. This is very common in wear surfaces where a very high concentration of dislocations is observed with sliding, rolling, or rubbing surfaces.

Outside the actual contact area (wear surface area), the microstructure of the copper reveals a considerably lower concentration of dislocations as is evidenced in the photomicrograph in figure 7-5. Figure 7-5 is obtained on the copper surface just outside the wear scar area. The figure shows that the dislocation concentration is considerably less than it is in figure 7-4. The cell structure is generated by the dislocations. However, in figure 7-5 there is a complete absence of the high density that is observed in figure 7-4.

It is interesting to observe that the photomicrographs of figures 7-4 and 7-5 were obtained with copper sliding on a steel surface at the cool temperatures provided by liquid methane. Had these experiments been conducted at room temperature, the high concentration of dislocations observed in figure 7-4 may not have been seen because, in sliding at essentially ambient temperature, the interface temperature and the bulk surface temperature of the copper may be sufficiently high so as to exceed the recrystallization temperature in the copper, and the material would essentially anneal itself. If during the sliding process the surfaces receive

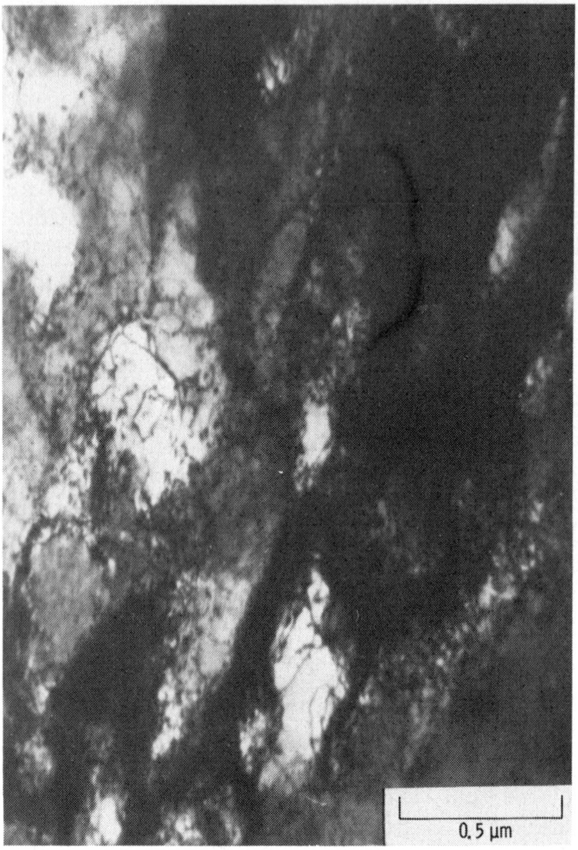

Figure 7-4. – Electron micrograph of copper subjected to sliding against 440C steel in liquid methane. Microstructure is representative of material from 200 micrometers below wear scar surface (ref. 2).

sufficient energy to maintain the system above the quantum of energy necessary for recrystallization, a continuous annealing temperature at the interface during the sliding process would inhibit the buildup of dislocations. If, however, the mechanical parameters in the surfaces were such as to reduce the quantum of energy below that necessary for recrystallization, then the photomicrograph in figure 7-4 would still be observed. The same effect can be achieved by changing the ambient temperature, which is what was done in the experiments conducted in liquid methane. The higher the ambient temperature at which the experiment is conducted, the more likely it is that the recrystallization temperature is reached and an annealing effect results in the surface.

The photomicrographs of figures 7-4 and 7-5 are for copper. The sliding velocity employed in the experiments (for which the photomicrographs in figs. 7-4 and 7-5 were taken) was 6.2 meters per second, which is a relatively

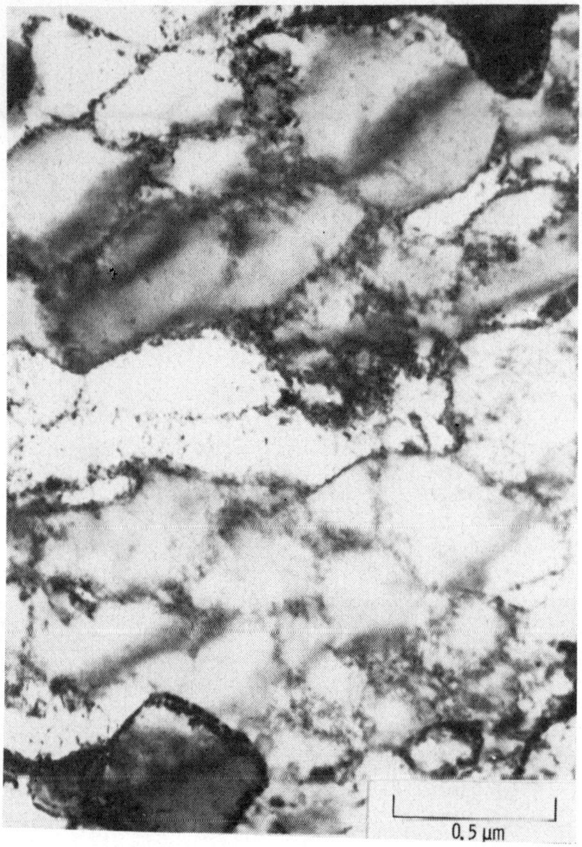

Figure 7-5.—Microstructure of copper adjacent to wear scar surface after sliding against 440C steel in liquid methane under 1-kilogram load at 6.2 meters per second. This microstructure is outside wear scar surface area (ref. 2).

high sliding velocity, under a fairly heavy load of 1 kilogram despite the fact that the materials were immersed in liquid methane. Under those conditions, the interfacial temperatures were not sufficient to bring about recrystallization, and the increase in dislocation concentration was observed in figure 7-4 in the wear contact zone. If, however, the sliding velocity or the load is increased so as to increase the quantum of energy placed at the interface, then the recrystallization temperatures can be achieved and the surface can be annealed. This was accomplished by increasing the sliding velocity from 6.2 to 12.4 meters per second, thereby increasing the quantum of energy placed on the surface of the copper in contact with the steel (ref. 2).

Increasing the interfacial energy caused the surface of the copper to undergo recrystallization. The evidence of this recrystallization is shown in a cross section of the copper wear surface in the photomicrographs in figure

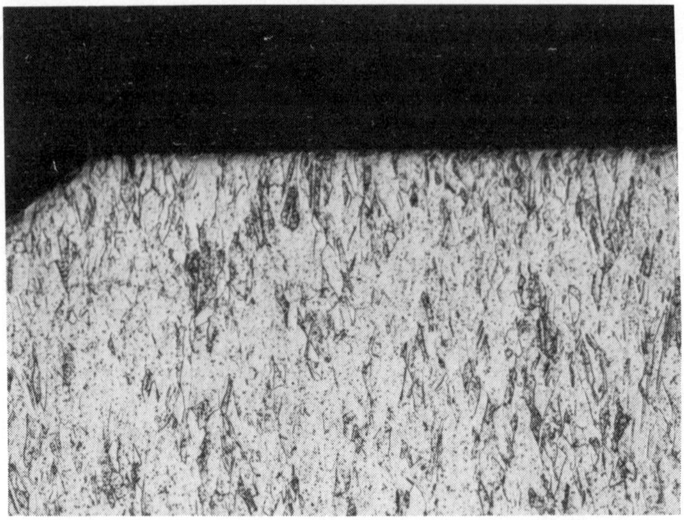

(*a*) *Entrance region.*

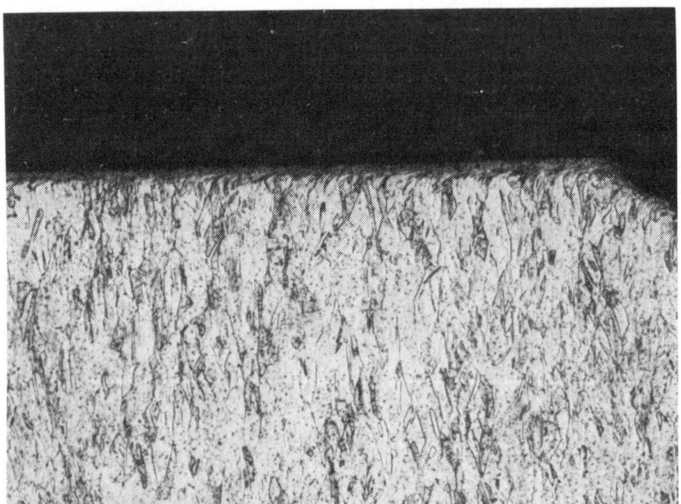

(*b*) *Exit region.*

Figure 7-6. – Metallographic sections cut parallel to sliding direction showing recrystallized layer on copper wear surface after sliding against 440C in liquid methane under 1-kilogram load at 12.4 meters per second (ref. 2).

7-6. Figure 7-6(a) shows the entrance region for the copper in contact with the 440C. Figure 7-6(b) shows the exit region with a very thin layer of recrystallized material at the surface.

This recrystallization process is kind of a self-healing process for materials. With metals, for example, increasing the amount of strain in the material increases the disclocation concentration (the defects) of the material. The recrystallization process produces an annealing effect. With annealing, if the time is sufficiently long at temperature, the crystallites can undergo grain growth at the surface of the material. In most tribological surfaces, however, the grain size is found to be very finely reduced in size as a result of the constant, continuous input of energy at the surface. There is insufficient time for grain growth to occur even though one is above the recrystallization temperature. In a tribological system, the metal does not experience the long time elements in conventional metallurgical areas where sufficient time may be allowed for grain growth to occur.

The defects (the high dislocation concentration in the wear surfaces) go into the generation of grain boundaries which are another form of defect in crystalline solids. The grain boundaries are nothing more than high concentrations of dislocations accommodating mismatches in adjacent orientations, as was discussed earlier in this text.

Correlation of Strain and Dislocation Density

Iron is probably one of the most common base materials used in alloys for tribological applications. Consequently, it is of considerable interest to those working in the field of tribology. Ruff has conducted wear experiments for steel sliding on iron to gain some insight into the amount of strain that develops in the surface and surficial regions of the iron, both unlubricated and lubricated (ref. 3). Steel was made to slide on the iron surface under a relatively light load of 50 grams with the surfaces either dry or lubricated with an oil. The strains measured in these two situations are presented in the data of figure 7-7.

In figure 7-7 the percent strain is plotted as a function of the distance from the wear track center in micrometers (μm). The upper curve in figure 7-7 is for the dry case of the steel sliding on the surface. The amount of strain is measured with surface analytical tools such as electron channeling and dislocation concentration measurements. An examination of the dry sliding case shows that the highest concentration of strain occurs at the surface of the material in the wear track center and that the amount of strain drops off relatively rapidly at some distance out from the center of the wear contact region. At various depths below the surface there is a considerable amount of strain, but it dissipates rather rapidly. At a 9-micrometer depth there is approximately 15-percent strain in the material at the wear track center, but it drops off with distance from the center. At 40 micrometers below the surface there is very little strain observed in the material; most of the strain dissipates by the time this depth is reached. The total wear track width in this instance is 160 micrometers.

When this same experiment is repeated in the presence of an oil, the total track width measured is considerably less, only 90 micrometers. Also, in addition to the width of the wear track being markedly less, the amount of strain and the strain pattern are somewhat different. The shape of the curve

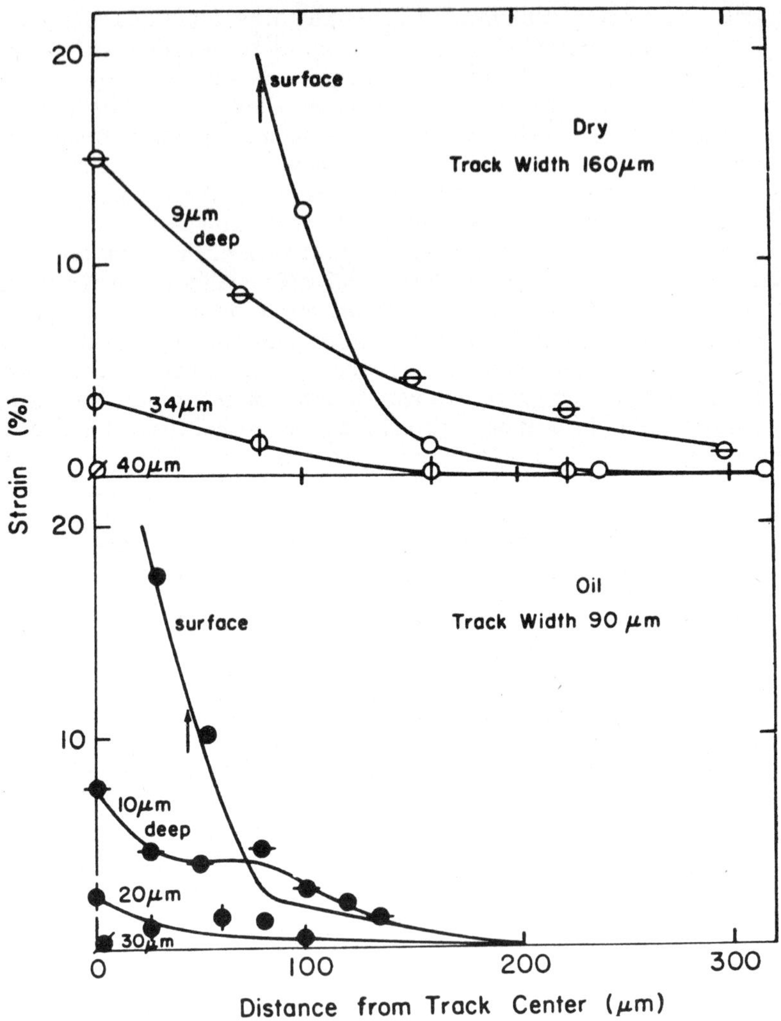

Figure 7-7. – Strain determinations at and beneath original surface for dry and oil-lubricated wear tests. Steel on iron; sliding distance, 17 meters; load, 50 grams (*ref. 3*).

for strain at the surface is similar to that observed for the unlubricated case; that is, the slope of the curve for strain with distance from the center of the wear contact zone is almost identical to that observed in the unlubricated case. It is simply shifted to the left on the plot. The maximum amount of strain achieved is comparable in both situations; there is simply less of it in the lubricated case.

When one examines the amount of strain subsurface, however, it readily becomes apparent from the data that the amount of strain drops off rather drastically and much more rapidly in the lubricated case subsurface than it does in the unlubricated case. For example, 20 micrometers beneath the surface the strain has dropped to approximately 3 percent in the wear track center. In contrast, for the unlubricated case, at 34 micrometers there was still at least this amount of strain in the material. At 30 micrometers beneath the surface, the strain is nearly dissipated in the iron for the lubricated case, while a depth of 40 micrometers was necessary to achieve the same effect in the unlubricated case. Thus, the total depth of strain for the unlubricated case is somewhat greater than it is for the lubricated case.

This greater amount of strain subsurface in the unlubricated case can be related to adhesion. If adhesion occurs at the interface, it is stronger in dry than in lubricated. Consequently, when the surfaces are moved tangentially over one another, a greater amount of force is required to overcome the adhesive bonds at the interface and to move the steel across the iron surface. To shear the interfacial junctions, more resistance is offered by the material, and that resistance is taken up in the form of strain or plastic deformation from the surface to the subsurface regions. While adhesion may occur in the lubricated case, the amount is much less. Consequently, the same depth of deformation or strain does not occur as in the unlubricated case because of a difference in adhesive behavior. In this particular set of experiments, the maximum amount of strain occurred at the surface, and this was the point of maximum shear stress because of the high degree of strain experienced at that location.

The relationship between strain and dislocation density is fairly straightforward in metallurgical systems. It is readily accepted that as the amount of strain increases the amount of dislocation density also increases. But in tribological systems, that concept required proof. In some abrasive wear experiments, Popov et al. demonstrated a relationship between dislocation density and the amount of strain that is observed in metals (ref. 4). They conducted some abrasive wear experiments where they measured the transformation of the surfaces in elemental iron and alloys during the abrasion process, and they correlated the amount of strain with the dislocation density. Some of their results are presented in figure 7-8 where both dislocation density and strain are plotted as functions of depth of the abraded wear surface. The amounts of strain are indicated in the lower curve for the iron and steel in dry sliding and oil lubricated. In the abrasive wear process, the amount of strain for the lubricated iron and steel is comparable to that of the elemental iron in the dry case. The maximum percent strain is seen at the surface. The strain for the iron is nearly 50 percent, while for the lubricated iron and steel it is approximately 35 percent. A considerable amount of strain occurs in the surface, however, as a result of the abrasive wear process. That strain drops off radically, however, as one moves subsurface. The data in figure 7-8 show that approximately 10 micrometers beneath the surface there is a sharp decrease in the amount of strain. A correlation between the percent of strain and dislocation density apparently exists for steel (fig. 7-8); the sharp decrease in strain with depth correlates with the sharp decrease in dislocation density with depth. At a depth of approximately 10 micrometers, both the strain and dislocation densities have dropped off considerably.

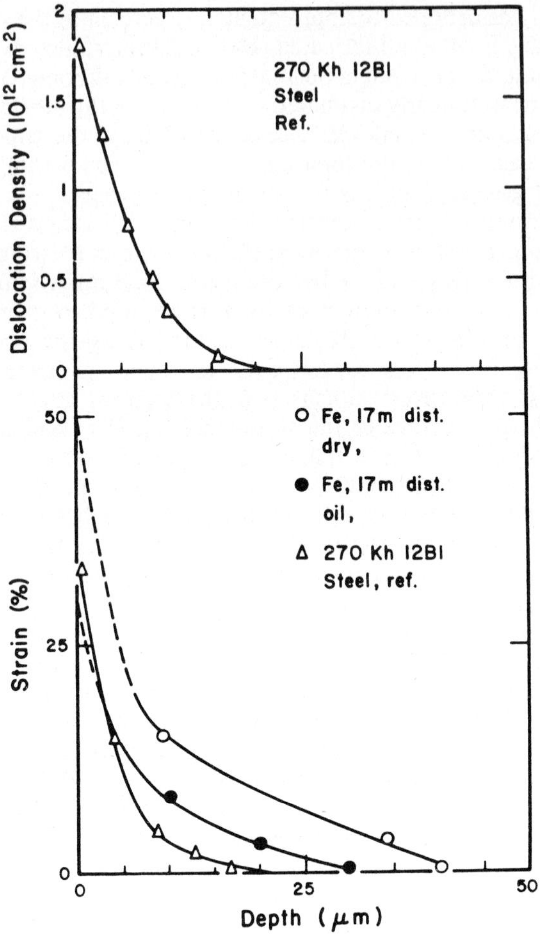

Figure 7-8.—Dislocation density and strain comparison (ref. 4).

Degrees of Wear

Wear is the removal of material from one or both of two solid surfaces in solid-state contact. It occurs when solid surfaces are in sliding, rolling, or rubbing motion relative to one another. Wear can occur on an extremely fine scale as might be the case in lightly loaded, well-lubricated systems where penetration of the boundary lubricating film is only intermittent and where solid-state contact occurs infrequently. Transfer and wear may occur on such a minute scale that it requires sensitive analytical surface tools to detect the presence of the wear or transfer of material from one surface to another.

On the other hand, the wear process can be extremely gross and thus result in the bulk removal of material from one surface or both in sliding, rubbing, or rolling contact with the result that the removed material is visible to the naked eye. Wear can take place gradually over a prolonged period of time with small amounts of material removed, or wear can take place rather drastically with the instantaneous loss of material from a surface. The gradual wear process is more typical of the kind of wear occurring in mild abrasion and corrosion processes; for example, it might be experienced with a polishing agent rubbed across a surface. Jeweler's rouge (iron oxide) rubbing on a surface produces a very fine form of abrasive wear that results in a loss of material from a surface. In practical tribological systems, this can occur when dirt is ingested between the surfaces of solids in contact. In addition to the mild abrasive wear condition, the mild corrosion situation that exists where the surfaces interact with the environment produces surface reaction products (compounds having their own individual identity). These materials can be removed from the surface. As a result of their removal, material from one (or both) of the solid surfaces in contact is lost from the system. This also can be a very long-term wear process. If, however, the corrosion process is accelerated (the reaction environment becomes more reactive or the concentration of reactant in the environment increases or the temperature or pressure of the system increases), the corrosion can become severe rather than mild. In such a situation, the loss of material due to corrosive wear may be enhanced considerably.

Types of Wear

A number of forms of wear have been categorized by various investigators through the years. It is generally held that the most common types of wear are (1) adhesive, (2) abrasive, (3) corrosive, (4) erosive, (5) cavitation, (6) fatigue, and (7) fretting. Each of these forms will be discussed individually. It is important, however, to recognize that, in wear, material is removed from surfaces. The form of that removal can take place in different manners according to the mechanisms operating in the system. Of the aforementioned wear mechanisms, one or more may be operating in one particular mechanical system.

With the wide variety of wear mechanisms that can operate in a practical tribological device, there are various forms of wear debris that can be generated as a result of the wear process. The wear particles generated in adhesive wear, for example, can take the nature of material plucked out of a surface. This occurs because of subsurface fracture where the adhesive bond at the interface is stronger than the cohesive bond in the weaker of the two materials. The particles can take the shape of flat platelets, as is experienced in fatigue or in some instances erosive wear, or they can take the form of individual lump-shaped particles of varying sizes, as might occur in abrasive and fretting wear. The wear particles can also occur in the form of flakes, as might occur in cavitation. In corrosive wear, the wear particles may not be particles at all but instead may be powders that form as a result of the environment interacting with the solid surface to form surface compounds which can be washed from the surface by lubricating fluids. The reaction products can also be gaseous where the environment interacts with a solid surface to produce volatile compounds.

In recent years, a number of investigators have observed, in a variety of different wear situations, the generation of spherical wear particles. Since these particles have been observed in a number of wear mechanisms, obviously they are not unique to any one wear mechanism. In some instances, these spherical particles have been examined in detail using the SEM. They have been found to be hollow spheres, which may be the end product of a flake or laminar piece of wear particle that has curled into a spherical shape (ref. 5). In figure 7-9, two spherical wear particles are shown. The photomicrograph on the left shows a fractured spherical wear particle. The outside is an extremely smooth sphere, and the inside contains furrows and ridges that are relatively smooth. The photomicrograph on the right shows a spherical wear particle that is not fractured but retains its spherical shape. It is cradled in what appears to be a larger particle that has already fractured into smaller pieces.

The spherical particles of figure 7-9 were obtained from grinding, which involves the abrasive wear process. The particles can vary from 1 to 30 micrometers in diameter. The spherical shape combined with the convoluted surface features on the inside of many of the particles indicates that melting followed by rapid cooling probably occurred. Since in the grinding procedures the interface temperatures can reach extremely high levels, it would not be surprising to find that the abrasive wear particles have undergone melting. The liquid wear particle then takes a spherical shape and gives the smallest surface to volume ratio. As a result, spherical wear particles are found in the grinding operation. Other investigators, however, have observed this form of wear debris in the fretting, cavitation, erosion, and fatigue processes.

To reduce wear in practical tribological devices, it is important to understand and appreciate the various wear mechanisms. If the mechanism of wear operating in a particular practical system can be identified, this can do much to assist in reducing wear.

Adhesive Wear

Nature of Adhesion

When two solid surfaces are placed in solid-state contact, bonding across the interface generally occurs. This bonding can be relatively weak when it is simply a physical attraction of one body for another (wherein the bond forces are of van der Waal's nature and relatively easy to rupture). It can also be extremely strong when it develops across the interface (where, e.g., chemical bonding occurs). When a load is applied to the two surfaces in contact the likelihood of strong bonding occurring is enhanced. This is true whether a surface film is present or not. Even with the most effective lubrication, contact occurs through the films if the loads or forces applied to the solid surfaces in contact are sufficiently high. When two atomically

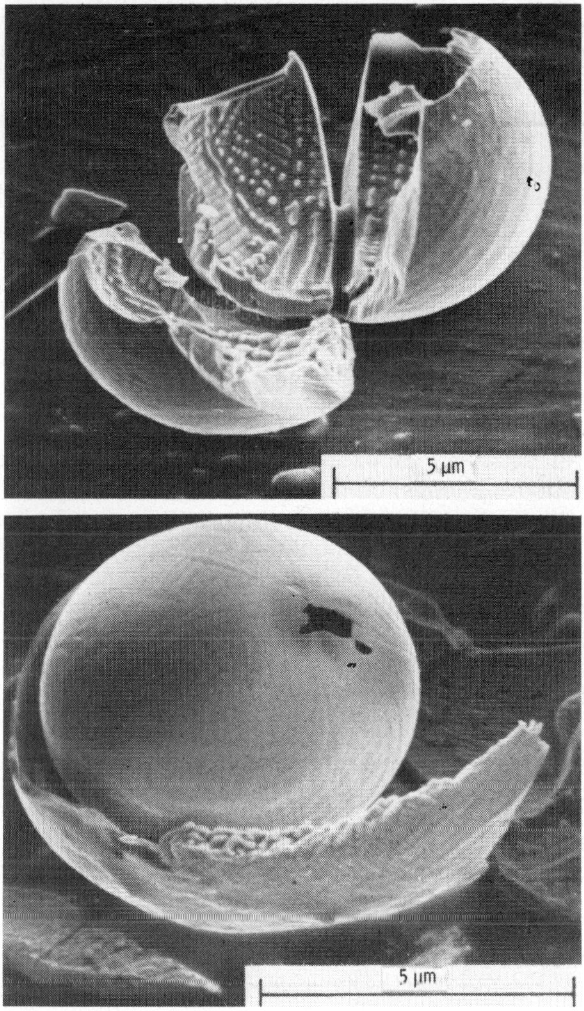

Figure 7-9. – Scanning electron micrographs of hollow spheres (ref. 5).

clean surfaces are placed in contact, the attractive forces of the solids for each other are sufficient to produce extremely strong bonding because of the electronic nature of the solid surfaces in the clean state. When the surfaces are clean, the entire real area of contact undergoes adhesion. Where the surfaces are covered with a film, whether it is simply a contaminating surface film or a lubricating film, the amount of adhesion in the real area of contact is some percent of the total real area of contact depending on the effectiveness of the surface film in preventing solid-state contact through the film. At extremely light loads, there may be a total absence of solid-state contact (for nascent surfaces); at heavier loads, however, contact through the surface films may occur.

The materials in contact, as well as the contaminating film, have a strong influence on whether adhesion occurs through contaminating films. For example, with aluminum in an ordinary air atmosphere, the surface is covered by a film of aluminum oxide. If a steel surface is pressed against aluminum under very light loads, the aluminum oxide may not be disturbed, and the only adhesion is the relatively weak adhesion between the steel and the aluminum oxide. If, however, the load is increased so that the elastic limit of the aluminum is reached and the aluminum undergoes plastic deformation under the oxide, the oxide is relatively brittle and breaks up; this condition exposes nascent aluminum to the steel at the interface. When the nascent aluminum contacts the steel, whether it be the elements of the steel itself or the oxide present on the steel surface, the bonding is extremely strong because elemental aluminum metal bonds very strongly to oxygen of the iron oxide or to the elemental iron itself. In either case, the adhesive forces measured across the interface are relatively high.

In any of these situations where adhesion occurs at the interface and tangential motion is imposed on the surfaces to cause one to move relative to the other, shear must take place to rupture the adhesive bonds at the interface to accomplish the tangential motion. In practice, the bonds very seldom rupture at the interface; instead, the bonds of the cohesively weaker of the two materials rupture. The cohesively weaker material transfers to the cohesively stronger material. This results in adhesive wear or the loss of material from one surface as a result of the adhesion process. In some systems, adhesion and adhesive wear can occur without tangential motion. A classic example of this is in making and breaking electrical contacts. The contacts are brought together normally to conduct electric current. When the surfaces are separated, they are separated in a normal direction. There is generally no tangential motion intentionally placed on the two surfaces in contact. Under such conditions, transfer can occur from one surface to another. For example, many have observed the ignition points in an automobile (before electronic ignition) where the mechanical contacts could be seen and their surfaces examined. Very frequently the mode of failure of the electrical contact was adhesive wear or the transfer of one material to the surface of the other. Generally, a pit was observed in one surface and a mound of material was observed to have transferred to the other from the contact. This is adhesive wear. In general, imposing an electrical current across an interface in a tribological system tends to provide more energy at the interface to promote adhesive wear and transfer.

Where the two solid surfaces placed in contact consisted of two different materials (indicated diagrammatically in fig. 7-10 for a pin in contact with a flat), adhesion can occur on touch. If the hemispherical rider is simply lifted off the surface, transfer can take place (material A to material B or material B to material A, depending on which is the cohesively weaker). If tangential motion is imposed between the two surfaces in contact, then a transfer film of material is laid down with the sliding process as indicated by the material accumulated on the solid surface (fig. 7-10). As the rider moves along the surface, it lays a film of transferred material because as the rider material A moves along the surface it makes and breaks new contacts with the material

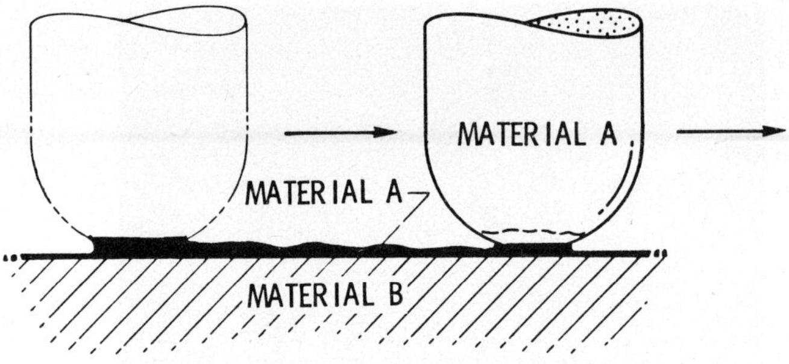

Figure 7-10. – Adhesive wear mechanism.

B; if these adhesive contacts are stronger than the cohesive bonds in material A, material A lays on the surface of material B with the sliding process. A transfer film can be seen on material B. A typical example of this occurrence would be for copper metal riders sliding on a steel disk surface. Generally, the copper is observed to transfer to the steel surface, and various analytical tools have been used to identify the presence of the copper. A very good tool which has revealed such transfer is the AES analysis. As little as 0.01 of a monolayer of copper on a steel surface can readily be detected with a single passage of a copper slider across the steel surface.

Where one wishes to study the transfer behavior of two dissimilar materials in contact, Auger spectroscopy is an ideal tool because it can detect all the elements present on the solid surface (except hydrogen and helium) in less than 0.1 second. Where two dissimilar wear resistant materials are in solid-state contact, transfer can be detected with the surface sensitive Auger analysis.

Where the surface transfer films are thicker, other analytical tools can be used to follow the transfer process. SEM is extremely useful for studying dissimilar materials in contact and observing the adhesive wear process. The physical nature of the transfer or the morphology of the transferred particles can be observed directly in the scanning electron microscope. To insure that one is observing adhesive wear and the transfer of material from one surface to another, an energy dispersive X ray analysis assists when that is incorporated directly into the scanning electron microscope.

An example of the usefulness of the scanning electron microscope coupled with the energy dispersive X-ray analysis for studying adhesive wear is presented in figure 7-11. In figure 7-11(a) is a scanning electron micrograph of an iron surface after an aluminum pin slid across the surface. The transfer of aluminum to the iron is readily apparent as stringers or globules of material (fig. 7-11(a)). To be certain that the material being observed on the surface is aluminum, an X-ray dispersive analysis map was made of the surface (fig. 7-11(b)). A comparison of figure 7-11(a) with 7-11(b) indicates that the material in figure 7-11(a) is transferred wear; this is confirmed in the X-ray map in figure 7-11(b). Thus, the aluminum has undergone transfer to the iron surface with a net loss of aluminum from the aluminum body.

(*a*) *Scanning electron micrograph.*

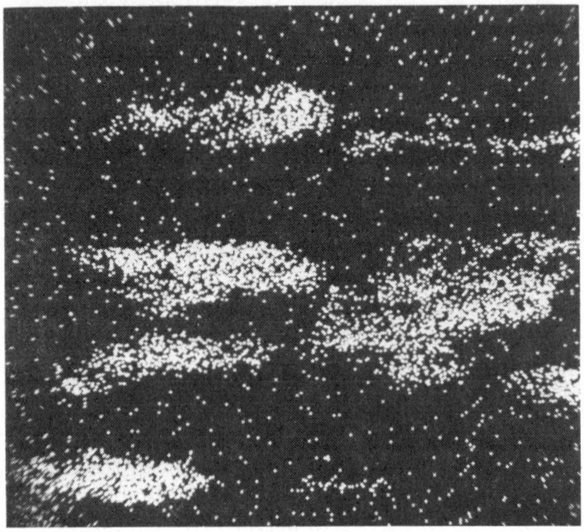

(*b*) *X-ray dispersive analysis map.*

Figure 7-11.—Aluminum transferred to iron after dry sliding. Sliding velocity, 0.15 centimeter per minute; load, 200 grams; temperature, 23° C; single pass.

The adhesive wear process manifested in figure 7-11 is a series of adhesions with shearing in the aluminum taking place repeatedly as the surface of the aluminum moves across that of the iron. When the same surface is passed over by the aluminum a second and a third time, a greater amount of aluminum is observed to transfer to the iron surface. With the first pass, it is adhesive bonding of aluminum to the iron with shear taking place in the cohesive bonds in the aluminum. On subsequent passes, however, an increasing amount of aluminum to aluminum bonding occurs; that is, aluminum from the aluminum body bonds to the transferred aluminum film because of a continuous increase in the coverage of the surface of the iron by aluminum. Ultimately, one is sliding aluminum on aluminum and the general adhesion and friction behavior is that of aluminum in contact with itself.

Aluminum and iron are extremely chemically active elements and strong bonding occurs between them. Therefore, it is not at all surprising that one might experience adhesion and transfer of aluminum to an iron surface. An obvious question is whether all metals behave like iron and aluminum when dissimilar metals are placed in solid-state contact. If elements are selected which do not react readily with other elements or with the environment, one finds that the adhesion process still takes place. For example, the members of the noble metals family, particularly gold, are relatively inert (interacting or reacting very weakly). Likewise, the members of the platinum metals family, while serving as good catalysts for catalzying a number of types of chemical reactions, are relatively weakly reactive; that is, they do not react with a strong chemistry or thermodynamics of reactions with either environmental constituents or other materials. Notwithstanding the foregoing, however, in sliding contact of noble metals with members of the platinum metals family, adhesion and transfer of these metals are observed regularly.

Of considerable practical interest are the metals rhodium and gold, since they are used in electrical contacts for surfaces and surface films. In such applications, the adhesion is extremely important because, if adhesive transfer occurs, it can destroy the electrical contact and cause a marked increase in electrical noise in these contacts. An examination of the sliding of gold across a rhodium surface shows that adhesion occurs for these relatively inert metals and transfer from the surface of one to the other readily occurs. Figure 7-12 presents an Auger spectrum for gold having slid on a

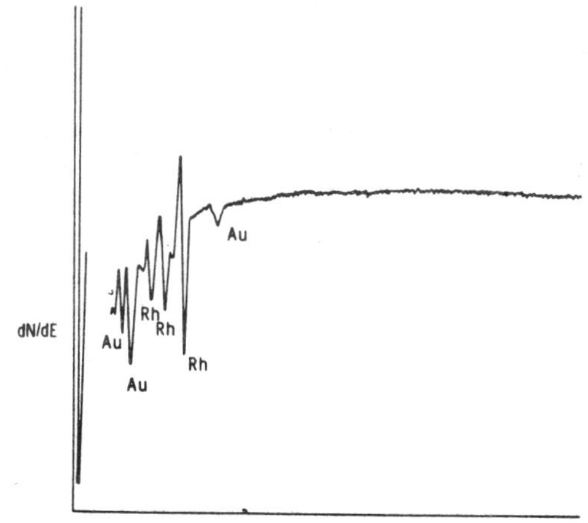

Figure 7-12. – Auger emission spectrum for rhodium (111) single-crystal surface after single pass of gold (111) crystal across it. Sliding velocity, 0.7 millimeter per minute; load, 10 grams; pressure, 1.3×10^{-8} pascal; temperature, 23° C.

rhodium surface. The Auger spectrum is for the rhodium surface. The spectrum presented in figure 7-12 shows Auger peaks for both rhodium and gold. This spectrum indicates that, with a single pass of the gold across the rhodium surface under extremely light loads of only 10 grams the adhesion of the gold to the rhodium was sufficiently strong to result in the transfer of gold to the rhodium surface. Thus, while both gold and rhodium may be relatively inert when they are placed in solid-state contact, adhesion occurs between them. Also, with tangential motion, visible transfer is observed. The general principle that the cohesively weaker of the two materials transfers to the cohesively stronger is borne out in the data of figure 7-12. Gold is the cohesively weaker of the two metals in contact, and since the gold is observed to have tansferred to the rhodium surface, the interfacial bond obviously is stronger than the cohesive bonds in the gold.

Adhesion followed by transfer is the general rule for dissimilar metals in contact when they are in a clean state. Nearly all metals in the clean state placed in solid-state contact show transfer of the cohesively weaker material to the cohesively stronger material.

This indicates that the adhesive bond at the interface is stronger than the cohesive bond in the cohesive weaker of the two materials. This, in turn, reflects that the interfacial bonds may be extremely strong even when no strong affinity of the elements exist at the interface. For example, some dissimilar metals in contact are completely immiscible in one another in the solid state; that is, there is no solubility of one element in the other. Copper and lead is one case, silver and iron another. Despite these insolubilities, strong adhesive bonds for these elements form and result in transfer of copper to lead as well as silver to iron. Surface energetics, therefore, are entirely different than bulk energetics; bulk properties should, therefore, not be used to predict surface behavior in tribological systems.

Dynamic Wear

In many tribological systems, it is extremely important to be able to measure the wear dynamically as it is taking place because changes in the rate of wear can take place with progressive wear. An example might be the system of aluminum sliding on iron mentioned earlier. Initially, aluminum is in contact with iron, and the shear process involves the shear of aluminum after the adhesion of aluminum to iron. With repeated passes over the surfaces, however, the aluminum builds up on the iron surface as a result of the development of a transfer film. At some time, aluminum is sliding on aluminum. Under such conditions, a change in the rate of wear might be experienced.

As was mentioned, Auger spectroscopy is a useful tool for studying such transfer where dissimilar materials are in contact. A cylindrical mirror Auger (CMA) analysis has been incorporated directly into a friction and wear experiment with the result that adhesive wear can be monitored continuously. In a pin-on disk experiment with an aluminum pin sliding across a steel surface, the transfer of aluminum to the steel was monitored continuously in the Auger spectra. In the CMA all the elements can be detected on the surface in less than 0.1 second. Thus, dynamic experiments can be conducted, and the chemistry of the surface can be monitored continuously. Since it is possible to deflect the beam with CMA, the chemistry in the wear track can be observed. Also, the beam can be deflected outside the wear track to observe the chemistry at that point. The effects of wear on the solid surface from a chemical point of view (both qualitatively and semiquantitatively) can thus be compared.

Auger spectra obtained on an oscilloscope with the CMA in a dynamic friction experiment for aluminum sliding on a steel surface are presented in figure 7-13. Figure 7-13(a) is an Auger spectra of the steel surface before sliding contact showing the low energy iron peak and the carbon Auger peak. The iron peak is due to the iron in the steel, and the carbon peak has a number of sources; one is the carbon present in the steel, and another is the carbon monoxide and carbon dioxide which may be adsorbed on the surface of the steel and its oxides. After one rotative pass of the aluminum slider across the surface, a bulge begins to appear in the iron peak as indicated in the Auger spectra in figure 7-13(b). The aluminum Auger peak appears adjacent to the low energy iron peak. Two things have occurred: (1) a bulge has appeared in the iron peak (which reflects the presence of aluminum), and (2) a diminution has occurred in the intensity of the carbon peak. The decrease in intensity of the carbon peak results from the aluminum on the surface blurring the carbon. After the aluminum slider slides across the steel

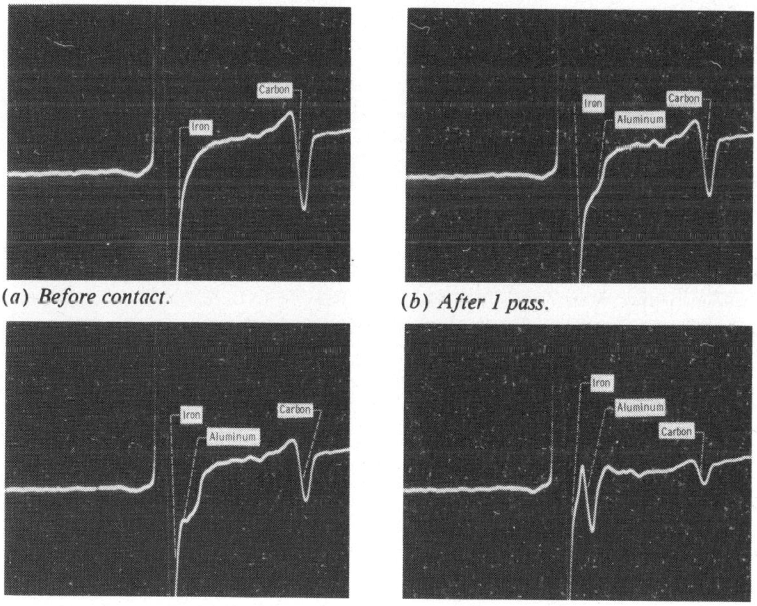

(*a*) *Before contact.* (*b*) *After 1 pass.*

(*c*) *After 10 passes.* (*d*) *After 20 passes.* CS-62854

Figure 7-13.—Auger analysis of steel disk surface before and during sliding contact with aluminum rider. Sliding velocity, 20 centimeters per minute; load, 500 grams; temperature, 23° C.

surface 10 times, the Auger spectra in figure 7-13(c) is obtained. In figure 7-13(c), the bulge in the iron peak (part of the aluminum peak) becomes much more pronounced, and a further diminution in the intensity or size of the carbon peak occurs. The carbon peak decreases in intensity as the carbon on the surface is covered with aluminum. After 20 passes, the Auger spectrum in figure 7-13(d) is obtained. Now, in addition to the iron peak, there is a distinct, separate aluminum peak associated with the aluminum which has transferred from the aluminum slider to the steel disk surface. In addition, a marked diminution is observed in the peak height of the carbon. The carbon has decreased in intensity markedly as a result of the buildup or transfer of aluminum to the steel surface.

The mechanism of transfer then can be related to the measured relative peak intensities of the aluminum to the carbon. The ratio of aluminum to carbon in the Auger spectrum gives some insight into the mode of transfer of aluminum to the steel surface. In figure 7-14, the ratio of aluminum to carbon Auger peak intensities are plotted as a function of the number of rider passes across the surface. For the first 12 passes, a relatively modest slope is seen from pass 4 through pass 12. It reflects a relatively gradual transfer. With, however, a transfer film of aluminum developing on the steel surface after 12 passes, a marked increase in the concentration of

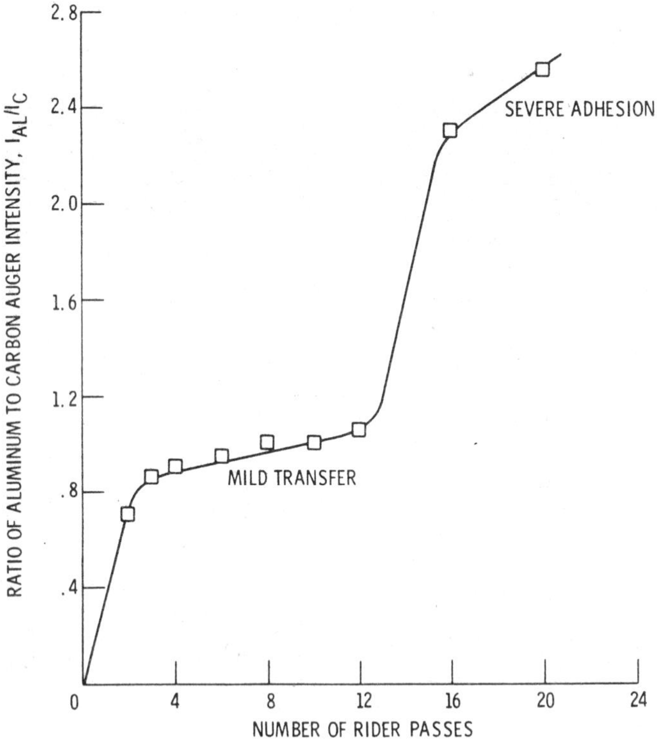

Figure 7-14. – Detection of aluminum adhesive transfer to steel surface during sliding friction experiment. Sliding velocity, 20 centimeters per minute; load, 500 grams; temperature, 23° C.

aluminum is observed on the surface at 16 passes; this indicates that the aluminum is now adhering to aluminum. The cohesive bonding of aluminum to aluminum is stronger than aluminum to steel. As a result, there is a change from mild adhesion to severe adhesion, and there is a transition from a mild wear condition (mild transfer) to one of severe adhesion and severe wear.

Many researchers in this field distinguish between mild and severe wear. It is particularly true in additive chemistry where additives are placed in oils and the measure of their effectiveness is their ability to maintain mild wear conditions. When loads on the surfaces become sufficiently high so as to result in severe wear, this transition in load is used as an indicator of the effectiveness of an antiwear additive in preventing wear and seizure from occurring for materials in contact. The transition then is really one from mild to severe wear (from mild to severe adhesion). In the cases of dissimilar metals in contact (such as the aluminum in contact with the steel in figs. 7-13 and 7-14), it is a transition from adhesion for dissimilar materials in contact to cohesion of like materials in contact when a transfer film has developed on the surface.

Fracture Cracks

Adhesive wear can take many forms as a result of the properties of materials in solid-state contact. In all cases, a necessary prerequisite for adhesive wear is adhesion at the interface. In some instances, transfer may occur in a smooth fashion as it does for the aluminum transferring to the steel surface. In other cases, the adhesion can produce a marked disruption of a solid surface. An example of the latter is observed when a copper polycrystalline slider moves across a copper bicrystal surface. Adhesion of copper to copper occurs across the interface. With tangential motion, the strong bonds at the interface result initially in the energy being dissipated in a form other than shear. Rather, separation takes place along slip bands in the material, and fracture cracks or cavities form in the material. These are demonstrated in the two grains of a copper bicrystal surface where a polycrystalline copper slider slid across the surface. Scanning electronmicrographs revealing the nature of the fracture cracks or cleavage cracks are presented in figure 7-15.

Figure 7-15 shows two photomicrographs, one taken in the wear track on the (111) grain of copper and the other on the (210) grain of copper. The magnification is the same in both. The size of the fracture crack in the (210) surface is much larger than on the (111) surface. On the (111) surface, there are a series of smaller cracks. A careful examination of the scanning electronmicrograph of these cracks reveals the extremely smooth nature of the bottom side of the crack. What has happened is that adhesion has occurred across the interface. As one attempts tangential motion, fracture occurs in the cohesively weakest zone. With the single crystal surface, the cohesively weakest zone is along slip or cleavage planes in the copper single crystal. Cracks or voids appear because the surface literally opens up; the material is pried open as the attempted tangential motion occurs. At some

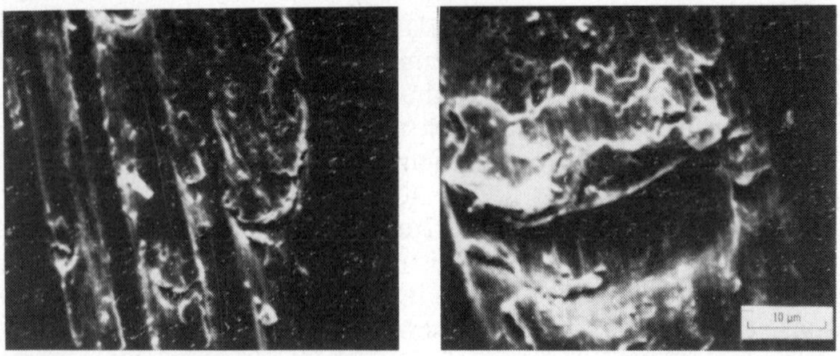

Figure 7-15.—Wear tracks on bicrystal grains. Copper slider; load, 100 grams; speed, 1.4 millimeters per minute.

point, the tangential force is sufficient to overcome the adhesive bond at the interface and fracture of the adhesive junction (or near the adhesive junction) occurs, leaving a curl of material above the surface. This is best seen in the upper portion of the photomicrograph for the (210) grain where material has been pulled above the plane of the flat surface, and it represents the region where the adhesive junction was fractured with tangential motion.

The topography of the surfaces under identical sliding conditions is markedly different in the two orientations. Generally, lower adhesive wear is observed for the high atomic density, lower surface energy (111) plane in the face-centered-cubic system than it is for other planes in that same system. Thus, the photomicrographs in figure 7-15 reveal the anisotropic wear behavior of copper.

The mechanism operating in the development of the fracture cracks as a result of adhesion in figure 7-15 are shown schematically in figure 7-16. In this figure, a rider or slider specimen slides on a flat disk surface (e.g., the

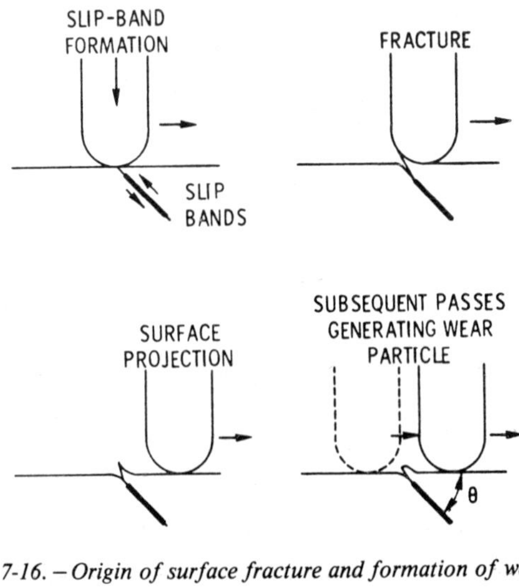

Figure 7-16.—Origin of surface fracture and formation of wear particle.

bicrystal surface of fig. 7-15). Adhesion occurs across the interface at the point of contact. When tangential motion is imposed to cause the slider to move across the surface, the adhesive bonds at the interface are stronger than the cohesive bonds across the slip planes, which happen to also be the cleavage planes in the face-centered-cubic metal copper. With the imposition of tangential motion, the slider tends to move forward pulling with it copper and causing separation across the slip bands in the copper. With further tangential motion, the force applied to the slider is sufficient to overcome the adhesive bond near the interface. Shear occurs and the junction is broken with the slider moving on. As it does, however, it leaves behind a small wake or curl of material setting above the surface. Subsequent passes of the slider across the surface cause shearing of the particle standing above the surface with the resulting generation of an adhesive wear particle. In this instance, the particle has been generated in a slightly different fashion than was observed and discussed in reference to figures 7-11 to 7-14. There, adhesion of the aluminum to the steel occurred on touch contact; with tangential motion, shearing occurred in the metal and resulted in adhesive transfer and ultimately in adhesive wear.

In figures 7-15 and 7-16, the adhesive wear process takes place by a slightly different two-step mechanism. The first step of the mechanism is adhesion, with tangential motion producing the generation of a curl or a discontinuity in the surface plane. The second step of the process involves shearing the curl; it is the shear process that generates the actual free wear particle. These two slightly different mechanisms produce wear as a result of the adhesion process.

Chemical Effects

The chemistry of solid surfaces and the changes therein influence the adhesive wear behavior of materials in solid-state contact. Probably one of the most frequently used materials in practical systems, as already mentioned, is steel. On a steel surface, in a normal air environment, one finds the normal oxides of iron (FeO, Fe_3O_4, and Fe_2O_3) in varying proportions, depending on the conditions of the surface, the environment, the ambient temperature, and other factors. In sliding, rubbing, or rolling contact for steel surfaces, the three oxides of iron play an important role in adhesive wear behavior. Fe_2O_3 is abrasive in nature, while FeO and Fe_3O_4 are not. Intimate contact of iron with itself results in the generation of adhesive wear particles. The role of the oxides, generally, is to inhibit the formation of the nascent metal junctions or bonds and thereby inhibit adhesive wear. In general, the normally occurring oxides present on a steel surface do a very effective job of inhibiting adhesion and adhesive transfer. However, the effectiveness of these surface films is in part a function of the particular mechanical parameters which are imposed on the system for the two solid surfaces in contact. Thus, such parameters as load pressing the two surfaces together and the sliding speed with which one surface rubs against another play a role in the adhesive wear behavior.

place is much greater than where the polymer slides along (with) the grinding grooves. Thus, for polymers in contact with rough metal surfaces, the wear process may be a combination of adhesion and/or cutting (abrasive) wear of the polymeric material. In such situations, the mechanical or physical features of the metal surface, such as surface finish or surface roughness, can play a significant part in the observed wear behavior of polymers. This is demonstrated in figure 7-18 for a polyethylene polymer sliding on various roughnesses of a stainless steel surface (unpublished data presented by M. A. Swikert at NASA Lewis Research Center Inspection in 1973). An examination of figure 7-18 reveals that for extremely smooth surfaces the wear rate of the polymer (where adhesion is at a maximum) is fairly high. As the amount of surface roughness increases, the wear to the polymer surface decreases. The initial high wear associated with the smooth surface may be a direct result of the adhesion forces binding the polymer to the metal surface. With increasing roughness, the real area of contact is interrupted by the surface topography and a decrease in the wear is observed. There is obviously (fig. 7-18) an optimum at approximately 15 rms metal surface roughness for a minimum in wear for the polyethylene. With an increase in roughness beyond 15 rms, the wear again begins to increase because of the skiving or cutting action referred to earlier. Thus, one can identify in figure 7-18 a very strong adhesive component of polymer to metal surface at the very smooth end of the scale (at 5 rms) and a combination of adhesion and cutting (skiving) action for very rough surfaces (70 rms).

At 15 rms where a minumum is observed, it is reasonable to assume that there is very little cutting action and, at the same time, the intimate contact of the polymer with the metal surfaces has been broken up by the surface roughness. Consequently, a reduction in wear is observed.

Other mechanical parameters also influence the wear of polymers. For example, the load on the polymers influences the real area of contact and also the adhesion of the polymer to, for example, metal surfaces. Changes in such mechanical parameters as load and speed also change the properties of the polymer at the interface. Most polymers have relatively low melting temperatures, and the mechanical parameters, such as load and speed, can bring about localized surface melting, which markedly changes the wear behavior particularly where adhesion plays an important role in the wear

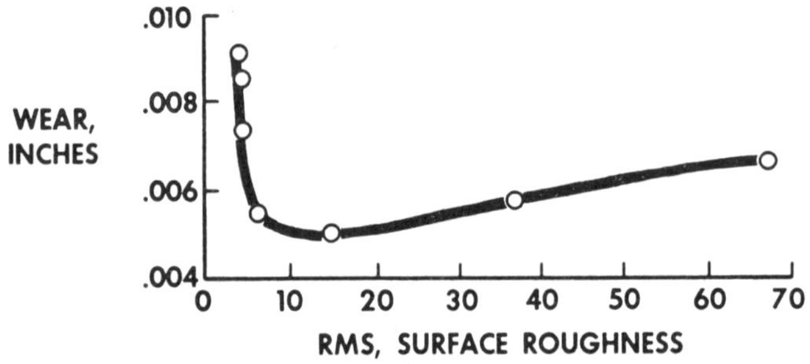

Figure 7-18. – Effect of stainless steel surface roughness on wear of polyethylene.

process of the polymers. Also, the chemical nature of the metals in contact influences the adhesive behavior of the polymers and, ultimately, the adhesive wear. Some metals interact much more strongly with polymers than do others. Metals that are notoriously reactive (e.g., titanium and aluminum) influence the adhesive wear behavior of polymers because they influence the transfer of polymer to the metal surface as a result of adhesive interactions.

Almost all polymers contain carbon, and titanium forms a very strong chemical bond with carbon. Many polymers contain oxygen in addition to carbon, and nearly all polymers contain hydrogen. Titanium has a strong affinity for both of these elements as well. Thus, there is a strong tendency for adhesive interactions by chemical bonding of many polymers to metals such as titanium. When such interactions take place, the polymer frequently transfers to the metal, and adhesive wear of the polymer occurs. Changing the load and speed effects the adhesive wear of the polymeric material. This effect is demonstrated in the surface profiles (fig. 7-19) obtained on ultra molecular weight polyethylene after it slid 2000 times against the titanium alloy in the absence of any lubricant (ref. 7). The surface profile in the upper trace represents the highest load examined. Decreasing load is shown in the figures. With a decrease in load, there is a decrease in the amount of polymeric material removed from the surface under equivalent conditions of total number of passes of sliding and material combinations. The only change was the decrease in load, and this load decrease produced a marked decrease in the amount of adhesive wear that was observed on the polymeric surface. Thus, load alters the adhesive wear behavior.

As was mentioned earlier, a considerable amount of chemistry takes place when metals interact with polymers, and surface analytical tools such as

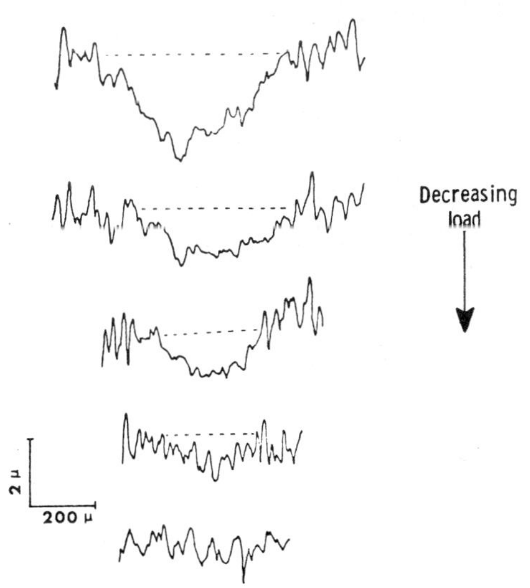

Figure 7-19. – Cross section of wear scar on ultra-high-molecular-weight polyethylene after 2000 passes of titanium alloy slider with no liquid present (ref. 7).

Auger spectroscopy are extremely useful in studying these chemical interactions. Auger can be used to identify the presence on the surface of a metal of elemental species from the polymer where transfer has occurred. Pepper (ref. 8) conducted a series of experiments with polyvinylchloride sliding against various metal and alloy surfaces, and he used Auger spectroscopy analysis to monitor the transfer of the polymer to the metal surface. In general, the polymer transfers to all metal surfaces.in sliding contact with the first pass of the polymeric material across the metal surface. Some of his results are presented in figure 7-20. Figure 7-20(a) is an Auger spectrum for an S-Monel disk. S-Monel is primarily a copper-nickel alloy containing silicon. The Auger spectrum in figure 7-20(a) is for the disk surface after sputter cleaning and prior to rubbing with a polyvinylchloride rider specimen. Examining the disk with Auger analysis indicates (fig. 7-20(a)) only the peaks associated with nickel and copper. Sliding with a single pass of the polyvinylchloride rider across the surface produces the Auger spectrum in figure 7-20(b) where, in addition to the nickel and copper peaks, chlorine and carbon peaks are also present. The presence of chlorine and carbon on the surface is due to the transfer of the polyvinylchloride to the S-Monel surface. A single pass of the slider across the surface is sufficient to transfer the polymer to the S-Monel. Two of the principle elements in the polyvinylchloride are carbon and chlorine. Both are readily detected by an Auger spectroscopy analysis. Thus, an Auger spectroscopy analysis is useful in studying the polymer transfer where very thin films of polymers are laid down in the interaction of polymers with metals.

Similar experiments were conducted by Pepper with PTFE. It was found that PTFE would also transfer to the S-Monel surface and that the film was uniform and continuous with no buildup in the multiple tranversals across the surface. The film thickness is estimated from the Auger analysis to be two to four atomic layers thick.

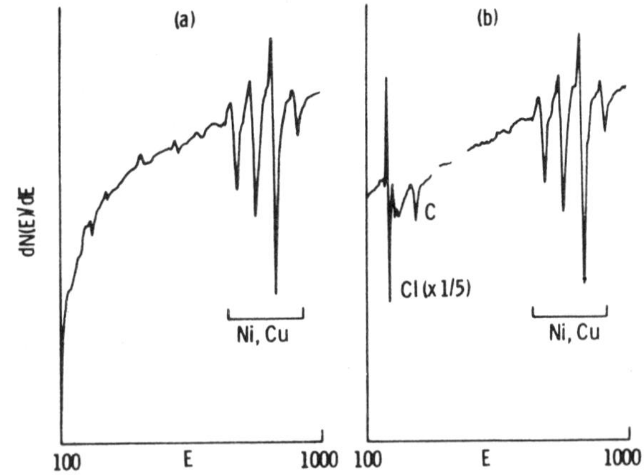

(a) Sputtered disk.

(b) After sliding PVC for one revolution.

Figure 7-20. – Auger spectrum from S-Monel disk. Beam current, 5 microamperes (ref. 8).

When analytical tools such as Auger spectroscopy are used to study the transfer of polymer films to metal surfaces, extreme care must be taken because many elements are susceptible to electron beam induced desorption; that is, the electron beam from the Auger analyzer has sufficient energy to produce desorption of the polymer from the metal surface. For example, if the electron beam is concentrated on one spot on the solid surface of PTFE transfer films on metal surfaces, a complete loss of fluorine from the surface occurs. The incoming electrons cause carbon to fluorine bond scission and the consequent liberation of fluorine. Thus, analyses are usually conducted on these surfaces by continuously moving the surface under the electron beam so as to be constantly sampling a new surface area and thereby getting a representative composition of the surface film. With the polyvinylchloride rubbing the metal surface, it appears that the polyvinylchloride actually undergoes decomposition during transfer. Hence, with a transfer film of approximately a monolayer in coverage, chemibsorbed chlorine appears to be present on the surface (fig. 7-20(b)).

With polychlorotrifluoroethylene (commonly referred to as KEL-F), the transfer film appears to be one of chain fragments of the polymer itself rather than the polymer's basic structure; hence, an analysis of the polymer gives a different Auger spectra than does an analysis of the film of KEL-F transferred to the metal surface as a result of sliding.

The observed behavior for these polymers appears to correlate with thermal decomposition characteristics of the polymers; this indicates that the polymers may undergo decomposition when they are sliding in contact with metal surfaces as a result of frictional heating. In other words, with the less thermodynamically stable polymers (such as the polyvinylchloride and KEL-F) the frictional heat at the interface becomes sufficiently high to cause the decomposition of these polymeric materials. With the PTFE, however, the decomposition temperature is sufficiently high so that the polymer transfers as polymer to the solid surface and does not decompose in the transfer process. In the study of adhesive wear behavior of polymers in contact with metals, care must be taken not to interpret the wear behavior of polymers in terms of chemistry of polymeric materials in general. For example, it is fairly well known that polymers in sliding, rubbing, or rolling contact undergo adhesive wear to metal surfaces. In the process of such contact, a number of changes can take place in the polymer at the interface including degradation of the polymer and local surface melting.

When a polymer is in contact with a metal surface, however, the chemistry of interaction can play a part in the mode of decomposition. This breakdown chemistry may be entirely different from that observed using conventional chemical techniques. For example, with ordinary pyrolysis of polymeric materials, certain percentages of the basic monomer from which the polymer must form are seen in the decomposition process. This is observed with polymers such as the KEL-F and PTFE; pyrolysis of these polymers in vacuum yields a percentage of the original monomer from which the polymers were formed. Some pyrolysis data for these two polymers and other polymeric compositions are presented in table 7-I (ref. 9). An examination of these data reveals that the temperature range for the

TABLE 7-I. – YIELD OF MONOMER IN THE PYROLYSIS OF SOME ORGANIC POLYMERS IN VACUUM[a] (IN PERCENT OF TOTAL VOLATILIZED)

Polymer	Temperature range (°C)	Yield of monomer (%)
Polymethylene	335–450	0·03
Polyethylene	393–444	0·03
Polypropylene	328–410	0·17
Poly(methylacrylate)	292–399	0·70
Hydrogenated polystyrene	335–391	1
Poly(propylene oxide)atactic	270–550†	2·80
Poly(propylene)isotactic	295–355	3·55
Poly(ethylene oxide)	324–363	3·90
Polyisobutylene	288–425	18·10
Poly(chlorotrifluoroethylene)	347–415	25·80
Poly(β-deuterostyrene)	345–384	39·70
Polystyrene	366–375	40·60
Poly(*m*-methylstyrene)	309–399	44·40
Poly(α-deuterostyrene)	334–387	68·40
Poly(α,β,β-trifluorostyrene)	333–382	72·00
Poly(methylmethacrylate)	246–354	91·40
Poly(tetrafluoroethylene)	504–517	96·60
Poly(α-methylstyrene)	259–349	100
Poly(oxymethylene)	below 200	100

[a]Reference 9.

[b]This value should probably be 330° C.

decomposition of most of the polymers is from approximately 250° to about 500° C. The actual useful range of polymers is therefore rather limited before decomposition occurs. The data in table 7-I indicate the temperature ranges for which pyrolysis begins and they also present the yield of monomer generated from pyrolysis of the polymer. For some polymers the amount of monomer yielded by the basic polymer structure is relatively low. For example, with polymethylene the percent is less than 0.03 percent.

In contrast, materials of interest in tribology (e.g., PTFE) yield a very high percentage of monomer on pyrolysis. If PTFE is decomposed or pyrolyzed in the temperature range of roughly 500° to 517° C, a 96.6-percent yield of monomer is obtained (table 7-I). This decomposition by ordinary pyrolysis is not, however, what is observed in the adhesive wear process of polymers to metal surfaces. The field ion microscopy studies with PTFE in contact with tungsten referred to in chapter 2 indicate that the polymer transfers by the mechanism of end-caps of the polymer chain adhering to the metal surface with carbon to metal bonding. Basically, CF_2 groups are attached to the metal surface by carbon to metal bonding. Thus, with the solid-state contact of PTFE to clean metal surfaces, the decomposition of the polymer as a result of metal polymer contact is by the generation or liberation of polymer end-caps from the polymer chain CF_2

groups as opposed to the generation of monomers, which would be C_2F_4 groups.

Similar results to those obtained in the FIM have been obtained in AES analysis of the solid surfaces. As mentioned earlier, when electron beam induced desorption of the polymer from the metal surface occurs, fluorine is liberated and only carbon remains on the surface. The bonding is carbon to metal. The fluorine sticking above the surface is open to electron beam induced desorption. While the fluorine to carbon bond is 51 percent ionic in character and is the strongest bond existing in organic structures, it still has vulnerability to the electron beam. The carbon to metal bond, however, is a much stronger bond and is not open to electron beam induced desorption. The Auger analysis of the metal surface reveals the presence of carbon on the surface but, after a time, an absence of fluorine because of electron beam induced desorption. This means that the carbon bonds much more strongly to the metal surface than the fluorine bonds to the carbon.

These observations show that the polymer metal chemistry in sliding or in adhesive contact is different from the quiescent pyrolysis chemistry. While conventional pyrolysis studies may be helpful in giving the upper temperature limits of usefulness for many polymers, they do not indicate the mechanism of decomposition of polymer that predicts adhesive wear of the polymer or loss of the polymer from the solid polymer body. It is difficult in tribological systems to gain insight into the actual temperatures the polymer is experiencing at the interface because the polymers are inherently good insulators and the temperatures at the interface may get extremely high in a very thin film; such high temperatures may not be felt subsurface to that thin film.

In addition to polymers in sliding, rubbing, and rolling contact with metals, there are situations where polymers are in solid-state contact with themselves. In such situations there is a question as to the nature of the adhesive behavior and the transfer of materials in these polymer to polymer systems. Adhesive wear occurs for polymers in sliding contact with polymers just as it does for polymers in sliding contact with metals. With polymers in sliding contact with metals, the polymer transfers to the metal surface, the shear takes place in the polymer, and the sacrifical element is generally the polymer (with some exceptions which have been described earlier). With polymers in contact with polymers, however, other factors must be considered to determine the nature of the adhesive transfer. When the cohesive binding in materials was discussed.(ch. 1), it was indicated that for metals in sliding contact with metals (where the metal surfaces are clean) adhesion occurs at the interface and the adhesive wear takes place in the cohesively weaker of the two materials; that is, the bond at the interface is stronger than the cohesive bond in the cohesively weaker of the two metals, and transfer of the cohesively weaker to the cohesively stronger occurs. The example cited earlier was aluminum in sliding contact with iron.

Comparable behavior exists for polymers in contact with polymers. Adhesive and cohesive bonds are extremely important. Even with polymers in contact with metals, it is certainly obvious that the cohesively weaker of the two materials is the polymer; for that reason polymers transfer to the

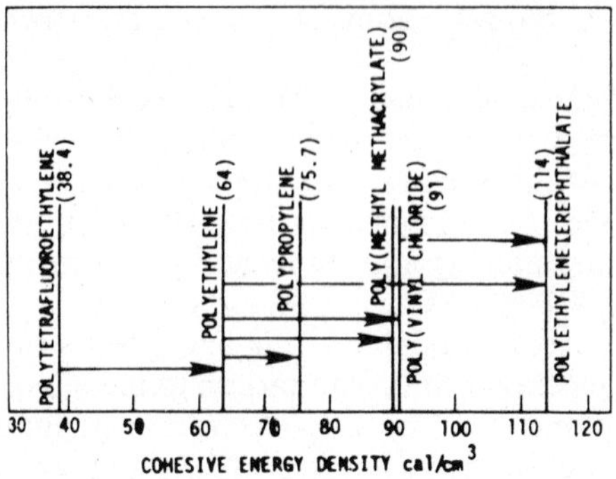

Figure 7-21.—Material transfer direction for various combinations from material of low cohesive energy density to material of higher cohesive energy density (ref. 10).

metal surface. The same principle applies for polymers in contact with polymers. Then, if adhesion at the interface results in strong bonding, transfer of the cohesively weaker polymer should occur to be cohesively stronger as a result of adhesion. The adhesive wear mechanism should, therefore, result in wear to the cohesively weaker of the two polymers. This is observed experimentally in the data of figure 7-21 obtained from the work of Jain and Bahadur (ref. 10). Their results indicate that the lower cohesive energy density material transfers to the higher cohesive energy density material for the various polymers indicated in the figure (including PTFE, polyethylene, polypropylyene, polymethylmethacrolate, polyvinylchloride, and polyethylenetrethalate).

Jain and Bahadur used infrared spectroscopy and differential thermal analysis to study the transfer of polymers from one surface to another. They found that the thickness of the transfer film was a function of sliding variables as determined by quantitative infrared spectroscopy. As might be anticipated, the thickness of the film increased with increasing sliding speed and time. This is analogous to what was observed with the aluminum in contact with iron where the amount of transfer increased with increasing time. There are analogies between metal to metal couples and polymer to polymer couples with respect to adhesive wear.

Carbon and Graphite

In addition to metals and polymers, other material couples have been observed undergoing adhesive wear. For example, carbon and carbon bodies are widely used in practical tribological systems such as mechanical seals where carbon bodies slide against metals. Very seldom is the carbon in rubbing, sliding, or rolling contact with itself. Carbon bodies can contain various amounts of graphite; the degree of graphitization in the carbon depends a great deal on the particular application for which the carbon body is intended. For metal surfaces containing normal residual surface oxides, the carbon or graphite body is generally observed to adhere; that is, adhesion takes place just as it does with metals and polymers to the metal surface. With tangential motion or rolling activity, the carbon is observed to transfer to the metal surface. With repeated passes of sliding, rolling, or rubbing across the same surface, a carbon transfer film develops on the metal surface. Thus, adhesion and adhesive transfer ultimately lead to adhesive wear of the carbon. In the initial stages of sliding, rolling, or rubbing contact, while the carbon film is building up on the metal surface, the wear rate is relatively high. This was also observed with dissimilar metals in contact where one metal is transferring to another surface. After the transfer film of carbon has been established on the metal surface, the wear rate decreases appreciably.

For fully graphitized carbon bodies or natural graphite, the orientation of the graphite has a pronounced effect on the observed wear behavior of the materials. Thus, orientation influences the adhesive wear process. This is indicated for natural graphite brush specimens oriented in two directions relative to a steel surface. In one case, the brushes are cut so that the basal planes on the graphite are parallel to the sliding or rotating surface. In the second case, the solid body is cut so that the cleavage or basal planes are oriented normal to the interface. Marked differences in the adhesive wear are observed in these two conditions as indicated by the data of figure 7-22 (ref. 11). In this figure the basal orientation parallel to the sliding interface produces a minimum in wear compared with the edge or prismatic planes rubbing at the interface. With the basal planes (cleavage planes) of the graphite parallel to the sliding interface, a thin film of graphite is rapidly laid down on the metal surface, and graphite is then sliding on graphite; basal plane is sliding over basal plane with easy shear accomplished between basal planes. As a consequence, the rate of wear reduces very rapidly

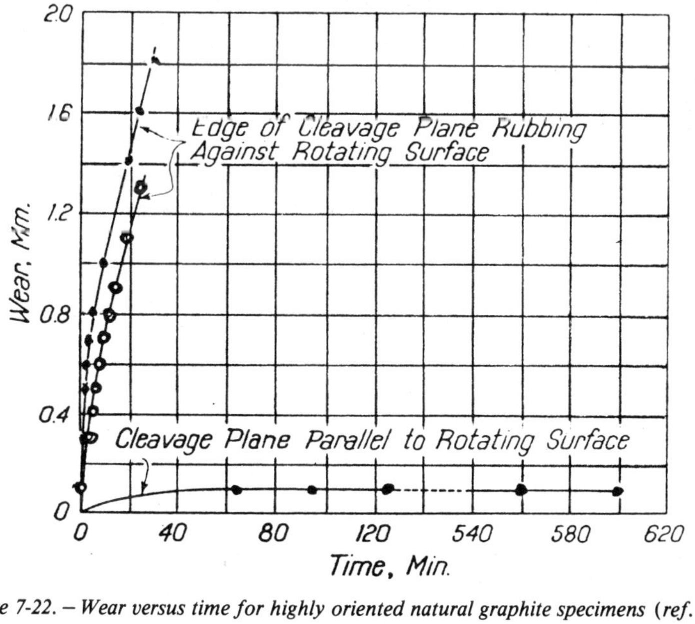

Figure 7-22. – Wear versus time for highly oriented natural graphite specimens (ref. 11).

because it takes a very short time to develop the oriented transfer film with the basal planes protecting the surface and with shear taking place between the transferred basal planes and those remaining in the carbon body. With the edges of the cleavage planes rubbing at the interface, however, the basal planes or large surface area planes are now normal to the surface. Wear can progress much more rapidly because adhesion of the prismatic orientation is much stronger to the metal surface than the basal orientation is, and the ease of shear between adjacent planes no longer exists because the planes are now oriented normal to the sliding surface. Therefore, the wear rate is much higher for the prismatic orientation (with the cleavage plane rubbing against the rotating surface).

The behavior observed in figure 7-22 for graphite is characteristic of materials with a lamellar crystal structure. For example, Lancaster has observed similar results for molybdenum disulfide in sliding contact with metal surfaces (ref. 12). With molybdenum disulfide, the basal plane is the low shear strength plane and, when a transfer film of molybdenum disulfide occurs from a solid body to a metal surface, shear takes place between adjacent basal planes with a minimum of adhesive wear of the molybdenum disulfide body. When, however, the planes are oriented normal to the surface, the edge sites of the molybdenum disulfide (the prismatic orientations) are very hard and the wear rate is much higher. In fact, in many instances, the edge planes of the molybdenum disulfide are sufficiently hard and brittle and have sufficient strength to produce a shearing or cutting of the metal surface.

The data in figure 7-22 apply to those situations where carbon or graphitic carbon bodies are used in sliding, rubbing, or rolling contact with metals. There are, however, situations where solid carbon is used in fiber form as additives in other materials (e.g., resins) for the purposes of strengthening and imparting to those resins good mechanical properties. Many of these solid bodies have been considered for use in practical tribological systems.

Fiber orientation in such matrix systems has a pronounced influence on the adhesive wear behavior of the resin. The resin transfers to the metal surface just as the carbon does. With carbon fibers present, however, the mechanism for wear of the resin is altered by the presence of the fiber as well as fiber orientation. When the carbon fibers are oriented such that they are parallel (rather than perpendicular) to the sliding interface, the adhesive wear rates are much higher than when the fibers are oriented perpendicular to the surface (ref. 13). The effect of fiber orientation is shown in figure 7-23. The highest wear rates in figure 7-23 are observed for two cases: (1) carbon fibers oriented parallel to the direction of sliding (B) and (2) where there is a total absence of carbon fibers in the resin matrix (A). When the fibers are oriented normal to the direction of sliding (C), wear is reduced. Minimum wear, however, is obtained when the carbon fibers are standing on end (like bristles on a brush) normal to the solid surface (D).

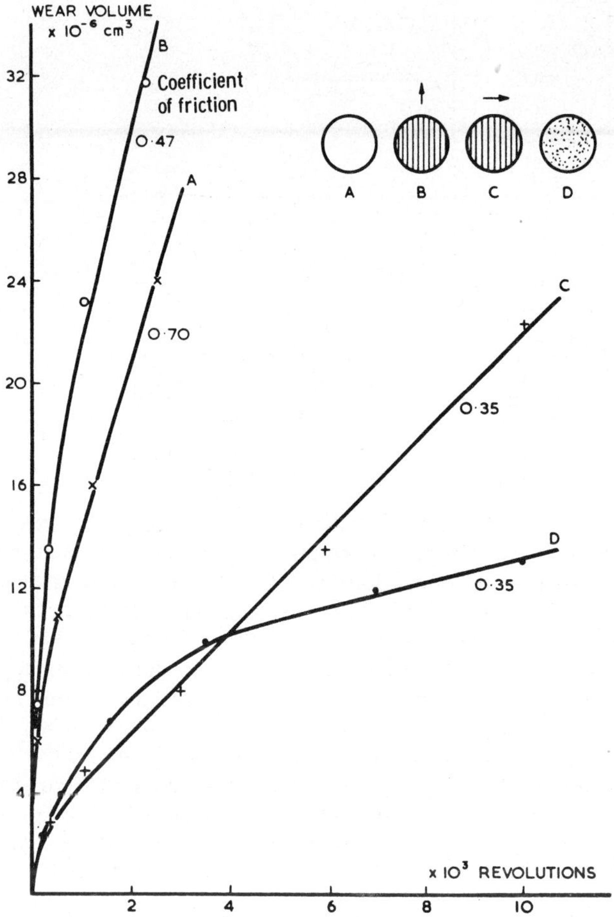

Figure 7-23.—Effect of orientation on wear of carbon fiber reinforced polyester resin (25 wt%) sliding against hardened tool steel (ref. 13).

Abrasive Wear

Nature of Abrasion

Another form of wear frequently encountered in practical tribological systems, which is of considerable significance and importance with respect to the influence of surfaces in wear behavior, is abrasive wear. Abrasive wear occurs in practical mechanical operations such as grinding, cutting, and machining. When dirt or debris becomes entrained in operating tribological systems and where it is sufficiently hard, it can produce scoring or abrasion of surfaces in sliding, rolling, or rubbing contact.

The fundamental mechanism involved in the abrasive process is shown schematically in figure 7-24. There are actually two modes by which abrasion and abrasive wear can take place. In one mode, the mode

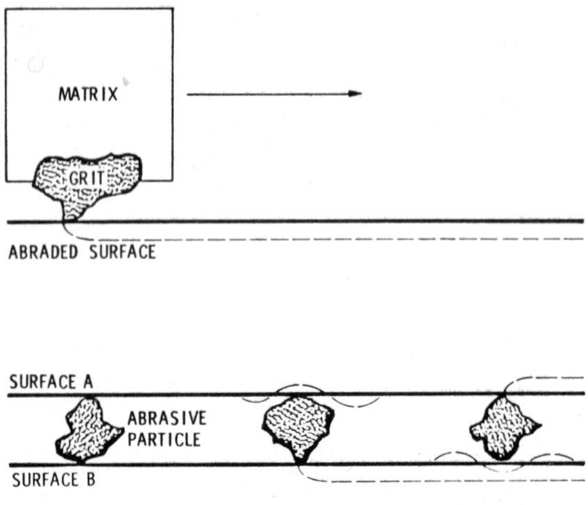

Figure 7-24. – Abrasive wear mechanism.

frequently encountered in industrial material removal processes such as cutting, grinding, and machining, an abrasive grit is embedded in a matrix (top of fig. 7-24). A practical analogy of this might be a grinding wheel which contains a number of grits of abrasive material embedded in the resin matrix, and the individual grit comes in contact with the workpiece or the surface to be abraided. The abrasive grit begins to cut or skive a groove in the workpiece surface and to remove material from the workpiece. Generally, the grit material is a material of high cohesive strength. The most common abrasive material used industrially is silicon carbide. Silicon carbide is a very hard, cohesively strong material having a high modulus of elasticity and resists plastic deformation. It, consequently, can remove (or cut or deform) other surfaces with which it comes in contact. Most metals are susceptible to cutting or deformation by silicon carbide.

If the grit is dragged tangentially across the surface (top of fig. 7-24), it will skive or cut a piece of material from the solid surface. When the surfaces are in dry contact, the material can be removed as a chip, flake, or curl of metal from the workpiece where the workpiece is a metal or alloy. When the surfaces are lubricated very effectively, however, and the grit becomes dull or worn so that sharp surfaces are not exposed, instead of an actual cutting operation with removal of metal, deformation of the surface can take place where the material is plastically deformed by the abrasive grit particle. Basically, two processes can occur in a conventional device. In a practical application of the single particle grit of figure 7-24 (e.g., a grinding wheel), there are numbers of particles or individual grits that act as cutting surfaces, some of which have sharp edges that do cutting and others that, because they have become dull as a result of the grinding process, deform the material rather than cut it. Both processes frequently occur simultaneously in the cutting or grinding operation. The goal is to maximize the removal process and minimize the deformation process, because, with deformation, material is simply moved from one area of the surface to another as opposed to removing it from the workpiece.

Adhesion also plays a role in the abrasion process. Very frequently, when grits are in contact with various metals in the clean state, metals adhere to the grit surface; adhesive transfer (as a result of adhesion at the interface between the metal and the silicon carbide) occurs, with transfer of the metal to silicon carbide. When this happens on a very large scale with very chemically active metals such as titanium and aluminum, the work wheel or the cutting surfaces become charged with metal. At that point, the cutting surface must be redressed or cleaned to remove the transferred metal in order to renew the effectiveness of the cutting or grinding edges of the individual grits. Miyoshi et al. have made detailed studies of the mechanism of adhesive transfer (and the role of adhesion in such transfer) in the grinding or cutting process (refs. 14 to 16).

A second mode or mechanism by which abrasive wear can occur is the situation described earlier where a high cohesive strength material (such as silicon dioxide, aluminum oxide, or silicon carbide) would become trapped as individual particles between two surfaces (as indicated schematically in the bottom of fig. 7-24 for surfaces A and B). Trapped between surfaces A and B is a relatively hard, high atomic density, high cohesive energy material that skives or cuts material from both surfaces. This particle acts, then, as a cutting surface for removing material. This basic mechanism is also involved in practical finishing operations such as polishing surfaces with abrasive powders. Aluminum oxide, magnesium oxide, or silicon carbide powders are used to polish surfaces. The surface to be finished may be surface A or surface B and the abrasive particles are applied to the surface and rubbed across the surface to remove material. Generally, the removal process occurs by a successive application of smaller and smaller abrasive particles until a fine polish of the surface results.

Polishing with abrasive particles is beneficial or desirable since it produces industrial finishes of practical usefulness. That same abrasion process, however, can be extremely detrimental when it takes place in operating machinery—for example, when abrasive grits become trapped between gear teeth and cause wear of the gear teeth or in a bearing where trapped abrasive particles can remove metal from the surfaces and cause the premature failure of mechanical components. Generally, however, the abrasive wear process when it occurs with fine particles trapped in bearings, gears, or mechanical seals is a gradual wearing process as opposed to the catastrophic one encountered in an adhesion wear mechanism. The foregoing discussions have established that the surfaces of solids (other than a few exceptions) are not flat and smooth as indicated in figure 7-24 but rather they contain surface irregularities or asperities. Frequently when one applies a very fine abrasive particle for polishing purposes, the function of the abrasive particle is to remove the tips of the asperties and produce a smoothing of the surface by that mechanism. Various investigators have examined surfaces which have been surface profiled before and after the abrasive process (ref. 17). Figure 7-25 shows a comparison of an initial unworn surface and the surface after wearing in a practical tribological system.

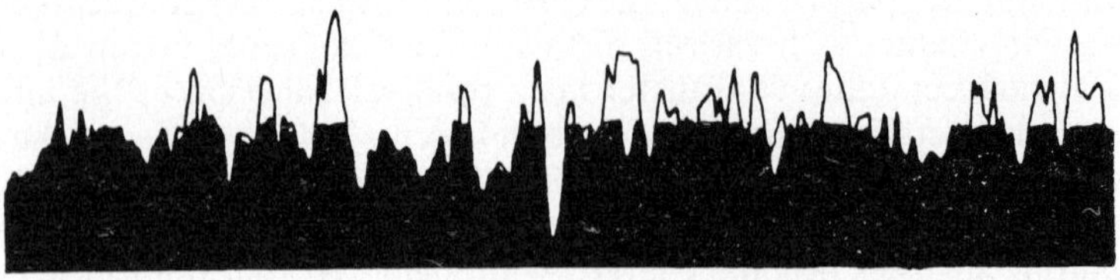

Figure 7-25. – Comparison of initial unworn and final worn profiles (ref. 17).

Two profiles are superimposed in figure 7-25. The black regions in the figure are the final worn surface profiles. The areas where there is no black, the peaks and spikes, are the regions indicating the initial profile. A comparison of these two profiles reveals a flattening or an abrading of the asperities.

Hardness Effects

Given a specific particle size and material for abrasive wear of a solid surface, the properties of the workpiece or solid surface that is being abraded have a very marked influence on how much abrasive wear takes place and how rapidly the process occurs for different materials. Through the years, a number of investigators have examined various properties of workpiece materials to gain some insight into those properties which exhibit an influence on the abrasive wear process. Khruschov (in Russia) measured the effect of hardness of various elemental metals on the resistance to abrasive wear and found, in general, a correlation between abrasive wear resistance and hardness (ref. 18). The harder the metal, the more resistant it is to abrasive wear, according to Khruschov. Some of his results are reproduced in figure 7-26 where resistance to wear is plotted as a function of hardness in kilograms per square millimeter.

The measurements presented in figure 7-26 are for bulk hardness as they relate to wear resistance. Other investigators have also done detailed studies on the effect of hardness and its relationship to wear resistance (ref. 19). In the course of these studies, it was found that there is a difference in wear behavior when one measures hardness of the bulk material as opposed to the surface microhardness. When various steels are hardened and tempered, differences in wear resistance are found, based on the tempering temperatures used to prepare the steels. Some of these results are presented in figure 7-27, which presents wear resistance for carbon steels as a function of hardness (as influenced by tempering). The steels contain various percents of carbon from 0.04 to 1.23 percent. Variations in carbon content in steel do not seem to alter markedly the resistance to abrasive wear of the steels. In figure 7-27(a), the wear resistance is plotted as a function of bulk hardness in DPH for three ranges of heat treatment: (1) annealed, (2) tempered (300° to 600° C), and (3) tempered (20° to 200° C). These data show that, with changes in the tempering temperatures, a marked change in

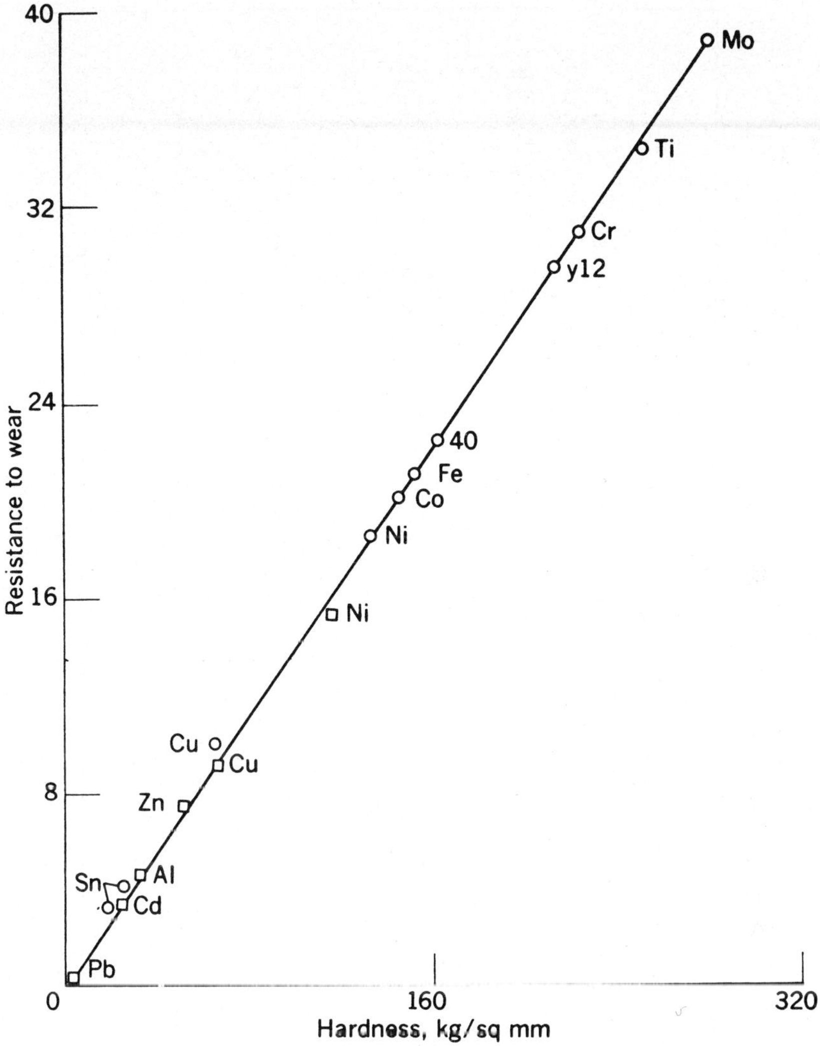

Figure 7-26. – Resistance to wear as function of hardness (ref. 18).

resistance to wear occurs. High temperature tempering produces a lower wear resistance than does low temperature tempering.

Figure 7-27(b) indicates the influence of wear resistance for the same steels as a function of surface microhardness in DPH; a comparison with the bulk hardness data of figure 7-27(a) shows that the wear resistance increases with increases in surface microhardness. It should be indicated in figures 7-27(a) and (b) that one is dealing with the reciprocal of the wear. As a consequence, the wear is actually decreasing in figures 7-27(a) and (b) with an increase in both bulk and surface microhardnesses. The interesting result to observe from figures 7-27(a) and (b) is that the wear rate is different for

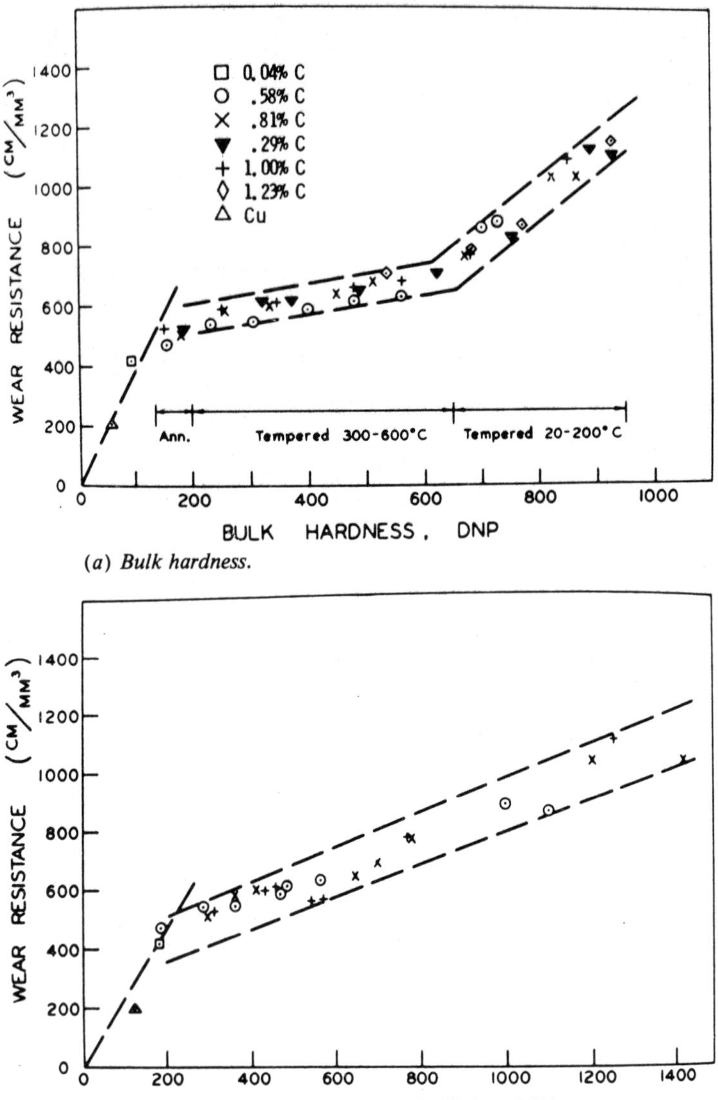

(a) *Bulk hardness.*

(b) *Surface microhardness.*

Figure 7-27. – Wear resistance of hardened and tempered carbon steels as function of hardness (ref. 19).

the two conditions. The correlation developed for surface microhardness is different from that observed for bulk hardness when compared to wear resistance of carbon steels.

The data of figure 7-27 show that the simple, straightforward relationship established for elemental metals in figure 7-26 by Khruschov must be modified when dealing with alloys that can be heat treated to produce various degrees of hardness. These changes in heat treatment cause changes in the resistance of the materials to wear. And these changes must be considered when working with alloy systems.

Surface Effects

In the study of the abrasive wear process, the properties of the abrasive are as important as the properties of the workpiece or the material being removed. For example, the nature of the abrasive and the chemical composition or chemistry of the abrasive influence the interactions of the abrasive with the workpiece material. Silicon carbide interacts differently (from a chemical point of view) with iron than diamond or aluminum oxide do. In addition, there are physical characteristics of the abrasive which influence the abrasive interaction between the abrasive and the material surfaces. Abrasives, in addition to simply removing material from a solid surface, leave the surface that has been abraded in a damaged state; that is, as a result of the abrasion process, a considerable amount of strain can be put into surfaces that are prone to plastic deformation. Even where materials are not very likely to deform plastically and are relatively brittle (such as ceramic materials and cast irons), the interaction of the abrasive with the solid surface can produce surface cracking. The energy supplied at the interface in the abrasion process is being dissipated in the formation of cracks as opposed to the generation of dislocations associated with strain in those materials that are sufficiently ductile to undergo deformation.

Analytical surface tools such as etch pitting techniques, SEM, and X-ray Laue techniques can be used to study the depth of damage that occurs in materials as a result of the sliding, rolling, or rubbing interaction encountered with the abrasion of surfaces. Various physical aspects of the abrasive material influence the depth of damage in materials. One of these is the particle size of the abrasive. There seems to be a correlation between the size of the abrasive particle and the depth of damage observed. Studies have been conducted with semiconductors such as silicon and germanium in single crystal form that had been abraded by abrasives of various particle sizes. These surfaces were examined by etch pitting techniques to determine the deformation subsurface as a result of the abrasion process.

Correlations have been found between the particle size of the abrasive and the depth of damage observed in the abraded material (ref. 20). Some results obtained for the abrasion of germanium and silicon single crystals where the single crystals were of a (111) orientation (i.e., the natural cleavage and slip planes in these crystal systems) are presented in figure 7-28. In this figure the depth of damage is plotted as a function of the nominal particle size of the abrasive. The data show that, as the particle size of the abrasive increases, the depth of damage in the surface (i.e., the residual damage after the abrasion process has taken place) is deeper with the larger particle. Thus, depth of damage for germanium and silicon is a direct function of the particle size of the abrasive in the abrasive wear process.

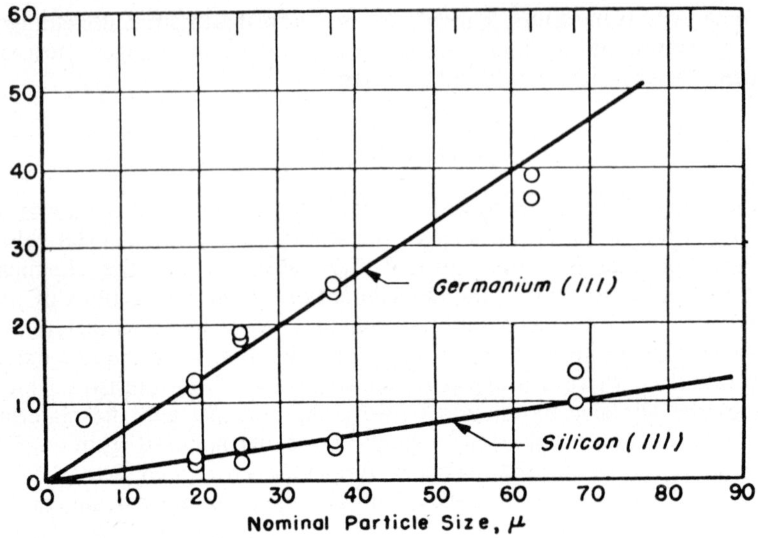

Figure 7-28.—Depth of damage on germanium and silicon as function of abrasive particle size (ref. 20).

Germanium is cohesively much weaker than silicon. As a consequence, the depth of damage observed in germanium is much greater than it is in silicon for the equivalent particle size of an abrasive. The cohesively weaker material experiences a greater depth of damage than the cohesively stronger one. When working with elemental materials, such as germanium and silicon in figure 7-28 and elemental metals in figure 7-26, direct relationships are readily apparent. In most practical devices encountering abrasive wear, however, the surfaces are not simple elemental metals but are frequently alloys or complex materials. In such situations, the question of the influence of various constituents on the abrasion process arises. How, for example, does an alloying element in an alloy affect the abrasion wear of a surface?

Abrasion of alloys has shown that alloy constituents do influence the surface abrasive interaction. Surface enrichment of one alloying element can be brought about by the strain associated with the abrasion process. Not only strain but also increased surface temperature can have an influence if the abrasion process is conducted in a dry state. Equilibrium segregation was discussed in chapter 5. Equilibrium segregation is also observed in the abrasive process because, in the interaction of the abrasive surface with the workpiece, a large quantity of energy must be dissipated at the interface. The energy arises from the interaction of the abrasive grit with the workpiece under high loads (and frequently high speed), and this puts a considerable amount of energy into the interface that must be dissipated very rapidly. That energy generally results in a considerable amount of heating and strain in the surfaces. The strain and increased surface temperature can bring about equilibrium segregation or segregation to the surface of alloy constituents where a particular alloy system is prone to segregation.

Studies in the abrasion of brass (a relatively simple material in common usage) by diamond reveals that, with a cutting of the surface by an abrasive grit of diamond, the chemistry of the brass changes (ref. 21). This effect is indicated in the Auger spectroscopy data in figure 7-29. In this figure two Auger spectra are presented: figure 7-29(a) is for an electropolished β-brass surface, and figure 7-29(b) is the bottom of the abraded track on the surface of the β-brass after a diamond grit has passed across the surface. In figure 7-29(a), the elements present on the surface are phosphorus, sulfur, carbon, oxygen, copper, and zinc. After abrasion of the surface, the same elements are present on the solid surface. However, the relative concentrations have changed. The concentration of sulfur and phosphorus, which can be present as contaminants from the environment, has decreased appreciably in the base of the wear track. Likewise, the carbon, which is present as adsorbed carbon monoxide and carbon dioxide as a contaminant on the surface of the β-brass, has diminished in size and intensity. The oxygen peak has increased in intensity probably because when the surfaces are abraded oxygen comes in contact with fresh, clean metal surfaces to generate an oxide. That oxide is thicker with clean metal than it would be with the other contaminants present on the solid surface. The other observation to be made in a comparison of figure 7-29(b) with that of figure 7-29(a) is that the concentration of zinc on the solid surface has increased appreciably after abrasion. This increase indicates that the ratio of zinc to copper in the brass surface prior to abrasion is considerably less than it is after the abrasion

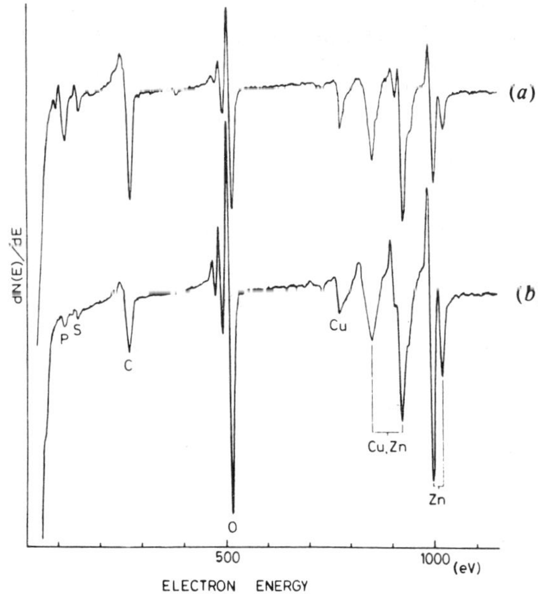

Figure 7-29.—Auger spectra obtained by point analysis from electropolished β-brass and bottom of track (ref. 21).

process has taken place. There is a higher concentration of zinc in the case of the freshly abraded alloy.

The electropolished surface may be depleted in zinc because the electropolishing agent preferentially attacks the zinc during electropolishing. Nonetheless, the chemistry of the surface which has been generated by the abrasion process is different from that of the polished surface.

The data in figure 7-29 raise an interesting point concerning the use of analytical surface tools and the examination of tribological surfaces. The most common technique used today for generating clean surfaces (i.e., the technique used to remove adsorbates and oxides from material surfaces) is ion bombardment of the surface with argon or some other inert gas to sputter away surface contaminants. The sputtering process itself is, however, somewhat selective when one is working with an alloy in that some alloy constituents have higher sputtering yields than others. For example, copper has a relatively high sputtering yield while carbon has a very low one. Thus, if one has a mixture of high and low sputtering yield substances in the form of an alloy, there may be a disparity in the yields during the sputtering process of the alloys in an attempt to clean the surface for analysis of the chemistry or composition of the alloy itself. Thus, the analytical tool for analyzing the surface may introduce, in the process of preparing that surface, modifications in the basic chemistry of the system that can alter the observed behavior.

The abrasion of surfaces in a good vacuum environment may be one way of mechanically generating a clean surface without obtaining the undesirable side effects (observed with sputtering) of having a surface enrichment of one alloying element as a result of differences in sputtering yields. Thus, the abrasion wear process may have potential for use as a cleaning technique.

So far in our discussion of the abrasive wear process, the assumption has been made that the abrasive removes material from the workpiece during their interaction. Sometimes, however, material is not removed from the workpiece and the abrasive becomes attached to the workpiece as a result of chemical bonding or deformation. For example, when diamond paste is used to polish an aluminum surface or other soft material, the polishing process results in particles of the diamond becoming embedded in the aluminum, or other soft metal surfaces. The diamond is lost from the abrasion process by being buried in the solid surface. This is shown in the scanning electron micrograph in figure 7-30, which indicates the presence of a diamond abrasive particle from the diamond polishing paste buried in the surface of aluminum as a result of an attempt to use diamond paste to polish the aluminum. The diamond particle can then act as an abrasive against other surfaces.

The presence of diamond, aluminum oxide, or silicon carbide abrasives in the surface of soft metals is not uncommon. Various analytical techniques have been used to detect and identify the abrasive materials embedded or buried in the soft materials.

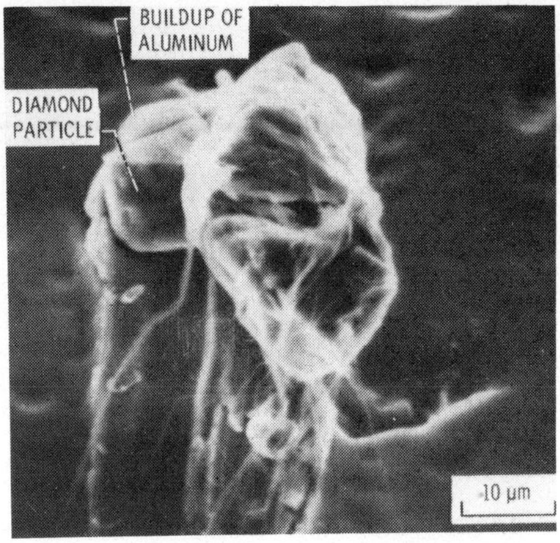

Figure 7-30. – Termination of score caused by hard diamond particle.

In figure 7-31 ion microprobe analyzer (IMA) spectra are presented for four metal surfaces (gold, silver, palladium, and platinum) after they have been abraded by silicon carbide abrasive paper for finishing purposes. An examination of the four spectra reveals that, in each case, the contact of silicon carbide with the metal resulted in interactions and embedment of silicon carbide in the metal surface. Silicon is seen in the IMA spectra for all four metals. The silicon concentration is higher in some than in others, but it definitely is present in all four metals.

Similar observations have been made with other abrasives (such as aluminum oxide) in contact with soft metals such as gold, silver, and copper. The surfaces become charged with the abrasive material. The results of using surface analysis to characterize the abrasive wear process indicate that extreme care should be taken in abrasive polishing soft metals because of the danger of embedding abrasive materials in the surface of these solids.

In addition to the physical parameters which influence the abrasive wear process, chemical reactions and other interactions that take place in material systems influence the abrasion of surfaces. If the chemistry of the abrasives themselves is considered, the different compounds used as abrasives can behave in entirely different manners in contact with different solid surfaces. For example, there are many carbides that are used in practical cutting, grinding, and machining operations. The application involved usually determines which of these carbides is preferred for cutting, grinding, or machining. However, there are basic differences in the wear of the abrasive, which is the tool. For example, among the carbides, there are marked differences in wear characteristics. Probably one of the preferred and most frequently used carbide materials is titanium carbide. There is even current interest in attempts to use it as a coating for the surfaces of ordinary steels to provide better wear resistance.

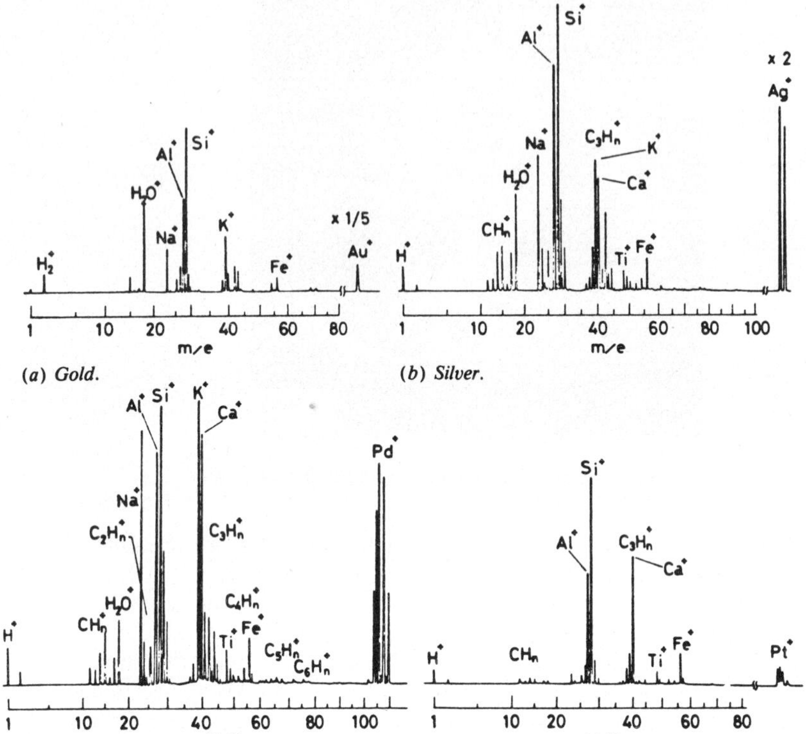

(*a*) *Gold.*

(*b*) *Silver.*

(*c*) *Palladium.*

(*d*) *Platinum.*

Figure 7-31. – IMA spectra of metals prepared by SiC abrasive paper finishing.

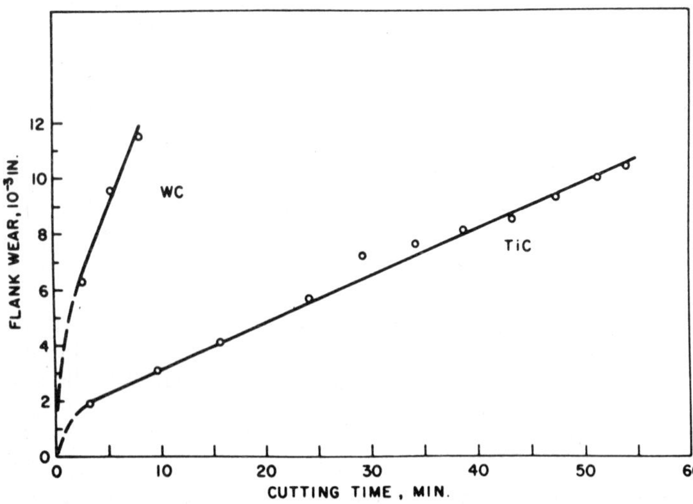

Figure 7-32. – Flank wear comparison of steel-finishing grades of titanium carbide and tungsten carbide.

An example of the differences in the wear behavior of carbides that can exist as cutting tools is demonstrated by the data in figure 7-32 where tool wear (commonly referred to as flank wear) is plotted as a function of cutting time. The data show that titanium carbide exhibits a much lower wear rate than does tungsten carbide for equivalent conditions.

Alloying

The chemistry of the cutting or grinding surface is important in tribological systems, but so, too, is the chemistry of the workpiece. Alloying, which was discussed earlier, can change the surface chemistry of the material being cut. Likewise, the alloy chemistry can alter mechanical properties so that the net effect of cutting, grinding or polishing is altered by the presence of alloying elements in elemental materials.

Miyoshi et al. have done some basic studies on the influence of alloy constituents in simple binary systems on abrasive wear resistance (ref. 15). During these studies, it was found that there are optimum concentrations of alloy constituents needed to produce a minimum in abrasive wear or, conversely stated, to produce a maximum in wear resistance. Several series of binary alloys were examined. One series, which is of particular interest to tribologists, is the iron-chromium series. Miyoshi et al. added various weight percents of chromium and iron and then examined the wear resistance of the iron-chromium alloys in abrasive contact with silicon carbide. The results of some of these experiments are presented in figure 7-33.

The data results in figure 7-33 are extremely interesting. They indicate that, with the addition of chromium to iron, there is a decrease in the groove height generated in the surface of the alloy. With as little as 1 weight percent chromium in iron, the height of the groove decreases appreciably. Further additions to 5 and 14 percent produce further reductions in the groove height generated in the surface as a result of the silicon carbide particle sliding across the surface. The concentration of chromium in iron and its effectiveness in reducing abrasion seems to optimize at approximately 14 weight percent of chromium in iron. At 19 weight percent the size of the groove increases again from that observed at 14 percent. The minimum in groove height appears to have occurred at 14 weight percent. The results of the data in figure 7-33 indicate that there are optimum concentrations for minimizing the amount of deformation that the surface experiences as a result of interactions with abrasive particles.

Polymers

Abrasive wear occurs to materials other than metals and alloys. It takes place in the wear of nonmetallic substances such as carbons, ceramics, and polymers. The polymers are extremely susceptible to abrasive wear because of the relatively weak cohesive strength of polymers compared to most other solids that polymers contact in tribological devices.

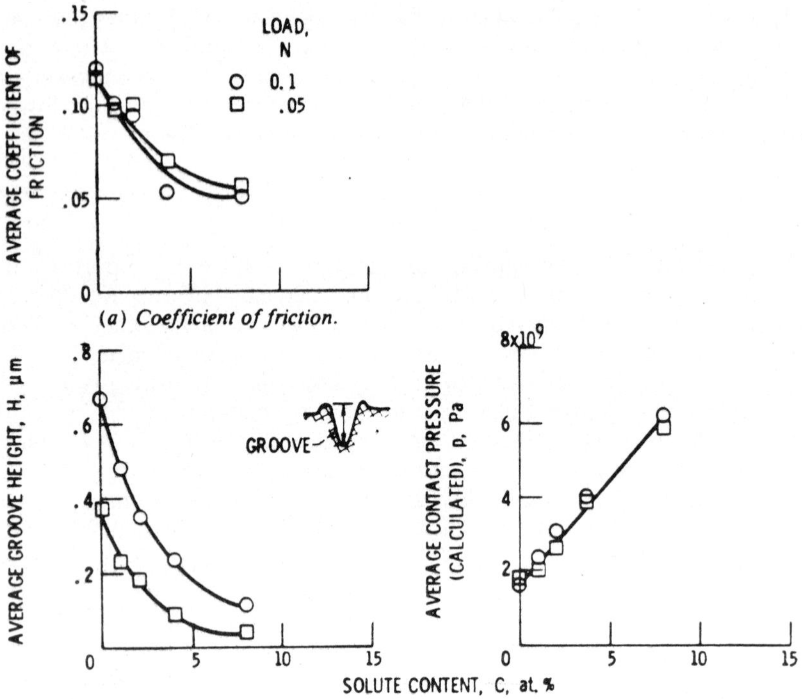

Figure 7-33. – Coefficients of friction, groove heights, and contact pressure for titanium-iron alloy and pure iron as function of solute content. Single-pass sliding of 0.25-millimeter-radius silicon carbide rider in mineral oil; sliding velocity, 3×10^{-3} meter per minute; room temperature.

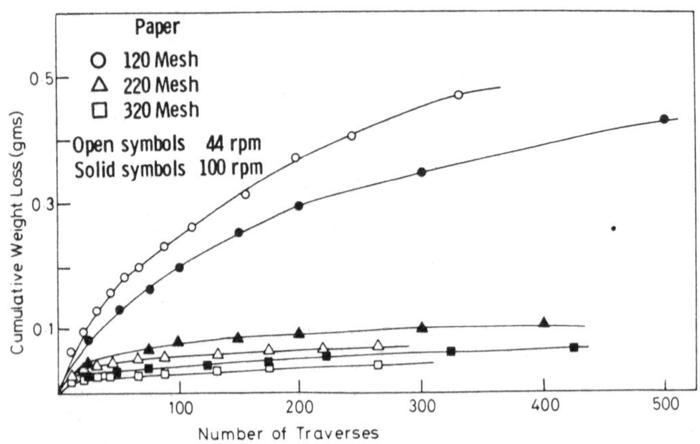

Figure 7-34. – Cumulative weight loss as function of number of traversals for PTFE (ref. 22).

In general, the polymers are very easily abraded by other surfaces, and there is a correlation between the abrasive wear of polymers and the various properties of the abrasives. For example, the subject of abrasive particle size was discussed earlier in reference to the abrasion and damage of the semiconductor surfaces silicon and germanium. A similar relationship exists for polymers contacted by abrasives. The particle size has an influence on the abrasive wear characteristics of the polymer. The larger the abrasive particle, the greater the amount of material removed from the polymer surface. Evidence for this is presented in figure 7-34 where the cumulative weight loss is plotted as a function of the number of traverses across the surface of PTFE with various abrasive papers (from 120 to 320 mesh) and at the rotative speeds of 44 and 100 rpm (ref. 22).

The larger the abrasive grit size, the greater the amount of cumulative weight loss (abrasive wear) to the polymer PTFE. The greatest amount of wear is experienced with the 120 mesh (largest particle size), intermediate wear is experienced with the 220 mesh (intermediate particle size), and the least amount of wear is experienced with the 320 mesh (smallest particle size). In the interaction of polymers with other solid surfaces, if abrasive wear occurs, it generally occurs to the polymers because of their relatively weak cohesive nature compared to that of the other solids. There are those relatively unique situations wherein the other surface (the cohesively stronger surface in contact with the polymer) may undergo abrasive wear as a result of the interaction of the polymer with the solid surface. These situations arise where the abrasion process is not simply one of simple abrasion but where other mechanisms (e.g., adhesion) are involved. Very frequently in the abrasive wear process, a careful examination of the process from a fundamental point of view reveals that some other mechanism (e.g., adhesion) is playing a role in the observed wear behavior.

A series of experiments was conducted in our laboratory with PTFE in sliding contact with annealed aluminum. The experiments were conducted with a pin-on disk configuration where the PTFE pin was a well characterized polymer and the disk was an aluminum disk that had been annealed. After the specimens were placed inside a vacuum chamber no attempt was made to clean the aluminum surface before conducting the experiment. The surface of the disk was monitored with AES analysis to identify the chemistry of the system.

With the sliding process and under a fairly heavy load of 500 grams on the PTFE rider, no polymer film developed on the aluminum disk surface as was discussed in the ADHESIVE WEAR section of this chapter. Instead, aluminum transferred to the polymer surface. The transferred aluminum became work hardened with continuous sliding, because of the continuous contact of the rider with the disk surface. Consequently, the work-hardened aluminum acted as a cutting tool to cut the aluminum disk surface. Auger analysis of the wear track revealed the generation of a clean aluminum surface as a result of the cutting action of the aluminum debris entrapped in the surface of the polymer with a complete loss of the oxide on the aluminum that was present before sliding. In other words, before sliding, an Auger analysis of the aluminum surface revealed the presence of both aluminum and oxygen peaks. With sliding of the PTFE on the surface, it was anticipated that a polymer transfer film would be observed on the aluminum disk. In contrast, however, a diminution in the oxygen peak of the Auger spectra was observed as a result of the cleaning action on the aluminum disk surface. Ultimately, the only peak that was seen in the Auger spectrum was that of aluminum. No transfer of carbon or fluorine to the aluminum disk was observed as a result of the polymer and metal interacting. The Auger spectrometer was removed from the system and this same experiment was repeated with a window in the system to permit photographing the entire process. A photograph taken while the PTFE was in sliding contact with the aluminum disk is shown in figure 7-35. In figure 7-35(a), the debris observed on the surface of the polished disk is not PTFE wear debris but rather aluminum chips that were generated as a result of the polymer being in sliding contact with the disk surface. A careful examination of the disk surface near the disk rider contact point reveals a small curl to the left of the disk rider contact point. This curl was generated as a result of the polymer sliding on the disk surface. All the debris present and observed in the photomicrograph is aluminum that was scraped from the surface as a result of the polymer sliding on it. The disk surface with the debris present and the wear track generated are shown in more detail in the photomicrograph in figure 7-35(b). An actual wearing of the disk surface took place (fig. 7-35(b)). Instead of the adhesive wear process taking place where the polymer transferred to the metal, it took place where the polymer abraded the metal by transferring it to the polymer with subsequent abrasion. The presence of the metal on the polymer surface was revealed with SEM (fig. 7-36).

In figure 7-36, the PTFE rider of figure 7-35 is turned on end and the wear spot is examined in the scanning electron microscope at two magnifications. There is a discrete individual aluminum particle buried in the matrix of the polymer. The embedded aluminum fragment is a piece of work-hardened aluminum that has transferred from the aluminum disk to the polymer surface. In the rubbing process it becomes work hardened and

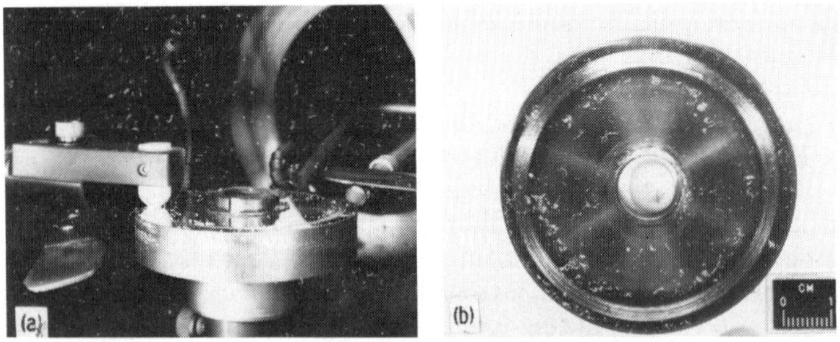

(*a*) *Rider and disk with Auger spectrometer removed.*

(*b*) *Top view of disk.*

Figure 7-35. – PTFE rider on aluminum disk. Surface of disk is covered with aluminum chips and PTFE rider has aluminum chips embedded in it. Load, 500 grams; sliding velocity, 1 centimeter per second; 26 revolutions in vacuum.

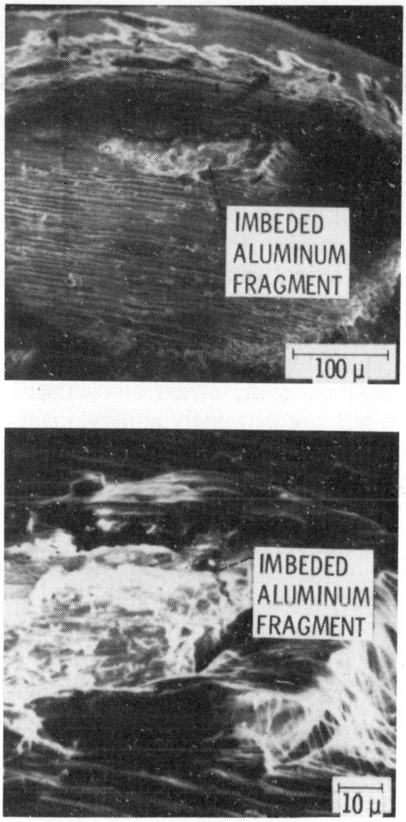

Figure 7-36.—PTFE rider wear scar showing lodged metal fragment. Run on aluminum (110) surface; single pass; load, 200 grams.

acts as a cutting tool (cuts the aluminum on the aluminum disk). The action of the aluminum wear particle is much like that of a particle embedded in a grinding wheel. The PTFE rider acts as the matrix (much as a resin in the grinding wheel does) and the work hardened aluminum particle acts as the abrasive grit (much as a grit of silicon carbide, aluminum oxide or diamond) might act in a grinding or cutting wheel.

Corrosive Wear

Nature of Corrosive Wear in Tribology

Another form of wear frequently encountered in mechanical systems is corrosive wear. Corrosive wear occurs when the environment interacts with the surfaces in solid-state contact to produce reaction products that affect the wear characteristics of the materials in sliding, rubbing, or rolling contact. The environmental interaction can take place by a number of mechanisms. Where the mechanical components are in dry sliding, oxygen from the normal air environment or the gaseous species present in the environment can interact with the solid surface. The reactive constituents that form can bring about corrosive wear. In addition, the presence of excessive quantities of extreme pressure or antiwear additives or other chemical agents in lubrication systems can bring about corrosive wear at an interface. Even with conventional liquid lubricants, high temperatures can cause excessive reaction of dissolved oxygen in the oils with the solid surfaces. This can result in corrosive wear of the elements if they are susceptible to corrosive or oxidative attack.

The magnitude of corrosive wear depends on a number of factors. First, it is influenced by the nature of the materials in solid-state contact—that is, their resistance to corrosive attack. What types of surface films are formed on these materials and are those films impervious to the environment? Some materials have surfaces that are extremely impervious to one environment but are readily attacked by another. A typical example might be a nickel surface. Nickel oxidizes and provides a protective surface film of nickel oxide in a normal air environment. Nickel, therefore, has a fairly respectable oxidation resistance. However, in a sulfur containing environment, the nickel surface is readily attacked by the sulfur. This leads to the second factor which influences the corrosive activity on a solid surface—namely, the environmental constituents. Some environmental constituents are much more reactive on the surface than are others. For example, a halogenated gas that readily decomposes to liberate halogen atoms can be extremely corrosive. When, however, a halogen atom is contained in a fluorinated hydrocarbon, the liberation of the halogen may be much more difficult to achieve and the gas may be much more resistant to decomposition. An example might be the interaction of a metal surface with carbon tetrachloride, which has one carbon atom and four chlorine atoms. Elevating the temperature slightly in a carbon tetrachloride environment can be extremely corrosive to ferrous base alloys. If, however, the chlorine is contained in monochlorotrifluoroethylene, the molecular structure is much more stable and, as a consequence, is much more resistant to decomposition on the solid surface and less prone to corrode ferrous base alloys.

Also of extreme importance is the concentration of the reactants (the surface area of the surfaces in solid-state contact and the concentration of the environmental constituent above that solid surface). The higher the concentration of a reactive gaseous species in an air environment, the more likely it is to corrode the solid surface. Still another parameter of importance is the surface temperatures. The higher the surface temperatures as a result of the rubbing process the more likely the environmental constituent interacts with the solid surface to form reaction products.

Extreme Pressure and Antiwear Additive Action

Corrosive wear is not entirely detrimental. In some situations it is actually desirable. For example, the whole concept of extreme pressure and antiwear additives is based on the premise that the additive interacts with the solid surface to form reaction products much like it does in corrosive wear. In the antiwear or antiseizure additive situation, however, the corrosion is one of a controlled nature. For example, typically organometallics containing sulfur, phosphorus, or chlorine are intended to decompose and then react at the solid surfaces to form inorganic chloride, sulfide, or phosphide films. These films in turn prevent metal to metal contact and avoid seizure in mechanical systems. The object is to form the protective reaction products but to limit the concentrations formed on the solid surface to a thin film at the contacting asperities only and thereby minimize the loss due to corrosion. It is a controlled corrosion process. Even the friction coefficients for materials in contact influence the corrosive wear. For example, the higher the friction coefficient, the greater the amount of energy dissipated at the interface and the higher the surface temperture; this, in turn, promotes surface reactivity. The pressure in the environment is another factor. The higher the ambient pressure of the gaseous species or liquid which is interacting with the solid surface, the more likely the reaction will take place (because of increased concentration of the reactive species).

Thus, corrosive wear can be undesirable when it occurs to excess, or it can be desirable when it limits other wear mechanisms (such as adhesive wear). When the reactivity is excessive, it is referred to as corrosive wear. When the reactivity is controlled, it is considered an effective lubricant. The fluorinated hyrdrocarbons (such as the refrigerants) are extremely stable molecularly. They are organic compounds containing a high concentration of fluorine, and the carbon to fluorine bond is the most stable in organic chemistry. Thus, these organic compounds provide a relatively inert environment in a conventional sense. With surfaces in sliding, rolling, or rubbing contact in an environment of a fluorinated hydrocarbon, another halogen (e.g., chlorine, bromine, or iodine) in addition to fluorine may be present. At the interface, these particular gaseous molecules can break down to form inorganic halides, chlorides, bromides, or iodides on the surface of metals in sliding, rubbing, or rolling contact. These inorganic compounds are very effective lubricants. Many of them have hexagonal crystal structures with easy shear along basal planes, making them inherently good solid film lubricants. These compounds have been explored for years as potential lubricants for metal surfaces. It was found that they can operate very effectively to lubricate solid surfaces at elevated temperatures. There is, however, a temperature where the material ceases to be an effective lubricant and instead becomes corrosive to the solids. In other words, there is a region where the additive acts as a good, effective boundary lubricant because the liberation of the halogen species produces surface inorganic compounds that provide a lubricating film. At high temperatures the reactivity becomes excessive because the breakdown of the molecular structure occurs at too high a rate. The breakdown of the molecular structure is catalyzed by the friction process and actually occurs on the friction surface; the ambient temperature alone is insufficient to cause decomposition or degradation of the organic molecular structure.

One such compound that is effective in this regard is dichlorodifluoromethane. At relatively modest temperatures above the ambient, it reacts with ferrous based materials such as steel surfaces to provide an effective surface film which gives very low friction (values frequently found in hydrodynamic lubrication) and provides very low wear for surfaces in sliding contact. As the ambient temperature is increased, the increase in the energy at the interface, that is, the energy of the environment plus the energy associated with the rubbing process, is sufficiently large to cause excessive reactivity. The dichlorodifluoromethane then becomes excessively reactive (reacts at too high a rate) and forms an excessive quantity of surface chlorides—principally, $FeCl_2$ and $FeCl_3$—ultimately resulting in corrosive wear.

The mechanism of corrosive wear depicted schematically in figure 7-37 shows the environment interacting with a solid surface. If that environment is dichlorodifluoromethane, wear results such as those observed in figure 7-38 are seen. The wear is relatively low at temperatures to 300° C (actually, orders of magnitude below that which would be observed for the metals in sliding contact in the absence of the dichlorodifluoromethane). Above 300° C, however, the wear rate increases rather markedly. This increase in wear is associated with excessive reactivity of the surface with the chlorine of the dichlorodifluoromethane, and this results in the formation of excessive iron chlorides on the solid surface.

The experiments of figure 7-38 were conducted in pin-on disks studies where a hemispherical slider was rubbing a disk of the same material—namely, a tool steel. The nature of the surfaces as a result of the reaction with the gaseous environment is shown in the photomicrographs in figure 7-39. There are three sets of pin-on disk specimens in figure 7-39. At 25° C, the surfaces show little or no evidence of the formation of surface compounds. Basically, the bright metallic nature of the metal surface is visible. At 300° C, however, the surface discolors because of the iron chlorides forming on the surface from the reaction of the rider and the disk with the gaseous environment. At 300° C, effective lubrication is still being obtained with the gas. At 600° C, however, the chemical reaction taking place on the solid surface is excessive, and the quantity of iron chlorides

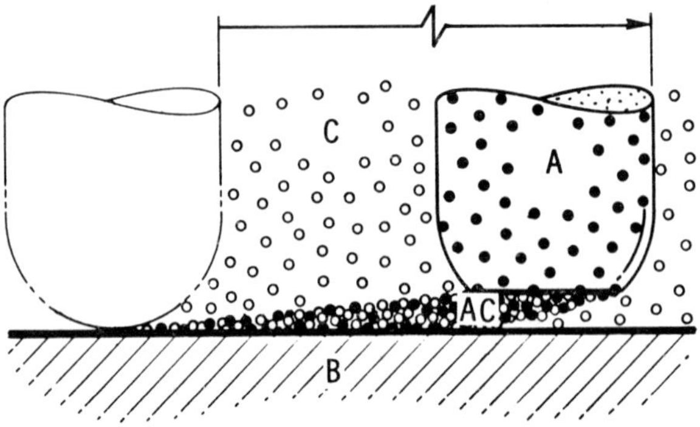

Figure 7-37. – Corrosive wear mechanism.

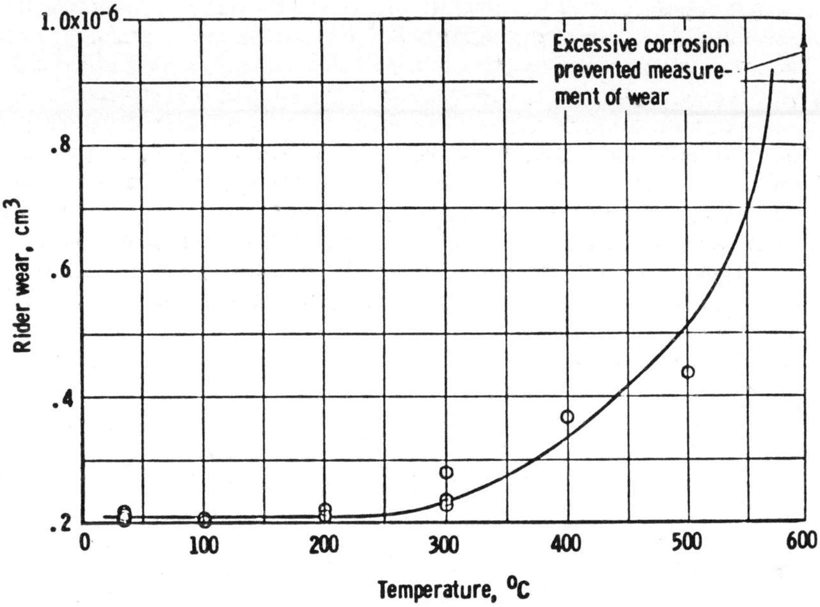

Figure 7-38. Wear of tool steel on tool steel at various temperatures with CF_2Cl_2 plus 1 percent SF_6 as lubricant. Sliding velocity, 60 centimeters per second; load, 1200 grams; duration, 1 hour.

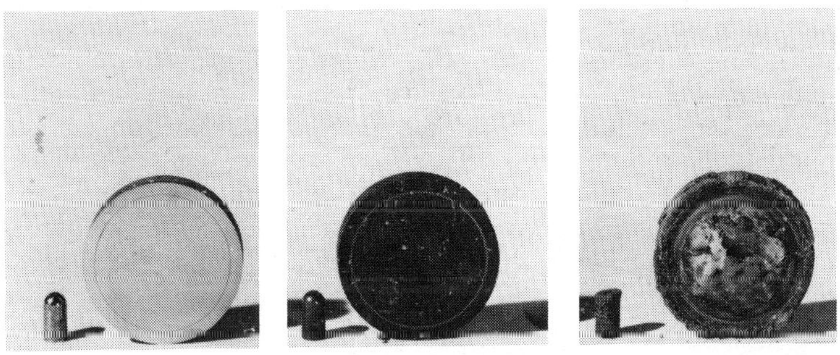

Figure 7-39. – Corrosion on slider specimens after runs at various temperatures with CF_2Cl_2 plus 1 percent SF_6 as lubricant. Disk and rider specimens, M-1 tool steel; duration of run, 1 hour.

formed is so great that very marked corrosion of the specimen surfaces results. Examining the photograph of the rider specimen in figure 7-39 indicates that the hemispherical tip of the rider has been worn away by corrosive wear; that is, the corrosive attack was so severe that the hemisphere of the rider was completely consumed. Thus, the gas dichlorodifluoromethane is an effective lubricant at 25° C for the steel on steel surfaces. At 600° C, the same gaseous species was extremely corrosive to the surface. The same reaction products are formed at both temperatures—the difference being one of quantity. The quantity is excessive at 600° C and is optimum at 25° C.

The same type of reactivity can occur with normal antiwear additives and extreme pressure additives in conventional lubricating oils when there is an excessive concentration of the additive or when the molecular structure of the additive is so weak that it decomposes readily at the metal interface. Excessive reactivity can occur in the form of copious quantities of a reaction product such as a chloride, sulfide, or phosphide that is corrosive wear. When the concentration is correct and the molecular structure is sufficiently stable, the reaction may be optimized at the interface to give sufficient amounts of reaction product to provide effective boundary lubrication but yet not such large quantities so as to produce the corrosive wear observed in figure 7-39.

Effect of Oxide

To an extent, one can tailor the molecular structure so that its ability to break down on the surface is optimal for the particular system in which it is employed. Even in an ordinary air environment the interaction of oxygen with the solid surface can be important. In the absence of the formation of surface oxides, extremely high friction and adhesive wear of metals can occur. In fact, in a vacuum system with the absence of surface oxides, complete seizure of metal surfaces takes place. The oxygen present on the solid surface is, in effect, a lubricant. In fact, the best lubricant we have is oxygen of the environment because it reduces friction coefficients of materials in solid-state contact from a state of complete seizure (with infinite friction coefficients) to a point where most metals exhibit friction coefficients between 0.5 and 1.5. The best liquid lubricants for lubricating solid surfaces only reduce the friction coefficient from the range 0.5 to 1.5 to about 0.1. Thus, the normal occurring oxides are, in a sense, lubricants of solid surfaces, particularly metals in the dry state.

The concentration of the oxide on the surface is extremely important. When the oxide is relatively thin, it can provide a protective surface film and it can reduce friction, adhesion, and adhesive wear. When, however, the ambient temperature is increased or mechanical parameters are altered to provide a large quantum of energy at the interface, excessive quantities of oxides can form on a solid surface and the ordinary oxidation process can lead to corrosive wear of materials in solid-state contact. For example, typical copper-based alloys operated at elevated tempertures undergo oxidative corrosion because of the excessive oxide formation on the solid surfaces. Many metallic alloy systems have inherently good resistance to oxidation. Protective thin surface films can form on the solid surfaces, which are impervious to film growth and provide effective lubrication. In stainless steels, the surface oxides are relatively thin and in some instances (e.g., the 400 series stainless steels) they are very effective in preventing adhesion, high friction, and adhesive wear. They are less effective, however, with the 300 series stainless steels which are much more likely to deform plastically and expose nascent metal to contact at the interface.

The oxides in alloy systems are, however, relatively complex and the oxide chemistry is extremely important in adhesion, friction, and wear. Some oxides are much better than others at reducing adhesion, friction, and wear. Where one has an alloy system, the oxides can vary on the solid surface as a function of condition such as, for example, ambient temperature. Mechanical parameters such as the load under which the surfaces are in contact and even the nature of the mechanical activity at the solid surface can alter the composition of the surface oxides that are generated. It should be remembered, however, that the formation of surface oxides is, in a strict sense, corrosion of the solid surface. It is a consumption of the solid surface in the formation of a substance other than the parent material. The solid surface may be either metal or alloy.

XPS or ESCA (as it was formerly called) is very effective for studying oxidation and corrosive wear of metal surfaces. It has been used by the specialist in corrosion and also by the tribologist for studying the formation of reaction products on solid surfaces. By studying the composition of the oxides on the surface and relating them to friction and wear, one can associate specific forms of oxides with desirable friction and wear behavior.

The use of XPS to study the oxidation of alloys frequently used in tribological systems provides some very useful information. For example, in the oxidation of Inconels (basically nickel-chromium alloys containing some iron), the nature and the form of oxides on the solid surface are very strongly influenced by temperature as well as by alloy chemistry. For example, with the Inconel series, oxidation at 100° C results in the formation of a duplex oxide layer of nickel oxide and chromium oxide (NiO and Cr_2O_3). Thus, one finds oxides of both elemental constituents on the surface. If, however, the temperature of the surface is increased to 280° C, not only does the surface contain NiO and Cr_2O_3, but it also contains a spinel of $NiFe_2O_4$ near the surface as well as Cr_2O_3 near the metal interface. The spinel is formed from the interaction of oxygen with the small concentration of iron that is present in the alloy.

The oxygen concentration affects the relative concentration of nickel oxide produced during oxidation. The chromium oxide phase, however, is unaffected by oxygen concentration; this suggests that the growth of the protective layer on Inconel is limited by the lower outward diffusion rate of chromiun ions.

A commonly used bearing steel in the aerospace industry is 440-C stainless steel. The oxidation process of this alloy is very much a function of the ambient temperatures. Auger analysis shows iron oxides on the surface under one set of conditions and, in another set of temperature conditions, almost exclusively Cr_2O_3.

As mentioned earlier, in addition to the ambient temperature, the temperature of the surfaces and the nature of the mechanical activity on the solid surface can influence the way the surface is oxidized. Some studies on the oxidation of a 5 percent iron, 95 percent nickel alloy indicate the effect of mechanical activity on the oxidation process. XPS data were obtained for the oxidized 5 percent iron, 95 percent nickel alloy after the surface had been abraded with silicon carbide paper (number 600) and then after the same type of surface had been polished with diamond paste (0.25 μm). The oxidation of these two surfaces was then studied (ref. 23) to see whether the differences in the mechanical activity indicated by the differences in the abrasion of the surfaces were important. If the differences in mechanical activity in the solid surface are simply one of degree in the abrasion process, the surface chemistry in the oxidation process should not change. The XPS data obtained were reduced and the results are presented in figure 7-40.

In figure 7-40, the elemental concentration of the oxides and the metal are presented for two conditions: (1) for the abraded surface and (2) for the surface that was polished with the diamond paste. In figure 7-40(a), there is a very high concentration of nickel metal exposed on the surface where the surfaces are abraded by the silicon carbide paper (which cuts more deeply

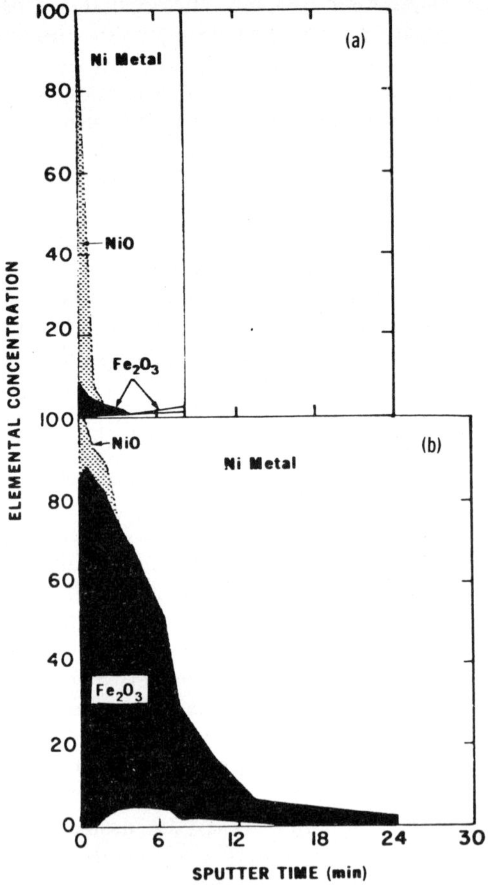

(*a*) *Surface previously abraded with #600 silicon paper.*

(*b*) *Surface previously polished with 0.25-micrometer diamond paste.*

Figure 7-40. – Effect of abrasion of 5-percent-iron–95-percent-nickel alloy surface on oxide film composiition (ref. 23).

into the surface than the diamond paste). A considerble amount of nickel metal is, therefore, exposed. In addition, there is a high concentration of nickel oxide present on the surface and some small concentration of the higher oxide of iron (Fe_2O_3). If the surface layer is removed by sputtering, the amount of nickel oxide in the surface drops off drastically and one is into the bulk of the material and sees basically nickel with very little or no iron in the XPS spectra. Hence, the principle surface film on the alloy surface that has been abraded with the silicon carbide paper is a very thin film of nickel oxide containing a small concentration of Fe_2O_3. The oxide is readily removed by a small amount of sputtering; this indicates that the film present on the solid surface is quite thin. However, if the surface is polished very slowly with diamond paste and then analyzed with XPS, the initial surfaces are seen to contain a fairly uniform mixture of nickel oxide plus Fe_2O_3. There is a much higher concentration of iron oxide on the surface where the surface has been abraded by the fine diamond paste particles. The composition of the film is a more even mixture of the two oxides (the oxide of iron and the oxide of nickel). Sputtering the surface to remove surface layers and doing a depth profile analysis with XPS reveal that the thickness of the film formed by the diamond paste polishing is much greater. It runs much more deeply into the surface than does the surface film generated by the abrasion of the 600 grit silicon carbide paper. Thus, there are marked differences in oxidation of the same alloy surface as a function of the way in which the surfaces are prepared. In considering the oxidation of metal surfaces and alloys, therefore, one must be concerned with the concentration of the reactants, the nature of the metal surface interacting, and the elemental constituents in the alloys and their reactivity toward the environmental constituents. However, one must also be concerned with the mechanical activity on the solid surface, since markedly different surface chemistry can be obtained from the differences in the mechanical activity on the solid surfaces. The differences shown in figure 7-40 are relatively minimal. The abrasion occurring in figures 7-40(a) and (b) is simply the difference in particle size and the chemistry of the two abrasives.

Controlled corrosion of a solid surface is extremely important in areas outside the lubrication of the solid surface in tribological devices. For example, very frequently it is desirable to produce a very smooth surface for mechanical components in practical tribological systems. One of the most effective ways of producing such a surface and eliminating or reducing the concentration of surface asperities is to use an electropolishing technique. Electropolishing is nothing more than controlled corrosion of the solid surface. One concentrates (by electrical means) a chemical material at the asperity tips to consume the asperity tips and thereby smooth or flatten the solid surface. In electropolishing, of course, the reactant is generally a liquid containing an acid, base, or salt material that interacts or reacts with the metal surface to form compounds which then go into solution. This controlled corrosion is very useful and beneficial since it can smooth the surface and reduce the sharp asperities that are or can be very detrimental to lubricating films and to the breakdown of these films.

An example of the changes in surface topography that can be achieved by using controlled corrosion associated with electropolishing is indicated by the two surface profile traces (ref. 24) in figure 7-41. One profile is for a ground surface (fig. 7-41(a)) and the other is for an electrochemically polished surface (fig. 7-41(b)).

A comparison of figure 7-41(a) with 7-41(b) indicates that the surface topography can be altered markedly by using controlled corrosive wear of a solid surface. The electrolyte reacts with the solid surface with the aid of electrical current to speed up the process. The tips of the asperities interact with the reacting solution to form reaction products that are soluble in the electrolyte and that dissolve in the solution. The driving force for the process, of course, is the localization of the current applied to the surface at the tips of the asperities. There is an irregular distribution of the current concentration on the surface; it is localized on the asperity regions that are the high spots in the solid surface and it causes the reaction to occur much more rapidly in those areas than in the low spots or regions that are not protruding markedly above the surface. The net result of this is a smoothening of the surface. This is seen by comparing the surface profile trace of figure 7-41(a) with that of figure 7-41(b).

This same type of surface smoothening has been observed in tribological devices where reaction takes place from the environment with the solid surfaces. In the absence of electrical current, the energy at the interface is supplied by rubbing and the friction associated with it. The friction process accelerates the reactivity of the environmental constituents with the solid surface to produce the smoothening effect seen in figure 7-41(b). This has been observed for liquid lubrication of solid surfaces where the liquid lubricant contains additives that are reactive with the solid surface. It has also been observed with the gaseous lubrication of solid surfaces, which were discussed earlier in this section.

(*a*) *Surface after grinding;* R_a, *0.340 micrometer.*

(*b*) *Surface after electrochemical polishing time of 5 minutes;* R_a, *0.160 micrometer.*

Figure 7-41. – Profile traces from one sample showing effect of electrochemical polishing. Horizontal magnification, 80; vertical magnification, 8000 (ref. 24).

Erosive Wear

Yet another mechanism of the loss of materials from solid surfaces involves the erosion of surfaces by the impingement or impacting of liquids, gases, or solids. The basic mechanism involved is depicted schematically in figure 7-42. In this figure the surface of material B is impacted by a particle of a second material A. Particle A strikes the surface of material B with some velocity. As a result of the impact in solid-state contact of particle A with the solid surface, material is removed from the surface of material B. The particle itself can vary in composition as well as in form. For example, it may be an atom or molecule of a gaseous species. Basically, the sputtering process of striking a metal surface by an argon ion is, in a real sense, an erosive wear process. The argon ion strikes the surface and knocks material from the surface. In the case of sputtering, the argon ion, which is in a gaseous form, is the incoming particle that strikes the solid surface and removes material from that surface. The same process can occur when liquid droplets strike a surface and remove material. For example, rain droplets hitting a surface with various velocities can remove material from a solid surface and ultimately cause erosive wear. Probably the most frequently discussed mechanism of erosive wear is where a solid particle strikes a solid surface and material is lost from that solid surface as a result of the impact in solid-state contact. From a practical point of view, the erosion process and erosive wear are extremely important.

From the mechanism displayed in figure 7-42, it becomes obvious that, with repeated impingements of particles on a solid surface, the surface after some period of time can be roughened. Examining an eroded metal surface, for example, shows that the surface topography has been altered. A very highly polished metal surface may become very rough as a result of the impingement. A surface scanning electron micrograph of an aluminum surface that was bombarded by silicon oxide particles is shown in figure 7-43. In this figure the aluminum surface was highly polished before it was impinged by silicon oxide particles. After erosion of the surface, a

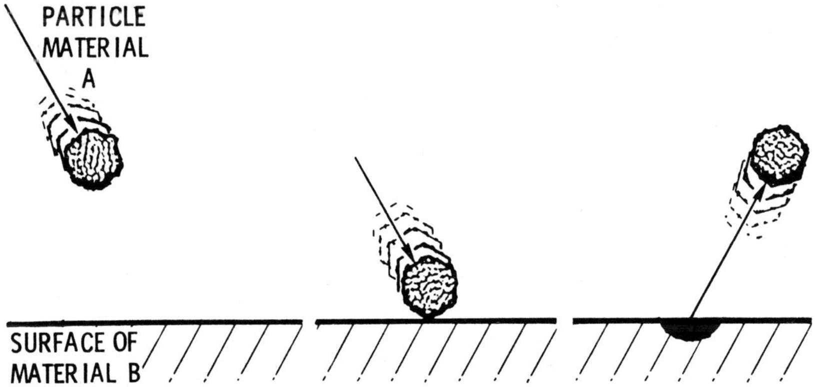

Figure 7-42.—Erosive wear mechanism.

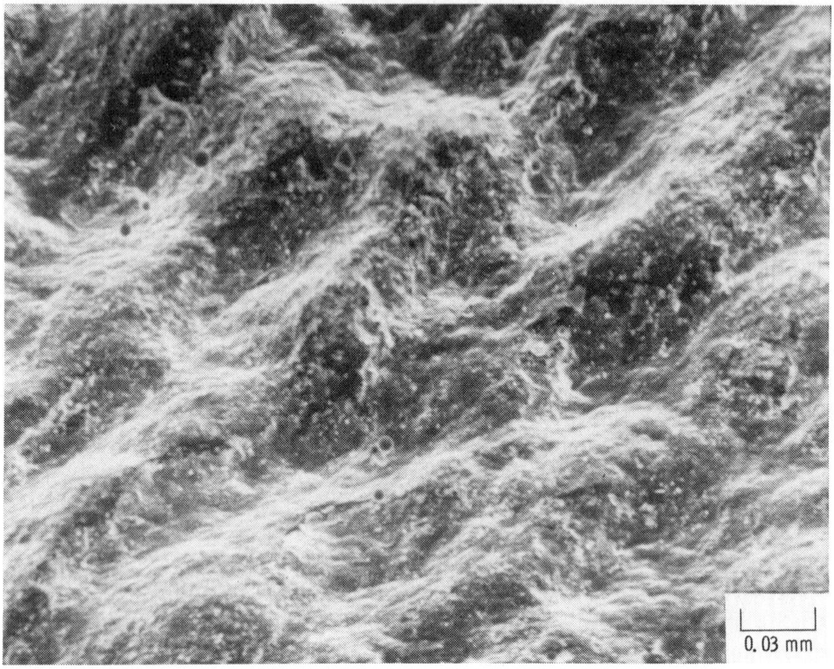

Figure 7-43. – Scanning electron micrographs of aluminum surface eroded by SiO_2 particles.

considerable amount of aluminum was removed from the surface by the incoming silicon oxide particles. If the process is repeated many times the surface is left in a very rough hill and valley condition.

In the erosion process, the nature of the particles striking the solid surface can vary just as the velocity or impact energy with which the solid particle strikes the surface can vary. The greater the velocity of the incoming particle, the greater the energy associated with it and the greater the energy imparted to the surface being impacted. This can have some interesting effects on the nature of the solid surface being impacted. For example, in some experiments conducted by Yust et al. it has been shown that, with ceramic surfaces, the impingement by silicon carbide particles at sufficiently high velocities can result in localized surface melting of the ceramic surface (ref. 25). This is shown in the scanning electron photomicrograph (fig. 7-44) obtained by Yust et al. for a ceramic surface that was impacted by silicon carbide particles. Note the long fingerlike projections extending downward in the photomicrograph; these projections indicate localized melting on the ceramic surface as the result of the erosive impact of the silicon carbide particles with the surface.

In trying to understand the mechanism of erosive wear, various investigators have attempted to relate the basic properties of materials to their erosive wear behavior. Are there basic fundamental properties of materials that have some influence on the rate at which material is removed by the erosive process? During the course of various investigations, some

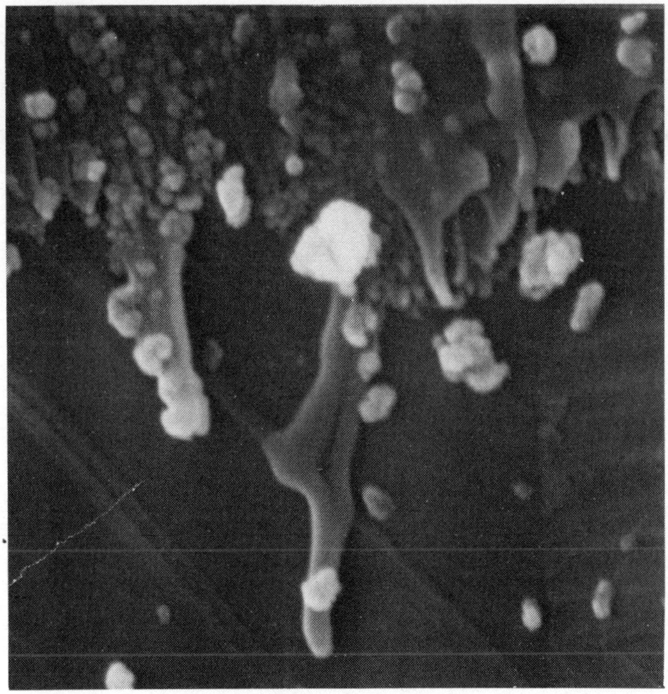

Figure 7-44. – Erosive melting of ceramic material surface (ref. 25).

basic properities were found to be related to the erosive nature of the materials being impacted. One such property has been, for metals, the binding energy in the metal. The cohesively strongly bonded metals appear to be more resistant to erosive wear than do the cohesively weakly bonded metals.

Vijh has made correlations for the erosive wear rate with various impacting particles and the metal to metal binding energies or cohesive energies for the metals (ref. 26). For a variety of different impacting particles, there is a basic or fundamental relationship between material loss from the metal surface and the cohesive binding energy of the metal. Figure 7-45 shows a plot of the volume erosive wear rate as a function of the binding energies for some elemental metals where the surface of the metal is impacted by quartz sand. From the log log plot in figure 7-45, it is apparent that there is a correlation between the erosive wear rate for elemental metals and their cohesive binding energies. The cohesively stronger metals exhibit lower erosive wear than do the cohesively weaker metals when impacted by quartz sand.

The same basic relationship was found by Vijh when other particles (silicon carbide) impacted the solid elemental metal surfaces (ref. 26). Figure 7-46 shows a plot of the volume of erosive wear rate as a function of the cohesive binding energies or metal to metal bond energies for elemental

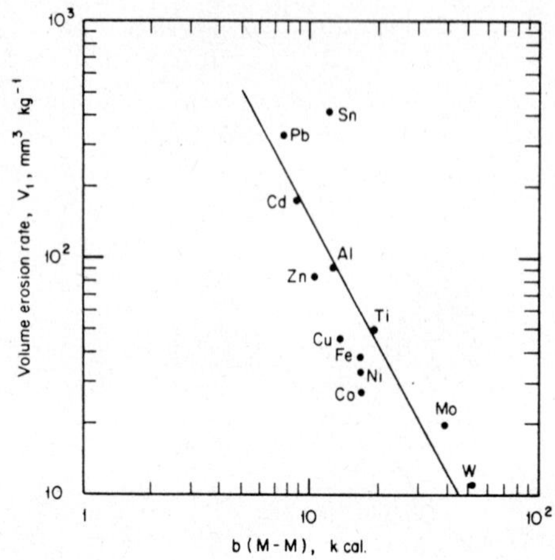

Figure 7-45. – Volume erosion rate V_1 (impact erosion by quartz sand) as function of metal-metal bond energies (ref. 26).

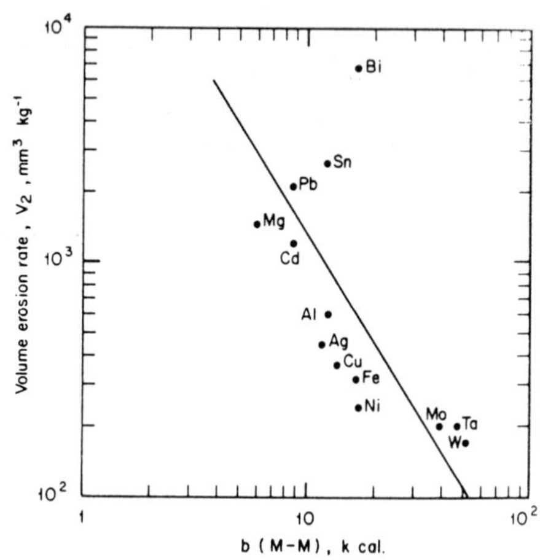

Figure 7-46. – Volume erosion rate V_2 (impact erosion by silicon carbide particles) as function of metal-metal bond energies (ref. 26).

metals. Just as was shown in figure 7-45 for quartz particles, figure 7-46 shows a direct relationship between the erosive wear and the cohesive binding energies in the metal when silicon carbide particles strike various elemental metal surfaces. Thus, the same basic relationship appears to exist whether the particles are quartz or silicon carbide that are impacting the surface; that is, the stronger the cohesive binding energy in the metals, the lower the erosive wear rate of that metal being impacted by solid particles.

There is no doubt but that there are other basic properties of metals that can be related to their resistance to erosive wear. Many of these properties are also related to the cohesive binding energies of the metals. For example, hardness, modulus of elasticity, and other such properties can be related to cohesive binding energies, and these, in turn, can be related to the erosive wear resistance. Cohesive binding energy is fundamentally a very useful property for that reason. Because it does relate to many other measured properties of metals, the use of that particular parameter in making correlations is extremely good. It is a very basic and fundamental property of the material.

Considerable research effort has been expended in the study of the rain erosion of solid surfaces. A large group of materials has been studied in an attempt to understand and correlate erosion of solid surfaces as a function of various materials in various classes and the properties of those materials as they relate to the resistance of the solid surfaces to erosion by rain droplets.

Langbein has attempted to show the relation between erosion in the basic classes of various materials as a function of the time for erosion (ref. 27). Some of his results are summarized in figure 7-47 for various classes of materials: glasses, polymers, ceramics, and metals. The range of the wear rates, erosive wear, as a function of time is presented for these various classes of materials. An examination of figure 7-47 reveals that the highest rate of erosive wear by rain particles is experienced with glasses, followed by polymers, ceramics, and then metals.

The results of figure 7-47 are rather interesting in that a variety of properties are exhibited by the classes of materials, and these properties, in general, do not seem to be related to the erosion resistance of the classes of materials. For example, glasses are relatively hard, brittle materials, and one might understand that hard brittle materials might fracture readily on the erosion by rain resulting in a relatively poor erosion resistance.

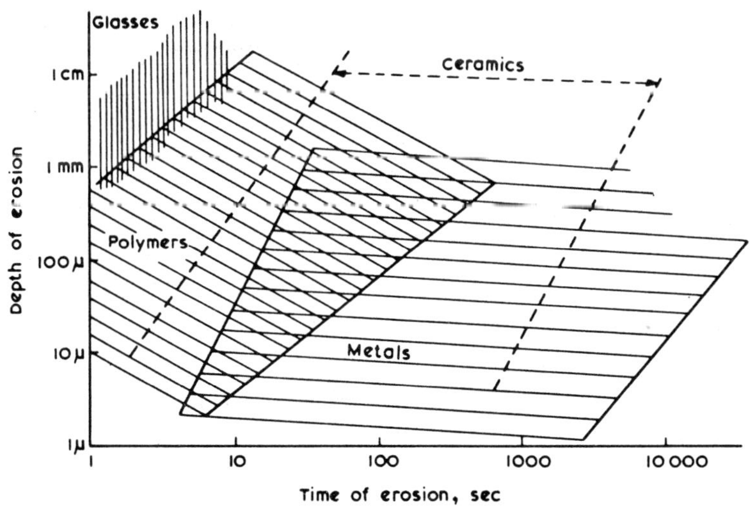

Figure 7-47. – Comparison of rain erosion properties of polymers with glasses, metals, and ceramics (ref. 27).

Polymers, on the other hand, exhibit properties at the other end of spectrum. They are extremely prone to deform rather readily and seem to exhibit properties that make them less likely to erode. Metals lie some place in between; ceramics are in a class analogous to that of the glasses. Even though ceramics are analogous in their mechanical behavior to glasses, their resistance to rain erosion apparently is significantly better than that of the glasses and many polymers. In some instances, they behave analogously to ductile metals in their resistance to rain erosion. With respect to liquid impingment on solid surfaces, there does not seem to be a fundamental correlation between the mechanical properties of the classes of materials involved and their resistance to rain erosion.

Even more fundamental properties such as cohesive energies or adhesive energies in the solids do not seem to relate to the results shown in figure 7-47. Polymers are relatively weak materials with respect to cohesive binding energies. Yet, they do have superior erosion resistance to materials such as glasses which are cohesively much more strongly bonded.

Fatigue Wear

Another mechanism for generating wear particles or losing material from solid surfaces in solid-state contact is associated with the fatique process where there are repeated applications of stress (by cyclic rotation across the solid surface or reciprocal sliding motion). Under such conditions, the generation of cracks, both surface and subsurface, leads to the formation of wear particles and the loss of material from solid surfaces.

The basic mechanism involved in fatique wear is shown schematically in figure 7-48. The process can occur in such elements as rolling element bearings where a ball or roller repeatedly rolls over the same surface and, after some time, a crack is observed to initiate either subsurface or in the

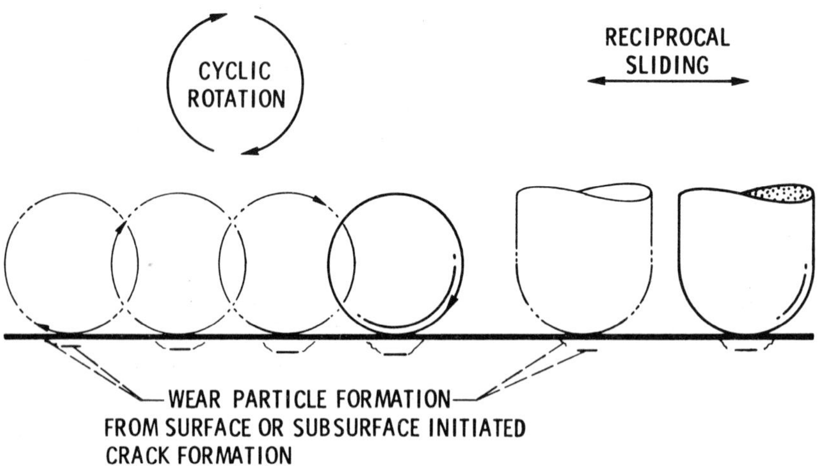

Figure 7-48. – Fatigue wear mechanism.

surface region. The subsurface crack gradually progresses to the surface and liberates a wear particle; however, the surface crack moves downward into the bulk, connects with other cracks, and generates wear particles. Generally the fatigue process occurs after many cycles of stress.

The development of subsurface cracks is shown in figure 7-3. Initiation of surface cracks, from reciprocal sliding, can be seen in the photomicrograph in figure 7-49. This photomicrograph is for a 200 000 molecular weight polyethylene oxide polymer that was in sliding contact with itself in reciprocal sliding in an inert atmosphere. The reciprocal sliding leads to the formation of surface initiated cracks in the polymer which progress from the surface into the subsurface. The photomicrograph in figure 7-50 shows two cracks, one on the upper part of the wear track and one in the center of the wear track. These ultimately join. Another crack just to the right of that particular crack is propagating in the direction of the larger crack. Ultimately, these cracks join and can lead to liberating particles of the polymer out of the bulk surface. Thus, in the fatigue process, both subsurface and surface initiated cracks develop that lead to the formation of wear debris. In relatively homogeneous materials, the initiation of the fatigue cracks can be assisted by dislocation mechanisms. For example, in the surface initiated crack, crack growth can progress subsurface with the aid of dislocation coalescence to form voids in the material. These voids

Figure 7-49. — Disk wear scar of 200 000-molecular-weight polyethylene oxide polymer in sliding contact with itself. Sliding velocity, 5.0 centimeters per minute; load, 25 grams; temperature, 23° C, duration, 1 hour, argon atmosphere.

assist in crack growth. This particular mechanism is displayed schematically in figure 7-50(a) where a surface fatigue crack develops in a surface and moves subsurface along a slip or cleavage plane. The coalescence of dislocations to form voids along the slip plane or cleavage plane as with sliding, rubbing, or rolling contact promotes the generation of dislocations subsurface and the coalescence of these dislocations. In other words, the accumulation of dislocations along the slip or cleavage plane results in the generation and growth of the surface originated fatigue crack.

Dislocations can also act in the development of subsurface initiated cracks because of their coalescence with repeated stress cycles applied to the solid surface, particularly where strain occurs in the material. This is indicated in figure 7-50(b) where applying surface stresses produces the region of maximum subsurface shear stress as well as the coalescence of dislocations along the slip or cleavage plane. Repeated cycles then cause the voids that are formed to grow and to increase in number until they coalesce to form a crack such as that in figure 7-3. Further stress cycles allow the crack to grow until it reaches the solid surface (as was observed with lithium fluoride in fig. 7-3.)

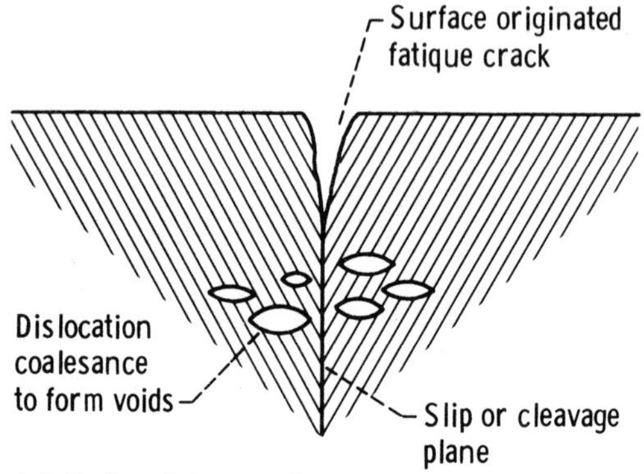

(*a*) *Surface fatigue crack.*

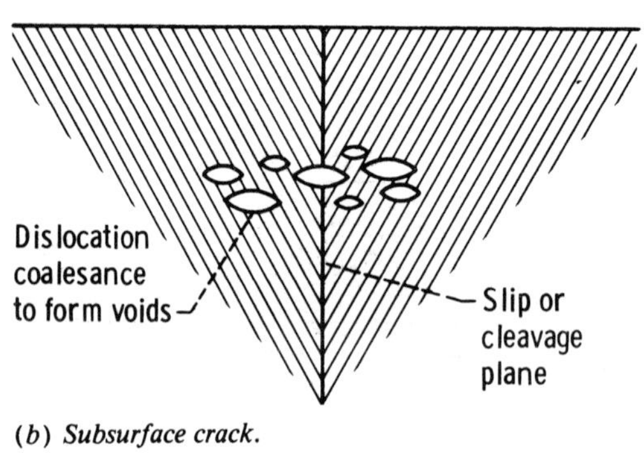

(*b*) *Subsurface crack.*

Figure 7-50. – Crack initiation in sliding, rolling, or rubbing.

Figure 7-51 is a schematic showing the mechanism of crack propagation (ref. 28). Once a primary crack has been initiated in the solid as a result of the fatigue process (fig. 7-51(a)), it can grow in the material as indicated in figure 7-51(b). In addition to the mechanisms displayed in figure 7-50, the secondary cracks that originate from the primary crack can develop as indicated in figures 7-51(c) and (d). Repeated stress cycles with the formation of the secondary crack can result ultimately in the formation of wear particles as indicated in figure 7-51(d). This results in liberating a wear particle from the solid surface and the formation of a void as indicated in figure 7-51(e). When this occurs repeatedly, the surface is ultimately left with a series of pits or voids (fig. 7-51(f)).

Fatigue crack propagation (as indicated in fig. 7-52) is influenced by a number of factors, one of which is the environment. The environment is extremely important in the behavior of fatigue crack growth in practical tribological systems. Even such things as variations in the relative humidity of the atmosphere can produce marked differences in the rate at which the crack grows in a material as a function of the number of repeated stress cycles.

In figure 7-52 the fatigue crack propagation is shown as a function of the amount of relative humidity in the air. Crack depth is plotted as a function of stress cycles in the repeated stressing of an aluminum alloy. There are three curves in figure 7-52: one for ordinary room air and two for dry air (moisture content for the one is 4 to 10 ppm and for the other it is 40 to 100 ppm). These data were obtained by Endo and Goto (ref. 29). The crack depth develops most rapidly in ordinary room air which has the highest concentration of moisture with the least number of stress cycles. The depth of the crack propagates very rapidly subsurface. The crack growth occurs at

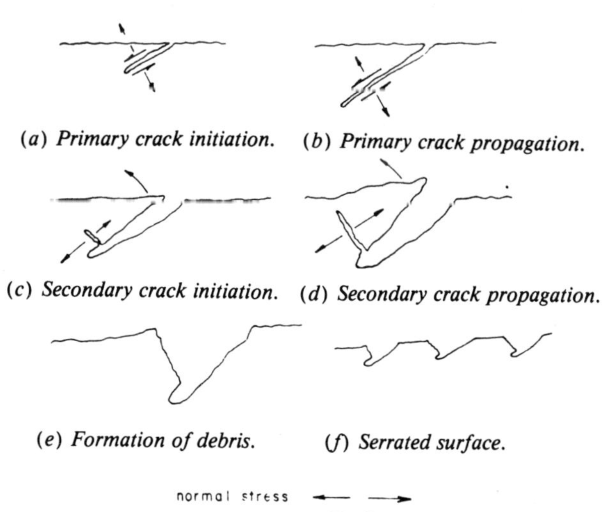

Figure 7-51. – Schematic diagram showing mechanism of crack propagation.

a lower rate in the dry air containing 40 to 100 ppm of moisture. A greater number of stress cycles is needed to generate cracks in the dry air containing only 4 to 10 ppm of moisture. In other words, the drier the air, the lower the rate of crack propagation subsurface in the material. The data in figure 7-52, then, indicate that fatigue crack growth is sensitive to the environment and that a small concentration of water vapor in the environment plays a role in the crack growth behavior and ultimately in the fatigue wear that is observed. This is because the greater the fatigue growth rate, the more rapidly fatigue wear particles are generated in the system.

If the fatigue processes are extremely sensitive to such things as parts per million of water vapor in an ordinary air environment, then it is reasonable to assume that the fatigue crack growth rate is even more sensitive to some of the surface active agents that are present in lubricants. This is a relatively virgin area of research, and very carefully controlled experiments are sorely needed to identify the role of surface active elements dissolved in conventional lubricants in fatigue crack growth of materials.

In homogeneous materials, fatigue cracks can generate and propagate as a result of dislocation coalescence along grain boundaries, slip bands, and other generated irregularities in the material. The generation of cracks as a result of the fatigue process for inhomogeneous materials (i.e., two-phase or second-phase materials) is even more likely because the second-phase particles can act as sites for the initiation of dislocations and the growth of cavities. This is particularly true, for example, of materials that contain a hard second phase, such as carbides in steels. It is well established in the metallurgical literature that carbides serve as sites for the coalescence of dislocations and the generation of cracks or voids in the material. The basic process for initiating and developing cracks or voids can be displayed schematically as in figure 7-53 where the small sphere represents an obstacle or a second-phase particle (assumed to be a carbide in the steel). Then the

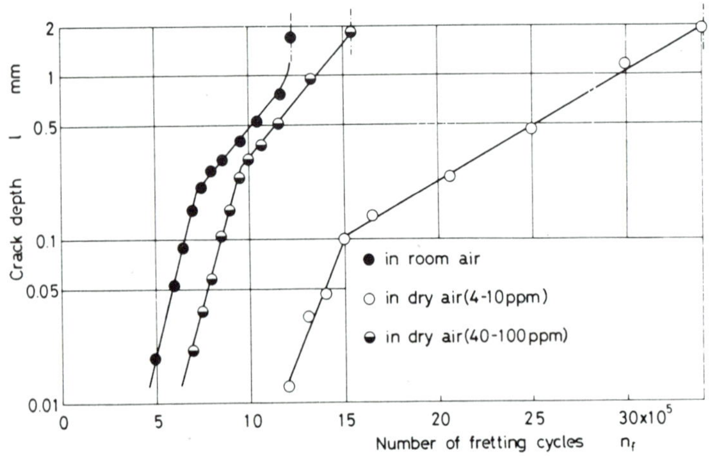

Figure 7-52.—Fatigue crack propagation curves for aluminum alloy in room air and in dry air (ref. 29).

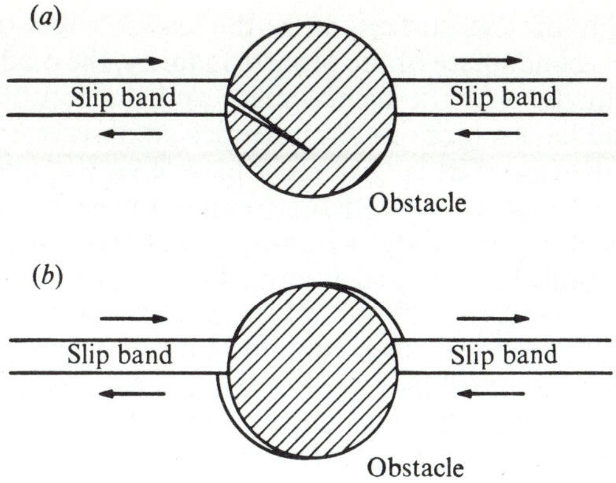

Figure 7-53. – Initiation of cavities at second phase in nonbrittle solid.

slip bands exist along the centerline of the included carbide or hard-phase particle. With strain of the material, slip is initiated along the slip band and dislocations move in the slip band to the location of the included hard-phase particle or obstacle to dislocation motion. The dislocations coalesce along the hard- or second-phase particle. On either side of the particle the dislocations coalesce, and cavities develop along the hard second-phase particles. These cavities can then grow to connect with other cavities that have formed along other second-phase particles until ultimately they lead to large cracks developing either subsurface or surface in second-phase, inhomogeneous nonbrittle solids such as the steels. These eventually lead to the formation of fatigue wear particles.

Fretting

Another form of wear that can be encountered in tribological devices is fretting, which generally occurs when surfaces are in oscillatory or reciprocating motion where the amplitude of the oscillation is relatively small. In fretting, the wear appears to be somewhat analogous to a corrosive wear mechanism in that a large and copious quantity of debris (reaction product debris) is generated. Actually, however, the fretting wear is a two-step mechanism. Initially, the rubbing (in reciprocating or oscillatory motion) of two solid surfaces (e.g., metals) in solid-state contact generally produces a failure because of the adhesive wear in the materials. Adhesive wear particles are generated at the interface, and they then become subsequently oxidized in the environment because a large quantity of energy is stored in the wear particles that are generated. Since the particles are very prone to oxidize heavily and very rapidly, copious quantities of oxide debris are generally found on the surfaces. For example, with steels and iron based alloys, large amounts of Fe_2O_3 are generally found in the fretting contact zone. The debris that is observed (the iron oxide, Fe_2O_3) is the end product of the fretting wear.

Since interaction of the surface with the environment is extremely important for the second phase of the process (namely, the oxidation stage), it might be anticipated that the environment and the constituents in the environment would have a pronounced influence on observed fretting wear behavior. This is the case. The environment plays a very strong role on the wear observed for surfaces which undergo fretting. Figure 7-54 presents two wear curves for an aluminum alloy in fretting experiments. In the figure the weight loss of the specimen is plotted as a function of the number of fretting cycles in two environments: in dry air, which contains only 4 to 10 ppm of water vapor, and in room air, which contains a much higher concentration of water vapor. An examination of figure 7-54 indicates that the amount of fretting wear that occurs in the moist atomsphere (room air) is considerably higher than that observed in the dry atmosphere.

It should be noted, from a careful examination of figure 7-54, that in dry air there is what appears to be a negative weight loss of the specimen. Actually, the specimen gains weight initially. The reason for this initial weight gain is that, for the aluminum surface, in the first step of the process adhesion takes place and metal is transferred back and forth from one surface to the other. The absence of moisture means there is insufficient oxygen to completely oxidize the adhesive wear particles generated. Wear particles are generated and readhere to the surfaces. (Some oxidation takes place because the experiments are done in air, but even in dry air, aluminum is extremely prone to oxidize very rapidly.) So, with the oxidation process occurring on a limited scale, adhesion of the wear particles to the parent surface still occurs. And, because of the uptake of the oxygen, there is a net weight gain in the specimens. The amount of oxide buried among the wear particles trapped on the surface has caused an increase in the weight of the sample after a time. With a number of fretting cycles, however, some of this material breaks free of the surface and an increase is observed in the

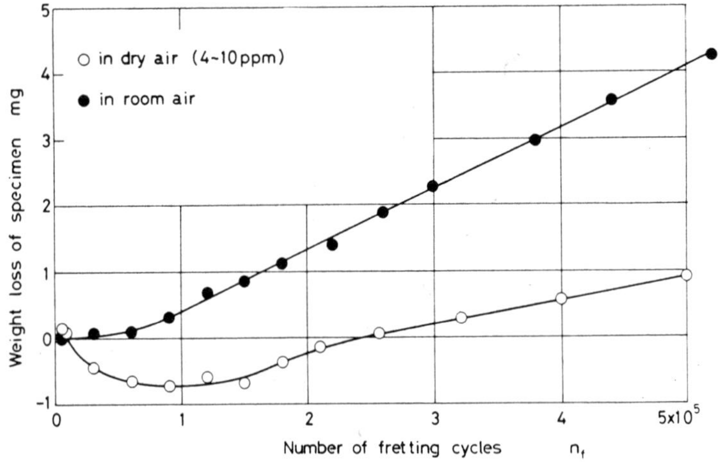

Figure 7-54. – Wear curves for aluminum alloy in room air and in dry air (ref. 29).

amount of debris generated; the result is a net weight loss from the sample surfaces.

With moist air (room air), the oxygen and water vapor present in the air are sufficient to cause a considerable amount of oxidation of the wear particles generated as a result of the adhesive wear process. These oxidized particles do not readhere to the surfaces and are liberated from the surface for an almost immediate net weight loss to the samples.

Cavitation

The last form of wear to be discussed in this chapter is cavitation. Cavitation wear is the loss of material surfaces as a result of a stream of liquid containing entrapped gas bubbles impinging the solid surface. The impinging liquid brings the gaseous bubbles to the surface at relatively high velocities. The bubbles strike the solid surface and collapse on impact; their collapse imposes shock waves on the solid surface and results in the liberation of material from the surface as indicated schematically in figure 7-55. Cavitation wear is a frequent problem in such practical systems as hydraulic devices where the hyrdaulic fluids contain entrapped gases that contact solid surfaces under relatively high velocities. Cavitation wear produces a roughening of the surface much like an etchant would do, and it produces an etched effect on the solid surface. It is frequently observed, for example, on impellers and propellers where the liquids have impacted near the hub and the vanes of the impeller or propeller to produce a cavitation wear effect. The wear is obvious because an etch appearing surface is evident as a result of the gaseous bubbles impacting the solid surface.

Unfortunately, there is a scarcity of fundamental data on the nature of cavitation wear involving the impact of gaseous bubbles on solid surfaccs. It would be extremely helpful if more research were conducted in this area so that some of the fundamental parameters (such as the velocity of incoming

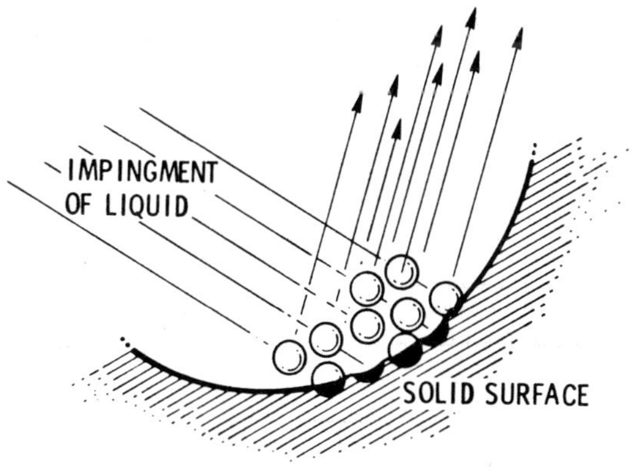

Figure 7-55. – Cavitation mechanism.

gaseous bubbles, the influence of the liquid media, and the effect of material properties of the solid surface on cavitation damage) could be made available to aid in understanding the mechanism of cavitation wear.

References

1. Davies, R. M.: The Determination of Static and Dynamic Yield Stresses Using a Steel Ball. Proc. Roy. Soc. (London), series A, vol. 197, no. 1050, June 1949, pp. 416-432.
2. Bill, R. C.; and Wisander, D.: Recrystallization as a Controlling Process and the Wear of Some Face-Centered Cubic Metals. Wear, vol. 41, 1977, pp. 351-363.
3. Ruff, A. W.: The Studies of Deformation at Sliding Wear Tracks in Iron. NBSIR 76-992, National Bureau of Standards, 1976.
4. Popov, V. S.; and Titukh, Y. I.: X-Ray Investigation of Transformation in the Surface of Alloys During Abrasive Wear. Metal Sci. Heat Treat., vol. 17, no. 1, Jan. 1975, pp. 23-26.
5. Jones, W. R., Jr.: Spherical Artifacts on Therograms. Wear, vol. 37, 1976, pp. 193-195.
6. Mailander, R.; and Dies, K.: Contributions to Investigation in the Process Taking Place During Wear. Arch. Eisenhuettenw., vol. 16, 1943, p. 385.
7. Rhee, S. H.; and Dumbleton, J. H.: The Application of the Zero Wear Model to Joint Prothesis. Wear, vol. 36, 1976, pp. 207-224.
8. Pepper, S. V.: Auger Analysis of Films Formed on Metals in Sliding Contact with Halogenated Polymers. J. Appl. Phys., vol. 45, no. 7, July 1974, pp. 2947-2956.
9. Madorsky, S. L.: Thermal Degradation of Organic Polymers. Interscience Publishers, 1964, p. 301.
10. Jain, V. K.; and Bahadur, S.: Metal Transfer in Polymer-Polymer Sliding. Wear, vol. 46, 1978, pp. 177-188.
11. Campbell, W. E.: Discussion of the Vapor Lubrication of Graphite in Relation to Carbon Brush Wear by Robert H. Savage. Mechanical Wear, J. T. Bruwell, ed., American Society of Mechanical Engineers, 1950, ch. 6, pp. 103-107.
12. Lancaster, J. K.: The Influence of Substrate Hardness on the Formation and Endurance of Molybdenum Disulphide Films. Wear, vol. 10, 1967, pp. 103-117.
13. Lancaster, J. K.: Composite Self-Lubricating Bearing Materials. Proc. Inst. Mech. Eng., vol. 182, pt. 1, no. 2, 1967-1968, pp. 33-43.
14. Miyoshi, K.; and Buckley, D. H.: Friction and Wear Behavior of Single-Crystal Silicon Carbide in Sliding Contact with Various Metals. Trans. ASLE, vol. 22, no. 3, July 1979, pp. 245-256.
15. Miyoshi, K.; and Buckley, D. H.: Friction and Wear Characteristics of Iron-Chromium Alloys in Contact with Themselves and Silicon Carbide. NASA TP-1387, 1979.
16. Miyoshi, K.; and Buckley, D. H.: The Friction and Wear of Metals and Binary Alloys in Contact with an Abrasive Grit of Single-Crystal Silicon Carbide. ASLE Preprint No. 79-LC-5C-1, Oct. 1979.
17. Stout, K. J.; King, T. G.; and Whitehouse, D. J.: Analytical Techniques in Surface Topography and Their Application to a Running-in Experiment. Wear, vol. 43, 1977, pp. 99-115.
18. Khruschov, M. M.: Resistance of Metals to Wear by Abrasion, as Related to Hardness. Proceedings of the Conference on Lubrication and Wear, Institute of Mechanical Engineers (London), 1957, pp. 655-659.
19. Mutton, P. J.; and Watson, J. D.: Some Effects of Microstructure on the Abrasion Resistance of Metals. Wear, vol. 48, 1978, pp. 385-398.
20. Faust, J. W., Jr.: Cleanness Factors in Mechanical Processes and Etching of Semiconductors. Symposium on Cleaning of Electronic Device Components and Materials, American Society for Testing Materials. Spec. Tech. Publ. 246, 1959, pp. 66-63. (See also Buck, T. M.: Damaged Surface Layers: Semiconductors. Surface Chemistry of Metals and Semiconductors. H. C. Gatos, ed., John Wiley and Sons, Inc., 1960, pp. 107-135.)

21. Takahashi, N.; and Okada, K.: Microscopic Changes in Surface Topography and Elemental Concentration on the Track Formed by the Intermittent Motion of Friction. Wear, vol. 38, 1976, pp. 177-180.
22. Cortellucci, R.; et al.: Abrasion of Plastics. Wear, vol. 47, 1978, pp. 397-405.
23. McIntyre, N. S.; and Zetaruk, D. G.; and Owen, D.: XPS Study of the Initial Growth of Oxide Films on Inconel 600 Alloy. Appl. Surface Sci., vol. 2, no. 1, Nov. 1978, pp. 55-73.
24. Hryniewicz, T.; Karpinski, T.; and Lukianowicz, C.: The Evaluation of Electrolytically Polished Surfaces. Wear, vol. 45, 1977, pp. 335-343.
25. Yust, C. S.; and Crouse, R. S.: Melting at Particle Impact Sites During Erosion of Ceramics. Wear, vol. 51, 1978, pp. 193-196.
26. Vijh, A. K.: Resistance of Metals to Erosion by Solid Particles in Relation to the Solid State Cohesion of Metals. Wear, vol. 39, 1976, pp. 173-175.
27. Langbein, R. G.: Rain Erosion of Polymers. Proceedings of the Rain Erosion Conference. A. A. Fyall and R. B. King, eds., Royal Aircraft Establishment (Farnborough, England), 1965, pp. 81-89.
28. Bailey, J. A.: Surface Damage During Machining of Annealed 18% Nickel Maraging Steel, Part I—Unlubricated Conditions. Wear, vol. 42, 1977, pp. 277-296.
29. Endo, K.; and Goto H.: Effect of Environment on Fretting Fatigue. Wear, vol. 48, 1978, pp. 347-367.

CHAPTER 8

Lubrication of Solid Surfaces

Surface analysis is extremely important and useful in studying the behavior of lubricants on solid surfaces and their role in reducing adhesion, friction, and wear of those surfaces. There are many different types of lubricants that are used to reduce the adhesion, friction, and wear of surfaces in solid-state contact. They include gases, liquids, and solids. Some of these materials have been discussed briefly in previous chapters. As was mentioned earlier, even oxygen in the ordindary environment is, in a strict sense of the word, a lubricant because it reduces adhesion, friction, and wear, and this is a measure of the ability of a materials to lubricate a solid surface. The greater the reduction of adhesion, friction, and wear of the solid surfaces in contact, the more effective the lubricating material is.

Very frequently, in sensitive systems, adhesion can be used to measure the effectiveness of a lubricant on a solid surface. However, the most common technique is to measure the friction coefficient for surfaces in solid-state contact and use this as an indicator of the effectiveness of a lubricant. For example, with most effective conventional (liquid) lubricants, a friction coefficient of 0.1 or below is considered to show good boundary lubrication for surfaces in solid-state contact. This, then, serves as a rule of thumb for indicating the effectiveness of various lubricants. Surface wear is another measure of the efficiency of lubricants to protect surface films. Frequently, inhibiting adhesive wear is a very critical requirement of a lubricant in many component systems.

In addition to using lubricants in tribological systems to reduce adhesion, friction, and wear, they are used in the metal and alloy processing industries to process metals and alloys. Metal operations include forging, drawing, milling, grinding, cutting, and shaping. In such systems, highly sophisticated techniques are not necessary for determining the efficiency of a lubricant. In forging, for example, the efficiency of a lubricant is measured by studying the change in dimensions or the profile of the surfaces of hollow disks (ref. 1). The disks are pressed between two platens in a forge with the lubricant applied to both surfaces of the disk, and then the surfaces are examined before their removal to determine the change in the shapes of the disks as a result of the deformation process.

The method for qualifying good from poor lubricants is based on the profiles of the disks. The differences between a good lubricant and a poor one are demonstrated in the schematics of figure 8-1 for hollow disk profiles. Figure 8-1(a) shows the disk profile that is obtained when the lubricant provides effective surface lubrication between the platens and the disk material. The profile of the disk and its hole is not markedly altered. When the lubricant is poor between the platens and the disk, adhesion and strong bonding of the disk to the platens occur. Adhesion and strong bonding tend, with applied forces, to restrict the deformation of the disks along the surface of the platen at the platen disk interface, and this causes the disk to bulge. The deformation is taken up and maximized in the central region of the disk as opposed to being fairly uniform throughout the thickness of the disk. Since the disk is restricted in its ability to deform at the disk platen interface, because the lubricant is unable to reduce the friction and provide a separating film at the interface between the disk and platen, the disk can not deform uniformly throughout its thickness.

Although the test used in figure 8-1 is relatively crude, it can give qualitative results concerning the efficiency of lubricants. Solid surfaces are extremely sensitive to small concentrations of lubricants on their surface, and extremely thin films can be very effective as lubricants in reducing adhesion, friction, and wear. This is especially true for extremely active metallic surfaces where the surface energy is reduced by the presence of even fractions of monolayers of various species on the solid surface.

Considerable research has been conducted on the influence of various adsorbates on the surfaces of clean metals. The effects that these adsorbates have on the friction and adhesion behavior for the metals in contact has been measured. It was found, for example, that with gaseous species such as

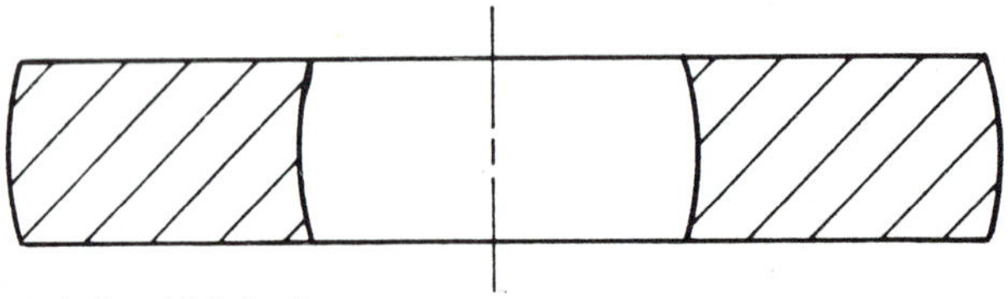

(a) Good lubrication.

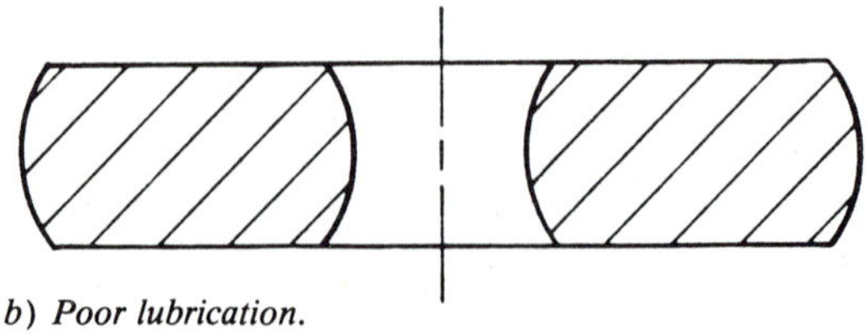

(b) Poor lubrication.

Figure 8-1.—Hollow disk profiles (ref. 1).

oxygen adsorbed to the surface fractions of monolayers are sufficient to produce very marked reductions in friction coefficients as shown in figure 8-2 for oxygen adsorbed on a tungsten surface sliding on tungsten. The upper curve (fig. 8-2) is oxygen atoms adsorbed as a function of surface exposure to oxygen. A complete monolayer of oxygen is indicated in the straight horizontal line in the upper portion of figure 8-2. The data in the figure indicate that relatively small exposures to oxygen bring about a very rapid covering of the surface of tungsten with an oxygen layer.

Measurements of friction coefficients with various amounts of surface coverage indicate that there are marked differences in the friction coefficients measured with various coverages of the tungsten surface by oxygen (fig. 8-2). The tungsten single crystal surfaces were of the (100) orientation for both the disk and rider, and the sliding velocity was extremely low, 0.001 centimeter per second with a 50-gram load, temperature of 20° C, and vacuum of 10^{-10} torr. For clean surfaces of tungsten (100) in contact with tungsten (100), a friction coefficient of 3.0 was measured. With as little as a

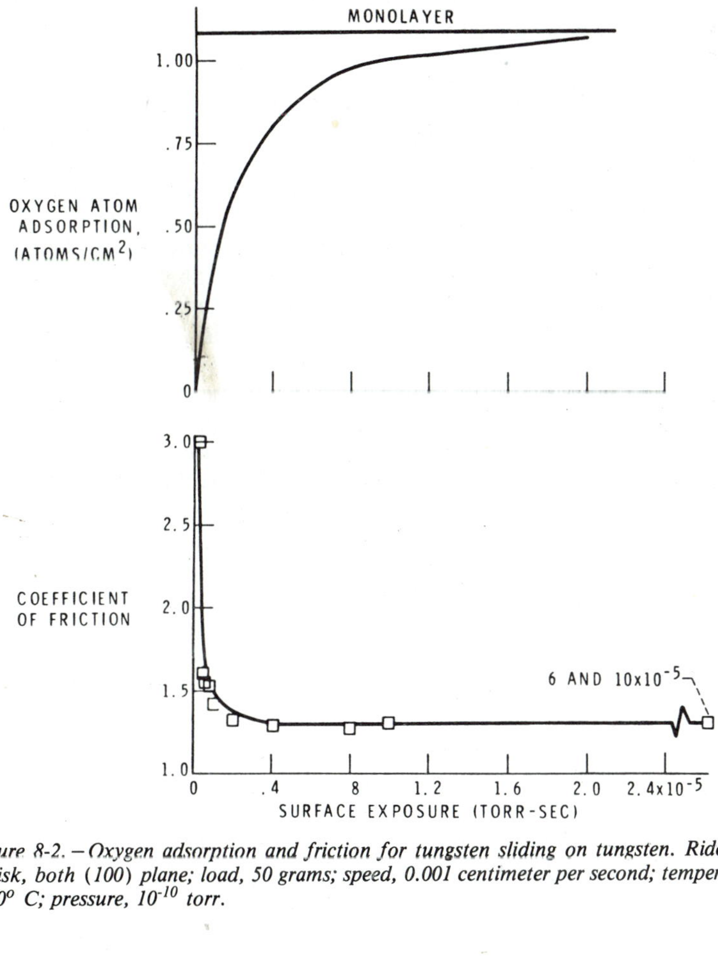

Figure 8-2. – Oxygen adsorption and friction for tungsten sliding on tungsten. Rider and disk, both (100) plane; load, 50 grams; speed, 0.001 centimeter per second; temperature, 20° C; pressure, 10^{-10} torr.

quarter of a monolayer of oxygen on the surface, however, the friction coefficient dropped to about 1.5 (representing 100-percent reduction).

The data in figure 8-2 show that oxygen is a very effective lubricant in reducing friction for tungsten in contact with itself. Also, further additions of oxygen to the surface beyond the quarter monolayer do not result in a marked change in the friction coefficient. It only decreases to about 1.3. By the time, however, the surface has been covered with a half monolayer of oxygen, the friction coefficient is nearly stabilized. Further additions of oxygen to the surface up to a full monolayer produce no further changes in observed friction. Thus, not even a full monolayer of oxygen is necessary to cause a very notable reduction in friction and to stabilize the friction properties of tungsten in contact with itself.

The data in figure 8-2 indicate the extreme importance of using surface analytical tools to characterize tribological systems. The extreme sensitivity of the surface (e.g., of metals such as tungsten) to even fractions of a monolayer of adsorbed species indicates that one must characterize the surfaces with which they are working very carefully if understanding the role of lubricants in the reduction of adhesion, friction, and wear of solid surfaces is to be achieved.

Molecular Structure

Almost anything on a surface of a clean solid that is very active such as a metal (except for the inert or noble gases) causes some change in the adhesion, friction, and wear properties. Almost any adsorbate, then, in a strict sense (with the exception of the noble gases) is a lubricant. However, the effectiveness of the material on the surface to provide lubrication is, to a large degree, a function of the structure of the particular species that is being placed on the surface. As was mentioned earlier, gases, liquids, and solids can act as lubricants. The real issue is one of structure. Certain structures are much more effective than others in lubricating solid surfaces.

Certain basic properties of elemental structures have been found to be related to the ability of the materials to reduce adhesion, friction, and wear and to provide effective boundary lubrication. Probably the most widely used materials in the lubrication of surfaces are liquids that have the straight chain hydrocarbon character. How does the chain length in a straight hydrocarbon influence the lubricating properties of that hydrocarbon? If one starts at the extreme low end of hydrocarbon chain lengths, for example, with a single carbon atom, the species is essentially methane and is in the gaseous state. As one progresses up the chain (increasing the carbon chain length), a transition occurs at room temperature from gas to liquid. Ultimately, if one proceeds high enough, one encounters solids. The entire range can be covered by changing the carbon chain length and, accordingly, the molecular weight of the material.

The carbon chain length has been found to influence the friction coefficient. Measurements have been made for various hydrocarbons adsorbed on a clean tungsten surface of three different orientations: the (210) surface, a (110) surface, and a (100) surface. The results are presented in figure 8-3 where the coefficient of friction is plotted as a function of the number of carbon atoms in the chain starting with methane and progressing through the carbon chain length to decane (i.e., from 1 to 10 carbon atoms). Figure 8-3 shows a decrease in friction coefficient with an increasing carbon chain length. Another observation from the data in figure 8-3 is that the orientation of the surface has an influence on the friction coefficient. The (110) surface exhibits lower friction than the (210) or the (100) surfaces do. The data in figure 8-3 were obtained in a vacuum environment with tungsten surfaces that were carefully cleaned before conducting the experiments. The system was well controlled.

The study of the effect of molecular chain length of lubricating species on the friction properties of solid surfaces, however, dates back many years. Sir William Hardy in some very classic experiments in the 1920's measured the friction properties for a number of different surfaces lubricated by a variety of lubricating species and found a number of fundamental relationships (ref. 2). The relationship of the carbon chain length to the friction reducing properties of the lubricating species was measured by him on conventional contaminated surfaces. He established that the chain length influences the friction properties for simple hydrocarbons and that this same type of basic relationship also exists for organic alcohols and acids. He found that the fundamental relationship between molecular weight (size of

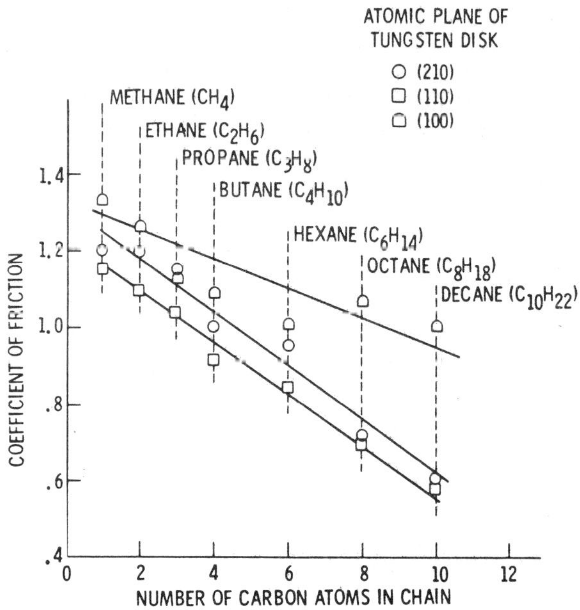

Figure 8-3. – Coefficient of friction for single-crystal tungsten rider (100) sliding on single crystal disk orientations (100), (110), and (210) with chemisorbed monolayers of hydrocarbons. Gases, homologous series methane through decane; load, 50 grams; sliding velocity, 0.001 centimeter per second; temperature, 20° C; ambient pressure, 10^{-10} torr (1.33×10^{-8} N/m^2).

the molecule of the organic acid or alcohol) and friction held for various surfaces including glass, steel, and bismuth. Figure 8-3 shows that it also applies to tungsten. Thus, given a particular organic structure, changing the chain length results in a decrease in the friction properties of a surface in the presence of that lubricating species.

Effect of Unsaturation

Many practical lubricants contain not only saturated straight hydrocarbons but also species that are unsaturated. Consequently, there is the question as to the influence of the degree of saturation of bonds in the carbon chain on the lubricating characteristics of a solid surface.

The degree of bond saturation decreases with the two carbon atom species ethane, ethylene, and acetylene. There is a greater degree of bond unsaturation in going from ethane to acetylene. There is a single bond in the carbon to carbon atom for ethane, double bond for ethylene, and triple bond for acetylene. Adhesion and friction measurements with iron surfaces covered with these particular species adsorbed on the solid surface indicate that the higher the degree of unsaturation the lower the adhesion and friction for metals in solid-state contact; that is, the friction measured for acetylene is much less than it is for ethane with the results for ethylene being between those of ethane and acetylene.

The argument proposed for reducing friction and adhesion with an increasing degree of unsaturation indicates when the fluid contacts the clean iron surface the unsaturated carbon to carbon bonds break very readily. The triple bond breaks more readily than the double bond, and the double more readily than the single. On bond cleavage within the molecules, the dangling bonds bond directly to the metal surface. Each carbon atom has a coordination number of four and can bond to its four nearest neighbors. With acetylene there is only a single hydrogen attached to the carbon, and on bond breaking there are three carbon atoms available for bonding to the metal surface. Actually, there are six atoms, because there are three associated with each carbon atom. Therefore, six bonds are available in acetylene for bonding to the metal surface and to occupy six iron sites on the solid surface; this ties up the binding energies associated with six iron atoms. With ethylene, bond breakage results in only four bonds, because there are two hyrdogens on each carbon atom in the ethylene structure. Only four iron atoms are tied up on the surface by the ethylene. In ethane there are only two, because ethane has, on carbon to carbon bond scission, only two bonds available for interaction with the iron surface since three hydrogen atoms are attached to each carbon atom.

The effect on lubrication of bond saturation and unsaturation has been observed with larger molecular weight species as well. For example, cetane contains fourteen carbon atoms in the chain length. An analogous chain length material that is unsaturated is 1-cetene. Lubricating properties of these two solids have been measured as well as some other properties (ref. 3). Some properties of 1-cetane and cetene are presented in figure 8-4 and in tables 8-I and 8-II.

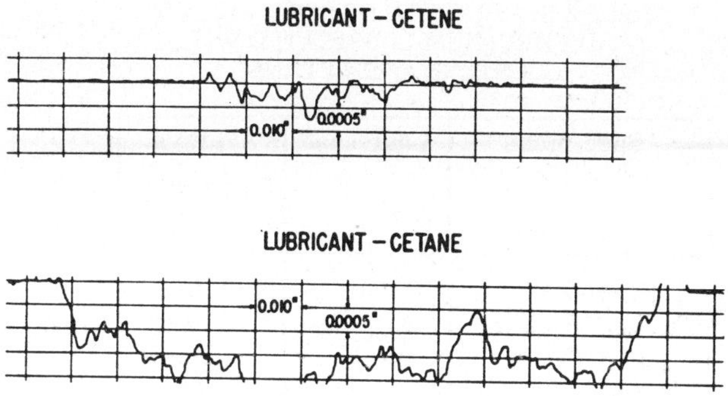

Figure 8-4. – Profiles of wear tracks on 410 stainless steel (ref. 3).

Table 8-I presents the physical properties (including melting point, boiling point, viscosity, density, and molecular weights) of the lubricating fluids 1-cetene and cetane. A comparison of the data in table 8-I shows that, from the viewpoint of physical properties, there is very little basic difference in the fundamental properties of 1-cetene and cetane.

TABLE 8-I. – PHYSICAL PROPERTIES OF LUBRICATING FLUIDS[a]

	1-Cetene[b]	Cetane[c]
Melting point, °C	2.2	18
Boiling point, °C	274	287
Viscosity, cps at 25° C	3.31	3.73
Density	0.7825	0.7733
Molecular weight	224.43	226.44

[a]Reference 3.
[b]$C_{14}H_{29}CH{=}CH_2$.
[c]$C_{14}H_{29}CH_2CH_3$.

Table 8-II presents the friction coefficients of steel, nickel, and chromium lubricated with 1-cetene and cetane. Again, with the exception of chromium, there is no great difference in the friction properties with the cetane and 1-cetene. With chromium, however, one does observe a relatively marked difference in the friction behavior of the species; the unsaturated 1-cetene gives a lower friction coefficient just as was observed with the acetylene in the lubrication of iron. Surface profile traces across the wear track of the 410 stainless steel surface lubricated with 1-cetene and cetane (fig. 8-4) indicate that the amount of wear that takes place to the surface is considerably greater on the 410 stainless steel with cetane. The 1-cetene is a much better boundary lubricant for reducing wear.

TABLE 8-II. – FRICTION OF STEEL, NICKEL, AND CHROMIUM LUBRICATED WITH 1-CENTENE AND CETANE[a]

Specimens	Lubricant	Average coefficient of friction
Cold-rolled steel	Cetane	0.31
	1-Cetene	.33
Nickel	Cetane	>.68
	1-Cetene	.60
Chromium	Cetane	>.68
	1-Cetene	.25

[a]Reference 3.

Effect of Halogen Additives

Figure 8-3 shows the effect of molecular weight on lubricating properties of organic molecules; figure 8-4 shows the effect of the degree of saturation in hydrocarbons on their lubricating ability. Most practical lubrication systems contain hyrdrocarbons that have substituted in their molecular structures surface active species that can interact with the surface and provide more effective lubrication than the simple hydrocarbon by itself. For example, chlorine or other halogens are added to hydrocarbon structures to provide more effective lubrication. The active halogen atom interacts more strongly with the solid surface than simple adsorption, which may be associated with the hydrocarbon adsorbing on the surface. This is especially useful in the presence of halogens (e.g., sulfur or phosphorus) and is particularly useful under heavy load or high-speed mechanical applications where there is a need for very strong bonding of the lubricating molecule to the solid surface.

This author has conducted studies with a benzene structure containing various halogens. The benzene structure was fluorinated, chlorinated, brominated, or iodinated, and the friction properties of these particular structures were measured where the only change in the molecular structure of the lubricant was the addition of a different halogen in each case. Friction properties were measured for these various materials to determine the influence of the different halogens on the friction behavior of the benzene molecular structure in the lubrication of solid surfaces. The sliding friction experiments were conducted with single crystals of gold sliding across iron single crystal surfaces that had been saturated with the halogenated benzene compounds. The friction coefficients measured for these particular structures are presented in figure 8-5.

Figure 8-5 plots friction coefficient as a function of load (from 1 to 30 g of the gold against the iron surface). The friction coefficients for

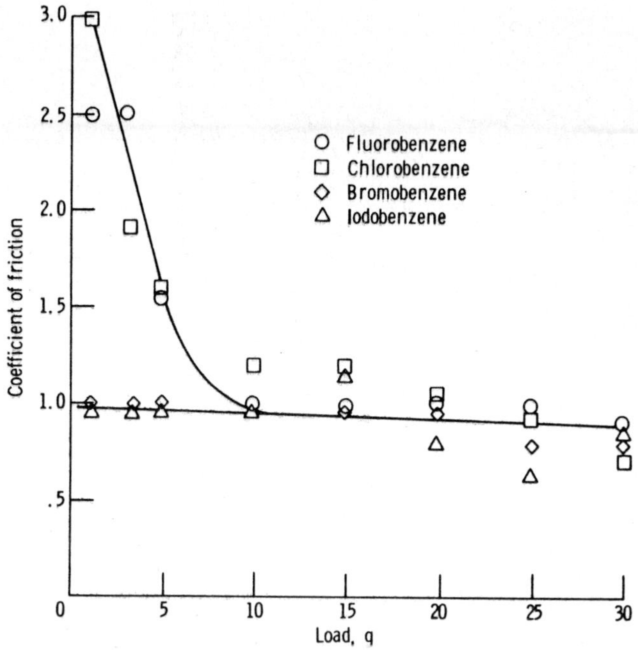

Figure 8-5. – Coefficient of friction for gold (111) single crystal sliding on iron (011) single crystal with iron surface saturated with halogenated benzene compounds.

bromobenzene and iodobenzene were essentially the same over the entire load range, and the friction was nearly independent of load. With fluorobenzene and chlorobenzene, however, the friction coefficients were extremely high initially but decreased with increasing load. The bromine and iodine are much more labile, or easy to decompose, in bromobenzene and iodobenzene. This allows more rapid interaction with the metal surface than is observed with fluorine and chlorine in fluorobenzene and chlorobenzene. It may be for this reason that the friction is lower at the very light loads with bromobenzene and iodobenzene. AES analysis is very useful in the study of the lubrication of the solid surfaces to determine the differences in film surface characteristics.

Figure 8-6 shows two Auger spectra for an iron surface, one lubricated with the chlorobenzene and the other with the iodobenzene of figure 8-5. The differences in the friction behavior can be explained by the differences in the surface chemistry observed. In the case of the chlorobenzene, chlorine and carbon are seen in addition to iron in the Auger spectrum. The chlorobenzene is adsorbed on the solid surface as a molecular structure with carbon and chlorine being present. Since iron can be seen in the Auger spectrum, the surface is not completely covered by chlorobenzene. The chlorine is so tightly bonded to the benzene that it does not bond or adsorb as strongly to the iron surface as do bromobenzene and iodobenzene.

Figure 8-6(b) shows an Auger spectrum for iodobenzene bonded to the iron surface. The Auger spectrum for the iodobenzene shows carbon only.

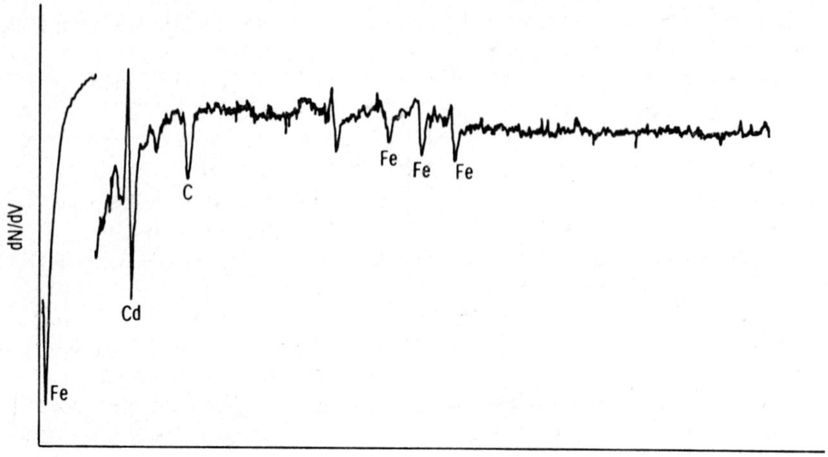

(*a*) *Chlorobenzene.*

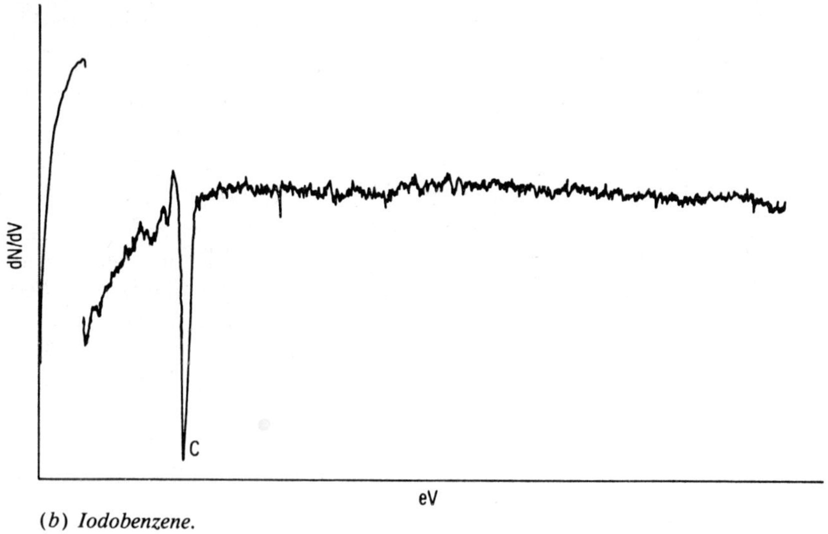

(*b*) *Iodobenzene.*

Figure 8-6.—Auger spectrum for iron (011) surface saturated with chlorobenzene and iodobenzene.

This spectrum indicates that the iodobenzene is decomposed, leaving the benzyl structure or benzene structure on the solid surface; the iodine has completely disappeared from the surface of the system. Also, figure 8-6(b) shows that the iron peaks are completely covered by the carbon. Iron is not visible in the Auger spectra in figure 8-6(b) as it is in the Auger spectra in figure 8-6(a); this indicates that the iodobenzene is much more effective than chlorobenzene in providing a protective surface film by shielding the iron. It is for this reason that, for the light loads, the friction coefficient is lower with iodobenzene than it is for chlorobenzene. Since the iodobenzenes and bromobenzenes are more labile, they provide a more effective surface film than does chlorobenzene, which remains in its molecular form and adsorbs to the surface in that form. In that form, however, it is not as effective as a lubricant because it does not completely shield or protect the iron surface. Auger spectroscopy is sensitive to a depth of four or five atomic layers. The data in figure 8-6, then, indicate that, in the iodobenzene case, the carbon layer on the solid surface is at least four or five atomic layers deep, because there is a complete absence of iron peaks in the spectrum. Thus, not only the molecular chain length and degree of saturation but also the presence of other active elements in the basic hydrocarbon structure can alter its lubricating behavior on solid surfaces in contact.

Figure 8-3 shows that as the molecular weight or hydrocarbon chain length increased the friction coefficient decreased. One can move through the various states of matter from gas to liquid to solid with increases in molecular weight. If this is true, one would expect to encounter a continuous decrease in friction coefficient with changes in the state of matter. Thus, one would expect lower friction with the liquid than with the gas, and likewise a lower friction coefficient for a solid than for a liquid of the same molecular structure. This is true as the chain length of the species increases, but it is also true when the same particular material that may already be existing in one particular state of matter is simply altered in its state. For example, many of the organic, straight chain acids are extremely good boundary lubricants on solid surfaces. A goodly number of these acids are solids. However, raising the temperature can cause localized melting and thus convert the solid on the surface to a liquid. With this change of state, there is an accompanying change in the friction behavior of materials. This change is reflected in the data in figure 8-7 for the static coefficient of friction as a function of temperature for palmitic acid on a quartz surface. This work

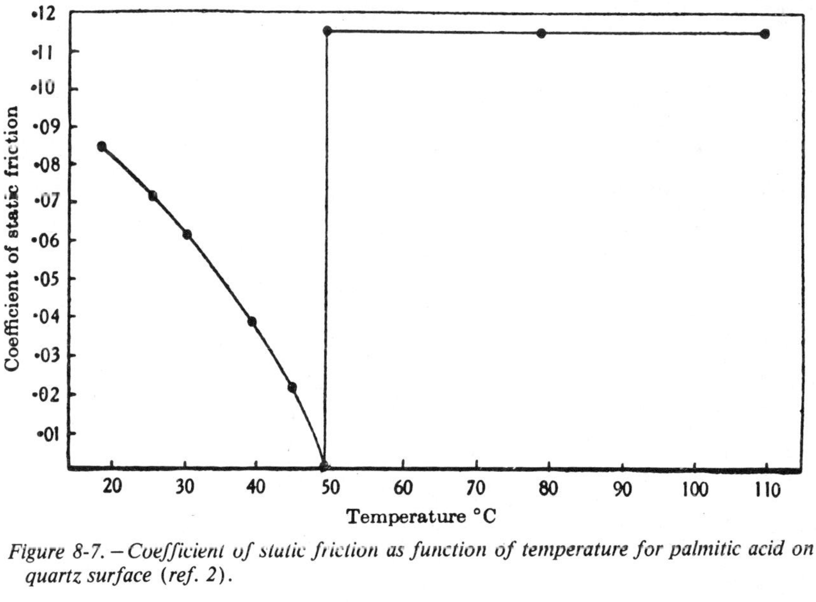

Figure 8-7. – Coefficient of static friction as function of temperature for palmitic acid on quartz surface (ref. 2).

was done by Hardy (ref. 2). One sees that as the temperature is increased the friction coefficient decreases in the 20° to 50° C region where there is a solid palmitic acid film on the surface. When 50° C (melting point of palmitic acid) is reached, however, with a conversion of the acid from the solid to the liquid state there is a notable jump in the friction coefficient from a very low value to a high value; the value remains high with further increases in temperature to 110° C. The results of figure 8-7 indicate, then, that palmitic acid in the solid form exhibits lower friction behavior than it does in the liquid form.

Subsurface Effects on Lubricant Behavior

Surface Temperature

Most of the lubricants used to reduce the adhesion, friction, and wear of materials are liquids with organic structures. Consequently, they have relatively low temperatures of decomposition. One must be concerned about the temperatures of surfaces and how they affect lubricant behavior. When rubbing metal surfaces extemely high flash temperatures (as high as 1000° C above the ambient) have been measured and these temperatures can exist in relatively lightly loaded sytems (ref. 4). A lubricant on the surface can reduce these surface temperatures somewhat but, even with effective lubrication, the temperature of the surface can still be relatively high and thus detrimental to the lubricant species. This temperature effect is indicated in figure 8-8 where the surface temperature is plotted as a function

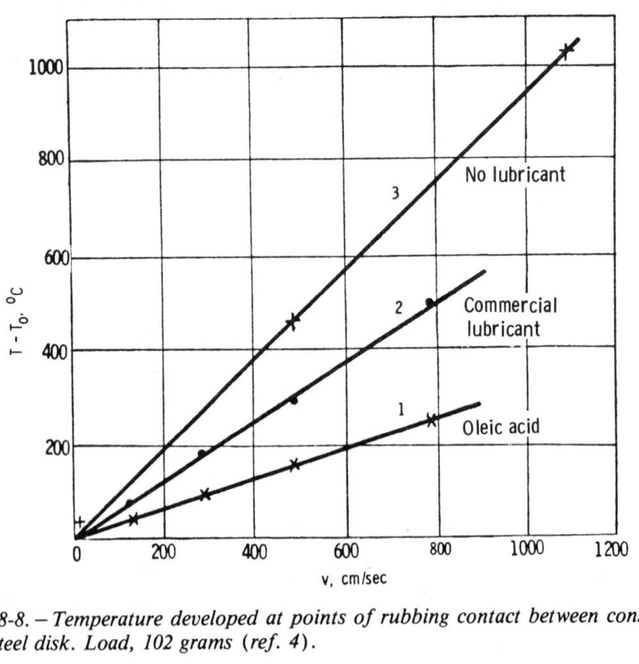

Figure 8-8. – Temperature developed at points of rubbing contact between constantan pin and steel disk. Load, 102 grams (ref. 4).

of the sliding velocity for a constantan pin sliding on a steel disk with a load of 102 grams and three different surface conditions. Curve 3 is for the surfaces in dry sliding, no lubricant on the surface; curve 2 is for the surfaces lubricated with a commercial lubricant; and curve 1 is for the surfaces lubricated with oleic acid. While a lubricant on the surface reduces the surface temperatures, the temperatures can still reach fairly high values. For example, with the commercial lubricant, temperatures as high as 500° C can be achieved, and this temperature is sufficiently high to cause degradation of most organic molecular structures. Thus, in lubricating solid surfaces, one must remember these temperature effects, particularly in boundary conditions where the supply of liquid lubricant is minimal and the lubricant is not available to act as a heat sink to carry away heat generated at the interface.

Mechanical Condition of Surfaces

In addition to concern for the temperature of the surfaces and their effect on lubricants, the lubricant performance is also influenced by other factors—for example, the mechanical condition of the solid surface. Most solid surfaces that are used in practical systems have been finished by grinding, machining, or polishing. As a result, the outermost layer consists of a highly refined structure that is in a highly strained state. This type of surface, for example, has a different type of reactivity toward lubricants than a surface which has been annealed and etched. The effect of the mechanical nature of properties of solid surfaces was recognized many years ago by Sir William Hardy (ref. 2). He observed the lubricated friction behavior for different materials, both burnished and etched surfaces. The burnished surface represents that which has been highly worked as might be a practical tribological surface. He compared his results to surfaces that were annealed and etched and found differences in the friction behavior. Some of his results are presented in table 8-III.

Table 8-III shows the results of using benzene, pyridene, ethyl alcohol, butyl xylene, octyl alcohol, and cyclohexanol to lubricate a bismuth surface. Bismuth was used by Sir William Hardy because, with bismuth, it was

TABLE 8-III. – SURFACE CONDITIONS[a]

	Burnished	Eteched	Ratio
Benzene	0.34	0.39	0.87
Pyridene	.33	.4	.83
Ethyl alcohol	.32	.39	.82
Butyl xylene	.27	.37	.72
Octyl alcohol	.25	.36	.7
Cyclohexanol	.20	.33	.6

[a]Reference 2.

easy to generate a highly deformed surface layer (which at the time was referred to as the Beilby layer). Researchers at that time thought that the very highly refined crystallite size in the surface layers of severely worked surfaces was really an amorphous material; as a result, it was referred to as the Beilby layer. Subsequent experimentation, however, has revealed that the Beilby layer is not an amorphous material but rather a very refined structure of fine-grained size. Notwithstanding the foregoing, the data in table 8-III indicate that the highly worked surface or burnished surface of bismuth exhibited lower friction coefficients than the etched and annealed surfaces. The worked surface is much more reactive because the energy state of the solid surface is much higher. Strained surfaces are always much more chemically active than unstrained surfaces. It might be anticipated, therefore, that a greater amount of reactivity or interaction of the lubricant with the solid surface would take place, and this would result in a more effective film forming on the surface of the burnished solids than that being formed for the etched solids. In each case in table 8-III, for a particular lubricant, the friction is lower for the burnished surface than it is for the etched surface. This reflects the influence of surface energetics, the more energetic burnished surface gives a stronger interaction with the lubricant than the etched and annealed surfaces. If one were studying the adhesion, friction, and wear behavior of these surfaces in the clean state, the opposite effect would be observed. There would be stronger solid to solid interactions with the burnished surfaces than there would be with the etched annealed surfaces, again because of the higher energetics of the worked surface.

Surface Chemistry

The combination of the higher surface temperatures experienced in tribological systems than in static surfaces and the strain associated with the rubbing, rolling, or sliding process can cause changes in the chemistry of the surface when a lubricant interacts with a solid surface. With alloys that are lubricated, for example, the chemistry may vary depending on the conditions to which the surfaces have been subjected. For example, figure 8-9 presents two Auger spectra for a 302 stainless steel surface that has been lubricated with a degassed mineral oil. In figure 8-9(a) is the surface of the basically as-received disk material. Figure 8-9(b) shows that same surface heated to 400° C for 10 minutes prior to conducting the friction experiment. Auger spectroscopy traces in the wear track revealed differences in the surface chemistry as a result of simply heating the one specimen surface and not the other. In figure 8-9(a) the Auger spectrum contains iron, oxygen, and carbon. Despite the fact that this is a 300 series stainless steel, there is an absence of chromium in the Auger spectra. When the sample has been heated to 400° C for just 10 minutes, however, in addition to iron and oxygen peaks, one finds chromium peaks. Thus, chromium appears on the surface for the sample that has been heated. For the sample that had been operated at room temperature there is no chromium in the Auger spectrum; this indicates an absence of chromium in the surface film. Thus, the surface

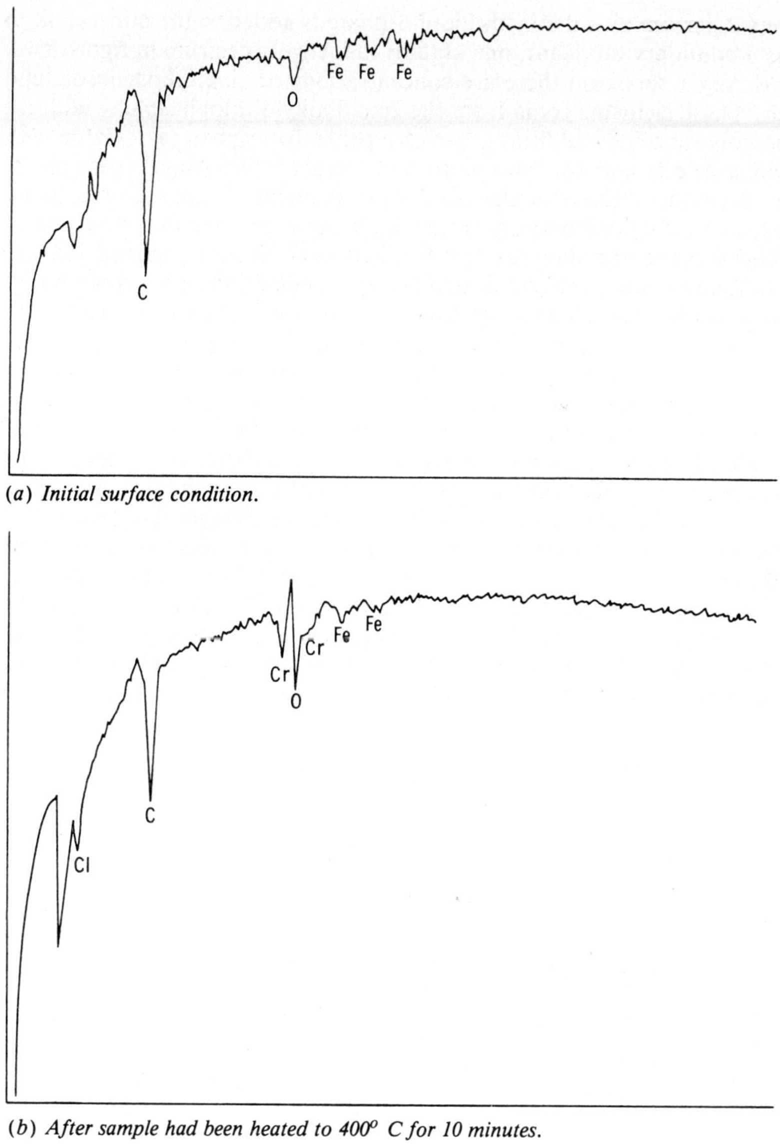

(*a*) *Initial surface condition.*

(*b*) *After sample had been heated to 400° C for 10 minutes.*

Figure 8-9.—Auger emission spectroscopy traces of wear track on 302 stainless steel disk lubricated with degassed mineral oil.

chemistry is changed by simply heating one surface and not the other. This is not a change in the lubricant, because the lubricant, a mineral oil, is essentially the same in both cases. The change, which is in the surface chemistry of the alloy, has been brought about by rubbing and heating the one surface.

Changing the chemistry of the lubricant can also change the chemistry of the metal surface. When, for example, a small concentration of approximately 2 percent zinc dialkyldithiophosphate is added to the mineral oil to act as a boundary lubricant, one obtains the Auger spectrum in figure 8-10. In this Auger spectrum there are concentrations of zinc, phosphorus, and sulfur. These elements come from the zinc dialkyldithiophosphate additive in the mineral oil. In addition, there are peaks for carbon (associated with the mineral oil) and for chromium and oxygen. The Auger spectrum in figure 8-10 was obtained under conditions identical to those used in figure 8-9(a), so a direct comparison can be made between these two Auger spectra. One sees that the alloy surface chemistry is different in figure 8-10 from that in figure 8-9(a). The addition of the zinc dialkyldithiophosphate brings about a surface enrichment in chromium of the 302 stainless steel comparable to what was achieved in figure 8-9(b). The samples in figure 8-10 were operated at room temperature just as they were in figure 8-9(a), but the alloy surface chemistry is not characteristic of that of figure 8-9(a). Instead, it is characteristic of that in figure 8-9(b); it indicates the complexity of lubricating solid surfaces and the complexity of the chemistry involved. The data also show the usefulness and importance of the availability of surface analytical tools in following the surface chemistry changes that take place.

In figures 8-9 and 8-10 the elements iron and chromium present in the alloy can react readily with the lubricant to form surface films, and there may be competitive reactions occurring at the surface for the lubricating species. Some of the species in the lubricant may have a stronger affinity for one element in the alloy than for the other. Where the elemental metal or the principal base element in an alloy can form more than one compound

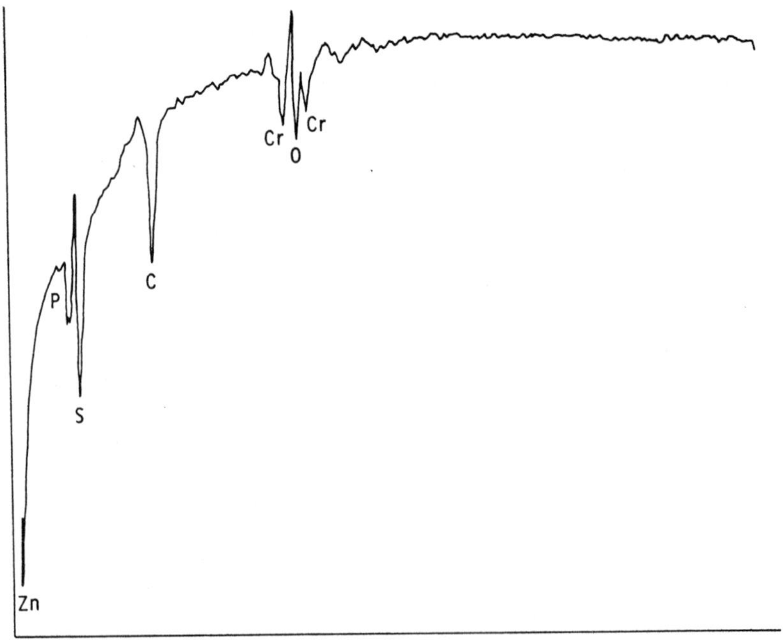

Figure 8-10.—Auger emission spectrum for 302 stainless steel wear surface after sliding with zinc dialkyldithiophosphate additive in mineral oil.

with a particular lubricant, changes in surface chemistry can take place; that is, the lubricant can undergo a change in its chemistry simply because the metal surface with which it is interacting may produce more than one particular compound form. An example of such behavior has been observed in the exposure of a conventional 52100 bearing steel with tricresyl phosphate. The immersion of the bearing steel to tricresyl phosphate for various periods of time has shown, in AES analysis, that the chemistry of the surface compound changes with time. This is indicated in the data of figure 8-11 which presents the Auger spectra obtained by Shafrin and Murday (ref. 5). Shafrin and Murday immersed the 52100 in tricresyl phosphate for 3, 15, and 21 days. The three Auger spectra for these experiments are presented in figure 8-11. The spectra for the short period of 3 days is characteristic of that associated with the phosphide of iron. An interpretive analysis of the presence of the phosphorus in the Auger spectrum and the electron energy associated therewith would reflect the formation of an iron phosphide on the surface. In contrast, however, for the samples that had been immersed in the TCP for 21 days, an analysis of the surface film reveals the presence of an iron phosphate. Thus, in one case, the short time immersion, the compound on the surface is iron phosphide, and for the longer time immersion, the surface film is iron phosphate.

The results of figure 8-11 are extremely interesting because they indicate that even where one has a fixed lubricant like tricresyl phosphate lubricating a particular solid surface such things as simply the change in the time with which the surface is in contact with the lubricant can cause a change in the compounds that are formed on the solid surface. Such a change is important since the friction and wear properties of iron phosphide and iron phosphate are different. These would then be reflected in friction and wear measurements.

Thus, extreme care must be taken in labeling the particular types of compounds that are formed on surfaces with various lubricants because such things as time and the alloy chemistry can change the nature of the surface

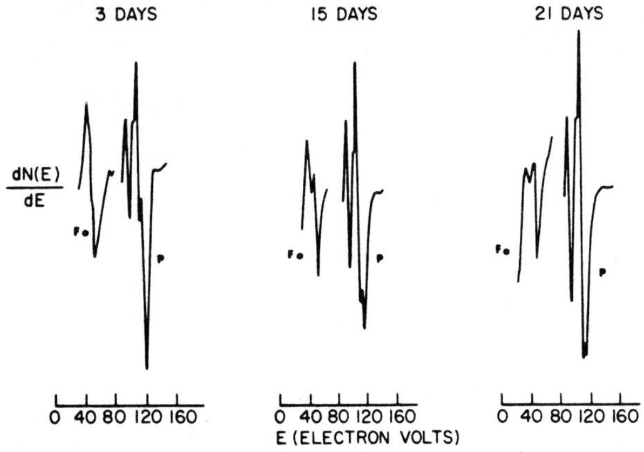

Figure 8-11.—Fe_{MMM} and P_{LMM} Auger electron line shapes for samples of 52100 steel (solvent cleaned) exposed to TCP at 100° C for 3, 15, and 21 days.

films formed; this is indicated by the data in figures 8-9 to 8-11. There is a very complex interaction of the alloy chemistry with that of the surface lubricants even when the alloy chemistry is relatively simple as it is in figure 8-11 (principally iron interacting with the lubricant).

Environmental Effects

In any practical lubrication system there are three basic components: (1) the solid surface to be lubricated, (2) the lubricant, where a conventional lubricant is to be used, and (3) the environment. The environment plays a very heavy role in the friction, adhesion, and wear behavior of solid surfaces in contact, and it also affects the lubricant itself. Furthermore, it can alter or affect the lubricant–solid surface interactions. For example, the influence of adsorbed oxygen from the environment on adhesion and friction of tungsten has already been discussed in this chapter. Hence, there is no question but that the environment interacting with the solid surfaces in the clean state has a pronounced effect on the adhesion, friction, and wear behavior. In addition, however, the environment can alter the lubricant properties. For example, many oils contain large concentrations of dissolved oxygen and, where the environment contains water vapor, lubricants may contain entrapped water as well. The entrapped oxygen and water vapor in lubricants can, in fact, act as antiwear additives; that is, they can react at the solid surface just as oxygen did on the tungsten (discussed earlier in this chapter) to form protective surface films. The same thing can happen when oxygen is dissolved in an oil. Some oils dissolve as much as 50 times their volume in gas, such as oxygen.

It was indicated that the burnishing of solid surfaces and its associated strain produces a surface which is much more reactive than is an annealed etched surface (table 8-III). If the energetics of the surface vary with the burnishing or annealing, then, for different orientations on the solid surface with their characteristically different surface energies, there might be a difference in the interactions of lubricants with such solid surfaces. Such differences are observed in practice.

Metals are probably of greatest interest with respect to the interaction of lubricants with solid surfaces. Studies have been conducted to examine the adsorption of various organic species on different planes of particular metals to determine the differences in the adsorption characteristics. The study, for example, of the adsorption of benzene on various orientations of nickel indicates differences in the structure adsorbed on the solid surface. The coupling of the use of LEED and high resolution electron energy loss spectroscopy gives insight into the adsorption on the different planes of nickel (ref. 6). The data obtained in experiments by Bertolini, Dalmai-Imelik, and Rousseau are presented in figure 8-12 for the nickel (001) and nickel (111) surfaces. Benzene is adsorbed to both surfaces, and a comparison of the energy loss spectra in figure 8-12 reveals differences for the two atomic planes of nickel. The authors of the data in figure 8-12 also used LEED to characterize the structure of the species adsorbed on the surface. Combining the LEED data with the data in figure 8-12 reveals that the

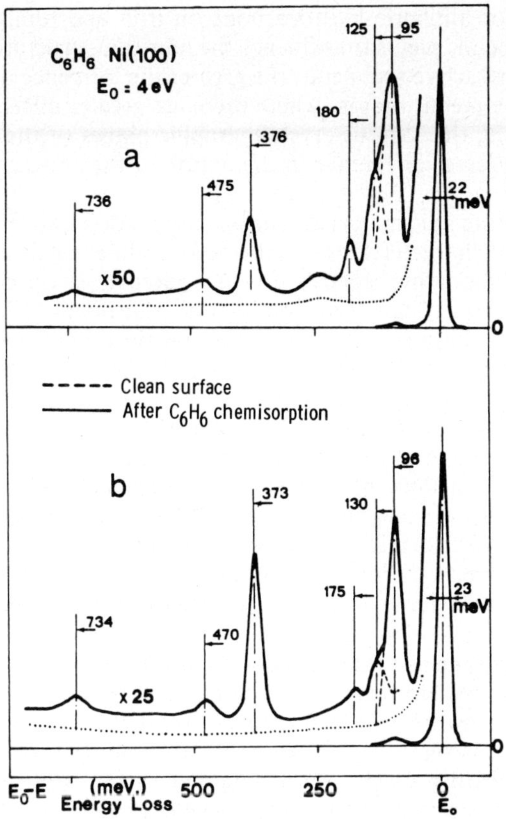

(*a*) *Ni(100); incident energy, 4 eV.*
(*b*) *Ni(111); incident energy, 1.8 eV.*

Figure 8-12. – Energy loss spectra of low energy electrons specularly reflected from Ni(100) face and Ni(111) face for clean surfaces and after C_6H_6 chemisorption at room temperature (ref. 6).

chemisorption of benzene on the nickel surfaces produces an ordered structure on the nickel (100) surface having a C (4×4) structure and a $(2\sqrt{3}\times 2\sqrt{3})$R 30° on the nickel (111) surface. The benzene, when it adsorbs on the nickel surface adsorbs with the structure retaining its aromatic character. It involves the 8 nickel atoms on the surface of the (100) face of nickel and the 12 on the (111) face of nickel. The interaction takes place by the φ electrons of the ring. Significant shifts of the carbon-hydrogen bonding and the carbon-hydrogen stretching vibrations in the energy loss spectra show a weakening of the C-H bonds because of the formation of chemisorption bonds in the coupling of the hydrogen atoms with the nickel substrate.

Such differences in adsorption characteristics for the aromatic structure benzene on two different nickel orientations could produce differences in friction and adhesion characteristics to the nickel surfaces such as observed by this author for aliphatic hydrocarbons on iron and tungsten. Surface orientation for metals seems to influence the lubricant structure of the solid surface. The more active the metal, the greater the influence of the surface orientation of the metal; that is, where there are greater differences in surface energy among the various crystallographic planes of the metal, there appears to be a greater difference in the nature of the structure formed on the solid surface.

With some nonmetallic materials, surface orientation has very little effect on the adsorption characteristics of lubricants. While metals are extremely active toward lubricating species in the gaseous environment, other materials (inorganic crystals, semiconductors, and nonmetallics) very frequently do not exhibit a strong affinity for constituents of the gaseous environment. Diamond, for example, does not chemisorb oxygen, nitrogen, ammonia, or hydrogen sulfide over a wide range of temperatures and pressures. The sticking coefficient, for example, for oxygen on all diamond surfaces at room temperature is about 10^{-7}. Annealing diamond at 450° C and 10^{-4} torr oxygen does not increase the oxygen Auger signal significantly. When, however, one heats the surface to 1300° C in an atmosphere of oxygen at 5×10^{-6} torr for about a half minute, the diamond surface begins to graphitize. Continued exposure to oxygen at 5×10^{-3} torr and 1300° C for an additional half minute results in a graphitized surface (ref. 7).

Sulfur from hydrogen sulfide does not chemisorb on the diamond surface at room temperature. However, using an electron beam on a solid surface can promote sulfurization of a diamond surface. Experiments by Lurie and Wilson have demonstrated that the electron beam can be used to induce adsorption of lubricating species on a diamond surface (ref. 7). However, when adsorption is stimulated by an external energy source, such as electron or ion beam, the resulting films formed on the solid surfaces are not really effective lubricants, because they are relatively unstable and can be induced to desorb very readily. When various species are adsorbed, there are few differences in their abilities to interact with various planes because of the relative inertness of the various planes of the diamond-type structure. For

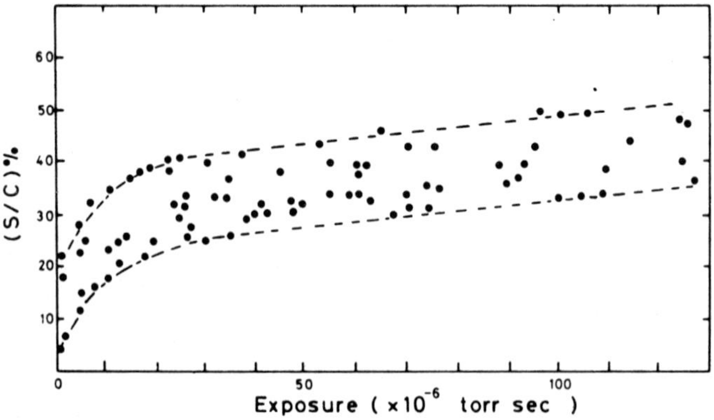

Figure 8-13. – Electron stimulated adsorption of sulfur on (111), (110), and (100) surfaces of diamond.

all practical purposes, no distinction or difference can be made along the various planes of diamond when exposed to the oxygen, nitrogen, ammonia, and hydrogen sulfide gases.

Some data are presented in figure 8-13 for the exposure of diamond surfaces to hydrogen sulfide. The (111), (110), and (100) surfaces of diamond were exposed to hydrogen sulfide. Plotted on the ordinate are the sulfur to carbon ratios as a function of exposure to hydrogen sulfide. All the concentration uptake of sulfur on the solid surface is within a band for all three orientations, and very little difference exists among the various orientations.

Surface Concentration

It was indicated earlier in this book that small concentrations of the lubricating species on the solid surface can markedly alter the adhesion, friction, and wear behavior. In chapter 1, static friction results were presented for fractions of a monolayer of coverage on iron, copper, and steel surfaces with oxygen and chlorine. Experimental data showed that fractions of a monolayer were sufficient to reduce appreciably the static friction coefficient for materials in solid-state contact. A considerable amount of the early literature on lubricants and their interaction with solid surfaces indicates that monolayers (e.g., organic acids on solid surfaces) are sufficient to provide effective boundary lubricating films.

One, however, does not have to resort to using large molecular weight species such as organic acids to feel the full effect of the presence of lubricating species on a solid surface relative to its ability to reduce adhesion, friction, and wear. Sulfur-containing additives are widely used in antiwear and extreme pressure lubrication applications because of their ability to provide effective surface films on ferrous based surfaces. Iron interacts readily with sulfur to form iron sulfide, which reduces adhesion and adhesive wear. Measurements (using LEED) on clean iron surfaces show that the adhesive forces for clean iron to itself are related to the amount of sulfur present on an iron surface. In contrast to the diamond in figure 8-13, iron readily adsorbs hydrogen sulfide. In fact, the adsorption takes place with such energetics that the hydrogen sulfide is dissociated into hydrogen and sulfur; hydrogen leaves the surface, and the sulfur remains combined with the iron.

Adhesion data are presented in figure 8-14 for an (011) single crystal surface of iron exposed to hydrogen sulfide. LEED and AES analysis were used to monitor the surface chemistry. There are essentially three concentrations of hydrogen sulfide present on the surface of iron in the figure 8-14 data. The data are plotted as force of adhesion versus normal load. The upper curve is for an iron (2×2) structure, the middle curve for an iron (1×2) structure, and the lower curve for a full monolayer. Thus, as the concentration of sulfur on the iron (011) surface is increased (moving from top curve to bottom curve), the adhesive force of the iron surface decreases in direct relation to the quantity of sulfur present on the surface. Adsorption of the hydrogen sulfide to a clean iron surface is instantaneous and, as has been

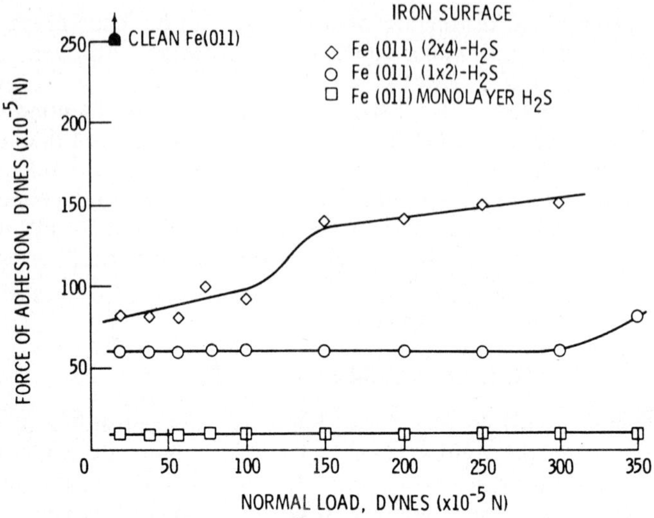

Figure 8-14. – Influence of hydrogen sulfide adsorption on adhesion of iron (011) surfaces. Diameter of contacting flat, 3.0 millimeters; contact time, 10 seconds.

mentioned, it occurs so energetically that there is dissociation of the hydrogen sulfide. In figure 8-14, a data point for a clean iron (011) surface is presented. A comparison of that data point with the data points obtained with even fractions of a monolayer of surface coverage indicate that sulfur is very effective in reducing the adhesion of iron. Both of the upper curves in figure 8-14 are for surface coverages of less than a monolayer.

The data of figure 8-2 coupled with those of figure 8-14 indicate the surface sensitivity of solids to even fractions of monolayers of lubricating species. The exposure of a surface to the presence of surface active species that can adsorb or react with the surface and provide a protective surface film is a function of simple adsorbed gases; it also occurs in well lubricated systems where additives are present in the oils. The interaction of the additives with the solid surfaces to provide a protective lubricating film is a function of the concentration of the additive in the particular lubricating media. Organometallic materials have considerable promise for lubrication applications, particularly those organometallics which contain low shear strength metals that on degradation of the lubricating species can form a metallic film on the solid surface in contact to inhibit adhesion, friction and wear.

One of the organometallics that has been investigated is cadmium in the form of dimethyl cadmium. The dimethyl cadmium at a sliding or rubbing interface decomposes to liberate methyl groups and cadmium metal; the latter can deposit on the metal surface and provide a protective coating of cadmium on the components in contact. Experiments have been conducted to determine the effect of concentration of this additive in ordinary mineral oil on friction behavior. Some results obtained in these experiments are presented in figure 8-15 where the coefficient of friction is plotted as a function of weight percent of dimethyl cadmium (in ordinary mineral oil)

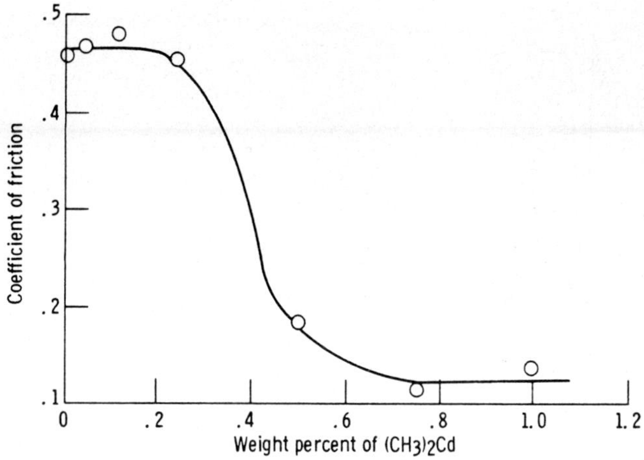

Figure 8-15.—Coefficient of friction for 302 stainless steel lubricated with various concentrations of dimethyl cadmium in mineral oil. Rider, 1045 steel; load, 1100 grams; sliding velocity, 150 meters per minute; temperature, 23° C.

lubricating steel surfaces. The friction coefficient is relatively high (approximately 0.45) at very low concentrations of the dimethyl cadmium in the mineral oil. When a concentration of approximately 0.5 weight percent is reached, there is a marked decrease in the friction coefficient for 1045 steel in sliding contact with 302 stainless steel. The marked decrease in friction coefficient (from 0.45 to approximately 0.1) at a concentration of approximately 0.75 weight percent of dimethyl cadmium in mineral oil can be related to the concentration effect of the additive in the mineral oil. A certain fixed concentration is necessary to achieve effective boundary lubrication. Below that concentration, ineffective lubrication is obtained; for all practical purposes it is as if the additive were not even present (as shown in fig. 8-15 by the data point at 0 concentration).

Auger spectroscopy spectra for two of the concentrations in figure 8-15 were obtained from the 302 stainless steel surface. One concentration was 0.25 weight percent where the friction is relatively high. An Auger emission spectrum was also obtained at a concentration of 0.5 weight percent where the friction coefficient had dropped from about 0.45 to less than 0.2. The Auger spectrum obtained are presented in figure 8-16. In figure 8-16(a) is the Auger spectrum for the dimethyl cadmium concentration of 0.25 weight percent. The spectrum contains essentially two basic elements carbon and oxygen. The Auger spectrum for the dimethyl cadmium concentration of 0.5 weight percent reveals cadmium and oxygen. Thus, the surface chemistry, as determined by Auger spectroscopy analysis, is different for the two concentrations. At the lower concentration, only carbon and oxygen are present on the solid surface, and they are not effective boundary lubricants. The friction coefficient is relatively high, approximately 0.45. With a higher concentration of dimethyl cadmium in the mineral oil, the friction coefficient drops, and that drop is associated with the formation of a cadmium film on the metal surface. Such a film is evidenced by the presence of cadmium in the Auger spectrum obtained from that surface.

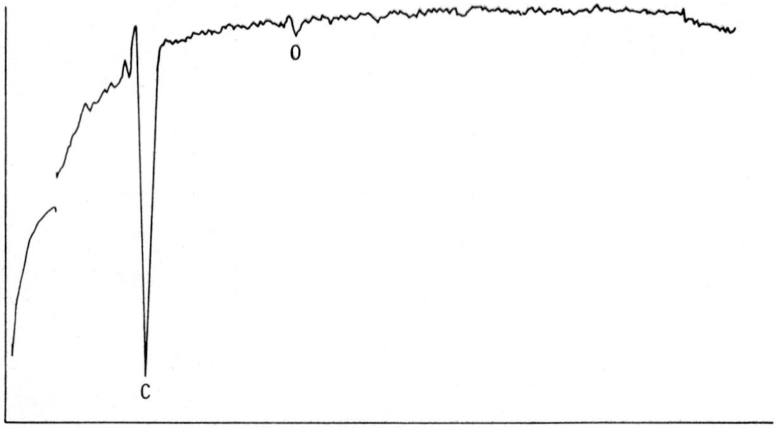

(*a*) *Dimethyl cadmium concentration, 0.25 weight percent.*

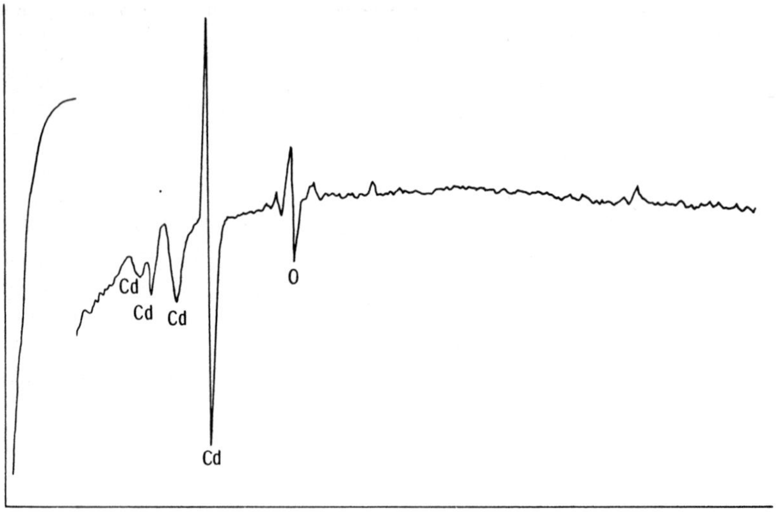

(*b*) *Dimethyl cadmium concentration, 0.50 weight percent.*

Figure 8-16.—Auger emission spectra for 302 stainless steel lubricated with dimethyl cadmium in mineral oil.

From the data in figures 8-2 and 8-14 to 8-16 it is apparent that concentration is extremely important with respect to the lubricating effects of various films on solid surfaces. First, extremely thin films may be effective in reducing adhesion and friction, and second, an optimum concentration is necessary to gain effective boundary lubrication with certain surface additives, particularly where those additives are present in a carrier medium such as a conventional oil.

In figures 8-4 to 8-6, the effect of the molecular structure of the lubricating species on the adhesion, friction, and wear properties of materials in solid-state contact was discussed with reference to the effect of particular groups and chain lengths on tribological behavior. In addition, the changes in structure can change the concentration of a species present on the solid surface. Simple substitutions of an atom in the molecular structure can alter the accommodation of that particular structure (increasing or decreasing it) on the solid surface. This, ultimately, has an effect on the adhesion and friction behavior. For example, the basic ethylene structure which has already been discussed can contain additional atoms in the structure. For example, ethylene oxide has oxygen present in it and vinyl chloride is basically the ethylene structure with the addition of a chlorine atom to the molecular structure. Substituting oxygen for chlorine or vice versa in the basic ethylene structure can produce marked differences in the accommodation of ethylene on a metal surface. For example, some LEED studies (in conjunction with AES analysis) indicate that the accommodation of these two particular molecules, ethylene oxide and vinyl chloride, on iron surfaces is different. For equal exposures, a higher concentration of ethylene oxide can be accommodated. This is indicated in the LEED photographs for the patterns obtained on the iron (011) surface that has been exposed to 1000 langmuirs of ethylene oxide and vinyl chloride (fig. 8-17). The pattern for the ethylene oxide adsorbed on the iron surface gives the hexagonal array of diffraction spots on the surface; the pattern indicates the presence of a close-packed structure on the surface. The presence of the ethylene oxide completely masks the iron LEED diffraction spots. On the surface which has adsorbed the vinyl chloride, the iron diffraction spots are still visible. The four brightest spots in the rectangular array are the diffraction spots for the iron (011) surface. The additional lighter diffraction spots between the iron diffraction spots located through the center of the pattern are associated with the adsorbed vinyl chloride on the solid surface.

The iron surface is completely masked by the ethylene oxide because of the close packing of the ethylene oxide structure on the solid surface. With vinyl chloride adsorbed, the iron LEED diffraction spots are still present. These results indicate that for equivalent exposure of the two different species (both containing the same basic molecular structure (ethylene) but with oxygen substituted for chlorine or vice versa) there is a greater accommodation on the iron surface for ethylene oxide than there is for vinyl

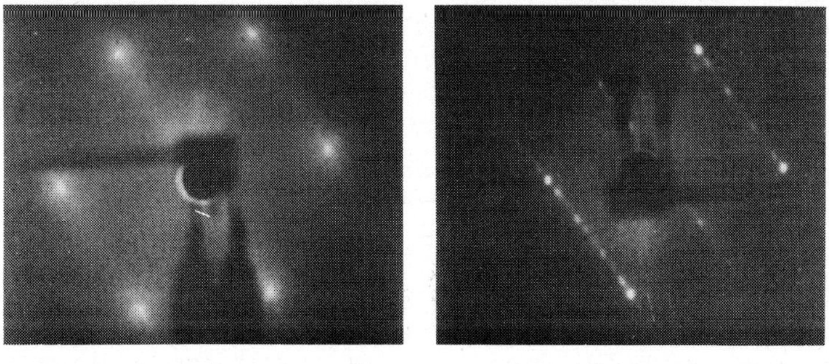

Figure 8-17.—LEED patterns obtained with two polymer forming hydrocarbons on iron (011) surface. Exposure, 1000 langmuirs.

chloride. Adhesion studies made with these two surfaces indicate that the adhesive bond forces to the surface containing ethylene oxide are considerably less than those measured for the iron surface containing vinyl chloride; that is, with vinyl chloride, there is exposed iron available for interaction across an interface. These iron surfaces are, however, completely covered when ethylene oxide is adsorbed on the surface. Thus, there is less metal to metal bonding across the interface and consequently lower adhesive forces with ethylene oxide. Molecular structure not only affects the basic tribological properties of the material from the point of view of the structure itself but also from the relative ability to interact with an environmental species. The amount taken up by the surface is a function of the molecular structure even when the basic molecular structure is the same with substitution of atomic species in the structure as the only difference.

Most of the data presented in this chapter have been for straight chain lubricating species on solid surfaces. Among the practical lubricants employed today, there are a host of organic aromatic compounds that are used for lubricating purposes, particularly at high temperatures. Many high temperature lubricants such as the polyphenyl ethers and other such lubricants are aromatic in nature. They contain ring structures. There is question as to which is the most effective in providing a lubricating film for surfaces in solid-state contact, a straight chain hydrocarbon or an aromatic structure (given equivalent bonding at the solid surface). The simplest aromatic structure to examine would be benzene where there are six carbon atoms in a ring with hydrogens attached; a simple experiment would be to examine the bonding of that to a metal surface. When one does this, most of the experimental results in the literature indicate that the benzene ring lies flat on the surface so that the six carbon atoms are available for bonding directly to the iron. Thus, six iron atoms on the metal surface of the iron would be bound or tied to the carbon as a result of the adsorption of the benzene on a clean iron surface. The same would be true for other metals such as platinum, rhodium, nickel, and tungsten. If one were to try to achieve a similar type of comparison with a straight chain hydrocarbon, the use of hexane (which consists of six carbon atoms in a straight chain) would not provide an effective comparison. The reason for this is that the straight chain hydrocarbon bonds at the end of the chain to the metal surface, with the bulk of the chain length setting above the surface. Thus, the interaction is between only one carbon atom in a chain and the metal surface; the balance of the chain remaining above the solid surface. Acetylene, though it contains only two carbon atoms, bonds to six iron atoms on a solid surface, and it is very effective in shielding the iron surface.

Earlier in this chapter the effectiveness of acetylene was compared to the effectiveness of ethylene and ethane. It was indicated that acetylene was much more effective in providing a shield for an iron surface than was either ethane or ethylene because of the availability of the triple bond which would, on the solid surface, bond to six metal atoms.

Adsorption studies have been conducted for acetylene and benzene on a platinum (100) surface. The results, which are compared using Auger spectroscopy, field emission spectroscopy, flash dehydrogenation, and LEED, indicate that both acetylene and benzene follow Langmuir adsorption characteristics in bonding to the platinum surface. Both acetylene and benzene cause a decrease in work function proportional to the coverage and correspond to a di-pole moment of 1.5 D for benzene and 0.5 for acetylene. Ultraviolet photoemission spectroscopy indicates π-bonding for the benzene with acetylene showing a broadening and chemical shift of the π of the metal and increased splitting of the orbital energies. The results, in general, indicate a π-electron bonding to the metal surface of both acetylene and benzene.

A comparison of the adsorption of these two species to the surface of platinum is presented in figure 8-18. In this figure the carbon to platinum ratio on the surface is plotted as a function of exposure in langmuirs to acetylene and benzene (ref. 8). The data indicate that there is a greater concentration of carbon on the surface with the adsorption of benzene than there is with the absorption of acetylene. Adhesion measurements with acetylene and benzene adsorbed to an iron surface indicate lower adhesion coefficients for benzene on the surface than are measured with acetylene on the solid surface. The aromatic structure of benzene provides more effective surface protection than does acetylene.

While in some systems the carbon atoms of the ring structure can bond directly to the metal surface, this may not always be the case. One example is where the ring structure contains very active atoms such as where halogens are added to the benzene ring. When the ring contains a halogen, bonding to the surface may be by the halogen atom which is much more surface active than the carbon. This would apply to such halogen atoms in the benzene molecular structure such as chlorine, bromine, or iodine. It does not apply, however, to fluorine. Fluorine appears to be so strongly bonded to carbon, both in the aromatic and aliphatic chains, that its interaction with metal surfaces is retarded by the strong interactions with the carbon; hence, bonding to metal surfaces can still occur by carbon atoms to metal atoms with fluorinated hydrocarbons.

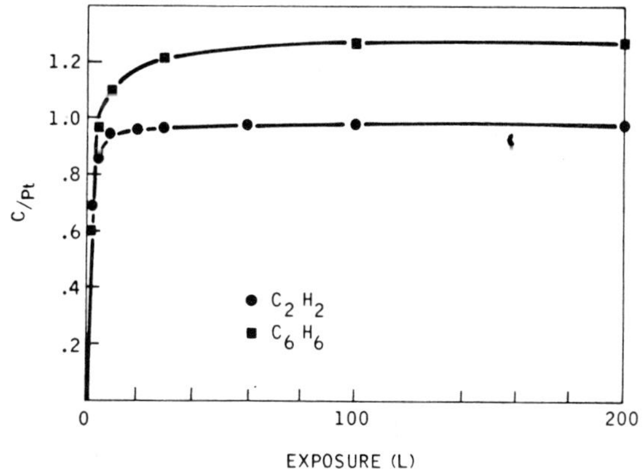

Figure 8-18.—Amount of acetylene and benzene adsorbed on Pt(100) as function of exposure. Coverage is expressed in terms of ratio of carbon to platinum atom density on surface; C/Pt = 1 corresponds to 1.3 × 10^15 atoms per square centimeter (ref. 7).

Environmental Effects on Lubricant Concentration

The interaction of the environment and environmental gaseous species with a lubricant and the solid surface can markedly alter lubrication behavior. Analytical surface tools are very useful in studying changes in the nature of solid surfaces and their tribological behavior as a result of environmental influences in both the concentration of reactant found on the surface and also its composition. Probably one of the lubricants of greatest interest is sulfur, and molecular structures which contain sulfur as a surface reactant provide effective boundary lubrication. Studies of sulfide compounds and their interaction with metal surfaces have indicated, over the years, that the presence of environmental constituents such as oxygen and water vapor influence the behavior of sulfur on the solid surface of metals (particularly ferrous surfaces). For example, simple thermogravimetric studies of the solid surfaces indicate that the interaction of sulfur containing fluids with iron surfaces is influenced by the oxygen in the environment. The data in figure 8-19 show this effect (ref. 9).

In figure 8-19, the uptake of sulphide by the iron surface is plotted on the ordinate as a function of the exposure time at 235° C in air and nitrogen. The di-n-butyl disulfide is adsorbed more readily on the solid surface when air is present. Thus, there appears to be a greater concentration of reactant on the iron surface in the presence of air.

In wear studies of mineral oils containing sulfur additives such as dibenzyldisulfide, the effect of load in an air environment also produces a change in the nature of the wear observed. For example, figure 8-20 shows the end of a rider specimen of 304 stainless steel that had been lubricated with

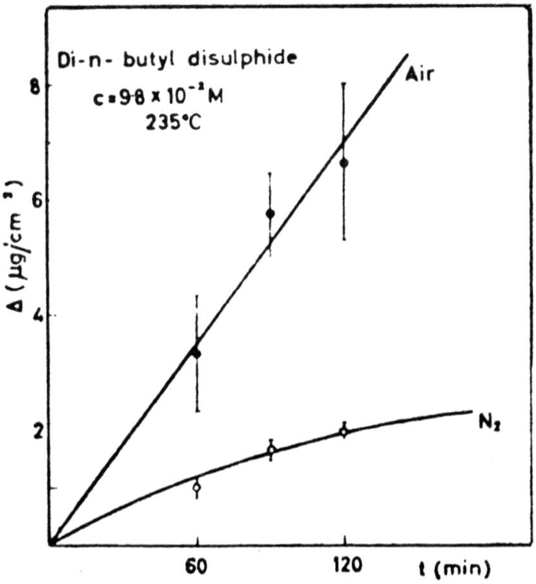

Figure 8-19. – Influence of oxygen in surface reaction with di-n-butyl disulfide (ref. 8).

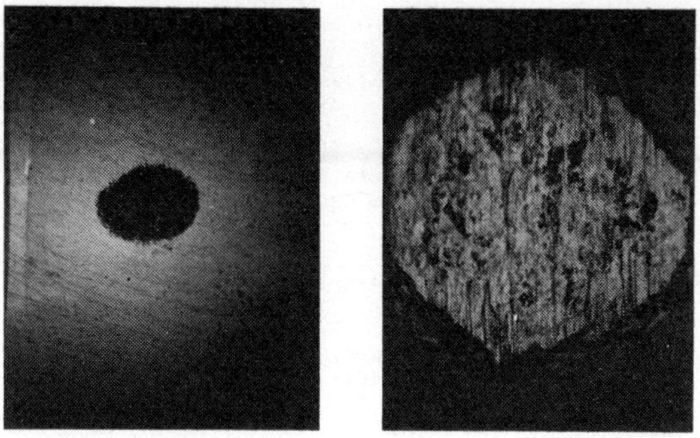

LOAD 5 LB; μ = 0.099 (LOW WEAR) LOAD 31 LB; μ = 0.29 (HIGH WEAR)

CS-78070

Figure 8-20. – Wear of 304 stainless steel rider. Mineral oil with 1 percent dibenzyldisulfide.

mineral oil containing 1 percent dibenzyldisulfide. In one instance, the load was relatively light (approximately 2.1 kg), and in the second, the load was increased to 0.4 kilogram. The wear scars obtained under otherwise equivalent conditions are presented in figure 8-20. At the light load, there is low wear of the surface; at the heavy load, there is very heavy (severe) wear with a marked change in friction. The friction is less than 0.1 in the lightly loaded condition and in excess of 0.29 in the heavily loaded condition.

Analytical tools can assist in understanding these two results: (1) an increase in the weight of the iron samples in figure 8-19 when the sulfur-containing compound is exposed to iron in the presence of air, and (2) greater wear at the heavier load in figure 8-20 for the stainless steel surfaces. XPS (X-ray photoelectron spectroscopy) is especially useful in conducting such analyses. For example, the surfaces shown in figure 8-20 were examined with XPS. Three areas or regions were examined. The wear scars as well as the outside wear contact zone in each of the photomicrographs in figure 8-20 were examined by XPS. The XPS results for the presence of sulfur and oxygen on the solid surface are presented in figure 8-21. Figure 8-21(a) shows three XPS traces: one for severe wear scar (photomicrograph in fig. 8-20 at high load), one for mild wear scar (smaller wear scar in fig. 8-20 at low load), and one for the unworn surface outside the wear scars of figure 8-20. An examination of the XPS spectra for sulfur containing compounds reveals that there is a very small concentration of iron sulfide in the left trace in figure 8-21(a). This particular trace is obtained directly from the solid surface after running. There was no attempt to remove surface layers which would include the mineral oil containing the additive dibenzyldisulfide. The principle XPS peaks detected were carbon from the mineral oil in the solid surface with small peaks of FeS and $FeSO_4$ for the severe wear condition; hardly any sulfur peaks were detected on the mild and unworn surfaces. If, however, the surface was sputtered for 30 seconds with argon bombardment to remove the carbon associated with the mineral oil, then the XPS spectra in figure 8-21(a) on the right side was obtained. For the severe wear scar, a very high concentration of iron sulfide was

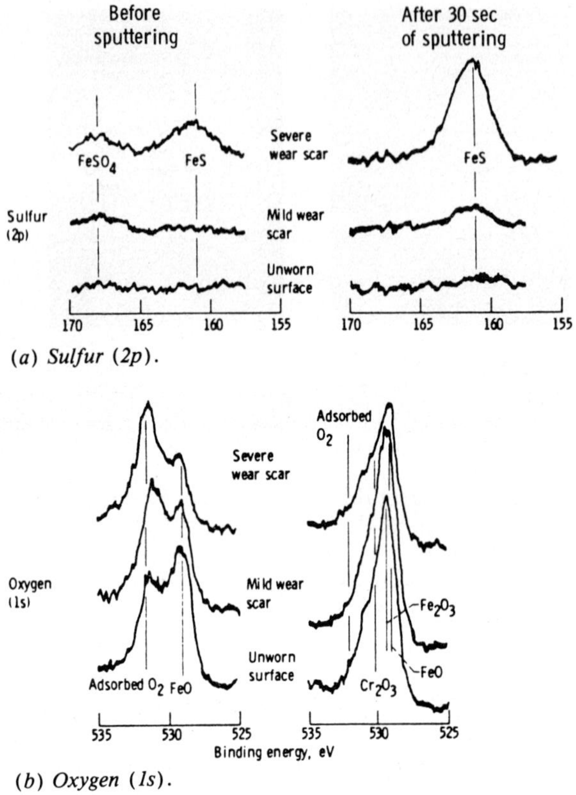

Figure 8-21. – XPS features from wear scars and unworn surface before and after sputtering.

found. There was very little iron sulfide in the mild wear scar and in the unworn surface. Classically, in the absence of oxygen, one might anticipate the formation of iron sulfide by the reaction of the sulfur containing additive dibenzyldisulfide with the solid surface with the resultant reaction product being iron sulfide. The results obtained in figure 8-21(a) are, therefore, not surprising—particularly those obtained after sputtering.

The fact that the severe wear scar contained iron sulfide, which is supposedly an antiwear additve, raises the obvious question as to what accounts for mild wear if iron sulfide does not, since iron sulfide was the anticipated reaction product which would be found in the mild wear regime not in the severe wear scar. It was anticipated that, in the severe wear scar, there would be very little sulfur present on the surface and principally nascent iron would be exposed as a result of the severe interaction of metal to metal contacts. The next likely candidate to be present on the solid surface would be oxygen-containing compounds, because the mineral oil does contain dissolved oxygen from the air environment and air is a part of the system.

An examination of the iron and the oxide peaks for the solid surface indicates that oxygen plays a very important role. The oxygen from the environment enters into the reaction products formed on the solid surface. In figure 8-21(b) are XPS spectra for the oxides and oxygen present on the surface before and after sputtering. In the traces on the left, the adsorbed oxygen (O_2) is present as well as the peaks for iron oxide (FeO). On the right side, after sputtering, are the O_2 peaks and the oxides of iron and chromium. A careful examination of the severe wear scar, the mild wear scar, and the unworn surface reveals that in the mild wear case the principle film on the solid surface is not sulfide but rather oxide. There is a higher concentration of oxide on the metal surface experienced in mild wear than on either the unworn surface or the surface containing the severe wear scar. Thus, it appears that the oxides present on the surface (the thicker oxides in the mild wear regime) are the effective antiwear additive that prevents severe wear from occurring.

Sulfur seems to promote the oxidation of ferrous based surfaces. This may account then for the increase in weight observed in figure 8-19 where the di-n-butyl disulfide is exposed to an iron surface in the presence of air. It also explains the wear behavior observed in figure 8-20, which indicates that oxides play a very important role in reducing wear of solid surfaces in solid-state contact. It is interesting to note that, with a simple iron surface exposed to sulfur, a sulfide film forms on the solid surface. If that surface is maintained in a vacuum so that the only thing present on the solid surface is iron sulfide, and that film at room temperature is then exposed to oxygen, the oxygen completely displaces the sulfur from the surface. An Auger spectroscopy analysis of these films reveals ultimately a complete displacement of the sulfur from the solid surface by oxygen leaving nothing but an oxide film. Thus, it may be that in many instances in practical lubrication devices, the role of sulfur in the molecular structure is really to enhance the formation of oxides on the surface which, in and of themselves, are good resistors of adhesion.

If one changes the lubricant from a standard mineral oil to some other base fluid using the same additive, differences in behavior in concentration of oxygen and sulfur on the surface are observed. For example, Coy has conducted studies using sulfur containing additives in various oils and, using Auger spectroscopy analysis, he found that the concentration of oxygen and sulfur on a solid surface (and the ratio of the two) varies depending on the oil in which the additive is present (ref. 10). Some of his results are presented in figure 8-22 where the sulfur to oxygen ratio is plotted as a function of sputtering time for various oils in which the sulfur containing additive finds itself. Figure 8-22 presents results for four different oils; it can be seen that the ratio of sulfur to oxygen varies appreciably for the four different oils. One basically is getting a depth profile analysis of the surface layers. This ratio changes with profiling, and the ratio at all times is different for the different oils. These results indicate that the system and the interactions (of environment, lubricant, and solid surface) are a lot more complex than one might anticipate. The environmental effects can not be discounted when considering the lubricating effects of various species on solid surfaces. In addition, one must consider the particular base fluid in which the active lubricating species is present, because a change in the base fluid may produce a change in the surface chemistry at the solid surface with the reacting species (fig. 8-22).

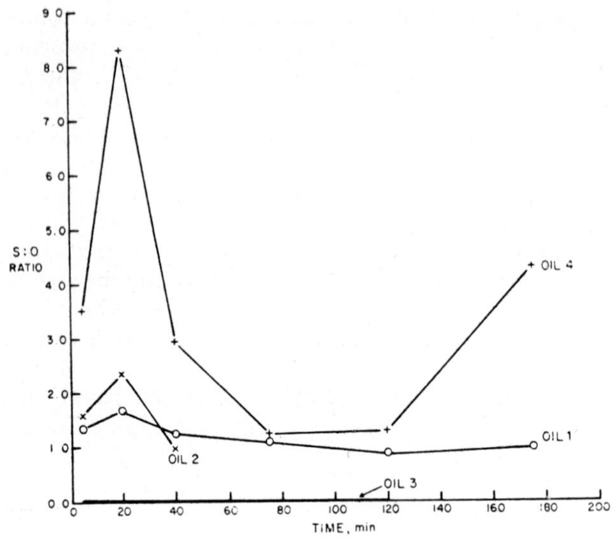

Figure 8-22. – Ratio of sulfur to oxygen (ref. 9).

Mechanical Effects on Lubricant Behavior

The exposure of conventional surfaces to lubricants generally results in adsorption of the lubricant to the solid surface, if the lubricant is at all surface active. In most practical lubrication systems, the surfaces of the solids to be lubricated contain adsorbed gases. In the case of metals, the surfaces have oxide films and, consequently, the bonding of the lubricant to the solid surface is by a mechanism generally of physical adsorption with some exceptions (where chemisorption actually occurs). This is true for hydrocarbon lubricants as well as for synthetically formulated species. This refers to the lubricant where the structure is a simple carbon-hydrogen type of structure without the presence of surface active atoms in the molecular species such as oxygen or chlorine. When the surfaces are involved in practical tribological systems and the mechanical activity of rubbing, sliding, or rolling at the contacting interface produces disruption of the surface films (the adsorbed films or the oxides), direct interaction of the lubricant species with the solid surface can take place. Under such conditions, the clean surface exposed as a result of mechanical activity at the solid surface is extremely reactive—particularly in the case of metals and alloys, which are the most commonly used materials in tribological devices. Under such conditions, hydrocarbons can chemibsorb and even chemically react with the metal surface. Furthermore, decomposition of the organic molecule can take place by the activity of the surface metal atoms, and this is reflected, for example, in the reactions which give rise to the formation of friction polymers.

Generally the first step in this process is the formation of radicals by the breaking of the carbon to carbon bond in the molecular structure with subsequent interaction of these radicals to form large molecular weight species which condense out onto the solid surface. These films that form as a result of the interaction of metal with straight hydrocarbons (whether they are aromatic, alphatic, naturally occurring, or synthetic structures) are not very effective in modern day machinery in preventing seizure of mechanical components under very heavy loads. The straight hydrocarbons have limited lubricating abilities. They are fine for light loads in low speed conditions. However, under heavy loads, high speed sliding conditions, or rolling conditions, the simple hydrocarbon structure is insufficient to provide effective boundary lubrication where surfaces are in solid-state contact. Consequently, the basic structure of the molecule that is used for lubricating purposes is frequently modified. This is one approach. For example, oxygen or chlorine may be added to the basic molecular structure of the lubricating species producing oxygenated or chlorinated hydrocarbons that then can act as a lubricant on the solid surface; the oxygen or the chlorine reacts with the surface to provide a protective inorganic surface film. One does not, under such conditions, rely exclusively on the lubricating characteristics or the film forming characteristics of the organic portion of the molecule. Rather, one relies on the formation of inorganic compounds which are thermodynamically much more stable than the organic compounds on the solid surface and which provide better surface protection to minimize adhesion and friction. The ultimate goal is to minimize the solid-state contact of nascent surfaces. Inorganic compounds are much more resistant to desorption, decomposition, and dissociation than are the simple organic structures present on solid surfaces. Hence, one simply substitutes an inorganic film (with its greater tenacity) for an organic film.

Another approach that can be taken to provide more effective lubrication of solid surfaces using conventional lubricants, such as straight mineral oils and general hydrocarbon structures, is to use additives that can interact with the solid surface. They find their way to the solid surface by migration through the oil or liquid and interact at the solid surfaces to provide a protective surface film. The film forms by decomposition of the compounds that are present in the oil; this liberates active surface species which interact with the solid surface to form inorganic compounds that provide effective boundary lubrication. There are a host of materials that have been used for these purposes. Generally, they are subdivided into two classes: one improves the basic lubricating properties of the base fluid or the oil carrier, and the second refers to extreme pressure additives that interact under very high load and high speed conditions (e.g., in gears to provide very stable surface films).

Some representative compounds of lubrication improvers and extreme pressure additives that provide lubricant films to withstand extreme pressure conditions are presented in table 8-IV (ref. 11). Some of the representative classes of materials that are used as additives are presented together with their constitutional or structural formulas. These include such materials as the esters like butyl stearate and acids and alcohols such as oleic acid and cetyl alcohol. Extreme pressure additives include tricresylphosphate, dibenzyldisulfide, and zinc dialkyldithiophosphate. These

TABLE 8-IV. – CLASSIFICATION AND CONSTITUTIONAL FORMULA OF ADDITIVES[a]

	Additives	Constitutional formula
Oiliness improvers	Rapeseed oil	CH_2OOCR_1 / $CH\,OOCR_2$ / CH_2OOCR_3
	Butyl stearate	$C_{17}H_{35}COOC_4H_9$
	Stearic acid	$C_{17}H_{35}COOH$
	Oleic acid	$CH_3(CH_2)_7CH=CH(CH_2)_7COOH$
	Cetyl alcohol	$CH_3(CH_2)_{14}CH_2OH$
	Oley! alcohol	$CH_3(CH_2)_7CH=CH(CH_2)_7CH_2OH$
Extreme pressure additives	Sulfurized fatty oil	Sulfide of fatty oil
	Tricresyl phosphate	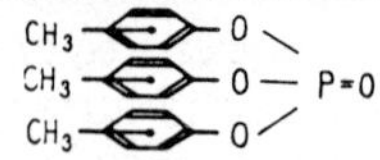
	Tributyl phosphite	$(C_4H_9O)_3P$
	Dibenzyl disulfide	
	Zinc-dialkyl dithio phosphate	$(RO)_2P(=S)-S-Z_n-S-P(=S)(OR)_2$ (R; $C_4 \sim C_8$)
	Chlorinated paraffin	Chloride of $CH_3(CH_2)_nCH_3$ (Cl; 40%)
	Pentachloro metylstearate	Chloride of $C_{17}H_{35}COOCH_3$ (5 Cl per 1 mol)

[a]Reference 11.

are probably the three principle extreme pressure additives employed commercially today to provide effective boundary lubrication where extreme loads and/or speeds are involved in mechanical devices. The extreme pressure lubricating properties and the reactivity or reaction charactersistics of tricresylphosphate with metal surfaces were discussed in reference to figure 8-11. Dibenzyldisulfide was discussed in reference to the formation of protective surface films and the influence of environment in figure 8-21. Zinc dialkyldithiophosphate (table 8-IV) is probably the most widely used extreme pressure and antiwear additive. A considerable amount of research has been conducted on it and dibenzyldisulfide.

For many years, it was common for researchers examining the reaction characteristics or interaction characteristics of lubricant additives with solid surfaces to use static immersion tests where the metal specimens were submerged in a bath of oil containing the additive. The oil was then heated to some temperature for a time with subsequent examination of the metal surface to determine what type of interaction of the additive present in the oil took place with the metal solid surface. Many shortcomings exist in this approach in that, in actual practical tribological devices, there is either sliding, rubbing, or rolling contact at the surface. This mechanical activity gives rise to the input of a considerable quantum of energy at the surface. This energy can produce a number of changes in the nature of the solid surfaces and in the lubricants that interact with that solid surface. The quantum of energy involved is sufficient to produce gross metallurgical changes in materials; consequently, it is sufficient to produce changes in the nature of the lubricant interaction with the solid surface. Therefore, very frequently the use of static immersion reaction studies of additives and oils with solid surfaces can produce misleading results. Evidence for that is presented by the data in figure 8-23 (ref. 12). Figure 8-23(a) shows the XPS spectra for a steel pin before any surface condition or treatment of the pin. The sulfur peak present in the solid surface is revealed in the XPS spectra.

Figure 8-23(b) shows the XPS spectra for the steel pin after the pin is immersed in oil containing the sulfur additive. An increase in the sulfur 2p intensity is observed. Figure 8-23(c) shows the XPS spectra (in the sulfur 2p binding energy region) for the steel surface after rubbing in a wear test under the same oil in which static immersion tests were conducted in figure

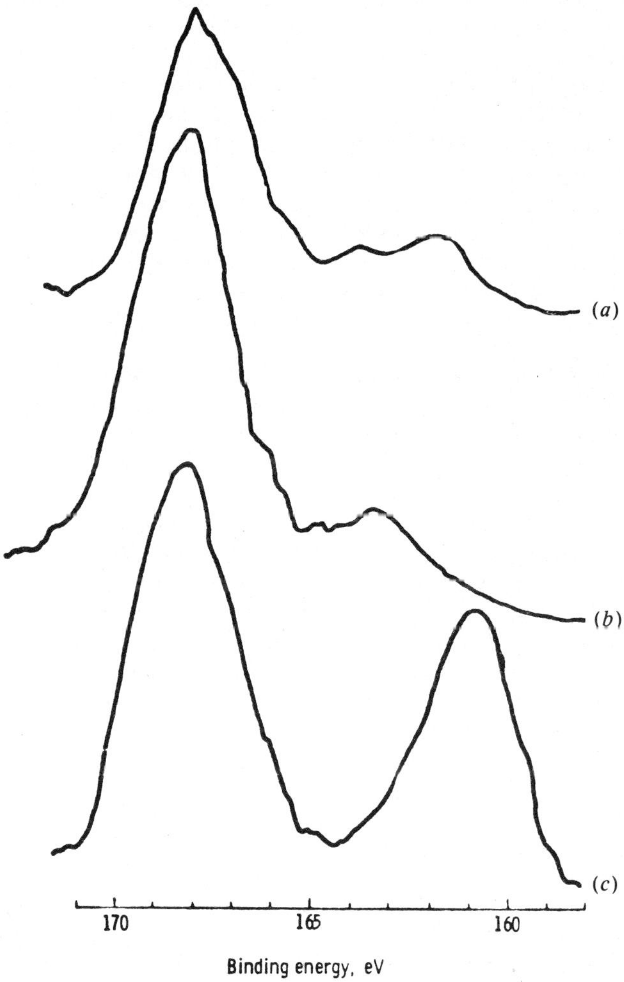

Figure 8-23. – Sulfur (2p) binding energy region on steel pins.

8-23(b). Comparing figure 8-23(b) with figure 8-23(c) shows a marked difference in the lower two spectra. A new additional peak at approximately 161 electron volts appears in the spectra that is not present in the spectra for static immersion. Thus, the sulfur compounds that formed in the surface as a result of the wear rubbing process are different from those on the surface as a result of static immersion. The mechanical activity associated with wear produces an alteration or change in the surface chemistry from that observed with simple immersion.

Baldwin, who obtained these data, indicates that the film present on the solid surface after wear testing is a metal sulfide (ref. 12). This surface sulfide then provides antiwear or extreme pressure lubricating properties to the solid surface. The sulfide is not present in the simple immersion data of figure 8-23(b). As one would anticipate, in order for the clean steel surface to be exposed to the lubricant additive and to accomplish the formation of a sulfide, nascent metal must be exposed. The only way that this can be accomplished is by cleaning the surface with such techniques as sputter cleaning or, as in figure 8-23, by rubbing the solid surfaces and exposing nascent metal or steel. With static immersion experiments, the steel surfaces are covered with oxides and adsorbates and the sulfur of the additive in the oil does not have an opportunity to interact directly with the metal surface. It is shielded; there is a barrier film of oxide and adsorbed layers which prevent the sulfur from coming into direct contact with the metal and thus prevent any reaction to provide a protective surface layer. For this chemistry to be accomplished, clean metal must be exposed.

Bird and Galvin also conducted experiments on surfaces lubricated with sulfur containing extreme pressure additives; they analyzed the surfaces with XPS and found that the results obtained from static immersion tests were not at all the same as those obtained from experiments conducted under mechanical activity (ref. 13).

With dibenzyldisulfide as an additive in white oil, Bird and Galvin used XPS to examine the sulfur 2p spectra for steel rollers that had been lubricated with the rollers being simply immersed and for the rollers in solid-state contact under various loading conditions. The results of some of their experiments are presented in the XPS spectra of figure 8-24. In figure 8-24 the sulfur 2p spectra, just the same as was presented in figure 8-23, are presented from the XPS data for the roller surfaces under various conditions: (1) simply immersed in the oil containing dibenzyldisulfide, (2) for the rollers in contact at various loads (loads indicated to the right of the spectra in fig. 8-24), and (3) in a cutting experiment. A careful examination of the spectra in figure 8-24 reveals marked differences in the XPS sulfur reaction products formed on the solid surface as a result of the nature of the mechanical activity taking place on the solid surface.

The first observation to be made from the data in figure 8-24 is that, in the absence of mechanical activity, the XPS spectrum is certainly entirely different from that obtained in the presence of mechanical activity. This result indicates, again, the importance of examining the surfaces in sliding or rubbing contact where mechanical activity is taking place as opposed to using static immersion to predict the nature of the surface films formed.

Not only does the spectrum change with the introduction of mechanical activity at the solid surfaces, but it also changes with changing mechanical conditions, such as increasing the load. For example, increasing the load in figure 8-24 produces a change in the spectrum. In the cutting experiments, while the nature of the mechanical activity is different from the rollers in contact, the basic surface chemistry appears to be similar to that obtained with the rollers in contact. The only difference again seems to be one of quantity of the relative reaction products formed as opposed to the actual nature or chemistry of the compounds themselves. The real difference in figure 8-24 exists between the presence and absence of any kind of mechanical activity at the solid surface.

The XPS data in figures 8-20 to 8-24 indicate that, with mechanical activity imposed at an interface with sulfur-containing additives, metallic sulfides and sulfates as well as oxides can form on the solid surface. The actual composition of the surface under certain mechanical conditions can be exclusively oxide (where mild wear is encountered) or exclusively sulfide, or,

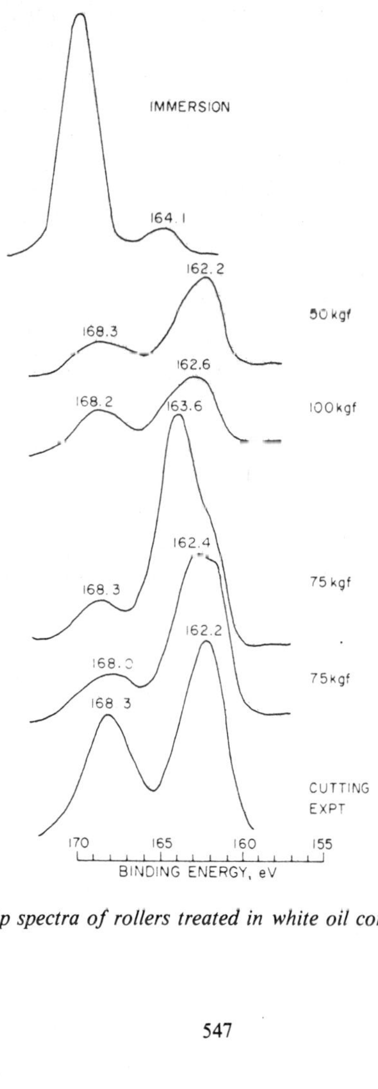

Figure 8-24. – Sulfur 2p spectra of rollers treated in white oil containing dibenzyldisulfide (ref. 12).

in addition, it can be a mixture of films, oxide, and sulfide (in some instances even sulfates). The nature of the chemistry of the film appears to be very heavily dependent on the mechanical activity. The load, speed, and temperature of the surface influence the ultimate chemistry seen on the surface with the surface analytical tool (XPS).

Analytical surface tools such as AES analysis and XPS are extremely useful in identifying the chemistry of the solid surfaces, particularly since we now know that the nature of mechanical activity at the solid surface can alter markedly the surface chemistry from that observed in static immersion studies. It is almost imperative, then, that one use some tool to identify the chemistry of the suface film generated, since that can vary in composition so widely as a function of the nature of the mechanical action that is taking place. When analytical surface tools are used, however, great care must be taken because of the effects of the instrument itself. For example, in an analytical surface tool such as AES analysis, a primary beam of electrons from 1500 to 3000 volts is directed at the solid surface. The primary incoming beam of electrons can produce a change in surface chemistry with delicate molecules such as some of the organic lubricating species that are employed in practical lubrication devices. Therefore, great care must be taken in interpreting data obtained from solid surfaces. For example, electron-induced desorption can occur because the primary beam of electrons from the Auger analysis impinges on the solid surface; that is, the energy associated with the incoming beam can be of sufficient intensity to liberate species directly from the solid surface. A complete removal of some species from a solid surface can occur by this technique and can produce misleading results when interpretating the data obtained. It is a caveat of which the experimenter who is using these surface analytical tools must be aware. He must not only know the capabilities of the instrumentation he is employing but also its limitations and its effect on the solid surface films that may be present.

For example, figure 8-25(a) presents an Auger spectrum as a functon of time for the zinc (from zinc dialkyldithiophosphate) lubricating a tungsten disk surface. Zinc is found to be present on the solid surface after a rubbing experiment has been conducted; that is, zinc is actually present in the wear track. The Auger spectrum obtained immediately after rubbing is shown to the far left near the ordinate of figure 8-25(a). The concentration of the zinc in the film on the surface is relatively high; this indicates that the zinc dialkyldithiophosphate has reacted with the solid surface and liberated zinc, which is present in the wear track of that solid surface. However, an examination of one particular location of the wear spot of the solid surface shows that the zinc decreases in intensity with time until after 15 minutes (fig. 8-25(a)) when a nearly complete loss of the zinc from the Auger spectra occurs. Now, if one were not aware that the electron beam could produce this type of desorption from the solid surface, one might be led to believe that there is a relatively minimal concentraion of zinc on the solid surface when in fact the zinc concentration in figure 8-25(a) is relatively large.

This electron-induced desorption is somewhat a function of the metal surface on which the additive finds itself. For most metals, this same kind of electron-induced desorption occurs for the zinc of zinc dialkyldithiophosphate (indicated in fig. 8-25(b)). In a similar type of experiment where the zinc dialkyldithiophosphate is present on a molybdenum disk surface, the zinc also desorbs. A concentration of zinc on the molybdenum surface is plotted in figure 8-25(b). Just as in figure 8-25(a) for the tungsten surface, a decrease in zinc intensity, with time, is observed until there is very little zinc left on the solid surface after 30 minutes of exposure to the electron beam.

Although it takes a greater period of time in figure 8-25(b) than in figure

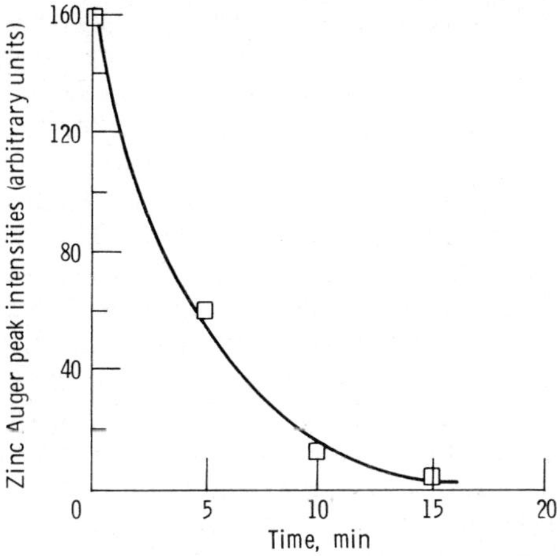

(*a*) *From wear track on tungsten disk; load, 300 grams.*

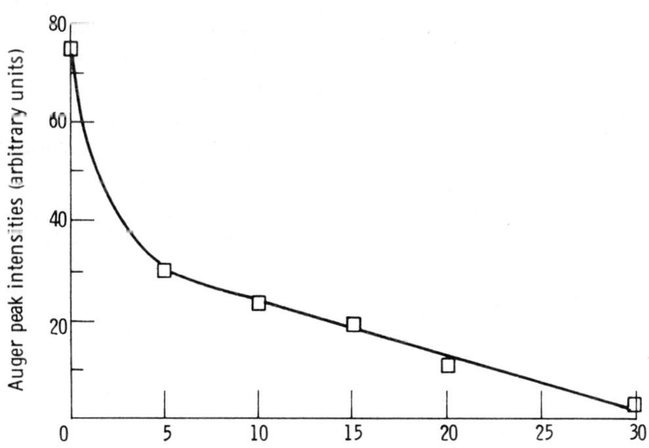

(*b*) *From wear track on molybdenum disk; load, 1600 grams.*

Figure 8-25.—Auger electron beam induced desorption of zinc. Lubricant, thin film of 10 percent zinc dialkyldithiosulfide in hexane; rider, aluminum; temperature, 23° C; sliding velocity, 30 centimeters per minute; Auger electron beam, 10 microamperes and 1360 volts.

8-25(a) to achieve similar results, ultimately the same effect is observed; the zinc is essentially removed from the solid surface by the electron beam of the Auger analyzer. This is a limitation of the Auger analyzer that must be recognized and appreciated when interpreting and analyzing surface films formed in tribological studies. While this may appear, at first, to be a disadvantage or limitation of the instrument, it also has its advantages; that is, the fact that the surface species are subject to an electron beam induced desorption can be used to aid in understanding the nature of the surface chemistry. Zinc dialkyldithiophosphate contains (in addition to zinc) sulphur and phosphorus, which also are found on the solid surfaces of the tungsten in figure 8-25(a) and the molybdenum in figure 8-25(b).

A careful examination of the solid surfaces and the Auger spectra for the sulphur and the phosphorus do not reveal electron beam induced desorption of the sulphur or phosphorus. The sulphur and phosphorus are stable on the solid surfaces. The electron beam energies involved in figures 8-25(a) and (b) from the Auger spectrometer are insufficient to produce a desorption of either the sulphur or phosphorus. This indicates that the sulphur and phosphorus are very strongly bonded to the metal surface and are impervious to the influences of the electron beam used in these experiments. It is only the zinc that is sensitive to the electron beam and undergoes desorption. One can use this to advantage when interpreting the nature of the bonding that may take place on the solid surface. If the sulphur and phosphorus do not change in their concentration on the solid surface as a function of exposure to the electron beam while the zinc does, it is apparent that the zinc is less strongly bonded to the metal surface because it comes off at the energies involved while the sulphur and phosphorus do not. Thus, the Auger electron beam induced desorption can be used to gain some qualitative indication of relative binding energies to the solid surface. One can vary the primary beam energies associated with the Auger spectrometer. By doing this, and varying the amount of energy (the quantum of energy) released at the solid surface, one can get some relative indication of the strength of the bonds formed from the various surface active elements present in an additive like zinc dialkyldithiophosphate.

The apparent limitation of the Auger spectrometer in analyzing lubricating films (indicated by the data of figs. 8-25(a) and (b)) can be overcome by simply moving the beam location on the solid surface either by using the instrumentation itself to accomplish that effect or by moving the specimen surface relative to the beam. This is one advantage of incorporating surface analytical tools directly into a tribological experiment.

In a pin on disk experiment, if one is examining the wear track and looking at the nature of reaction film formed in the wear track, where electron beam induced desorption occurs, one can simply move the disk relative to the Auger beam and analyze a different location on the wear track. Or one can continuously move the wear track under the beam and do an averaging of the intensity of the beam over the entire wear surface area. This latter approach is probably a much better one since it gives an average of the film thickness over the entire wear surface area as opposed to a specific spot location thickness.

Pepper examined the effect of electron beam induced desorption on metal surfaces lubricated with such polymeric materials such as polyvinylchloride. He found that the chlorine in polyvinylchloride (just as the zinc in the zinc dialkyldithiophosphate) was sensitive to electron beam induced desorption from the Auger spectrometer (ref. 14). He overcame this problem by moving the disk relative to the Auger spectrometer. Some of his results are presented in the data of figure 8-26.

Figure 8-26 presents the Auger spectra for the chlorine and carbon peaks associated with the polyvinylchloride. With no disk movement (zero velocity) there is a relatively small chlorine peak at 180 electron volts (fig. 8-26(a)). The carbon peak is of relatively large intensity at 271 electron volts. If the disk rotates very slowly under the beam so that one is continuously monitoring a different location of disk surface, the full intensity of the chlorine Auger peak is revealed in the spectra of figure 8-26(b). Note the marked difference in the size of the chlorine peaks in figures 8-26(a) and (b). Also note that the carbon peak intensity is decreased somewhat in figure 8-26(b) from that observed in figure 8-26(a); the difference is the result of shielding by the chlorine. When the greater amount of chlorine is present in the Auger spectrum, its presence on the surface masks some of the Auger electrons being emitted from the carbon. Consequently, a somewhat lower carbon peak intensity is observed in figure 8-26(b). One gets a much better representation of the chemistry of the solid surface when doing an Auger spectra with the surfaces moving, where electron beam induced desorption is a problem.

Again, however, the data in figure 8-26 indicate to the experimenter that the chlorine is relatively weakly bonded to the solid surface. Actually, it would appear from the spectra of figure 8-26, that the carbon is bonded to the solid surface and the chlorine is simply attached to the carbon. That would also account for the shielding of the carbon and the lower intensity of carbon observed in the Auger spectra in figure 8-26(b) relative to that of figure 8-26(a).

Similar results have been observed by Pepper in the analysis of polytetrafluoroethylene films on metal surfaces where solid bodies of polytetrafluoroethylene were in sliding or rubbing contact with metal

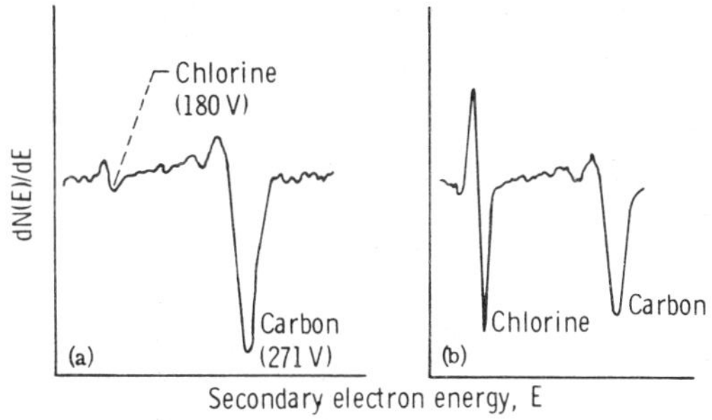

(*a*) *Before sliding.* (*b*) *After sliding.*

Figure 8-26. – Chlorine and carbon part of Auger spectrum of steel disk on which polyvinylchloride was deposited from solution in tetrahydrofuran. Sliding velocity, 0.03 centimeter per second; beam current, 1 microampere.

surfaces. A transfer film of polytetrafluoroethylene to the metal surface is observed and, again, an Auger analysis reveals the presence of polytetrafluoroethylene even with the single pass of the slider across the surface. However, the beam from the Auger analyzer induces electron beam desorption of the fluorine. This condition indicates that the bonding to the metal surface may be by the carbon atom, since with both polyvinylchloride and polytetrafluoroethylene the halogen atom is subject to electron beam induced desorption and comes off the solid surface.

It is reasonable to conclude from the data in figure 8-26 and the additional data obtained by Pepper with polytetrafluoroethylene that the bonding mechanism is by the carbon. If the halogens reacted directly with the metal surface, the films formed on the solid surface would be very strongly bonded, because of the chemical nature of the metal fluorides and chlorides formed. The metal fluorides and chlorides are inorganic compounds having good stability. They would be expected to remain on the surface at the energies associated with the electron beam intensities employed in the data obtained in figure 8-26. It would therefore appear that the halogens are not bonded to the metal surface but remain bonded to the carbon atom while the carbon is bonded to the solid surface.

References

1. Avitzur, B.; and Kohser, R. A.: Disk and Strip Forging for the Determination of Friction and Flow Strength Values. ASLE Trans., vol. 21, no. 2, Apr. 1978, pp. 143-151.
2. Hardy, W. B.: Collected Scientific Papers. The University Press (Cambridge, England), 1936. See Papers No. 37 (1919), No. 39 (1920), No. 40 (1920), No. 41 (1922), No. 42 (1922), No. 43 (1923), No. 44 (1925), No. 46 (1925), No. 50 (1926), and No. 57 (1928).
3. Owens, R. S.; and Barnes, W. J.: The Use of Unsaturated Hydrocarbons as Boundary Lubricants for Stainless Steels. ASLE Trans., vol. 10, no. 1, Jan. 1967, pp. 77-84.
4. Bowden, F. P.; and Tabor, D.: Friction and Lubrication of Solids. Oxford Clarendon Press (London), 1950, p. 40.
5. Shafrin, E. G.; and Murday, J. S.: Auger Compositional Analysis of Ball Bearing Steels Reacted with Tricresyl Phosphate, ASLE Trans., vol. 21, no. 4, Oct. 1978, pp. 329-336.
6. Bertolini, J. C.; Dalmai-Imelik, G.; and Rousseau, J.: Benzene Adsorption on Nickel (100) and (111) Faces Studied by LEED and High Resolution Electron Energy Loss Spectroscopy. Surface Sci., vol. 67, no. 2, Oct. 1977, pp. 478-488.
7. Lurie, P. G.; and Wilson, J. M.: The Diamond Surface. I: The Structure of the Clean Surface and the Interaction with Gases and Metals. Surface Sci., vol. 65, 1977, pp. 453-475.
8. Fischer, T. E.; Kelemen, S. R.; and Bonzell, H. P.: Adsorption of Acetylene and Benzene on the Platinum (100) Surface. Surface Sci., vol. 64, 1977, pp. 157-175.
9. Llopis, J.; et al.: Surface Reactions of Iron with Hydrocarbon Solutions of Organic Sulphides. Corrosion Sci., vol. 4, no. 1, Mar. 1964, pp. 27-49.
10. Coy, R. C.: Discussion of the Paper—Element Concentration Analysis of Films Generated on a Phosphor Bronze Pin Worn Against Steel under Conditions of Boundary Lubrication by M. H. Jones. ASLE Trans., vol. 21, no. 2, Apr. 1978, pp. 107-108.
11. Kawai, N.; et al.: The Frictional Characteristics of Mineral Oils in Sheet Metal Drawing. Bull. Jap. Soc. Mech. Eng., vol. 15, no. 83, 1972, pp. 635-641.
12. Baldwin, Bernard A.: Chemical Characterization of Wear Surfaces Using X-Ray Photoelectron Spectroscopy. Lubr. Eng., vol. 32, no. 3, Mar. 1976, pp. 125-130.
13. Bird, R. J.; and Galvin, G. D.: The Application of Photoelectron Spectroscopy to the Study of E. P. Films on Lubricated Surfaces. Wear, vol. 37, 1976, pp. 132-167.
14. Pepper, S. V.: Sliding of Polyvinylchloride on Metals Studied by Auger Electron Spectroscopy. NASA TN D-7533, 1974.

CHAPTER 9

Effect of Surface Films on the Mechanical Behavior of Solid Surfaces

It has been recognized for many years that films on solid surfaces can influence the mechanical behavior of the surficial layers of the solid; that is, the presence of the surface films somehow alters the nature of the mechanical properties of the solid surface. There is a very fundamental interrelationship, then, between the presence of surface films and tribological responses, since the responses are dependent on mechanical behavior of the solid surfaces.

Various investigators through the years have observed the effect of mechanical properties of solid surfaces and the influence on those properties by the presence of surface films. Joffe in the 1920's observed that halide crystals, such as potassium chloride, could be deformed very readily in a moist atmosphere, or underwater, while in dry air they were relatively brittle and fractured fairly readily (ref. 1). This increase in ductility was due to the presence of the water on the solid surface.

In the 1930's, Roscoe observed that the presence of oxides on certain metals, such as cadmium, produced a surface hardening effect (ref. 2).

In the 1920's and 1930's, the Russian researcher, Rehbinder, observed that the presence of certain organic acids on the surfaces of solids resulted in a surface softening or a reduction in the mechanical properties of solids (ref. 3).

In 1950, Kramer observed a similar effect in the presence of surface acids. In Kramer's work, the effect on mechanical behavior was related to the formation of surface compounds. The explanation for the formation of compounds in the absence of normal surface chemistry for the development of such films was that the surfaces, with deformation, liberated electrons (called exoelectrons). These exoelectrons provided the necessary energies at the surface for the formation of surface compounds, which then altered the mechanical behavior of the solid surfaces (ref. 4). Some of these effects are shown schematically in figure 9-1.

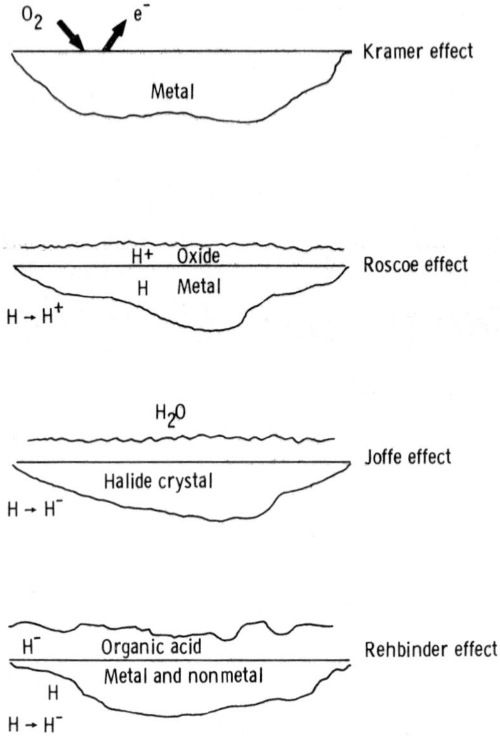

Figure 9-1. – Effect of surface films on mechanical behavior.

Kramer Effect

The Kramer effect was postulated by Kramer to explain the change in the rate of work hardening of aluminum and copper crystals when tested in paraffin oils containing different concentrations of stearic acid. Kramer suggested that the fatty acids, such as stearic acid, react at the metal surface to form a metal soap, and that these metal soaps dissolved, or desorbed, into the environment. The rate of dissolution of the fatty acids in the environment was a function of the rate of reaction at the surface. However, an examination of the aluminum and copper in the presence of the paraffin oil where the metal stearate was added directly to the oil showed no such effect. And furthermore, no effect was observed with gold crystals. It is known that gold soaps are not normally formed because the free energy required is too large. Despite this, however, Kramer did observe the change in work hardening for gold in the presence of the paraffinic oil containing the stearic acid. He then suggested that this was due to the presence of electrons on the surface which were emitted during the deformation process of a metal. In other words, during the deformation, the metal emitted electrons (exoelectrons) from the solid surface. These exoelectrons then can interact with the stearic acid to generate or promote the formation of a metal stearate. Even with normal surface chemistry, one would not consider or anticipate the formation of metal stearates. Thus, in the case of gold, a stearate would form despite the lack of a normally observed gold stearate.

For many years other investigators have observed the liberation of exoelectrons from the surfaces of metals during deformation, working, cutting, and grinding of metal surfaces. The liberation of these exoelectrons has been a research curiosity. Most investigators believe that the deformation of the metal surface causes the liberation of the exoelectrons from the solid surface.

Ferrante, however, conducted some experiments in a vacuum environment with magnesium single crystal surfaces (both strained and completely annealed) and found that, in the clean state, the surfaces emitted no exoelectrons. The deformed surface and the clean surface (which was annealed) exhibited the same basic characteristics; the cleaning process or deformation process did not liberate exoelectrons. And straining one surface did not yield exoelectrons when the metal surfaces were in the clean state. Hence, deformation, in and of itself for a clean magnesium surface, does not give rise to exoelectrons. Ferrante found, however, that if he admitted a small concentration of oxygen into the system, on both the strained and the annealed crystal surfaces, an equivalent quantity of exoelectrons was emitted from the solid surfaces. With the admission of further amounts of oxygen, exoelectron emission increased. Ultimately, after some time, when the surface was covered with an oxygen layer, the exoelectron emission from the solid surface ceased (ref. 5).

In figure 9-2, the exoelectron emission measured by Ferrante is presented as a function of time where both the oxygen pressure and the exoelectrons given off by the surfaces were measured. With the initial admission of oxygen to the system there was an immediate, large liberation of exoelectrons from the surface, as indicated by the sharp spike or peak in

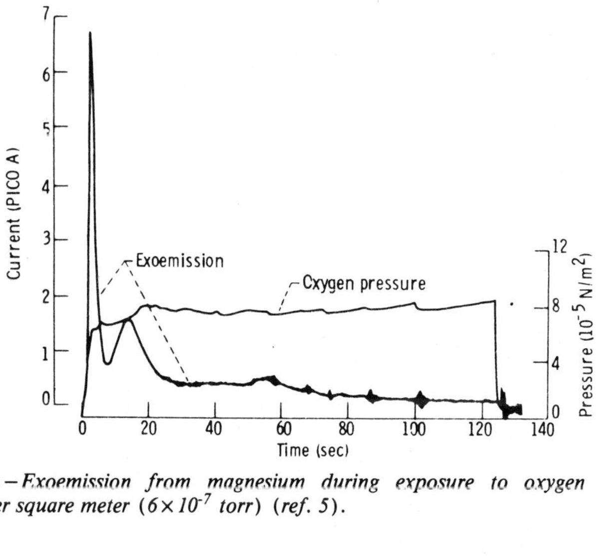

Figure 9-2. – Exoemission from magnesium during exposure to oxygen at 8×10^{-5} newton per square meter (6×10^{-7} torr) (ref. 5).

figure 9-2 near time zero. Exoemission decreased quickly. And with the continued admission of oxygen to the system, there was a second burst of exoelectrons from the solid surface that formed a smaller peak. With further additions of oxygen to the system, however, there were no changes in the exoelectrons liberated from the solid surface; in fact, the exoelectron emission began to decay with further exposure to oxygen as a function of time.

The initial, large peak associated with exoelectron emission in figure 9-2 is due to the adsorption of the oxygen on the surface of the magnesium. The chemical reaction to form the surface film liberates energy. That energy is in the form of exoelectrons emitted from the surface. Subsequently, the oxygen present on the surface rearranges to form the compound magnesium oxide, and with that, a second spurt of electrons is emitted from the surface which accounts for the second exoelectron peak observed in figure 9-2. The surface is now covered with an oxide, and the exoelectron emission from the solid surface subsides.

The Auger spectra obtained from the surface while the oxygen was admitted is presented in figure 9-3. The lower spectra is for the clean surface, showing an absence of oxygen, and the upper spectrum is for the magnesium surface once it has been covered with oxygen (as indicated in fig. 9-2).

The results of figures 9-2 and 9-3 indicate that the source of exoelectrons

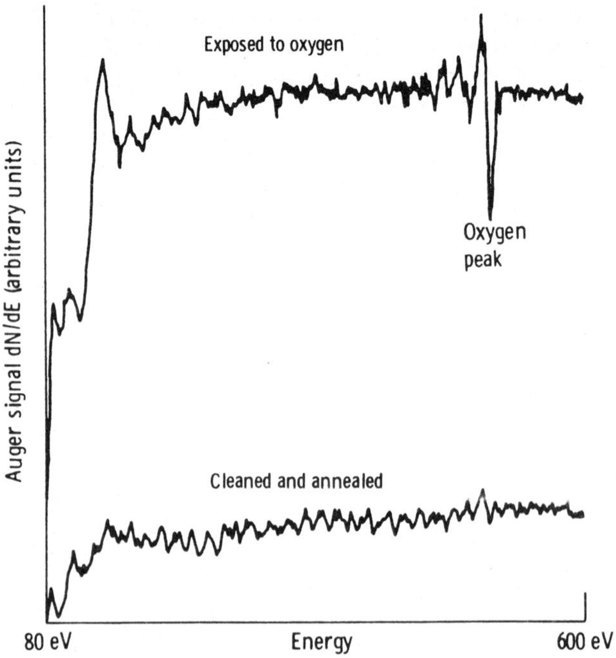

Figure 9-3.—AES spectrum of magnesium surface following cleaning and annealing at 300° C and following exposure to oxygen at 6.5×10^{-3} newton second per square meter (50 langmuirs).

from the solid surface is not the actual deformation of the solid surface but rather the interaction of the surface with the environment. In the case of the magnesium (figs. 9-2 and 9-3), the environment is oxygen, and on chemical interaction (first chemisorption and subsequently chemical reaction) there is an emission or liberation of energy in the release of exoelectrons from the solid surface.

The environment could conceivably be something other than oxygen and the same type of process may still take place. The emission of the electrons from the surface is not, as has been believed through the years by Kramer and many other investigators, a result of the deformation or the mechanical working of the solid surface and the generation of clean surface; instead, it is a result of a chemical interaction.

Roscoe Effect

A second effect, shown schematically in figure 9-1, is the Roscoe effect. Roscoe observed that when single crystals of cadmium contain an oxide layer the hardness of the metal is altered by the presence of the oxide (ref. 2). There was actually an increase in the hardness of the metal surface layers (an increase in the microhardness) in the presence of cadmium oxide. This effect has been referred to as the Roscoe effect and is manifested in tribological systems in sliding, rolling, and rubbing contact.

Much like cadminum, zinc forms an oxide which produces a surface hardening effect (ref. 6). Sliding friction experiments with zinc single crystals on the basal plane with either freshly cleaved surfaces or in an oxidized state reveal the presence of complete plastic deformation in the surface that has been unoxidized (the freshly cleaved zinc basal orientation (ref. 7)). The same sliding conditions with an oxidized (unoxidized) zinc surface results in the formation of twins. A different mechanism actually prevails at the sliding interface, and deformation is severely restricted in the presence of the oxide film.

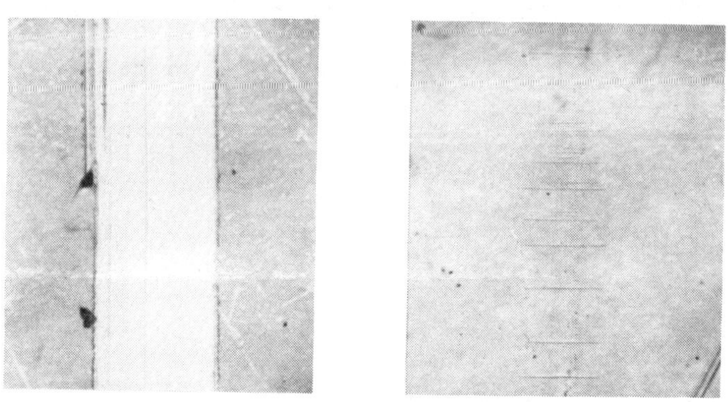

UNOXIDIZED SURFACE OXIDIZED SURFACE

Figure 9-4.—Deformation tracks developed on zinc (0001) surface in hexadecane. Load, 200 grams.

Evidence for this behavior is presented in the two photomicrographs in figure 9-4. In the left photomicrograph, a wear track is presented that was generated on the zinc single crystal after cleaving, in the absence of any oxides, with the cleaved crystal under hexadecane. The photomicrograph on the right is for an oxidized zinc surface. Note the presence of the series of fine lines which look like lamellae running normal to the track for the oxidized surface in the lower photomicrograph. These lines are actually, under high magnification, twins that develop on the zinc surface as a result of the deformation process. Note that there is not the wide wear track generated on the zinc surface in the presence of the oxide that is observed for the unoxidized surface. There is a marked reduction in the amount of deformation that takes place on the solid surface when the surface has been oxidized, because oxidation inhibits plastic deformation and strain in the metal with sliding contact. The fact that the presence of surface films increases the hardness of the materials is not restricted to oxides although Roscoe's initial observations were for cadmium oxide on cadmium and, in figure 9-4, the zinc oxide ZnO on zinc single crystal surfaces. Other species present on the solid surface can also impose a hardening effect.

Probably one of the most ductile materials used in tribological systems is gold, which is extremely ductile and prone to deform plastically. Thus, it is an ideal material to examine with respect to the Roscoe effect. Gold is not oxidized; thus, it is not possible to achieve surface hardening by an oxidation mechanism. Gold does, however, form certain halides. Gold chloride does provide surface hardening. In fact, the hardening effect of chlorine on the surface of gold is extremely pronounced. In wear experiments, the deformation track generated on the surface as a result of sliding actually undergoes fracture with cracks developing in the surfaces when the gold chloride film is present on the surface. This is not observed in the absence of a chloride film. In the absence of a chloride film, gold

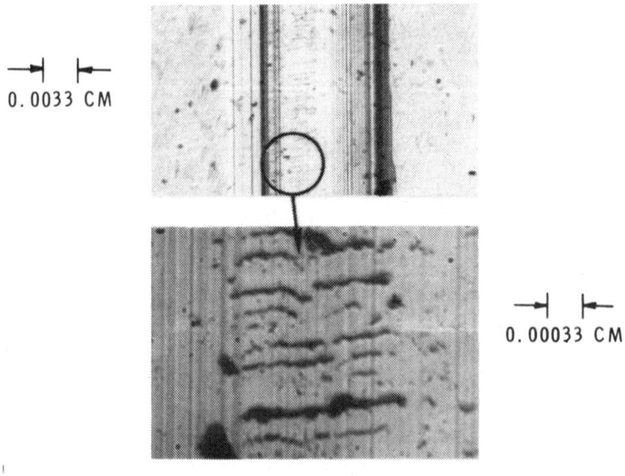

Figure 9-5.—Cracks developed in single crystal gold surface with chloride film present. Load, 200 grams; speed, 0.005 millimeter per second; temperature, 20° C; sliding direction, [110].

behaves in a very ductile manner and deforms entirely plastically with no evidence for fracture cracking.

A wear surface of a sliding wear track that was generated on the gold surface with a gold chloride film present is shown in the photomicrographs in figure 9-5. In figure 9-5 the wear track is shown at low magnification; an increase in magnification of a portion of the wear track is shown in the lower photomicrograph. A careful examination of the lower photomicrograph reveals that the surface of the track in the wear contact region contains a series of fracture cracks that initiates at the surface and moves subsurface in the material; the cracks run normal to the direction of sliding. In the absence of the chloride film on the surface, these cracks are not observed under identical sliding conditions; this indicates that the surface has been hardened by the presence of the chloride, which makes it much more likely to undergo fracture cracking than does the normally ductile gold.

Joffe Effect

Another surface mechanism presented in figure 9-1 is the Joffe effect. This effect occurs, for example, when a water film (aqueous film) is present on the surfaces of such materials as halide crystals. When water is present on potassium chloride, what would normally be a brittle material becomes extremely ductile. For example, if a bar of potassium chloride crystal is bent in dry air, the bar fractures and cracks into pieces. If that same attempted bending is accomplished in a moist atmosphere, the bar undergoes some deformation and can be bent a little before fracture occurs. If the bar is submerged in an aqueous solution, it can be bent completely plastically like a piece of taffy with no evidence of fracture in the potassium chloride. Thus, the initiation of fracture cracks in the surface is arrested by the presence of the surface active species (in this instance, water) on the solid surface. The water arrests the formation of fracture cracks which are initiated at the surface. It also inhibits the movement of cracks where the cracks are already present. This is actually a change in the hardness of the material—essentially an increase in the ductility of the halide crystal as a result of the surface film being present on the solid. While this phenomena was observed initially with alkaline halide crystals, it occurs for other materials as well. For example, both magnesium oxide and aluminum oxide are sensitive to water vapor on their solid surfaces.

An example of the Joffe effect can be seen in the data of Westbrook and Jorgensen presented in figure 9-6 (ref. 8). The hardness is plotted (fig. 9-6) as a function of indentation time for magnesium oxide at room temperature in two environments: dry toluene and moist air. An examination of the data in figure 9-6 reveals that water on the surface of the magnesium oxide appreciably reduces hardness. This change in hardness in the presence of the water vapor is not a marginal change but an extremely marked one. The differences in hardness become even greater with an increase in indentation time allowing for plastic behavior of the magnesium oxide under the indenter. Thus, where the indenter has been allowed to stand in contact with

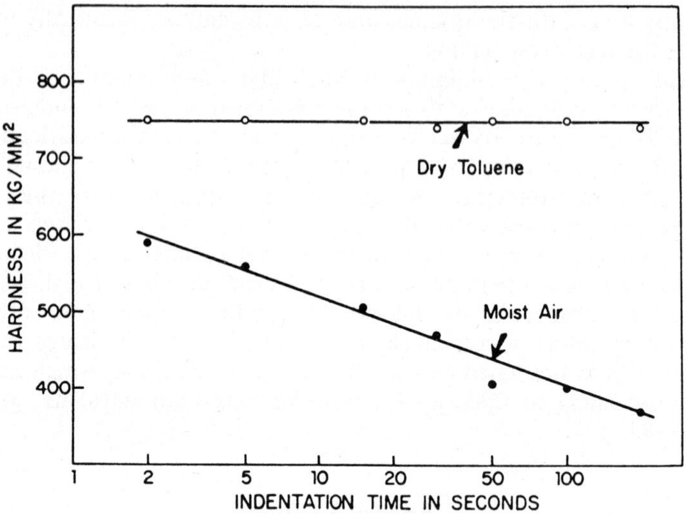

Figure 9-6. – Illustration of time dependence of microhardness of cleaved MgO in moist air. Temperature, 25° C (ref. 8).

the solid surface for more than 100 seconds, there is nearly a 100-percent reduction in the hardness of the magnesium oxide. The magnesium oxide becomes more plastic and deforms in a plasticlike manner in the presence of water vapor on its solid surface. In dry toluene, however, impressing the indenter on the surface results in fracture cracking without deformation.

The data in figure 9-6 indicate the significance of the Joffe effect in altering the mechanical behavior of solid surfaces in contact. Since the hardness of the magnesium oxide is influenced by moisture, it is only reasonable to assume that tribological behavior will, in turn, also be influenced by this change in surface mechanical behavior. The Joffe effect produces a surface softening (increase in ductility of the material) as well as reducing the formation and growth of fracture cracks and the velocity with which the cracks move in materials. The medium above the surface which acts on the solid surface to affect the solid material need not be water. Other species can accomplish the same effect. For example, oxygen has been observed, in the fracture cracking of steels, to arrest the crack growth in steels. This effect is shown by the data in figure 9-7 for the crack growth of a high strength steel as a function of the environment (ref. 9).

In figure 9-7, the crack length is plotted as a function of time that the material is exposed to two different environments: H_2 and $H_2 + 0.6$ percent O_2. The initial environment is pure hydrogen; in pure hydrogen, the crack grows with time. If one admits a small concentration of oxygen (0.6 percent) into the hydrogen environment, the growth of the crack is arrested. It only takes a short time for the crack to stop growing, but ultimately the growth rate is completely arrested in the atmosphere containing a small concentration of oxygen. If the admission of oxygen into the environment is turned off and the system is again exposed to pure hydrogen, an increase in crack length again occurs. Then, after a time, if the hydrogen is again

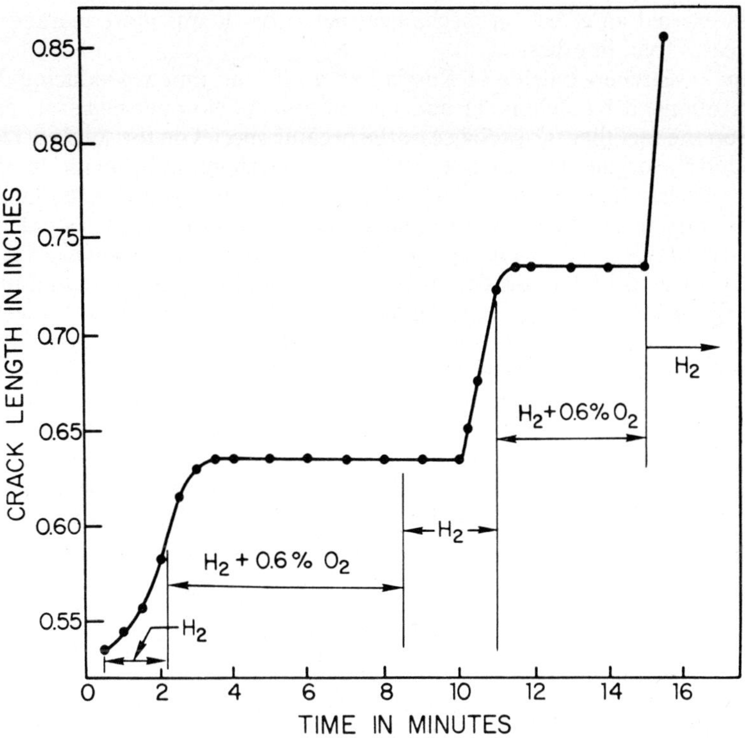

Figure 9-7. – Subcritical crack growth in high strength steel as function of environment (ref. 9).

supplemented with a small concentration of oxygen, crack growth is arrested just as before. Consequently, as shown by the data in figure 9-7, certain environmental species can arrest or inhibit the growth of cracks in materials. In the case of the high strength steel (fig. 9-7), oxygen was the effective surface active specie that inhibited crack growth. With other materials, the environment may be different but the same effect can be accomplished.

Rehbinder Effect

Another surface environment that affects the mechanical behavior of solids is the Rehbinder effect, which is named after Paul Rehbinder, the Russian scientist who first observed the phenomena (ref. 3). Rehbinder observed that if certain organic acids in vaseline oil were present on solid surfaces, the ductility of the surface was increased. The organic acids on the surfaces of metals (and nonmetals) increase the ductility or deformability of the solid surface and thus reduce the hardness in the surficial layers. The extensive research conducted by Rehbinder and his students determined that, for a host of materials and surface active species, the surface active species exerted an effect on mechanical behavior. It was more marked in some cases than in others.

Some researchers outside of Russia had a difficult time reproducing the results obtained by Rehbinder and his colleagues. No one, however, can question the fact that the presence of the organic species on the solid surface (particularly organic acids) does affect the mechanical behavior in the surface layers. It is not restricted to the presence of organic acids. Other species, such as alcohols, can produce similar effects. A classic example of the effect of adsorbed surface species on the deformability of solid surfaces is demonstrated by Grosskreutz who stripped a 3000-angstrom aluminum oxide film from aluminum metal and conducted tensile experiments on the aluminum oxide in a vacuum environment and in an ordinary air environment with moisture present. Grosskreutz found a difference in the plastic deformation characteristics of the same aluminum oxide film in the two environments (ref. 10). The film in the air environment was much more prone to plastic deformation as it strained more readily with less stress applied than did the specimen examined in the vacuum environment. In figure 9-8 tensile stress is plotted as a function of strain. The amount of

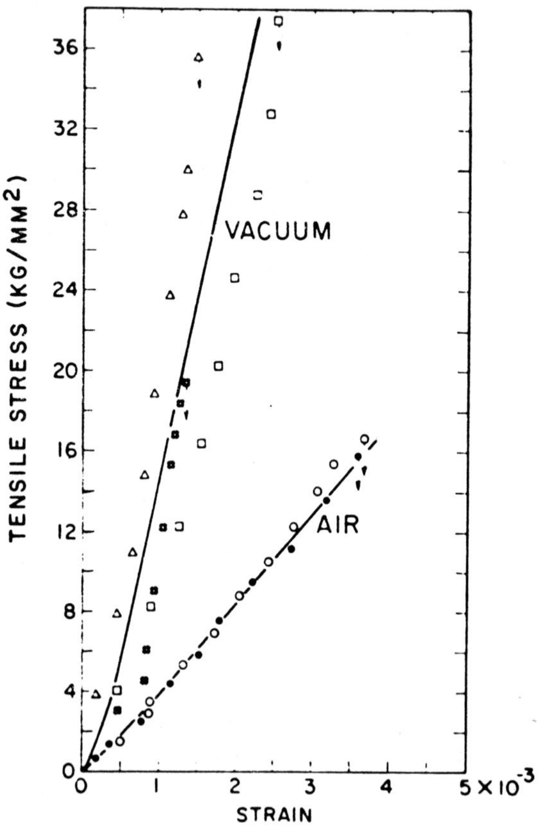

Figure 9-8.—Effects of gaseous environment on stress-strain curves of stripped 3000 Å anodic films of alumina pulled in air and in vacuum (10^{-6} torr) (ref. 10).

strain that occurred at a particular stress is much greater in the presence of air than it is in a vacuum environment; this indicates again the effect of surface adsorbates on the deformability or the strain behavior of the surfaces of materials.

Rehbinder found that the presence of organic species in solid surfaces reduced the shear strength and hardness of metals as well as nonmetals, including organic solids. He found that certain organic solid single crystals were very sensitive to the presence of hydrocarbons. Their surface strength depended on the nature of the hydrocarbons adsorbed on the solid surface of the organic crystal. During some of his experimentation, he examined the cleavage behavior of such organic crystals as naphthalene and found it to be sensitive to the presence of surface active hydrocarbons. The surface energy was reduced in varying amounts by the particular form of the hydrocarbon present on the solid surface (ref. 11). It was particularly sensitive to the chain length of the hydrocarbon adsorbed on the surface. Some of the experimental results of Rehbinder et al. are presented in figure 9-9 for measurements of free surface energy on the cleavage plane of naphthalene monocrystals in hydrocarbons and alcohols of the saturated series.

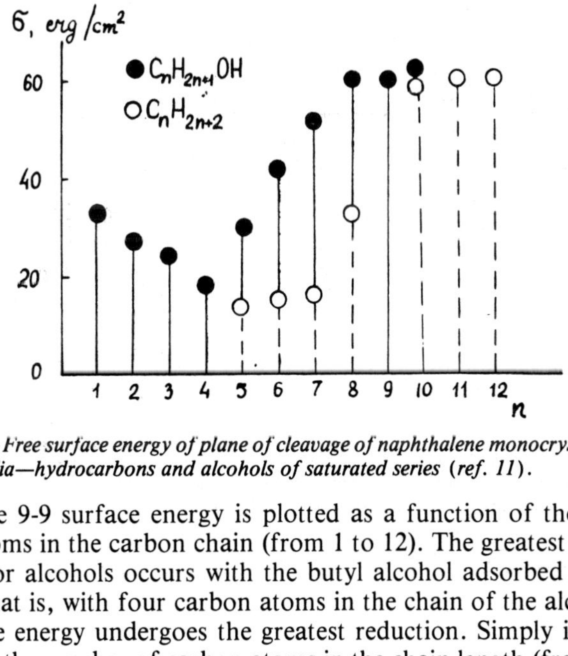

Figure 9-9. – Free surface energy of plane of cleavage of naphthalene monocrystals in various liquid media—hydrocarbons and alcohols of saturated series (ref. 11).

In figure 9-9 surface energy is plotted as a function of the number of carbon atoms in the carbon chain (from 1 to 12). The greatest reduction in strength for alcohols occurs with the butyl alcohol adsorbed on the solid surface; that is, with four carbon atoms in the chain of the alcohol group, the surface energy undergoes the greatest reduction. Simply increasing or decreasing the number of carbon atoms in the chain length (from this value of 4) results in an increase in the strength of the material.

Summation of Surface Film Effects

Some of the surface influenced mechanical properties of materials are summarized from a tribological point of view in figures 9-10 and 9-11. In

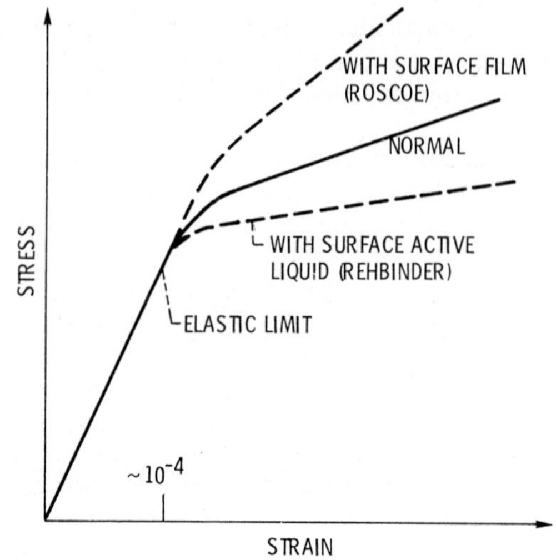

Figure 9-10. – Schematic illustration of principal extrinsic surface effects.

figure 9-10 is a typical stress-strain curve for a material in the normal state and with the surface active liquid present on the surface (to manifest the Rehbinder effect) and also with an oxide film present on the solid surface (to manifest the Roscoe effect). With the surface film present that provides surface hardening or increased strengthening with the Roscoe effect, one observes increases in stress for a given strain (fig. 9-10). With the application to the solid surface of a surface active species such as an organic acid which imparts a softening to the surface, one observes the stress strain curve for the solid surface with the surface active liquid present. Between those two is the normal surface.

If the friction and wear characteristics of surfaces exhibiting these three sets of surface conditions are measured, one observes differences in both friction and wear results. In figure 9-11 are friction and wear data as functions of load for zinc single crystals in three environments: dry sliding as cleaved (comparable to the normal condition of fig. 9-10), oxidized surface (comparable to the Roscoe effect of fig. 9-10), and in the presence of 5 percent hydrochloric acid and water (comparable to the surface active liquid of fig. 9-10). An examination of the track widths for a ruby ball sliding across a zinc single crystal surface reveals that the track width is narrowest with the oxide film present; that is, where there is maximum surface hardening, there is a minimum in the size of the track width generated. Thus, the wear track width data of figure 9-11 correlates with the stress strain data of figure 9-10. For the dry sliding case, the intermediate case, one finds that the track width is greater than that for the oxidized surface (i.e., the normal case in fig. 9-10). With the surface active liquid present on the solid, 5 percent hydrochloric acid in water, one observes the maximum track width formed in the sliding wear process. This is comparable to the stress strain curve in figure 9-10 for the surface active liquid present. Thus, the track width data of figure 11 correlate completely

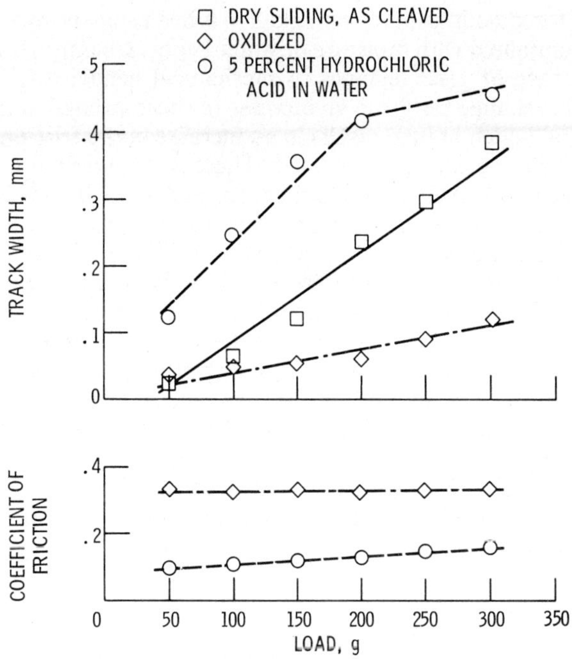

Figure 9-11. – Width of wear track and coefficient of friction produced with ruby ball sliding on zinc single crystal (0001) surface in [10$\bar{1}$0] direction. Sliding velocity, 1.4 millimeters per minute; temperature, 23° C; dry argon atmosphere.

with the stress strain data of figure 9-10, the normal condition being intermediate between the two extremes.

The coefficient of friction for the surface with the surface active liquid present gives the lowest friction because, in the sliding process, there is much less resistance to shear at the interface in the presence of the surface active liquid than there is in the presence of the surface oxide. The aqueous media can act as a lubricant to reduce the friction coefficient for the ruby in sliding contact with the zinc surface. There is some analogous behavior for materials in the presence of surface active liquids with respect to the mechanical properties of the surface layers. In general, the Rehbinder effect involves principally surface active liquids (such as the basic homologous series of hydrocarbons, their alcohols, and surface active materials such as water) and some other organic species including organic acids. The Rehbinder effect has shown increases in plasticity with the presence of the surface active material. However, not all materials behave in this manner in the presence of, say, surface hydrocarbons. For example, the examination of soda lime glass in air and in vacuum where the air contains moisture indicates markedly different behavior than might be anticipated. For most metals in a vacuum environment in the absence of surface films, the friction is higher in vacuum for the metals than it is in air. This is true for even ceramic materials such as aluminum oxide. With glass, however, opposite effects are obtained. In the vacuum environment, the friction force measured for glass in contact with glass in a clean state is lower than that for glass contaminated with moisture or water vapor. Chemisorbed water vapor on the surface of glass changes its mechanical behavior. With respect to friction, this change results in an increase in shear strength and resistance to deformation which in turn results in an increase in the friction properties of the glass in sliding contact with itself. There is stronger bonding because of the hydrogen-hydroxyl bridges that are formed at the interface. Thus, when these active surface agents are removed from the solid surface, there is relatively little interaction between the oxides of the two solid surfaces, the silicon dioxide of both solid surfaces. It has also been observed with some hydrocarbons on glasses that the presence of hydrocarbons, just as with water vapor, increases the hardness of the glass. This is the opposite effect to that which Rehbinder observed on many other materials. An example of an increase in the hardness of glass is shown in the data of figure 9-12 where the hardness is plotted as a function of the number of carbon atoms in the chain where two homologous series are employed, the hydrocarbons and the alcohols (ref. 12). From the data it can be observed that with both n-alkanes and n-alcohols there is a change in hardness with an increase in

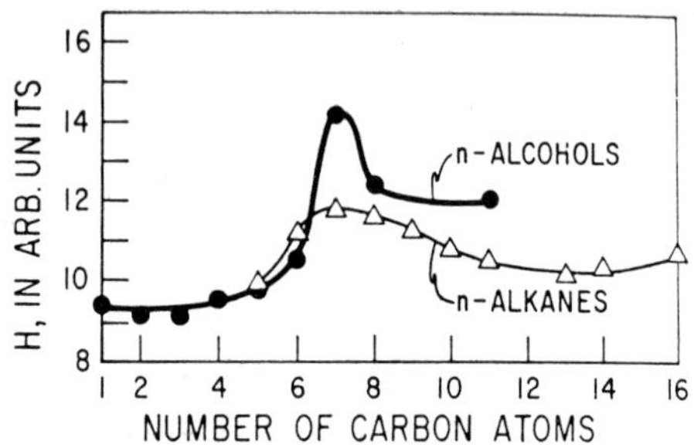

Figure 9-12.—Pendulum hardness of soda-lime glass in homologous series of hydrocarbons and alcohols (ref. 12).

the number of carbon atoms in the chain length. At approximately 6 carbon atoms for the alkanes an increase in the hardness is observed. With the alcohols, at 7 carbon atoms in the chain length, a very marked increase in the hardness is observed for the soda glass. That increase in hardness again tapers off, however, when the chain length is further increased. The results of figure 9-12 would indicate than that an opposite effect to the normal Rehbinder effect is observed in hardness measurements on soda lime glass. There are certain species present on soda lime glass that can cause an increase in its hardness; this increase is very analogous to the increase in hardness of metals produced by their oxides in the Roscoe effect.

References

1. Joffe, Adam F.: The Physics of Crystals. L. B. Loeb, ed., McGraw-Hill Book Co., Inc., 1928.

2. Roscoe, R.: The Plastic Deformation of Cadmium Single Crystals. Phil. Mag., vol. 21, 1926, pp. 399-406.
3. Rehbinder, P. A.; and Likhtman, V. I.: Effect of Surface-Active Media on Strains and Rupture in Solids. Proceedings of the Second International Congress on Surface Activity, London, no. 3, 1957, pp. 563-580.
4. Kramer, I. R.: The Effect of Surface-Active Agents on the Mechanical Behavior of Aluminum Single Crystals. Trans. AIME, vol. 221, no. 5, Oct. 1961, pp. 989-993.
5. Ferrante, J.: Exoelectron Emission from a Clean, Annealed Magnesium Single Crystal During Oxygen Adsorption. Trans. ASLE, vol. 20, no. 4, Oct. 1976, pp. 328-332.
6. Harper, S.; and Cottrell, A. H.: Surface Effects and the Plasticity of Zinc Crystals. Proc. Phys. Soc. (London), Series B, vol. 63, part 5, 1950, pp. 331-338.
7. Buckley, D. H.: Effect of Surface Films on Deformation of Zinc Single-Crystal Surface During Sliding. Trans. ASLE, vol. 15, no. 2, Apr. 1972, pp. 96-102.
8. Westbrook, J. H.; and Jorgensen, P. J.: Indentation Creep of Solids. Trans. AIME, vol. 223, no. 2, Feb. 1965, pp. 425-428. See also Westbrook, J. H.: Environment Sensitive Mechanical Behavior. A. R. C. Westwood and N. C. Stoloff, eds., Gordon and Breach, 1966, pp. 247-268.
9. Hancock, G. G.; and Johnson, H. H.: Hydrogen, Oxygen, and Subcritical Crack Growth in a High Strength Steel. Trans. AIME, vol. 236, no. 4, Apr. 1966, pp. 513-516.
10. Grosskreutz, J. C.: The Effect of Oxide Films on Dislocation-Surface Interactions in Aluminum. Surface Sci., vol. 8, 1967, pp. 173-190.
11. Rehbinder, P. A.; and Shchukin, E. D.: Surface Phenomena in Solids During Deformation and Fracture Processes. Progress in Surface Science, vol. 3, Sidney G. Davidson, ed., Pergamon Press, 1973, pp. 97-188.
12. Einsberger, F. M.: Properties of Glass Surfaces. Annual Review of Material Science, vol. 2, R. A. Huggins, ed., 1972, pp. 529-572.

CHAPTER 10

Solid Film Coatings

The concept of lubrication brings to the minds of most of the uninitiated the terms oils and greases. In addition to oils and greases, however, there are a host of solids that are used as lubricants. A solid lubricant is any solid which, when interposed between two surfaces in contact, reduces adhesion, friction, and wear between those surfaces. Thus, in a strict sense, the normal oxides present on the surfaces of most metals are solid lubricants since they prevent cold welding or gross seizure of contacting surfaces. There are a variety of other solids that lubricate well beyond the effectiveness of ordinary oxides. These include graphite, molybdenum disulfide, the dichalconides (of tungsten, molybdenum, tantalum, and niobium), metal halides, other metal sulfides, and metal phosphides. In addition, there are polymers that are used as solid components in lubrication systems; they are used as bearing retainers, gears, bushings, and in numerous other mechanical components. Useful solid polymers include polytetrafluoroethylene (Teflon or PTFE), nylon, and the polymides.

Solid lubricants are used as films in mechanical components as well as being incorporated into conventional greases. The present world consumption of greases containing molybdenum disulfide is well over 10^8 pounds per year. These greases contain from 1 to 10 percent molybdenum disulfide to improve their load carrying capacity and reduce wear. The use of solid lubricants extends from aerospace applications to ordinary household use. Solid lubricants are used to lubricate precision gears, bearings, latches, and even depolyable booms in spacecraft. Some come in aerosol cans for application on practical devices, and others require sophisticated application methods such as plasma physics deposition techniques. Hard fats such as tallow and lard were probably the first solid lubricants used. For centuries solids were used for centuries to lubricate the wheels of wagons because fluids could not be retained between the axle and the wheel. Shaft seals to contain fluid lubricants are a relatively recent invention.

Graphite and molybdenum disulfide, two of the most widely used solid lubricants today, were first used for writing rather than for lubricating. This was probably a result of their color and relative ease of transfer to surfaces. Perhaps the first use of graphite as a solid lubricant was in high temperature metal working operations such as the extrusion of steel. The first documentation of the use of molybdenum disulfide as a solid lubricant dates back several centuries. Widespread use of solid lubricants did not occur until about 1947. The literature indicates that up to 1947 there were only about 10 references to the use of solids as lubricants. From 1947 to the present day, however, the number has increased to nearly 1000. Interest in solid lubricants is so widespread that international conferences were held on this subject in 1971 and 1978.

In 1923, Dickerson and Pauling clarified the crystal structure of naturally occurring molybdenum disulfide and attributed its ease of shear and low friction properties to its crystal structure. In 1925, Bragg characterized the structure of graphite and associated its easy shear with the flaky nature of the layer lattice structure. These observations give insight into the structural effects of solid lubricants. Low shear strength solids that adhere well to solid surfaces are generally effective solid lubricants. Detailed high pressure physics studies in the 1930's revealed a number of other organic solids with shear strengths such that they became candidate lubricants. This work was extended in the 1940's. Both crystal structure and sheer properties were considered when selecting solid lubricants in the 1950's.

The object in lubrication with solids is to impose a low shear strength solid film between two surfaces in contact. If the surfaces to be lubricated are relatively hard, as is generally the case with bearings, gears, seals, and other mechanical components, the load is supported by the substrate with minimal real area of contact. When tangential motion is initiated between the surfaces, shear takes place in the lubricant film, and the friction forces are accordingly less than they would be in the absence of the solid film interposed at the interface. In lubrication, reducing wear is frequently as important as achieving low friction. Adhesive wear occurs when the adhesive junction is stronger than cohesion in the weaker of the two materials in contact. With tangential motion under such conditions, subsurface fracture or shear can occur and result in transfer from one surface to another. A solid lubricant reduces the strength of the solid to solid junction and, therefore, reduces adhesive wear. Abrasive wear occurs, as was mentioned earlier in the text, when hard particles move under a load across the surface of a softer substrate. Materials are then removed by the cutting action of the harder substance. The ease of shear in solid lubricants reduces the abrasion by reducing the probability of abrasive particle formation.

Corrosive wear, as was also mentioned earlier, occurs when constituents of the environment interact successively with contacting surfaces. Protective solid films, such as gold, applied to surfaces can inhibit this type of wear. It is apparent that one of the requirements for a solid to function as a solid film lubricant is that it have a relatively low shear strength. The shear strength of a number of solids that are potentially useful as lubricants has been reported. In general, within a particular class of material, the friction coefficient correlates with the shear strength of the solid. In figure 10-1, the shear strengths of cadmium chloride, cadmium bromide, and cadmium iodide are presented as functions of applied pressure (ref. 1). Since the shear strength increases with the increase in applied pressure, it might be anticipated that the friction would increase with applied load when these materials are present on solid surfaces. Friction coefficients for the three halides of cadmium are presented in figure 10-1. At a fixed load, the values presented above the curves indicate that the lower the shear strength of the solid, the lower the friction coefficient.

In addition to having a low shear strength, the solid must be able to adhere to the surface to be lubricated. Adherence to the surface may be chemical or mechanical in nature. Thus, when extreme pressure additives are placed in oils, these additives react chemically to form compounds with the metal of the surface to be lubricated. Many addititives contain chlorine, phosphorus, or sulfur so that the interaction of the additives with the surface results, for example, in the formation of the chlorides, phosphates, or sulfides of iron. These compounds prevent adhesion of the surfaces at high loads where oils may be squeezed out of the contact zone. Adherence of the solid film lubricant to the surface can frequently be simply mechanical in nature. For example, with PTFE in sliding or rubbing contact with a metal surface, the metal asperity shears the polymer and leaves particles buried in the valleys between asperities. The surface topography can be sufficiently regular to provide mechanical interlocking, a phenomenon exemplified by passing a file over a solid. The file becomes charged with particles that are simply mechanically bound to the file surfaces preventing further filing action. Even with solids such as PTFE (with low shear strength and low surface energy) chemical bonding to a metal surface can occur if the metal is atomically clean. Adhesion studies in the field ion microscope (FIM) indicate that such bonding of PTFE to metal surfaces is actually achieved. This can occur on simple touch contact of the

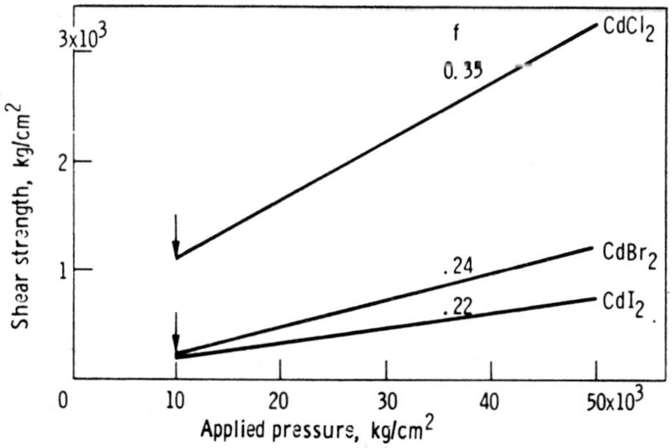

Figure 10-1. – Shear properties of layer-lattice cadmium halides.

polymer to the metal surface. With layer lattice compounds such as graphite and molybdenum disulfide, it has been suggested the mechanism responsible for adhesion is that the sharp edges of the crystallites inherent in the crystal geometry embed in the surface to be lubricated.

When the adhesion of a solid lubricant is not good, techniques must be used to achieve bonding to the surface. These include burnishing, surface pretreatments, and the use of binders. Unlike liquid lubricants that are mobile and can be continuously moved into the contact zone, solids are generally immobile. They have a finite life in the contact zone that is frequently a function of how well they adhere to the surface to be lubricated.

In many practical applications, solid lubricants are used because high temperatures are involved and there is a need for good thermal stability. Solid lubricants such as graphite and molybdenum disulfide are stable in air up to 400° C. Graphite begins to oxidize appreciably above that temperature, principally to carbon dioxide. Molybdenum disulfide oxidizes to molybdenum trioxide (MoO_3). In vacuum, molybdenum disulfide is stable to 750° C. Above 400° C, certain ceramic coatings provide effective boundary lubrication. For example, enamels containing fluorides of the alkaline earth metals lubricate to 1000° C.

Other properties important to solid lubricants are chemical inertness, good ductility, high melting point, good electrical conductivity (for certain applications), and corrosion preventive ability. The solid lubricant must also be capable of being prepared in high purity. Frequently, small amounts of impurity can adversely affect the lubricating characteristics of a solid. Silica, for example, is an impurity found in molybdenum disulfide. Figure 10-2 indicates the effects of various concentrations of silica on the friction

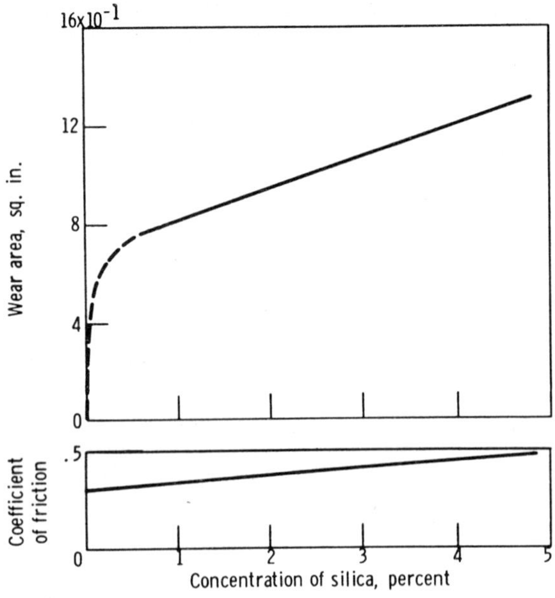

Figure 10-2. – Effect of silica on pure MoS_2 film on steel.

TABLE 10-I. – CLASSES OF SOLID LUBRICANTS AND EXAMPLES

Class	Example
I. Inorganic compounds	
(a) Laminar solids	Graphite, MoS_2, CF_n
(b) Nonlaminar solids	PbO, CaF_2
(c) Soft metals	Pb, Sn, In, Au, Ag, Cd
II. Organic compounds	
(a) Fats, soaps, waxes	Tallow, stearic acid
(b) Polymers	PTFE, polyimide
(c) Thermally stable	Phthalocyinine

characteristics of molybdenum disulfide used to lubricate steel. While the change in the friction coefficient with increasing concentration of silica in molybdenum disulfide to 5 percent is very small, the wear is markedly affected by the presence of the silica, the silica being abrasive to the steel surface (fig. 10-2).

Solid lubricants can be classified according to their properties. Classes of solid lubricants and some examples of each are indicated in table 10-I. The solids that are most widely used as lubricants are the layer laminar (layer lattice) solids. Of these, graphite and molybdenum sulfide have received the most attention.

Graphite and Molybdenum Disulfide

Graphite as used in lubrication has the laminar planar hexagonal crystal structure shown schematically in figure 10-3. This structural characteristic of graphite provides it with many anisotropic properties. Since each carbon atom can be bonded to four other carbon atoms, the carbon to carbon bonding within each plane is relatively strong. In the absence of contaminants between layers graphite does not lubricate; the contaminants, which are principally water and hydrocarbons, desorb fairly readily. For

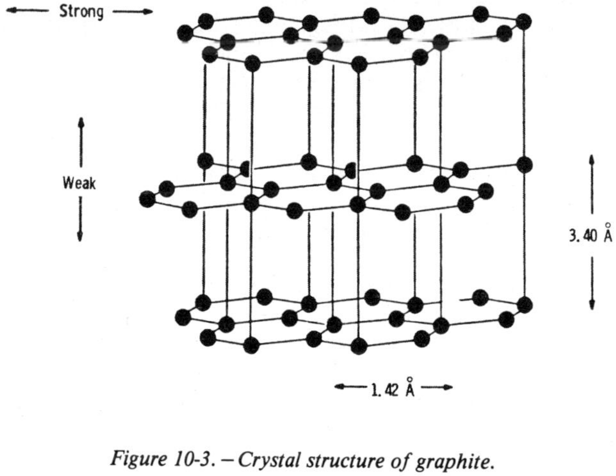

Figure 10-3. – Crystal structure of graphite.

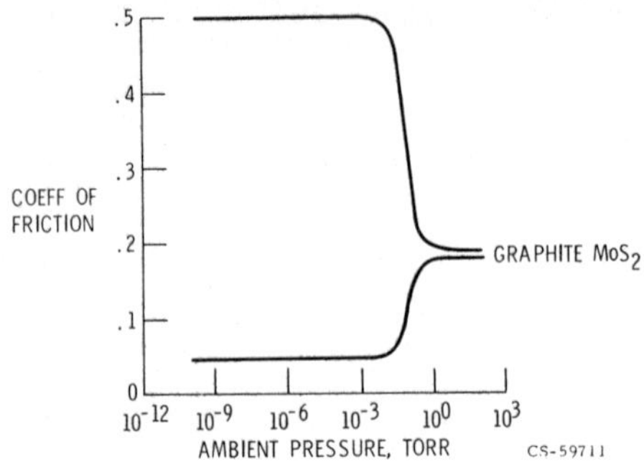

Figure 10-4. – Effect of ambient pressure on friction of graphite and MoS_2.

this reason graphite does not lubricate in a hard vacuum. This effect is demonstrated by the data in figure 10-4 where friction coefficient is plotted as a function of ambient pressure for both graphite and MoS_2. In a vacuum of 10^{-10} torr graphite exhibits a relatively high friction coefficient (0.5). As the pressure is increased (toward atmospheric), at approximately 10^{0} torr, the friction coefficient begins to drop drastically with the admission to the system of air, which adsorbs to the surface of the graphite. The friction coefficient of graphite drops to about 0.2 near atmospheric pressure. In contrast to graphite, however, the friction characteristics of molybdenum disulfide improve in a vacuum environment (fig. 10-4). As contaminants present on the surface of molybdenum disulfide are desorbed in a vacuum system, there is also a decrease in friction coefficient. The friction coefficient begins to decrease at pressures of approximately 10^{-2} torr and reduces to about 0.04 at pressures of 10^{-10} torr. Thus, while graphite is a poor lubricant for vacuum applications, molybdenum disulfide is an extremely good one, exhibiting friction coefficients superior to those of effective lubricating oils which characteristically exhibit friction coefficients of approximately 0.1.

The fact that a solid has a layer lattice (laminar) crystal structure does not insure lubricating qualities. Although both boron nitride and mica have layer lattice structures, neither is a solid lubricant. Their adherence to the surface to be lubricated is poor. When graphite is used as a lubricant, usually it is in contact with a metal surface. There is evidence that a residual metal oxide film must be present on the metal surface in order to develop a graphite film; this was discussed earlier in this text. In the absence of these oxides, graphite does not transfer to the metal surface; both friction and wear are higher in the absence of a transfer film than they are in the presence of such a film.

Molybdenum disulfide, like graphite, has a hexagonal crystal structure in its lubricating form, as indicated in figures 10-5 and 10-6. The distance between the adjacent sulphur layers is greater than the thickness of the layers themselves. This accounts for the easy shear and good lubricating characteristics of molybdenum disulfide. Unlike graphite, molybdenum disulfide does not depend on the presence of adsorbates for its lubricating characteristics. In fact, molybdenum disulfide lubricates better in the

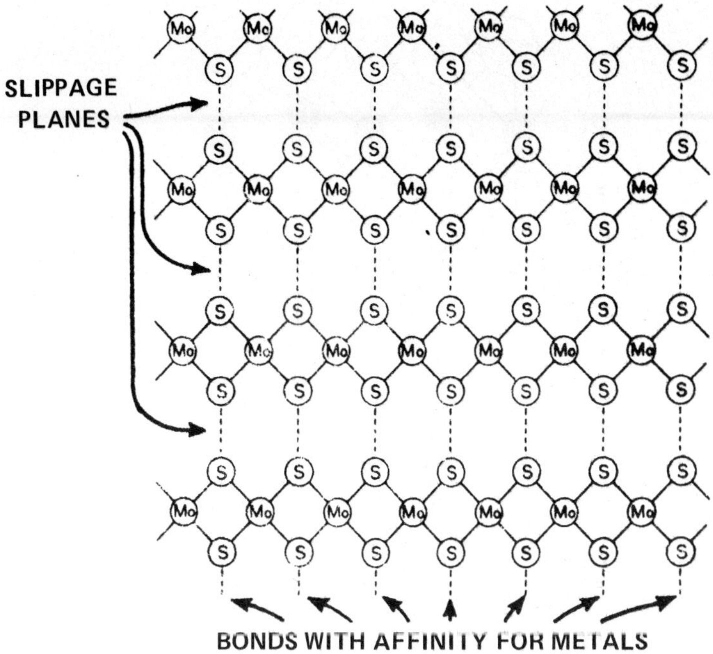

Schematic sectional diagram of MoS_2 lattice structure

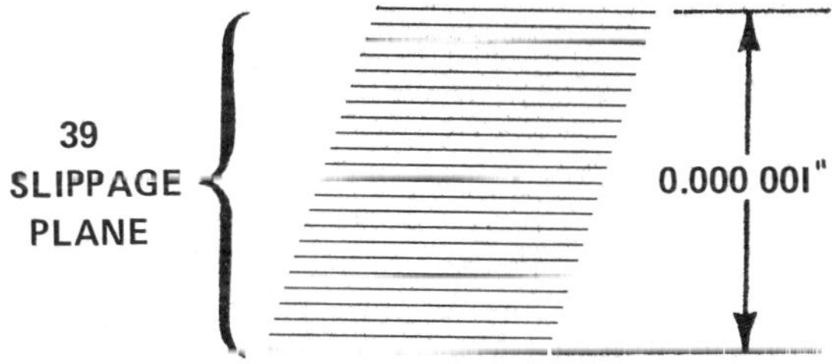

Slippage planes in a particle of Moly-Sulfide.

Figure 10-5. – MoS_2 crystal lattice.

absence of these adsorb surface films (fig. 10-4), as has already been discussed. Much like cadmium iodide, there exist weak van der Waals forces between sulphur atoms and adjacent layers. The presence of absorbates such as water vapor increases the friction coefficient obtained with molybdenum disulfide. Wear also is observed to increase with an increase in the concentration of water in the environment.

Because of the crystal structure of graphite and molybdenum disulfide, both solids orient rapidly when rubbed on surfaces. The basal planes become essentially parallel to the substrate surface which facilitates easy

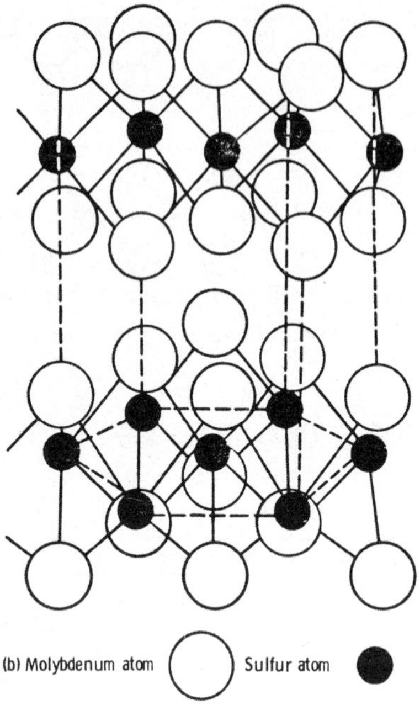

Figure 10-6. – Crystal structure of molybdenum disulfide.

shear and, accordingly, low friction. Both friction and wear with these solids are affected by orientation. With the basal plane of graphite normal to the interface, the rate of wear is high. When the basal plane is parallel to the interface, the rate of wear decreases for graphite in contact with metal surfaces such as copper.

The friction coefficient obtained with molybdenum disulfide varies considerably with orientation. With a basal plane normal to the substrate surface, a friction coefficient of 0.26 is obtained on a steel surface. When the basal plane is oriented parallel to the sliding interface, the friction coefficient decreases to 0.10. Frequently, when a molybdenum disulfide coating is first rubbed on the surface the friction coefficient decreases after a number of passes have been made. This decrease is associated with the transition from a randomly oriented molybdenum disulfide crystallite to orientation with the crystallites nearly parallel to the surface.

The effect of mechanical parameters on the lubricating effectiveness of graphite and molybdenum disulfide are markedly different. With an increase in sliding speed, for example, the friction coefficient for graphite in air increases, while that for molybdenum disulfide decreases. With graphite, an increase in the friction coefficient is observed when the load is increased; in contrast, an increase in load generally results in a decrease in the friction coefficient observed with molybdenum disulfide as indicated by the data in figure 10-7 (ref. 2).

Graphite is a good lubricant at room temperature and at 500° C but not at intermediate temperatures. Its high temperature lubricating

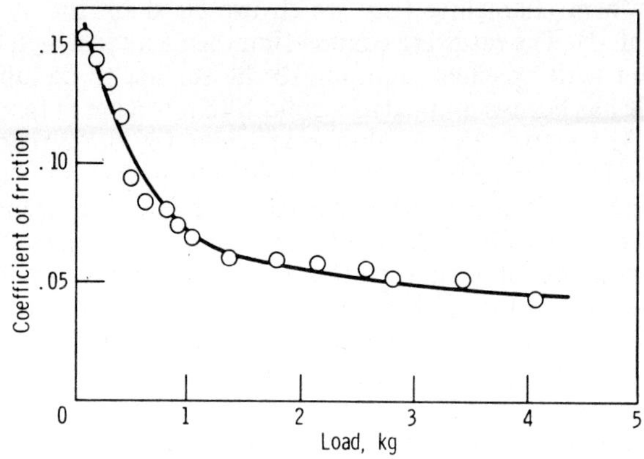

Figure 10-7. – Variation of friction coefficient with load for MoS_2 film on Cr (ref. 2).

characteristics gave rise to its use in metal forming. The explanation for this behavior is that, at room temperature, environmental contaminants separate graphite lamellae. These materials desorb above room temperature but, when the temperature is sufficiently high, the oxidation rate of the metal substrate is such that the oxides aid in the lubrication process. With molybdenum disulfide, since contaminants are detrimental to lubricating effectiveness, the temperature increase from 20° to 100° C yields a reduction in friction coefficient. Beyond that temperature, friction characteristics remain relatively unchanged until severe oxidation of molybdenum disulfide occurs. Oxidation of molybdenum disulfide is low in air below 370° C but increases rapidly above that temperature. Lubricating effectiveness of molybdenum disulfide is not lost as long as there remains some unoxidized molybdenum disulfide on the surface. Particle size also has an influence on the oxidation rate as might be anticipated. The oxidation rate from X-ray diffraction studies indicates that, as might be anticipated, the smaller the particle size the higher the oxidation rate.

Other Types of Solid Lubricants

The dichalconides of molybdenum, tungsten, tantalum, and niobium are also solid lubricants. Tungsten disulfide resists oxidation better than molybdenum disulfide and molybdenum diselenide. Because it has better electrical conductivity than many of the other dichalconides, it is used in electrical contacts.

Metal halides that act as solid lubricants include cadmium iodide, cadmium chloride, cadmium bromide, cobalt chloride, lead iodide, and mercuric iodide. A problem encountered with these materials is the corrosiveness of their structures. Many inorganics have low shear strengths and therefore act as solid lubricants even though they do not have a layer lattice structure. Lead oxide, cadmium oxide, and boric oxide are examples. Lead oxide and calcium fluoride have been used with other oxides and fluorides to form enameling frits which are fired on the surfaces to be lubricated (ref. 3). The resulting enamel film then acts as a high temperature solid lubricant with excellent adhesion to the surface to be lubricated.

Boric oxide has been examined as a solid film lubricant at temperatures to 650° C. In figure 10-8, the friction coefficient for boric oxide (B_2O_3) is presented as a function of temperature from 400° to 600° C. At 400° C, the friction coefficient obtained with boric oxide is high. Near its melting point, the friction coefficient of boric oxide decreases to less than 0.10, a value frequently encountered with effective liquids as boundary lubricants at room temperature. In figure 10-8, specimens with two geometries were used in making the friction measurements. Both gave essentially the same result. The viscosity of boric oxide at 450° C is approximately 10^6 poise while at 600° C it is only 2×10^2 poise. This marked reduction in viscosity results in a decrease in the force necessary for viscous shear and consequently a reduction in the friction coefficient.

In addition to nonlamellar solids and the lamellar solids, there are soft metals that can act as very effective solid film lubricants. When thin films of these soft metals (such as lead, cadmium, tin, indium, silver, or gold) are applied to hard substrates, the films provide effective boundary lubrication. The load between the surfaces is supported by the hard substrate, and with tangential motion, shear takes place in the soft, low shear strength metal

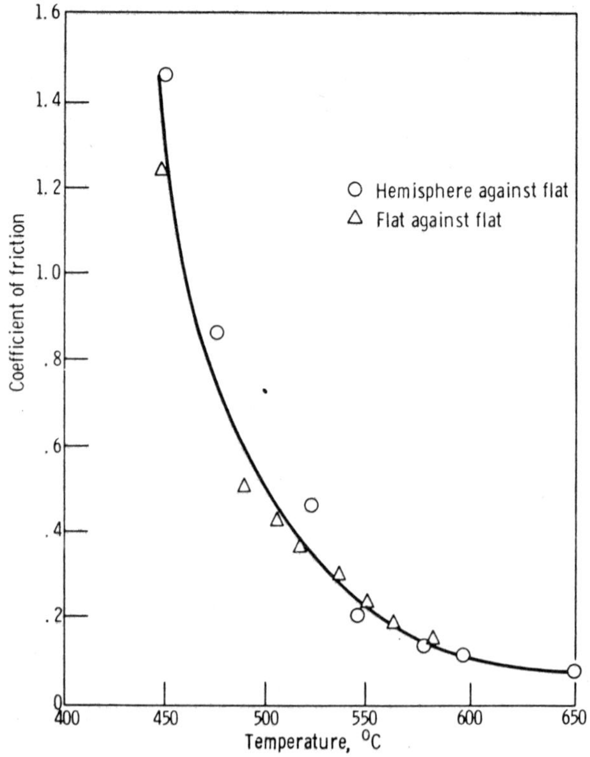

Figure 10-8. – Variation of friction coefficient for boric oxide with temperature (ref. 4).

film. Such films, while generally exhibiting higher friction characteristics than those experienced by lubricating with the layer laminar solids such as molybdenum disulfide, are still very useful where both lubrication and corrosion protection are required.

With soft metal films, film thickness is extremely important. The data of figure 10-9 were obtained with a hard steel slider on a hard steel plate coated with indium (ref. 5). With thin films, the friction coefficient was high because some shearing of the steel as well as of indium is occurring. At a thickness of 10^{-4} to 10^{-5} centimeter, the film gives the most effective lubrication. When the film thickness is increased, it contributes to friction and the coefficient of friction also increases.

In addition to soft metal films (laminar and nonlaminar solid lubricants), other materials (fats, soaps, waxes, and organic acids) have been used to a great extent as solid film lubricants. A large quantity of metallic soaps are used as thickners—for example, in greases. Tallow and lard, the earliest solid lubricants, are still used in some specialized areas today as are long chain paraffinic acids, alcohols, and esters. For example, stearates of calcium, magnesium, and sodium are widely used as dry powders in wire drawing. With the long chain acids, alcohols, and esters, the functional groups attach to the metal surface with the aliphatic chain extending up from the metal surface. This orientation leads to two phenomena: first, it provides close molecular packing on the metal surface and thereby minimizes metal to metal contact; second, the friction decreases with an increase in chain length because the separation of the two surfaces to be lubricated increases with increasing chain length.

While organic solids give low friction values, they generally cannot be used at high loads or at temperatures above their melting points. The effect on friction of the molecular weight or chain length of a straight chain paraffinic soap on the steel surface is shown in figure 10-10 (ref. 6). There is a direct relationship between the static friction coefficient and the molecular weight of the lubricant or, as it were, the chain length of the paraffinic hydrocarbon.

Polymeric materials are also widely used as solid film lubricants.

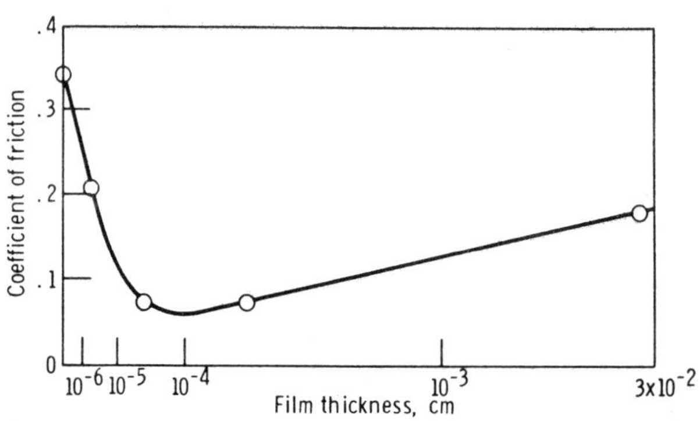

Figure 10-9. – Variation of friction coefficient with thickness of indium plate on hard steel (ref. 5).

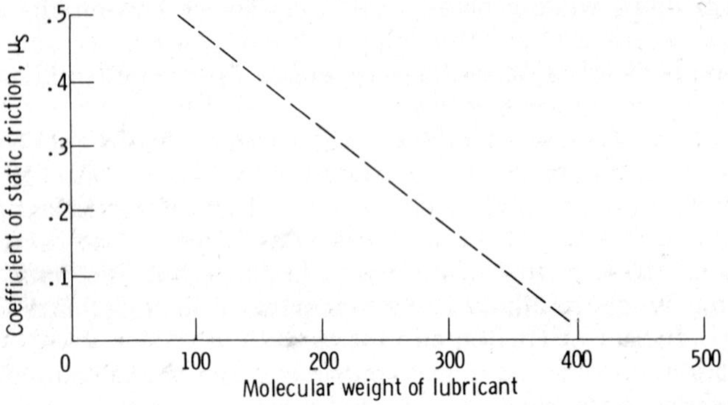

Figure 10-10. – *Coefficient of friction as function of molecular weight of straight-chain paraffin soaps on steel.*

Probably next to the layer lattice or laminar solids polymers have seen the most extensive use as solid lubricants. Materials such as PTFE and the polyimides are used as solid bodies; in addition, these materials are used as films. Since polyimides have good mechanical strength and rigidity, they can be fabricated into bearing retainers, gears, seals, and valve seats. Materials such as PTFE, however, require the incorporation of fillers to improve dimensional stability. Glass fibers or metals, in either powder or fiber form, are used for this purpose.

The interfacial frictional energy generated can result in polymer degradation and fillers influence degradation as indicated by the data of figure 10-11. In figure 10-11(a), the concentration of PTFE degradation fragments is seen to increase with sliding velocity for the glass-filled PTFE.

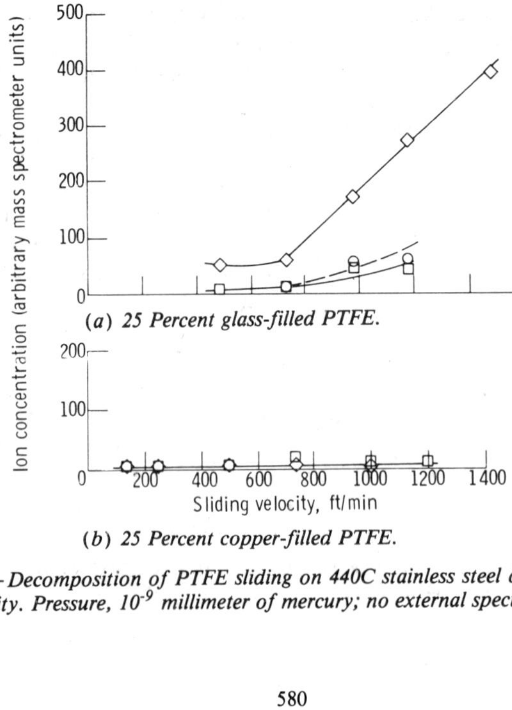

Figure 10-11. – *Decomposition of PTFE sliding on 440C stainless steel disk as function of sliding velocity. Pressure,* 10^{-9} *millimeter of mercury; no external specimen heating.*

When copper is used as the filler, the amount of degradation does not increase.

Remarkably low friction coefficients can be obtained with polymers such as PTFE. Friction coefficients from 0.01 to 0.04 have been achieved from temperatures as low as that of liquid hydrogen to the decomposition temperature of the polymer. There are also some dyes, such as the phenanthrenes and the phthalocyanines, that have been examined as high temperature lubricants. Phthalocyanines (metal free or copper) have been the most extensively used. In general, their lubricating properties are inferior to molybdenum disulfide but are, under certain conditions, superior to graphite. Their biggest single use has been as a thickener in high temperature greases. Bonding the phthalocyanines to metal surfaces is by the formation of chelates.

Solid lubricants are used and applied to surfaces in a variety of ways. One of the principle uses of solid lubricants is as dispersions in oils and greases. Until recently, colloidal graphite was one of the most popular additives; molybdenum disulfide is now much more widely used. In studies with diester greases, the presence of molybdenum disulfide increases the oxidation rate of the grease; this is especially the case when the molybdenum disulfide is of small particle size. Solid lubricants are sometimes applied as powders. While this technique has proven effective in metal forming and extrusion, it is ineffective when the loose powders can be easily pushed out of the contact zone. To overcome this problem, films, such as molybdenum disulfide, are developed by mechanical burnishing. The loose powder is applied to the surface by a brush or polishing cloth. Some adhesion is accomplished.

Another technique is to use resins to make compacts of the solid lubricant material. The compact is then made to contact the surface. If the compact is kept in continuous contact, a solid lubricant film is constantly being regenerated. This technique has been effectively used in metallic retainers to lubricate ball bearings. Some of the lubricants can also be formed by chemical and electrochemical methods. Metal oxides and sulfides chemically formed on metal surfaces reduce friction as evidenced by the data in table 10-II. Both oxides and sulfides reduce the static friction coefficients of steel, copper, and brass. Halogen containing gases such as monochlorotrifluoromethane, dichlorodifluoroethane, and sulfur hexofluoride have been effectively used to generate high temperature solid lubricant films by reaction. While these gases may be thermally stable at ambient temperature, frictional heating at the contacting metal interface is sufficent to decompose the molecule locally and thus liberate the reactant to form a metal halide, which acts as a solid lubricant.

When dry surface films are desired, the most common method of film application is to use binders, frequently resins (either air drying or heat cured). Phenolics, epoxies, silicones and polyimides are good binders. Although the ratio of lubricant to binder is typically 1 to 1 by volume, it varies with lubricant, resin, and lubricant particle size. The mixture is usually sprayed on the component. When the resin requires heating to cure it, the film is baked. In some applications, organic resins are undesirable and inorganic binders are used—for example, sodium silicate. An aqueous

TABLE 10-II. – FRICTION-REDUCING EFFECT OF SOLID FILMS ON STEEL, BRASS, AND COPPER

No.	Metal combination	Treatment of surfaces	Static coefficient of friction	
			Untreated surfaces	Treated surfaces
1.	Steel-steel	Oxide	0.78	0.27
2.	Steel-steel	Sulphide	0.78	0.39
3.	Brass-brass	Sulphide	0.88	0.57
4.	Copper-copper	Sulphide	1.21	0.74
5.	Copper-copper	Oxide	1.21	0.76
6.	Steel-steel	White oil	0.78	0.32
7.	Steel-steel	Oxide, white oil	0.78	0.19
8.	Steel-steel	Sulphide, white oil	0.78	0.16

solution of the lubricant and binder is sprayed on the surface, and the water is evaporated by heating. This type of coating is particularly useful in vacuum applications.

Defining Solid Film Lubricants by Using Plasma Physics

In recent years, a number of new techniques have been developed which involve using plasma physics to deposit the solid film lubricants. These techniques include ion implantation, ion plating, and sputtering. The sputtering process is used to deposit hard face coatings (for improved wear resistance) and soft, low shear strength films (for use as solid film lubricants). The next sections deal with these plasma deposition techniques. Special attention is being given to these techniques because they are becoming more and more popular as a way of depositing adherent, dense, and reliable solid film lubricant coatings.

Ion Implantation

Ion implantation is really not a technique for developing coatings on surfaces but rather a way of treating the surface layers of an already existing solid component so as to alter its adhesion, friction, and wear behavior by burying ions (which have inherently low friction and wear characteristics) in the solid surface layers. The implantation of ions of a desired species is accomplished, as it were, by impacting the surface layers and thereby altering the chemistry of the solid surface (since they are available at and near the solid surface). In ion implantation, ions are produced by a variety of means in an ion source. These generated ions are accelerated by an electric field toward the part to be implanted—in tribology, the tribological surface. The energies associated with the incoming ions are usually between 10 000 and 200 000 electron volts. Since these ions most often have a single electronic charge, these valves correspond to the ionic voltages required. A lower limit must be set for the energies because sufficient energy is required

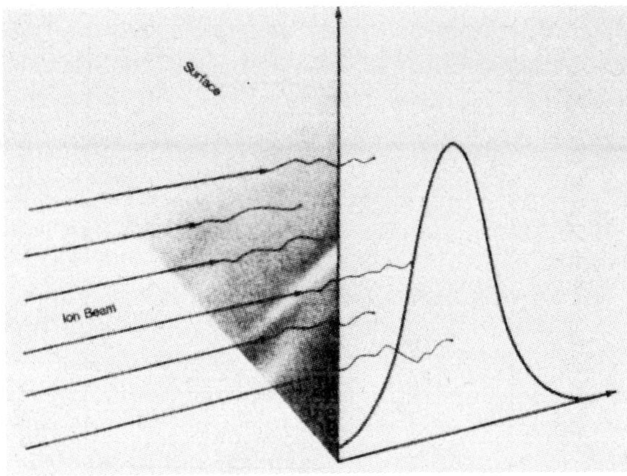

Figure 10-12. – Schematic diagram illustrating penetration of ions implanted in metal surface and resulting depth distribution of implanted atoms (ref. 7).

for the ions to penetrate thin oxide layers normally present on metallic surfaces. The process of ion implantation is generally carried out in a relatively modest vacuum, typically of the order of 10^{-5} or 10^{-6} torr.

The ions penetrate the solid surface as indicated schematically in figure 10-12 (ref. 7). The ions become buried in the surface to various depths depending on the distribution of their energies. All the ions coming to or arriving at the surface are not at the same energy level. The variation in energies which the ions carry with them determines the depth to which they become buried in the surface; the distribution of the implanted species varies in the surface layers based on the energies of the incoming ions. Figure 10-12 shows the distribution of energies as indicated by the depth of implantation.

Ions in the concentration range of, typically, 10^{15} to 10^{18} ions per square centimeter are employed. The gaseous species used to generate the ions include a variety of materials that are known to impart good adhesion, friction, and wear characteristics to solid surfaces. These might include such elements as nitrogen, oxygen, boron, and even metallic ions such as lead. Dearnaley has used ion implantation as a way of depositing implanted species which would reduce friction and wear characteristics in materials in solid-state contact (ref. 7). They have studied stainless steels in sliding contact with steel disk surfaces that have been implanted with nitrogen ions, and they have measured the effect of the implanted ions on the wear behavior of the stainless steel in contact with the steel surface (ref. 7). Some of their results are presented in figure 10-13. Figure 10-13 shows the wear rate in cubic centimeters per centimeter of sliding distance plotted as a function of load in kilograms for the unimplanted and the implanted steel disk surfaces. A concentration of 2×10^{17} nitrogen ions per square centimeter was used at an applied voltage of 50 000 volts. Examining these data reveals that implanting nitrogen ions in the surface layers gives improved wear resistance to the surface.

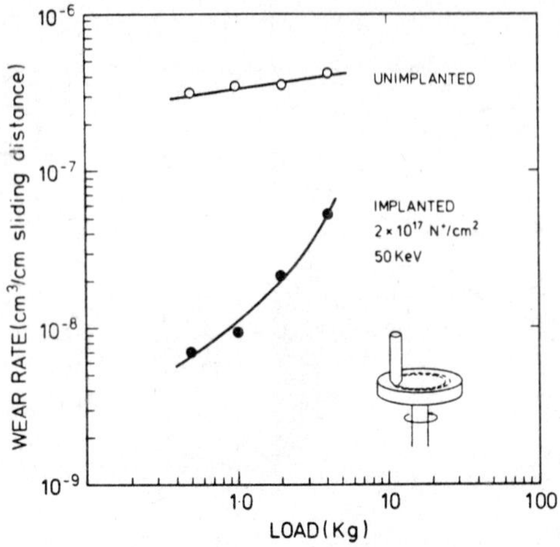

Figure 10-13. – Volumetric wear between stainless steel pin and steel disk implanted with nitrogen ions (2×10^{17} ions/cm^2 at 50 keV) as function of applied load (ref. 7).

Ion Plating

A technique that is rapidly expanding in its use for applying thin solid film lubricants, particularly metals, is the system known as ion plating. Ion plating involves a plasma physics approach to the deposition of films. In contrast to ion implantation, which operates at relatively high energies (above 10 keV), ordinary ion plating operates typically in the low voltage range (1 to 5 keV). There is much less energy involved in the ion plating process, ion plating is much simpler to use, and the equipment required is nowhere as costly as that for ion implantation. Also, it has an outstanding versatility for applying thin, uniform, dense, and adherent solid film lubricants to all types of surfaces. Complex geometric surfaces can be ion plated very effectively with very thin films in a controlled manner.

The tribological coatings applied by ion plating are superior to those applied by conventional electro-deposition techniques. The mechanism for ion plating was originally developed by Maddox (ref. 8). The basic system in its mode of operation is shown schematically in figure 10-14. In ion plating, the surface to be coated, the substrate, is mounted on a pedestal inside a vacuum chamber. A vacuum system is required, but a relatively crude vacuum, much like ion implantation, of the order of 10^{-5} or 10^{-6} torr is satisfactory. An evaporator filament source is also positioned inside the vacuum chamber just beneath the substrate to be coated. Around this substrate is a grounded shield. A variable leak valve is also provided inside the vacuum chamber to bleed gases into the chamber. A high voltage power supply is required to apply the voltage to the surface to be coated, and a filament power supply is used to supply current to the evaporator or filament source.

The vacuum chamber is evacuated and the chamber is backfilled with argon gas or some other inert gas to a typical pressure of 15 to 20×10^{-3} torr. A high voltage is applied to the system between the substrate surface and the evaporant filament source; the evaporant filament source is the anode, and

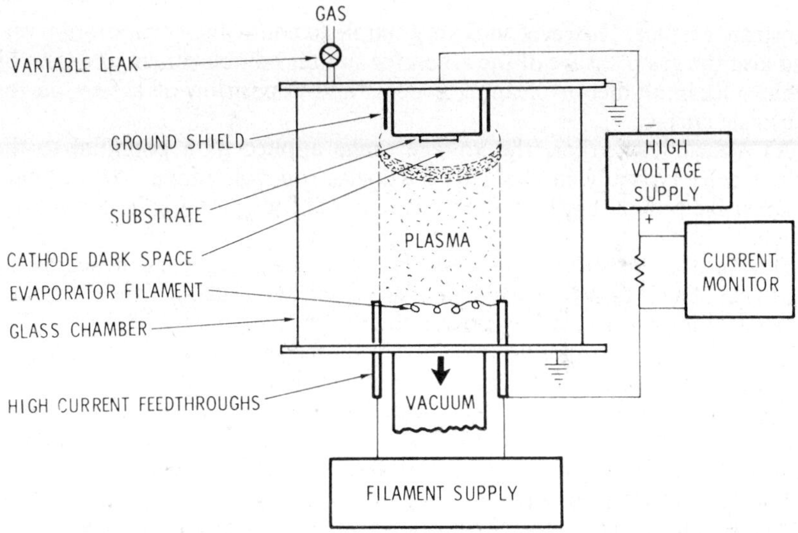

Figure 10-14. – Simple ion plating setup using dc discharge and thermal evaporation as material source.

the substrate is the cathode (negatively charged surface). With the generation of positively charged argon ions inside the vacuum chamber, the substrate is bombarded with argon ions with a conventional or straightforward argon ion bombardment technique. The argon ion bombardment of the surface causes the sputter removal of contaminants from the substrate surface including adsorbates and oxide layers (with metal or alloy substrates). While the substrate is being cleaned by the argon plasma and after the cleaning has been completed, the material that is to be deposited on the substrate is contained on the evaporant filament source; that is, some of the material to be deposited on the substrate is applied to the evaporant filament. The evaporator filament is heated by the filament power supply until the filament is hot enough to evaporate the material to be deposited. This material evaporates from the evaporator filament into the argon plasma, which is located between the filament and the substrate. Metals have lower ionization potentials than argon gas. Thus, when a metal evaporates from the evaporator filament source, the metal immediately becomes ionized in the argon plasma. Since there is a potential between the substrate and the anode, the ions move to the substrate. They have energy when they arrive at the substrate surface.

While the sputter cleaning process is going on with the argon ion bombardment, the metallic ions are also being deposited on the substrate. By balancing the argon pressure and the evaporation rate from the filament, one can deposit the coating material while continuously cleaning the substrate. A careful balance must be achieved between the two, however; otherwise, if the removal rate during the sputtering bombardment for cleaning purposes with argon ions greatly exceeds the deposition rate of the metallic ions, the material may be removed from the surface faster than it is deposited. The net result is that no coating is actually applied to the substrate surface. However, it is very simple to control the evaporation rate and also the gas pressure of the argon for sputter bombardment and thereby achieve a simultaneous cleaning process and deposition of a film on the substrate surface.

As mentioned earlier, the voltage range applied as a potential to the substrate is typically in the 1 to 5 kiloelectron volt range. An ordinary inexpensive power supply is all that is required for operating in this voltage range.

During the deposition process, a dark space appears beneath the substrate. The location of the dark space is fairly critical. This is controlled by controlling the ionization process and the gas pressure in the system. The basic system for ion plating is relatively simple; it has only one electrode, the substrate, a source for evaporating material, and a gas bleedin valve all in an ordinary unsophisticated vacuum chamber. The filament power supply and the high voltage supply are relatively inexpensive components. Thus, one can build an ion plating system relatively inexpensively. This is one of its attractive features.

A question arises as to what goes on in the ion plating process at the surface. The actual events occurring at the surface during the ion plating process are shown schematically in figure 10-15 from the work of Maddox. The substrates depicted in figure 10-15 are the same as that shown in figure 10-14 with the high voltage being applied just as it was in figure 10-14. The evaporating filament or the anode is the evaporator filament in figure 10-14. An examination of figure 10-15 reveals a number of events taking place simultaneously during the ion plating process. During this process, the coating material M is leaving or evaporating from the evaporating filament. Metastable atoms of the coating material M that do not have sufficient energy to escape the filament source return to the parent surface. Likewise, since atoms are being liberated to the space and are being generated as ions, there is an excess of electrons in the vicinity of the evaporating filament, and these come to the parent surface to generate additional ions.

At the substrate surface, both gaseous atoms and ions strike the surface. Since the ionization of the plasma is not 100 percent effective, in the plasma there are ions as well as atoms of the species that are used for sputter cleaning the surface and for maintaining a plasma during the ion plating process. In addition, both atoms and ions are arriving at the substrate surface of the cathode. Controversary exists over the degree or the effectiveness of ionization in the ion plating process; that is, when the atoms leave the filament source and become ions in the plasma, just exactly what percent of the metal atoms is ionized is not certain. The percent has been indicated as being from 1 to 15 percent, depending on the particular source and the investigator who made the measurements. Nonetheless, a sufficient percentage of the metal atoms is ionized so that the metallic material which is to be deposited on the substrate is carried to all points on the solid surface as indicated by the equipotential lines surrounding the specimen. This indicates that the incoming species basically coat the entire specimen. The ionized material and the ionized plasma have sufficient momentum to carry along with them the neutrals to the solid surface.

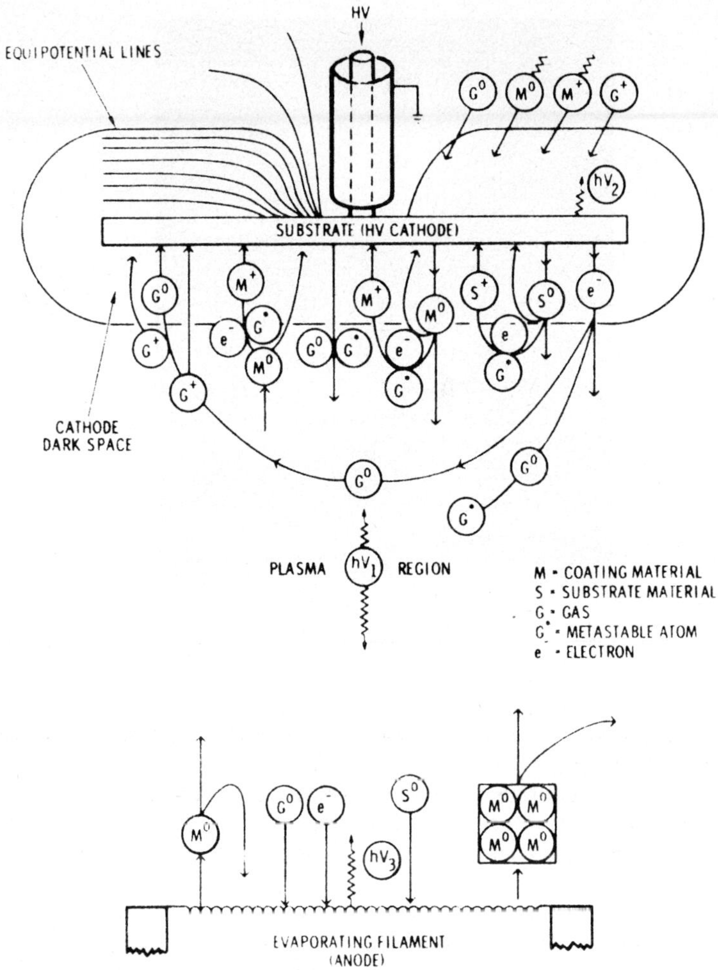

Figure 10-15. – Schematic representation of processes which occur in dc gas discharge with thermal vaporization source.

The plasma, as generated during the actual ion plating process, is shown in the photograph in figure 10-16. The cathode is the disk specimen or substrate to be coated. The dark space is the space or region shown schematically in figures 10-14 and 10-15. The glow discharge region is the region between the anode and the cathode just short of the dark space, and the anode is the filament that is being heated. It contains the material that is to be deposited on the substrate or disk surface of figure 10-16. Thus, in figure 10-16, the surface of the disk is being argon ion and atom bombarded for cleaning purposes. While that process is going on, the filament is heated to incadescence. The plating material is then evaporated from the filament

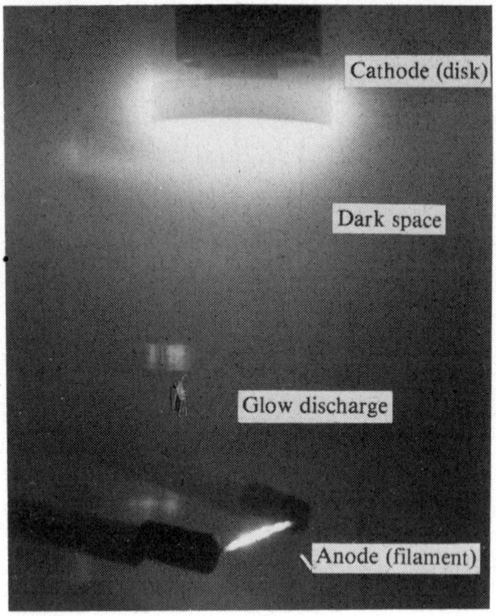

Figure 10-16. – Glow discharge during ion plating.

source into the glow discharge. The evaporant is ionized to generate positively charged ions which are carried to the disk surface to be coated along with the neutrals.

A variety of techniques (involving filament or evaporant source) have been used in ion plating to generate the metallic ions needed in the plasma deposition process associated with ion plating. Some of these are depicted schematically in figure 10-17. One approach is to use a metal bearing gas where gas containing the metal ions is admitted to the system and then these metal ions are directed at the substrate surface. Another approach is to use arcs where one has two electrodes of the material to be deposited. This is especially useful where one is depositing very high melting point materials that are difficult to evaporate using ordinary resistance heated tungsten filaments. If the melting point of the material is very high, such as the melting point of rhenium which is close to that of tungsten, it would be very difficult to use a tungsten susceptor to achieve the evaporation of rhenium. However, electrodes could be made of rhenium, and then an arc could be generated between the two electrodes to cause vaporization of the rhenium and that could be directed at the solid surface. Another approach would be to sputter the material to be deposited into the plasma. This is a very useful technique because there is no limit to the materials that could be sputtered and placed into the plasma by the sputtering process.

The most commonly used techniques, and one of the first to be used, is ordinary resistance heating (fig. 10-17). Probably this is still the most popular technique for routine applications. Yet another technique is the focused electron beam technique where one has high melting point materials

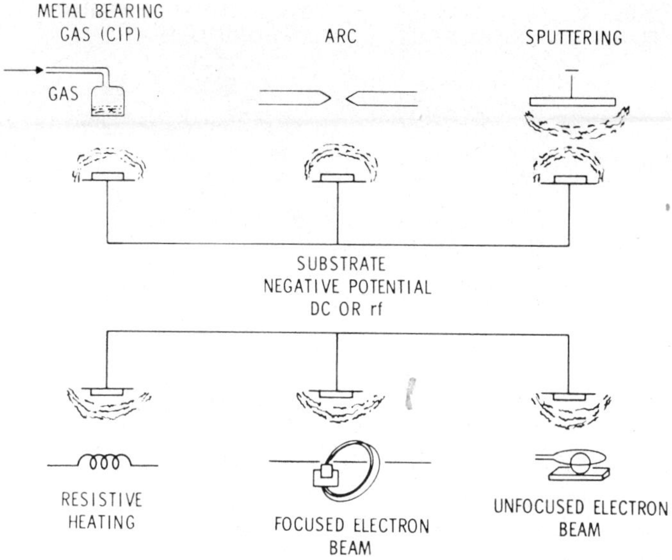

Figure 10-17. – Various techniques for providing vapor to be deposited in ion plating.

that he wants to evaporate into the plasma. The focused electron beam is very effective for accomplishing this. Also, there is, as indicated in figure 10-17, the unfocussed electron beam approach to achieving vaporization of material and its movement into the plasma for deposition on the surface.

One of the advantages of using ion plating for the deposition of coatings is that, because one is using a plasma, very complex geometric surfaces can be coated uniformly with dense films of materials. This is particularly useful in practical tribological systems where the mechanical components involved are frequently of complex geometric configuration; that is, they are not simple flat surfaces. For example, bearing cages (retainers), gears, seals, etc., do not have simple geometries that are very readily amenable to straightforward line of sight types of deposition. In this respect, ion plating is very useful.

The effectiveness of plasma deposition for coating complex surfaces is shown schematically in figure 10-18, where it is contrasted with conventional vapor deposition. If one wishes to coat a bearing retainer, for example, the conventional vacuum deposition allows only the coating of the surfaces that are in line of sight with the evaporating specie. This is shown in figure 10-18(a) as being the edge of a bearing retainer. The reason for this is that conventional vacuum deposition, as indicated in figure 10-18(a), is a line of sight technique so only the surface that the evaporant sees is actually coated. With ion plating, however, since one has a positively charged ion moving into and through the plasma to the substrate to be coated, the charge effects on the substrate and on the gaseous plasma allow very uniform coatings of very complex geometric surfaces. Thus, with ion plating, one is able to coat very complex tribological surfaces such as bearings, gears, and seals with ease. The reason for this is demonstrated in

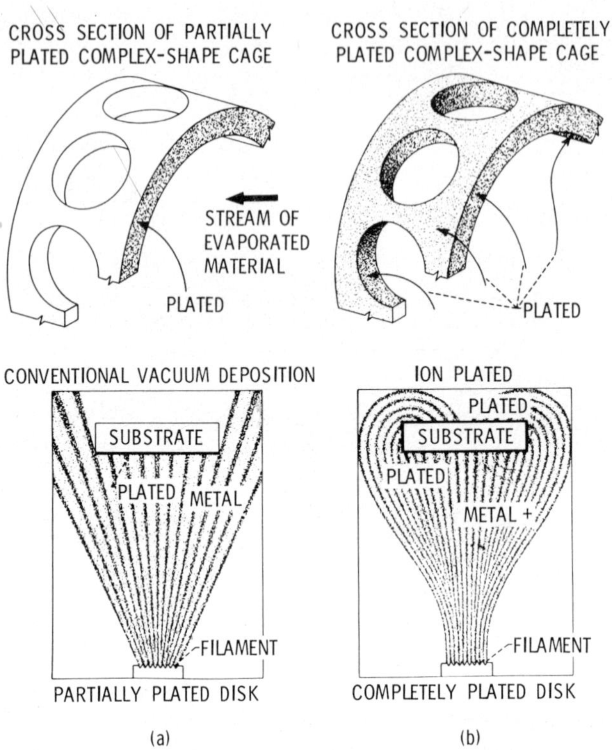

Figure 10-18. – Comparison of conventional vacuum deposition with ion plating techniques.

the schematic drawing of figure 10-18(b). The figure indicates that the surface is coated very uniformly. The substrate is negatively charged, and the metallic ions are positively charged; the metallic ions are drawn to all points on the solid surface to be coated.

Yet another distinct advantage of the ion plating process over other methods for applying solid film lubricants to substrates is the adherence or adhesion of the coating to the substrate. With the application of most solid film lubricants, as was discussed earlier in this chapter, materials must be added to the solid film lubricant to achieve bonding to the substrate. This involves the use, for example, with inorganic solid film lubricants, of such things as binders to achieve adhesion of the lubricant coating to the substrate. Even when metallic films are electroplated, there is a sharp interface between the substrate and the electrodeposited coating. Frequently, in tribological mechanical components this region (interface between the coating and the actual substrate) serves as a site for the degradation of the coating. It is usually the weakest region in the system, and generally yields under the stresses associated with operation; in practical applications, it results in the early failure or the loss of coating material from a substrate.

With ion plating, since there is a potential at the substrate, the metallic ions coming to the surface do so with considerable kinetic energy.

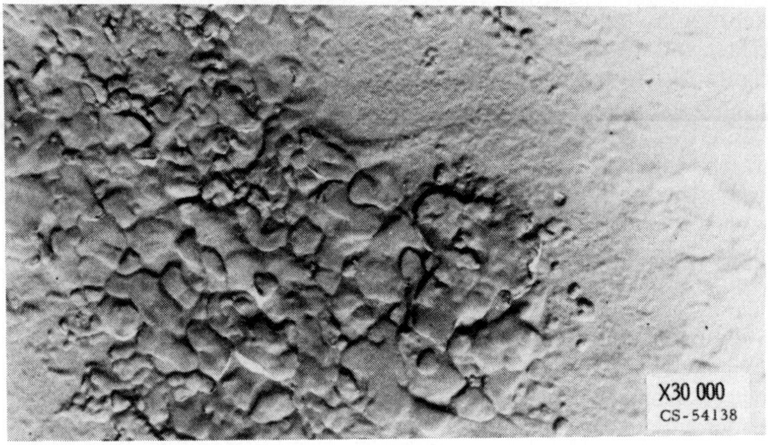

Figure 10-19.—Electron micrograph of cross section of nickel film applied on tungsten surface by ion plating. Magnification, 30 000.

Consequently, they do not simply deposit on the surface but become buried in the surficial layers; they form what is referred to as a graded or diffused interface. There is no sharp line when one looks at the film in cross section. There is a relatively uniform decrease in the concentration of deposited material as one traverses, in cross section, from surface to subsurface. There is no sharp demarcation between coating and substrate. This type of interface is ideal for lubrication applications, because in many practical tribological components the extreme stresses associated with the mechanical activity (motion) at the surface can disrupt lubricant films where the films do not strongly adhere to the substrate surface.

Figures 10-19 and 10-20 indicate the nature of this diffused interface for a

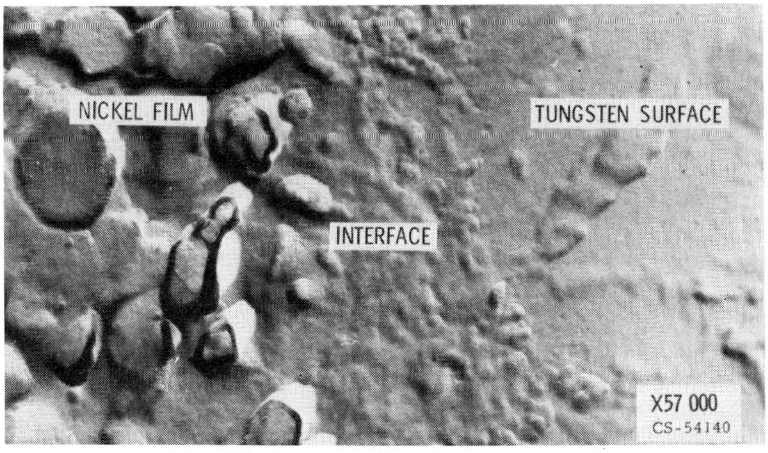

Figure 10-20.—Electron micrograph of cross section of nickel film applied on tungsten surface by ion plating. Magnification, 57 000.

nickel film ion plated onto a tungsten surface. In both of these figures, the interface and the coating and substrate material are shown in cross section. In figure 10-19 it is difficult to see and identify the end of the nickel material and the beginning of the tungsten substrate surface. The photomicrograph in figure 10-20 shows, at an increased magnification, the nickel coating (or film), the tungsten substrate, and the interfacial region. Note that in the interface there is no sharp line of demarcation between the nickel film and the tungsten substrate surface. The nickel was only brought out by etching to show the differences between the nickel and tungsten. The attack by the etchant is different for the nickel than it is for the tungsten. Were it not for the use of a selectively attacking chemical agent to etch, it would be difficult or impossible to detect any kind of interfacial region. This is an ideal coating because it can withstand high surface contact stresses without being displaced from the solid surface. Similar interfaces (diffuse or graded interfaces) have been observed on a wide variety of substrates and coating materials. This indicates that the ion plating process is such that the interface is not sharp and that the maximum amount of adhesion can be obtained with ion plating because of the formation of the diffuse or graded interface (refs. 9 and 10).

There is a question, when using a plasma deposition process such as ion plating, of the ability of the process to give a uniform deposition of the film on all parts of a solid surface to be coated. There are some variations in the coating thickness that exist in the ion plating process because of the geometry facing the dark space in the plasma. However, these differences in coating thickness as functions of specimen geometry can be minimized by optimizing various parameters associated with the ion plating process. One such parameter is gas pressure. By carefully controlling the gas pressure in the system, one can obtain an optimum uniformity of coating thickness on complex geometric surfaces.

The effect on film thickness, and its variation, with geometry and gas pressure, is demonstrated in figure 10-21. In figure 10-21 the film thickness is plotted as a function of the gas pressure or the plasma pressure for a flat plate that has been coated on the front side (side facing the dark space of the plasma) as well as the back side. It can be seen from the two curves that, at a lowest gas pressure of approximately 5 millimeters of mercury and at the highest pressure of about 25 millimeters of mercury, there is a minimum in the difference in coating thickness on the front and back sides of the plate. The curves show that by varying the parameters in the system one can vary the thickness and uniformity of the coating by using the ion plating process.

There are limitations, however, despite the fact that the coating process is capable of applying films to very complex geometric surfaces. For example, where one has a tube with a very small diameter hole, it is extremely difficult to uniformly coat the entire inside of the tube by ion plating, because the plasma becomes essentially choked in the throat of the tube; this choking inhibits the deposition of the coating through the entire length of the tube. This is, however, a function of the length to diameter (L/D) ratio of the tube. With certain L/D ratios, if sufficient plasma is able to enter the tube, tubes can be coated very uniformly by ion plating. For

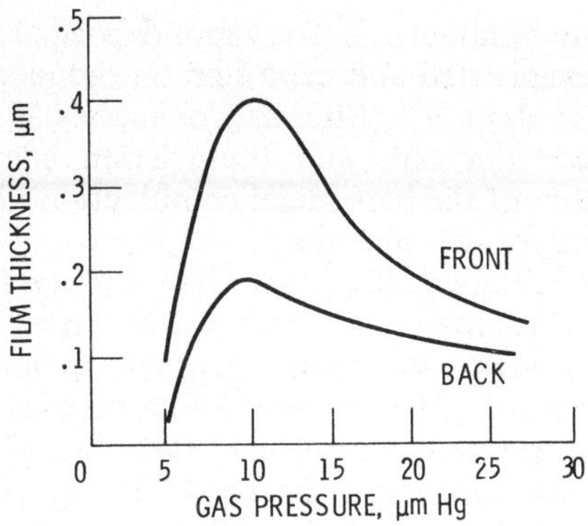

Figure 10-21. – Film thickness on front and back surfaces of flat plate specimen as function of argon pressure.

example, lead has been very effectively applied to Teflon tubing by the ion plating process.

The tribological properties of ion plated films are markedly superior to those films applied by other deposition techniques where the film materials are identical. For example, because of the graded interface achieved with ion plating, the deposition of soft metal films to substrates provides very good adherence. This good adherence is reflected in the endurance life of the coating on the solid surface. Some friction data are presented in figure 10-22, and they reflect this effect for vapor deposited and ion plated gold films applied to a nickel-chrome alloy with a niobium rider sliding on the surface (ref. 11). The friction coefficient for the bare metal combination without the application of a solid film of gold for lubrication purposes is

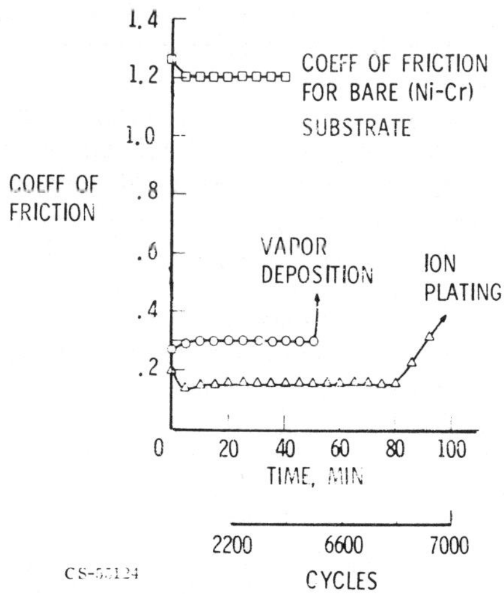

Figure 10-22. – Coefficient of friction of niobium sliding on nickel-chromium alloy with gold deposited by vapor deposition and ion plating about 2000 Å thick. Load, 250 grams; speed, 5 feet per minute; ambient temperature; pressure, 10^{-11} torr (ref. 11).

seen in figure 10-22 to be about 1.2. The vapor deposited film, that is, where the gold is simply evaporated and deposited on the metal surface, gives a friction coefficient of about 0.3. After approximately 50 minutes, however, the vapor deposited film fails and the friction coefficient rises very drastically toward that of the bare metal combination, the bare niobium is in contact with the nickel-chrome alloy.

With the ion plated gold film, one first observes a lower friction coefficient (less than 0.2) than that obtained with the vapor deposited film. Also, the friction coefficient remains low for a much longer period, approximately 80 minutes. The ion plated film then begins to fail slowly, unlike the vapor deposited film which fails instantaneously (with a corresponding instantaneous rise in friction). The ion plated film, because of the diffuse or graded interface, fails gradually, and the friction coefficient increases from the value obtained for the effectively lubricated solid surface to that obtained for the bare metal over a prolonged period which may take many hours. The reason for this result is that the gold is buried in the surface of the substrate; hence, as one wears through the lubricant film, the contact is with a substrate containing buried gold, and this contributes to a reduced friction force. As this film is worn through and one penetrates further and further into the substrate, the friction coefficient of the bare metal combinations is ultimately achieved. With the vapor deposited film, however, this occurs very rapidly because the film separates at the interface from the substrate and is wiped away from the solid surface very rapidly; this condition causes an almost instantaneous increase in the friction force.

The tenacity of the ion plated films to substrates can be further demonstrated in other mechanical tests to evaluate the coating behavior and its adherence. One such test is an ordinary tensile test. If, for example, ordinary tensile specimens are pulled to elongation and fracture, the tensile specimen undergoes plastic deformation. If the material pulled to tensile fracture is a relatively ductile material, there is a considerable amount of plastic flow that occurs before fracture. Applying films, by ion plating and other techniques, to ductile metals such as nickel and then pulling to tensile fracture is a good way to measure the adherence of coatings to substrates. If the coating flakes off with the tensile extension of the specimen (and the plastic deformation that ensues), then the coating is relatively poorly adhered to the solid substrate. If, however, the coating remains intact with the substrate, then adhesion is very good.

Some tensile experiments were conducted with nickel specimens coated with gold, and the results of some of these tensile tests are shown in the photomicrographs of figure 10-23 where there are three tensile specimens (ref. 11). In the upper photograph is the original tensile specimen prior to pulling to elongation and fracture. The middle photograph is that of an uncoated nickel specimen that has been pulled through elongation to fracture; the elongation is approximately 39 percent prior to fracture. Careful examination of the surface of the nickel specimen reveals a modeling or a roughening of the texture of the surface of the tensile specimen over the entire length that has undergone deformation. If a tensile specimen is coated with a gold film and pulled to tensile fracture, the results

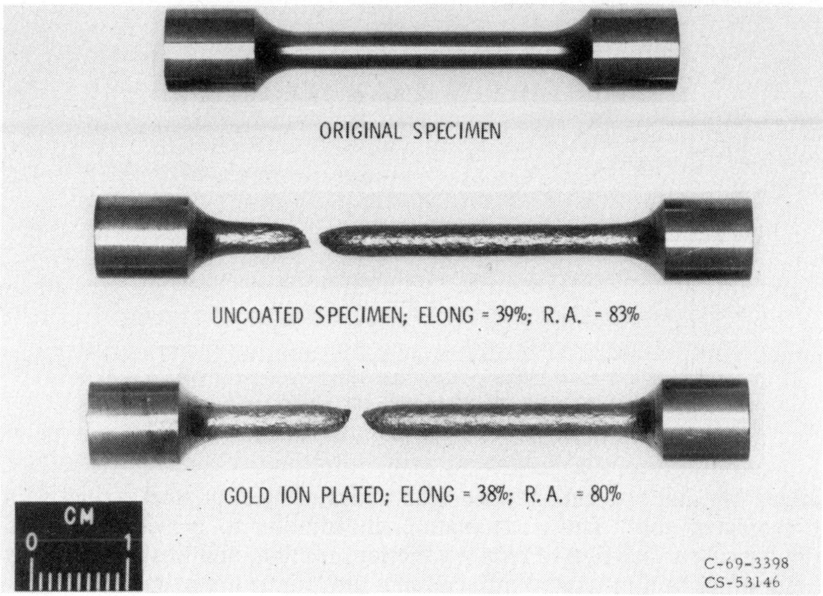

Figure 10-23. – Comparison of uncoated and gold-ion-plated nickel tensile specimens after fracture (ref. 11).

obtained are as shown in the lower photomicrograph in figure 10-23. In this instance, the gold coating on the surface flows plastically with the nickel substrate to the point of fracture. In the necking region of the tensile specimen where fracture has occurred, the coating remains intact until actual fracture. Thus, in the slope region of the tensile specimen, where necking occurred before fracture, the gold is still completely covering the nickel substrate; this indicates the strong adhesion of the gold to the substrate. The gold behaves as the substrate, flows with it, and takes on the character of the substrate material as opposed to its own individual identity. This is the ideal type of adhesion for tribological applications. It appears in bearings and gears where high contact stresses and deformations can cause the separation of coatings at a sharp interface.

Ion plated gold has also been found to influence other mechanical properties of substrate materials. One such property is fatigue. Fatigue is extremely important in tribology; it is important in both bearings and gears. Some experiments have shown that the fatigue life of components can be extended by applying ion plated films to the surface (ref. 12). Evidence for this is presented in figure 10-24 where contact stress is plotted as a function of cycles to failure for uncoated steel, steel coated with electroplated gold, and steel coated with ion plated gold. The results indicate that the presence of the electroplated gold has no effect on the fatigue life. The coating fails as if it were the plain, uncoated steel. But with the application of the ion plated gold, an increase in fatigue life (at a given stress) is observed. Conversely, an increase in the stress to produce equivalent fatigue life is indicated. That is, one can operate at higher stress levels to achieve the same

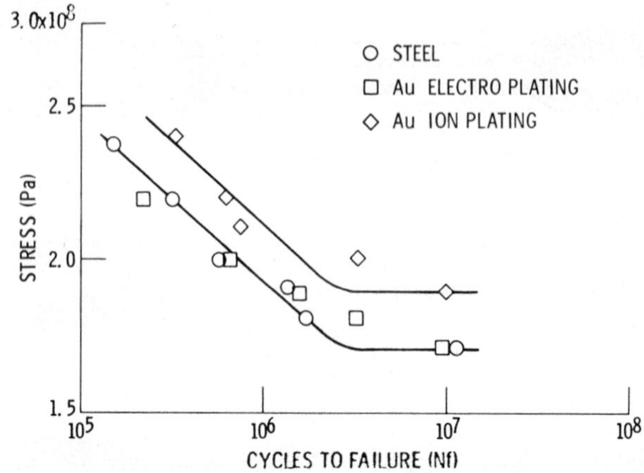

Figure 10-24. – Effect of ion plating on fatigue of low carbon steel (ref. 12).

fatigue life that was encountered with uncoated steel or steel coated with electroplated gold. Thus, ion plating, in addition to providing a good tribological surface film to reduce adhesion, friction, and wear, is also very useful (in certain instances) in increasing the fatigue life of materials.

Sputtering

Ion plating is extremely useful for depositing films where a diffuse or graded interface is desired for maximum adhesion. However, it has its limitations. One of them is that compounds cannot be deposited by that technique without resorting to hybrid systems that employ something other than simply ion plating, because the ion plating process requires the evaporation of the material to be deposited. For example, many compounds in a vacuum environment dissociate before evaporating. A classic example of such behavior is shown by molybdenum disulfide, a solid film lubricant of interest to most tribologists.

Since molybdenum disulfide, when heated in a vacuum environment, dissociates into molybdenum and elemental sulfur before it evaporates, ion plating cannot be used to deposit molybdenum disulfide. Fortunately, however, there are other techniques available for depositing such inorganic compounds that are of interest in tribology.

One such technique is sputtering. There are two types of sputtering, depending on the power source used for depositing the coating materials: direct current (DC) sputtering and radiofrequency (RF) sputtering. The most commonly used technique today is the RF sputtering. It is extremely versatile and allows almost any material to be deposited on almost any substrate. The basic components of the sputtering system are not too different from those used in ion plating. RF sputtering requires a relatively crude vacuum system that can achieve pressures of the order of 10^{-6} torr. Therefore, a system using diffusion and mechanical pumps with cold trapping can be used. A schematic representation of a device used for RF sputtering and a photomicrograph showing the system in operation are presented in figure 10-25. An examination of the schematic in figure 10-25 indicates that the RF sputtering process requires an RF power supply; for

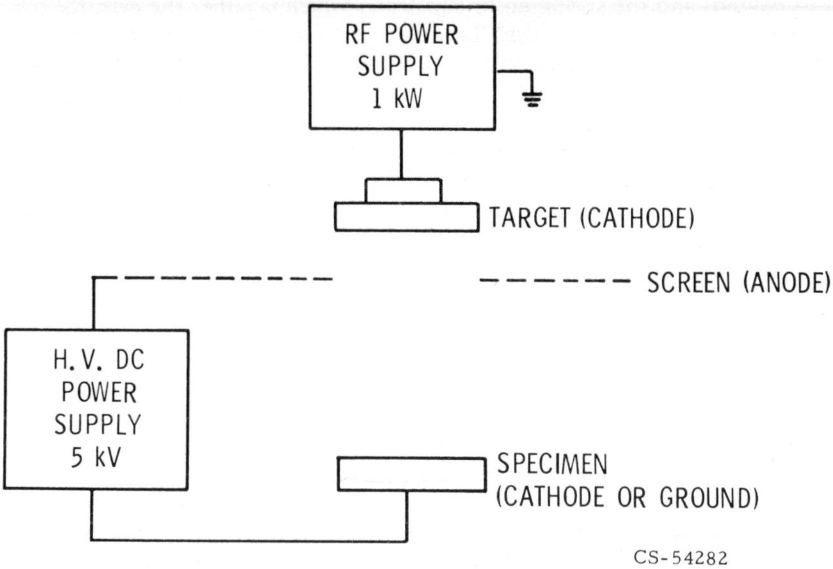

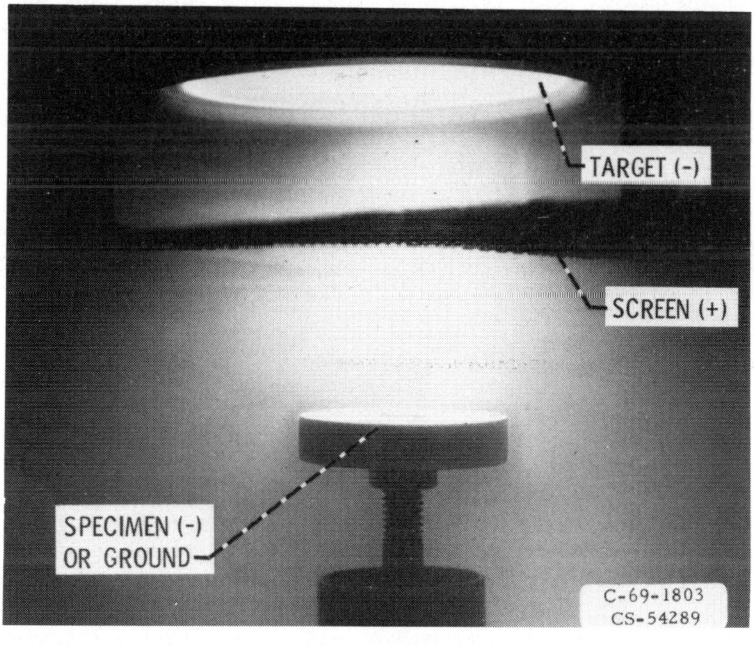

Figure 10-25. – Sputtering apparatus.

DC sputtering, a DC power supply would be used in place of the RF. A target is attached to the power supply. The target would be the cathode, and it would contain the material to be deposited as a film on the solid substrate. A screen, or anode, would be placed between the target material (material to be coated) and the specimen or substrate, which is either the cathode or ground. The specimen can then be attached to a high voltage DC power supply for setting the screen anode potential. This provides, then, a bias during the sputtering process.

In this particular system, with the admission of argon gas into the system, the surface to be coated can be sputter cleaned, just as it is in ion plating, by generating argon ions and bombarding the specimen surface to be coated with these argon ions. The photograph in figure 10-25 shows the system in operation. The specimen, a disk, is approximately 6.5 centimeters in diameter and may be at either negative potential or ground. The screen is at a positive potential, the target at a negative potential. At some point, the argon ions (positively charged) strike the target material (negatively charged) and dislodge the film material (e.g., molybdenum disulfide) from the solid surface; this dislodged material is carried to the specimen (cathode) surface where it is deposited. As shown by figure 10-25, the sputtering system is relatively simple. The glow in the photograph is due to the plasma.

It is possible by using a pedestal arrangement, as indicated in figure 10-25, to coat complex geometric surfaces fairly uniformly with solid film lubricating materials, such as molybdenum disulfide. Normally in the electronics industry flat surfaces are coated by setting them on a large, flat substrate surface. However, with the pedestal arrangement, bearing components have been effectively coated with solid films (e.g., molybdenum disulfide (ref. 13)). One such example is shown in the photomicrographs of figure 10-26. Figure 10-26 shows two sets (one

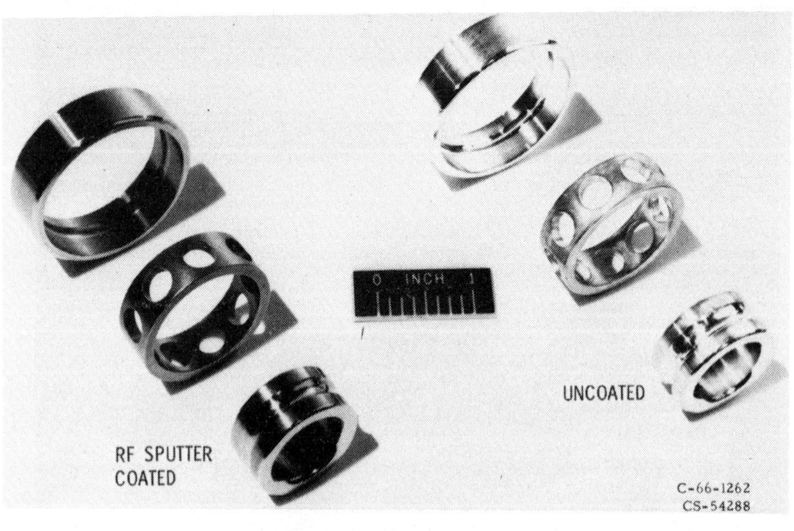

Figure 10-26. – Ball bearing assembly completely coated with MoS_2 film by radiofrequency sputtering (ref. 13).

uncoated and the other RF sputter coated with molybdenum disulfide) of three bearing components: a bearing outer race, a bearing inner race, and a bearing cage. On the left are the surfaces RF coated with molybdenum disulfide; the coating is uniform over the entire surface area of the bearing races as well as in the cage. Even the pockets of the cage have been uniformly coated with the molybdenum disulfide. The optimum film thickness for bearing applications is approximately 2000 to 3000 angstroms for the molybdenum disulfide film. In addition to bearing surfaces, other geometric surfaces of interest to the tribologist have been very effectively coated by RF sputtering. These include gears, seals, electrical contacts, and other components.

Just as with the ion plating process, sputtering provides good adhesion of the coating to the substrate material. This is accomplished because the surface is argon ion bombarded and thus cleaned to remove adsorbates and oxides before the deposition of the coating material (whether molybdenum disulfide or some other compound for lubricating or corrosion protection purposes). Since the surface is clean, it is in a highly energetic state. In such a state, there is a strong tendency for interaction of the clean metal surface with the incoming depositing species. As a consequence, good adhesion results.

In the sputtering process, the adherence is not such as obtained with ion plating. A sharp interface is obtained in the sputter deposition process. One does not have the potential associated with the ion plating process that allows burying the coating material in the substrate. The coating does leave a line of demarcation between coating and substrate. The adhesion is, however, very good, far superior to that encountered with techniques other than ion plating. This is demonstrated in the data of figure 10-27 (ref. 13) where tensile specimens were pulled to tensile fracture. This same process was done earlier in reference to the data of figure 10-23 for ion plated films.

In figure 10-27, molybdenum disulfide was applied to nickel and Inconel specimen surfaces. The specimens were pulled to tensile fracture. An

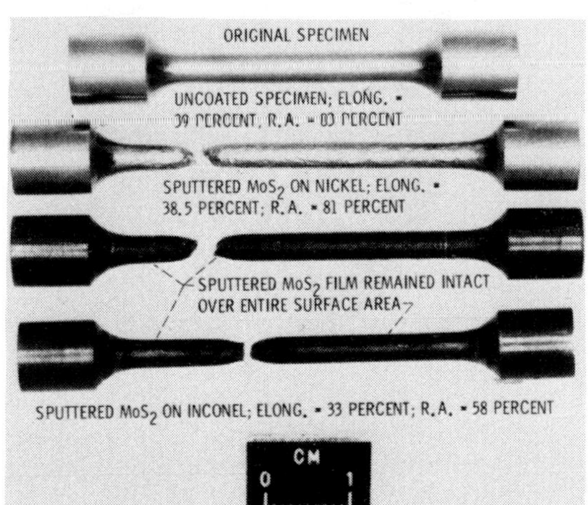

Figure 10-27. – Comparison of uncoated and radiofrequency sputtered MoS_2 on nickel and Inconel tensile specimens after fracture (ref. 13).

examination of the specimens after fracture indicated that the molybdenum disulfide remained intact over the entire surface area that had undergone plastic deformation. The nickel specimen underwent plastic flow just as was observed in figure 10-23 with the ion plated specimens. The surface became modeled and deformed. Despite this deformation and destruction of the smooth surface topography of the original tensile specimen, the molybdenum disulfide remained intact on the solid surface, right down to the point of tensile fracture. The black coloration indicated in figure 10-27 reflects the presence of the molybdenum disulfide on the surface in the necked region of both the nickel-coated and the Inconel-coated specimens.

The versatility of the RF sputter deposition process, as was mentioned earlier, is extremely great. Almost any material can be used as a target and bombarded with argon ions for deposition on almost any substrate material. To demonstrate this, films of PTFE (or teflon) were deposited on paper surfaces. Figure 10-28 is a photograph showing a piece of filter paper that was coated with Teflon by RF sputter deposition. The filter paper was backed by a metal disk. The hexagon-shaped white area in the center of the photograph is where a nut held the paper covered disk specimen to the electrode in the apparatus shown in figure 10-25.

Thus, one can apply such materials as polymers to organic substrates such as paper in the RF sputtering process. Normally in practical tribological systems, one would not be applying PTFE to paper. However, PTFE is a very effective solid film lubricant in thin film form, and one may

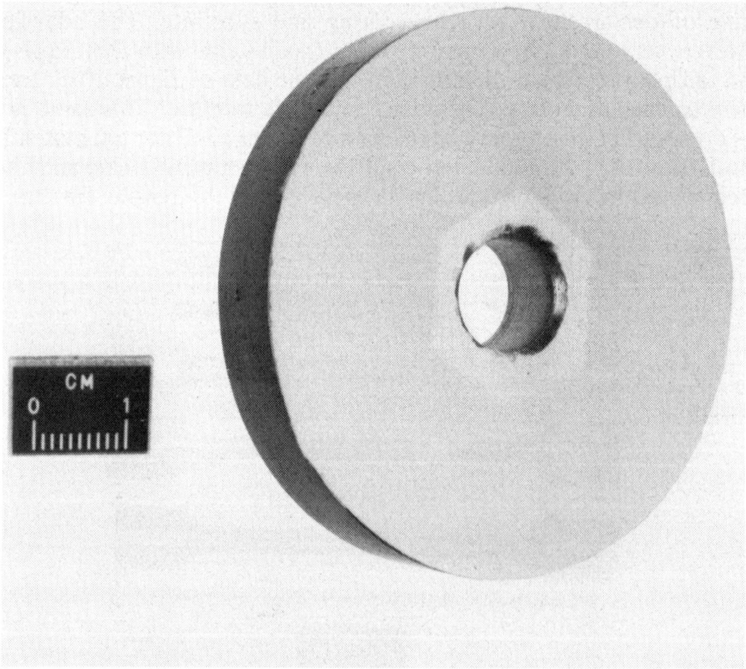

Figure 10-28. – Sputtered Teflon on paper.

wish to apply that material to other substrates, including metals and ceramics, which may be components of practical tribological devices.

The superior adhesion of molybdenum disulfide to substrate surfaces applied by the RF sputtering process can be shown in endurance evaluation of the coating materials. In addition to the sputter deposition of films, other techniques have been used in the past to apply solid film lubricants such as molybdenum disulfide to substrates. These techniques, including burnishing and using resin binders as well as other organic and inorganic binders, have been used to achieve adhesion of the molybdenum disulfide to substrate surfaces. RF sputtered coatings have been compared in endurance experiments with coatings applied by these other techniques (ref. 13). Some representative data obtained on cycles to failure for coatings applied by three different techniques are presented in the data of figure 10-29. In figure 10-29, molybdenum disulfide has been applied by (1) burnishing, (2) using a resin binder to a thickness of 13×10^4 angstroms (normal recommended thickness for molybdenum disulfide films applied by this technique), and (3) DC sputtering to a thickness of approximately 2000 angstroms. From the cycles to failure, it can be seen in the data of figure 10-29 that the greatest endurance life was achieved with the 2000-angstrom sputtered molybdenum disulfide film. After 5.8×10^5 cycles there was no evidence of film failure whereas the resin-bonded film in a much greater thickness failed earlier; the burnished film had the shortest endurance life of the three films examined. The data in figure 10-29 indicate the effectiveness of the good adhesion of molybdenum disulfide to the substrate surface; they also show that using the pure molybdenum disulfide (without binders) promotes or enhances the endurance life of solid film lubricants.

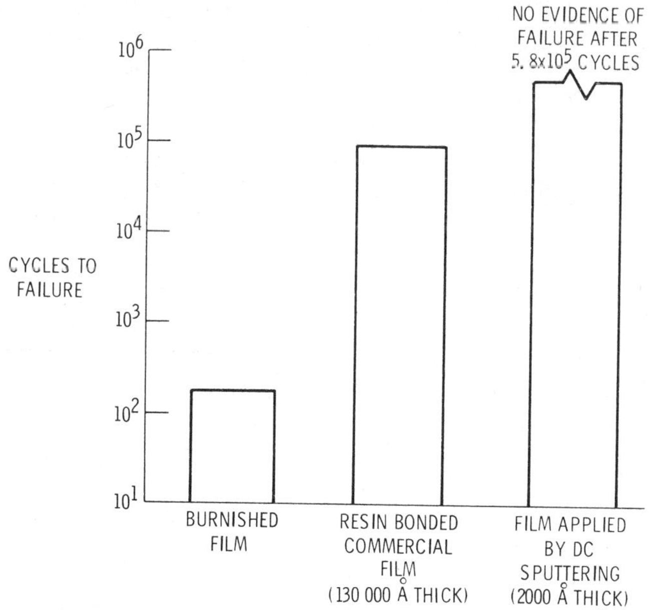

Figure 10-29. – Endurance lives of MoS_2 films applied by various techniques (ref. 13).

The endurance lives of films applied to solid surfaces for lubrication purposes can be increased even further by using other techniques such as applying transition layers between the soft lubricating film and the substrate material. Spalvins found that applying certain hard-face coatings directly to a substrate by sputtering and then applying a soft lubricating film (such as molybdenum disulfide) over those coatings can give a longer endurance life (ref. 14). This is demonstrated by the data in figure 10-30 for a molybdenum disulfide film (2000 Å) applied directly by sputter deposition to a 440C bearing steel and also applied over a chrome silicide, which was first deposited on the 440C steel. With the chrome silicide layer between the substrate and the molybdenum disulfide coating (2000 Å thick), a much longer endurance life was achieved. The exact mechanism for the improved performance of the molybdenum disulfide lubricant with the chrome silicide present at the interface is not fully understood. It is, however, highly specific in that certain interface materials or coatings improve the adhesion of lubricant films while others do not.

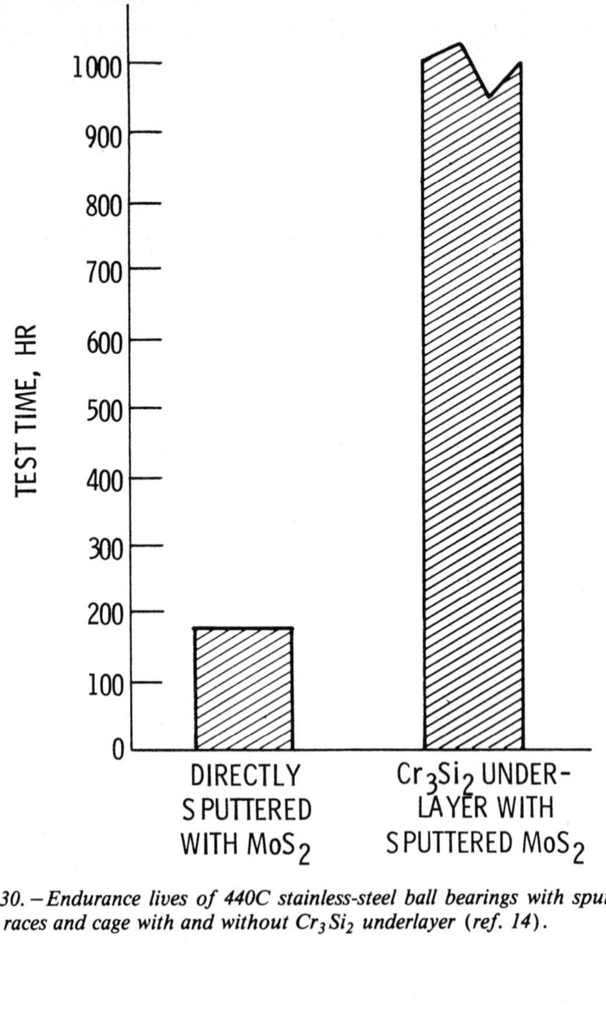

Figure 10-30.—Endurance lives of 440C stainless-steel ball bearings with sputtered MoS_2 films on races and cage with and without Cr_3Si_2 underlayer (ref. 14).

Many parameters involved in the sputter deposition process can be varied: (1) the gaseous pressure at which the surfaces are coated, that is, the argon plasma pressure, (2) the temperature of the substrate, (3) the gas used for bombarding the surface, (4) the target, (5) the composition of the gas, (6) the bleeding in of additional active gases to the vacuum system while the sputtering is taking place, and (7) the use of different power levels. All of these can vary the nature of the resultant deposited film.

One variable in the sputtering deposition process which has been found to have a marked influence on the coating characteristics resulting from sputtering has been the substrate temperature. Certain materials deposited by sputtering techniques are sensitive to substrate temperature. One such coating material is molybdenum disulfide. If molybdenum disulfide is sputter deposited at a sufficiently low temperature, the coating resulting from the sputtering process is completely ineffective as a lubricant. In fact, very high friction coefficients are measured for the coating material, and it does not reflect any of the characteristics normally associated with the lubricating properties of molybdenum disulfide. If the temperature of the substrate is sufficiently high, however, the friction properties of the molybdenum disulfide coating are characteristic of molybdenum disulfide and good lubricating properties are observed.

The reasons for this strange behavior are seen in the data in figure 10-31 where the friction coefficient is plotted as a function of substrate temperature for molybdenum disulfide films applied by sputter deposition

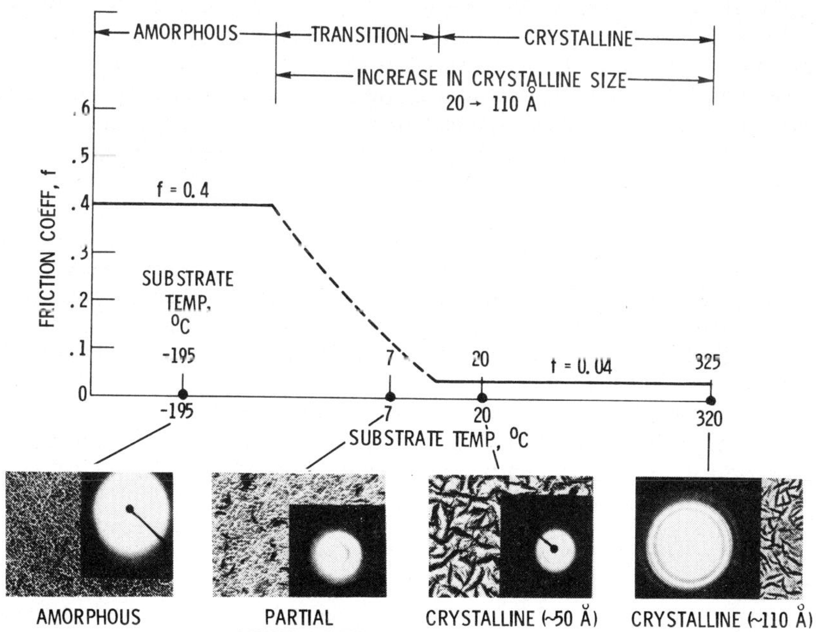

Figure 10-31.—Substrate temperature effects on MoS_2 film morphology and friction coefficient (ref. 14).

on a substrate (ref. 14). The substrate was held at temperatures from -195° to 320° C. At the very low -195° C, the molybdenum disulfide film exhibited a friction coefficient of 0.4. Careful examination of the friction properties of the molybdenum disulfide indicated that the high friction behavior continued to temperatures greater than -195° C. A friction coefficient of 0.4 is not characteristic of the good lubricating properties of molybdenum disulfide. Electron diffraction techniques and SEM studies of the solid surface indicated that the film exhibiting a friction coefficient of 0.4 was amorphous as reflected in the electron diffraction pattern presented on the left side of figure 10-31. The pattern indicates an amorphous nature by a complete absence of diffracting rings which are normally associated with the crystalline nature of a compound. If the temperature of the substrate is increased, however, a temperature region is reached where the friction coefficient for the molybdenum disulfide begins to decrease from 0.4. Ultimately, at about 10° to 15° C, the friction coefficient reaches a very low value of approximately 0.04. (The friction undergoes a tenfold decrease from 0.4 to 0.04 over this range of temperatures.) This decrease in the friction coefficient is associated with a transition from the amorphous to crystalline form of molybdenum disulfide.

Examination with electron diffraction of a molybdenum disulfide film at about 7° C indicates the photomicrograph and the electron diffraction pattern second from the left in figure 10-31. These show a structure which is different from that obtained at -195° C. The electron diffraction pattern begins to show rings, which indicate partial crystallinity. In constrast, at 20° C a completely crystalline pattern is obtained with a particle size (from the transmission electron micrograph) of about 50 angstroms. At 7° C there is only partial crystallinity and the size of the molybdenum dilsulfide crystallites is approximately 20 angstroms. Thus, from 7° to 20° C, one moves from partial crystallinity to complete crystallinity in molybdenum disulfide. The completely crystalline film at 20° C reflects the friction characteristics of 0.04 that are associated with the effect of lubrication of molybdenum disulfide. A further increase in the temperature of the substrate to 325° C increases the number of diffraction rings observed for the molybdenum disulfide as indicated by the electron diffraction pattern in figure 10-31. However, no further change in the friction coefficient is observed. But there is a change in the crystallite size to approximately 110 angstroms.

The data in figure 10-31, then, indicate that the substrate temperature is important in the nature of the film obtained. While the films obtained at -195° and 20° C are essentially molybdenum disulfide, the film at 20° C is crystalline while that at -195° C is amorphous. A tenfold difference in friction behavior is observed with this difference in morphology; the amorphous material is not a lubricant ($f=0.4$) and the crystalline material is ($f=0.04$).

Another variable which can affect the lubricating or tribological properties of deposited films in the sputter deposition process is the dc bias voltage (shown schematically in fig. 10-25(a)). Variations in coating composition can be achieved (ref. 14) depending on the voltage applied to the system. This is indicated in the friction data of figure 10-32 for a molybdenum disulfide coating. In figure 10-32 the friction coefficient is plotted as a function of dc bias voltage over a voltage range of 0 to 600 volts. The friction coefficient for the unlubricated substrate material—namely, 440C stainless steel—is also presented in figure 10-32. At a relatively low bias voltage of 0 to 200 volts dc, a friction coefficient in the range associated with effective lubrication of molybdenum disulfide is achieved. Above 200 volts dc, however, the friction coefficient begins to increase markedly and, at 400 volts bias, the friction coefficient is characteristic of that of the 440C substrate; this indicates that the molybdenum disulfide film has failed as a lubricant. Thus, in addition to substrate temperature, the bias voltage used in the deposition of tribological coatings is also extremely important.

Since the molybdenum disulfide that is sputter deposited is applied to a clean substrate, and the molybdenum disulfide itself has been carefully outgassed in the process of evacuating the vacuum system, the coatings that result from the sputter deposition process are extremely sensitive to environmental contaminants when the coatings are exposed to the environment. This effect, or the effect of environment, is seen in coating performance and is reflected in the friction data measured for molybdenum disulfide sputter deposited films where the pressure ranges from essentially atmospheric to 10^{-9} torr. Friction coefficients as functions of ambient

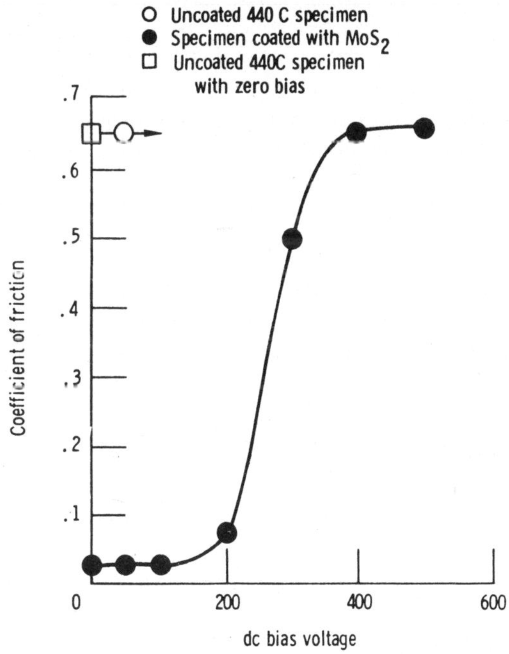

Figure 10-32. – Effect of negative dc bias on coefficient of sliding friction for sputtered MoS_2. Load, 250 grams; speed, 40 rpm; substrate and rider, 440C; pressure, 1×10^{-3} torr (ref. 14).

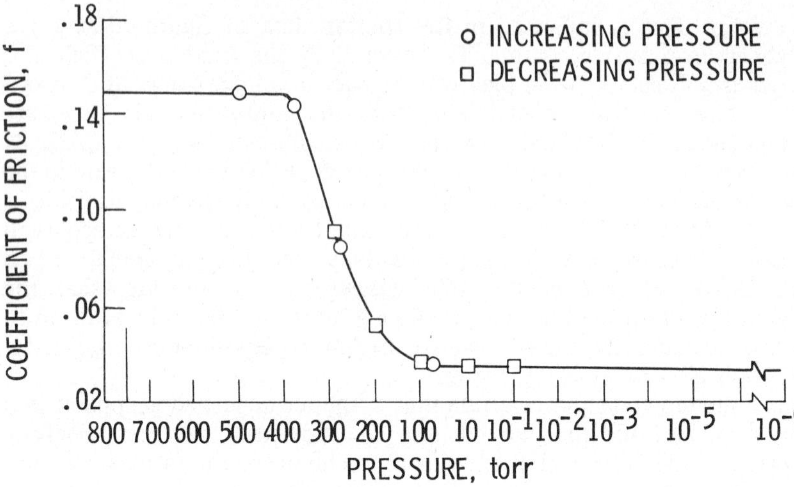

Figure 10-33. – Effect of pressure on coefficient of sliding friction for sputtered MoS_2. Load, 250 grams; speed, 40 rpm; substrate and rider, nickel; room temperature.

pressure are plotted in figure 10-33. An examination of the figure reveals that at very low pressures of 100 torr and less the friction coefficient is approximately 0.04. As the pressure is increased toward atmospheric from 100 torr, an increase in friction coefficient is observed; the friction coefficient ultimately rises to about 0.14 near atmospheric pressure. The contaminants in the environment (namely, water and oxygen as well as some hydrocarbons) tend to contaminate the surface of the molybdenum disulfide, reduce its characteristic lubricating effectiveness, and increase the fricton coefficient. If the molybdenum disulfide remains pure and uncontaminated by environmental constituents, it gives its best friction performance.

In addition to applying coatings with good lubricating characteristics, RF and DC sputtering can be used to deposit coatings which are very hard and tenaciously bonded to the surface to improve wear resistance. There are, for example, many industrial processes which require cutting and shearing edges to remain sharp with the minimum amount of wear. Improved performance of these cutting edges and endurance life can be achieved if very hard coatings are applied to the edges. Coatings such as carbides, silicides, and borides have been found to be very effective in providing good wear resistance to substrates. One of the problems, however, encountered in trying to use hard face coatings on substrate surfaces is the poor adhesion of the coating to the substrate. Under the severe mechanical stresses involved with practical components the coatings usually fail by spalling off the surface. Poor adhesion and a mismatch of properties of the coating to the substrate cause the spalling.

Investigations of various parameters to explore ways of maximizing the adhesion of hard face coatings to substrates have been undertaken (refs. 14 to 18). Various properties of hard face coating materials and substrates and various parameters appear to have importance from such investigations. They seem to indicate that one can optimize the adherence and the performance of hard face coatings on substrates by careful attention to variables in the system. For example, the thickness of the coating on the substrate is extremely critical. One desires a sufficiently thick coating to provide wear resistance. However, the coating should not be so thick that it takes on its own characteristic properties. Since these properties could be entirely different from the substrate, they might cause a mismatch of mechanical properties, and this could result in the coating spalling from the surface. Thus, there is an optimum coating thickness for maximum tribological effectiveness (ref. 14).

This is demonstrated by the data in figure 10-34 for the very common, hard face coating material titanium carbide. The friction coefficient is plotted for steel sliding on steel that is coated with titanium carbide. Friction is plotted at three different loads. The data (fig. 10-34) show that as the load is increased a thicker coating is required to achieve the low friction coefficients associated with titanium carbide—namely, a friction coefficient slightly in excess of 0.2.

There is an optimum coating thickness. If the coating is too thick, spalling of the coating occurs. The chemistry of a hard face coating is extremely important in the performance of the coating when applied to a substrate material. Using surface analytical tools such as X-ray photoelectron spectroscopy (XPS) has been very helpful in identifying the chemistry and relating that chemistry to tribological performance (refs. 16 to 18). For example, coatings have been applied using a bias voltage. A bias voltage in the sputtering process for the deposition of such coatings as titanium carbide has been found to produce an improved endurance life of the titanium carbide coatings. Better performance is achieved with bias

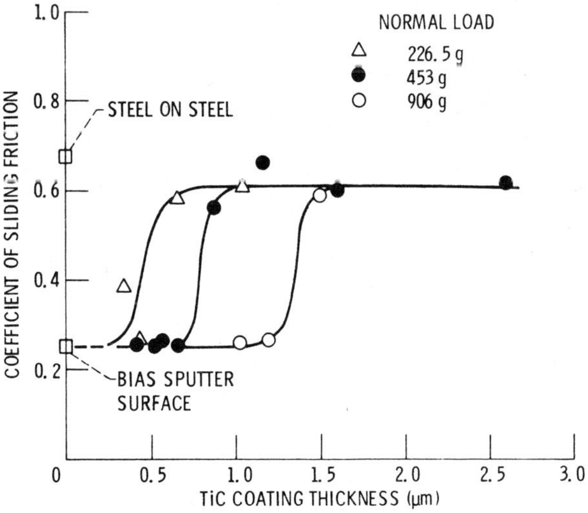

Figure 10-34. – Coefficient of sliding friction after 20 cycles for steel rider on sputtered TiC film as function of film thickness. Sliding speed, 1.5 centimeters per second (ref. 14).

rather than without bias. The chemical explanation for this performance is shown by the data in figure 10-35. XPS data are presented for a titanium carbide film deposited on a substrate with or without bias (ref. 19); the bias was -500 volts. An examination of the carbon and titanium peaks associated with titanium carbide reveals that using the bias produces a much improved concentration of titanium carbide in the film. Without the bias, there is a mixed oxide carbide system; with the dc bias, a purer titanium carbide structure is obtained on the solid surface. Thus, biasing improves the coating materials from a compound point of view.

In the discussion of molybdenum disulfide applied to substrates by sputter deposition, it was found desirable to clean the substrate surface very effectively before applying the film. Better adhesion of molybdenum disulfide to the substrate occurred in the presence of a clean substrate. With the deposition of hard face coatings, however, an entirely different result has been observed. With the hard face coatings, very frequently it is desirable to deliberately form or preform compounds on the surface of the substrate before applying the hard face coatings by sputter techniques. This is analogous to what was observed with the deposition of molybdenum disulfide over a chrome silicide film. The presence of the chrome silicide (fig. 10-30) improved the endurance charactersitics of the molybdenum disulfide film. Likewise, it has been found that the application of an oxidized substrate or the oxidation of substrates to form oxides and the use of these oxides as intermediate layers between the substrate and the hard face coating tend to improve the tribological performance of the coating materials.

This improved performance is indicated in some wear surface profiles

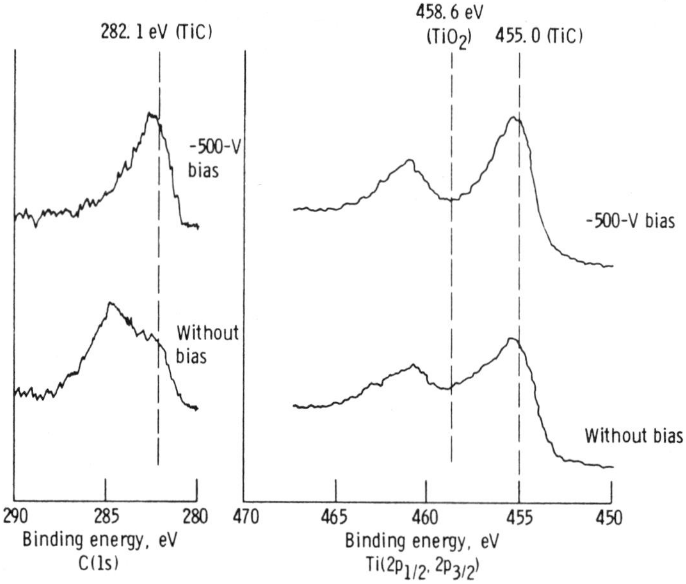

Figure 10-35. – Representative C(1s) and Ti(2p) XPS peaks from biased and unbiased TiC sputtered coatings (ref. 19).

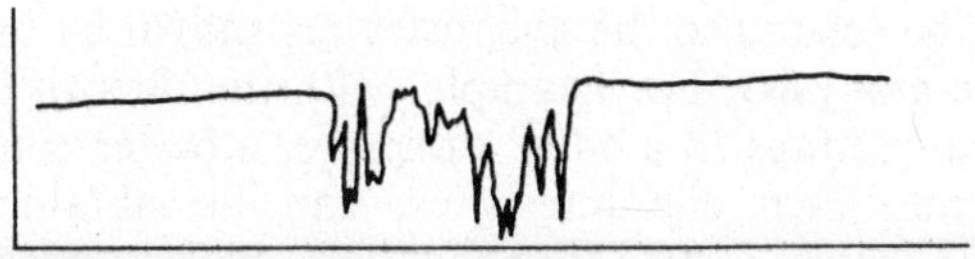

(*a*) *Sputter etched at – 1200 volts.*

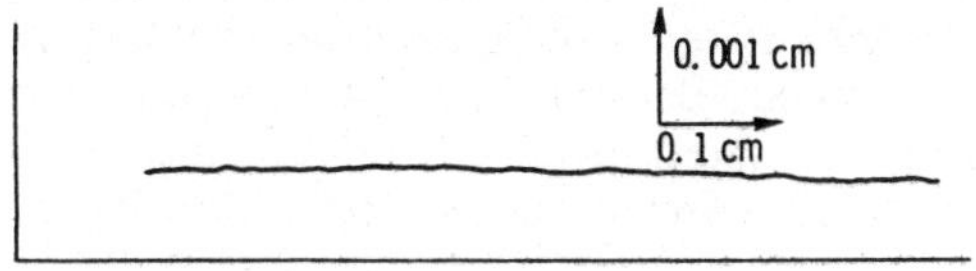

(*b*) *Preoxidized at 340° C for 60 hours.*

Figure 10-36. – Surface profile tracings of disk wear track for 440C disk radiofrequency sputter coated with TiC (–500-V bias). Load, 5 newtons; sliding time, 60 minutes; nitrogen atmosphere (ref. 19).

that have been obtained from surfaces that were sputter coated with titanium carbide. Results are presented in figure 10-36. In figure 10-36, a dc substrate surface was cleaned by sputter etching at -1200 volts and a wear experiment was run on the coating in a nitrogen atmosphere (ref. 16). The etched surface wore very heavily as indicated by the wear surface profile trace of figure 10-36(a). A considerable amount of material has been removed from the surface as a result of the sliding process. If, however, the substrate surface (in this instance, a 440C steel) is preoxidized at 340° C for approximately 60 hours to form an oxide layer and then the titanium carbide is sputter deposited on the oxide layer, the wear surface profile of figure 10-36(b) is observed. In figure 10-36(b), little or no wear is detected on the coating surface (the same carbide as in fig. 10-36(b)). Thus, marked differences in the wear resistance of the coating are achieved by sandwiching an oxide layer between the titanium carbide and the 440C substrate. This performance is observed despite the fact that the outermost layer, that which the mating component sees, is essentially titanium carbide in both cases.

Generally, in the sputtering process, an inert gas is used as the species to bombard and clean the substrate surface and to bombard the target material. The reason for this is that one does not want an active atom or a molecule striking the target surface because such an atom or molecule would interact with the target material and possibly form other compounds which might be undesirable. Therefore, generally inert gases such as argon, zenon, or krypton are used. The chemistry of the coating, however, can be varied by introducing active gases into the inert gas plasma in small concentrations; these active gases can interact with the coating material or the substrate during the cleaning process. One can vary the chemistry of the coating as well as that of the substrate before depositing the coating material by this technique. This can result in variations in surface chemistry and improved performance of hard faced coatings on substrates.

Bleed-in gases such as nitrogen and acetylene have been used by Brainard and Wheeler to apply hard faced coatings such as titanium carbide and titanium nitride to substrates (ref. 16). XPS has shown that the improved performance can be related to the chemistry (as shown by using both XPS and depth profile analysis). For example, with the deposition of titanium nitride hard faced coatings to a 440C substrate, a better coating, from the standpoint of endurance characteristics, can be obtained if a small concentration of nitrogen is bled into the argon plasma during deposition. The presence of the nitrogen promotes the nitriding of the iron on the 440C substrate and gives rise to an iron nitride on the surface to which the titanium nitride is applied. Between the titanium nitride coating and the iron nitride substrate a mixed layer of titanium nitride and iron nitride occurs. This mixed layer helps to act as a transition layer from the 440C

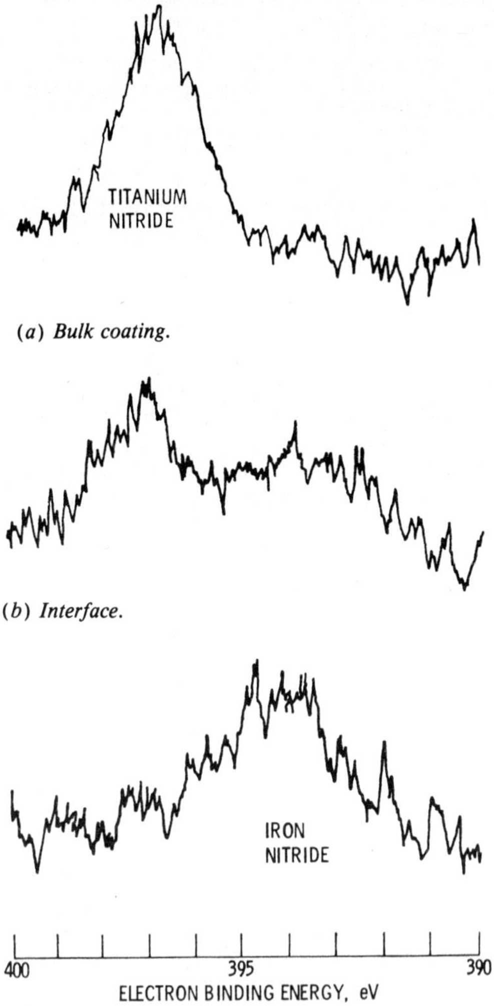

(*a*) *Bulk coating.*

(*b*) *Interface.*

(*c*) *Substrate.*

Figure 10-37. – Nitrogen XPS peaks for 440C disk exposed to and radiofrequency sputtered coated in argon plasma containing nitrogen partial pressure of 1.3 $\times 10^{-2}$ *newton per square meter. Bias, –500 volts (ref. 20).*

substrate to the titanium nitride coating. There is a gradual change in the properties from the characteristic properties of the 440C substrate to the properties of the titanium nitride. This transition region helps to improve adhesion and resistance to spalling of the hard face coating.

Some XPS data of the titanium nitride coating and depth profile analysis of that coating are presented in figure 10-37 (ref. 16). The XPS data in figure 10-37(a) for the coated 440C specimen show the titanium nitride peak on the outermost layer. Depth profiling or removal of surface layers by argon to the interface region was then carried out. At the interface between the titanium nitride and the 440C is a mixed film of titanium nitride and iron nitride (fig.10-37(b)). Further depth profiling to the substrate reveals the XPS spectra of figure 10-37(c) where an iron nitride is found on the 440C substrate surface. Thus, introducing a small concentration of nitrogen into the argon plasma helps to achieve a transition layer of iron nitride and titanium nitride in the surficial layers.

Just as with the soft molybdenum disulfide coatings, sputter deposition of hard face coatings is sensitive to variable parameters. For example, the presence or absence of a bias or the presence or absence of partial pressures of nitrogen can vary the performance and the morphology of the coating on the substrate very markedly. For example, some of these variables alter the orientation of the film material from a crystallographic point of view.

Examining the titanium carbide deposited on iron substrates indicates that varying parameters such as biasing and introducing partial pressures of nitrogen can result in a variation in the orientation viewed on the surface of the specimen coated with titanium carbide by sputtering. This is indicated in the data of figure 10-38 where there is a variation of titanium carbide (111) reflections for coatings applied by sputter deposition employing three variable conditions. In one set of conditions, there is a -500 volt bias with a

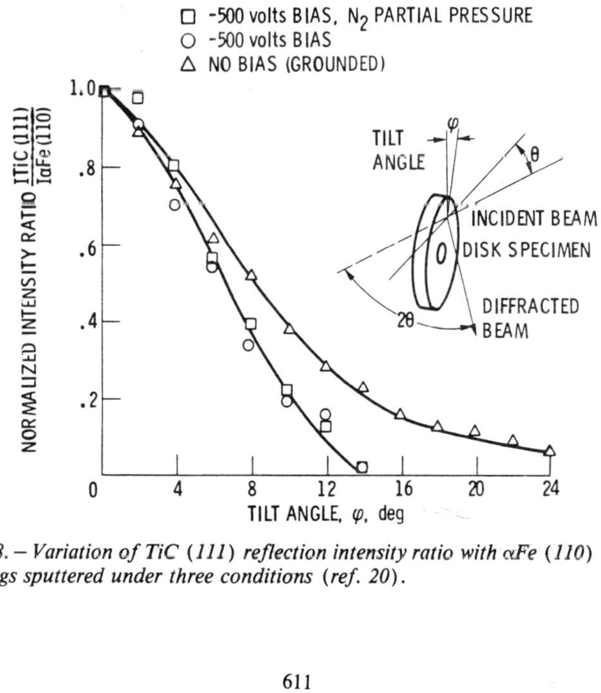

Figure 10-38. – Variation of TiC (111) reflection intensity ratio with αFe (110) (substrate) for coatings sputtered under three conditions (ref. 20).

nitrogen partial pressure. In the second set of conditions, there is a -500 volt bias but no nitrogen partial pressure. In a third set of conditions, there was no bias and the speciman was essentially grounded. Examining the data of figure 10-38 indicates that the (111) reflections were essentially the same for the -500 volt bias specimens with and without the nitrogen partial pressure. The reflections from the (111) surfaces, however, were different in the absence of the biasing. Thus, biasing appears to influence the crystallographic orientation observed in the coating surface layers, whereas the presence or the absence of the nitrogen partial pressure does not seem to markedly influence crystallographic orientation. An X-ray diffraction analysis, however, of the titanium carbide peaks indicates a shift in the titanium carbide peaks with and without the presence of the nitrogen partial pressure (fig. 10-39). In figure 10-39 the X-ray diffraction angles are presented for titanium carbide (111) orientation and silver (111) orientation. The silver is included in both samples for reference purposes. An

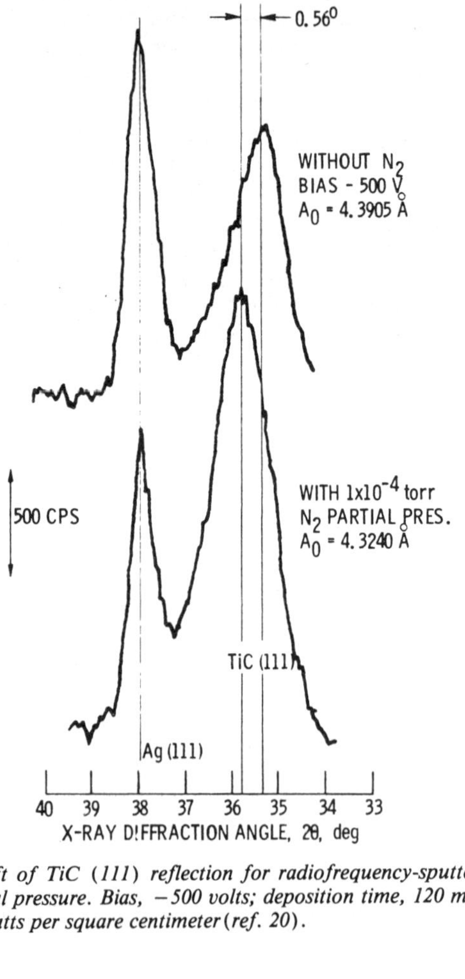

Figure 10-39.—Shift of TiC (111) reflection for radiofrequency-sputtered TiC with and without N_2 partial pressure. Bias, −500 volts; deposition time, 120 minutes; copper Kα radiation, 1.63 watts per square centimeter (ref. 20).

examination of the carbide peaks indicates that there is a shift in the titanium carbide peak of approximately 0.56° for the two coatings. The one without the nitrogen partial pressure is shifted from that containing nitrogen partial pressure.

Surface analytical tools including AES and XPS have proved to be very useful in analyzing the chemistry of hard faced coatings applied to substrates, particularly with the application of depth profiling. A coating is applied to the substrate and then analyzed by AES or XPS at the outermost layer. The coating is then systematically removed by argon sputter bombardment or argon ion depth profiling, and the coating is analyzed as one moves from the surface through the interface and ultimately to the substrate. Such an analysis of the chemistry from a standpoint of elemental analysis using AES and compound formation with the use of XPS can give considerable insight into the reasons coatings behave in a particular manner.

Brainard and Wheeler conducted a detailed study of the refractory compounds of molybdenum (carbides, borides, and silicides) as coatings on substrates to determine the effectiveness of such coatings as hard faced wear resistant materials (refs. 20 and 21). During their investigation, Brainard and Wheeler employed XPS to analyze the coatings together with a depth profile analysis of these molybdenum compounds. The results of some of their investigations are presented in figure 10-40. Figure 10-40 presents schematically the composition (on the left side) of the molybdenum compound which would be essentially on the outermost surface. Moving from left to right, one goes in depth profile analysis through the coating material (the carbide, boride, or silicide), through the interface between the coating and the interface composition, and through the interface composition until finally reaching the substrate region. An analysis of such regions with XPS indicates that the coatings become keyed into the substrate by the interface oxides. For example, in the case of molybdenum carbide, molybdenum oxide becomes tied in with the molybdenum carbide coating and the iron oxide that is present on the substrate. The iron oxide, of course, is tied to the iron substrate. This locking or keying in helps to promote adhesion of the coating to the substrate.

The data of figure 10-40 show that the oxides of the iron present on the surface are a lower oxide of iron—namely, Fe_3O_4—despite the fact that initial oxides present on the surface may be Fe_2O_2. The deposition of the coating on a clean surface results frequently in a substitution reaction or the reduction of the higher oxides of iron to a lower oxide form with the formation of the oxide of the coating material in the interfacial region. In the case of molybdenum carbide, it is molybdenum oxide. With molybdenum boride, there are two oxides of the coating material, molybdenum oxide and boric oxide, that form in the interfacial region. A similar oxide may form with the carbon of molybdenum carbide; however, since the product would be gaseous, it is probably liberated to the system and pumped away. Again, with molybdenum silicide, as was observed with molybdenum boride, both molybdenum and silicon oxides are found trapped in the interfacial region between the molybdenum silicide coating and the iron oxide layer. It is hypothesized that iron sacrifices some of its

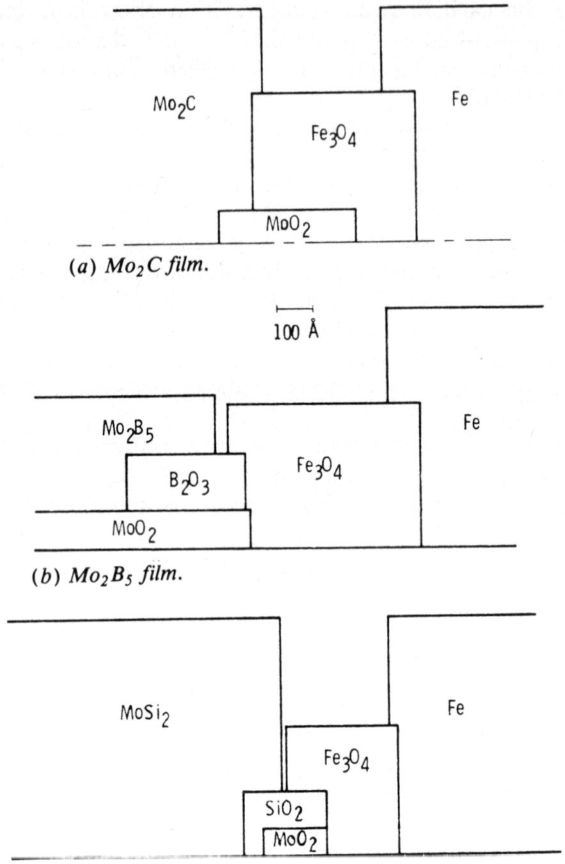

(c) *MoSi$_2$ film.*

Figure 10-40. – Schematic drawing of interfacial region of Mo_2C, Mo_2B_5, and $MoSi_2$ radiofrequency sputtered coatings on oxidized 440C substrates. Bias, −300 volts (ref. 22).

oxygen to the coating material to form the keying interfacial compounds—in this particular instance, oxides of the coating material. The data of figure 10-40 indicate the extreme usefulness of surface analytical tools in characterizing surface thin film coatings applied for improved tribological performance.

Chemical Vapor Deposition

Still another technique for applying tribological coatings to substrates is chemical vapor deposition (CVD). CVD had been used for a long time to apply coatings to substrates for various purposes. In recent years, Hintermann, et al. have begun to use it to deposit tribological coatings (ref. 22). The basic mechanism of CVD is shown schematically in figure 10-41

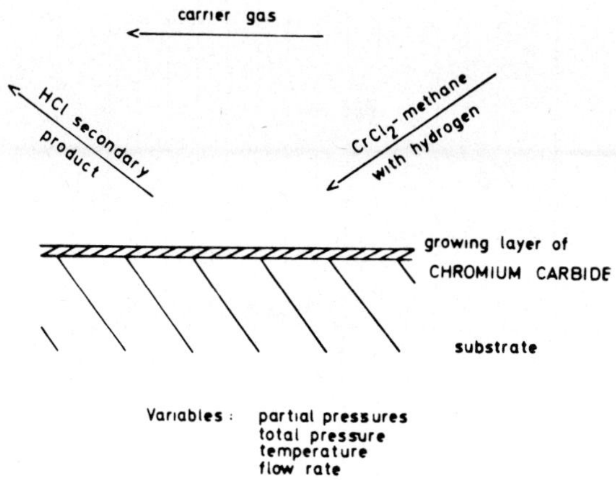

Figure 10-41. – Schematic drawing of CVD reaction showing formation of chromium carbide (ref. 23).

for depositing a chrome carbide layer on a substrate. A carrier gas is admitted into a vacuum system. The carrier gas may contain methane with chrome chloride and hydrogen. The substrate is heated and the gaseous mixture is allowed to come in contact with the substrate surface. Upon contact with the substrate surface the chrome chloride is reduced to chrome carbide by the interaction of the chrome chloride with the methane, and a chrome carbide film is formed on the surface. The chlorine is liberated with the hydrogen to form hydrogen chloride; the hydrogen chloride is pumped from the system as a byproduct of the reaction of the chrome chloride with methane and hydrogen. The result is a chrome carbide film left on the surface. Very frequently in CVD high substrate temperatures are required to initiate the decomposition of the gaseous species on the solid surface for the formation of the desired surface film. Also, the balance of the constituents (the ratios of the relative constituents in the gaseous plasma) is extremely critical to achieve the proper surface chemistry. Variations of stoichometry of gases can result in liberating products other than the desired. In CVD, more use of analytical surface tools such as XPS is needed to characterize the real chemistry of the CVD films. Some of the variables, as indicated in figure 10-41, are the partial pressure of the gases in the system, the total system pressure, the temperature of the substrate, and the flow rates of the gases. For industrial processes, controlling all these to the required levels may be extremely difficult. This makes the use of analytical surface tools and analysis of CVD coatings all the more imperative.

Some researchers have used certain analytical tools in characterizing CVD films. Tungsten carbide coatings, for example, deposited by CVD have been analyzed using the microprobe (ref. 23). A typical microprobe scan of a tungsten carbide coating applied to a steel substrate with a nickel inner layer is shown in figure 10-42. The microprobe analysis shows iron from the steel substrate with an intermediate peak for nickel with the tungsten from tungsten carbide coating being present beyond the nickel layer. Note that there is an overlap of the nickel with the iron substrate and

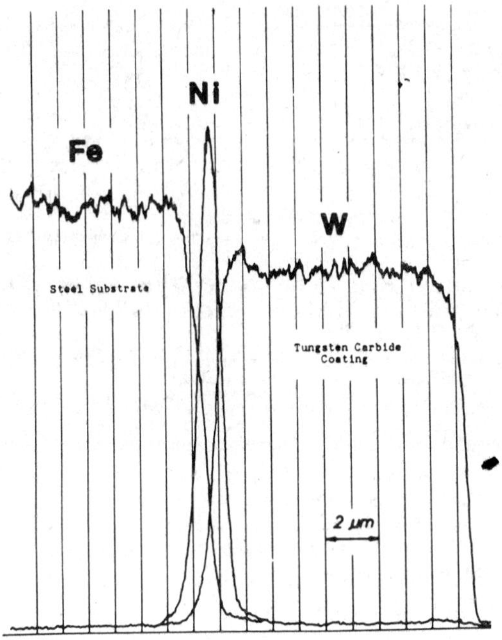

Figure 10-42. – Microprobe scan across steel/Ni/W_2C interface. Tungsten carbide coating deposited at 450° C (ref. 24).

with the tungsten carbide coating material; these result from diffusion of the nickel into both the substrate and the tungsten carbide coating at the high temperatures involved in the CVD process. In this case, the deposition process was accomplished at 450° C. This particular temperature is relatively low for CVD, and in many CVD processes the temperatures employed on the substrate are much higher than those indicated in figure 10-42. Again, these substrate temperatures are one of the limiting factors in using CVD for coatings in certain applications.

References

1. Bridgman, P.W.: The Physics of High Pressure. MacMillan Press, 1931.
2. Haltner, A.J.; and Oliver, C.S.: The Frictional Properties of Some Solid Lubricant Films Under High Load. J. Chem. Eng. Data, vol. 6, no. 1, Jan. 1961, pp. 128-130.
3. Sliney, H.E.: Plasma-Sprayed Metal-Glass and Metal-Glass Fluoride Coatings for Lubrication to 900° C. Trans. ASLE, vol. 17, no. 3, July 1974, pp. 182-189.
4. Peterson, M.B.; Florek, J.J.; and Murray, S.F.: Consideration of Lubricants for Temperatures Above 100° F. Trans. ASLE, vol. 2, no. 2, Apr. 1960, pp. 225-234.
5. Bowden, F.P; and Tabor, D.: The Friction and Lubrication of Solids. Vol. 2, Oxford Clarendon Press, 1964, p. 115.
6. Hardy, W. B.: Collected Scientific Papers. Cambridge University Press, 1936.
7. Dearnaley, G.: The Ion Implantation of Metals and Engineering Materials. Trans. Inst. Metal Finish., vol. 56, no. 1, 1978, pp. 25-31.
8. Mattox, D.M.: Interface Formation and Adhesion of Deposited Thin Films. SC-R-65-852, Sandia Corp., 1965.

9. Spalvins, T.; Przybyszewski, J.S.; and Buckley, D.H.: Deposition of Thin Films by Ion Plating on Surfaces Having Various Configurations. NASA TN D-3707, 1966.
10. Spalvins, T.: Bonding of Metal Lubricant Films by Ion Plating. Lubr. Eng., vol. 27, no. 2, Feb. 1971, pp. 40-46.
11. Spalvins, T.: Characteristics of Ion Plating Films Including Mechanical Properties and Lubrication. Sputtering and Ion Plating, NASA SP-5111, 1972, pp. 41-57.
12. Spalvins, T.: Coatings for Wear and Lubrication. NASA TM-78841, 1978.
13. Spalvins, T.: Friction Characteristics of Sputtered Solid Film Lubricants. NASA TM X-52819, 1970.
14. Spalvins, Talivaldis: Coatings for Wear and Lubrication. Thin Solid Films, vol. 53, 1978, pp. 285-300.
15. Brainard, W.A.; and Wheeler, D.R.: Use of Nitrogen-Argon Plasma to Improve Adherence of Sputtered Titanium Carbide Coatings on Steel. J. Vac. Sci. Technol., vol. 16, no. 1, Jan.-Feb. 1979, pp. 31-36.
16. Brainard, William A.; and Wheeler, Donald R.: Adherence of Sputtered Titanium Carbides. Thin Solid Films, vol. 63, 1979, pp. 363-368.
17. Wheeler, D.R.; and Brainard, W.A.: X-ray Photoelectron Spectroscopy Study of Radiofrequency Sputtered Refractory Compound Steel Interfaces. NASA TP-1161, 1978.
18. Brainard, W.A.; and Wheeler, D.R.: Friction and Wear of Radiofrequency-Sputtered Borides, Silicides and Carbides. NASA TP-1156, 1978.
19. Brainard, W.A.; and Wheeler, D.R.: X-ray Photoelectron Spectroscopy Study of Radiofrequency Sputtered Titanium Carbide, Molybdenum Carbide and Titanium Boride Coatings and their Friction Properties. NASA TP-1033, 1977.
20. Wheeler, D. R.; and Brainard, W. A.: X-ray Photoelectron Spectroscopy Study of Radiofrequency Sputtered Chromium Boride, Molybdenum Disilicide, and Molybdenum Disulfide Coatings and their Friction Properties. NASA TN D-8482, 1977.
21. Brainard, W.A.: The Friction and Wear Properties of Sputtered Hard Refractory Compounds. NASA TM-78895, 1978.
22. Hintermann, H.E.; Perry, A.J.; and Horvath, E.: Chemical Vapour Deposition Applied in Tribology. Wear, vol. 47, 1978, pp. 407-415.
23. Archer, N.J.; and Yee, K.K.: Chemical Vapour Deposited Tungsten Carbide Wear-Resistant Coatings Formed at Low Temperatures. Wear, vol. 48, 1978, pp. 237-250.

Author Index

This index gives the author, the page on which the author/reference is cited, the number of the reference in parentheses, and the page number in bold type on which the reference is listed.

As an example,

Page on which the reference listing appears
Reference number
Page cited

Bowden, F. P.: 2(2), **16**;

Subject Index